AF292275

Developments in Mathematics

Volume 66

Series Editors

Krishnaswami Alladi, Department of Mathematics, University of Florida, Gainesville, FL, USA

Pham Huu Tiep, Department of Mathematics, Rutgers University, Piscataway, NJ, USA

Loring W. Tu, Department of Mathematics, Tufts University, Medford, MA, USA

Aims and Scope

The Developments in Mathematics (DEVM) book series is devoted to publishing well-written monographs within the broad spectrum of pure and applied mathematics. Ideally, each book should be self-contained and fairly comprehensive in treating a particular subject. Topics in the forefront of mathematical research that present new results and/or a unique and engaging approach with a potential relationship to other fields are most welcome. High-quality edited volumes conveying current state-of-the-art research will occasionally also be considered for publication. The DEVM series appeals to a variety of audiences including researchers, postdocs, and advanced graduate students.

More information about this series at http://www.springer.com/series/5834

Teresa W. Haynes • Stephen T. Hedetniemi
Michael A. Henning

Editors

Structures of Domination in Graphs

 Springer

Editors

Teresa W. Haynes
Department of Mathematics and Statistics
East Tennessee State University
Johnson City, TN, USA

Department of Mathematics
and Applied Mathematics
University of Johannesburg
Johannesburg, South Africa

Michael A. Henning
Department of Mathematics
and Applied Mathematics
University of Johannesburg
Johannesburg, South Africa

Stephen T. Hedetniemi
School of Computing
Clemson University
Clemson, SC, USA

ISSN 1389-2177 ISSN 2197-795X (electronic)
Developments in Mathematics
ISBN 978-3-030-58891-5 ISBN 978-3-030-58892-2 (eBook)
https://doi.org/10.1007/978-3-030-58892-2

This Springer imprint is published by the registered company Springer Nature Switzerland AG
The registered company address is: Gewerbestrasse 11, 6330 Cham, Switzerland

Preface

While concepts related to domination in graphs can be traced back to the mid-1800s in connection to various chessboard problems, domination was first defined as a graph theoretical concept in 1958. Domination in graphs has experienced rapid growth from its introduction, resulting in over 1200 papers published on domination in graphs by the late 1990s.

Noting the need for a comprehensive survey of the literature on domination in graphs, in 1998 Haynes, Hedetniemi, and Slater published the first two books on domination, *Fundamentals of Domination in Graphs* and *Domination in Graphs: Advanced Topics*. We refer to these books as Books I and II.

The explosive growth of this field since 1998 has continued, and today more than 4,000 papers have been published on domination in graphs, and the material in Books I and II is now more than 20 years old. Thus, the authors feel it is time for an update on the developments in domination theory since 1998. We also want to give a comprehensive treatment of only the major topics in domination. This coverage of domination, including the major results and updates, will be in the form of three books, which we call Books III, IV, and V.

Book III, *Domination in Graphs: Core Concepts*, is written by the authors and concentrates, as the title suggests, on the three main types of domination in graphs: domination, independent domination, and total domination. It contains major results on these basic domination numbers, including proofs of selected results that illustrate many of the proof techniques used in domination theory.

For the companion books, Books IV and V, we invited leading researchers in domination to contribute chapters.

Book IV concentrates on the most-studied types of domination that are not covered in Book III. Although well over 70 types of domination have been defined, Book IV focuses on those that have received the most attention in the literature, and contains chapters on paired domination, connected domination, restrained domination, multiple domination, distance domination, dominating functions, fractional dominating parameters, Roman domination, rainbow domination, locating-domination, eternal and secure domination, global domination, stratified domination, and power domination.

The present volume, Book V, is divided into three parts. The first part focuses on several domination-related concepts: broadcast domination, alliances, domatic numbers, dominator colorings, irredundance in graphs, private neighbor concepts, game domination, varieties of Roman domination, and spectral graph theory. The second part covers domination in (i) hypergraphs, (ii) chessboards, and (iii) digraphs and tournaments. The third part focuses on the development of algorithms and complexity of (i) signed, minus, and majority domination, (ii) power domination, and (iii) alliances in graphs. The third part also includes a chapter on self-stabilizing domination algorithms.

The authors of the chapters in Book V provide a survey of known results with a sampling of proof techniques in their areas of expertise. To avoid excessive repetition of definitions and notation, Chapter 1 provides a glossary of commonly used terms.

This book is intended as a reference resource for researchers and is written to reach the following audiences: first, established researchers in the field of domination who want an updated, comprehensive coverage of domination theory; second, researchers in graph theory who wish to become acquainted with newer topics in domination, along with major developments in the field and some of the proof techniques used; and third, graduate students with interests in graph theory, who might find the theory and many real-world applications of domination of interest for master's and doctoral theses topics. We also believe that Book V provides a good focus for use in a seminar on either domination theory or domination algorithms and complexity, including the new algorithm paradigm of self-stabilizing domination algorithms.

We wish to thank the authors who contributed chapters to this book as well as the reviewers of the chapters.

Johnson City, TN, USA
Clemson, SC, USA
Johannesburg, South Africa

Teresa W. Haynes
Stephen T. Hedetniemi
Michael A. Henning

Contents

Part II   Domination in Selected Graph Families

Part III   Algorithms and Complexity

Glossary of Common Terms

Teresa W. Haynes, Stephen T. Hedetniemi, and Michael A. Henning

1 Introduction

It is difficult to say when the study of domination in graphs began, but for the sake of this glossary let us say that it began in 1962 with the publication of Oystein Ore's book *Theory of Graphs* [15]. In *Chapter 13 Dominating Sets, Covering Sets and Independent Sets* of [15], we see for the first time the name *dominating set*, defined as follows: "A subset D of V is a *dominating set* for G when every vertex not in D is the endpoint of some edge from a vertex in D." Ore then defines the *domination number*, denoted $\delta(G)$, of a graph G, as "the smallest number of vertices in any minimal dominating set." So, at this point, and for the first time, domination has a "name" and a "number."

Of course, prior to this Claude Berge [3], in his book *Theory of Graphs and its Applications*, which was first published in France in 1958 by Dunod, Paris,

T. W. Haynes (✉)
Department of Mathematics and Statistics, East Tennessee State University, Johnson City, TN, USA

Department of Mathematics and Applied Mathematics, University of Johannesburg, Johannesburg, South Africa
e-mail: haynes@etsu.edu

S. T. Hedetniemi
School of Computing, Clemson University, Clemson, SC, USA
e-mail: hedet@clemson.edu

M. A. Henning
Department of Mathematics and Applied Mathematics, University of Johannesburg, Johannesburg, South Africa
e-mail: mahenning@uj.ac.za

had previously defined the same concept, but had, in *Chapter 4 The Fundamental Numbers of the Theory of Graphs* of [3], given it the name "the coefficient of external stability."

Before Berge, Dénes König, in his 1936 book *Theorie der Endlichen und Unendlichen Graphen* [13], had defined essentially the same concept, but in *VII Kapitel, Basisproblem für gerichtete Graphen*, König gave it the name "punktbasis," which we would today say is an independent dominating set.

And even before König, in the books by Dudeney in 1908 [8] and W. W. Rouse Ball in 1905 [2], one can find the concepts of domination, independent domination, and total domination discussed in connection with various chessboard problems. And it was Ball who, in turn, credited such people as W. Ahrens in 1910 [1], C. F. de Jaenisch in 1862 [7], Franz Nauck in 1850 [14], and Max Bezzel in 1848 [4] for their contributions to these types of chessboard problems involving dominating sets of chess pieces.

But it was Ore who gave the name *domination* and this name took root. Not long thereafter, Cockayne and Hedetniemi [6] gave the notation $\gamma(G)$ for the domination number of a graph, and this also took root and is the notation adopted here.

Since the subsequent chapters in this book will deal with domination parameters, there will be much overlap in the terminology and notation used. One purpose of this chapter is to present definitions common to many of the chapters in order to prevent terms being defined repeatedly and to avoid other redundancy. Also, since graph theory terminology and notation sometimes vary, in this glossary we clarify the terminology that will be adopted in subsequent chapters.

We proceed as follows. In Section 2.1, we present basic graph theory definitions. We discuss common types of graphs in Section 2.2. Some fundamental graph constructions are given in Section 2.3. In Section 3.1 and Section 3.2, we present parameters related to connectivity and distance in graphs, respectively. The covering, packing, independence, and matching numbers are defined in Section 3.3. Finally in Section 3.4, we define selected domination-type parameters that will occur frequently throughout the book.

For more details and terminology, the reader is referred to the two books *Fundamentals of Domination in Graphs* [10] and *Domination in Graphs, Advanced Topics* [11], written and edited by Haynes, Hedetniemi, and Slater and the book *Total Domination in Graphs* by Henning and Yeo [12]. An annotated glossary, from which many of the definitions in this chapter are taken, was produced by Gera, Haynes, Hedetniemi, and Henning in 2018 [9].

2 Basic Terminology

In this section, we give basic definitions, common types of graphs, and fundamental graph constructions.

2.1 Basic Graph Theory Definitions

Before we proceed with our glossary of parameters, we need to define a few basic terms, which are used in the definitions in the following subsections. For an integer $k \geq 1$, we use the standard notation $[k] = \{1, \ldots, k\}$ and $[k]_0 = [k] \cup \{0\} = \{0, 1, \ldots, k\}$.

A (finite, undirected) *graph* $G = (V, E)$ consists of a finite nonempty set of *vertices* $V = V(G)$ together with a set $E = E(G)$ of unordered pairs of distinct vertices called *edges*. Each edge $e = \{u, v\}$ in E is denoted with any of e, uv, vu, and $\{u, v\}$. We say that a graph G has *order* $n = |V|$ and *size* $m = |E|$.

Two vertices u and v in G are *adjacent* if they are joined by an edge e, that is, u and v are adjacent if $e = uv \in E(G)$. In this case, we say that each of u and v is *incident* with the edge e. Further, we say that the edge *e joins* the vertices u and v. Two edges are *adjacent* if they have a vertex in common. Two vertices in a graph G are *independent* if they are not adjacent. A set of pairwise independent vertices in G is an *independent set* of G. Similarly, two edges are *independent* if they are not adjacent.

A *neighbor* of a vertex v in G is a vertex u that is adjacent to v. The *open neighborhood* of a vertex v in G is the set of neighbors of v, denoted $N_G(v)$. Thus, $N_G(v) = \{u \in V : uv \in E\}$. The *closed neighborhood of* v is the set $N_G[v] = \{v\} \cup N_G(v)$. For a set of vertices $S \subseteq V$, the *open neighborhood* of S is the set $N_G(S) = \bigcup_{v \in S} N_G(v)$ and its *closed neighborhood* is the set $N_G[S] = N_G(S) \cup S$. If the graph G is clear from the context, we omit it in the above expressions. For example, we write $N(v)$, $N[v]$, $N(S)$, and $N[S]$ rather than $N_G(v)$, $N_G[v]$, $N_G(S)$, and $N_G[S]$, respectively.

For a set of vertices $S \subseteq V$ and a vertex v belonging to the set S, the *S-private neighborhood* of v is defined by $\mathrm{pn}[v, S] = \{w \in V : N_G[w] \cap S = \{v\}\}$, while its *open S-private neighborhood* is defined by $\mathrm{pn}(v, S) = \{w \in V : N_G(w) \cap S = \{v\}\}$. As remarked in [12], the sets $\mathrm{pn}[v, S] \setminus S$ and $\mathrm{pn}(v, S) \setminus S$ are equivalent and we define the *S-external private neighborhood* of v to be this set, abbreviated $\mathrm{epn}[v, S]$ or $\mathrm{epn}(v, S)$. The *S-internal private neighborhood* of v is defined by $\mathrm{ipn}[v, S] = \mathrm{pn}[v, S] \cap S$ and its *open S-internal private neighborhood* is defined by $\mathrm{ipn}(v, S) = \mathrm{pn}(v, S) \cap S$. We define an *S-external private neighbor* of v to be a vertex in $\mathrm{epn}(v, S)$ and an *S-internal private neighbor* of v to be a vertex in $\mathrm{ipn}(v, S)$.

The *degree* $d_G(v)$ of a vertex v is the number of neighbors v has in G, that is, $d_G(v) = |N_G(v)|$. Again if the graph G is clear from the context, we use $d(v)$ rather than $d_G(v)$. We remark that some books use $\deg(v)$ and $\deg v$ to denote the degree of v. We leave it to the authors to choose which of these notations to adopt in their chapters. For a subset of vertices $S \subseteq V$, the *degree of* v *in* S, denoted $d_S(v)$, is the number of vertices in S adjacent to the vertex v; that is, $d_S(v) = |N_G(v) \cap S|$. In particular, if $S = V$, then $d_S(v) = d_G(v)$. The *degree sequence* of a graph G with vertex set $V = \{v_1, v_2, \ldots, v_n\}$ is the sequence $d_1, d_2, \ldots, d_n$, where $d_i = d(v_i)$ for $i \in [n]$. Often the degree sequence, $d_1, d_2, \ldots, d_n$ is written in non-increasing order, and so $d_1 \geq d_2 \geq \cdots \geq d_n$.

An *isolated vertex* is a vertex of degree 0 in G. A graph is *isolate-free* if it does not contain an isolated vertex. The minimum degree among the vertices of G is denoted by $\delta(G)$ and the maximum degree by $\Delta(G)$. A *leaf* is a vertex of degree 1, while its neighbor is a *support vertex*. A *strong support vertex* is a (support) vertex with at least two leaf neighbors.

For subsets X and Y of vertices of G, we denote the set of edges that join a vertex of X and a vertex of Y in G by $[X, Y]$.

Two graphs G and H are *isomorphic*, denoted $G \cong H$, if there exists a bijection $\phi : V(G) \to V(H)$ such that two vertices u and v are adjacent in G if and only if the two vertices $\phi(u)$ and $\phi(v)$ are adjacent in H. A *parameter* of a graph G is a numerical value (usually a non-negative integer) that can be associated with a graph such that whenever two graphs are isomorphic, they have the same associated parameter value.

By a *partition* of the vertex set V of a graph G, we mean a family $\pi = \{V_1, V_2, \ldots, V_k\}$ of nonempty pairwise disjoint sets whose union equals V, that is, for all $1 \le i < j \le k$, $V_i \cap V_j = \varnothing$ and

$$\bigcup_{i=1}^{k} V_i = V.$$

For such a partition π, we will say that π has *order k*.

A *walk* in a graph G from a vertex u to a vertex v is a finite, alternating sequence of vertices and edges, starting with the vertex u and ending with the vertex v, in which each edge of the sequence joins the vertex that precedes it in the sequence to the vertex that follows it in the sequence. A *trail* is a walk containing no repeated edges, and a *path* is a walk containing no repeated vertices. We will mainly be concerned with paths. A path between two vertices u and v is called a (u, v)-*path* or a u-v *path* or a u, v-*path* in the literature. The *length* of a walk equals the number of edges in the walk. A graph G is *connected* if for any two vertices u and v in G, there is a (u, v)-path.

A *cycle* is a path in which the first and last vertices are the same and all other vertices are distinct. A *chord* of a cycle C is an edge between two nonconsecutive vertices of C.

The *distance* $d(u, v)$ between two vertices u and v, in a connected graph G, equals the minimum length of a (u, v)-path in G. A shortest, or minimum length, path between two vertices u and v is called a (u, v)-*geodesic*; a v-*geodesic* is any shortest path from v to another vertex; a *geodesic* is any shortest path in a graph. The *diameter* of G is the maximum length of a geodesic in G.

A graph $G' = (V', E')$ is a *subgraph* of a graph $G = (V, E)$ if $V' \subseteq V$ and $E' \subseteq E$. A subgraph G' of a graph G is called a *spanning subgraph* of G if $V' = V$. If $G = (V, E)$ and $S \subseteq V$, then the *subgraph of G induced by S* is the graph $G[S]$, whose vertex set is S and whose edges are all the edges in E both of whose vertices are in S.

Let F be an arbitrary graph. A graph G is said to be *F-free* if G does not contain F as an induced subgraph.

If $G = (V, E)$ and $S \subseteq V$, the subgraph obtained from G by deleting all vertices in S and all edges incident with one or two vertices in S is denoted by $G - S$; that is, $G - S = G[V \smallsetminus S]$. If $S = \{v\}$, we simply denote $G - \{v\}$ by $G - v$. The *contraction* of an edge $e = xy$ in a graph G is the graph obtained from G by deleting the vertices x and y and all edges incident to x or y and adding a new vertex and edges joining this new vertex to all vertices that were adjacent to x or y in G.

A *component* of a graph is a maximal connected subgraph. An *odd (even) component* is a component of odd (even) order. Let $oc(G)$ equal the number of odd components of G. A vertex $v \in V$ is a *cutvertex* if the graph $G - v$ has more components than G. An edge $e = uv$ is a *bridge* if the graph $G - e$ obtained by deleting e from G has more components than G.

2.2   Common Types of Graphs

A graph of order $n = 1$ is called a *trivial graph*, while a graph with at least two vertices is called a *nontrivial graph*. A graph of size $m = 0$ is an *empty graph*, while a graph with at least one edge is a *nonempty graph*. Recall that a *connected graph* is a graph for which there is a path between every pair of its vertices.

A *k-regular graph* is a graph in which every vertex has degree k for some $k \geq 0$. A *regular graph* is a graph that is k-regular. A 3-regular graph is also called a *cubic graph*.

A graph of order n that is itself a cycle is denoted by C_n, and a graph that is itself a path is denoted by P_n. Note that a cycle is a 2-regular graph.

A *forest* is an *acyclic* graph, that is, a graph with no cycles. A *tree* is a connected acyclic graph. Equivalently, a tree is a connected graph having size one less than its order. Hence, if T is a tree of order n and size m, then T is connected and $m = n - 1$. Note that every component of a forest is a tree, and a forest in which every component is a path is called a *linear forest*.

If G is a vertex disjoint union of k copies of a graph F, we write $G = kF$.

A *complete graph* is a graph in which every two vertices are adjacent. A complete graph of order n is denoted by K_n. A *triangle* is a subgraph isomorphic to K_3 or C_3, since $K_3 \cong C_3$.

A graph G is *bipartite* if its vertex set can be partitioned into two independent sets X and Y. The sets X and Y are called the *partite sets* of G. A *complete bipartite graph*, denoted $K_{r,s}$, is a bipartite graph with partite sets X and Y, where $|X| = r$, $|Y| = s$, and every vertex in X is adjacent to every vertex in Y. The graph $K_{r,s}$ has order $r + s$, size rs, $\delta(K_{r,s}) = \min\{r, s\}$ and $\Delta(K_{r,s}) = \max\{r, s\}$.

A *star* is a nontrivial tree with at most one vertex that is not a leaf. Thus, a star is a complete bipartite graph $K_{1,k}$ for some $k \geq 1$. A *claw* is an induced copy of the graph $K_{1,3}$. Thus, a *claw-free graph* is a $K_{1,3}$-free graph.

For $r, s \geq 1$, a *double star* $S(r, s)$ is a tree with exactly two (adjacent) vertices that are not leaves, one of which has r leaf neighbors and the other s leaf neighbors. Equivalently, a *double star* is a tree having diameter equal to 3.

A *diamond* is an induced copy of the graph $K_4 - e$, which is obtained from a copy of the complete graph of order 4 by deleting any edge e.

A graph G can be *embedded* on a surface S if its vertices can be placed on S and all of its edges can be drawn between the vertices on S in such a way that no two edges intersect. A graph G is *planar* if it can be embedded in the plane; a *plane graph* is a graph that has been embedded in the plane.

A *rooted tree* T is a tree having a *distinguished vertex* labeled r, called the *root*. Let T be a rooted tree with root r. For each vertex v, let $P(v)$ be the unique (r, v)-path in T. The *parent* of a vertex v is its neighbor on $P(v)$, while the other neighbors of v are called its *children*. The set of children of v is denoted by $C(v)$. Note that the root r is the only vertex of T with no parent. A *descendant* of v is any vertex $u \neq v$ such that $P(u)$ contains v, while an *ancestor* of v is a vertex $u \neq v$ that belongs to $P(v)$ in T. In particular, every child of v is a descendant of v, while the parent of v is an ancestor of v. A *grandchild* of v is a descendant of v at distance 2 from v. We let $D(v)$ denote the set of descendants of v, and we define $D[v] = D(v) \cup \{v\}$. The *maximal subtree* at v, denoted T_v, is the subtree of T induced by $D[v]$. The *depth* of a vertex v in T equals $d(r, v)$, and the *height* of v, denoted $\mathrm{ht}(v)$, is the maximum distance from v to a descendant of v. Thus, $\mathrm{ht}(v) = \max\{d(v, w) : w \text{ is a descendant of } v\}$.

For classes of graphs not defined here, we refer the reader to the definitive encyclopedia on graph classes, *Graph Classes: A Survey* [5] by Brandstädt, Le, and Spinrad.

2.3   *Graph Constructions*

Given a graph $G = (V, E)$, the *complement* of G is the graph $\overline{G} = (V, \overline{E})$, where $uv \in \overline{E}$ if and only if $uv \notin E$. Thus, the complement $\overline{G}$ of G is formed by taking the vertex set of G and joining two vertices by an edge whenever they are not joined in G.

By a *graph product* $G \otimes H$ on graphs G and H, we mean a graph whose vertex set is the Cartesian product of the vertex sets of G and H (that is, $V(G \otimes H) = V(G) \times V(H)$) and whose edge set is determined entirely by the adjacency relations of G and H. Exactly how it is determined depends on what kind of graph product we are considering.

The *Cartesian product* $G \,\square\, H$ of two graphs G and H is the graph with vertex set $V(G) \times V(H)$, where two vertices (u_1, v_1) and (u_2, v_2) are adjacent if and only if either $u_1 = u_2$ and $v_1 v_2 \in E(H)$ or $v_1 = v_2$ and $u_1 u_2 \in E(G)$.

The *direct product* (also known as the *cross product, tensor product, categorical product*, and *conjunction*) $G \times H$ of two graphs G and H is the graph with vertex set $V(G) \times V(H)$, where two vertices (u_1, v_1) and (u_2, v_2) are adjacent in $G \times H$ if and only if $u_1 u_2 \in E(G)$ and $v_1 v_2 \in E(H)$.

Given a graph $G = (V, E)$ and an edge $uv \in E$, the *subdivision* of edge uv consists of (i) deleting the edge uv from E, (ii) adding a new vertex w to V, and (iii) adding the new edges uw and wv to E. In this case, we say that the edge uv has been *subdivided.* The *subdivision graph* $S(G)$ is the graph obtained from G by subdividing every edge of G exactly once.

Given a graph $G = (V, E)$, the *line graph* $L(G) = (E, E(L(G)))$ is the graph whose vertices correspond 1-to-1 with the edges in E, and two vertices are adjacent in $L(G)$ if and only if the corresponding edges in G have a vertex in common, that is, if and only if the corresponding two edges are adjacent.

The *corona* $G \circ K_1$ of a graph G, also denoted $\mathrm{cor}(G)$ in the literature, is the graph obtained from G by adding, for each vertex $v \in V$, a new vertex v' and the edge vv'. The edge vv' is called a *pendant edge*. The *k-corona* $G \circ P_k$ of G is the graph of order $(k+1)|V(G)|$ obtained from G by attaching a path of length k to each vertex of G so that the resulting paths are vertex-disjoint. In particular, the 2-*corona* $G \circ P_2$ of G is the graph of order $3|V(G)|$ obtained from G by attaching a path of length 2 to each vertex of G so that the resulting paths are vertex-disjoint. The *generalized corona* $G \circ H$ is the graph obtained by adding a copy of H for each vertex v of G and joining v to every vertex of H. Thus, a generalized corona $G \circ H$, where $H = K_1$, is the ordinary corona $G \circ K_1$. We note that whether $G \circ P_k$ is intended to denote a k-corona or a generalized corona will be clear from context or specifically stated by the author.

3 Graph Parameters

In this section, we present common graph parameters that may appear in this book.

3.1 Connectivity and Subgraph Numbers

In this subsection, we present parameters related to connectivity in graphs.

(a) *blocks* $\mathrm{bl}(G)$, number of blocks in G. A *block* of a graph G is a maximal nonseparable subgraph of G, that is, a maximal subgraph having no cutvertices.
(b) *bridges* $\mathrm{br}(G)$, number of bridges in G.
(c) *circumference* $\mathrm{cir}(G)$, maximum length or order of a cycle in G.
(d) *clique number* $\omega(G)$, maximum order of a complete subgraph of G.
(e) *components* $c(G)$, number of maximal connected subgraphs of G.
(f) A *vertex cut* of a connected graph G is a subset S of the vertex set of G with the property that $G - S$ is disconnected (has more than one component). A vertex cut S is a *k-vertex cut* if $|S| = k$.
(g) *vertex connectivity* $\kappa(G)$, minimum cardinality of a vertex cut of G if G is not the complete graph, and $\kappa(K_n) = n - 1$. A graph G is *k-vertex-connected* (or *k-connected*) if $\kappa(G) \geq k$ for some integer $k \geq 0$. Thus, $\kappa(G)$ is the smallest number of vertices whose deletion from G produces a disconnected graph or the trivial graph K_1. A nontrivial graph has connectivity 0 if and only if it is disconnected.

(h) An *edge cut* of a nontrivial connected graph G is a nonempty subset F of the edge set of G with the property that $G - F$ is disconnected (has more than one component). Thus, the deletion of an edge cut from the connected graph G results in a disconnected graph. An edge cut F is a *k-edge cut* if $|F| = k$.

(i) *edge connectivity* $\lambda(G)$, minimum cardinality of an edge cut of G if G is nontrivial, while $\lambda(K_1) = 0$. A graph G is *k-edge-connected* if $\lambda(G) \geq k$ for some integer $k \geq 0$. Thus, $\lambda(G)$ is the smallest number of edges whose deletion from G produces a disconnected graph or the trivial graph K_1. Hence, $\lambda(G) = 0$ if and only if G is disconnected or trivial.

(j) *girth* of G, denoted *girth(G)* or $g(G)$ in the literature, the length of a shortest cycle in G.

3.2 Distance Numbers

This subsection contains the definitions of parameters, which are defined in terms of the distances $d(u, v)$ between vertices u and v in a graph.

(a) *eccentricity* $\mathrm{ecc}(v) = \max\{d(v, w) : w \in V(G)\}$.

(b) *diameter* $\mathrm{diam}(G)$, maximum distance among all pairs of vertices of G. Equivalently, the diameter of G is the maximum length of a geodesic in G. Thus, the diameter of G is the maximum eccentricity taken over all vertices of G. Two vertices u and v in G for which $d(u, v) = \mathrm{diam}(G)$ are called *antipodal* or *peripheral vertices* of G. A *diametral path* in G is a geodesic whose length equals the diameter of G.

(c) The *periphery* of a graph G is the subgraph of G induced by its peripheral vertices.

(d) *radius* $\mathrm{rad}(G) = \min\{\mathrm{ecc}(v) : v \in V(G)\}$.

(e) The *center* of a graph G, denoted $C(G)$, is the subgraph of G induced by the vertices in G whose eccentricity equals the radius of G. A vertex $v \in C(G)$ is called a *central vertex* of G.

3.3 Covering, Packing, Independence, and Matching Numbers

As previously defined, a set S is *independent* if no two vertices of S are adjacent.

A set M of edges is called a *matching* if no two edges of M are adjacent, and a matching of maximum cardinality is a *maximum matching*. Given a matching M, we denote by $V[M]$ the set of vertices in G incident with an edge in M. A matching M of G is a *perfect matching* if $V[M] = V(G)$. Thus, if G has a perfect matching M, then G has even order $n = 2k$ for some $k \geq 1$ and $|M| = k$.

A vertex and an edge are said to *cover* each other in a graph G if they are incident in G. A *vertex cover* in G is a set of vertices that covers all the edges of G, while an *edge cover* in G is a set of edges that covers all the vertices of G. Thus, a vertex cover in G is a set of vertices that contains at least one vertex of every edge in G.

A subset S of vertices in G is a *packing* if the closed neighborhoods of vertices in S are pairwise disjoint. Equivalently, S is a packing in G if for every $u, v \in S$, $d(u, v) > 2$. Thus, if S is a packing in G, then $|N_G[v] \cap S| \le 1$ for every vertex $v \in V(G)$. A packing is also called a *2-packing* in the literature. More generally, for $k \ge 2$, a set S is a *k-packing* in G if for $u, v \in S$, $d(u, v) > k$.

A subset S of vertices in G is an *open packing* if the open neighborhoods of vertices in S are pairwise disjoint. Thus, if S is an open packing in G, then $|N_G(v) \cap S| \le 1$ for every vertex $v \in V(G)$.

All of the parameters in this subsection have to do with sets that are independent or cover other sets. These include some of the most basic of all parameters in graph theory.

(a) *vertex independence numbers* $i(G)$ and $\alpha(G)$, minimum and maximum cardinalities of a maximal independent set in G. The lower vertex independence number, $i(G)$, is also called the *independent domination number* of G, while the upper vertex independence number, $\alpha(G)$, is also called the *independence number* of G. (While the notation $i(G)$ is fairly standard for the independent domination number, we remark that the independence number is also denoted by $\beta_0(G)$ in the literature.)

(b) *vertex covering numbers* $\beta(G)$ and $\beta^+(G)$, minimum and maximum cardinalities of a minimal vertex cover in G. (We remark that the vertex covering number is also denoted by $\tau(G)$ or by $\alpha(G)$ in the literature.)

(c) *edge covering numbers* $\beta'(G)$ and $\beta'^+(G)$, minimum and maximum cardinalities of a minimal edge cover in G.

(d) *k-packing numbers* $\rho_k(G)$, maximum cardinality of a k-packing in G for $k \ge 2$. When $k = 2$, the k-packing number $\rho_k(G)$ is called the *packing number* of G, denoted by $\rho(G)$. Thus, $\rho(G)$ is the maximum cardinality of a packing in G.

(e) *open packing numbers* $\rho^o(G)$, maximum cardinality of an open packing in G.

(f) *matching numbers* $\alpha'^-(G)$ and $\alpha'(G)$, minimum and maximum cardinalities of a maximal matching in G. The upper matching number, $\alpha'(G)$, is also called the *matching number* of G. Recall that a *perfect matching* is a matching in which every vertex is incident with an edge of the matching. Thus, if a graph G of order n has a perfect matching, then $\alpha'(G) = \frac{1}{2}n$. It should be noted that by a well-known theorem of Gallai, that if G is a graph of order n with no isolated vertices, then $\alpha(G) + \beta(G) = n = \alpha'(G) + \beta'(G)$. (We remark that the matching number is also denoted by $\beta_1(G)$ in the literature.)

3.4 Domination Numbers

A *dominating set* in a graph $G = (V, E)$ is a set S of vertices of G such that every vertex in $\overline{S} = V \setminus S$ has a neighbor in S. Thus, if S is a dominating set of G, then

$N_G[S] = V$ and every vertex in $\overline{S}$ is therefore adjacent to at least one vertex in S. For subsets X and Y of vertices of G, if $Y \subseteq N_G[X]$, then the set X *dominates* the set Y in G. In particular, if X dominates $V(G)$, then X is a dominating set of G.

The many variations of dominating sets in a graph G are based on (i) conditions that are placed on the subgraph $G[S]$ induced by a dominating set S, (ii) conditions that are placed on the vertices in $\overline{S}$, or (iii) conditions that are placed on the edges between vertices in S and vertices in $\overline{S}$. We mention only the major domination numbers here.

A *total dominating set*, abbreviated TD-set, in a graph G with no isolated vertex is a set S of vertices of G such that every vertex in V is adjacent to at least one vertex in S. Thus, a subset $S \subseteq V$ is a TD-set in G if $N_G(S) = V$. Every graph without isolated vertices has a TD-set, since $S = V$ is such a set. If X and Y are subsets of vertices in G, then the set X *totally dominates* the set Y in G if $Y \subseteq N_G(X)$. In particular, if X totally dominates $V(G)$, then X is a TD-set in G.

A *paired dominating set*, abbreviated PD-set, of G is a set S of vertices of G such that every vertex is adjacent to some vertex in S and the subgraph $G[S]$ induced by S contains a perfect matching M. Two vertices joined by an edge of M are said to be *paired* and are also called *partners* in S.

A *connected dominating set*, abbreviated CD-set, in a graph G is a dominating set S of vertices of G such that $G[S]$ is connected.

(a) *domination numbers* $\gamma(G)$ and $\Gamma(G)$, minimum and maximum cardinalities of a minimal dominating set in G. The parameters $\gamma(G)$ and $\Gamma(G)$ are referred to as the *domination number* and *upper domination number* of G, respectively. A dominating set of G of cardinality $\gamma(G)$ is called a γ-*set* of G, while a minimal dominating set of cardinality $\Gamma(G)$ is called a Γ-set of G.

(b) *independent domination number* $i(G)$, minimum cardinality of a dominating set in G that is also independent. An independent dominating set of G of cardinality $i(G)$ is called an i-set of G. We note that the maximum order of a minimal independent dominating set equals the vertex independence number $\alpha(G)$.

(c) *total domination numbers* $\gamma_t(G)$ and $\Gamma_t(G)$, minimum and maximum cardinalities of a minimal total dominating set of G. The parameters $\gamma_t(G)$ and $\Gamma_t(G)$ are referred to as the *total domination number* and *upper total domination number* of G, respectively. A TD-set of G of cardinality $\gamma_t(G)$ is called a γ_t-*set* of G, while a minimal TD-set of cardinality $\Gamma_t(G)$ is called a Γ_t-set of G.

(d) *paired domination numbers* $\gamma_{\mathrm{pr}}(G)$ and $\Gamma_{\mathrm{pr}}(G)$, minimum and maximum cardinalities of a minimal PD-set of G. The parameters $\gamma_{\mathrm{pr}}(G)$ and $\Gamma_{\mathrm{pr}}(G)$ are referred to as the *paired domination number* and *upper paired domination number* of G, respectively. A PD-set of G of cardinality $\gamma_{\mathrm{pr}}(G)$ is called a γ_{pr}-*set* of G, while a minimal PD-set of cardinality $\Gamma_{\mathrm{pr}}(G)$ is called a Γ_{pr}-set of G.

(e) *connected domination numbers* $\gamma_c(G)$ and $\Gamma_c(G)$, minimum and maximum cardinalities of a minimal CD-set of G. The parameters $\gamma_c(G)$ and $\Gamma_c(G)$ are referred to as the *connected domination number* and *upper connected domination number* of G, respectively. A CD-set of G of cardinality $\gamma_c(G)$ is called a γ_c-*set* of G, while a minimal CD-set of cardinality $\Gamma_c(G)$ is called a Γ_c-set of G.

References

1. W. Ahrens, *Mathematische Unterhaltungen und Spiele* (Druck und Verlag von B. G. Teubner, Berlin, 1910), pp. 311–312
2. W.W.R. Ball, *Mathematical Recreations and Essays*, 4th edn. (Macmillan, London, 1905)
3. C. Berge, *The Theory of Graphs and its Applications* (Methuen, London, 1962)
4. M. Bezzel, Schachfreund Berliner Schachzeitung **3**, 363 (1848)
5. A. Brandstädt, V.B. Le, J.P. Spinrad, *Graph Classes: A Survey*. SIAM Monographs on Discrete Mathematics and Applications (SIAM, Philadelphia, 1999)
6. E.J. Cockayne, S.T. Hedetniemi, Towards a theory of domination in graphs. Networks **7**, 247–261 (1977)
7. C.F. de Jaenisch, *Applications de l'Analyse Mathematique au Jeu des Echecs* (Petrograd, Moscow, 1862)
8. H.E. Dudeney, *The Canterbury Puzzles and Other Curious Problems* (E. P. Dutton and Company, New York, 1908)
9. R. Gera, T.W. Haynes, S.T. Hedetniemi, M.A. Henning, An annotated glossary of graph theory parameters with conjectures, in *Graph Theory, Favorite Conjectures and Open Problems, Volume 2*, ed. by R. Gera, T.W. Haynes, S.T. Hedetniemi (Springer, Berlin, 2018), pp. 177–281
10. T.W. Haynes, S.T. Hedetniemi, P.J. Slater, *Fundamentals of Domination in Graphs* (Marcel Dekker, New York, 1998)
11. T.W. Haynes, S.T. Hedetniemi, P.J. Slater, eds., *Domination in Graphs, Advanced Topics* (Marcel Dekker, New York, 1998)
12. M.A. Henning, A. Yeo, *Total Domination in Graphs*. Springer Monographs in Mathematics (Springer, New York, 2013). ISBN: 978-1-4614-6524-9 (Print) 978-1-4614-6525-6 (eBook)
13. D. König, *Theorie der Endlichen und Unendlichen Graphen*. Akademische Verlagsgesellschaft M. B. H., Leipzig, 1936 (Chelsea, New York, 1950)
14. F. Nauck, Briefwechsel mit allen für alle. Illustrirte Zeitung **15**, 182 (1850)
15. O. Ore, *Theory of Graphs*. American Mathematical Society Colloquium Publications, vol. 38 (American Mathematical Society, Providence, 1962)
16. A.M. Yaglom, I.M. Yaglom, *Challenging Mathematical Problems with Elementary Solutions, Combinatorial Analysis and Probability Theory*, vol. I (Holden-Day, San Francisco, 1964)

Part I
Related Parameters

Broadcast Domination in Graphs

Michael A. Henning, Gary MacGillivray, and Feiran Yang

AMS Subject Classification: 05C65, 05C69

1 Introduction

The concept of broadcast domination was birthed by combining the concepts of distance and domination in graphs and applying them to modeling the problem of positioning broadcasting radio transmitters, where each transmitter may have a different effective radiated power. To formally define broadcast domination, we

The research of the author Michael A. Henning supported in part by the University of Johannesburg. The research of the author Gary MacGillivray supported by the Natural Sciences and Engineering Research Council of Canada.

M. A. Henning
Department of Mathematics and Applied Mathematics, University of Johannesburg, Johannesburg, South Africa
e-mail: mahenning@uj.ac.za

G. MacGillivray (✉)
Department of Mathematics and Statistics, University of Victoria, V8W 2Y2, P.O. Box 1700 STN CSC, Victoria, BC, Canada
e-mail: gmacgill@math.uvic.ca

F. Yang
Department of Mathematics and Applied Mathematics, University of Johannesburg, Auckland Park, 2006 South Africa

Department of Mathematics and Statistics, University of Victoria, V8W 2Y2, P.O. Box 1700 STN CSC, Victoria, BC, Canada
e-mail: fyang@uvic.ca

recall the fundamental concepts of distance and domination in graph theory. The *distance* between two vertices u and v in a graph G, denoted by $d_G(u, v)$, or simply $d(u, v)$ if the graph G is clear from context, is the length of a shortest (u, v)-path in G. The *eccentricity* $\mathrm{ecc}_G(v)$ of a vertex v in G is the maximum distance of a vertex from v in G. The maximum eccentricity among the vertices of G is the *diameter* of G, denoted by $\mathrm{diam}(G)$, while the minimum eccentricity among the vertices of G is the *radius* of G, denoted by $\mathrm{rad}(G)$. A *central vertex* of G is a vertex whose eccentricity equals $\mathrm{rad}(G)$. A tree is either *central* or *bicentral*, depending on whether it has one or two central vertices. A *diametrical path* in G is a shortest path whose length is equal to $\mathrm{diam}(G)$. We note that the two vertices at the end of a diametrical path have maximum eccentricity in G.

A *dominating set* in a graph G is a set S of vertices of G such that every vertex outside S is adjacent to at least one vertex in S. The *domination number* of G, denoted by $\gamma(G)$, is the minimum cardinality of a dominating set in G.

A *neighbor* of a vertex v in G is a vertex adjacent to v. The *open neighborhood* of a vertex v in G, denoted by $N_G(v)$, is the set of all neighbors of v in G, while the *closed neighborhood* of v is the set $N_G[v] = N_G(v) \cup \{v\}$. We denote the *degree* of a vertex v in G by $d_G(v) = |N_G(v)|$. The minimum and maximum degrees among all vertices of G are denoted by $\delta(G)$ and $\Delta(G)$, respectively.

For an integer $k \geq 1$, the *closed k-neighborhood* of v in G, denoted by $N_k[v;G]$, is the set of all vertices within distance k from v, that is, $N_k[v;G] = \{u : d_G(u, v) \leq k\}$. The *open k-neighborhood* of v, denoted by $N_k(v;G)$, is the set of all vertices different from v and at distance at most k from v in G, that is, $N_k(v;G) = N_k[v;G] \setminus \{v\}$.

If the graph G is clear from context, we omit the subscript G. For example, we simply write $N(v)$, $N[v]$, $N_k(v)$, and $N_k[v]$ rather than $N_G(v)$, $N_G[v]$, $N_k(v;G)$, and $N_k[v;G]$, respectively. When $k = 1$, the set $N_k[v] = N[v]$ and the set $N_k(v) = N(v)$. In what follows, for an integer $k \geq 1$, we use the standard notation $[k] = \{1, \ldots, k\}$ and $[k]_0 = [k] \cup \{0\} = \{0, 1, \ldots, k\}$.

For a graph $G = (V, E)$ with a vertex set V and an edge set E, a function $f : V \to \{0, 1, 2, \ldots, \mathrm{diam}(G)\}$ is called a *broadcast* on G. For each vertex v in G, the value $f(v)$ is called the *strength* (or the *weight*) of the broadcast from v. For each vertex $u \in V$, if there exists a vertex v in G (possibly, $u = v$) such that $f(v) > 0$ and $d(u, v) \leq f(v)$, then f is called a *dominating broadcast* on G. A vertex v with $f(v) > 0$ can be thought of as the site from which the broadcast is transmitted with strength $f(v)$, and such a vertex is called an *f-broadcast vertex* or simply a *broadcast vertex* if the function f is clear from context. The *ball of radius r around v* is defined as $N_r[v] = \{u \in V : d(u, v) \leq r\}$. Thus, the ball $N_{f(v)}[v]$ is the set of vertices that *hear* the broadcast from v. Vertices u with $f(u) = 0$ do not broadcast. For $X \subseteq V$, we define

$$f(X) = \sum_{v \in X} f(v).$$

The *cost* of the dominating broadcast f is the quantity $f(V)$, which is the sum of the strengths of the broadcasts over all vertices in G. The minimum cost of a dominating broadcast is the *broadcast domination number* of G, denoted by $\gamma_b(G)$.

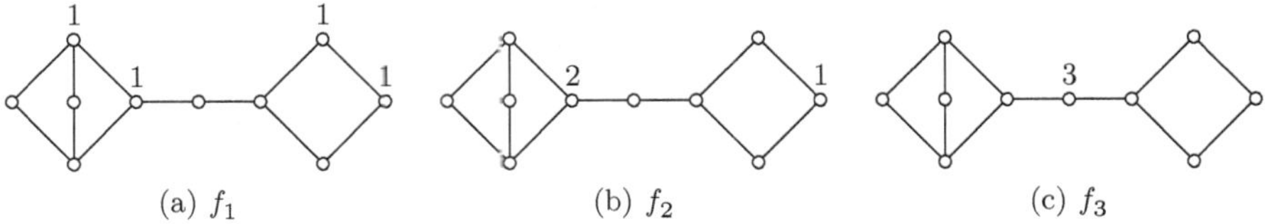

(a) f_1 (b) f_2 (c) f_3

Fig. 1 Three broadcast dominating functions of a graph G

An *optimal broadcast* is a broadcast with cost equal to $\gamma_b(G)$. For the graph G shown in Figure 1, three broadcast dominating functions are illustrated in Figure 1(a), 1(b) and 1(c). The cost of f_1, f_2, and f_3 is 4, 3, and 3, respectively. For this graph G, we have $\gamma_b(G) = 3$ and both f_2 and f_3 are optimal broadcasts.

Broadcast domination in graphs was first introduced and studied in 2001 by Erwin [21, 22]. Erwin observed that if a dominating broadcast f satisfies $f(v) \in \{0, 1\}$ for all $v \in V$, then f is the characteristic function of a dominating set and hence has cost at most $\gamma(G)$. Furthermore, he observed that a broadcast $f : V \rightarrow \{0, 1, \ldots, \mathrm{diam}(G)\}$ that assigns the strength $\mathrm{rad}(G)$ to a central vertex of a connected graph G and the strength 0 to all remaining vertices of G has cost $f(V) = \mathrm{rad}(G)$. If $G = K_1$, then $\gamma_b(G) = 1 = \gamma(G)$, while $\mathrm{rad}(G) = 0$. Hence, we assume that $G \neq K_1$ and therefore has order at least 2. Thus, the broadcast domination number of a graph G is at most its domination number and at most its radius. We state this formally as follows.

Observation 1. ([21, 22]) *If G is a connected graph of order at least 2, then*

$$\gamma_b(G) \leq \min\{\gamma(G), \mathrm{rad}(G)\}.$$

Graphs for which the broadcast domination number is equal to the radius are called *radial*. In view of Observation 1, we can replace $\mathrm{diam}(G)$ by $\mathrm{rad}(G)$ in the definition of a dominating broadcast in a graph G. Erwin [21, 22] showed that if the domination number or the radius of a graph is at most 3, then the broadcast domination number is determined.

Proposition 2. ([21, 22]) *If G is a connected graph of order at least 2 and $k = \min\{\gamma(G), \mathrm{rad}(G)\}$ where $k \in [3]$, then $\gamma_b(G) = k$.*

In 2006, Dunbar, Erwin, Haynes, Hedetniemi, and Hedetniemi [20] defined a key concept called efficient broadcast. A dominating broadcast is *efficient* if no vertex hears a broadcast from two different vertices. If f is not an efficient dominating broadcast in a graph $G = (V, E)$, then there exists a vertex v such that $d(v, x) \leq f(x)$ and $d(v, y) \leq f(y)$, where x and y are broadcasting vertices in G. In this case, we can reassign the value 0 to both x and y, assign the value $f(w) + f(x) + f(y)$ to a vertex w that is within distance $f(y)$ from x and also within distance $f(x)$ from y, and leave the value of all other vertices unchanged under f. The cost of the new broadcast is equal to the cost of the original broadcast. This process can be repeated until an efficient broadcast is found. This yields the following result.

Theorem 3. ([20]) *Every graph G has an optimal dominating broadcast that is efficient.*

As first observed by Herke [31], the broadcast domination number of a connected graph is equal to the minimum broadcast domination number among its spanning trees.

Observation 4. ([31]) *If G is a connected graph, then*

$$\gamma_b(G) = \min\{\gamma_b(T) \mid T \text{ is a spanning tree of } G\}.$$

2　The Dual of Broadcast Domination

Graph theoretic minimization (respectively, maximization) problems expressed as linear programming problems have dual maximization (respectively, minimization) problems. Much of the early work on linear programming duality problems for domination type parameters is done by Slater. A survey of these results can be found in the 1998 survey paper of Slater [44]. The dual concept of coverings and packings is also well studied in graph theory. For a survey on the combinatorics underlying set packing and set covering problems, we refer the reader to the 2001 monograph by Cornuéjols [17].

In this section, we discuss the dual (in the sense of linear programming) of broadcast domination, namely *multipacking*. The term multipacking was first introduced in the Master's thesis of Teshima [47] in 2012. Here, broadcast domination was considered as a linear programming problem, and the linear programming dual was used to give the definition of a multipacking. A *multipacking* is a set $S \subseteq V$ in a graph $G = (V, E)$ such that for every vertex $v \in V$ and for every integer $r \geq 1$, the ball of radius r around v contains at most r vertices of S, that is, there are at most r vertices in S at distance at most r from v in G. We note that in this definition of a multipacking, we may restrict our attention to $r \in [\mathrm{diam}(G)]$. By our earlier observations, we can in fact restrict the integer r to belong to the set $[\mathrm{rad}(G)]$. The *multipacking number* of G is the maximum cardinality of a multipacking of G and is denoted by $\mathrm{mp}(G)$. We define next the multipacking number in terms of the dual of the linear programming problem for broadcast domination.

Let $G = (V, E)$ be a graph with $V = \{v_1, v_2, \ldots, v_n\}$. The definition of $\gamma_b(G)$ leads to a 0–1 integer program, which we now describe. For each vertex v_i and integer $k \in [\mathrm{rad}(G)]$, let x_{ik} be an indicator variable giving the truth value of the statement "the strength of the broadcast f at vertex v_i equals k," that is,

$$x_{ik} = \begin{cases} 1 \text{ if } f(v_i) = k \\ 0 \text{ otherwise.} \end{cases}$$

The formulation of the primal integer program for broadcast domination is given by

Broadcast Domination γ_b

Minimize $\displaystyle\sum_{k=1}^{\mathrm{rad}(G)} \sum_{i=1}^{n} k x_{ik},$

    subject to

$$\sum_{d(v_i,v_j)\leq k} x_{ik} \geq 1 \text{ for all vertices } v_i \text{ and } v_j,$$

    $x_{ik} \in \{0, 1\}$ for each vertex v_i and integer $k \in [\mathrm{rad}(G)]$.

Multipacking Number mp(G)

Maximize $\displaystyle\sum_{k=1}^{n} y_j,$

    subject to

$$\sum_{d(v_i,v_j)\leq k} y_j \leq k \text{ for all vertices } v_i \text{ and } v_j \text{ and integer } k \in [\mathrm{rad}(G)],$$

    $y_k \in \{0, 1\}$ for each $k \in [n]$.

Since broadcast domination and multipacking are dual problems, we have the following observation.

Observation 5. *For every graph G, we have* $\mathrm{mp}(G) \leq \gamma_b(G)$.

The graph G shown in Figure 2 satisfies $\mathrm{mp}(G) = 3$, where the darkened vertices form a multipacking of maximum cardinality in G. As observed earlier, $\gamma_b(G) = 3$, and so for this example, we have $\mathrm{mp}(G) = \gamma_b(G)$.

In 2014, Hartnell and Mynhardt [26] provided the following lower bound on the multipacking number of a graph.

Theorem 6. ([26]) *If G is a connected graph, then* $\mathrm{mp}(G) \geq \left\lceil \frac{1}{3}(\mathrm{diam}(G) + 1) \right\rceil$.

Proof.. Let $P: v_0, v_1, \ldots, v_d$ be a diametrical path of G, where $d = \mathrm{diam}(G)$. Let $V_i = \{v \in V : d(v, v_0) = i\}$ for all $i \in [d]$, and let $M = \{v_i : i \equiv 0 \ (\mathrm{mod}\ 3)\}$. We note that $|M| = \lceil \frac{1}{3}(d + 1) \rceil$. By our choice of the set M, every vertex $v \in V(P)$ satisfies $|N_r[v] \cap M| \leq r$ for all integers $r \geq 1$. We now consider an arbitrary vertex $w \in V$. We note that $w \in V_j$ for some $j \equiv [d]_0$. Since $v_j \in V_j$ and $M \subseteq V(P)$, we note that $N_r[w] \cap M \subseteq N_r[v_j] \cap M$, implying that $|N_r[w] \cap M| \leq r$ for all integers $r \geq 1$. Since $w \in V$ is arbitrary, this implies that the set M is a multipacking in G. Thus, $\mathrm{mp}(G) \geq |M| = \lceil \frac{1}{3}(d + 1) \rceil = \lceil \frac{1}{3}(\mathrm{diam}(G) + 1) \rceil$. $\qquad\square$

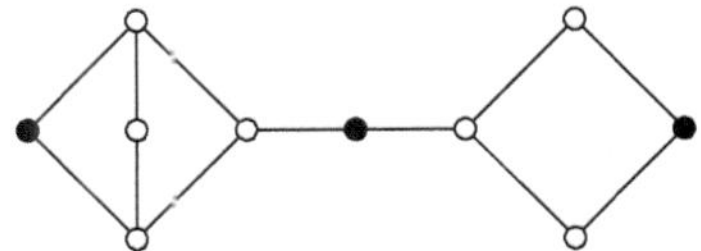

Fig. 2 A graph G with $\mathrm{mp}(G) = 3$

As an immediate consequence of Observation 5 and Theorem 6, we have the following lower bound on the broadcast domination number first observed by Erwin [21, 22].

Corollary 7. ([21, 22]) *If G is a connected graph, then $\gamma_b(G) \geq \left\lceil \frac{1}{3}(\mathrm{diam}(G) + 1) \right\rceil$.*

We note that if G is a path P_n on $n \geq 2$ vertices, then $\gamma(G) = \lceil \frac{1}{3}n \rceil = \lceil \frac{1}{3}(\mathrm{diam}(G) + 1) \rceil$. Hence, by Observations 1 and 5 and Theorem 6, we have that the lower bound of Theorem 6 is tight. Furthermore, we have the following result.

Proposition 8. *For every integer $n \geq 2$,*

$$\mathrm{mp}(P_n) = \gamma_b(P_n) = \gamma(P_n) = \left\lceil \frac{n}{3} \right\rceil.$$

By Observation 4, for $n \geq 3$, we have $\gamma_b(C_n) = \gamma_b(P_n)$, and so by Proposition 8, $\gamma_b(C_n) = \lceil \frac{n}{3} \rceil$. However, $\mathrm{mp}(C_n) = \lfloor \frac{n}{3} \rfloor$ for all $n \geq 3$. Thus, for cycles, we have the following result.

Proposition 9. ([47]) *For every integer $n \geq 3$, $\mathrm{mp}(C_n) = \gamma_b(C_n)$ if and only if $n \equiv 0$ (mod 3). For n (mod 3) $\in \{1, 2\}$, we have $\mathrm{mp}(C_n) = \gamma_b(C_n) - 1$.*

By Theorem 6, if G is a connected graph, then $3\mathrm{mp}(G) \geq \mathrm{diam}(G) + 1$. By definition, $\mathrm{diam}(G) \geq \mathrm{rad}(G)$. By Observation 1, $\mathrm{rad}(G) \geq \gamma_b(G)$. Hence, $3\mathrm{mp}(G) \geq \mathrm{diam}(G) + 1 \geq \mathrm{rad}(G) + 1 \geq \gamma_b(G) + 1$, or, equivalently, $\gamma_b(G) \leq 3\mathrm{mp}(G) - 1$. Hence, as a consequence of our earlier results, we have the following upper bound on the broadcast domination number in terms of its multipacking number.

Corollary 10. ([26]) *If G is a connected graph, then $\gamma_b(G) \leq 3\mathrm{mp}(G) - 1$.*

If the multipacking number of a graph G is at least 2, then Hartnell and Mynhardt [26] improved the upper bound in Corollary 10 slightly.

Theorem 11. ([26]) *If G is a connected graph with $\mathrm{mp}(G) \geq 2$, then $\gamma_b(G) \leq 3\mathrm{mp}(G) - 2$.*

As a consequence of Corollary 10, we have the following upper bound on the ratio $\gamma_b(G)/\mathrm{mp}(G)$.

Corollary 12. ([26]) *If G is a connected graph, then $\dfrac{\gamma_b(G)}{\mathrm{mp}(G)} < 3$.*

In 2012, Teshima [47] proved that the graph G shown in Figure 3 satisfies $\gamma_b(G) = 4$ and $\mathrm{mp}(G) = 2$. Assigning a strength 2 to each of the two vertices of degree 2 in G as illustrated in Figure 3, and a strength of 0 to the remaining vertices of G produces an optimal broadcast of G. An example of a multipacking of maximum cardinality in G is given by the set of two darkened vertices of G illustrated in Figure 3. This example serves to show the existence of a graph G for which the ratio $\gamma_b(G)/\mathrm{mp}(G) = 2$.

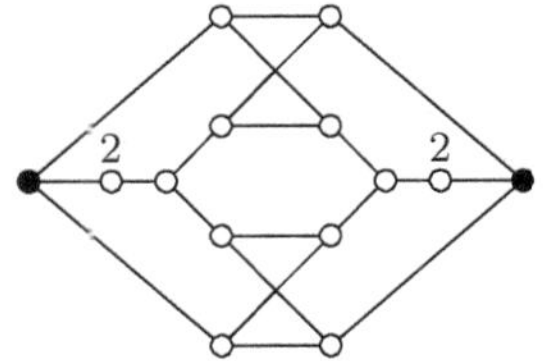

Fig. 3 A graph G with $\gamma_b(G) = 4$ and $\mathrm{mp}(G) = 2$

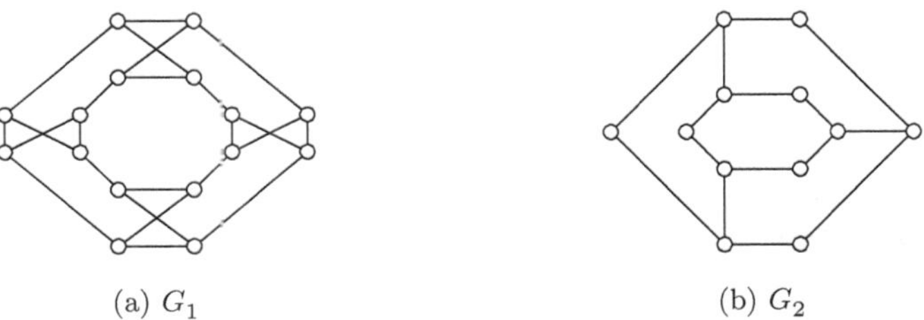

(a) G_1 (b) G_2

Fig. 4 Two graphs satisfying $\gamma_b(G) = 4$ and $\mathrm{mp}(G) = 2$

To date, no graph G has been found satisfying $\gamma_b(G)/\mathrm{mp}(G) > 2$. Beaudou, Brewster, and Foucaud [4] posed the following conjecture.

Conjecture 1. ([4]) *If G is a connected graph, then $\gamma_b(G) \le 2\mathrm{mp}(G)$.*

There are a few known examples of connected graphs G which achieve the conjectured bound, that is, $\gamma_b(G) = 2\mathrm{mp}(G)$. For example, if G is a cycle C_4 or C_5, then $\gamma_b(G) = 2$ and $\mathrm{mp}(G) = 1$, and so $\gamma_b(G) = 2\mathrm{mp}(G)$. As observed earlier, if G is the graph shown in Figure 3, then $\gamma_b(G) = 4$ and $\mathrm{mp}(G) = 2$, and so $\gamma_b(G) = 2\mathrm{mp}(G)$. Two additional examples of graphs G with $\gamma_b(G) = 4$ and $\mathrm{mp}(G) = 2$ are the graphs $G = G_1$ and $G = G_2$ shown in Figure 4(a) and 4(b), respectively. Graph G_1 is attributed to C. R. Dougherty in [4, Figure 3(c)] as private communication, while graph G_2 is given in [4].

In 2014, Hartnell and Mynhardt [26] gave a construction of a graph G_k such that $\gamma_b(G_k) - \mathrm{mp}(G_k) = k$ for any given integer $k \ge 1$, showing that the difference $\gamma_b - \mathrm{mp}$ can be arbitrarily large. In order to explain their construction, let H be the graph obtained from three vertex-disjoint copies F_1, F_2, and F_3 of $K_{2,4}$ as follows. Let u_i be a vertex of degree 2 in F_i for $i \in [2]$, and let v_1 and v_2 be two vertices of degree 2 in F_3. Let H be obtained from the disjoint union of F_1, F_2, and F_3 by joining v_i to u_i for $i \in [2]$. Let x be a vertex of degree 2 in F_1 different from u_1, and let y be a vertex of degree 2 in F_2 different from u_2. The graph H is illustrated in Figure 5.

Let M be a multipacking of maximum cardinality in H. Each induced subgraph F_i of H contains at most one vertex of M, implying that $\mathrm{mp}(H) = |M| \le 3$. By Theorem 6, $\mathrm{mp}(H) \ge \left\lceil \frac{1}{3}(\mathrm{diam}(H) + 1) \right\rceil = \left\lceil \frac{1}{3}(8 + 1) \right\rceil = 3$. Consequently, $\mathrm{mp}(H) = 3$. An example of a multipacking of maximum cardinality in H is given by the set

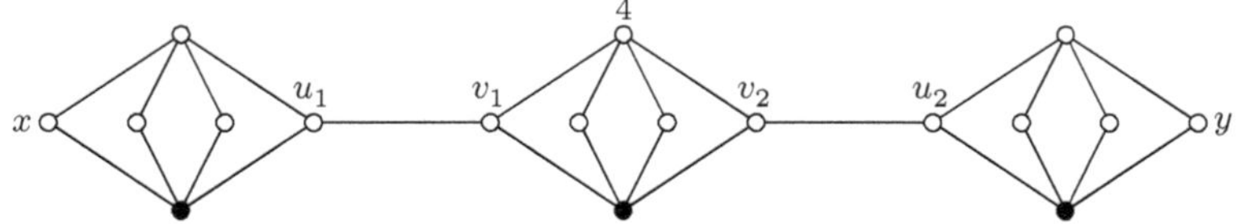

Fig. 5 A graph H with $\gamma_b(H) = 4$ and $\mathrm{mp}(H) = 3$

of three darkened vertices of H illustrated in Figure 5. By Observation 5, we have $\gamma_b(H) \geq \mathrm{mp}(H) = 3$. If $\gamma_b(H) = 3$, then since $\mathrm{rad}(H) = 4$, every optimal broadcast in H must contain at least two broadcast vertices (of positive strength), one of which therefore has strength 1 and the other strength 2. But then at least one of the vertices x and y hears no broadcast, a contradiction. Hence, $\gamma_b(H) \geq 4$. Since $\mathrm{rad}(H) = 4$ and $\gamma(H) = 6$, by Observation 1, we have $\gamma_b(H) \leq 4$. Consequently, $\gamma_b(H) = 4$.

We now return to the general construction given by Hartnell and Mynhardt [26]. For $k = 1$, let $G_k = H$. For $k \geq 2$, let $H_1, H_2, \ldots, H_k$ be k vertex-disjoint copies of the graph H, where x_i and y_i are the vertices in H_i named x and y in H. Let G_k be constructed from the disjoint union of the graphs $H_1, H_2, \ldots, H_k$ by adding the edges $y_i x_{i+1}$ for $i \in [k-1]$. As shown in [26], $\gamma_b(G_k) = 4k$ and $\mathrm{mp}(G_k) = 3k$. This yields the following result.

Theorem 13. ([26]) *For every integer $k \geq 1$, there exists a connected graph G_k such that $\gamma_b(G_k) = 4k$ and $\mathrm{mp}(G_k) = 3k$. Thus, the following hold in the graph G_k.*

(a) $\gamma_b(G_k) - \mathrm{mp}(G_k) = k$.
(b) $\gamma_b(G_k)/\mathrm{mp}(G_k) = \frac{4}{3}$.

Recall that in Theorem 11, if G is a connected graph with $\mathrm{mp}(G) \geq 2$, then $\gamma_b(G) \leq 3\mathrm{mp}(G) - 2$. Hartnell and Mynhardt [26] asked whether the factor 3 in this bound can be improved. In 2019, Beaudou, Brewster, and Foucaud [4] answered their question in the affirmative, resulting in a significant improvement of this upper bound on the broadcast domination number in terms of its multipacking number.

Theorem 14. ([4]) *If G is a connected graph, then $\gamma_b(G) \leq 2\mathrm{mp}(G) + 3$.*

Hartnell and Mynhardt [26] were the first to observe that Conjecture 1 is true when $\mathrm{mp}(G) \leq 2$. The conjecture is shown in [4] to hold for all graphs with multipacking number at most 4.

Theorem 15. ([4]) *If G is a connected graph and $\mathrm{mp}(G) \leq 4$, then $\gamma_b(G) \leq 2\mathrm{mp}(G)$.*

By Observation 5, for every graph G, we have $\mathrm{mp}(G) \leq \gamma_b(G)$. In 2017, Mynhardt and Teshima [47] proved that equality holds here for the class of trees, thereby extending a classic result due to Meir and Moon [37] that the domination number equals the 2-packing number for trees.

Theorem 16. ([47]) *For every tree T, we have $\gamma_b(T) = \mathrm{mp}(T)$.*

For any integer programming problem, a natural variation of the problem can be obtained by considering the LP relaxation. Since broadcast domination

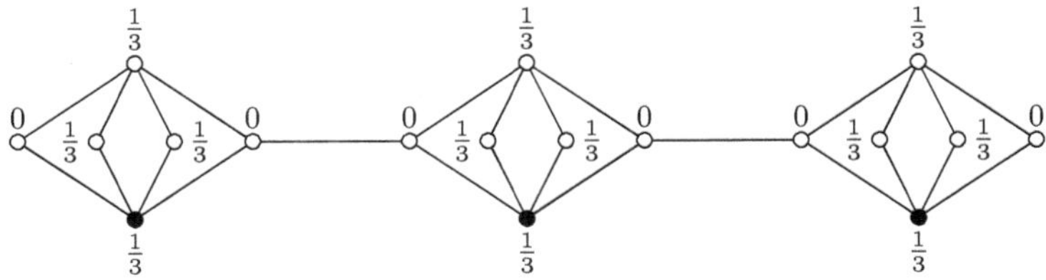

Fig. 6 A graph H with $\mathrm{mp}_f(H) = 4$ and $\mathrm{mp}(H) = 3$

and multipacking can be regarded as integer programming problems, Brewster, Mynhardt, and Teshima [11] used this idea to study fractional broadcast domination and fractional multipacking. Here, the broadcast strength of a vertex can be a fraction, and a vertex can be considered to be fractionally in a multipacking. For example, we can assign 1/3 strength to all vertices in C_4, for a total cost of 4/3, resulting in a fractional dominating broadcast where each vertex hears a total strength at least one. On the other hand, we can pack 1/3 for each vertex in C_4 and it will give a multipacking of size 4/3. We denote the fractional broadcast domination number as $\gamma_{b,f}(G)$ and the fractional multipacking number as $\mathrm{mp}_f(G)$. The duality theorem of linear programming yields the result below.

Theorem 17. ([11]) *If G is a connected graph, then*

$$\mathrm{mp}(G) \leq \mathrm{mp}_f(G) = \gamma_{b,f}(G) \leq \gamma_b(G).$$

The difference $\mathrm{mp}_f(G) - \mathrm{mp}(G)$ can be arbitrarily large. The graph H shown in Figure 5 has fractional multipacking number at least 4 since we can pack 1/3 on the degree 2 and 4 vertices with the exception of x and y, which are not packed. The resulting fractional multipacking is shown in Figure 6. Thus, $\mathrm{mp}_f(H) \geq 4$. As observed earlier, $\gamma_b(G) = 4$, implying by Theorem 17 that $\mathrm{mp}_f(H) \leq 4$. Consequently, $\mathrm{mp}_f(H) = 4$.

Using the previous construction G_k given by Hartnell and Mynhardt [26], we can readily deduce the following result.

Theorem 18. *For every integer $k \geq 1$, there exists a connected graph G_k such that* $\mathrm{mp}_f(G_k) = 4k$ *and* $\mathrm{mp}(G_k) = 3k$.

3 Broadcast Domination in Trees

Broadcasts in trees have a special structure, which was exploited in the thesis by Herke [31] in 2007 and in the papers by Herke and Mynhardt [32] in 2009 and Cockayne, Herke, and Mynhardt [16] in 2011. In order to determine the broadcast domination number of a tree, the above authors introduced the concept of a *shadow tree* of a tree, defined as follows.

Suppose $P: v_0 v_1 v_2 \ldots v_d$ is a diametrical path in a tree T. The shadow tree is constructed in two steps. First, consider the forest $F = T - E(P)$ obtained from T by deleting all edges on the path P. For each vertex v_k of P, let Q_k be a longest path in F emanating from v_k. Let Q_k start at v_k and end at the vertex b_k (possibly, $v_k = b_k$). We note, for example, that Q_0 is the trivial path consisting of the vertex $v_0 = b_0$, and Q_d is the trivial path consisting of the vertex $v_d = b_k$. For example, consider the tree T in Figure 7, where the vertices of the diametrical path $P: v_0 v_1 v_2 \ldots v_d$ and the vertices b_1, b_4, and b_5 are labeled as shown. We note that in this example, $v_i = b_i$ for $i \in \{0, 2, 3, 6, 7\}$.

In the first step of the construction of a shadow tree, we reduce the tree T to the subtree, T_{reduced}, of T induced by the vertices belonging to the set

$$V(P) \cup \left(\bigcup_{k=1}^{d-1} V(Q_k) \right).$$

For the tree T in Figure 7, the resulting reduced tree T_{reduced} is shown in Figure 8.

In the second step of the construction of a shadow tree, if $d(v_k, b_k) \geq d(v_k, b_i)$ for some $k \in [d]$ and $i \in [d] \setminus \{k\}$, then we remove the vertices in $V(Q_i) \setminus \{v_i\}$ from the tree T_{reduced}. We repeat this process until no such indices k and i exist. The resulting tree is a *shadow tree* of T, denoted by T_{shadow}. The *shadow* of vertex b_k is the set of vertices $\{v \in V(T_{\text{shadow}}) : d(v_k, b_k) \geq d(v_k, v)\}$. In our example, in the

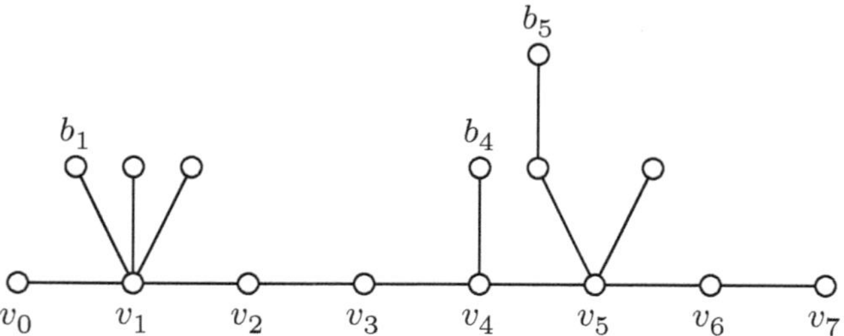

Fig. 7　A tree T

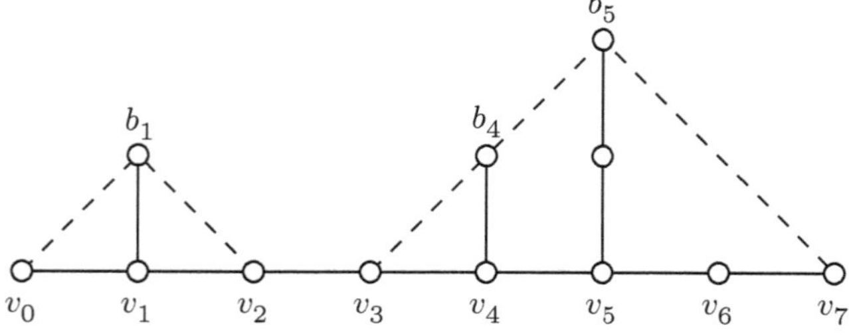

Fig. 8　A reduced tree T_{reduced} of T

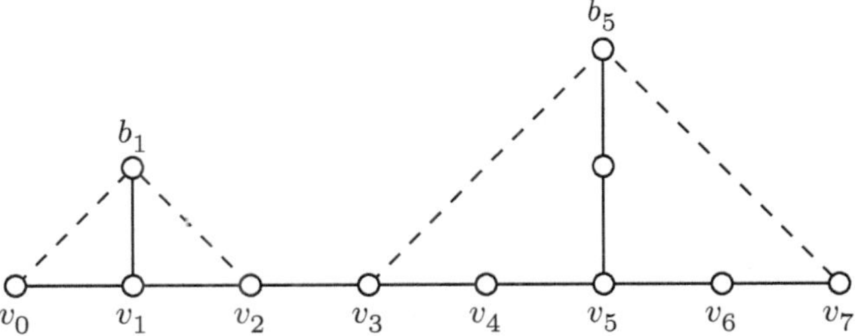

Fig. 9 A shadow tree T_{shadow} of T

reduced tree T_{reduced} shown in Figure 8, we have $d(v_5, b_5) \geq d(v_5, b_4)$. According to the second step of our construction, we remove the vertex b_4 from the tree T_{reduced}. The resulting shadow tree T_{shadow} of T is shown in Figure 9.

Herke and Mynhardt [32] showed that the broadcast domination number of a tree equals the broadcast domination number of its shadow tree.

Theorem 19. ([32]) *For a tree T and its shadow tree S_T, $\gamma_b(T) = \gamma_b(S_T)$.*

By Theorem 19, it therefore suffices for us to consider only the shadow tree S_T of a tree T to determine the broadcast domination number of T.

Herke and Mynhardt [32] also introduced the important definitions of split-sets and split-edges. Let T be a tree with diametrical path P. A *split-set* is a set of edges on P whose removal splits T into components such that for each component T_i has even positive diameter and $T_i \cap P$ is a diametrical path of T_i. A *split-edge* is an edge that is contained in some split-set. For example, in Figure 9, $v_2 v_3$ is a split-edge. On the other hand, the edge $v_3 v_4$ is not a split edge since its removal creates a subtree with diametrical path from b_5 to v_7. In general, all the edges that have two ends in some shadow (visually in Figure 9, the only edge that is not in some shadow is $v_2 v_3$) cannot be a split-edge. Herke and Mynhardt [32] showed that the broadcast domination number is a function of the largest size of a split-set.

Theorem 20. ([32]) *If M is split-set with maximum cardinality m of a tree T, then*

$$\gamma_b(T) = \left\lceil \frac{1}{2}(\text{diam}(T) - m) \right\rceil.$$

Recall that by Observation 1, if G is a connected graph of order at least 2, then $\gamma_b(G) \leq \text{rad}(G)$. Graphs G satisfying $\gamma_b(G) = \text{rad}(G)$ are called *radial graphs*, which form an important class of graphs with respect to broadcast domination. Several characterizations of radial graphs are given in the literature. A characterization of radial trees is given by Herke and Mynhardt [32].

Theorem 21. ([32]) *A tree T is radial if and only if it has no non-empty split-set.*

As an application of shadow graphs, Herke and Mynhardt [32] gave an upper bound for the broadcast domination number of a tree in terms of its order.

Theorem 22. ([32]) *If T is a tree of order n, then $\gamma_b(T) \le \left\lceil \frac{n}{3} \right\rceil$.*

Proof.. We proceed by induction on the order $n \ge 1$. The result is immediate for $n \in [3]$. This establishes the base case. Let $n \ge 4$ and assume that every tree T' of order $n' < n$ satisfies $\gamma_b(T) \le \lceil \frac{1}{3}n' \rceil$. Let T be a tree of order n. Suppose that T has two adjacent vertices u_1 and u_2 of degree 2. Let v_i be the neighbor of u_i different from u_{3-i} for $i \in [2]$. Thus, $Q: v_1 u_1 u_2 v_2$ is a path in T. There exists an edge $e \in E(Q)$ such that $T - e$ has two components T_1 and T_2 of orders n_1 and n_2, respectively, where $n_1 \equiv 0 \pmod 3$. Thus, $n_1 = 3t$ for some $t \ge 1$. Applying the inductive hypothesis to T_1 and T_2, we have $\gamma_b(T_1) \le t$ and $\gamma_b(T_2) \le \lceil \frac{1}{3}n_2 \rceil = \lceil \frac{1}{3}(n - 3t) \rceil$. Hence,

$$\gamma_b(T) \le \gamma_b(T_1) + \gamma_b(T_2) \le t + \left\lceil \frac{1}{3}(n - 3t) \right\rceil = \left\lceil \frac{1}{3}n \right\rceil.$$

Hence, we may assume that T does not have two adjacent vertices of degree 2, otherwise the desired result follows. Let $\mathrm{rad}(T) = k$, and let $P: v_1 v_2 \ldots v_d$ be a diametrical path in a tree T, and so $\mathrm{diam}(T) = d - 1$. By Observation 1, $\gamma_b(T) \le \mathrm{rad}(T) = k$.

Recall that a tree is central if it contains exactly one central vertex (whose eccentricity equals the radius of the tree), while a tree is bicentral if it has two central vertices. Suppose, firstly, that T is central. In this case, $d = 2k + 1$. By our assumption that T has no adjacent vertices of degree 2, the pigeonhole principle shows that at least $\lceil \frac{1}{2}(2k - 1) \rceil = k - 1$ of the vertices in $V(P) \setminus \{v_1, v_d\}$ are adjacent to vertices not on P, implying that $n \ge (2k + 1) + (k - 1) = 3k$. Therefore, in this case, $\gamma_b(T) \le k \le \frac{1}{3}n$. Suppose, secondly, that T is bicentral. In this case, $d = 2k$. Analogously, as before, at least $\lceil \frac{1}{2}(2k - 2) \rceil = k - 1$ of the vertices in $V(P) \setminus \{v_1, v_d\}$ are adjacent to vertices not on P, implying that $n \ge 2k + (k - 1) = 3k - 1$. Therefore, in this case, $\gamma_b(T) \le k \le \lceil \frac{1}{3}n \rceil$. This completes the proof by induction. $\square$

By Observation 4, Theorem 22 gives an upper bound on the broadcast domination number of a graph.

Corollary 23. ([32]) *If G is a connected graph of order n, then $\gamma_b(G) \le \left\lceil \frac{n}{3} \right\rceil$.*

This bound of Corollary 23 is tight for paths and cycles.

4 Broadcast Domination in Graph Products

In this section, we present selected results on broadcast domination in graph products. By a *graph product* $G \otimes H$ on graphs G and H, we mean the graph that has vertex set the Cartesian product of the vertex sets of G and H, that is,

$$V(G \otimes H) = V(G) \times V(H) = \{(g, h) \mid g \in V(G) \text{ and } h \in V(H)\},$$

and has an edge set that is determined entirely by the adjacency relations of G and H. Exactly how it is determined depends on what kind of graph product we are considering. In this section, we consider four such graph products, namely the Cartesian product ($\square$), the direct product ($\times$), the strong product ($\boxtimes$) of graphs, and the lexicographic product ($\bullet$).

Two vertices (g_1, h_1) and (g_2, h_2) in the *Cartesian product* $G \square H$ of graphs G and H are adjacent if either $g_1 = g_2$ and $h_1 h_2$ is an edge in H or $h_1 = h_2$ and $g_1 g_2$ is an edge in G.

Two vertices (g_1, h_1) and (g_2, h_2) in the *direct product graph* $G \times H$ of graphs G and H are adjacent if $g_1 g_2 \in E(G)$ and $h_1 h_2 \in E(H)$.

Two vertices (g_1, h_1) and (g_2, h_2) in the *strong direct product* $G \boxtimes H$ of G and H are adjacent if and only if $u = v$ and $u'v' \in E(H)$ or $u' = v'$ and $uv \in E(G)$ or $uv \in E(G)$ and $u'v' \in E(H)$.

Two vertices (g_1, h_1) and (g_2, h_2) in the *lexicographic product* $G \bullet H$ of G and H are adjacent if and only if either $g_1 g_2 \in E(G)$ or $g_1 = g_2$ and $h_1 h_2 \in E(H)$.

In 2009, Braser and Spacaman [10] studied broadcast domination in the Cartesian product, the direct product, and the strong product of graphs and established the following upper bounds on the broadcast domination number in the respective product graphs.

Theorem 24. ([10]) *For all graphs G and H,*

$$\gamma_b(G \square H) \le \frac{3}{2}(\gamma_b(G) + \gamma_b(H)).$$

Theorem 25. ([10]) *For all graphs G and H,*

$$\gamma_b(G \boxtimes H) \le \frac{3}{2}\max\{\gamma_b(G), \gamma_b(H)\}.$$

Theorem 26. ([10]) *For all graphs G and H,*

$$\gamma_b(G \times H) \le \begin{cases} 3\max\{\gamma_b(G), \gamma_b(H)\} & \text{if } \mathrm{rad}(G) \ne \mathrm{rad}(H) \\ 3\min\{\gamma_b(G), \gamma_b(H)\} + 1 & \text{otherwise.} \end{cases}$$

Dunbar, Erwin, Haynes, Hedetniemi, and Hedetniemi [20] presented results on broadcast domination in $m \times n$ grid graphs $G_{n,m}$, or equivalently in the Cartesian product $P_m \square P_n$ of paths P_m and P_n. They showed that it suffices to have one broadcast vertex in the center of the grid, broadcasting with strength $\mathrm{rad}(G_{n,m}) = \lfloor \frac{m}{2} \rfloor + \lfloor \frac{n}{2} \rfloor$. This is illustrated in Figure 10(a) in the case of a 4×4 grid.

Theorem 27. ([20]) *For integers $m \ge 1$ and $n \ge 1$,*

$$\gamma_b(P_m \square P_n) = \mathrm{rad}(P_m \square P_n) = \left\lfloor \frac{m}{2} \right\rfloor + \left\lfloor \frac{n}{2} \right\rfloor.$$

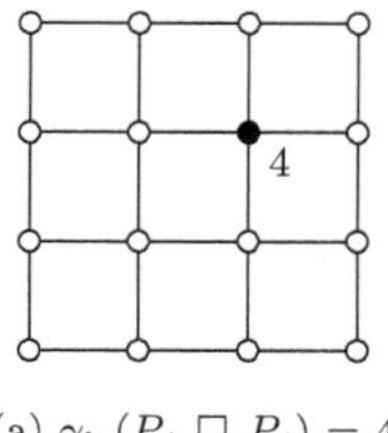

(a) $\gamma_b (P_4 \,\square\, P_4) = 4$

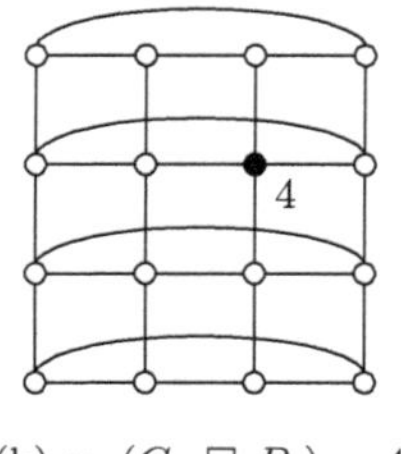

(b) $\gamma_b (C_4 \,\square\, P_4) = 4$

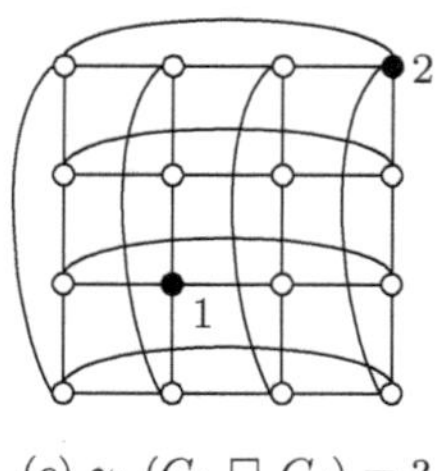

(c) $\gamma_b (C_4 \,\square\, C_4) = 3$

Fig. 10 Broadcast domination in $P_4 \square P_4$, $C_4 \square P_4$, and $C_4 \square C_4$

We remark that this result was also established by Braser and Spacaman [10].

In 2019, Beaudou and Brewster [3] later extended the result in $m \times n$ grids to multipacking.

Theorem 28. ([3]) *For $m \geq 4$ and $n \geq 4$,*

$$\mathrm{mp}(P_n \square P_m) = \gamma_b(P_n \square P_m),$$

with the exception of $P_4 \square P_6$, where $\mathrm{mp}(P_4 \square P_6) = 4$ *and* $\gamma_b(P_4 \square P_6) = 5$.

In 2015, Koh and Soh [33] determined the broadcast domination number of the Cartesian product of a cycle and a path. We illustrate this in Figure 10(b) in the case of the Cartesian product $C_4 \square P_4$.

Theorem 29. ([33]) *For integers $m \geq 3$ and $n \geq 2$,*

$$\gamma_b(C_m \square P_n) = \begin{cases} \frac{m}{2} & \text{if } n = 2 \text{ and } m \text{ is even,} \\ \lfloor \frac{m}{2} \rfloor + \lfloor \frac{n}{2} \rfloor & \text{otherwise.} \end{cases}$$

Braser and Spacaman [10] gave results on broadcast domination in the Cartesian products $C_m \square C_n$ of cycles C_m and C_n, also called the *torus* in the literature.

Theorem 30. ([10]) *For $m \geq 3$ and $n \geq 3$,*

$$\gamma_b(C_m \square C_n) = \begin{cases} \mathrm{rad}(C_m \square C_n) - 1 & \text{if both } m \text{ and } n \text{ are even,} \\ \mathrm{rad}(C_m \square C_n) & \text{otherwise.} \end{cases}$$

Using a differen approach to that used in [10], Soh and Koh [46] determined the broadcast domination number of the torus $C_m \square C_n$. We illustrate this in Figure 10(c) in the case of the torus $C_4 \square C_4$. We remark that the broadcast domination number of the grid $P_m \square P_n$ and the torus $C_m \square C_n$ is the same, except when both m and n are even.

Theorem 31. ([46]) *For $m \geq 3$ and $n \geq 3$, we have* $\gamma_b(C_m \square C_n) = \lceil \frac{m+n}{2} \rceil - 1$.

In 2014, Soh and Koh [45] studied broadcast domination in the strong product, the direct product, and the lexicographic product of two paths.

Theorem 32. ([45]) *For integers* $m \geq n \geq 1$,

$$\gamma_b(P_m \boxtimes P_n) = \left\lceil \frac{1}{2}\left(m - \left\lfloor \frac{m}{\max\{p,3\}} \right\rfloor\right)\right\rceil,$$

where $p = 2\left\lceil \frac{n-1}{2} \right\rceil + 1$.

The broadcast domination number of the strong product $P_4 \boxtimes P_4$ equals 2 and is illustrated in Figure 11(a).

Theorem 33. ([45]) *For integers* $m \geq n \geq 1$,

$$\gamma_b(P_m \times P_n) = \begin{cases} m & \text{if } n = 1, \\ n \cdot \left(\frac{m+1}{n+1}\right) & \text{if } n \geq 2, \text{ both } m \text{ and } n \text{ are odd with } \frac{m+1}{n+1} \\ & \text{an integer}, \\ 2 \cdot \left\lceil \frac{1}{2}\left(m - \frac{m}{n+1}\right)\right\rceil & \text{otherwise.} \end{cases}$$

The broadcast domination number of the direct product $P_4 \times P_4$ equals 2 and is illustrated in Figure 11(b).

Theorem 34. ([45]) *The broadcast domination number of the lexicographic product of* P_m *and* P_n *has*

$$\gamma_b(P_m \bullet P_n) = \begin{cases} \max\{\left\lceil \frac{m}{3} \right\rceil, \left\lceil \frac{n}{3} \right\rceil\} & \text{if } m = 1 \text{ or } n \in \{1,2,3\}, \\ \max\{\left\lceil \frac{2m}{5} \right\rceil, 2\} & \text{if } m \geq 2 \text{ and } n \geq 4. \end{cases}$$

The broadcast domination number of the lexicographic product $P_4 \bullet P_4$ equals 2 and is illustrated in Figure 11(c), where the darkened vertex has a strength of 2.

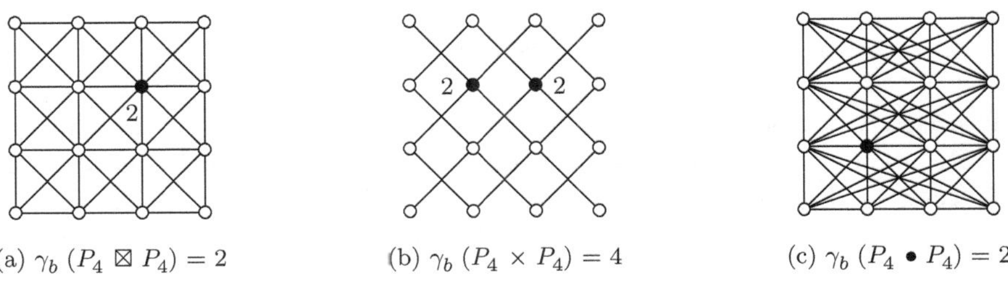

(a) $\gamma_b (P_4 \boxtimes P_4) = 2$ (b) $\gamma_b (P_4 \times P_4) = 4$ (c) $\gamma_b (P_4 \bullet P_4) = 2$

Fig. 11 Broadcast domination in $P_4 \boxtimes P_4$, $P_4 \times P_4$, and $P_4 \bullet P_4$

5   Irredundant Broadcasts

The upper domination number $\Gamma(G)$ of a graph G is the maximum cardinality of a minimal dominating set of G. The concept of the upper broadcast domination number was first defined by Erwin [21] and also studied, for example, in [1, 20, 38]. A dominating broadcast is *minimal* if reducing the strength of any vertex that is broadcasting results in a broadcast that is no longer dominating. Thus, a dominating broadcast f on a graph G is a minimal dominating broadcast if no broadcast $g < f$ is dominating. The maximum cost of a minimal dominating broadcast is the *upper broadcast domination number* of G, denoted as $\Gamma_b(G)$. Thus, for a graph $G = (V, E)$,

$$\Gamma_b(G) = \max\{f(V) \mid f \text{ is a minimal dominating broadcast of } G\}.$$

Ahmadi, Fricke, Schroder, Hedetniemi, and Laskar [1] point out that if you broadcast with strength $\mathrm{diam}(G)$ from a vertex v with $\mathrm{ecc}(v) = \mathrm{diam}(G)$, then you have a minimal dominating broadcast. Hence, we have the following lower bound on the upper broadcast domination number.

Theorem 35. ([1]) *Every graph G satisfies* $\mathrm{diam}(G) \le \Gamma_b(G)$.

A set $X \subseteq V(G)$ is *irredundant* if each $x \in X$ dominates a vertex y that is not dominated by any other vertex in X. We note that a maximal irredundant set is not necessarily a dominating set. The concept of an irredundance broadcast was first introduced in 2015 by Ahmadi et al. [1]. Essential for their definition of an irredundant broadcast, they use an important property given by Erwin [21] that makes a dominating broadcast a minimal dominating broadcast. Thereafter, they define an irredundant broadcast f to be *maximal irredundant* if no broadcast $g > f$ is irredundant and observe that any minimal dominating broadcast is maximal irredundant. The *lower* and *upper broadcast irredundance numbers* of G are given by

$$\mathrm{ir}_b(G) = \min\{f(V) \mid f \text{ is a maximal irredundant broadcast of } G\}$$

and

$$\mathrm{IR}_b(G) = \max\{f(V) \mid f \text{ is an irredundant broadcast of } G\},$$

respectively. We note that if the strength of any vertex in an optimal dominating broadcast is reduced, this will decrease the number of vertices that hears the broadcast (as all vertices in G can no longer hear the broadcast), so every optimal dominating broadcast is maximal irredundant. As a result, we have that $\mathrm{ir}_b(G) \le \gamma_b(G)$. More generally, by the properties established by Erwin [21], and by the above definitions, we have the following inequality chain.

Theorem 36. ([1]) *Every graph G satisfies the inequality chain given by*

$$\mathrm{ir}_b(G) \leq \gamma_b(G) \leq \gamma(G) \leq \Gamma(G) \leq \Gamma_b(G) \leq \mathrm{IR}_b(G).$$

Conditions under which an irredundant broadcast is maximal irredundant were determined by Mynhardt and Roux [38]. Their main result is that the ratio γ_b/ir_b is bounded.

Theorem 37. ([38]) *For every graph G, we have* $\gamma_b(G) \leq \frac{5}{4}\mathrm{ir}_b(G)$.

However, Mynhardt and Roux [38] give constructions illustrating that the ratio IR_b/Γ_b is unbounded for general graphs.

In 2017, Bouchemakh and Fergani [8] continued the study of the upper broadcast number and established the following upper bound for the upper broadcast domination number.

Theorem 38. ([8]) *If G is a connected graph of order n, then* $\Gamma_b(G) \leq n - \delta(G)$ *and the bound is sharp on paths, stars, and complete graphs.*

Bouchemakh and Fergani also studied the upper broadcast number on grids. Recall that the Cartesian product of graphs G and H is denoted by $G \square H$.

Theorem 39. ([8]) *For integers* $m \geq n \geq 2$, *we have* $\Gamma_b(P_m \square P_n) = m(n - 1)$.

6 Independent Domination Broadcasts

In 2006, Dunbar, Erwin, Haynes, Hedetniemi, and Hedetniemi [20] defined the concept of an independent broadcast. A broadcast f on a connected graph G is an *independent broadcast* if every pair of vertices u and v for which $f(u)>0$ and $f(v)>0$, we have $d(u, v) > \max\{f(u), f(v)\}$. Equivalently, an independent broadcast on G is a broadcast f of G such that for every vertex x of G, $f(v)>0$ implies that $f(u)=0$ for every vertex u of G within distance at most $f(v)$ from v. Thus, if f is an independent broadcast, then no broadcast vertex can hear a broadcast from any other broadcast vertex. As observed in [20], an independent broadcast need not be a dominating broadcast. The *broadcast independence number* $\alpha_b(G)$ of G is the maximum cost of an independent broadcast of G. The *lower broadcast independence number* $i_b(G)$ of G equals the minimum cost of a maximal independent broadcast of G. Thus,

$$i_b(G) = \min\{f(V) \mid f \text{ is a maximal independent broadcast of } G\}, \text{ and}$$
$$\alpha_b(G) = \max\{f(V) \mid f \text{ is an independent broadcast of } G\}.$$

An independent broadcast of G of cardinality $\alpha_b(G)$ is called an α_b-broadcast of G. In 2014, Bouchmakh and Zemir [9] studied broadcast independence on grids

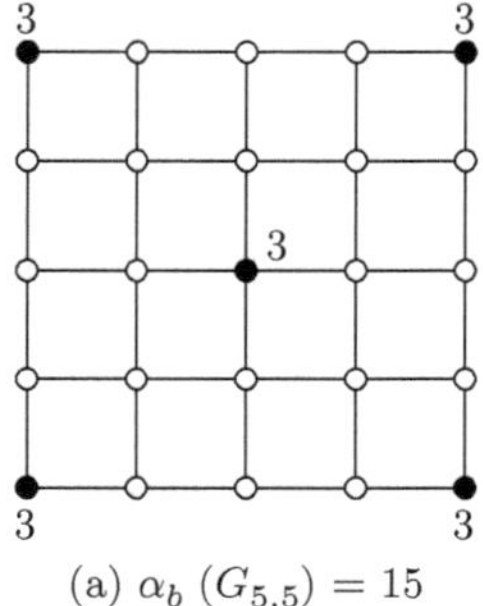
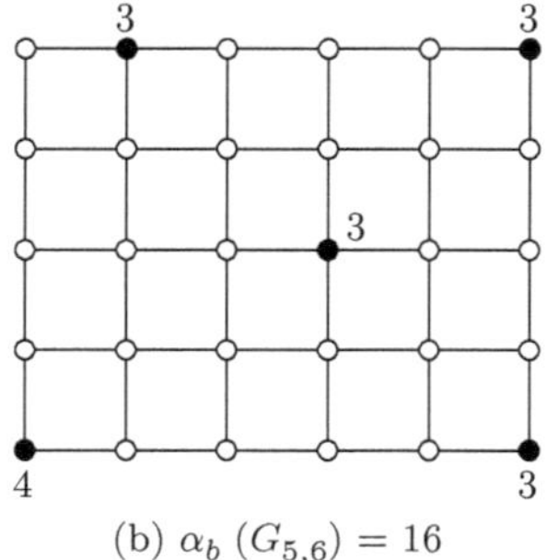

(a) $\alpha_b\,(G_{5,5}) = 15$ (b) $\alpha_b\,(G_{5,6}) = 16$

Fig. 12 α_b-Broadcasts of $G_{5,5}$ and $G_{5,6}$

and gave bounds for the broadcast independence number on $2 \times n$, $3 \times n$, and $4 \times n$ grids. Recall that the $m \times n$ grid graph is denoted by $G_{n,m}$, and so $G_{n,m} = P_m \square P_n$.

Theorem 40. ([9]) *The following hold.*

(a) *For $n \geq 2$, $\alpha_b(G_{2,n}) = 2(n-1)$.*
(b) *For $n \geq 3$, $\alpha_b(G_{3,n}) = 2n$.*
(c) *For $n \geq 4$, $\alpha_b(G_{4,n}) = 2(n+1)$.*
(d) *$\alpha_b(G_{5,5}) = 15$ and $\alpha_b(G_{5,6}) = 16$.*

An α_b-broadcast of $G_{5,5}$ and $G_{5,6}$ is illustrated in Figure 12(a) and 12(b), respectively, where the broadcast vertices are darkened and their strengths are given.

Theorem 41. ([9]) *For integers $n \geq m \geq 5$ where $(m, n) \notin \{(5, 5), (5, 6)\}$,*

$$\alpha_b(G_{m,n}) = \left\lceil \frac{mn}{2} \right\rceil.$$

Later, Ahmane, Bouchmakh, and Sopena studied the broadcast independence number for caterpillars [2]. In 2019, Bessy and Rautenbach [5] studied the relationship between the broadcast independence number and the independence number, where the independence number $\alpha(G)$ of G is the maximum cardinality of an independent set in G. Their main result is that the broadcast independence number and the independence number are within a constant factor from each other.

Theorem 42. ([5]) *For every connected graph G, we have $\alpha(G) \leq \alpha_b(G) \leq 4\alpha(G)$.*

Bessy and Rautenbach [5] also characterize all extremal graphs satisfying equality in the bound of Theorem 42. Imposing a girth condition on the graph, Bessy and Rautenbach [6] improve the upper bound in Theorem 42.

Theorem 43. ([5]) *Let G be a connected graph of girth g and minimum degree δ. If $g \geq 6$ and $\delta \geq 3$ or $g \geq 4$ and $\delta \geq 5$, then $\alpha_b(G) \leq 2\alpha(G)$.*

Theorem 44. ([5]) *For every positive integer k, there is a connected graph G of girth at least k and minimum degree at least k such that*

$$\alpha_b(G) \leq 2\left(1 - \frac{1}{k}\right)\alpha(G).$$

The results of Theorems 43 and 44 imply that lower bounds on the girth and the minimum degree of a connected graph G can lower the fraction $\alpha_b(G)/\alpha(G)$ from 4 to below 2, but not any further.

7 *k*-Broadcast Domination

Another variation of broadcast domination is the k-broadcast domination, which was first studied in 2018 by Henning, MacGillivray, and Yang [29] (also see [49]). In a k-broadcast, instead of requiring every vertex to hear at least one broadcast, this time we restrict every vertex to hear at least k broadcasts. The following motivation is given in [29]. Consider a large city partitioned into many neighborhoods, each of which has a radio tower to transmit emergency information. It is desirable to have some redundancy in the system so that everyone can hear a broadcast even if some of the towers are not functioning. The goal is to design a broadcast protocol with the property that every neighborhood receives some number k of broadcasts, and which does not require every tower to be used. Formally, let $f : V \to \{0, 1, \ldots, \mathrm{diam}(G)\}$ be a broadcast on a graph $G = (V, E)$, and let V_f^+ be the set of all vertices in G with positive strength under f, that is,

$$V_f^+ = \{v \in V \mid f(v) > 0\}.$$

If, for each vertex $u \in V$, there exist k different vertices $v_1, v_2, \ldots, v_k \in V_f^+$ such that for $i \in [k]$ we have $d(u, v_i) \leq f(v_i)$, then f is called a *dominating k-broadcast*. The *cost* of a dominating k-broadcast is the quantity $f(V)$. The minimum cost of a dominating k-broadcast is the *k-broadcast domination number* of G and is denoted by $\gamma_{b_k}(G)$. When $k = 1$, we note that the 1-broadcast domination number is the broadcast domination number, $\gamma_b(G)$, of G.

The integer programming formulation of k-broadcast domination is a bit more complicated than for 1-broadcast domination. Here, we need to introduce the ℓ-adjacency matrix and the ball matrix. Adopting the notation in [29], the *ℓ-adjacency matrix A^ℓ* is the $n \times n$ incidence matrix, where the rows correspond to vertices and the columns correspond to closed ℓ-neighborhoods. The (i, j)th entry of A^ℓ is 1 if the vertex v_i is contained in the closed ℓ-neighborhood of v_j, otherwise it is 0. Clearly, the closed neighborhood adjacency matrix is just A^1. We define the *ball matrix* to be

$$A^* = \begin{bmatrix} A & A^2 & \ldots & A^r \end{bmatrix},$$

where r is the radius of the graph. The columns of A^* are the characteristic vectors of closed ℓ-neighborhoods of vertices. For $i \in [n]$ and $\ell \in [r]$, let $x_{i\ell}$ be a Boolean variable representing the truth value of the statement "there is a broadcast from vertex v_i of strength ℓ." Let

$$x = [x_{11}, x_{21}, \ldots, x_{n1}, x_{12}, x_{22}, \ldots, x_{n2}, \ldots, x_{1r}, x_{2r}, \ldots, x_{nr}]^T.$$

We have $A^*x \geq [k, k, \ldots, k]^T$ if and only if every vertex hears at least k broadcasts. We must also add constraints to guarantee that each vertex broadcasts at most once. For each vertex v_i, the constraint $x_{i1} + x_{i2} + \cdots + x_{ir} \leq 1$, or equivalently $-x_{i1} - x_{i2} - \cdots - x_{ir} \geq -1$, guarantees that v_i broadcasts with at most one strength, and therefore at most once. The 0–1 integer program $B_k(G)$ for finding $\gamma_{b_k}(G)$ is described below:

$$\text{Minimize } \sum_{\ell=1}^{\mathrm{rad}(G)} \sum_{i=1}^{n} \ell x_{i\ell},$$

subject to
$$A^*x \geq [k, k, \ldots, k]^T,$$
$$x_{i1} + x_{i2} + \cdots + x_{ir} \leq 1 \text{ for each } i \in [n],$$
$$x_{i\ell} \in \{0, 1\} \text{ for } v_i \in V \text{ and } \ell \in [r].$$

Recall that the definition of multipacking arose from the dual of the linear programming relaxation of the broadcast domination integer program, and then consider it as an integer program. Analogously, we shall use the dual of the linear programming relaxation of $B_k(G)$ to formulate the definition of k-multipacking. Considering the dual of the linear programming relaxation of $B_k(G)$ as an integer program leads to the following integer program, $M_k(G)$:

$$\text{Maximize } \sum_{j=1}^{n} k \cdot c_j - r_j,$$

subject to
$$\left(\sum_{d(v_i,v_j) \leq \ell} c_j \right) - r_i \leq \ell \text{ for each } \ell \in [r] \text{ and each } i \in [n],$$
$$c_j, r_j \in \mathbb{N} \text{ for each vertex } v_j.$$

There are two values, c_j and r_j, associated with each vertex v_j. The variable c_j indicates how many times a vertex v_j is chosen in the k-multipacking (vertices can be chosen more than once). The variable r_j is the relaxation value on vertex v_j which allows the multipacking restriction on that specific vertex to be relaxed by r_j. Thus, a *k-multipacking* is a pair (c, r), where $c : V \rightarrow \mathbb{N}$ and $r : V \rightarrow \mathbb{N}$ such that for every vertex $v_i \in V$, we have

$$\sum_{d(v_i,v_j) \leq \ell} c(v_j) \leq \ell + r(v_i)$$

for each $\ell \in [r]$. The *value* of a k-multipacking (c, r) is the quantity

$$\sum_{v \in V}(k \cdot c(v) - r(v)).$$

The largest value of a k-multipacking of G is the *k-multipacking number* of G, denoted by $\mathrm{mp}_k(G)$. As remarked in [29], multipacking is the same as 1-multipacking. To see this, suppose we take a maximum multipacking of a graph G and consider whether it is an optimal solution to $M_1(G)$. If we want to increase the value of some variable c_j, we must also increase some r_i from 0 by at least the same amount, otherwise a constraint is violated at some vertex. Conversely, given a maximum 1-multipacking, if we want to decrease some positive r_i, then some c_j must be decreased by at least the same amount, otherwise a constraint will be violated. As a result, $\mathrm{mp}_1(G) = \mathrm{mp}(G)$. Thus, the k-multipacking problem is a generalization of multipacking.

The *fractional k-broadcast domination number of G*, $\gamma_{b_k, f}(G)$, is the optimum solution of the linear programming relaxation of $B_k(G)$, and the *fractional k-multipacking number of G*, $\mathrm{mp}_{k,f}(G)$, is the optimum solution of the linear programming relaxation of $M_k(G)$. By the strong duality theorem of linear programming, we have the following result.

Theorem 45. ([29]) *If G is a connected graph, then*

$$\mathrm{mp}_k(G) \leq \mathrm{mp}_{k,f}(G) = \gamma_{b_k,f}(G) \leq \gamma_{b_k}(G).$$

As observed in [29], by the definition of k-multipacking, we can always let c_i be 1 for all the vertices included in a maximum multipacking and r_i be 0 for all vertices, yielding $k \cdot \mathrm{mp}(G)$ as a lower bound on the k-multipacking number.

Observation 46. ([29]) *For a graph G, $\mathrm{mp}_k(G) \geq k \cdot \mathrm{mp}(G)$.*

As an example of a dominating 2-broadcast and a 2-multipacking, consider the tree shown in Figure 13(a). A dominating 2-broadcast is obtained by broadcasting with strength 1 from v_2 and strength 2 from v_1, as illustrated in Figure 13(b). This gives $\gamma_{b_2}(G) \leq 3$. On the other hand, we can assign the function values of $(c(v_i), r(v_i))$ as $(1, 0)$, $(0, 1)$, $(1, 0)$, and $(0, 0)$ at the vertices v_1, v_2, v_3, and v_4, respectively. This gives a 2-multipacking with value 3, thus $\mathrm{mp}_2(G) \geq 3$. By the duality theorem of linear programming, we have $3 \leq \mathrm{mp}_2(G) \leq \gamma_{b_2}(G) \leq 3$, implying that $\gamma_{b_2}(G) = \mathrm{mp}_2(G) = 3$.

The 2-broadcast domination number can differ from twice the broadcast domination number by an arbitrarily large constant difference as shown in [29].

Theorem 47. ([29]) *For any integer t, there exists a connected graph G with $\gamma_{b_2}(G) \leq 2 \cdot \gamma_b(G) - t$ and there exists a graph H with $\gamma_{b_2}(H) \geq 2 \cdot \gamma_b(H) - t$.*

The following upper bound on the 2-broadcast domination number of a tree in terms of its order is given in [29].

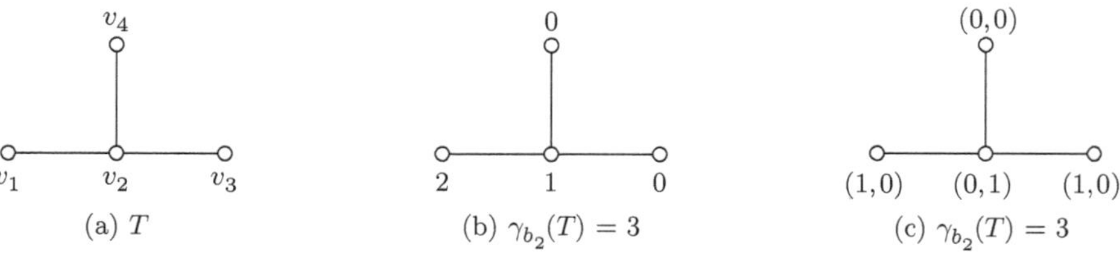

Fig. 13 A tree T with $\gamma_{b_2}(T) = \mathrm{mp}_2(T) = 3$

Theorem 48. ([29]) *If T is a tree of order $n \geq 2$, then $\gamma_{b_2}(T) \leq \lceil \frac{4n}{5} \rceil$.*

The proof of Theorem 48 is by induction on the order $n \geq 2$. The tree is carefully split into two subtrees, and various cases are considered in terms of different properties of the resulting subtrees. Since a dominating 2-broadcast in a spanning tree of a connected graph G is also a dominating 2-broadcast in G, as an immediate consequence of Theorem 48, we have the following result.

Corollary 49. ([29]) *If G is a connected graph of order $n \geq 2$, then $\gamma_{b_2}(G) \leq \lceil \frac{4n}{5} \rceil$.*

The authors in [29] believe that the bound in Theorem 48 could possibly be improved using a more detailed analysis and pose the following conjecture.

Conjecture 2. ([29]) *If T is a tree of order $n \geq 2$, then $\gamma_{b_2}(T) \leq \frac{1}{3}(2n + 4)$.*

If T is a path P_n where $n \equiv 1 \pmod 3$, then $\gamma_{b_2}(T) = (2n + 4)/3$. Thus, if Conjecture 2 is true, then the bound is tight.

8   Limited Broadcast Domination

In this section, we consider a limited version of the broadcast function. For a graph $G = (V, E)$ and an integer $k \geq 1$, a function $f : V \to \{0, 1, \ldots, k\}$ is called a *k-limited dominating broadcast*, abbreviated kLD-broadcast, in G if for each vertex $u \in V$, there exists a vertex v in G such that $f(v) > 0$ and $d(u, v) \leq f(v)$. The minimum cost of a kLD-broadcast is the *k-limited broadcast domination number* of G, denoted by $\gamma_{b,k}(G)$. A kLD-broadcast of cost $\gamma_{b,k}(G)$ is called a $\gamma_{b,k}$-*broadcast of G*; that is, a $\gamma_{b,k}$-broadcast of G is a minimum kLD-broadcast of G. The 1-limited broadcast domination number, $\gamma_{b,1}(G)$, of G is precisely the domination number, $\gamma(G)$. For $k \geq 1$, the function that assigns the weight 1 to the vertices of a minimum dominating set of G (of cardinality $\gamma(G)$) and the weight 0 to the remaining vertices of G is a kLD-broadcast of cost $\gamma(G)$, implying that $\gamma_{b,k}(G) \leq \gamma(G)$. By definition, every kLD-broadcast is a dominating broadcast, and so $\gamma_b(G) \leq \gamma_{b,k}(G)$. We state this formally as follows.

Observation 50. *For an integer $k \geq 1$ and for every graph G,*

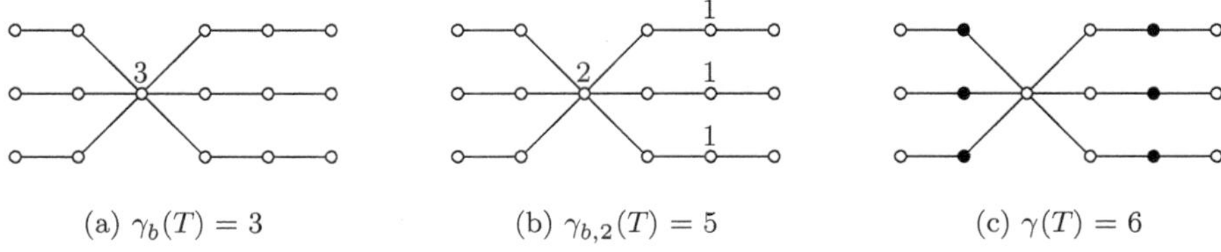

(a) $\gamma_b(T) = 3$ (b) $\gamma_{b,2}(T) = 5$ (c) $\gamma(T) = 6$

Fig. 14 A tree T with $\gamma(T) = \gamma_{b,2}(T) = \gamma(T) = 3$

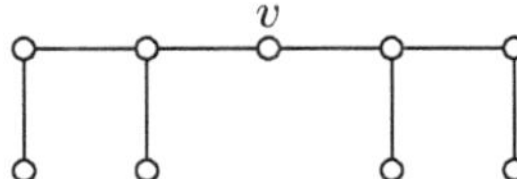

Fig. 15 The tree T_9 with gluing vertex v

$$\gamma_b(G) \leq \gamma_{b,k}(G) \leq \gamma(G) \quad \text{and} \quad \gamma_{b,k}(G) \leq \gamma_{b,k+1}(G).$$

The inequalities in the inequality chain of Observation 50 can be strict. For example, consider the tree T shown in Figure 14. In the case when $k = 2$, we have $\gamma_b(T) = \gamma_{b,3}(T) = 3$, $\gamma_{b,2}(T) = 5$, and $\gamma(T) = \gamma_{b,1}(T) = 6$. An optimal broadcast with cost equal to $\gamma_b(G) = 3$ is shown in Figure 14(a). A 2-limited dominating broadcast of cost $\gamma_{b,2}(T) = 5$ is shown in Figure 14(b), while a dominating set of T of cardinality $\gamma(T) = 6$ is shown in Figure 14(c).

We note that the optimal broadcast in Figure 14(b) is not efficient. More generally, the result of Theorem 3, that every graph has an optimal dominating broadcast that is efficient, does not apply to k-limited broadcasting for $k \geq 2$. This above example illustrates that, even for a tree, an optimal 2-limited dominating broadcast may not be efficient. However, every graph has efficient k-limited broadcasts for some values of k, e.g., $k = \text{rad}(G)$ and $k = \text{diam}(G)$. This raises the question: what is the smallest value of k for which G has an efficient k-limited broadcast?

Limited broadcast domination in graphs was first studied in 2013 by Rad and Khosvravi [42], where some fundamental properties were introduced. The first major results for limited broadcast domination were given in 2018 by Cáceres, Hernando, Mora, Pelayo, and Puertas [13, 14].

Theorem 51. ([13, 14]) *For $k \geq 2$, if G is a connected graph, then*

$$\gamma_{b,k}(G) = \min\{\gamma_{b,k}(T) \mid T \text{ is a spanning tree of } G\}.$$

In order to establish a tight upper bound for the 2-limited broadcast domination number of a general graph, Cáceres et al. [13] construct a family of trees $\mathcal{F}$ as follows. Let T_9 be the tree shown in Figure 15. We call the central vertex of degree 2 (that is not a support vertex) the *gluing vertex* of T_9. A tree T belongs to the family $\mathcal{T}$ if T is obtained from $k \geq 1$ vertex-disjoint copies of the tree T_9 shown in Figure 15 by adding $k - 1$ edges between the gluing vertices.

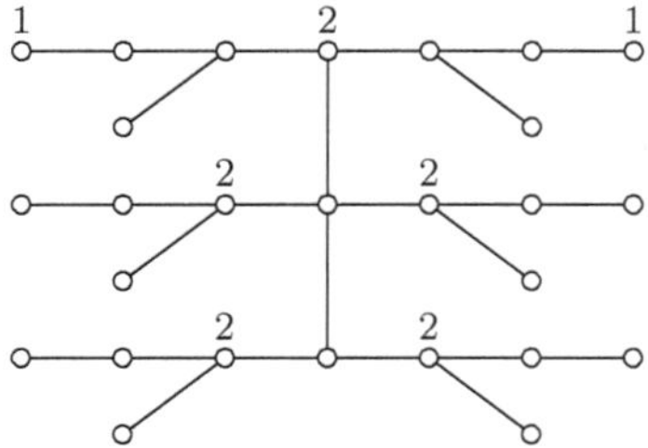

Fig. 16 A tree T in the family $\mathcal{T}$

An example of a tree T in the family $\mathcal{T}$ is shown in Figure 16. Furthermore, an example of a 2-limited dominating broadcast of cost $\gamma_{b,2}(T) = 12$ for the tree T is illustrated in Figure 16.

We are now in a position to state the result of Cáceres et al. [13].

Theorem 52. ([13]) *If T is a tree of order n, then $\gamma_{b,2}(G) \leq \left\lceil \frac{4}{9}n \right\rceil$, with equality if and only if $T \in \mathcal{F} \cup \{P_1, P_2, P_4\}$.*

As a consequence of Theorems 51 and 52, we have the following upper bound on the 2-limited broadcast domination number of a graph.

Theorem 53. ([13]) *If G is a connected graph of order n, then $\gamma_{b,2}(G) \leq \left\lceil \frac{4}{9}n \right\rceil$.*

Cáceres et al. [13] showed that the upper bound in Theorem 53 can be improved if the graph G contains a dominating path, that is, a path P such that every vertex not on P has a neighbor on P. In this case, we note that the graph G has a caterpillar as a spanning tree, where a caterpillar is a tree in which the removal of all leaves yields a path.

Theorem 54. ([13]) *If G is a graph of order n that contains a dominating path, then $\gamma_{b,2}(G) \leq \left\lceil \frac{2}{5}n \right\rceil$.*

In a subsequent paper, Cáceres et al. [14] generalized the upper bound on the 2-limited broadcast domination number of a tree given in Theorem 52 to the k-limited broadcast domination number for all $k \geq 2$.

Theorem 55. ([14]) *If T is a tree of order n and $k \geq 2$ is an integer such that $k < \mathrm{rad}(T)$, then*

$$\gamma_{b,k}(T) \leq \left\lceil \frac{k+2}{3(k+1)} \cdot n \right\rceil,$$

and this bound is tight.

As a consequence of Theorems 51 and 55, we obtain the following bound on the k-limited broadcast domination number of a graph.

Theorem 56. ([13]) *For $k \geq 2$, if G is a connected graph of order n, then $\gamma_{b,k}(G) \leq \left\lceil \frac{k+2}{3(k+1)} \cdot n \right\rceil$.*

Recently, Henning, MacGillivray, and Yang [30] studied 2-limited broadcast domination in subcubic graphs. A *subcubic graph* is a graph whose maximum degree is at most 3, while a *cubic graph* (also called a 3-*regular graph* in the literature) is a graph in which every vertex has degree 3. The following conjecture is posed in [30].

Conjecture 3. ([30]) *If G is a cubic graph of order n, then $\gamma_{b,2}(G) \leq \frac{1}{3}n$.*

Conjecture 3 is shown in [30] to be true if the cubic graph G is (C_4, C_6)-free, where a (C_4, C_6)-*free graph* is a graph that does not contain a 4-cycle or a 6-cycle as an induced subgraph.

Theorem 57. ([30]) *If G is a cubic graph of order n that is (C_4, C_6)-free, then $\gamma_{b,2}(G) \leq \frac{1}{3}n$.*

9   Algorithmic and Complexity Results

We consider in this section the problem of finding the broadcast domination number of an arbitrary graph. We state the decision problem formally as follows:

BROADCAST DOMINATION

Input: A graph G, and an integer $k \geq 1$.
Question: Is $\gamma_b(G) \leq k$?

The most interesting feature about dominating broadcasts is that the broadcast domination number $\gamma_b(G)$ can be computed in polynomial time for any graph, as shown by Heggernes and Lokshtanov [27] in 2006. This is quite counter-intuitive since computing the domination number of a graph is in general NP-hard.

Theorem 58. ([27]) *The broadcast domination number of a graph of order n can be computed in $O(n^6)$ time, implying that Broadcast Domination is solvable in polynomial time.*

To find an optimal dominating broadcast, Heggernes and Lokshtanov first considered a ball graph of the original graph. The *ball graph* of a dominating broadcast is a graph whose vertices are the broadcast neighborhoods of the original graph where two vertices of the ball graph are adjacent if the two broadcast neighborhoods contain a pair of adjacent vertices in the original graph. By Theorem 3, every graph G has an optimal efficient dominating broadcast, implying that there exists an optimal efficient broadcast whose ball is either a path or a cycle. The idea is to assume that for each vertex $v \in V$, the broadcast neighborhood of v is an end point of a ball graph which is a path. This finds all possible optimal dominating broadcasts which are paths. Next, the case when the ball graph is a cycle is considered. A broadcast neighborhood from the original graph is first removed, giving a path ball graph for the remaining subgraph. The running time of this process when the ball graph is a path is $O(n^4)$, and when the ball graph is a cycle, the running time is $O(n^6)$. Heggernes and Sæther [28] later conjectured that BROADCAST DOMINATION can be solved in $O(n^5)$ time in general.

In the literature, several algorithms have been given to find the broadcast number and the multipacking number for trees. In 2009, Dabney, Dean, and Hedetniemi [19] (also see [18]) gave a linear algorithm to find an optimal dominating broadcast for trees. The linearity of the algorithm is based on a complex data structure. Later, Brewster, MacGillivray, and Yang [12] (see also [49]) gave a simpler greedy algorithm which makes use of shadow trees, split-edges, and split-sets.

Theorem 59. ([18, 19]) *The broadcast domination number of a tree of order n can be computed in* $O(n)$ *time, implying that Broadcast Domination is solvable in linear time for trees.*

We consider next the following decision problem:

MULTIPACKING

Input: A graph G, and an integer $k \geq 1$.
Question: Is $mp(G) \geq k$?

In 2014, Mynhardt and Teshima [39] (also in [47]) showed that the multipacking number of a tree of order n can be computed in linear time. Brewster, MacGillivray, and Yang [12] (also see [49]) gave a simpler algorithm for finding an optimal multipacking on trees.

Theorem 60. ([39, 47]) *The multipacking number of a tree of order n can be computed in* $O(n)$ *time, implying that Multipacking is solvable in linear time for trees.*

A *block graph* is a graph in which every block is a complete graph. In particular, every tree is a block graph. Heggernes and Sæther [28] showed that BROADCAST DOMINATION can be solved efficiently on block graphs. As block graphs form a superclass of trees, their result extends the result given by Theorem 59. Due to the tree-like structure of a block graph, their algorithm is efficient and elegant.

Theorem 61. ([28]) *The broadcast domination number of a block graph of order n can be computed in* $O(n + m)$ *time.*

A graph G is an *interval graph* if there exists a one-to-one correspondence between its vertex set and a family of closed intervals in the real line, such that two vertices are adjacent if and only if their corresponding intervals intersect. Blair, Heggernes, Horton, and Manne [7] studied algorithmic and complexity results for broadcast domination in interval graphs, series-parallel graphs, and trees. Employing a dynamic programming method, they found optimal broadcasts for interval graphs and for series-parallel graphs.

Theorem 62. ([7]) *The broadcast domination number of an interval graph of order n can be computed in $O(n^3)$ time.*

Theorem 63. ([7]) *The broadcast domination number of a series-parallel graph of order n and radius r can be computed in $O(nr^4)$ time.*

The complexity result for broadcast domination in interval graphs given in Theorem 62 was subsequently improved by Chang and Peng [15]. Although their method is similar to that employed in [7], they use a better data structure resulting to improve the running time.

Theorem 64. ([15]) *The broadcast domination number of an interval graph of order n and size m can be computed in $O(n + m)$ time.*

A *chordal graph* is a graph in which every cycle of length at least 4 has a chord, where a chord of a cycle is an edge that is not part of the cycle but joins two vertices of the cycle. Equivalently, a chordal graph is a graph in which every induced cycle contains exactly three vertices. Chordal graphs can also be defined in terms of a perfect elimination ordering. A *perfect elimination ordering* in a graph G is an ordering of the vertices of the graph such that, for each vertex v, the set of neighbors of v that occur after v in the order form a clique. A graph is chordal if and only if it has a perfect elimination ordering. Heggernes and Sæther [28] gave the following complexity result to solve BROADCAST DOMINATION in chordal graphs.

Theorem 65. ([28]) *The broadcast domination number of a chordal graph of order n can be computed in $O(n^4)$ time.*

Furthermore, Heggernes and Sæther [28] conjectured that BROADCAST DOMINATION can be solved in $O(n^2)$ time for chordal graphs of order n. Although partial results have been obtained, their conjecture has yet to be fully settled. However, pleasing progress for the important subclass of chordal graphs, called strongly chordal graphs, has been made.

For $k \geq 3$, a graph G is called a *k-trampoline* (also called a *k-sun* in the literature) if it contains a k-clique with vertex set $\{v_1, v_2, \ldots, v_k\}$ and, for each pair $\{v_i, v_{i+1}\}$, there is a vertex w_i of degree 2 adjacent only to v_i and v_{i+1} in G for all $i \in [k]$, where addition is taken modulo k. Thus, a k-trampoline has order $2k$. A 3-trampoline is shown in Fig. 17. A graph G is a *strongly chordal graph* if it is chordal and does not contain a k-trampoline as an induced subgraph, for any k.

Brewster, MacGillivray, and Yang [12] (see also [49]) showed that for strongly chordal graphs, BROADCAST DOMINATION can be solved in $O(n^3)$ time. This

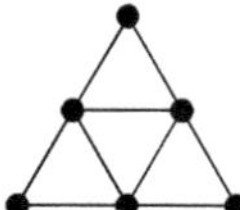

Fig. 17 A 3-trampoline

algorithm is different from previous algorithms in that it uses integer programming to find an optimal broadcast. As shown in Section 2, BROADCAST DOMINATION can be defined as a linear programming problem. However, if the solution to the linear program is fractional, it is not a solution of BROADCAST DOMINATION. In general, although the class of chordal graphs does not necessarily have an integer solution to the linear program, the subclass of strongly chordal graphs always has integer solutions. A matrix is *totally balanced* if it does not contain any cycle of length at least 3. It was proved in [12] that the constraint matrix of a strongly chordal graph is totally balanced. A matrix is called Γ-*free* if it does not contain

$$\Gamma = \begin{bmatrix} 1 & 1 \\ 1 & 0 \end{bmatrix}$$

as a submatrix. Lubiw [34, 35] proved that a totally balanced matrix can have a Γ-free ordering, and Farber [23] showed that the linear programming problem associated with Γ-free matrices always has an integer solution and it can be solved greedily. Combining all the results above, Brewster et al. [12] provided an efficient algorithm for the class of strongly chordal graphs.

Theorem 66. ([12, 49]) *The broadcast domination number of a strongly chordal graph of order n can be computed in* $O(n^3)$ *time.*

Brewster et al. [12] noted that every strongly chordal graph G satisfies $\gamma_b(G) = \mathrm{mp}(G)$ by the duality theorem of linear programming.

Corollary 67. ([12, 49]) *If G is strongly chordal graph, then* $\gamma_b(G) = \mathrm{mp}(G)$.

We consider next the following decision problem:

k-LIMITED BROADCAST DOMINATION

Input: A graph G, integers $k \geq 1$ and $\ell \geq 1$.
Question: Is $\gamma_{b,k}(G) \leq \ell$?

Cáceres et al. [13, 14] used a reduction from 3-SAT to prove that for $k \geq 2$, k-LIMITED BROADCAST DOMINATION is NP-complete.

Theorem 68. ([13, 14]) *For $k \geq 2$, k-Limited Broadcast Domination is NP-complete for general graphs.*

Cáceres et al. [13] considered trees and proved the following result.

Theorem 69. ([13]) 2-*Limited Broadcast Domination can be solved in linear time for trees.*

10   Concluding Comments

In this chapter, we have surveyed selected results on the broadcast domination in graphs. Other results on broadcast domination can be found, for example, in [24, 25, 36, 40, 41, 43, 48]. We close with a small list of conjectures and open problems. We repeat our earlier three conjectures.

Conjecture 1 ([4]). *If G is a connected graph, then $\gamma_b(G) \leq 2mp(G)$.*

Conjecture 2 ([29]). *If T is a tree of order $n \geq 2$, then $\gamma_{b_2}(T) \leq \frac{1}{3}(2n + 4)$.*

Conjecture 3 ([30]). *If G is a cubic graph of order n, then $\gamma_{b,2}(G) \leq \frac{1}{3}n$.*

We present next a list of open problems, taken from the "Stephen Hedetniemi treasure chest of intriguing ideas and open research questions."

Problem 1. Recall that by Theorem 35, every graph G satisfies $\text{diam}(G) \leq \Gamma_b(G)$. Is $\text{diam}(G) < \Gamma_b(G)$ possible?

Problem 2. What can you say about $\gamma_{b,3}(T)$ for trees, as compared with $\gamma_{b,2}(T)$?

Problem 3. What is the smallest value of k for which a graph G has an efficient k-limited broadcast?

Problem 4. In normal domination, every vertex has broadcast strength of 1. In k-limited broadcasting, a vertex can be assigned any integer value between 0 and k, that is, if f is a k-limited dominating broadcast, then $f : V \rightarrow \{0, 1, \ldots, k\}$. What if the only values allowed are 0 and k, that is, only one type of broadcast vertex is made, one of strength k, and so $f : V \rightarrow \{0, k\}$?

Problem 5. Does a broadcast domination chain analogous to that presented in Theorem 36 exist for k-limited broadcasting?

Problem 6. It would seem worthwhile to define a type of dominating broadcast f in a graph G in which for every vertex v in V, there exists a distinct broadcast vertex w in V such that $d(v, w) \leq f(w)$. This would enable a broadcast vertex v to compare its broadcast with that being given by another broadcast vertex. This is the total version of broadcast domination, called *broadcast total domination*.

Problem 7. Let us define a new concept called *connected dominating broadcasting*. Given a set B of broadcast vertices, we construct a corresponding broadcast network $N = (B, C)$, whose vertices are the broadcast vertices B, and two broadcast vertices are connected by an edge uv in C if $d(u, v) \leq \min\{f(u), f(v)\}$, that is u and v

can hear each others' broadcasts. What we want is that this broadcast network N is connected. The problem with normal broadcast domination is that the broadcast stations are all independent and totally disconnected. We want connected broadcast domination. This is a much more realistic model.

Problem 8. Even more interesting than connected or network broadcasting is the directed model, whereby a broadcast vertex v can hear a broadcast from a broadcast vertex u, but not necessarily conversely. In this case, you need a central, or originating, broadcast vertex v^* from which there is a directed path to all other broadcast vertices. Such a vertex v^* then originates a broadcast to all broadcast vertices along these directed edges, which in turn broadcast the originating message to all remaining non-broadcast vertices. That is, between any two broadcast vertices u and v, if $d(u, v) \leq f(v)$, then there is an arc from v to u, meaning that u can hear a broadcast from v. But it is possible that v cannot hear a broadcast from u. In this way, you can define the directed broadcast network existing among the broadcast vertices, and this is a directed graph. You want this broadcast network to be connected in the further sense that there is a central vertex from which a given broadcast message can be relayed to all broadcast vertices over the arcs in the network, which in turn can then broadcast this message to all non-broadcast vertices, i.e., the listeners.

Problem 9. Consider, as an example, the 7×7 grid graph. Place the value 4 in the center vertex, square $(4, 4)$. Place the value *1* at the four vertices of degree *4* at distance *2* from a vertex of degree *2*, as illustrated in Fig. 18(a). You have a dominating broadcast with a connected broadcast network that is a directed $K_{1,4}$. If you decrease the broadcast strength of the center vertex to 3, then you have a less expensive 3-limited dominating broadcast, but now the broadcast network is totally disconnected. The two broadcasts are illustrated in Figure 18(a) and 18(b), respectively.

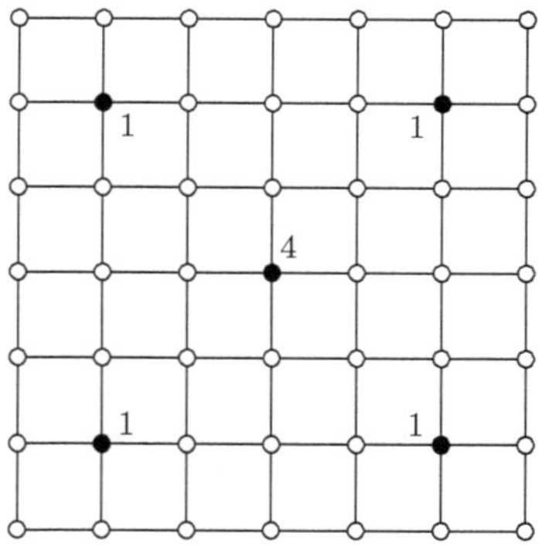

(a) A connected broadcast of cost 8

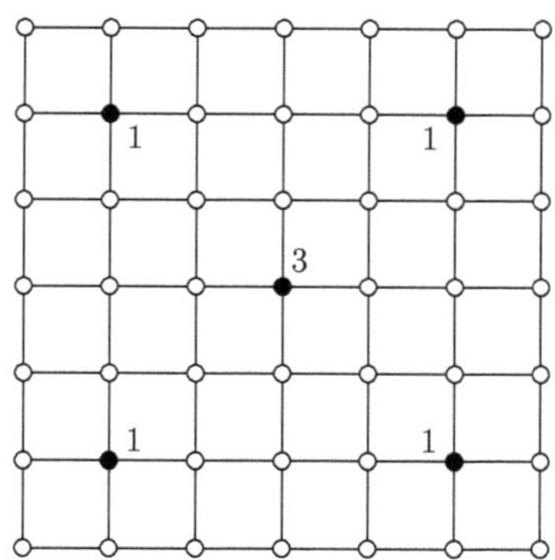

(b) A disconnected broadcast of cost 7

Fig. 18 Two broadcasts in the 7×7 grid graph

References

1. D. Ahmadi, G.H. Fricke, C. Schroder, S.T. Hedetniemi, R.C. Laskar, Broadcast irredundance in graphs. Congr. Numer. **224**, 17–31 (2015)
2. M. Ahmane, I. Bouchmakh, E. Sopena, On the broadcast independence number of caterpillars. Discrete Appl. Math. **244**, 20–35 (2018)
3. L. Beaudou, R. Brewster, On the multipacking number of grid graphs. Discrete Math. Theor. Comput. Sci. **21**, #23 (2019)
4. L. Beaudou, R. Brewster, F. Foucaud, Broadcast domination and multipacking: bounds and the integrality gap. Australas. J. Combin. **74**, 86–97 (2019)
5. S. Bessy, D. Rautenbach, Relating broadcast independence and independence. Discret. Math. **342**(12), 111589, 7 pp (2019)
6. S. Bessy, D. Rautenbach, Girth, minimum degree, independence, and broadcast independence. Commun. Comb. Optim. **4**(2), 131–139 (2019)
7. J.R.S. Blair, P. Heggernes, S. Horton, F. Manne, Broadcast domination algorithms for interval graphs, series-parallel graphs, and trees. Congr. Numer. **169**, 55–77 (2004)
8. I. Bouchemakh, N. Fergani, On the upper broadcast domination number. Ars Combin. **130**, 151–161 (2017)
9. I. Bouchemakh, M. Zemir, On the broadcast independence number of grid graph. Graphs Combin. **30**, 83–100 (2014)
10. B. Bresar, S. Spacapan, Broadcast domination of products of graphs. Ars Combin. **92**, 303–320 (2009)
11. R. Brewster, C.M. Mynhardt, L. Teshima, New bounds for the broadcast domination number of a graph. Cent. Eur. J. Math. **11**(7), 1334–1343 (2013)
12. R. Brewster, G. MacGillivray, F. Yang, Broadcast domination and multipacking in strongly chordal graphs. Discrete Appl. Math. **261**, 108–118 (2019)
13. J. Cáceres, C. Hernando, M. Mora, I.M. Pelayo, M.L. Puertas, Dominating 2-broadcast in graphs: complexity, bounds and extremal graphs. Appl. Anal. Discrete Math. **12**(1), 205–223 (2018)
14. J. Cáceres, C. Hernando, M. Mora, I.M. Pelayo, M.L. Puertas, General bounds on limited broadcast domination. Discrete Math. Theor. Comput. Sci. **20**, #13 (2018)
15. R.Y. Chang, S.L. Peng, A linear-time algorithm for broadcast domination problem on interval graphs, in *The 27th Workshop on Combinatorial Mathematics and Computation Theory* (2010), pp. 184–188
16. E.J. Cockayne, S. Herke, C.M. Mynhardt, Broadcasts and domination in trees. Discret. Math. **311**, 1235–1246 (2011)
17. G. Cornuéjols, *Combinatorial Optimization. Packing and Covering.* CBMS-NSF Regional Conference Series in Applied Mathematics, vol. 74 (Society for Industrial and Applied Mathematics (SIAM), Philadelphia, 2001), xii + 132 pp.
18. J. Dabney, A linear-time algorithm for broadcast domination in a tree. Master's Thesis, Clemson University, 2007
19. J. Dabney, B.C. Dean, S.T. Hedetniemi, A linear-time algorithm for broadcast domination in a tree. Networks **53**, 160–169 (2009)
20. J.E. Dunbar, D.J. Erwin, T.W. Haynes, S.M. Hedetniemi, S.T. Hedetniemi, Broadcasts in graphs. Discrete Appl. Math. **154**, 59–75 (2006)
21. D.J. Erwin, Cost domination in graphs. Ph.D. Thesis, Western Michigan University, 2001
22. D.J. Erwin, Dominating broadcasts in graphs. Bull. Inst. Combin. Appl. **42**, 89–105 (2004)
23. M. Farber, Domination, independent domination, and duality in strongly chordal graphs. Discrete Appl. Math. **7**, 115–130 (1984)
24. P.F. Faul, Adjunctions in broadcast domination with a cost function. Australas. J. Combin. **72**, 70–81 (2018)
25. L. Gemmrich, C.M. Mynhardt, Broadcasts in graphs: diametrical trees. Australas. J. Combin. **69**, 243–258 (2017)

26. B.L. Hartnell, C.M. Mynhardt, On the difference between broadcast and multipacking number of graphs. Utilitas Math. **94**, 19–29 (2014)
27. P. Heggernes, D. Lokshtanov, Optimal broadcast domination in polynomial time. Discret. Math. **306**, 3267–3280 (2006)
28. P. Heggernes, S.H. Sæther, Broadcast domination on block graphs in linear time, in *Computer Science—Theory and Applications*. Lecture Notes in Computer Science, vol. 7353 (Springer, Heidelberg, 2012), pp. 172–183
29. M.A. Henning, G. MacGillivray, F. Yang, k-Broadcast domination and k-multipacking. Discrete Appl. Math. **250**, 241–251 (2018)
30. M.A. Henning, G. MacGillivray, F. Yang, 2-Limited broadcast domination in subcubic graphs. Discrete Appl. Math. **285**, 691–706 (2020)
31. S. Herke, Dominating broadcasts in graphs. Master's Thesis, University of Victoria, 2007
32. S. Herke, C.M. Mynhardt, Radial trees. Discret. Math. **309**, 5950–5962 (2009)
33. K.M. Koh, K.W. Soh, Dominating broadcast labeling in Cartesian products of graphs. Electron. Notes Discrete Math. **48**, 197–204 (2015)
34. A. Lubiw, Γ-Free matrices. Master's Thesis. Department of Combinatorics and Optimization, University of Waterloo, 1982
35. A. Lubiw, Doubly Lexical orderings of matrices. SIAM J. Comput. **16**, 854–879 (1987)
36. S. Lunney, C.M. Mynhardt, More trees with equal broadcast and domination numbers. Australas. J. Combin. **61**, 251–272 (2015)
37. A. Meir, J.W. Moon, Relations between packing and cover numbers of a tree. Pacific J. Math. **61**, 225–233 (1975)
38. C.M. Mynhardt, A. Roux, Dominating and irredundant broadcasts in graphs. Discrete Appl. Math. **220**, 80–90 (2017)
39. C.M. Mynhardt, L.E. Teshima, Broadcasts and multipackings in trees. Utilitas Math. **104**, 227–242 (2017)
40. C.M. Mynhardt, J. Wodlinger, A class of trees with equal broadcast and domination numbers. Australas. J. Combin. **56**, 3–22 (2013)
41. C.M. Mynhardt, J. Wodlinger, Uniquely radial trees. J. Combin. Math. Combin. Comput. **93**, 131–153 (2015)
42. N.J. Rad, F. Khosravi, Limited dominating broadcast in graphs. Discrete Math. Algorithms Appl. **5**(4), 1350025, 9 pp (2013)
43. S.M. Seager, Dominating broadcasts of caterpillars. Ars Combin. **88**, 307–319 (2008)
44. P.J. Slater, LP-duality, complementarity, and generality of graphical subset parameters, in *Domination in Graphs*. Monographs Textbooks Pure Applied Mathematics, vol. **209** (Dekker, New York, 1998), pp. 1–29
45. K.W. Soh, K.M. Koh, Broadcast domination in graph products of paths. Australas. J. Combin. **59**, 342–351 (2014)
46. K.W. Soh, K.M. Koh, Broadcast domination in tori. Trans. Comb. **4**(4), 43–53 (2015)
47. L. Teshima, Broadcasts and multipackings in graphs. Master's Thesis, University of Victoria, 2012
48. F. Yang, New results on broadcast domination and multipacking. Master's Thesis, University of Victoria, 2015
49. F. Yang, Limited broadcast domination. Ph.D. Thesis, University of Victoria, 2019

Alliances and Related Domination Parameters

Teresa W. Haynes and Stephen T. Hedetniemi

1 Introduction

An alliance is generally thought of as a treaty or formal agreement between two or more parties, made in order to unite for a common cause. In 2002, P. Kristiansen, S. M. Hedetniemi, and S. T. Hedetniemi [69] introduced several types of alliances in graphs to model such agreements. The study of alliances in graphs has become a popular area of research with around 100 papers published since its inception in 2002. In fact, three survey papers have been published on the topic [39, 73, 99]. Since these recent overviews are readily available, the purpose of this chapter is not to give a comprehensive survey of alliances in graphs. Instead our goal is to provide selected results along with sample proof techniques used in studying alliances. Furthermore, since an alliance need not be a dominating set and this book is on domination in graphs, the main focus of this chapter will be on alliances that are also dominating sets. In particular, we select a specific dominating alliance (a global defensive alliance) to serve as an illustration of work in the field. Also, we present a brief overview of some precursors of alliances and two recently defined related parameters, namely cost effective sets and distribution sets. For algorithms and complexity of alliances, we refer the reader to Chapter 17 of this volume.

T. W. Haynes (✉)
Department of Mathematics and Statistics, East Tennessee State University, Johnson City, TN, USA

Department of Mathematics and Applied Mathematics, University of Johannesburg, Johannesburg, South Africa
e-mail: haynes@etsu.edu

S. T. Hedetniemi
School of Computing, Clemson University, Clemson, SC, USA
e-mail: hedet@clemson.edu

We begin with some terminology and a discussion of precursors of alliances in Section 2. Definitions, preliminary results, and examples of alliances are given in Section 3. In Section 4, we present a survey of core results on a type of dominating alliance, namely, a global defensive alliance. The related concepts of cost effective sets and distribution sets along with open problems and ideas for future work are discussed in Section 5.

We will use the following terminology throughout the chapter. Given a vertex set $S \subseteq V$, we let $\overline{S} = V \setminus S$ and we use the notation $d_S(v) = |N(v) \cap S|$ for the number of neighbors of v that are in S and $d_{\overline{S}}(v) = |N(v) \cap \overline{S}|$ for the number of neighbors of v in $\overline{S}$. Thus, $d(v) = d_S(v) + d_{\overline{S}}(v)$. The *boundary* of a set S is the set $\partial(S) = N(S) \cap \overline{S}$. Let the *generalized corona* $G \circ H$ be the graph obtained by adding a copy of H for each vertex v of G and joining v to every vertex of H.

2 Preliminary Definitions and Background

This section contains a discussion of parameters that can be considered as precursors to alliances. We present several types of sets, all of which are defined in terms of conditions on the degrees of vertices with respect to sets S, $\partial(S)$, and $\overline{S}$, such as $d_S(v)$, $d_{\overline{S}}(v)$, and $d(v)$, either for vertices $v \in S$ or vertices $v \in \overline{S}$. Each of these types of sets naturally arise in different real-world contexts. We present these as they occurred chronologically in the literature.

2.1 Unfriendly Sets and Satisfactory Partitions

Problems of partitioning the vertex set of a graph with constraints on the degrees of vertices in the sets can be traced to a problem of unfriendly partitions of graphs introduced by Borodin and Kostochka [9] in 1977.

Definition 2.1 *A partition $\{S, \overline{S}\}$ of the vertex set V of a graph G is unfriendly if for every vertex $v \in S$, $d_{\overline{S}}(v) \geq d_S(v)$ and for every vertex $u \in \overline{S}$, $d_S(u) \geq d_{\overline{S}}(u)$.*

Stated equivalently, a partition of the vertex set of a graph G into two sets is called *unfriendly* if every vertex $v \in V$ has at least as many neighbors in the opposite set as it has in its own set. In 1990, Aharoni, Milner, and Prikry [1] settled, in the affirmative, a conjecture by Cowan and Emerson that every graph has an *unfriendly partition* (see also Shelah and Milner [85]).

Theorem 2.2 ([1]) *Every nontrivial graph has an unfriendly partition.*

Proof. Among all partitions of the vertex set V of G into two nonempty sets, select $\{S, \overline{S}\}$ to be one that maximizes the number of edges having one end in S and the other in $\overline{S}$. We claim that $\{S, \overline{S}\}$ is an unfriendly partition. If not, then there must be at least one vertex having more neighbors in its own set than in the other set. Moving this vertex to the other set would therefore increase the number of edges between the two sets, a contradiction. $\qquad\square$

It can also be shown that if you start with any arbitrary partition of V into two nonempty sets and repeatedly find a vertex v in either set, which has more neighbors in the set containing it than neighbors in the other set, and then move v to the other set, this process will repeatedly increase the number of edges between the two sets. Therefore, after a finite number of such moves, this process will terminate with an unfriendly bipartition.

Corollary 2.3 *Every graph G without isolated vertices has a partition into two dominating sets.*

Proof Let $\{S, \overline{S}\}$ be an unfriendly partition of the vertex set of G. We show that both S and $\overline{S}$ are dominating sets of G. It suffices to consider only set S as the same argument used for S applies to $\overline{S}$. If S is not a dominating set, then there exists a vertex, say $x \in \overline{S}$ having no neighbors in S. But in this case, x has more neighbors (at least one) in $\overline{S}$ than it has in S (none), contradicting that $\{S, \overline{S}\}$ is an unfriendly partition. It follows that both S and $\overline{S}$ are dominating sets of G. $\qquad\square$

Corollary 2.3 is reminiscent of the following well-known theorem and corollary of Ore [72].

Theorem 2.4 (Ore [72]) *If G is a graph having no isolated vertices, then the complement $\overline{S}$ of any minimal dominating set S is a dominating set.*

Corollary 2.5 *Every graph G without isolated vertices has a partition into two dominating sets.*

Although Corollary 2.3 and Corollary 2.5 are the same, the reasoning behind them makes a difference. With Corollary 2.5, we know that at least one of the two dominating sets is a minimal dominating set, and furthermore, it can be a minimum dominating set of cardinality $\gamma(G)$. We do not have such an assurance in Corollary 2.3.

Our discussion of unfriendly partitions gives rise to the following definitions.

Definition 2.6 *A dominating set S is unfriendly if for every vertex $w \in \overline{S}$, $d_S(w) \geq d_{\overline{S}}(w)$, that is, w has at least as many neighbors in S as it has in $\overline{S}$. An unfriendly dominating set is very unfriendly if this inequality is strict, that is, $d_S(w) > d_{\overline{S}}(w)$.*

Definition 2.7 *A dominating set S is friendly if for every vertex $w \in \overline{S}$, $d_S(w) \leq d_{\overline{S}}(w)$, that is, w has at least as many neighbors in $\overline{S}$ as it has in S. A friendly dominating set is very friendly if this inequality is strict, that is, $d_S(w) < d_{\overline{S}}(w)$.*

This leads to the observation: every isolate-free graph G has a partition into an independent dominating set S and an unfriendly dominating set $\overline{S}$, since the complement $\overline{S}$ of every maximal independent set S is necessarily a (very) unfriendly dominating set. Notice that, by definition, the vertex set V is vacuously a friendly dominating set, so every graph has a friendly dominating set. We will revisit friendly and unfriendly dominating sets in Section 5.

In some sense dual to an unfriendly partition, a *satisfactory partition* is a partition of the vertex set of a graph into two sets such that each vertex has at least as many neighbors in the set containing it as it has in the opposite set. We note that this "friendly" version is equivalent to a partition of the vertex set into two strong defensive alliances. Satisfactory partitions have been studied in [43–45] and [81]. However, unlike unfriendly partitions, not every graph has a satisfactory partition. For example, complete graphs of odd order and complete bipartite graphs $K_{r,s}$, when r or s is odd, do not have satisfactory partitions. In fact, it is an NP-complete problem to decide if an arbitrary graph has a satisfactory partition [4].

2.2 (σ, ρ)-sets

In 1994, Telle [92, 93] introduced the idea of (σ, ρ)-sets S, in which σ is a nonnegative integer condition that must hold on the number of neighbors a vertex in S must have in S, and ρ is a nonnegative integer condition that must hold on the number of neighbors a vertex in $\overline{S}$ must have in S.

This was generalized by Haynes, Hedetniemi, and Slater in 1998 [50] as follows. There are four possible values under consideration, namely, $d_S(v)$ for $v \in S$, $d_{\overline{S}}(v)$ for $v \in S$, $d_S(v)$ for $v \in \overline{S}$, and $d_{\overline{S}}(v)$ for $v \in \overline{S}$. Table 1 illustrates how different domination parameters are defined using combinations of these four values. A blank

Table 1 Degree Conditions

S is	$v \in S, d_S(v)$	$v \in S, d_{\overline{S}}(v)$	$v \in \overline{S}, d_S(v)$	$v \in \overline{S}, d_{\overline{S}}(v)$
a D-set			≥ 1	
an independent set	$= 0$			
an ID-set	$= 0$		≥ 1	
a TD-set	≥ 1		≥ 1	
a PD-set			$= 1$	
a RD-set			≥ 1	≥ 1
a k-dominating set			$\geq k$	
a D-set and $\overline{S}$ is a D-set		≥ 1	≥ 1	
a $[1, k]$-dominating set			≥ 1 and $\leq k$	
an odd D-set	even		odd	
an open odd D-set	odd		odd	
an efficient D-set	$= 0$		$= 1$	
a 1-dependent D-set	≤ 1		≥ 1	

entry in the table implies that this condition is not relevant to the definition. Let D-set, TD-set, ID-set, PD-set, and RD-set denote dominating set, total dominating set, independent dominating set, perfect dominating set, and restrained dominating set, respectively.

2.3 Signed Domination

In 1995, Dunbar, Hedetniemi, Henning, and Slater [30] introduced the concept of *signed domination* in graphs as follows.

Definition 2.8 *A function* $f: V \rightarrow \{-1, 1\}$ *is called a signed dominating function if for every vertex* $v \in V$, $f(N[v]) = \Sigma_{u \in N[v]} f(u) \geq 1$.

In effect, a signed dominating function defines a partition of the vertex set of G into two sets S and $\overline{S}$, where $S = \{v : f(v) = 1\}$ and $\overline{S} = \{u : f(u) = -1\}$, such that for every vertex $v \in V$, $d_S(v) > d_{\overline{S}}(v)$. We note that in a signed dominating function the set S is a very unfriendly dominating set of the set $\overline{S}$, or equivalently, S is a global offensive alliance of G, which we will define in the next section.

2.4 Minus Domination

In 1999, Dunbar, Hedetniemi, Henning, and McRae [31] introduced a variation on signed domination called *minus domination*, as follows.

Definition 2.9 *A function* $f: V \rightarrow \{-1, 0, 1\}$ *is a minus dominating function if for every vertex* $v \in V$, $f(N[v]) = \Sigma_{u \in N[v]} f(u) \geq 1$.

In effect, a minus dominating function defines a partition $\{S^{-1}, S^0, S^1\}$ of the vertex set V, where $S^{-1} = \{v : f(v) = -1\}$, $S^0 = \{v : f(v) = 0\}$, and $S^1 = \{u : f(u) = 1\}$, such that for every vertex $v \in V$, $d_{S^1}(v) > d_{S^{-1}}(v)$.

2.5 Strong and Weak Dominating Sets

In 1996, Sampathkumar and Pushpa Latha [80] focused on the degrees of the vertices in a dominating set S and how they related to the degrees of vertices in $\overline{S}$.

Definition 2.10 *A dominating set* S *is said to be strong if for every vertex* $v \in \overline{S}$, *there exists a vertex* $u \in S \cap N(v)$ *such that* $d(u) \geq d(v)$. *Similarly, a dominating set* S *is said to be weak if for every vertex* $v \in \overline{S}$, *there exists a vertex* $u \in S \cap N(v)$ *such that* $d(u) \leq d(v)$.

2.6 α-Dominating Sets

In 2000, Dunbar, Hoffman, Laskar, and Markus [32] introduced the concept of α-domination, where α is a number $0 < \alpha \leq 1$.

Definition 2.11 *A set S of vertices in a graph G is an α-dominating set if for every vertex $v \in \overline{S}$, $\frac{d_S(v)}{d(v)} \geq \alpha$, where $0 < \alpha \leq 1$.*

Note that when $\alpha \geq 1/2$, the vertices in $\overline{S}$ corresponding to an α-dominating set S satisfy the unfriendly condition of Aharoni et al. that every vertex in $\overline{S}$ has at least as many neighbors in S as it has in $\overline{S}$. However, no unfriendly condition is required for the vertices in an α-dominating set S.

2.7 Communities

In 2000, Flake, Lawrence, and Giles [41] introduced the concept of a *community* in a graph as follows.

Definition 2.12 *A community is a vertex subset $C \subseteq V$ of a graph G, such that for all vertices $v \in C$, v has at least as many edges connecting to vertices in C as it does to vertices in $\overline{C}$.*

This definition can be rephrased as follows. A *community* is a set S of vertices having the property that for every vertex $v \in S$, $d_S(v) \geq d_{\overline{S}}(v)$.

3 Alliances

In 2002, concepts almost the same as communities were introduced by Kristiansen, Hedetniemi, and Hedetniemi [69], but in a completely different context, that of alliances in networks rather than communities. In 2004, the authors followed the proceedings [69] with a more detailed introduction to alliances in [57].

An alliance is generally thought of as a treaty or formal agreement between two or more parties or nations, made in order to unite for a common cause or for mutual support. For example, defensive alliances are formed during times of war, where the allies agree to join forces if one or more of them are attacked, and offensive alliances can be formed in times of peace, where allies might have to join forces in order to keep peace. In addition to alliances for national defense, applications of alliances are widespread in nature from social and business associations to political and scientific groupings. As mentioned in the introduction, the study of alliances in graphs has become a popular area of research with around 100 papers published since its introduction in 2002.

The popularity of alliances is further evidenced by three recent survey papers on the topic [39, 73, 99]. The first survey by Fernau and Rodríguez-Velázquez [39] in 2014 focuses mainly on defensive alliances. Yero and Rodríguez-Velázquez [99] wrote a second survey in 2017. In it, they note that graph parameters and types of alliances have been studied under many different names, and they provide a new general and unifying framework for a wide variety of alliances. In 2018, Ouazine, Slimani, and Tari [73] published the third survey on alliances. This survey gives another generalization of alliances and presents results for both defensive and offensive alliances.

In this section, we present definitions, examples, and preliminary results on alliances. Recall that for a set S, the boundary of S, denoted $\partial(S)$, is the set of vertices in $\overline{S}$ that have a neighbor in S.

Definition 3.1 *A nonempty set of vertices S of a graph G is a defensive alliance if for every $v \in S$, $|N[v] \cap S| \geq |N(v) \cap \overline{S}|$. The minimum cardinality of a defensive alliance of G is the defensive alliance number of G, denoted by $a(G)$.*

This can be stated equivalently as follows. A *defensive alliance* is a nonempty set S of vertices having the property that for every vertex $v \in S$, $d_S(v) \geq d_{\overline{S}}(v) - 1$. Conceptually, each vertex in S is in alliance with its neighbors in S for defense against possible attacks from neighbors in $\partial(S)$. For each vertex $v \in S$, an attack at v by the vertices in $\partial(S)$ that are adjacent to v can result in no worse than a draw (assuming strength in numbers). Thus, each vertex in S can be successfully defended against attacks from its neighbors in $\partial(S)$.

By changing focus from vertices in S to vertices in $\partial(S)$, Hedetniemi et al. [57] defined the following.

Definition 3.2 *A nonempty set of vertices S of a graph G is an offensive alliance if for every vertex $v \in \partial(S)$, $|N(v) \cap S| \geq |N[v] \cap \overline{S}|$. The minimum cardinality of an offensive alliance of G is the offensive alliance number of G, denoted by $a_o(G)$.*

Equivalently, an *offensive alliance* is a set S of vertices having the property that for every vertex $v \in \partial(S)$, $d_S(v) \geq d_{\overline{S}}(v) + 1$. In terms of application of an offensive alliance S, it is reasonable to think that each vertex in S is in alliance with its neighbors in S against its neighbors in $\partial(S)$. For the set S as a whole, since an attack by an offensive alliance S on the vertices of $\partial(S)$ can result in no worse than a "tie," the vertices in S can "successfully" attack any single vertex in $\partial(S)$.

For examples, we note that the offensive alliance and defensive alliance numbers are equal for a complete graph, that is, $a(K_n) = a_o(K_n) = \left\lceil \frac{n}{2} \right\rceil$. Note also that any vertex of degree 0 or 1 is a defensive alliance. It is shown in [57] that $a(G) = 1$ if and only if G has a vertex of degree 0 or 1, and it is shown in [37] that $a_o(G) = 1$ if and only if G is a star. The alliance numbers for paths and cycles follow.

Proposition 3.3 *For paths P_n and cycles C_n with $n \geq 3$,*

1. *([57]) $a(P_n) = 1$ and $a(C_n) = 2$,*
2. *([37]) $a_o(P_n) = \left\lfloor \frac{n}{2} \right\rfloor$ and $a_o(C_n) = \left\lceil \frac{n}{2} \right\rceil$.*

Paths and cycles provide examples where the offensive alliance number can be larger than the defensive alliance number. To see that these two numbers are incomparable, consider the complete bipartite graphs $K_{r,s}$, where $2 \le r \le s$. Then $a(K_{r,s}) = \lfloor \frac{r}{2} \rfloor + \lfloor \frac{s}{2} \rfloor$, while $a_o(K_{r,s}) = \lceil \frac{r+1}{2} \rceil$. Thus, the defensive alliance number is larger than the offensive alliance number for $K_{r,s}$ when $r \ge 4$.

We mention upper bounds on these two alliance numbers before defining more alliance numbers. For two sets of vertices A and B, we define an edge having one end in A and the other in B, an *AB-edge*.

Theorem 3.4 ([42]) *If G is a connected graph of order $n \ge 2$, then $a(G) \le \lceil \frac{n}{2} \rceil$.*

Proof The result is trivial if G has a vertex of degree at most 1. Among all balanced bipartitions $\{A, B\}$ of V, where $|A| - |B| \le 1$, let $\pi = \{A, B\}$ be one that minimizes the number of AB-edges. Without loss of generality, we can assume that $|A| = \lceil \frac{n}{2} \rceil$ and $B = \lfloor \frac{n}{2} \rfloor$. If A or B is a defensive alliance, then the result holds. Hence, assume that neither A nor B is a defensive alliance. Thus, there exist vertices $a \in A$ and $b \in B$ such that $|N[a] \cap A| < |N(a) \cap B|$ and $|N[b] \cap B| < |N(b) \cap A|$. But then swapping a and b will produce a balanced bipartition with fewer AB-edges, contradicting our choice of π. $\square$

As we have seen, Theorem 3.4 is sharp for complete graphs.

Theorem 3.5 ([37]) *If G is a graph of order $n \ge 2$, then $a_o(G) \le \frac{2n}{3}$.*

Proof Since the result is trivial if G has an isolated vertex, we may assume that the minimum degree of G is at least 1. Color the vertices of V with three colors, say Red, Green and Blue, so that the number of monochromatic edges is minimum. Let v be a vertex colored red. Then v has at least as many green (respectively, blue) neighbors as it has red ones, else v could be recolored green (respectively, blue), decreasing the number of monochromatic edges. Thus, the union of any two color classes is an offensive alliance, implying that $a_o \le \frac{2n}{3}$. $\square$

The authors of [37] note that the bound of Theorem 3.5 is tight for K_3, $K_{2,2,2}$, and the generalized corona $K_3 \circ K_2$.

A defensive (respectively, offensive) alliance S is called *strong* if the inequality is strict. We state the definitions formally as follows.

Definition 3.6 *A nonempty set of vertices S of a graph G is a*

1. *strong defensive alliance if for every $v \in S$, $|N[v] \cap S| > |N(v) \cap \overline{S}|$. The minimum cardinality of a strong defensive alliance of G is the strong defensive alliance number of G, denoted by $\hat{a}(G)$.*
2. *strong offensive alliance if for every vertex $v \in \partial(S)$, $|N(v) \cap S| > |N[v] \cap \overline{S}|$. The minimum cardinality of a strong offensive alliance of G is the strong offensive alliance number of G, denoted by $\hat{a}_o(G)$.*

In 2009, Shafique and Dutton [84] published a paper that had an interesting connection to the Aharoni, Milner, and Prikry theorem (Theorem 2.2) that every graph has an unfriendly partition. A partition $\{S, \overline{S}\}$ of a vertex set of a graph G

is called an *alliance-free partition*, if neither S nor $\overline{S}$ contains a strong defensive alliance as a subset. Shafique and Dutton [84] prove that a connected graph G has an alliance-free partition exactly when G has a block that is neither an odd clique nor an odd cycle. For more on alliance numbers, the reader is referred to [2, 3, 5, 8, 20, 37, 59, 63, 65–67, 70].

In 2009, Brigham, Dutton, Haynes, and Hedetniemi [16] studied alliances that are both defensive and offensive.

Definition 3.7 *A nonempty set of vertices S of a graph G is a powerful alliance if for every $v \in N[S]$, $|N[v] \cap S| \geq |N[v] \setminus S|$. The minimum cardinality of a powerful alliance of G is the powerful alliance number of G, denoted by $a_p(G)$.*

This can be stated equivalently as follows. A *powerful alliance* is a set S of vertices having the property that for every vertex $v \in S$, $d_S(v) \geq d_{\overline{S}}(v) - 1$ and for every vertex $v \in \partial(S)$, $d_S(v) \geq d_{\overline{S}}(v) + 1$. As examples, for the complete graph K_n, $a_p(K_n) = \left\lceil \frac{n}{2} \right\rceil$, and the values of the powerful alliance number of paths and cycles are given in the next result.

Proposition 3.8 ([16]) *For paths P_n and cycles C_n with $n \geq 3$,*

$$a_p(P_n) = \left\lfloor \frac{2n}{3} \right\rfloor \text{ and } a_p(C_n) = \left\lceil \frac{2n}{3} \right\rceil.$$

Powerful alliances are also studied in [14, 15, 46].

An alliance S of any type (defensive, offensive, or powerful) is called *global* if S is in addition a dominating set. We state the definitions formally as follows.

Definition 3.9 *Any alliance S is a global alliance if S is a dominating set.*

1. *The global defensive alliance number $\gamma_a(G)$ (respectively, global strong defensive alliance number $\gamma_{\hat{a}}(G)$, is the minimum cardinality of a global defensive alliance (respectively, global strong defensive alliance) of G.*
2. *The global offensive alliance number $\gamma_o(G)$ (respectively, global strong offensive alliance number $\gamma_{\hat{o}}(G)$) is the minimum cardinality of a global offensive alliance (respectively, global strong offensive alliance) of G.*
3. *The global powerful alliance number $\gamma_{a_p}(G)$ is the minimum cardinality of a global powerful alliance of G.*

The next result follows directly from the definitions.

Proposition 3.10 *For any graph G,*

$$\gamma(G) \leqslant \gamma_a(G) \leqslant \gamma_{\hat{a}}(G),$$

$$\gamma(G) \leqslant \gamma_o(G) \leqslant \gamma_{\hat{o}}(G), \text{ and}$$

$$\gamma(G) \leq \gamma_{a_p}(G).$$

As we mentioned in the introduction, we will survey the results on global defensive alliances as a sampling of global alliances. For more on global offensive alliances, see [10, 12, 23, 26, 29, 48, 62, 74, 89, 98].

Shafique and Dutton [82, 83] generalized alliances to k-alliances for an integer k. Formally, a nonempty set S of vertices of a graph G is a *defensive k-alliance* (respectively, an *offensive k-alliance*) if every vertex of S (respectively, the boundary of S) has at least k more neighbors in S than it has in $\overline{S}$. Note that for $k = -1$, a defensive k-alliance is the standard defensive alliance and for $k = 0$ it is a strong defensive alliance. Similarly, the case for $k = 1$ (respectively, $k = 2$) in a k-offensive alliance corresponds to the normal offensive alliance (respectively, strong offensive alliance). A set $S \subseteq V$ is a *powerful k-alliance* if it is both a defensive k-alliance and an offensive $(k + 2)$-alliance. Much of the research on alliances has been on these generalized k-alliances. See [6, 18, 21, 24, 25, 38, 40, 60, 77, 78, 86, 87, 90, 91, 94–97, 100], for example.

We conclude this section by mentioning that Haynes and Lachniet [49] defined the alliance partition number of a graph as follows. A partition of the vertex set of G into defensive alliances is called an *alliance partition*. The *alliance partition number* $\psi_a(G)$ is the maximum order of any alliance partition of G. For example, the alliance partition number of grid graphs $G_{r,c} = P_r \square P_c$ (the Cartesian product of path graphs on r and c vertices) was determined in [49]. For examples, see Figures 1 and 2, where the alliances in the partition are circled with dashed blue lines.

Theorem 3.11 ([49]) *For the grid graph $G_{r,c}$,*

$$\psi_a(G_{r,c}) = \begin{cases} \left\lceil \frac{c+1}{2} \right\rceil & \text{if } 1 = r \leq c \\[2ex] c & \text{if } 2 = r \leq c \\[2ex] c & \text{if } 3 = r \leq c \text{ and } c \text{ is odd} \\[2ex] c + 1 & \text{if } 3 = r \leq c \text{ and } c \text{ is even} \\[2ex] \left\lfloor \frac{r-2}{2} \right\rfloor \left\lfloor \frac{c-2}{2} \right\rfloor + r + c - 2 & \text{if } 4 \leq r \leq c. \end{cases}$$

We note that the alliance partition number is also studied in [35] and studied under a different name, the quorum coloring number, in [58]. As defined by Hedetniemi, Hedetniemi, Laskar, and Mulder [58] in 2013, the *quorum coloring number* is the maximum order k of a vertex partition $\pi = \{V_1, V_2, \ldots, V_k\}$ such that for every vertex $v \in V_i$, $|N[v] \cap V_i| \geq |N[v]|/2$, that is, at least half of the vertices in the closed neighborhood of every vertex v have the same color as v. They determined the quorum number, and hence the alliance partition number, of a hypercube Q_n as follows.

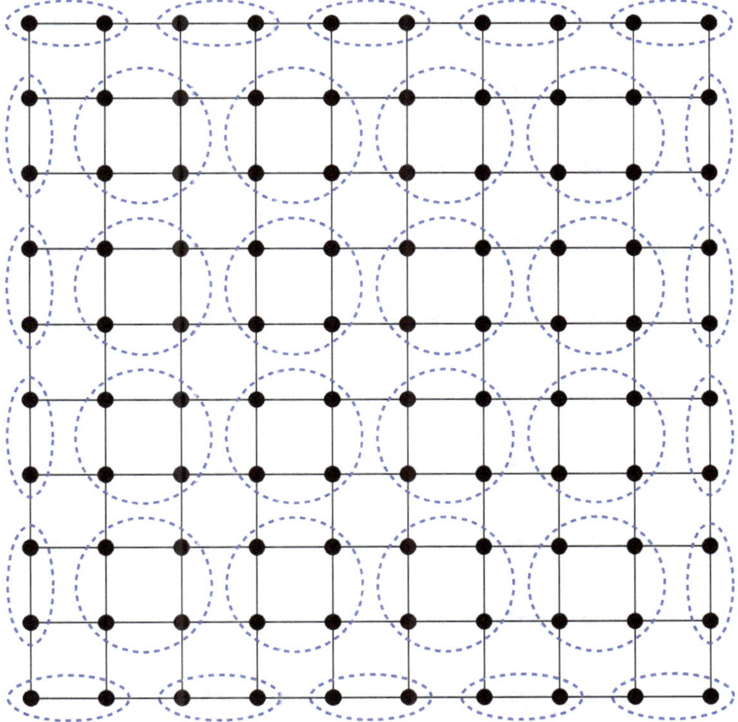

Fig. 1 $\psi_a(G_{10,10}) = 34$

Theorem 3.12 ([58]) *For the hypercube Q_n, $\psi_a(Q_n) = 2^{\lceil \frac{n}{2} \rceil}$.*

Sahbi and Chellali [79] showed that the decision problem associated with the quorum coloring number (alliance partition number) is NP-complete. Partitioning the vertex set of a graph into alliances for other types of alliances is also studied in [84, 90, 96, 100], for example.

4 Global Defensive Alliances

Global defensive alliances were defined in [57, 69] and first studied in [51] and [52]. As examples, we determine the global defensive alliance and global strong defensive alliance numbers of the Petersen graph P.

Proposition 4.1 *For the Petersen graph P, $\gamma_a(P) = 4$, while $\gamma_{\hat{a}}(P) = 5$.*

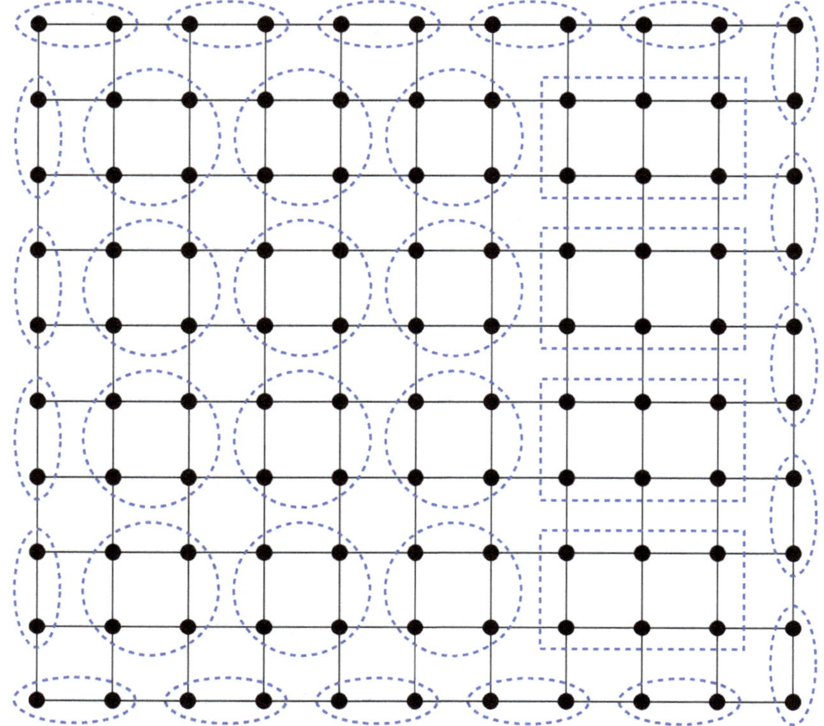

Fig. 2 $\psi_a(G_{10,11}) = 35$

Fig. 3 Petersen Graph P,
$\gamma_a(P) = 4$

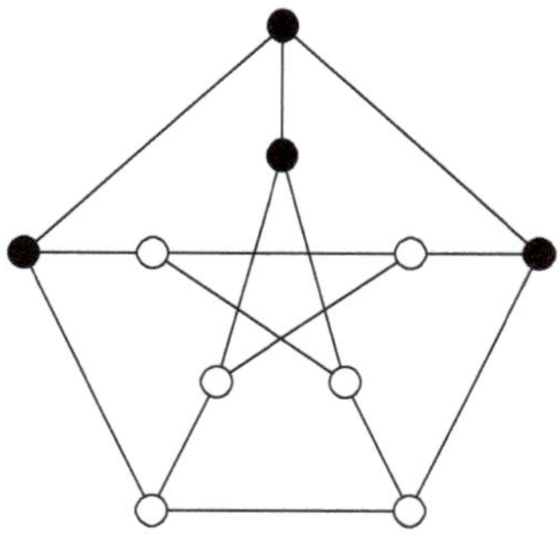

Proof Figure 3 demonstrates a global defensive alliance (the set of darkened vertices) of the Petersen graph P, while Figure 4 illustrates a global strong defensive alliance of P. Hence, $\gamma_a(P) \leq 4$ and $\gamma_{\hat{a}}(P) \leq 5$.

First, to see that $\gamma_a(P) \geq 4$, let S be a γ_a-set of P. Since P is 3-regular and for every vertex $v \in S$, $d_S(v) + 1 \geq d_{\overline{S}}(v)$, it follows that there are no isolated vertices in $P[S]$. That is, every vertex in S must have a neighbor in S. Furthermore, no set of three vertices having this property dominates P, implying that $\gamma_a(P) \geq 4$, and so $\gamma_a(P) = 4$.

Fig. 4 Petersen Graph P, $\gamma_{\hat{a}}(P) = 5$

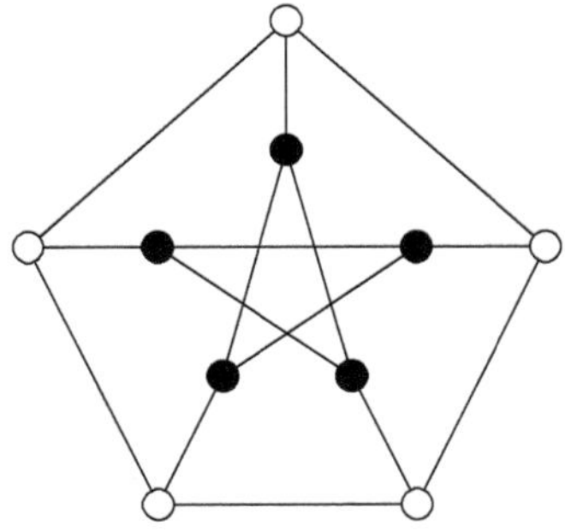

Table 2 Global defensive alliance numbers for families of graphs

Family of graphs G	$\gamma_a(G)$	$\gamma_{\hat{a}}(G)$
Complete graphs K_n	$\left\lfloor \frac{n+1}{2} \right\rfloor$	$\left\lceil \frac{n+1}{2} \right\rceil$
Complete bipartite graphs $K_{r,s}$, $2 \leq r \leq s$	$\left\lfloor \frac{r}{2} \right\rfloor + \left\lfloor \frac{s}{2} \right\rfloor$	$\left\lceil \frac{r}{2} \right\rceil + \left\lceil \frac{s}{2} \right\rceil$
Cycles C_n	$\left\lfloor \frac{n}{2} \right\rfloor + \left\lceil \frac{n}{4} \right\rceil - \left\lfloor \frac{n}{4} \right\rfloor$	$\left\lfloor \frac{n}{2} \right\rfloor + \left\lceil \frac{n}{4} \right\rceil - \left\lfloor \frac{n}{4} \right\rfloor$
Paths P_n, $n \geq 3$, $n \not\equiv 2 \ (mod\ 4)$	$\left\lfloor \frac{n}{2} \right\rfloor + \left\lceil \frac{n}{4} \right\rceil - \left\lfloor \frac{n}{4} \right\rfloor$	$\left\lfloor \frac{n}{2} \right\rfloor + \left\lceil \frac{n}{4} \right\rceil - \left\lfloor \frac{n}{4} \right\rfloor$
Paths P_n, $n \geq 3$, $n \equiv 2 \ (mod\ 4)$	$\left\lfloor \frac{n}{2} \right\rfloor + \left\lceil \frac{n}{4} \right\rceil - \left\lfloor \frac{n}{4} \right\rfloor - 1$	$\left\lfloor \frac{n}{2} \right\rfloor + \left\lceil \frac{n}{4} \right\rceil - \left\lfloor \frac{n}{4} \right\rfloor$
Double Star $S(r, s)$, $1 \leq r \leq s$	$\left\lfloor \frac{r-1}{2} \right\rfloor + \left\lfloor \frac{s-1}{2} \right\rfloor + 2$	$\left\lfloor \frac{r}{2} \right\rfloor + \left\lfloor \frac{s}{2} \right\rfloor + 2$

We note that $\gamma_a(P) = 4 \leq \gamma_{\hat{a}}(P)$. To see that $\gamma_{\hat{a}}(P) \geq 5$, suppose to the contrary that $\gamma_{\hat{a}}(P) = 4$ and let D be a $\gamma_{\hat{a}}$-set of P. Since P has no 4-cycle as a subgraph, at least one of the vertices, say x, in D has $d_D(x) = 1$ and so $d_D(x) + 1 = d_{\overline{D}}(x) = 2$, contradicting that D is a global strong defensive alliance. Hence, $\gamma_{\hat{a}}(P) \geq 5$, and so $\gamma_{\hat{a}}(P) = 5$. $\qquad\qquad\qquad\qquad\qquad\qquad\qquad\qquad\qquad\qquad\qquad\qquad\qquad\quad\square$

Values of the global defensive alliance and global strong defensive alliance numbers for several families of graphs are given [52]. We summarize these results in Table 2.

For the remainder of this section, we focus on bounds on the global (strong) defensive alliance numbers for general graphs and trees. Bounds for several other families of graphs have been studied in [19, 33, 34, 47, 61, 68, 71, 76, 88, 91, 101].

4.1 Bounds for General Graphs

From our previous discussion, $\gamma(G) \leqslant \gamma_a(G) \leqslant \gamma_{\hat{a}}(G) \leq n$ for any graph G of order n. Bullington, Eroh, and Winters [17] showed that for positive integers a, b, and c, where $2 \leq b$ and $c \leq \frac{1}{2}(ab + 2b - a\sqrt{b/a}$, there exists a graph G with $\gamma(G) = a$, $\gamma_a(G) = b$, and $\gamma_{\hat{a}}(G) = c$.

For examples showing that equality and strictness can occur in each of the inequalities, we consider a family of caterpillars. A *caterpillar* is a tree that reduces

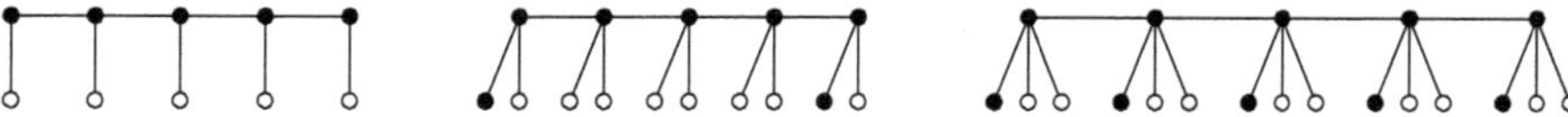

Fig. 5 Caterpillars $G_{5,1}$, $G_{5,2}$, and $G_{5,3}$

to a nonempty path, called the *spine*, upon the removal of all its leaves. For $k \geq 2$, let $G_{k,i}$ denote the caterpillar that has the path P_k as its spine and each vertex on the spine is adjacent to i leaves. In other words, $G_{k,i}$ is the generalized corona $P_k \circ \overline{K}_i$.

We note that $\gamma(G_{k,1}) = \gamma_a(G_{k,1}) = \gamma_{\hat{a}}(G_{k,1}) = k$ and that the vertices of the spine form a γ-set, a γ_a-set, and a $\gamma_{\hat{a}}$-set. Furthermore, $\gamma(G_{k,2}) = \gamma_a(G_{k,2}) = k < k + 2 = \gamma_{\hat{a}}(G_{k,2})$, while $\gamma(G_{k,3}) = k < k + 2 = \gamma_a(G_{k,3}) < 2k = \gamma_{\hat{a}}(G_{k,3})$. See Figure 5 for caterpillars $G_{5,1}$, $G_{5,2}$, and $G_{5,3}$, for examples, where the vertices on the spine form a γ-set and the darkened vertices form a $\gamma_{\hat{a}}$-set. The vertices on the spine also form a γ_a-set of $G_{k,1}$ and $G_{k,2}$, while for $G_{k,3}$, the spine vertices along with two leaves (one adjacent to each end of the spine) form a γ_a-set.

It is observed in [52] that a graph G has $\gamma_a(G) = n$ if and only if $G = \overline{K}_n$. In fact, Haynes, Hedetniemi, and Henning [52] prove the following sharp upper bounds.

Theorem 4.2 ([52]) *Let G be a graph with order n and minimum degree $\delta = \delta(G)$.*

1. $\gamma_a(G) \leq n - \left\lceil \frac{\delta}{2} \right\rceil$, and
2. $\gamma_{\hat{a}}(G) \leq n - \left\lfloor \frac{\delta}{2} \right\rfloor$.

Complete graphs achieve both the bounds of Theorem 4.2.

As we have seen, the domination number is a lower bound on the global defensive alliance number and hence for the global strong defensive alliance number. It was established in [52] that the total domination number $\gamma_t(G)$ is also a lower bound on the global (strong) defensive alliance number for graphs G with given minimum degree.

Theorem 4.3 ([52]) *Let G be a graph with minimum degree $\delta(G)$.*

1. *If $\delta(G) \geq 1$, then $\gamma_t(G) \leq \gamma_{\hat{a}}(G)$.*
2. *If $\delta(G) \geq 2$, then $\gamma_t(G) \leq \gamma_a(G)$.*

These bounds are sharp as can be seen with cycles, that is, for cycles C_n with $n \geq 3$, $\gamma_t(C_n) = \gamma_a(C_n) = \gamma_{\hat{a}}(C_n)$. The caterpillar $G_{k,3}$ (as previously defined) shows that strictness in each bound can occur as $\gamma(G_{k,3}) = \gamma_t(G_{k,3}) = k < k+2 = \gamma_a(G_{k,3}) < 2k = \gamma_{\hat{a}}(G_{k,3})$, for $k \geq 3$. The graph G in Figure 6 gives another example of strictness in the inequalities, where $\gamma(G) \neq \gamma_t(G)$. To see this, we note that the support vertices of G along with a single vertex from the triangle form a γ-set of G, while the support vertices along with their nonleaf neighbors form a γ_t-set of G. Further, the leaves of G along with the three vertices of the triangle form a γ_a-set of G, and the set of darkened vertices in Figure 6 is a $\gamma_{\hat{a}}$-set of G.

Favaron [36] considered relationships between the global (strong) defensive alliance numbers and the independent domination number. We give here the result

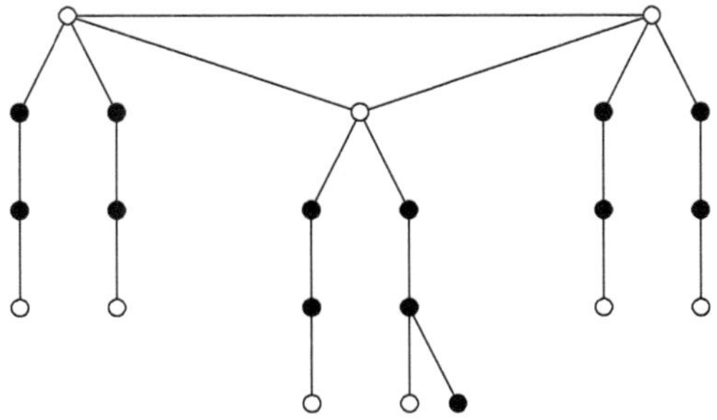

Fig. 6 Graph G with $\gamma(G)=7$, $\gamma_a(G)=10$, $\gamma_t(G)=12$, and $\gamma_{\hat{a}}(G)=13$

and the proof to the first part of the theorem but omit the second proof as it can be proven similarly.

Let $\mathcal{F}_1$ be the family of graphs G obtained from the complete graph K_k by attaching k leaves adjacent to each vertex of the K_k, that is, $G = K_k \circ \overline{K}_k$. Let $\mathcal{F}_2$ be the family of graphs G obtained from the complete graph K_k by attaching $k-1$ leaves adjacent to each vertex of the K_k, that is, $G = K_k \circ \overline{K}_{k-1}$.

Theorem 4.4 ([36]) *For every graph G,*

1. $i(G) \leq (\gamma_a(G))^2 - \gamma_a(G) + 1$ *with equality if and only if $G \in \mathcal{F}_1$.*
2. $i(G) \leq (\gamma_{\hat{a}}(G))^2 - 2\gamma_{\hat{a}}(G) + 2$ *with equality if and only if $G \in \mathcal{F}_2$.*

Proof Let S be a γ_a-set of G, A be a maximal independent set of $G[S]$, and B a maximal independent set of $G[\overline{S} \setminus N(A)]$. Then $A \cup B$ is a maximal independent set of G, and so $i(G) \leq |A| + |B|$. Since S is a defensive alliance, for each $v \in S$, $d_{\overline{S}}(v) \leq d_S(v) + 1$. Furthermore, since S is a dominating set,

$$|B| \leq |\overline{S} \setminus N(A)| \leq \sum_{v \in S \setminus A} d_{\overline{S}}(v) \leq \sum_{v \in S \setminus A} (d_S(v) + 1) \leq |S| - |A| + \sum_{v \in S \setminus A} d_S(v). \tag{1}$$

Thus,

$$i(G) \leq |A| + |B| \leq |S| + \sum_{v \in S \setminus A} d_S(v). \tag{2}$$

Since every vertex $v \in S$ has at most $|S| - 1$ neighbors in S, $i(G) \leq |S| + (|S| - |A|)(|S| - 1)$ with $|A| \geq 1$. Hence, $i(G) \leq |S|^2 - |A|(|S| - 1) \leq |S|^2 - |S| + 1 = (\gamma_a(G))^2 - \gamma_a(G) + 1$.

If $i(G) = (\gamma_a(G))^2 - \gamma_a(G) + 1$, then $|A| = 1$ and $d_S(v) = |S| - 1$ for every $v \in S \setminus A$, that is, S is a clique and A consists of a vertex $a \in S$. Moreover, for any $a \in S$, equality in (1) gives $|B| = |\overline{S} \setminus N(A)|$. Thus, $\overline{S} \setminus N(A)$ is independent, and $|\overline{S} \setminus N(A)| = \sum_{v \in S \setminus \{a\}} d_{\overline{S}}(v) = \sum_{v \in S \setminus \{a\}} (d_S(v) + 1)$. This implies that $N_{\overline{S}}(v)$ is independent and $d_{\overline{S}}(v) = d_S(v) + 1$ for all $v \in S$. Moreover, $N_{\overline{S}}(v) \cap N_{\overline{S}}(u) = \emptyset$ for all $u, v \in S$. It follows that $G \in \mathcal{F}_1$.

For the converse, let G be a graph in $\mathcal{F}_1$, that is, $G = K_k \circ \overline{K}_k$ for $k \geq 1$. Note that the vertices of the clique K_k form a minimum dominating set of G and a global alliance of G, so $\gamma_a(G) = k$. Let v be a vertex of the clique K_k in G, $L(G)$ be the set of leaves of G, and $L(v)$ the set of leaves adjacent to v in G. Then the set $(L(G) \setminus L(v)) \cup \{v\}$ is a minimum independent dominating set of G. Hence, $i(G) = |(L(G) \setminus L(v)) \cup \{v\}| = k^2 - k + 1 = (\gamma_a(G))^2 - \gamma_a(G) + 1$.

A similar argument establishes Part (2) of the theorem.   □

Sharp lower bounds on the global defensive alliance and global strong defensive alliance numbers in terms of the order of a graph are given in [52].

Theorem 4.5 ([52]) *If G is a graph of order n, then*

1. $\gamma_a(G) \geq \frac{\sqrt{4n+1}-1}{2}$, *and*
2. $\gamma_{\hat{a}}(G) \geq \sqrt{n}.$

Proof Let G be a graph of order n, and let S be a γ_a-set of G with $|S| = \gamma_a(G) = k$. Since S is a defensive alliance, each vertex $v \in S$ has at least $\left\lfloor \frac{d(v)}{2} \right\rfloor$ neighbors in S. Hence, $k = |S| \geq |\{v\}| + \left\lfloor \frac{d(v)}{2} \right\rfloor$. Moreover, $d_{\overline{S}}(v) \leq \left\lceil \frac{d(v)}{2} \right\rceil \leq k$. Since S is a dominating set, $|\overline{S}| = n - |S| = n - k \leq \sum_{v \in S} d_{\overline{S}}(v) \leq k^2$. Equivalently, $k^2 + k - n \geq 0$, and so $k \geq \frac{\sqrt{4n+1}-1}{2}$, proving part (1).

Taking into account that strict inequality must hold for a global strong defensive alliance, a similar argument gives the bound of (2).   □

Families of generalized coronas achieve sharpness for the bounds of Theorem 4.5. In particular, $K_k \circ \overline{K}_k$ for $k \geq 3$ has order $n = k^2 + k$ and $\gamma_a(K_k \circ \overline{K}_k) = k = \frac{\sqrt{4n+1}-1}{2}$, while $K_k \circ \overline{K}_{k-1}$ has order $n = k^2$ and $\gamma_{\hat{a}}(K_k \circ \overline{K}_{k-1}) = k = \sqrt{n}$.

Haynes, Hedetniemi, and Henning [52] also proved the lower bounds of $\frac{2n}{\Delta+3}$ and $\frac{2n}{\Delta+2}$ on the global defensive alliance number and the global strong defensive alliance number, respectively, of bipartite graphs with order n and maximum degree Δ. Rodríguez-Velázquez and Sigarreta [75] showed the bipartite condition was not necessary for the bound on the global defensive alliance number to hold and improved the bound of Theorem 4.5 for the global strong defensive alliance number as follows.

Theorem 4.6 ([75]) *If G is a graph of order n with maximum degree $\Delta = \Delta(G)$, then*

1. $\gamma_a(G) \geq \frac{2n}{\Delta+3}$, *and*
2. $\gamma_{\hat{a}}(G) \geq \frac{n}{\Delta/2+1}.$

Proof Let G be a graph of order n and maximum degree $\Delta = \Delta(G)$. Let S be a γ_a-set of G. Now each vertex $v \in S$ has at least $\left\lfloor \frac{d(v)}{2} \right\rfloor$ neighbors in S and has at most $\left\lceil \frac{d(v)}{2} \right\rceil$ neighbors in $\overline{S}$. That is, for each $v \in S$, $d_{\overline{S}}(v) \leq d_S(v) + 1$, and so

$$\sum_{v \in S} d_{\overline{S}}(v) \le \sum_{v \in S} d_S(v) + |S|. \tag{3}$$

Moreover, since S is a dominating set,

$$n - |S| \le \sum_{v \in S} d_{\overline{S}}(v). \tag{4}$$

By Equations 3 and 4, we have

$$2n - 3|S| \le \sum_{v \in S} (d_{\overline{S}}(v) + d_S(v)) \le \sum_{v \in S} d(v) \le \Delta|S|. \tag{5}$$

Thus, $|S| = \gamma_a(G) \ge \frac{2n}{\Delta+3}$.

Next let S be a $\gamma_{\hat{a}}$-set of G. Then,

$$d_{\overline{S}}(v) \le \left\lfloor \frac{d(v)}{2} \right\rfloor. \tag{6}$$

By Equations 4 and 6, $\gamma_{\hat{a}}(G) \ge \frac{n}{\Delta/2+1}$. □

Recall that $K_{r,s}$ denotes the complete bipartite graph with partite sets of cardinality r and s. To see that the bounds of Theorem 4.6 are sharp, we again consider generalized coronas. It is noted in [52] that $K_{k,k} \circ \overline{K}_{k+1}$ for $k \ge 1$ has $\Delta = 2k+1$ and $n = 2k + 2k(k+1) = 2k^2 + 4k$. Since the $2k$ vertices of the $K_{k,k}$ form a global defensive alliance, $\frac{2n}{\Delta+3} = 2k \le \gamma_a(K_{k,k} \circ \overline{K}_{k+1}) \le 2k$, and so the bound is sharp. Moreover, as noted in [75] and by Theorem 4.1, the Petersen graph P has order $n = 10$, $\Delta(P) = 3$, and $\gamma_{\hat{a}}(P) = 4 = \frac{n}{\Delta/2+1}$.

Rodríguez-Velázquez and Sigarreta [75] also determined lower bounds on global defensive alliance numbers of a graph in terms of its spectral radius λ (the largest eigenvalue of the adjacency matrix of the graph).

Theorem 4.7 ([75]) *If G is a graph of order n with spectral radius λ, then*

1. $\gamma_a(G) \ge \frac{n}{\lambda+2}$, *and*
2. $\gamma_{\hat{a}}(G) \ge \frac{n}{\lambda+1}$.

4.2 Bounds for Trees

We present bounds on the global (strong) defensive alliance numbers of trees. In particular, upper bounds are detailed in Section 4.2.1 and lower bounds in Section 4.2.2.

4.2.1   Upper Bounds

The upper bound of Theorem 4.2 on the global defensive alliance number for general graphs was improved for trees in [52]. The authors also characterized trees attaining the new bound. To present the new bound and characterization, we introduce some additional notation and define two families of trees. For a vertex v in a rooted tree T, let $C(v)$ and $D(v)$ denote the sets of children and descendants, respectively, of v, and let $D[v] = D(v) \cup \{v\}$.

Let $\mathcal{T}_1$ be the family of all trees defined as follows: Let $T \in \{P_5, K_{1,4}\}$ or let T be the tree obtained from $tK_{1,4}$ (the disjoint union of $t \geq 2$ copies of the star $K_{1,4}$) by adding $t - 1$ edges between leaves of these copies of $K_{1,4}$ in such a way that the center of each $K_{1,4}$ is adjacent to exactly three leaves in T.

Theorem 4.8 ([52]) *If T is a tree of order $n \geq 4$, then $\gamma_a(T) \leq \frac{3n}{5}$, with equality if and only if $T \in \mathcal{T}_1$.*

Proof We proceed by induction on the order $n \geq 4$ of T. If $n = 4$, then either T is the path P_4 or the star $K_{1,3}$, and so $\gamma_a(T) = 2 < 3n/5$. Suppose, then, that for all trees T' of order n', where $4 \leq n' < n$, $\gamma_a(T') \leq 3n'/5$, with equality if and only if $T \in \mathcal{T}_1$.

Let T be a tree of order $n \geq 4$. If T is a star, then $\gamma_a(K_{1,n-1}) = \left\lfloor \frac{n-1}{2} \right\rfloor + 1 \leq \frac{3n}{5}$ with equality if and only $n = 5$, that is, if and only if $T = K_{1,4} \in \mathcal{T}_1$. If T is a double star, then from the results in Table 2, $\gamma_a(T) < 3n/5$. If T is the path P_5, then $\gamma_a(T) = 3 = 3n/5$ and $T \in \mathcal{T}_1$. Hence, we may assume that $\operatorname{diam}(T) \geq 4$ and that $T \neq P_5$.

Among all support vertices of T of eccentricity $\operatorname{diam}(T) - 1$, let v be one of minimum degree. Root T at a vertex r, where r is at distance $\operatorname{diam}(T) - 1$ from v. Let u denote the parent of v and x the parent of u.

Let T' be the tree of order n' obtained from T by deleting v and its children, that is, $T' = T - D[v]$. Since $\operatorname{diam}(T) \geq 4$ and $T \neq P_5$, it follows from our choice of v that $n' \geq 4$. Applying the inductive hypothesis to T', $\gamma_a(T') \leq 3n'/5$. Let S' be a γ_a-set of T'. Let $|C(v)| = \ell_v$, and so $n = n' + \ell_v + 1$.

If $u \in S'$, then adding v and $\left\lfloor \frac{\ell_v - 1}{2} \right\rfloor$ children of v to S' forms a global defensive alliance of T, and so $\gamma_a(T) \leq |S'| + (\ell_v + 1)/2 \leq 3(n - \ell_v - 1)/5 + (\ell_v + 1)/2 < 3n/5$. Hence, we may assume that $u \notin S'$, else we have the desired result.

If $d(v) = 2$, then adding the child of v to S' produces a global defensive alliance of T, and so $\gamma_a(T) \leq |S'| + 1 \leq 3(n - 2)/5 + 1 < 3n/5$. Hence, we many assume that $\ell_v \geq 2$.

We consider two possibilities based on the $d(u)$. Assume first that $d(u) \geq 3$. If u has a child v' different from v that is a support vertex, then, by our choice of v, $|C(v')| \geq \ell_v \geq 2$. But then we can always choose S' to contain u and v', contradicting our assumption that $u \notin S'$. Hence, every child of u different from v must be a leaf. If u is adjacent to more than one leaf, then again we can choose $u \in S'$, a contradiction. Hence, $d(u) = 3$ and the child y (say) of u different from v is a leaf. Since $u \notin S'$, it follows that $y \in S'$. Deleting y from S' and adding u, v, and $(\ell_v - 1)/2$ children of v to S' yields a global defensive alliance of T, and so $\gamma_a(T) \le |S'| - 1 + 2 + (\ell_v - 1)/2 = |S'| + (\ell_v + 1)/2 < 3n/5$.

Second assume that $d(u) = 2$. If $\mathrm{diam}(T) = 4$, then x is adjacent to r, that is, x is a support vertex. By our choice of v, $d(x) \ge d(v) \ge 3$. Thus, $T - D[u]$ is a star $K_{1,k}$ with x as its center, where $k \ge \ell_v \ge 2$. The set $\{x, u, v\}$ together with $(\ell_v - 1)/2$ leaves adjacent to v and $(k - 1)/2$ leaves adjacent to x is a global defensive alliance of T, and so $\gamma_a(T) \le 3 + (\ell_v - 1)/2 + (k - 1)/2 = (k + \ell_v + 4)/2 = (n + 1)/2$. Since $n \ge 7$, $\gamma_a(T) < 3n/5$. Thus, we may assume that $\mathrm{diam}(T) \ge 5$.

Let $T^* = T - D[u]$ have order n^*. Since $\mathrm{diam}(T) \ge 5$, it follows from our choice of v that $n^* \ge 4$. Applying the inductive hypothesis to T^*, $\gamma_a(T^*) \le 3n^*/5$ with equality if and only if $T^* \in \mathcal{T}_1$.

Let S^* be a γ_a-set of T^*. Adding u and v along with $\left\lfloor \frac{\ell_v - 1}{2} \right\rfloor$ children of v to S^* gives a global defensive alliance of T. Hence, if $\ell_v = 2$, then $\gamma_a(T) \le |S^*| + 2 = 3(n - 4)/5 + 2 < 3n/5$; while if $\ell_v \ge 3$, then $\gamma_a(T) \le |S^*| + (\ell_v + 3)/2 \le 3(n - \ell_v - 2)/5 + (\ell_v + 3)/2 \le 3n/5$. Furthermore, if $\gamma_a(T) = 3n/5$, then $\ell_v = 3$ and $\gamma_a(T^*) = |S^*| = 3n^*/5$. By the inductive hypothesis, $T^* \in \mathcal{T}_1$. Moreover, by our choice of v, the support vertex adjacent to r has at least three leaf neighbors in T^*. Hence, T^* is not the path P_5. If $T^* = K_{1,4}$, then $T \in \mathcal{T}_1$. So we may assume that $T^* \ne K_{1,4}$. That is, T^* is a tree obtained from $t \ge 2$ copies of $K_{1,4}$ by adding $t - 1$ edges as described in the definition of the family $\mathcal{T}_1$. We note that a γ_a-set of T^* can be chosen to consist of the center, the nonleaf neighbor of the center, and one leaf neighbor from each of the t copies of $K_{1,4}$.

Suppose x is a central vertex of one of the copies of $K_{1,4}$ in T^*. Now S^* contains at least one child of x that is a leaf in T^*. Deleting this child of x from S^* and adding u, v, and one child of v forms a global defensive alliance of T. Thus, $\gamma_a(T) \le |S^*| - 3 - 1 = |S^*| + 2 = 3(n - 5)/5 + 2 < 3n/5$, a contradiction. Hence, x must be a leaf of one of the copies of $K_{1,4}$ in T^*. Let z be the center of the $K_{1,4}$ containing x in T^*, and let $N(z) = \{z_1, z_2, z_3, x\}$.

Now x may or may not be a leaf in T^*. If x is a leaf in T^*, then in T, z is adjacent to exactly two leaves, z_1 and z_2, say. Now let D^* be a γ_a-set of T^* that contains all the central vertices of the t copies of $K_{1,4}$ in T^*, exactly one leaf adjacent to each central vertex and all the leaves of $K_{1,4}$ that are incident to the $t - 1$ added edges when constructing T^*. In particular, $z, z_3 \in D^*$. We may assume that $x \in D^*$. Let $D = (D^* - \{x, z, z_3\}) \cup \{z_1, z_2, u, v, w\}$, where w is any child of v. Therefore, D is a global defensive alliance of T of cardinality $\gamma_a(T^*) + 2 < 3n/5$, a contradiction. Hence, x is not a leaf in T^*, that is, x is adjacent to a vertex (a leaf) in a copy of $K_{1,4}$ in T^*. Thus, z_1, z_2, and z_3 are leaves in T^*, and it follows that $T \in \mathcal{T}_1$. $\qquad \square$

A parallel result for the global strong defensive alliance number is also given [52]. Since its proof is similar to the one for Theorem 4.8, we omit it and only state the theorem here. Let $\mathcal{T}_2$ be the family of trees described as follows: Let T be the tree obtained from the disjoint union $t \ge 1$ copies of the star $K_{1,3}$ by adding $t - 1$ edges between leaves of these copies of $K_{1,3}$ in such a way that the center of each $K_{1,3}$ is adjacent to at least one leaf in T. Let $\mathcal{T}_2$ be the family of all such trees T.

Theorem 4.9 ([52]) *If T is a tree of order $n \ge 3$, then $\gamma_{\hat{a}}(T) \le \frac{3n}{4}$, with equality if and only if $T \in \mathcal{T}_2$.*

The global defensive alliance number can be much larger than the vertex independence number for general graphs. For example, the complete graph K_n has $\alpha(K_n) = 1 \leq \left\lfloor \frac{n+1}{2} \right\rfloor = \gamma_a(K_n)$. However, the following result shows that for trees T, $\gamma_a(T)$ is bounded above by $\alpha(T)$. To prove the theorem, we will use the following observation.

Observation 4.10 ([22]) *If T is a tree obtained from a tree T' by attaching a star $K_{1,p}$, for $p \geq 1$, with center u by adding edge uv for some vertex v of T', then $\alpha(T) = \alpha(T') + p$.*

Theorem 4.11 ([22]) *For any tree T, $\gamma_a(T) \leq \alpha(T)$, and this bound is sharp.*

Proof We proceed by induction on the order n of T. It is straightforward to check the result for trees of order $n = 1$ and $n = 2$. Let T be a tree of order $n \geq 3$ and assume that $\gamma_a(T') \leq \alpha(T')$ for every tree T' of order $n' < n$. If T is a star, then $\gamma_a(T) = \lceil n/2 \rceil \leq \alpha(T) = n - 1$, and the result holds.

Assume that T is not a star, and let v be a support vertex of T with exactly one nonleaf neighbor, say w. Let T be the tree obtained from T by removing v and all its leaf neighbors. Since T is not a star, T' has order at least two. Let S' be a γ_a-set of T'. We consider two cases based on the number ℓ_v of leaves adjacent to v in T.

Assume first that $\ell_v \geq 2$. If $w \in S'$, then S' can be extended to a global defensive alliance of T by adding v and $\left\lfloor \frac{\ell_v - 1}{2} \right\rfloor$ leaves adjacent to v. If $w \notin S'$, then adding v and $\left\lceil \frac{\ell_v - 1}{2} \right\rceil$ leaves from $N(v)$ gives a global defensive alliance of T. In either case, $\gamma_a(T) \leq \gamma_a(T') + \left\lceil \frac{\ell_v - 1}{2} \right\rceil + 1$. By Observation 4.10, $\alpha(T) = \alpha(T') + \ell_v$. Applying the inductive hypothesis to T', we obtain $\gamma_a(T) \leq \gamma_a(T') + \left\lceil \frac{\ell_v - 1}{2} \right\rceil + 1 \leq \alpha(T') + \left\lceil \frac{\ell_v - 1}{2} \right\rceil + 1 \leq \alpha(T) - \ell_v + \left\lceil \frac{\ell_v - 1}{2} \right\rceil + 1$, and therefore $\gamma_a(T) \leq \alpha(T)$.

Next assume that $\ell_v = 1$. Let v' be the leaf neighbor of v. Then S' can be extended to a global defensive alliance of T by adding v if $w \in S'$ or adding v' if $w \notin S'$. Thus, $\gamma_a(T) \leq \gamma_a(T') + 1$. By Observation 4.10, $\alpha(T) = \alpha(T') + 1$. Applying the inductive hypothesis to T', we obtain $\gamma_a(T) \leq \gamma_a(T') + 1 \leq \alpha(T') + 1 = \alpha(T)$.

That this bound is sharp may be seen by considering the tree H_k, formed from an odd path P_{2k+1}, for $k \geq 0$, labelled $1, 2, \ldots, 2k+1$, where for each odd labelled vertex v of the path, a new P_5 is added by identifying its center vertex with v. Then $\gamma_a(H_k) = \alpha(H_k) = 3(k+1)$. For example, see H_3 in Figure 7. $\qquad\square$

Fig. 7 The tree H_3

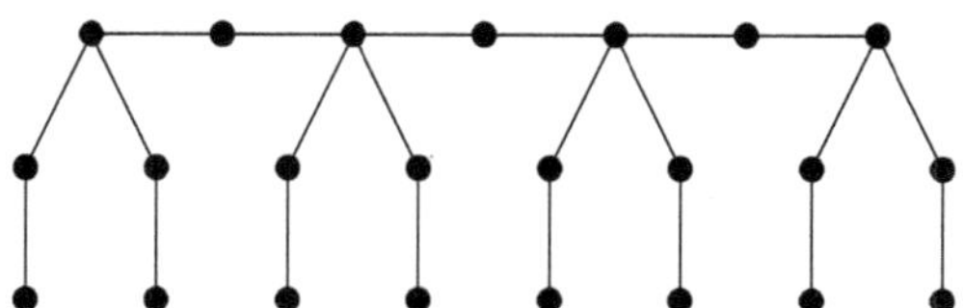

Since $\alpha(T) \leqslant (n + \ell - 1)/2$ for every nontrivial tree T with ℓ leaves [7], the next corollary is an improvement on the bound of Theorem 4.8 for $\ell \leqslant n/5$.

Corollary 4.12 ([22]) *For every nontrivial tree T with ℓ leaves, $\gamma_a(T) \leqslant (n + \ell - 1)/2$.*

The following bounds on the global strong defensive alliance number are also found in [22].

Theorem 4.13 ([22]) *If T is a tree of order $n \geq 3$ with s support vertices, then*

1. $\gamma_{\hat{a}}(T) \leq \frac{3\alpha(T)-1}{2}$, *and*
2. $\gamma_{\hat{a}}(T) \leq \alpha(T) + s - 1$.

4.2.2 Lower Bounds

From our previous discussion, $\gamma(T) \leq \gamma_a(T) \leq \gamma_{\hat{a}}(T)$ for all trees T. Trees T having $\gamma(T) = \gamma_{\hat{a}}(T)$ are characterized in [53].

The following sharp lower bounds for trees are determined in [52].

Theorem 4.14 ([52]) *If T is a tree of order n, then*

1. $\gamma_a(T) \geq (n+2)/4$, *and*
2. $\gamma_{\hat{a}}(T) \geq (n+2)/3$.

Proof For part (1), let S be a γ_a-set of T and $\gamma_a(T) = k$. Further, let $F = T[S]$, and so $V(F) = S$. Since F is a forest, $\sum_{v \in S} d_S(v) = 2|E(F)| \leq 2(|S|-1) = 2(k-1)$. Since S is a global defensive alliance, $d_{\overline{S}}(v) \leq d_S(v) + 1$ for all $v \in S$. Combining these inequalities with the fact that S is a dominating set, it follows that

$$n - k = |\overline{S}| \leq \sum_{v \in S} d_{\overline{S}}(v) \leq \sum_{v \in S}(d_S(v) + 1) \leq 2(k-1) + k = 3k - 2.$$

Hence, $k \geq (n+2)/4$.

For part (2), let S be a $\gamma_{\hat{a}}$-set of T and $\gamma_{\hat{a}}(T) = k$. Let $F = T[S]$. As in part (1), $\sum_{v \in S} d_S(v) = 2|E(F)| \leq 2(|S|-1) = 2(k-1)$. Since S is a global strong defensive alliance, $d_{\overline{S}}(v) \leq d_S(v)$ for every $v \in S$. Since S is a dominating set,

$$n - k = |\overline{S}| \leq \sum_{v \in S} d_{\overline{S}}(v) \leq \sum_{v \in S} d_S(v) \leq 2(k-1),$$

and so $k \geq (n+2)/3$. $\qquad\qquad\square$

We note that the tree T of order n obtained from a tree F of order n' by adding $d_F(v) + 1$ leaves adjacent to each vertex v of F has $\gamma_a(T) = n' = (n+2)/4$, attaining the bound of Theorem 4.14(1). Further, the tree T of order n obtained from a tree F of order n' by adding $d_F(v)$ leaves adjacent to each vertex v of F has $\gamma_{\hat{a}}(T) = n' = (n+2)/3$, attaining the bound of Theorem 4.14(2).

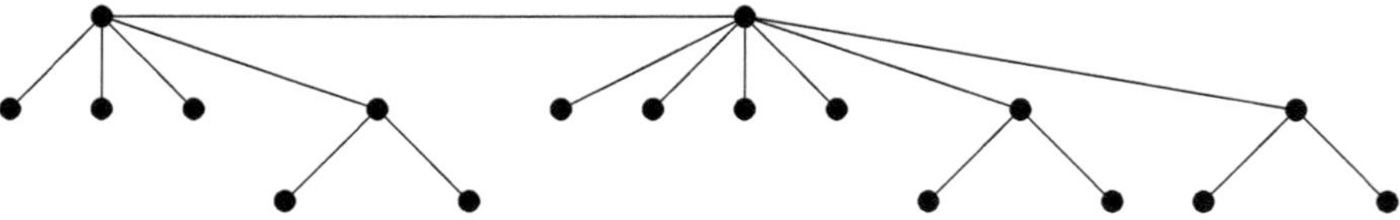

Fig. 8 Tree T_1

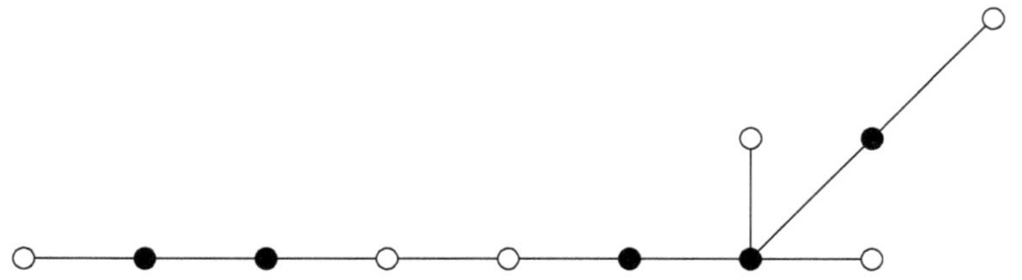

Fig. 9 Tree T_2

Rodríguez-Velázquez and Sigarreta [76] generalized the lower bounds of Theorems 4.14 as follows.

Theorem 4.15 ([76]) *Let T be a tree of order n.*

1. *If S is a γ_a-set of T and T[S] has c components, then $\gamma_a(T) \geq (n + 2c)/4$.*
2. *If S is a $\gamma_{\hat{a}}$-set of T and T[S] has c components, then $\gamma_{\hat{a}}(T) \geq (n + 2c)/3$.*

Note that if $T[S]$ is connected, then the bounds of Theorem 4.15 reduce to the bounds of Theorem 4.14. Bouzefrane, Chellali, and Haynes [13] also improved the lower bounds given in Theorem 4.14 as follows.

Theorem 4.16 ([13]) *Let T be a tree of order $n \geq 2$ with ℓ leaves and s support vertices. Then,*

1. $\gamma_a(T) \geq \frac{3n-\ell-s+4}{8}$, *and*
2. $\gamma_{\hat{a}}(T) \geq \frac{3n-\ell-s+4}{6}$.

We note that constructive characterizations are given in [13] for the trees attaining each of the bounds of Theorem 4.16. For examples of trees achieving the bounds, consider the trees T_1 and T_2 in Figures 8 and 9. The tree T_1 shown in Figure 8 has order $n = 18$, $\ell = 13$, and $s = 5$. The set of support vertices of T_1 is a γ_a-set of T_1, and so $\gamma_a(T_1) = 5 = (3n - \ell - s + 4)/8$. Also, the tree T_2 shown in Figure 9 has order $n = 11$, $\ell = 4$, and $s = 3$. The set of darkened vertices form a $\gamma_{\hat{a}}$-set of T_2, so $\gamma_{\hat{a}}(T_2) = 5 = (3n - \ell - s + 4)/6$.

In [36], Favaron considered relationships between the global (strong) defensive alliance numbers and the independent domination number of trees. We state the result but omit the proofs as they are similar to the proof of Theorem 4.4 for general graphs.

Let $\mathcal{H}_1$ be the family of trees T obtained from a tree H by attaching $d_H(u) + 1$ leaves to each vertex u of H. Let $\mathcal{H}$ be the family of trees H such that for every maximal independent set I of H, the number of components of the forest of $H - I$ is at most $|n(H)|/2$. Further, let $\mathcal{H}_2$ be the family of trees T obtained from a tree $H \in \mathcal{H}$ by attaching $d_H(u)$ leaves to each vertex u of H.

Theorem 4.17 ([36]) *For every tree T of order $n \geq 2$,*

1. $i(T) \leq 2\gamma_a(T) - 1$ *with equality if and only if $T \in \mathcal{H}_1$.*
2. $i(T) \leq 3\gamma_{\hat{a}}(T)/2 - 1$ *with equality if and only if $T \in \mathcal{H}_2$.*

In concluding this section, we note that relationships between global defensive alliance and global offensive alliance numbers are given in [11, 102].

5 Related Parameters and Future Work

We conclude this chapter with ideas for future research involving alliances. We begin by presenting two related concepts, namely cost effective sets and distribution sets, that were defined subsequent to the introduction of alliances. As they are relatively unstudied compared to alliances, we include them here. Additional avenues for future research are discussed in Section 5.1.3.

5.1 *Cost Effective and Distribution Sets*

Cost effective and distribution sets depend on degrees of vertices in the sets S and $\overline{S}$ and are similar to alliances.

5.1.1 Cost Effective Sets

In 2012 Haynes, Hedetniemi, Hedetniemi, McCoy, and Vasylieva [54] introduced cost effective sets and cost effective domination in graphs. This was a reworking of the 1990 concept of unfriendly partitions of Aharoni, Milner, and Prikry [1]. Cost effective sets were proposed to model applications, where services are provided to clients.

Consider the client–server model of human relationships, in which we let a set S represent a collection of servers, providing services to the vertices in $\partial(S)$ over the edges between S and $\partial(S)$. We say that a server, a vertex $u \in S$, is *cost effective* if it serves at least as many clients as other servers, that is, if $d_{\overline{S}}(v) \geq d_S(v)$.

Definition 5.1 *A subset S of vertices of a graph G is cost effective if for every vertex $v \in S$, $d_{\overline{S}}(v) \geq d_S(v)$. The cost effective number $CE(G)$ equals the maximum cardinality of a cost effective set in G, and the lower cost effective number $ce(G)$ equals the minimum cardinality of a maximal cost effective set in G.*

Notice that the property of being a cost effective set is hereditary, that is, every subset of a cost effective set is cost effective. Notice also that every independent set is a cost effective set.

Proposition 5.2 *For any graph G, $\alpha(G) \leq CE(G)$.*

Since a set can be maximal independent but not maximal cost effective, no inequality exists between $ce(G)$ and $i(G)$.

If the inequality is strict, that is, if $d_{\bar{S}}(v) > d_S(v)$ for a vertex $v \in S$, then v is said to be *very cost effective*.

Definition 5.3 *A subset S of vertices in a graph G is very cost effective if every vertex of S is very cost effective. The very cost effective number $VCE(G)$ equals the maximum cardinality of a very cost effective set in G, and the lower very cost effective number $vce(G)$ equals the minimum cardinality of a maximal very cost effective set in G.*

Cost effective domination numbers are defined as expected. The following results are given in [54].

Observation 5.4 ([54]) *For a connected graph G of order $n \geq 2$,*

$$\gamma_{ce}(G) \leq \left\lfloor \frac{n}{2} \right\rfloor.$$

Observation 5.5 ([54]) *Every independent dominating set S in an isolate-free graph G is a very cost effective dominating set.*

Corollary 5.6 ([54]) *For any isolate-free graph G,*

$$\gamma(G) \leq \gamma_{ce}(G) \leq \gamma_{vce}(G) \leq i(G) \leq \alpha(G) \leq \Gamma_{vce}(G) \leq \Gamma_{ce}(G) \leq \Gamma(G).$$

For more on cost effective sets, the reader is referred to [27, 55, 56, 64].

5.1.2 Distribution Centers

In 2018 Desormeaux, Haynes, Hedetniemi, and Moore [28] defined a distribution center in a graph to model a supply and demand situation. In business, a distribution center for products is a structure or a group of units used to store goods that are to be distributed to retailers, to wholesalers, or directly to consumers. Distribution centers are usually thought of as being demand driven.

Definition 5.7 *A nonempty set S of vertices in a graph G is a distribution center if for each vertex $v \in \partial(S)$, there exists a vertex $u \in S$ such that $u \in N(v)$ and $|N[u] \cap S| \geq |N[v] \cap \bar{S}|$. The minimum cardinality of a distribution center of a graph G is the distribution center number $dc(G)$.*

Equivalently, a nonempty set S of vertices in a graph G is a *distribution center* if for each vertex $v \in \partial(S)$, there exists a vertex $u \in S \cap N(v)$ with $d_S(u) \geq d_{\overline{S}}(v)$. For such vertices, we say that u supplies the demand of v, or equivalently, v is supplied by u.

One perspective of a distribution center S is to think of a vertex $v \in \partial(S)$ and its neighbors in $\overline{S}$ as needing some amount of resource units, one unit per vertex, while each vertex in S is able to supply one unit of the resource. Thus, a vertex in $\partial(S)$ makes a demand on the distribution center S and is supplied by one of its neighbors in S. Vertex v asks a vertex $u \in S \cap N(v)$ to deliver $d_{\overline{S}}(v) + 1$ units for itself and its neighbors in $\overline{S}$. This is possible only if the vertex u can receive from itself and its neighbors in S at least this demand, that is, $d_S(u) \geq d_{\overline{S}}(v)$. Hence, such a set S models a distribution center that is capable of providing *two-day delivery* to any vertex (customer) in $\partial(S)$: on day 1, each neighbor of $u \in S$ ships one unit of resource to u, and then, on day 2, vertex u ships $d_{\overline{S}}(v) + 1$ units of resource to its neighbor $v \in \partial(S)$.

Notice the contrast between a distribution center and an offensive alliance. With an offensive alliance S, the neighbors of a vertex $v \in \overline{S}$ that are in S can provide one unit of resource in one day. Thus, the total demand of $d_{\overline{S}}(v) + 1$ by vertex v can be met by the vertices in S, each sending one unit of resource to v. In this way, an offensive alliance is like a one-day distribution center.

One can think of a distribution center as a type of an alliance between the vertices of S to service the vertices in $\partial(S)$. Although distribution centers and offensive alliances are similar concepts, the corresponding parameters can easily be shown to be incomparable. To see this, note that for cycles C_n with $n \geq 5$, $a_o(C_n) = \left\lceil \frac{n}{2} \right\rceil > 2 = \mathrm{dc}(C_n)$. On the contrary, for the complete bipartite graph $K_{r,s}$ with $1 \leq r \leq s$, $a_o(K_{r,s}) = \left\lceil \frac{r+1}{2} \right\rceil < r = \mathrm{dc}(K_{r,s})$.

As with alliances, a distribution center that is also a dominating set is called a *global distribution center*.

Definition 5.8 *A set S of vertices of a graph G is a* global distribution center *if for each vertex $v \in \overline{S}$, there exists a vertex $u \in S \cap N(v)$ such that $d_S(u) \geq d_{\overline{S}}(v)$. The* global distribution center number $\gamma_{\mathrm{dc}}(G)$ *is the minimum cardinality of a global distribution center of G.*

Clearly, every global distribution center is a dominating set. Moreover, every graph G has a distribution center and a global distribution center since the set $V(G)$ is trivially both of these.

Observation 5.9 ([28]) *For any graph G of order n, $\gamma(G) \leq \gamma_{\mathrm{dc}}(G)$ and $\mathrm{dc}(G) \leq \gamma_{\mathrm{dc}}(G) \leq n$.*

We conclude our discussion of distribution centers by illustrating the distribution and global distribution numbers of the Petersen graph P in Figures 10 and 11, respectively, where the darkened vertices represent the appropriate sets.

Notice in this example that both the set of darkened vertices and the set of undarkened vertices are global distribution centers. In fact, in any *prism*, that is,

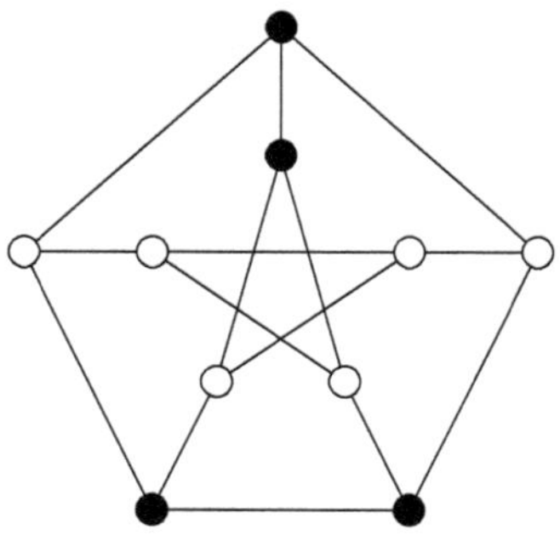

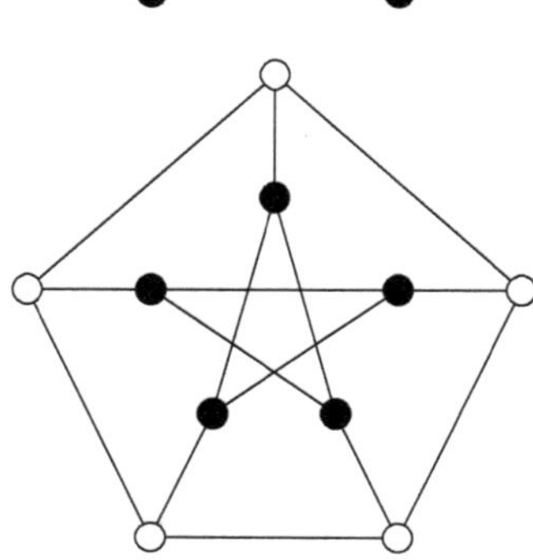

Fig. 10 Petersen Graph P, $\mathrm{dc}(P) = 4$

Fig. 11 Petersen Graph P, $\gamma_{\mathrm{dc}}(P) = 5$

a Cartesian product of the form $G \square K_2$, the set of vertices in each copy of G in $G \square K_2$ is a global distribution center. As another illustration, it can be seen that the set of vertices in any two (or more) consecutive rows (or columns) of a grid graph of the form $G_{m,n} = P_m \square P_n$ forms a distribution center in a grid graph.

5.1.3 Future Research

Much work remains to be done on cost effective sets and distribution sets in graphs, since they were only introduced in 2012 and 2017, respectively.

We note that Corollary 2.3 raises two optimization questions:

Problem 5.10 Over all possible unfriendly partitions $\{S, \overline{S}\}$ of a graph G, what is the smallest and largest cardinality of sets S and $\overline{S}$, or, equivalently, what is the largest difference $|S| - |\overline{S}|$?

Problem 5.11 Over all possible unfriendly partitions $\{S, \overline{S}\}$ of a graph G, what is the smallest and largest cardinality of the set of edges between S and $\overline{S}$?

The definitions given in this chapter suggest a broader avenue for future research.

For example, our discussion of unfriendly partitions suggests the definition of friendly and unfriendly sets as follows. A set S is *friendly* if for every vertex $v \in \partial(S)$, $d_S(v) \leq d_{\overline{S}}(v)$, and is *very friendly* if this inequality is strict, that is, $d_S(v) < d_{\overline{S}}(v)$.

If we reverse this inequality, we get the following:

Table 3 Degree Conditions

	$d_S(v) \leq d_{\overline{S}}(v)$	$d_S(v) \geq d_{\overline{S}}(v)$
for every $v \in S$	cost effective	internally strong
for every $v \in \partial(S)$	friendly	unfriendly

A set S is *unfriendly* if for every vertex $v \in \partial(S)$, $d_S(v) \geq d_{\overline{S}}(v)$, and is *very unfriendly* if this inequality is strict, that is, $d_S(v) > d_{\overline{S}}(v)$.

If we change focus from every vertex $v \in \overline{S}$ to every vertex $u \in S$, we get the following:

A set S is *cost effective* if for every vertex $u \in S$, $d_S(u) \leq d_{\overline{S}}(u)$, and is *very cost effective* if this inequality is strict, that is, $d_S(u) < d_{\overline{S}}(u)$.

If we reverse this inequality, we get the following:

A set S is *internally strong* if for every vertex $u \in S$, $d_S(u) \geq d_{\overline{S}}(u)$, and is *very internally strong* if this inequality is strict, that is, $d_S(u) > d_{\overline{S}}(u)$.

We conclude by summarizing these concepts in Table 3. As usual, if $\partial(S) = \overline{S}$, then the relevant table entries represent dominating sets. Thus, three new types of dominating sets are defined in Table 3.

References

1. R. Aharoni, E.C. Milner, K. Prikry, Unfriendly partitions of a graph. J. Combin. Theory Ser. B **50**, 1–10 (1990)
2. G. Araujo-Pardo, L. Barriére, Defensive alliances in circulant graphs. Ars Combin. **115**, 115–138 (2014)
3. G. Araujo-Pardo, L. Barriére, Defensive alliances in regular graphs. Ars Combin. **115**, 35–54 (2014)
4. C. Bazgan, Z. Tuza, D. Vanderpooten, The satisfactory partition problem. Discrete Appl. Math. **154**, 1236–1245 (2006)
5. C. Bazgan, H. Fernau, Z. Tuza, Aspects of upper defensive alliances. Discrete Appl. Math. **266**, 111–120 (2019)
6. S. Bermudo, J.A. Rodríguez-Velázquez, J.M. Sigarreta, I.G. Yero, On global offensive k-alliances in graphs. Appl. Math. Lett. **23**, 1454–1458 (2010)
7. M. Blidia, M. Chellali, O. Favaron, Independence and 2-domination in trees. Australas. J. Combin. **33**, 317–327 (2005)
8. B. Bliem, S. Woltran, Defensive alliances in graphs of bounded treewidth. Discrete Appl. Math. **251**, 334–339 (2018)
9. A.V. Borodin, A.V. Kostochka, On an upper bound of a graph's chromatic number, depending on the graph's degree and density. J. Combin. Theory Ser. B **23**, 247–250 (1977)
10. M. Bouzefrane, M. Chellali, On the global offensive alliance number of a tree. Opuscula Math. **29**, 223–228 (2009)
11. M. Bouzefrane, M. Chellali, A note on global alliances in trees. Opuscula Math. **31**, 153–158 (2011)
12. M. Bouzefrane, S. Ouatiki, On the global offensive alliance in unicycle graphs. AKCE Int. J. Graphs Comb. **15**, 72–78 (2018)
13. M. Bouzefrane, M. Chellali, T.W. Haynes, Global defensive alliances in trees. Util. Math. **82**, 241–252 (2010)

14. R.C. Brigham, R.D. Dutton, Bounds on powerful alliance numbers. Ars Combin. **88**, 135–159 (2008)
15. R.C. Brigham, R.D. Dutton, S.T. Hedetniemi, A sharp lower bound on the powerful alliance number of $C_m \times C_n$. Congr. Numer. **167**, 57–63 (2004)
16. R.C. Brigham, R.D. Dutton, T.W. Haynes, S.T. Hedetniemi, Powerful alliances in graphs. Discrete Math. **309**, 2140–2147 (2009)
17. G. Bullington, L. Eroh, S.J. Winters, Bounds concerning the alliance number. Math. Bohem. **134**, 387–398 (2009)
18. W. Carballosa, J.M. Rodríguez, J.M. Sigarreta, Y. Torres-Nuñez, Computing the alliance polynomial of a graph. Ars Combin. **135**, 163–185 (2017)
19. C. Chang, M. Chia, C. Hsu, D. Kuo, L. Lai, F. Wang, Global defensive alliances of trees and Cartesian product of paths and cycles. Discrete Appl. Math. **160**, 479–487 (2012)
20. M. Chellali, Offensive alliances in bipartite graphs. J. Combin. Math. Combin. Comput. **73**, 245–255 (2010)
21. M. Chellali, Trees with equal global offensive k-alliance and k-domination numbers. Opuscula Math. **30**, 249–254 (2010)
22. M. Chellali, T.W. Haynes, Global alliances and independence in trees. Discuss. Math. Graph Theory **27**, 19–27 (2007)
23. M. Chellali, L. Volkmann, Independence and global offensive alliance in graphs. Australas. J. Combin. **47**, 125–131 (2010)
24. M. Chellali, L. Volkmann, Global offensive k-alliance in bipartite graphs. Opuscula Math. **32**, 83–89 (2012)
25. M. Chellali, T.W. Haynes, B. Randerath, L. Volkmann, Bounds on the global offensive k-alliance number in graphs. Discuss. Math. Graph Theory **29**, 597–613 (2009)
26. M. Chellali, T.W. Haynes, L. Volkmann, Global offensive alliance numbers in graphs with emphasis on trees. Australas. J. Combin. **45**, 87–96 (2009)
27. M. Chellali, T.W. Haynes, S.T. Hedetniemi, Client-server and cost effective sets in graphs. AKCE Int. J. Graphs Comb. **15**, 211–218 (2018)
28. W.J. Desormeaux, T.W. Haynes, S.T. Hedetniemi, C. Moore, Distribution centers in graphs. Discrete Appl. Math. **243**, 186–193 (2018)
29. M.C. Dourado, L. Faria, M.A. Pizaña, D. Rautenbach, J.L. Szwarcfiter, On defensive alliances and strong global offensive alliances. Discrete Appl. Math. **163**, 136–141 (2014)
30. J.E. Dunbar, S.T. Hedetniemi, M.A. Henning, P.J. Slater, Signed domination in graphs, in *Graph Theory, Combinatorics and Applications* (Wiley, New York, 1995), pp. 311–322
31. J.E. Dunbar, S.T. Hedetniemi, M.A. Henning, A.A. McRae, Minus domination in graphs. Discrete Math. **199**(1–3), 35–47 (1999)
32. J.E. Dunbar, D.G. Hoffman, R.C. Laskar, L.R. Markus, α-domination. Discrete Math. **211**, 11–26 (2000)
33. R.G. Eballe, R.M. Aldema, E.M. Paluga, R.F. Rulete, F.P. Jamil, Global defensive alliances in the join, corona and composition of graphs. Ars Combin. **107**, 225–245 (2012)
34. R.I. Enciso, R.D. Dutton, Lower bounds for global alliances of planar graphs. Congr. Numer. **187**, 187–192 (2007)
35. L. Eroh, R. Gera, Alliance partition number in graphs. Ars Combin. **103**, 519–529 (2012)
36. O. Favaron, Global alliances and independent domination in some classes of graphs. Electron. J. Combin. **15**(1), Research Paper 123, 9 pp. (2008)
37. O. Favaron, G.H. Fricke, W. Goddard, S.M. Hedetniemi, S.T. Hedetniemi, P. Kristiansen, R. Laskar, R.D. Skaggs, Offensive alliances in graphs. Discuss. Math. Graph Theory **24**, 263–275 (2004)
38. H. Fernau, J.A. Rodríguez-Velázquez, J.M. Sigarreta, Offensive k-alliances in graphs. Discrete Appl. Math. **157**(1), 177–182 (2009)
39. H. Fernau, J. A. Rodríguez-Velázquez, A survey on alliances and related parameters in graphs. Electron. J. Graph Theory Appl. **2**, 70–86 (2014)
40. H. Fernau, J.A. Rodríguez-Velázquez, J.M. Sigarreta, Global powerful r-alliances and total k-domination in graphs. Util. Math. **98**, 127–147 (2015)

41. G.W. Flake, S. Lawrence, C.L. Giles, Efficient identification of web communities, in *Proceedings of the 6th ACM SIGKDD International Conference on Knowledge Discovery and Data Mining* (KDD-2000) (2000), pp. 150–160
42. G.H. Fricke, L.M. Lawson, T.W. Haynes, S.M. Hedetniemi, S.T. Hedetniemi, A note on defensive alliances in graphs. Bull. Inst. Combin. Appl. **38**, 37–41 (2003)
43. M.U. Gerber, D. Kobler. Partitioning a graph to satisfy all vertices. Eur. J. Oper. Res. **125**(2), 283–291 (2000)
44. M.U. Gerber, D. Kobler, Partitioning a graph to satisfy all vertices. Technical Report, Swiss Federal Institute of Technology, Lausanne, 2001
45. M.U. Gerber, D. Kobler. Classes of graphs that can be partitioned to satisfy all their vertices. Australas. J. Combin. **29**, 201–214 (2004)
46. A. Harutyunyan, A fast algorithm for powerful alliances in trees, in *Combinatorial Optimization and Applications. Part I.* Lecture Notes in Comput. Sci., vol. 6508 (Springer, Berlin, 2010), pp. 31–40
47. A. Harutyunyan, Some bounds on global alliances in trees. Discrete Appl. Math. **161**, 1739–1746 (2013)
48. A. Harutyunyan, Global offensive alliances in graphs and random graphs. Discrete Appl. Math. **164**, 522–526 (2014)
49. T.W. Haynes, J.A. Lachniet, The alliance partition number of grid graphs. AKCE Int. J. Graphs Comb. **4**, 51–59 (2007)
50. T.W. Haynes, S.T. Hedetniemi, P.J. Slater, *Fundamentals of Domination in Graphs* (Marcel Dekker, New York, 1998)
51. T.W. Haynes, S.T. Hedetniemi, M.A. Henning, Global defensive alliances, in *Proc. 17th Int. Symp. Comput. Inform. Sci., I, ISCIS XVII (Orlando, FL, 2002)* (CRC Press, Boca Raton, 2002), pp. 303–307
52. T.W. Haynes, S.T. Hedetniemi, M.A. Henning, Global defensive alliances in graphs. Electron. J. Combin. **10**(1), R47, 13 pp. (2003)
53. T.W. Haynes, S.T. Hedetniemi, M.A. Henning, A characterization of trees with equal domination and global strong alliance numbers. Util. Math. **66**, 105–119 (2004)
54. T.W. Haynes, S.M. Hedetniemi, S.T. Hedetniemi, T.L. McCoy, I. Vasylieva, Cost effective domination in graphs. Congr. Numer. **211**,197–209 (2012)
55. T. W. Haynes, S.T. Hedetniemi, I. Vasylieva, Very cost effective bipartitions in graphs. AKCE Int. J. Graphs Combin. **12**, 155–160 (2015)
56. T.W. Haynes, S.T. Hedetniemi, T.L. McCoy, T.K. Rodriguez, Bounds on cost effective domination numbers. Quaest. Math. **39**, 773–783 (2016)
57. S.M. Hedetniemi, S.T. Hedetniemi, P. Kristiansen, Alliances in graphs. J. Combin. Math. Combin. Comput. **48**, 157–177 (2004)
58. S.M. Hedetniemi, S.T. Hedetniemi, R. Laskar, H.M. Mulder, Quorum colorings of graphs. AKCE Int. J. Graphs Comb. **10**, 97–109 (2013)
59. C. Hegde, B. Sooryanarayana, Strong alliances in graphs. Commun. Comb. Optim. **4**, 1–13 (2019)
60. J.C. Hernández-Gómez, J.M. Sigarreta, Global dual k-alliance in cubic graphs. Adv. Appl. Discrete Math. **12**, 173–178 (2013)
61. C.J. Hsu, F.H. Wang, Y.L. Wang. Global defensive alliances in star graphs. Discrete Appl. Math. **157**, 1924–1931 (2009)
62. N. Jafari Rad, A note on the global offensive alliances in graphs. Discrete Appl. Math. **250**, 373–376 (2018)
63. L.H. Jamieson, S.T. Hedetniemi, A.A. McRae, The algorithmic complexity of alliances in graphs. J. Combin. Math. Combin. Comput. **68**, 137–150 (2009)
64. F.P. Jamil, H.M. Nuenay-Maglanque, Cost effective domination in the join, corona and composition of graphs. Eur. J. Pure Appl. Math. **12**, 978–998 (2019)
65. T. Kawaura, K. Kimura, N. Niki, Defensive alliances with prescribed and proscribed vertices. Congr. Numer. **220**, 219–226 (2014)

66. H. Kharazi, A. Mosleh Tehrani, On the defensive alliances in graph. Trans. Comb. **8**, 1–14 (2019)
67. K. Kimura, M. Koyama, A. Saito, Small alliances in a weighted graph. Discrete Appl. Math. **158**, 2071–2074 (2010)
68. M. Kiyomi, Y. Otachi, Alliances in graphs of bounded clique-width. Discrete Appl. Math. **223**, 91–97 (2017)
69. P. Kristiansen, S.M. Hedetniemi, S.T. Hedetniemi, Introduction to alliances in graphs, in *Proc. 17th Int. Symp. Comput. Inform. Sci., ISCIS XVII (Orlando, FL, 2002)* (CRC Press, Boca Raton, 2002), pp. 308–312
70. R. Lewoń, A. Malafiejska, M. Malafiejski, Global defensive sets in graphs. Discrete Math. **339**, 1861–1870 (2016)
71. C. Lin, J. Liu, Y. Wang, Global strong defensive alliances of Sierpiński-like graphs. Theory Comput. Syst. **53**, 365–385 (2013)
72. O. Ore, *Theory of Graphs*, vol. 38 (Amer. Math. Soc. Colloq. Publ., Providence, 1962)
73. K. Ouazine, H. Slimani, A. Tari, Alliances in graphs: parameters, properties and applications-a survey. AKCE Int. J. Graphs Comb. **15**, 115–154 (2018)
74. J.A. Rodríguez-Velázquez, J.M. Sigarreta, Global offensive alliances in graphs, in *CTW2006–Cologne-Twente Workshop on Graphs and Combinatorial Optimization*. Electron. Notes Discrete Math., vol. 25 (Elsevier Sci. B. V., Amsterdam, 2006), pp. 157–164
75. J.A. Rodríguez-Velázquez, J.M. Sigarreta, Spectral study of alliances in graphs. Discuss. Math. Graph Theory **27**, 143–157 (2007)
76. J.A. Rodríguez-Velázquez, J.M. Sigarreta, Global alliances in planar graphs. AKCE Int. J. Graphs Comb. **4**, 83–98 (2007)
77. J.A. Rodríguez-Velázquez, J.M. Sigarreta, Global defensive k-alliances in graphs. Discrete Appl. Math. **157**, 211–218 (2009)
78. J.A. Rodríguez-Velázquez, I.G. Yero, J.M. Sigarreta, Defensive k-alliances in graphs. Appl. Math. Lett. **22**, 96–100 (2009)
79. R. Sahbi, M. Chellali, On some open questions concerning quorum colorings of graphs. Discrete Appl. Math. **247**, 294–299 (2018)
80. E. Sampathkumar, L. Pushpa Latha, Strong weak domination and domination balance. Discrete Math. **161**, 235–242 (1996)
81. K.H. Shafique, R.D. Dutton, On satisfactory partitioning of graphs. Congr. Numer. **154**, 183–194 (2002)
82. K.H. Shafique, R.D. Dutton, Maximum alliance-free and minimum alliance-cover sets. Congr. Numer. **162**, 139–146 (2003)
83. K.H. Shafique, R.D. Dutton, A tight bound on the cardinalities of maximum alliance-free and minimum alliance-cover sets. J. Combin. Math. Combin. Comput. **56**, 139–145 (2006)
84. K.H. Shafique, R.D. Dutton, Partitioning a graph into alliance-free sets. Discrete Math. **309**, 3102–3105 (2009)
85. S. Shelah, E.C. Milner. Graphs with no unfriendly partitions, in *A Tribute to Paul Erdös*, ed. by A. Baker, B. Bollobás, A. Hajnal (Cambridge University Press, Cambridge, 1990), pp. 373–384
86. J.M. Sigarreta, Upper k-alliances in graphs. Int. J. Contemp. Math. Sci. **6**(41–44), 2121–2128 (2011)
87. J.M. Sigarreta, S. Bermudo, H. Fernau, On the complement graph and defensive k-alliances. Discrete Appl. Math. **157**(8), 1687–1695 (2009)
88. J.M. Sigarreta, J.A. Rodriőguez, On defensive alliances and line graphs. Appl. Math. Lett. **19**, 1345–1350 (2006)
89. J.M. Sigarreta, J.A. Rodríguez, On the global offensive alliance number of a graph. Discrete Appl. Math. **157**, 219–226 (2009)
90. J.M. Sigarreta, I.G. Yero, S.E. Bermudo, J.A. Rodríguez-Velázquez, Partitioning a graph into offensive k-alliances. Discrete Appl. Math. **159**, 224–231 (2011)
91. M. Tavakoli, S. Klavžar, Distribution of global defensive k-alliances over some graph products. CEJOR Cent. Eur. J. Oper. Res. **27**, 615–623 (2019)

92. J.A. Telle, Complexity of domination-type problems in graphs. Nordic J. Comput. **1**, 157–171 (1994)
93. J.A. Telle, Vertex partitioning problems: characterization, complexity and algorithms on partial k-trees. PhD thesis, Dept. Computer and Information Science, University of Oregon, 1994
94. L. Volkmann, Connected global offensive k-alliances in graphs. Discuss. Math. Graph Theory **31**, 699–707 (2011)
95. I.G. Yero, J.A. Rodríguez-Velázquez, Boundary defensive k-alliances in graphs. Discrete Appl. Math. **158**, 1205–1211 (2010)
96. I.G. Yero, J.A. Rodríguez-Velázquez, Partitioning a graph into global powerful k-alliances. Graphs Combin. **28**, 575–583 (2012)
97. I.G. Yero, J.A. Rodríguez-Velázquez, Boundary powerful k-alliances in graphs. Ars Combin. **111**, 495–504 (2013)
98. I.G. Yero, J.A. Rodríguez-Velázquez, Computing global offensive alliances in Cartesian product graphs. Discrete Appl. Math. **161**, 284–293 (2013)
99. I.G. Yero, J.A. Rodríguez-Velázquez, A survey on alliances in graphs: defensive alliances. Util. Math. **105**, 141–172 (2017)
100. I.G. Yero, S. Bermudo, J.A. Rodríguez-Velázquez, J.M. Sigarreta, Partitioning a graph into defensive k-alliances. Acta Math. Sin. (Engl. Ser.) **27**, 73–82 (2011)
101. I.G. Yero, M. Jakovac, D. Kuziak, On global (strong) defensive alliances in some product graphs. Commun. Comb. Optim. **2**, 21–33 (2017)
102. A. Yu, An inequality on global alliances for trees. Discrete Appl. Math. **185**, 227–229 (2015)

Fractional Domatic, Idomatic, and Total Domatic Numbers of a Graph

Wayne Goddard and Michael A. Henning

AMS Subject Classification: 05C69, 05C65

1 Introduction

In this chapter, we survey some results concerning the fractional domatic, fractional idomatic, and fractional total domatic numbers of a graph. First, we recall the fundamental concepts of a dominating set, an independent dominating set, and a total dominating set.

A *dominating set* of a graph G is a set S of vertices of G such that every vertex not in S has a neighbor in S. The *domination number* of G, denoted $\gamma(G)$, is the minimum cardinality of a dominating set.

An *independent dominating set* of G is a set that is both a dominating set and an independent set. The *independent domination number*, denoted $i(G)$, is the

The research of the author Michael A. Henning was supported in part by the University of Johannesburg.

W. Goddard (✉)
School of Computing and School of Mathematical and Statistical Sciences, Clemson University, Clemson, SC 29634, USA

Department of Mathematics and Applied Mathematics, University of Johannesburg, Auckland Park 2006, Johannesburg, South Africa
e-mail: goddard@clemson.edu

M. A. Henning
Department of Mathematics and Applied Mathematics, University of Johannesburg, Johannesburg, South Africa
e-mail: mahenning@uj.ac.za

minimum cardinality of an independent dominating set of G. An independent set of vertices in a graph G is a dominating set of G if and only if it is a maximal independent set. Thus, $i(G)$ is equivalently the minimum cardinality of a maximal independent set of vertices in G. A survey on independent domination in graphs can be found in [9].

A *total dominating set*, abbreviated TD-set, of a graph G with no isolated vertex is a set S of vertices such that every vertex in G is adjacent to a vertex in S. The *total domination number*, denoted by $\gamma_t(G)$, is the minimum cardinality of a TD-set of G. For a recent book on total domination in graphs, we refer the reader to [14].

The parameters studied in this chapter represent the fractional relaxation of the count of the maximum number of disjoint dominating, independent dominating, and total dominating sets. We discuss these in the next three sections. In Section 5, we discuss a common framework in hypergraphs and in Section 6 some generalizations.

2 The Fractional Domatic Number

In this section, we survey results on the fractional domatic number of a graph. First, we recall the concept of the domatic number of a graph. The maximum number of vertex-disjoint dominating sets in a graph G is called the *domatic number* of G. While this is often denoted by $d(G)$, we will use here the notation $dom(G)$. The domatic number was introduced in 1975 by Cockayne and Hedetniemi [4] and has since been the subject of a large number of publications; a rough estimate says that it occurs in more than 200 papers to date. Much of the early work on the domatic number of a graph and its variants was due to Zelinka. As he remarked in [31], the word "domatic" was created from the words "dominating" and "chromatic" since, although it is defined using the concept of domination in graphs, it is somewhat analogous to the chromatic number of a graph, where we partition the vertex set into classes having certain properties (in this case, each class is a dominating set).

We consider here a fractional analog of the parameter $dom(G)$. For a family $\mathcal{F}$ of subsets of $V(G)$, let $m(\mathcal{F})$ denote the maximum number of times a vertex appears in $\mathcal{F}$ (equivalently, the maximum degree of the hypergraph with $\mathcal{F}$ as the hyperedges). The *fractional domatic number* of a graph G, denoted FDOM(G), is defined as

$$\text{FDOM}(G) = \max \frac{|\mathcal{F}|}{m(\mathcal{F})},$$

where the maximum is taken over all families $\mathcal{F}$ of dominating sets of G. (For a discussion of why FDOM can be viewed as a fractional analog and why maximum can be used instead of supremum in the above formula, see Section 5.) The fractional domatic number seems to have been formally introduced in 2006 by Suomela [26], although the concept was studied in 2000 by Fujita, Yamashita, and Kameda [8].

We start with the following immediate lower and upper bounds on the fractional domatic number of a graph.

Theorem 1 *For every graph G, we have*

$$dom(G) \leq \mathrm{FDOM}(G) \leq \frac{n(G)}{\gamma(G)}.$$

Proof. To prove the lower bound, consider a family $\mathcal{F}$ that consists of a maximum number of vertex-disjoint dominating sets of G. In this case, $|\mathcal{F}| = dom(G)$ and $m(\mathcal{F}) = 1$, implying that $\mathrm{FDOM}(G) \geq |\mathcal{F}|/m(\mathcal{F}) = dom(G)$.

To prove the upper bound, let $\mathcal{F}$ be a family of dominating sets of G. Each set in the family $\mathcal{F}$ has size at least $\gamma(G)$. Thus,

$$\gamma(G) \cdot |\mathcal{F}| \leq \sum_{F \in \mathcal{F}} |F| \leq n(G) \cdot m(\mathcal{F}),$$

or, equivalently, $|\mathcal{F}|/m(\mathcal{F}) \leq n(G)/\gamma(G)$, whence the result. $\square$

We note that equality occurs in the upper bound of Theorem 1 for all complete graphs. We show next that equality also occurs in the upper bound for all cycles. We use the standard notation $[k] = \{1, \ldots, k\}$.

Proposition 2 *For $n \geq 3$ we have $\mathrm{FDOM}(C_n) = n/\gamma(C_n)$.*

Proof. Let G be the cycle $v_1 v_2 \ldots v_n v_1$. Let S be an arbitrary minimum dominating set of G; so $|S| = \gamma(G)$. Let $\mathcal{S} = \{j \mid j \in [n] \text{ and } v_j \in S\}$. For $i \in [n]$, let

$$S_i = \{v_{i+j} \mid j \in \mathcal{S}\},$$

where addition is taken modulo n. Each set S_i is a minimum dominating set of G. Let $\mathcal{F} = \{S_1, S_2, \ldots, S_n\}$. We note that each vertex of G appears in exactly $\gamma(G)$ of these sets, implying that $m(\mathcal{F}) = \gamma(G)$ and therefore that

$$\frac{|\mathcal{F}|}{m(\mathcal{F})} = \frac{n}{\gamma(G)}.$$

Hence, $\mathrm{FDOM}(G) \geq n/\gamma(G)$. The desired result now follows from the upper bound of Theorem 1. $\square$

The above proposition immediately generalizes to any circulant. Recall that $\gamma(C_n) = \lceil n/3 \rceil$. Hence, as a consequence of Proposition 2, the fractional domatic number of a cycle is determined as follows:

$$\mathrm{FDOM}(C_n) = \begin{cases} 3 & \text{if } n \equiv 0 \,(\mathrm{mod}\ 3) \\ \frac{3n}{n+2} & \text{if } n \equiv 1 \,(\mathrm{mod}\ 3) \\ \frac{3n}{n+1} & \text{if } n \equiv 2 \,(\mathrm{mod}\ 3). \end{cases}$$

It is immediate that a graph G of minimum degree δ has $dom(G) \le \delta + 1$. The same upper bound holds for the fractional domatic number:

Theorem 3 *If a graph G of order n has minimum degree δ, then*

$$\frac{n}{n-\delta} \le \text{FDOM}(G) \le \delta + 1.$$

Proof. Let v be a vertex of minimum degree in G, and so $d_G(v) = \delta$. Let $\mathcal{F}$ be a family of dominating sets of G. Each set in the family $\mathcal{F}$ must contain the vertex v or a neighbor of v in order to dominate v. By the pigeonhole principle, at least one vertex in the closed neighborhood of v appears in at least $|\mathcal{F}|/(\delta + 1)$ sets, and so $m(\mathcal{F}) \ge |\mathcal{F}|/(\delta + 1)$, or, equivalently, $|\mathcal{F}|/m(\mathcal{F}) \le \delta + 1$. This is true for every family $\mathcal{F}$ of dominating sets of G, implying that $\text{FDOM}(G) \le \delta + 1$.

To prove the lower bound, consider the collection $\mathcal{F}$ of all subsets F of the vertex set of exactly $n - \delta$ elements. Since every vertex belongs to F or has a neighbor in F, each such subset F is a dominating set of H; that is, $\mathcal{F}$ is a family of dominating sets of G. Furthermore, every vertex is in $\binom{n-1}{n-\delta-1}$ sets of $\mathcal{F}$, and so

$$\frac{|\mathcal{F}|}{m(\mathcal{F})} = \frac{\binom{n}{n-\delta}}{\binom{n-1}{n-\delta-1}} = \frac{n}{n-\delta}.$$

Hence, $\text{FDOM}(G) \ge n/(n-\delta)$. $\square$

Thus, for example, by Theorem 3, if a graph G contains an isolated vertex, then $dom(G) = \text{FDOM}(G) = 1$. However, if G has no isolated vertex and S is a maximal independent set in G, then both S and $V(G) \smallsetminus S$ are dominating sets, implying that $dom(G) \ge 2$. Thus, $\text{FDOM}(G) \ge 2$ if and only if G contains no isolated vertex.

It remains an open problem to characterize the graphs G achieving equality in the upper bound of Theorem 3. In the special case when the graph is a regular graph, Fujita, Yamashita, and Kameda proved in [8] the following result.

Theorem 4 ([8]) *If G is a δ-regular graph of order n, then $\text{FDOM}(G) = \delta + 1$ if and only $dom(G) = \delta + 1$.*

We note that the parameter FDOM is monotonic, in that it cannot decrease on the addition of edges. We next examine the fractional domatic number of the disjoint union or join of graphs. Let G and H be two graphs. The *join* of G and H, written $G \oplus H$, is the graph obtained from the disjoint union of G and H by joining each vertex of G to every vertex of H, while the *union* of G and H, written $G + H$, is the graph consisting of the disjoint union of G and H. As a consequence of Lemma 32 later, we get the following result.

Theorem 5 *For graphs G and H, it holds that*

$$\text{FDOM}(G + H) = \min(\text{FDOM}(G), \text{FDOM}(H)).$$

A *universal vertex* of a graph G is one adjacent to all other vertices in G.

Lemma 6 *For graphs G and H without a universal vertex with orders n_1 and n_2, respectively, the following hold:*

(a) $\mathrm{FDOM}(G \oplus H) = n_1$ *if* $n_1 = n_2$.
(b) $\min(n_1, n_2) < \mathrm{FDOM}(G \oplus H) \le (n_1 + n_2)/2$ *otherwise.*

Proof. The upper bound follows from Theorem 1, since the join of G and H has domination number 2. For the lower bound in (a), let $\mathcal{F}$ consist of n_1 disjoint pairs, each pair containing one vertex of G and one vertex of H. For the lower bound in (b), assume $n_1 < n_2$. Then, let $\mathcal{F}$ consist of (i) all pairs of one vertex in G and one vertex in H and (ii) the set $V(H)$. Then, $|\mathcal{F}| = n_1 n_2 + 1$ and $m(\mathcal{F}) = n_2$, whence the result. $\square$

Fujita, Yamashita, and Kameda proved the following surprising and beautiful result in [8].

Theorem 7 ([8]) *If G is a cubic graph, then G has a family $\mathcal{F}$ of five dominating sets of G such that $m(\mathcal{F}) \le 2$. Furthermore, such a family $\mathcal{F}$ can be constructed in polynomial time.*

As an immediate consequence of Theorem 7, we have the following lower bound on the fractional domatic number of a cubic graph.

Theorem 8 *If G is a cubic graph, then $\mathrm{FDOM}(G) \ge \frac{5}{2}$.*

As remarked by Fujita et al. [8], it is not true that if G is a cubic graph, then G has a family $\mathcal{F}$ of six dominating sets of G such that every vertex is in at most two of these. We note that if a cubic graph had such a property, then this would imply that $\mathrm{FDOM}(G) \ge 6/2 = 3$. However, that in turn would imply that $\gamma(G) \le n(G)/3$, but Kostochka and Stodolsky [19] showed there are cubic graphs G, where $\gamma(G) = 8n(G)/23 + o(1)$.

Abbas, Egerstedt, Liu, Thomas, and Whalen [1] generalized Theorem 7 to a larger class of graphs, namely the class of $K_{1,6}$-free graphs; that is, graphs with no induced subgraph isomorphic to $K_{1,6}$. Their study was motivated by a problem encountered both in the multiagent robotics and in the mobile sensor networks domains. As remarked in [1], the generalization to $K_{1,6}$-free graphs is of interest in multiagent robotics, because the class of $K_{1,6}$-free graphs includes the class of unit disk graphs, where for every vertex there is a disk of radius 1 centered at the vertex representing its transmission or interaction range (see [23]).

In order to state their result, we recall a 1989 result due to McCuaig and Shepherd [22]. Let $\mathcal{B}$ be the family of seven graphs shown in Figure 1. McCuaig and Shepherd [22] showed that if G is a connected graph of minimum degree at least 2 and G is not one of graphs in the family $\mathcal{B}$, then the domination number of G is at most two-fifths its order.

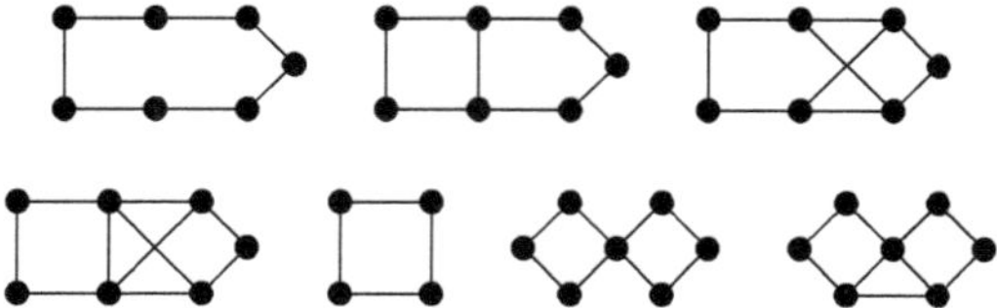

Fig. 1 The family $\mathcal{B}$ of seven exceptional graphs

We are now in a position to state the result of Abbas et al. [1]:

Theorem 9 ([1]) *If G is a $K_{1,6}$-free connected graph with minimum degree at least 2 and $G \notin \mathcal{B} \cup \{K_{2,3}\}$, then G has a family $\mathcal{F}$ of five dominating sets of G such that $m(\mathcal{F}) \leq 2$, implying that $\mathrm{FDOM}(G) \geq \frac{5}{2}$.*

We remark that the proof of Theorem 9 given by Abbas et al. [1] is algorithmic and gives a polynomial-time algorithm to find such a family $\mathcal{F}$.

We conclude this section with a comment about planar graphs. By Theorem 3, the maximum fractional domatic number of a planar graph is at most 6, since the minimum degree is at most 5. As observed in [12], an example of a planar graph with fractional domatic number equal to 6 is the icosahedron; this has six disjoint independent dominating sets of size 2, each consisting of a vertex and the unique vertex at distance 3 from it.

3 The Fractional Idomatic Number

In this section, we survey results on the fractional idomatic number of a graph. First, we recall the concept of the idomatic number of a graph. The maximum number of vertex disjoint independent dominating sets in a graph G is called the *idomatic number* of G denoted by $\mathrm{idom}(G)$. This terminology was introduced by Zelinka [29], but the parameter was originally defined by Cockayne and Hedetniemi [5]. In this section, we consider a fractional version of the idomatic number of a graph. The *fractional idomatic number* of graph G, denoted $\mathrm{FIDOM}(G)$, is defined as

$$\mathrm{FIDOM}(G) = \max \frac{|\mathcal{F}|}{m(\mathcal{F})},$$

where the maximum is taken over all families $\mathcal{F}$ of independent dominating sets of G and where, as before, $m(\mathcal{F})$ is the maximum number of times an element appears in $\mathcal{F}$. (We note that this parameter should be defined as the supremum, but as explained in Section 5, one can show that the supremum is always achieved.) The results in this section are mainly due to the authors [12].

Using analogous proofs to those presented in Theorems 1 and 3, one can establish the following immediate lower and upper bounds on the fractional idomatic number of a graph.

Theorem 10 ([12]) *The following hold in a graph G with minimum degree δ.*

(a) $\mathrm{idom}(G) \leq \mathrm{FIDOM}(G) \leq n(G)/i(G)$.
(b) $\mathrm{FDOM}(G) \leq \delta + 1$.

Examples of equality in the upper bound of Theorem 10(a) are the cycles of length a multiple of 3. We note that $\mathrm{FIDOM}(G) = 1$ only if the graph G has an isolate. As shown, for example, by Favaron [7], the independent domination number of a graph of order n can be as much as $n - o(n)$, even with prescribed minimum degree. For such a graph G, we have that $\mathrm{idom}(G) = 1$ and that $\mathrm{FIDOM}(G) = 1 + o(1)$. Even restricted to special classes of graphs, one can still see this behavior. Here is one such result. Recall that a graph is *claw-free* if it does not contain $K_{1,3}$ as an induced subgraph.

Proposition 11 ([12]) *There exist connected claw-free graphs G with arbitrarily large minimum degree for which* $\mathrm{FIDOM}(G) = 1 + o(1)$.

Proof. Let G be the claw-free graph G constructed as follows. For a and d positive integers, let G be obtained from a complete graph H of order ad as follows. Let X_1, $\ldots$, X_a be a partition of $V(H)$ into sets each of size d. For each resulting set X_i, we add a vertex x_i of degree d adjacent to every vertex of X_i for $i \in [a]$. The resultant split graph G is claw-free and has minimum degree d. Letting $X = \{x_1, \ldots, x_a\}$, we note that every independent dominating set of G contains at least $a - 1$ vertices of X, implying that

$$\mathrm{FIDOM}(G) \leq \frac{a}{a-1}.$$

Furthermore, for all $i \in [a]$ if $F_i = (X \setminus \{x_i\}) \cup \{x_i'\}$, where x_i' is an arbitrary vertex in X_i, then $\mathcal{F} = \{F_1, \ldots, F_a\}$ is a family of independent dominating sets of G satisfying $m(\mathcal{F}) = a - 1$, implying that

$$\mathrm{FIDOM}(G) \geq \frac{|\mathcal{F}|}{m(\mathcal{F})} = \frac{a}{a-1}.$$

Consequently, $\mathrm{FIDOM}(G) = a/(a-1)$. The desired result now follows by taking a and d arbitrarily large. $\qquad\square$

We next present a lower bound on the fractional idiomatic number in terms of dynamic colorings. An *r-dynamic coloring*, also called *r-hued coloring* in the literature, of a graph G is a proper coloring of the vertices of G such that every vertex v has at least $\min(d_G(v), r)$ colors in its neighborhood, where $d_G(v)$ is the degree of the vertex v in G. For more details on r-dynamic colorings, we refer the reader to Jahanbekam et al. [17].

Theorem 12 ([12]) *If G is a graph with minimum degree at least r that has an r-dynamic coloring using k colors, then*

$$\mathrm{FIDOM}(G) \geq \frac{k}{k-r},$$

and therefore $i(G) \leq (k-r)n/k$.

As a consequence of Theorem 12, we have the following lower bound on the fractional idiomatic number in terms of the chromatic number.

Theorem 13 ([12]) *If G is an isolate-free graph with chromatic number k, then*

$$\text{FIDOM}(G) \geq \frac{k}{k-1}.$$

By Theorem 10(b), the maximum fractional idomatic number of a planar graph is at most 6, since the minimum degree is at most 5. As noted earlier, an example of a planar graph with fractional idomatic number equal to 6 is the icosahedron; this has six disjoint independent dominating sets of size 2, each consisting of a vertex and the unique vertex at distance 3 from it.

We next consider lower bounds on the fractional idomatic number of a planar graph. As a consequence of Theorem 13 and a result by MacGillivray and Seyffarth [20] that provides an upper bound for the independent domination number in terms of the chromatic number, we obtain the following result.

Theorem 14 ([12]) *The following hold in a planar connected graph G of order n.*

(a) *If $n \geq 2$, then $\text{FIDOM}(G) \geq \frac{4}{3}$.*
(b) *If $n \geq 10$, then $i(G) \leq \frac{3}{4}n - 2$.*

Recall that the *corona $G \circ P_1$* of a graph G, also denoted cor(G) in the literature, is the graph obtained from G by adding a pendant edge to each vertex of G. As remarked in [12], the two bounds in Theorem 14 are sharp because of the corona $K_4 \circ P_1$ of K_4 illustrated in Figure 2.

If one considers planar graphs of minimum degree at least 2, then the lower bound on the fractional idiomatic number in Theorem 14(a) can be improved. The key is a result due to Kim, Lee, and Park [18], who showed that every connected planar graph has a 2-dynamic coloring using at most four colors, except for C_5. Therefore, by Theorem 12 (with $r = 2$ and $k = 4$), we have that every connected planar graph G, except possibly for C_5, satisfies $\text{FIDOM}(G) \geq 2$. However, since the 5-cycle has $\text{FIDOM}(C_5) = \frac{5}{2}$, the 5-cycle is no exception to the lower bound $\text{FIDOM}(G) \geq 2$. We therefore have the following lower bound on the fractional idiomatic number of a planar graph with minimum degree at least 2.

Theorem 15 ([12]) *If G is a planar graph with $\delta(G) \geq 2$, then $\text{FIDOM}(G) \geq 2$.*

Fig. 2 A planar graph G with $\text{FIDOM}(G) = \frac{4}{3}$

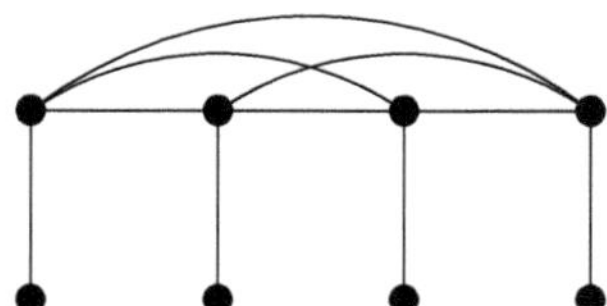

As an immediate consequence of Theorem 10(a) and Theorem 15, we have the following upper bound on the independent domination number of a planar graph with minimum degree at least 2.

Corollary 16 ([12]) *If G is a planar graph with $\delta(G) \geq 2$, then $i(G) \leq \frac{1}{2}n(G)$.*

The following construction, given in [12], shows that there exists an infinite family of planar graphs G with minimum degree two that satisfy $\mathrm{FIDOM}(G) = 2$. For $s \geq 2$, let H_s be the graph obtained from a 4-cycle $v_1 v_2 v_3 v_4 v_1$ by adding s new vertices whose neighbors are the pair $\{v_i, v_{i+2}\}$ for each $i \in [2]$. When $s = 4$, for example, the resulting graph H_s is illustrated in Figure 3. As observed in [12], it holds that $i(H_s) = \frac{1}{2}n(H_s)$ and $\mathrm{FIDOM}(H_s) = 2$. Thus, the bounds of Theorem 15 and Corollary 16 are tight.

As remarked in [12], there are numerous families of planar graphs G with minimum degree two that satisfy $\mathrm{FIDOM}(G) = 2$ and $i(G) < \frac{1}{2}n(G)$. It remains, however, an open problem to characterize the graphs achieving equality in the bounds of Theorem 15 and Corollary 16. It is also noted in [12] that there are numerous families of planar graphs G with minimum degree two that do not have two disjoint independent dominating sets, and therefore such graphs G satisfy $\mathrm{idom}(G) = 1$ and $\mathrm{FIDOM}(G) \geq 2$.

It remains an open problem to determine a best possible lower bound on the fractional idiomatic number of a planar graph with minimum degree 3. In this case, we believe that the upper bound of Theorem 15 can be improved from 2 to $\frac{5}{2}$. We pose this formally as a conjecture.

Conjecture 1 *If G is a planar graph with $\delta(G) \geq 3$, then $\mathrm{FIDOM}(G) \geq \frac{5}{2}$.*

We note that if $G = C_5 \,\square\, K_2$ is the 5-prism illustrated in Figure 4, then $i(G) = 4 = \frac{2}{5}n$ and $\mathrm{FIDOM}(G) = \frac{5}{2}$. Thus, if Conjecture 1 is true, the bound is best possible.

Fig. 3 The planar graph H_4 with $i(H_4) = \frac{1}{2}n(H_4)$ and $\mathrm{FIDOM}(H_4) = 2$

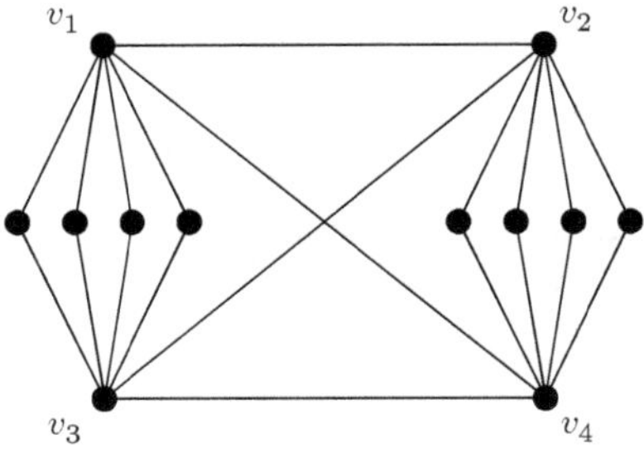

Fig. 4 The 5-prism $C_5 \,\square\, K_2$

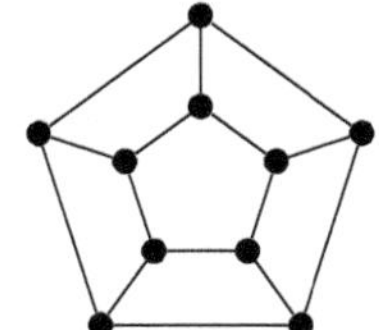

The fractional idiomatic number of a maximal outerplanar graph is easy to compute. If G is a maximal outerplanar graph, then G is 3-colorable and every color class is an independent dominating set, implying that $\mathrm{idom}(G) \geq 3$. Furthermore, since G has minimum degree 2, Theorem 10(b) implies that $\mathrm{FIDOM}(G) \leq 3$. Consequently, $\mathrm{idom}(G) = \mathrm{FIDOM}(G) = 3$ for a maximal outerplanar graph G, as observed in [12].

If G is a general outerplanar graph, then G is 3-chromatic and, by Theorem 13, we therefore have $\mathrm{FIDOM}(G) \geq \frac{3}{2}$. This, in turn, implies by Theorem 10(a) that $i(G) \leq \frac{2}{3}n(G)$. If G is obtained from K_3 by attaching $k \geq 1$ pendant edges to each vertex of the triangle, then the resulting graph G has order $n(G) = 3(k+1)$ and satisfies $i(G) = 2k+1$. Thus, for k sufficiently large, the bounds $i(G) \leq \frac{2}{3}n(G)$ and $\mathrm{FIDOM}(G) \geq \frac{3}{2}$ are asymptotically sharp. If, however, we impose a minimum degree condition, then we can improve the lower bound of $\frac{3}{2}$ on the fractional idiomatic number to a lower bound of 2, as shown by the following result in [12].

Theorem 17 ([12]) *If G is an outerplanar graph with minimum degree at least* 2, *then* $\mathrm{FIDOM}(G) \geq \mathrm{idom}(G) \geq 2$.

Let G and H be two graphs. As above, the *join* of G and H is written as $G \oplus H$, while their disjoint *union* is written as $G + H$. The *lexicographic product* of G and H, written $G[H]$, is the graph with vertex set $V(G) \times V(H)$, where two vertices (g_1, h_1) and (g_2, h_2) are adjacent in $G[H]$ if and only if $g_1 g_2 \in E(G)$ or $g_1 = g_2$ and $h_1 h_2 \in E(H)$. For these graph operations, the fractional idiomatic number behaves as follows.

Theorem 18 ([12]) *For graphs G and H, the following hold:*

(a) $\mathrm{FIDOM}(G + H) = \min(\mathrm{FIDOM}(G), \mathrm{FIDOM}(H))$.
(b) $\mathrm{FIDOM}(G \oplus H) = \mathrm{FIDOM}(G) + \mathrm{FIDOM}(H)$.
(c) $\mathrm{FIDOM}(G[H]) = \mathrm{FIDOM}(G) \times \mathrm{FIDOM}(H)$.

The following result is shown in [11].

Theorem 19 ([11]) *If G is a graph with minimum degree at least* 2 *and maximum degree at most* 3, *then* $\mathrm{idom}(G) \geq 2$.

As an immediate consequence of Theorem 19, every cubic graph G satisfies $\mathrm{idom}(G) \geq 2$, a result attributed to Berge. We note that there are many cubic graphs G satisfying $\mathrm{idom}(G) = 2$. We pose the following question.

Question 1 *Is it true that if G is a connected cubic graph different from $K_{3,3}$, then* $\mathrm{FIDOM}(G) \geq \frac{5}{2}$?

As observed earlier, if $G = C_5 \,\square\, K_2$ is the 5-prism shown in Figure 4, then $\mathrm{FIDOM}(G) = \frac{5}{2}$. Hence, if Question 1 is true, then the lower bound value 5/2 for the fractional idiomatic number would be best possible.

4 The Fractional Total Domatic Number

In this section, we survey results on the fractional total domatic number of a graph. First, we recall the concept of the total domatic number of a graph. The *total domatic number* of a graph G, denoted by *tdom(G)* and first defined by Cockayne, Dawes, and Hedetniemi [6], is the maximum number of total dominating sets into which the vertex set of G can be partitioned. The parameter *tdom(G)* is equivalent to the maximum number of colors in a (not necessarily proper) coloring of the vertices of a graph, where every color appears in every open neighborhood. Chen, Kim, Tait, and Verstraete [3] called this the *coupon coloring problem*. This parameter is now well studied. We refer the reader to Chapter 13 in the book [14] on total domination in graphs for a survey of results on the total domatic number and to [10] for a recent paper on this topic.

In this section, we consider a fractional version of the total domatic number of a graph. The *fractional total domatic number* of a graph G, denoted $FTD(G)$, is defined as

$$FTD(G) = \max \frac{|\mathcal{F}|}{m(\mathcal{F})},$$

where the maximum is taken over all families $\mathcal{F}$ of total dominating sets of G and where, as before, $m(\mathcal{F})$ is the maximum number of times an element appears in $\mathcal{F}$. (As before, the parameter should be defined as the supremum, but as shown in Section 5, the supremum is always achieved.)

The following trivial lower and upper bounds on the fractional total domatic number of a graph are established in [10].

Theorem 20 ([10]) *If G is an isolate-free graph, then*

$$tdom(G) \leq FTD(G) \leq \frac{n(G)}{\gamma_t(G)}.$$

Theorem 21 ([10]) *If a graph G of order n has minimum degree $\delta \geq 1$, then*

$$\frac{n}{n - \delta + 1} \leq FTD(G) \leq \delta.$$

Thus, for example, by Theorems 20 and 21, we get the following observation.

Proposition 22 ([10]) *The following hold in a graph G with minimum degree δ.*

(a) *If $\delta \geq 1$, then $FTD(G) = 1$.*
(b) *If $\delta \geq 2$, then $FTD(G) > 1$.*

We note that there are graphs G with arbitrarily large minimum degree with $FTD(G) < 1 + \varepsilon$ for any given $\varepsilon > 0$. Indeed, these are the graphs that Zelinka [30] provided as examples that have *tdom(G)* = 1 and arbitrarily large minimum degree.

A lower bound on the fractional total domatic number of a claw-free graph with minimum degree at least 2 is determined in [10].

Theorem 23 ([10]) *If G is a claw-free graph with $\delta \geq 2$, then $FTD(G) \geq \frac{3}{2}$.*

The lower bound of Theorem 23 is in a sense best possible in that the graphs K_3 and C_6 have fractional total domatic number exactly 3/2. However, asymptotically the bound should be improvable. As remarked in [10], perhaps it is true that if G is a connected, claw-free graph with $\delta \geq 2$, then $FTD(G) \geq 2 - o(1)$. Furthermore, if G is a connected, claw-free graph with $\delta \geq 3$, then maybe this guarantees that $FTD(G) \geq 2$. As shown in [10], there are arbitrarily large connected $K_{1,4}$-free graphs with fractional total domatic number exactly 3/2.

We note that the union of two disjoint dominating sets is a total dominating set. Thus, it is immediate that $tdom(G) \geq \lfloor dom(G)/2 \rfloor$. But, in the case that the ordinary domatic number is odd, one can say slightly more:

Theorem 24 ([10]) *If G is an isolate-free graph, then $FTD(G) \geq \frac{1}{2}dom(G)$.*

Proof. Let $dom(G) = k$, and let $D_1, \ldots, D_k$ be k disjoint dominating sets in the graph G. The family

$$\mathcal{F} = \{ D_i \cup D_j \mid 1 \leq i < j \leq k \}$$

is a family of $\binom{k}{2}$ total dominating sets of G. Since every vertex of G appears in at most $k - 1$ sets in the family $\mathcal{F}$, we note that $m(\mathcal{F}) = k - 1$. Thus,

$$FTD(G) \geq \frac{|\mathcal{F}|}{m(\mathcal{F})} = \frac{\binom{k}{2}}{k - 1} = \frac{k}{2}.$$

The desired result now follows recalling that $k = dom(G)$. $\square$

Equality in Theorem 24 occurs, for example, in complete graphs.

We consider now planar graphs. A *triangulated disc* is a (simple) planar graph all of whose faces are triangles, except possibly for the outer face. Matheson and Tarjan [21] showed that if G is a triangulated disc, then $dom(G) \geq 3$. Hence, as an immediate consequence of Theorem 24 and the Matheson–Tarjan result, we have the following lower bound on the fractional total domatic number of a triangulated disc.

Theorem 25 ([10]) *If G is a triangulated disc, then $FTD(G) \geq \frac{3}{2}$.*

As remarked in [10], the lower bound of Theorem 25 is tight as may be seen by considering the triangulated disc G illustrated in Figure 5, where the shaded area consists of any maximal planar graph (or, equivalently, triangulation). Let S be the set of three vertices on the outer face of G that have degree at least 4. If $\mathcal{F}$ is an arbitrary family of total dominating sets of G, then each set in the family $\mathcal{F}$ contains at least two vertices of S. By averaging, there is a vertex in S that belongs to at least $2|\mathcal{F}|/3$ sets in $\mathcal{F}$, implying that $m(\mathcal{F}) \geq 2|\mathcal{F}|/3$, or, equivalently, $|\mathcal{F}|/m(\mathcal{F}) \leq 3/2$.

Fig. 5 A triangulated disc G
with $FTD(G) = \frac{3}{2}$

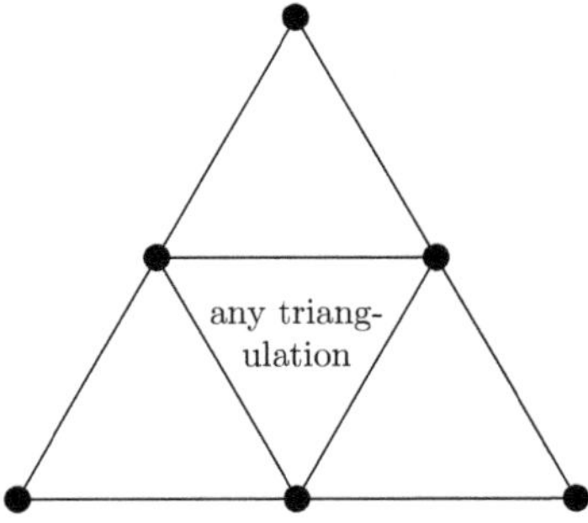

Since this is true for every family $\mathcal{F}$ of total dominating sets of G, this implies that $FTD(G) \leq 3/2$. Consequently, by Theorem 25, $FTD(G) = 3/2$.

We next consider triangulations, where by *triangulation* we mean a simple graph embedded in some orientable surface such that every region is a triangle. The following key lemma establishes an upper bound on the fractional total domatic number of a graph G in terms of its average degree, which we denote by $\overline{d}_{av}(G)$.

Lemma 26 ([10]) *If G is a triangulation of order at least 4, then*

$$FTD(G) \leq \overline{d}_{av}(G) - 1.$$

If G is a planar triangulation of order n, then $\overline{d}_{av}(G) = 6 - \frac{12}{n}$. Thus, as an immediate consequence of Lemma 26, we have the following upper bounds on the total domatic and fractional total domatic numbers of a planar graph.

Theorem 27 ([10]) *The following hold in a planar graph G.*

(a) $tdom(G) \leq 4$.
(b) $FTD(G) \leq 5 - \frac{12}{n}$.

As remarked in [10] there are planar graphs G with $tdom(G) = 4$. For example, if G is obtained from a truncated tetrahedron and adding a vertex inside each hexagonal face that is joined to all vertices on the boundary, then G is a planar graph G of order 16 satisfying $tdom(G) = 4$. Illustrated in Figure 6 (which corrects a figure in [10]) is a spanning subgraph thereof that still has four disjoint total dominating sets: the vertices labeled i form a total dominating set for each $i \in [4]$.

As shown in [10], there are planar graphs G for which $FTD(G) > 4$. We note that the result of Lemma 26 applies on all surfaces. In particular, since the average degree of a toroidal graph is at most 6, this yields the following upper bounds on the total domatic and fractional total domatic numbers of a toroidal graph.

Theorem 28 *If G is a toroidal graph, then $tdom(G) \leq FTD(G) \leq 5$.*

There are toroidal graphs G satisfying $tdom(G) = FTD(G) = 5$. The example provided in [10] is illustrated in Figure 7, where the top and bottom dotted lines

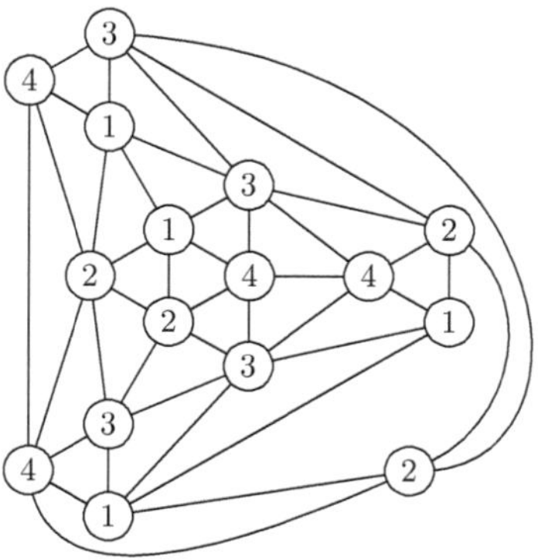

Fig. 6 A planar graph G with $tdom(G) = 4$

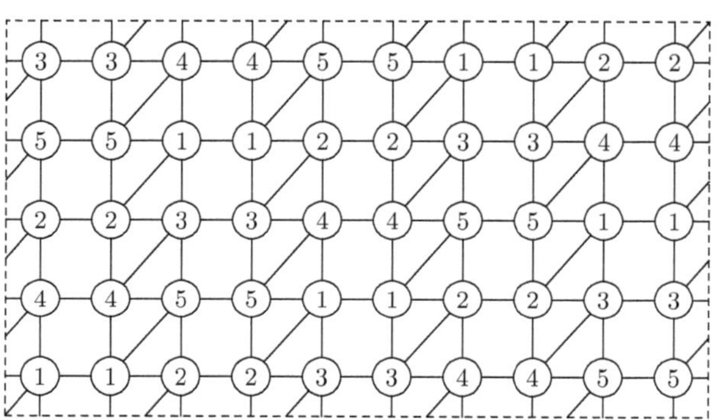

Fig. 7 A toroidal graph G with $tdom(G) = 5$

should be identified and similarly with the left and right dotted lines; the vertices labeled i form a total dominating set of G for each $i \in [5]$.

It remains generally an open problem to determine good lower bounds for triangulations. Since every planar triangulation is a triangulated disc, Theorem 25 implies that every planar triangulation G satisfies $FTD(G) \geq \frac{3}{2}$. However, we believe this lower bound can be improved. In this regard, the following conjectures are posed in [10].

Conjecture 2 ([10]) *If G is a planar triangulation of order at least* 4, *then* $tdom(G) \geq 2$.

Using the Four Color Theorem, it is shown in [10] that Conjecture 2 is true if every vertex has odd degree or if the dual of G is Hamiltonian. By characterizing the maximal outerplanar graphs H that have $tdom(H) < 2$, Nagy [24] showed that the conjecture is true if G is Hamiltonian. A stronger version of the above is the following conjecture.

Conjecture 3 ([10]) *Every planar triangulation with at least four vertices has a proper 4-coloring (C_1, C_2, C_3, C_4) such that $C_1 \cup C_2$ and $C_3 \cup C_4$ are total dominating sets.*

Fig. 8 The Heawood graph

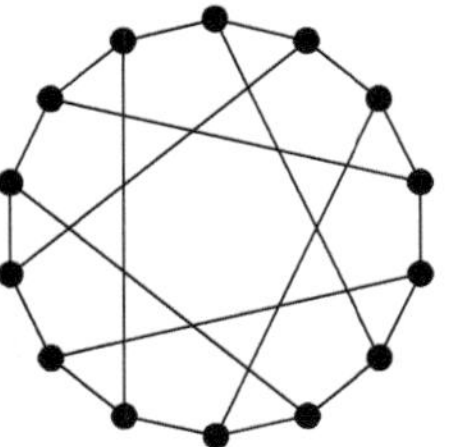

As with the parameter FDOM, the parameter *FTD* is monotonic. The fractional total domatic number of the disjoint union of graphs behaves as expected.

Theorem 29 ([10]) *For all graphs G and H, it holds that FTD(G + H) = min{FTD(G), FTD(H)}.*

We conclude this section with some comments on regular graphs. It remains a long-standing open problem to characterize those cubic graphs that do not have two disjoint total dominating sets; that is, the 3-regular graphs with $tdom(G) = 1$. It is well known that the Heawood graph, shown in Figure 8, is the smallest example of such a graph G without two disjoint total dominating sets.

Nevertheless, a natural question is whether the fractional total domatic number of a cubic graph is always at least 2. This question was answered in [16].

Theorem 30 ([16]) *If G is a connected cubic graph, then FTD(G) $\geq$ 2.*

Recall that Fujita, Yamashita, and Kameda proved in Theorem 7 the beautiful result that every connected cubic graph G has a family $\mathcal{F}$ of five dominating sets such that every vertex is in at most two of these. The following strengthening of Theorem 30 is conjectured in [10].

Conjecture 4 ([10]) *If G is a connected cubic graph, then G has a family of four total dominating sets of G such that every vertex is in at most two of these.*

For regular graphs with higher minimum degree, we have the following lower bound on the fractional total domatic number of a regular graph.

Theorem 31 ([16]) *For all k $\geq$ 3, if G is a k-regular graph, then*

$$FTD(G) > \frac{k}{1 + \ln(k)}.$$

5 Fractional Definitions and Hypergraphs

The three parameters explored in this chapter can be defined in terms of hypergraphs. The *fractional matching number* of a hypergraph H, denoted $v_*(H)$, is defined by the linear program

$$\max \sum_{e \in E(H)} w(e) \text{ such that } \forall e \in E : w(e) \geq 0 \text{ and } \forall v \in V(H) : \sum_{e \ni v} w(e) \leq 1.$$

The *matching number* of H, denoted $v(H)$, is the maximum number of disjoint hyperedges. Clearly, $v^*(H) \geq v(H)$. Also, by linear programming duality, the fractional matching number equals the fractional transversal/cover number.

As before, for a multiset $\mathcal{F}$ of $E(H)$, we define $m(\mathcal{F})$ as the maximum number of times a vertex of H appears in $\mathcal{F}$.

Since the linear program has a rational solution,

$$v^*(H) = \sup \frac{|\mathcal{F}|}{m(\mathcal{F})} = \max \frac{|\mathcal{F}|}{m(\mathcal{F})},$$

where the maximum and the supremum are over all such multisets $\mathcal{F}$. See Chapter 1 of [25] for a fuller discussion.

Consider a family $\mathcal{D}$ of subsets of the vertex set of a graph G. This set can naturally be thought of as a hypergraph $H_\mathcal{D}$. Then, the matching number of $H_\mathcal{D}$ is the maximum number of disjoint members of $\mathcal{D}$. If we let $\mathcal{D}$ be the set of dominating sets, we get the domatic number and fractional domatic number, discussed in Section 2. If we let $\mathcal{D}$ be the set of independent dominating sets, we get the idomatic number and fractional idomatic number, discussed in Section 3. If we let $\mathcal{D}$ be the set of total dominating sets, we get the total domatic number and fractional total domatic number, discussed in Section 4.

The results about the behavior of the three fractional parameters under disjoint union are a special case of a result in hypergraphs. Given two disjoint hypergraphs H_1 and H_2, we define their *direct sum* as the hypergraph with vertex set $V(H_1) \cup V(H_2)$ and edge set $\{e_1 \cup e_2 | e_1 \in E(H_1), e_2 \in E(H_2)\}$. Bujtás and Tuza [2] showed that the matching number of the direct sum of two hypergraphs equals the smaller of the two matching numbers. We note that the analogous result is true for the fractional matching number too.

Lemma 32 *If hypergraph H is the direct sum of hypergraphs H_1 and H_2, then $v^*(H) = \min(v^*(H_1), v^*(H_2))$.*

Proof. For $\ell \in \{1, 2\}$, let h_i be an optimal weighting of $E(H_i)$. Let $Y = \max(v^*(H_1), v^*(H_2))$. Then define the weighting h of the direct sum H by

$$h(e_1 \cup e_2) = h_1(e_1)h_2(e_2)/Y.$$

For each vertex $v \in V(H_1)$, we have

$$\sum_{e \ni v} h(e) = \sum_{e_1 \ni v} h_1(e_1) \sum_{e_2} h_2(e_2)/Y \leq \sum_{e_1 \ni v} h_1(e_1) \leq 1.$$

Similarly, the constraint is satisfied for $v \in V(H_2)$. And the total weight of h is $v^*(H_1)v^*(H_2)/Y = \min(v^*(H_1), v^*(H_2))$. It follows that

$$\nu^*(H) \geq \min(\nu^*(H_1), \nu^*(H_2)).$$

Conversely, let h be the optimal weighting for the direct sum H. Then, define the weighting h_1 on hypergraph H_1 by

$$h_1(e_1) = \sum_{e \supseteq e_1} h(e).$$

It follows readily that for each vertex $v \in V(H_1)$, we have $\sum_{e_1 \ni v} h_1(e_1) \leq 1$. That is, the weighting h_1 represents a fractional matching of H_1. Thus, $\nu^*(H_1) \geq \nu^*(H)$. Analogously, $\nu^*(H_2) \geq \nu^*(H)$. That is, $\nu^*(H) \leq \min(\nu^*(H_1), \nu^*(H_2))$.

The two inequalities combined give the desired result. $\qquad\square$

As a consequence the results on disjoint union follow for the fractional domatic, independent domatic, and total domatic numbers given earlier.

6 More on Hypergraphs

While the previous section described a general conversion from a particular domatic number of a graph to the matching number of an associated hypergraph, another hypergraph provides a more general setting for the fractional domatic and fractional total domatic number.

Recall that a subset T of vertices is a *transversal* (also called *vertex cover* or *hitting set*) in a hypergraph H if T has a nonempty intersection with every edge of H. The *transversal number* $\tau(H)$ of H is the minimum size of a transversal in H. We denote by $disj_\tau(H)$ the *disjoint transversal number* of a hypergraph H, which is the maximum number of disjoint transversals in H. Analogous to the fractional total domatic number, one can define the fractional disjoint transversal number. The *fractional disjoint transversal number* of H, denoted $FDT(H)$, is defined as

$$FDT(G) = \max \frac{|\mathcal{F}|}{m(\mathcal{F})},$$

where the maximum is taken over all families $\mathcal{F}$ of transversals of H. Analogous to earlier results, we have the following bounds on the fractional disjoint transversal number.

Theorem 33 ([10]) *For every isolate-free hypergraph H of order n,*

$$disj_\tau(H) \leq FDT(H) \leq \frac{n}{\tau(H)}.$$

For $k \geq 2$, if H is the complete k-uniform hypergraph of order n, then $\tau(H) = n - k + 1$, and so by Theorem 33, $FDT(H) \leq n/(n - k + 1)$. To prove that

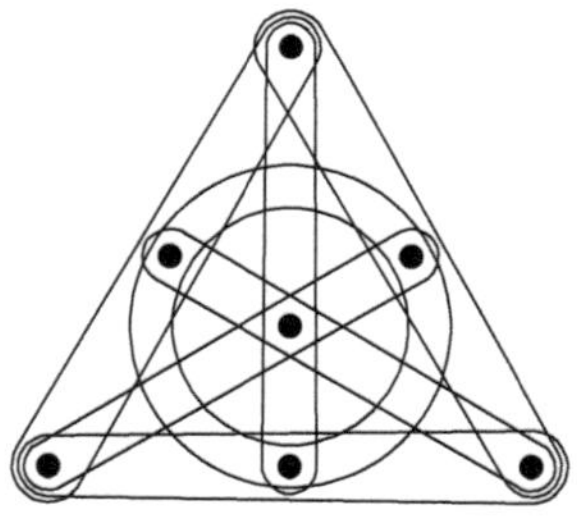

Fig. 9 The Fano plane F_7

$FDT(H) \geq n/(n-k+1)$, we consider the collection $\mathcal{F}$ of all $n-k+1$ element subsets of $V(H)$. The resulting family $\mathcal{F}$ is a family of $\binom{n}{n-k+1}$ transversals of H. Each vertex is in $\binom{n-1}{n-k}$ sets of $\mathcal{F}$, and so

$$FDT(H) \geq \frac{|\mathcal{F}|}{m(\mathcal{F})} = \frac{\binom{n}{n-k+1}}{\binom{n-1}{n-k}} = \frac{n}{n-k+1}.$$

Consequently, $FDT(H) = n/(n-k+1)$. Thus, the upper bound of Theorem 33 is achieved, for example, by the complete k-uniform hypergraph of order n.

As another example, if F_7 is the Fano plane, illustrated in Figure 9, then $\tau(F_7) = 3$, and so by Theorem 33, $FDT(F_7) \leq 7/3$. If we take $\mathcal{F}$ to be the family consisting of the seven edges of F_7, then $\mathcal{F}$ is a family of transversals of F_7. Each vertex belongs to exactly three sets of $\mathcal{F}$, and so $FDT(F_7) \geq |\mathcal{F}|/m(\mathcal{F}) = 7/3$. Consequently, $FDT(F_7) = 7/3$. Thus, the Fano plane achieves equality in the upper bound of Theorem 33. We note however that $disj_\tau(F_7) = 1$.

Analogous to Lemma 32, we have the following result on the fractional disjoint transversal number of the disjoint union of hypergraphs.

Theorem 34 ([10]) *If H is the disjoint union of isolate-free hypergraphs H_1 and H_2, then $FDT(H) = \min\{FDT(H_1), FDT(H_2)\}$.*

We describe next the interplay between the fractional (total) domatic number and the fractional disjoint transversal number. The *open neighborhood hypergraph*, abbreviated ONH, of a graph G is the hypergraph $\mathcal{ON}(G)$ whose vertex set is $V(G)$ and whose hyperedges are the open neighborhoods of vertices in G. Thus, if $H = \mathcal{ON}(G)$, then $V(H) = V(G)$ and $E(H) = \{N_G(x) \mid x \in V(G)\}$. As first observed by Thomassé and Yeo [27], a total dominating set in G is a transversal in $\mathcal{ON}(G)$ and conversely. Thus, the transversal number of $\mathcal{ON}(G)$ is precisely the total domination number $\gamma_t(G)$. Similarly, the *closed neighborhood hypergraph*, abbreviated CNH, of a graph G is the hypergraph $\mathcal{CN}(G)$ whose vertex set is $V(G)$ and whose hyperedges are the closed neighborhoods of vertices in G. Again, a dominating set in G is a transversal in $\mathcal{CN}(G)$ and conversely. We state this connection formally as follows.

Proposition 35 ([10]) *For every graph G, $FDOM(G) = FDT(\mathcal{CN}(G))$ and for every isolate-free graph G, $FTD(G) = FDT(\mathcal{ON}(G))$.*

As an example, if G is the Heawood graph, illustrated in Figure 8, then the ONH consists of two disjoint copies of the Fano plane F_7. Therefore, by Theorem 34 and Proposition 35, we have $FTD(G) = FDT(\mathcal{ON}(G)) = FDT(F_7 \cup F_7) = FDT(F_7) = 7/3$. In contrast, the Heawood graph does not have two disjoint total dominating sets; that is, $tdom(G) = 1$. Nevertheless, there is a general fractional lower bound.

Theorem 36 ([16]) *If H is a 3-regular 3-uniform hypergraph, then $FDT(H) \geq 2$, and this bound is tight.*

As regards tightness, it is shown in [13] that there are infinitely many (connected) 3-regular 3-uniform hypergraphs H satisfying $\tau(H) = \frac{1}{2}n(H)$, implying by Theorem 33 that each of these hypergraphs satisfies $FDT(H) \leq 2$. The lower bound in Theorem 36 shows that $FDT(H) \geq 2$. Consequently, $FDT(H) = 2$ for these hypergraphs.

As a consequence of a result due to Thomassen [28] and a relationship given in [15] between the total domatic number of a k-regular graph and 2-colorings of k-uniform k-regular hypergraphs, we have that if H is a 4-regular 4-uniform hypergraph, then $FDT(H) \geq disj_\tau(H) \geq 2$. However, it is conjectured in [16] that this lower can be improved when H is a 4-regular 4-uniform hypergraph.

Conjecture 5 ([16]) *If H is a 4-regular 4-uniform hypergraph, then $FDT(H) \geq \frac{7}{3}$.*

We note that if Conjecture 5 is true, then it implies that every 4-regular graph G satisfies $FTD(G) \geq \frac{7}{3}$.

Using probabilistic arguments, the following lower bound on the fractional disjoint transversal number of a k-regular k-uniform hypergraph was established in [16].

Theorem 37 ([16]) *For all $k \geq 3$, if H is a k-regular k-uniform hypergraph, then*

$$FDT(H) > \frac{k}{1 + \ln(k)}.$$

Furthermore, this bound is essentially best possible as there exist k-regular k-uniform hypergraphs H_k with $FDT(H_k) \leq \frac{k}{\ln(k)}(1 + o(1))$.

7   Conclusion

In this chapter, we survey results on the fractional relaxation of the count of the maximum number of disjoint dominating, independent dominating, and total dominating sets in a graph. We discuss a common framework in hypergraphs and show that the fractional domatic and fractional total domatic numbers of a graph can be placed in a more general hypergraph setting. We present the main results known to date on the fractional domatic parameters and list several outstanding open problems and conjectures that have yet to be settled.

References

1. W. Abbas, M. Egerstedt, C.-H. Liu, R. Thomas, P. Whalen, Deploying robots with two sensors in $K_{1,6}$-free graphs. J. Graph Theory **82**, 236–252 (2016)
2. C. Bujtás, Z. Tuza, C-perfect hypergraphs. J. Graph Theory **64**, 132–149 (2010)
3. B. Chen, J.H. Kim, M. Tait, J. Verstraete, On coupon colorings of graphs. Discrete Appl. Math. **193**, 94–101 (2015)
4. E.J. Cockayne, S.T. Hedetniemi, Optimal domination in graphs. IEEE Trans. Circuits Syst. **CAS-2**, 855–857 (1975)
5. E.J. Cockayne, S.T. Hedetniemi, Disjoint independent dominating sets in graphs. Discret. Math. **15**, 213–222 (1976)
6. E.J. Cockayne, R.M. Dawes, S.T. Hedetniemi, Total domination in graphs. Networks **10**, 211–219 (1980)
7. O. Favaron, Two relations between the parameters of independence and irredundance. Discret. Math. **70**, 17–20 (1988)
8. S. Fujita, M. Yamashita, T. Kameda, A study on r-configurations—a resource assignment problem on graphs. SIAM J. Discrete Math. **13**, 227–254 (2000)
9. W. Goddard, M.A. Henning, Independent domination in graphs: a survey and recent results. Discret. Math. **313**, 839–854 (2013)
10. W. Goddard, M.A. Henning, Thoroughly dispersed colorings. J. Graph Theory **88**, 174–191 (2018)
11. W. Goddard, M.A. Henning, Acyclic total dominating sets in cubic graphs. Appl. Anal. Discrete Math. **13**, 74–84 (2019)
12. W. Goddard, M.A. Henning, Independent domination, colorings and the fractional idomatic number of a graph. Appl. Math. Comput. **382**, 125340 (2020)
13. M.A. Henning, A. Yeo, Hypergraphs with large transversal number and with edge sizes at least three. J. Graph Theory **59**, 326–348 (2008)
14. M.A. Henning, A. Yeo, *Total Domination in Graphs*. Springer Monographs in Mathematics (Springer, Berlin, 2013). ISBN: 978-1-4614-6524-9 (Print), 978-1-4614-6525-6 (Online)
15. M.A. Henning, A. Yeo, 2-colorings in k-regular k-uniform hypergraphs. Eur. J. Comb. **34**, 1192–1202 (2013)
16. M.A. Henning, A. Yeo, A note on fractional disjoint transversals in hypergraphs. Discret. Math. **340**, 2349–2354 (2017)
17. S. Jahanbekam, J. Kim, O. Suil, D.B. West, On r-dynamic coloring of graphs. Discrete Appl. Math. **206**, 65–72 (2016)
18. S.-J. Kim, S.J. Lee, W.-J. Park, Dynamic coloring and list dynamic coloring of planar graphs. Discrete Appl. Math. **161**, 2207–2212 (2013)
19. A.V. Kostochka, B.Y. Stodolsky, On domination in connected cubic graphs. Discret. Math. **304**, 45–50 (2005)
20. G. MacGillivray, K. Seyffarth, Bounds for the independent domination number of graphs and planar graphs. J. Comb. Math. Comb. Comput. **49**, 33–55 (2004)
21. L.R. Matheson, R.E. Tarjan, Dominating sets in planar graphs. Eur. J. Comb. **17**, 565–568 (1996)
22. W. McCuaig, B. Shepherd, Domination in graphs with minimum degree two. J. Graph Theory **13**, 749–762 (1989)
23. M. Mesbahi, M. Egerstedt, *Graph Theoretic Methods in Multiagent Networks*. Princeton Series in Applied Mathematics (Princeton University Press, Princeton, 2010), xx+403 pp. ISBN: 978-0-691-14061-2
24. Z.L. Nagy, Coupon-coloring and total domination in Hamiltonian planar triangulations. Graphs Combin. **34**, 1385–1394 (2018)
25. E.R. Scheinerman, D.H. Ullman, *Fractional Graph Theory. A Rational Approach to the Theory of Graphs* (Wiley, New York, 1997), xviii+211 pp.

26. J. Suomela, Locality helps sleep scheduling. *Working Notes of the Workshop on World-Sensor-Web: Mobile Device-Centric Sensory Networks and Applications* (WSW, Boulder, 2006), p. 41–44
27. S. Thomassé, A. Yeo, Total domination of graphs and small transversals of hypergraphs. Combinatorica **27**, 473–48 (2007)
28. C. Thomassen, The even cycle problem for directed graphs. J. Amer. Math. Soc. **5**, 217–229 (1992)
29. B. Zelinka, Adomatic and idomatic numbers of graphs. Math. Slovaca **33**, 99–103 (1983)
30. B. Zelinka, Total domatic number and degrees of vertices of a graph. Math. Slovaca **39**, 7–11 (1989)
31. B. Zelinka, Domatic numbers of graphs and their variants: A survey, in *Domination in Graphs*. Monographs Textbooks Pure Applied Mathematics, vol. 209 (Dekker, New York, 1998), pp. 351–377

Dominator and Total Dominator Colorings in Graphs

Michael A. Henning

AMS Subject Classification: 05C65, 05C69

1   Introduction

A *dominating set* of a graph G is a set $S \subseteq V(G)$ such that every vertex in $V(G) \setminus S$ is adjacent to at least one vertex in S. The *domination number* of G, denoted by $\gamma(G)$, is the minimum cardinality of a dominating set of G. A dominating set of G of cardinality $\gamma(G)$ is called a *γ-set of G*.

A *total dominating set*, abbreviated a TD-set, of a graph G with no isolated vertex is a set $S \subseteq V(G)$ such that every vertex in $V(G)$ is adjacent to at least one vertex in S. The *total domination number* of G, denoted by $\gamma_t(G)$, is the minimum cardinality of a TD-set of G. A TD-set of G of cardinality $\gamma_t(G)$ is called a *γ_t-set of G*. Total domination is now well studied in graph theory. The literature on the subject of total domination in graphs has been surveyed and detailed in the book [19].

A *proper vertex coloring* of a graph G is an assignment of colors (elements of some set) to the vertices of G, one color to each vertex, so that adjacent vertices are assigned distinct colors. A proper vertex coloring whose colors are taken from a set of k colors, usually the set $[k] = \{1, 2, \ldots, k\}$, is called a *proper k-coloring*. In a given coloring of G, a *color class* of the coloring is a set consisting of all those vertices assigned the same color. The *vertex chromatic number* of G, denoted $\chi(G)$, is the smallest positive integer k for which G has a proper k-coloring. A *χ-coloring* of G is a proper k-coloring of G that uses $\chi(G)$ colors. In what follows, we simply

M. A. Henning (✉)
Department of Mathematics and Applied Mathematics, University of Johannesburg, Johannesburg, South Africa
e-mail: mahenning@uj.ac.za

call a proper vertex coloring a *coloring*, and we refer to the vertex chromatic number as the *chromatic number*.

In this chapter, we combine the concept of domination (total domination) in graphs with the concept of colorings in graphs and study dominator colorings (respectively, total dominator colorings) of a graph. In Section 3, we formally define dominator colorings in graphs, and in Section 4, we formally define the analogous concept of total dominator colorings in graphs. In these sections, we present selected results on the so-called dominator chromatic number and total dominator chromatic number of a graph.

2   Graph Theory Notation

For completeness, we include some graph theory terminology that we will use in this chapter. A vertex of degree 1 is called a *leaf*, and its unique neighbor is called a *support vertex*. Two vertices v and w are *neighbors* in a graph G if they are adjacent, that is, if $vw \in E(G)$. The *open neighborhood* of a vertex v in G is the set of neighbors of v, denoted $N_G(v)$, whereas the *closed neighborhood* of v is $N_G[v] = N_G(v) \cup \{v\}$. The *open neighborhood* of a set $S \subseteq V(G)$ is the set of all neighbors of vertices in S, denoted $N_G(S)$, whereas the *closed neighborhood* of S is $N_G[S] = N_G(S) \cup S$. The *S-private neighborhood* of a vertex $v \in S$ is defined by $\text{pn}_G(v, S) = \{w \in V(G) : N_G[w] \cap S = \{v\}\}$. Thus, $\text{pn}_G(v, S) = N_G[v] \setminus N_G[S \setminus \{v\}]$. We note that if $v \in \text{pn}_G(v, S)$, then the vertex v is isolated in the subgraph $G[S]$. A vertex outside the set S that belongs to the set $\text{pn}_G(v, S)$ is called an *S-external private neighbor* of v. If the graph G is clear from the context, we omit the subscript G in the above definitions. For example, we write $N[v]$ and $N[S]$ rather than $N_G[v]$ and $N_G[S]$, respectively.

We denote a *complete graph* on n vertices by K_n, and we denote a *path* and *cycle* on n vertices by P_n and C_n, respectively. We denote a *complete bipartite graph* with partite sets of cardinality m and n by $K_{m,n}$. A *star* is a graph $K_{1,n}$ for some $n \geq 1$. A *double star* is a tree with exactly two (adjacent) non-leaf vertices. If one of these vertices is adjacent to ℓ_1 leaves and the other to ℓ_2 leaves, then we denote the double star by $S(\ell_1, \ell_2)$. By a *nontrivial graph*, we mean a graph of order at least two. The *corona* cor(G) of a graph G, also denoted $G \circ K_1$ in the literature, is the graph obtained from G by attaching a leaf v' to every vertex v of G. The 2-*corona* $G \circ P_2$ of G is the graph of order $3|V(G)|$ obtained from G by attaching a path of length 2 to each vertex of G so that the resulting paths are vertex-disjoint.

Given a graph F, a graph G is *F-free* if it does not contain any induced subgraph isomorphic to F. If G is $K_{1,3}$-free, then G is said to be *claw-free*. A graph is *chordal* if it contains no induced cycle of length 4 or more. A graph is a *split graph* if its vertex set can be partitioned into a clique and an independent set. A *universal vertex* in a graph is a vertex that is adjacent to every other vertex in the graph. A *clique* in G is a complete subgraph of G. The *clique number* of G, denoted $\omega(G)$, is the maximum cardinality of a clique in G.

A set of vertices in a graph G is a *packing* if the vertices in S are pairwise at distance at least 3 apart, that is, if u and v are arbitrary distinct vertices in S, then $d(u, v) \geq 3$. Equivalently, S is a packing if the closed neighborhoods of vertices in S are pairwise disjoint. A subset S of vertices in a graph G is an *open packing* if the open neighborhoods of vertices in S are pairwise disjoint. Further the set S is a *perfect packing* (respectively, a *perfect open packing*) if every vertex belongs to exactly one of the closed (respectively, open) neighborhoods of vertices in S. The *packing number* $\rho(G)$ (respectively, the *open packing number* $\rho^o(G)$) is the maximum cardinality of a packing (respectively, open packing) in G.

A vertex and an edge are said to *cover* each other in a graph G if they are incident in G. A *vertex cover* in G is a set of vertices that covers all the edges of G. The *vertex covering number* $\tau(G)$ (also denoted by $\beta(G)$ or vc(G) in the literature) is the minimum cardinality of a vertex cover in G. The independence number $\alpha(G)$ of a graph G is the maximum cardinality of an independent set in G.

3 Dominator Colorings

A vertex in a graph G *dominates* itself and all vertices adjacent to it. Further, a vertex is a *dominator* of a set S if it dominates every vertex in S. A *dominator coloring* of a graph G is a proper coloring of G with the additional property that every vertex in $V(G)$ dominates all vertices in at least one color class, that is, each vertex of the graph belongs to a singleton color class or is adjacent to every vertex of some (other) color class. The *dominator chromatic number* $\chi_d(G)$ of G is the minimum number of color classes in a dominator coloring of G. A χ_d-*coloring* of G is a dominator coloring of G that uses $\chi_d(G)$ colors.

The concept of a dominator coloring in a graph was birthed in the late 1970s when Cockayne, Hedetniemi, and Hedetniemi [9] defined the domatic number of a graph involving partitions into dominating sets. In 2006, Hedetniemi, Hedetniemi, and McRae [14] further studied the concept of dominator colorings in graphs. (We remark that these two papers are cited as [4] and [13], respectively, in the 2006 paper by Gera, Horton, and Rasmussen [13].) On March 15, 2004, Hedetniemi, Hedetniemi, Laskar, McRae, and Wallis [15] submitted a paper on dominator partitions in graphs, but due to the backlog in the journal at the time, the paper only appeared 5 years later! In 2006, Gera et al. [13] published a paper on dominator colorings in graphs, and in 2007, Gera [11, 12] continued the study of dominator colorings.

Since every vertex is a dominator of itself, the coloring of G that assigns a unique color to each vertex is a trivial dominator coloring of G. Thus, every graph G has a dominator coloring, and therefore the dominator chromatic number $\chi_d(G)$ is well-defined. Since every dominator coloring of G is a coloring of G, we have the following observation.

Observation 1 *For every graph G, we have $\chi(G) \leq \chi_d(G)$.*

Fig. 1 A χ_d-coloring of a path P_{17}

The simplest example to show that strict inequality may occur in Observation 1 is to take G to be a path P_4 given by $v_1v_2v_3v_4$. We note that $\chi(G)=2$ and the unique 2-coloring of G has color classes $\{v_1, v_3\}$ and $\{v_2, v_4\}$. However, neither the vertex v_1 nor v_4 dominates any color class, implying that $\chi_d(G) \geq 3$. However, the 3-coloring of G with color classes $V_1 = \{v_1, v_3\}$, $V_2 = \{v_2\}$, and $V_3 = \{v_4\}$ is a dominator coloring of G, noting that the vertices v_1 and v_3 dominate the color class V_2, the vertex v_2 dominates both color classes V_1 and V_2, and the vertex v_4 dominates its own color class V_3. Thus, $\chi_d(G) \leq 3$. Consequently, $\chi_d(G)=3$. As shown by Theorems 2 and 3, the difference $\chi_d(G) - \chi(G)$ can be made arbitrarily large by taking, for example, G to be a path P_n or cycle C_n of sufficiently large order n.

We note that if G is a star $K_{1,k}$ where $k \geq 1$, then a proper 2-coloring of G is also a dominator coloring of G, and so $\chi(G) = \chi_d(G) = 2$. If $G = K_n$, then $\chi(G) = \chi_d(G) = n$. Hence, equality in Observation 1 is possible.

Gera et al. [13] determined the dominator chromatic number of a path P_n on n vertices. We note that $\chi_d(P_2) = \chi_d(P_3) = 2$. As observed earlier, $\chi_d(P_4) = 3$. It is a simple exercise to verify that $\chi_d(P_5) = 3$. If G is the path $P_n: v_1v_2 \ldots v_n$ where $n \geq 6$, let $f : V(G) \to \{1, 2, \ldots, 2 + \lceil \frac{n}{3} \rceil\}$ be the dominator coloring defined by

$$
f(v_i) = \begin{cases} 1 & \text{when } n \ (\mathrm{mod}\ 6) \in \{1, 3\} \\ 2 & \text{when } n \ (\mathrm{mod}\ 6) \in \{0, 4\} \\ \left\lceil \dfrac{i}{3} \right\rceil + 2 & \text{when } n \ (\mathrm{mod}\ 6) \in \{2, 5\}. \end{cases}
$$

However if $n \equiv 1 \ (\mathrm{mod}\ 3)$, then we redefine $f(v_n)$ to be the value $\lceil \frac{n}{3} \rceil + 2$. When $n = 16$, for example, the resulting dominator coloring is illustrated in Figure 1 where here $f(v_{16}) = \lceil \frac{16}{3} \rceil + 2 = 8$ (and where color 1 is blue, color 2 is white, color 3 is green, etc.).

Gera et al. [13] proved that the dominator coloring f defined above is a χ_d-coloring of the path P_n.

Theorem 2 ([13]) *For $n \geq 2$, we have*

$$
\chi_d(P_n) = \begin{cases} 1 + \lceil \frac{n}{3} \rceil & \text{if } n \in \{2, 3, 4, 5, 7\} \\ 2 + \lceil \frac{n}{3} \rceil & \text{otherwise.} \end{cases}
$$

(a) $\chi_d(C_5) = 3$ (b) $\chi_d(P_5) = 3$

Fig. 2 χ_d-coloring of C_5 and P_5

In 2007, Gera [12] determined the dominator chromatic number of a cycle C_n on n vertices.

Theorem 3 ([12]) *We have* $\chi_d(C_3) = 3$, $\chi_d(C_4) = 2$, *and* $\chi_d(C_5) = 3$, *while for* $n \geq 3$ *and* $n \notin \{4, 5\}$, *we have*

$$\chi_d(C_n) = \left\lceil \frac{n}{3} \right\rceil + 2.$$

We note that if H is a spanning proper subgraph of G, then a χ_d-coloring of G may not be a dominator coloring of H. As a simple example, the χ_d-coloring of the cycle C_5 shown in Figure 2(a) is not a dominator coloring of P_5, even though $\chi_d(C_5) = \chi_d(P_5) = 3$.

For disconnected graphs, we have the following upper and lower bounds on the dominator chromatic number.

Theorem 4 ([13]) *If G is a disconnected graph with components G_1, G_2, $\ldots$, G_k where $k \geq 2$, then*

$$k - 1 + \max \{\chi_d(G_i) \mid i \in [k]\} \leq \chi_d(G) \leq \sum_{i=1}^{k} \chi_d(G_i).$$

Proof Let C_i be a χ_d-coloring of G_i for all $i \in [k]$, where we can choose the colorings so that no two color classes uses the same color. Let C be the union of these k color classes, and so the restriction of C to the component G_i yields the χ_d-coloring C_i for all $i \in [k]$. The coloring C is a chromatic dominator coloring of G, and so

$$\chi_d(G) \leq |C| = \sum_{i=1}^{k} |C_i| = \sum_{i=1}^{k} \chi_d(G_i) = \sum_{i=1}^{k} \chi_d(G_i).$$

To prove the lower bound, consider a component of G with largest dominator chromatic number. Each of the remaining $k - 1$ components of G requires at least one additional color, since every vertex must be a dominator of some color class. Hence, $\chi_d(G) \geq k - 1 + \max \{\chi_d(G_i) \mid i \in [k]\}$. $\qquad\square$

That the lower bound of Theorem 4 is tight may be seen by taking G to be the vertex-disjoint union of $k \geq 2$ stars $K_{1,n}$, for some $n \geq 2$. Each component H of G has

$\chi_d(H) = 2$. Assigning to all leaves of G the same color and assigning to the central vertex of each of the k stars a unique color produce a dominator coloring of G using $k + 1 = (k - 1) + 2 = k - 1 + \max\{\chi_d(H) \mid H$ is a component of $G\}$ colors.

That the upper bound of Theorem 4 is tight may be seen by taking G to be the vertex-disjoint union of $k \geq 2$ copies of $K_{2,n}$ for some $n \geq 2$. Let $\mathcal{C}$ be a dominator coloring of G. Since every vertex must be a dominator of some color class, we note that each component of G has at least one color not used in any other color class. Suppose that some component H of G uses exactly one color, say color 1, not used in any other color class. If two or more vertices in H are colored 1, then no vertex in the color class associated with the color 1 is a dominator of any color class, a contradiction. Hence, exactly one vertex in H is colored 1. However, every vertex different from v and not adjacent to v in the component H is therefore not a dominator of any color class, a contradiction. Hence, each component of G has at least two colors unique to that component, implying that $\chi_d(G) \geq 2k = \sum_{i=1}^{k} \chi_d(G_i)$, where $G_1, G_2, \ldots, G_k$ denote the components of G.

3.1 Bounds on the Dominator Chromatic Number

By definition of a dominator coloring, we have the following observation.

Observation 5 *If v is an arbitrary vertex in a graph G, then in every dominator coloring of G, the closed neighborhood $N[v]$ of v contains a color class.*

Theorem 6 *If G is a graph, then $\chi_d(G) \geq \rho(G)$, with strict inequality if there is no perfect packing in G.*

Proof If S is a packing in G, then by Observation 5, the closed neighborhoods of vertices in S contain at least $|S|$ color classes, and so $\chi_d(G) \geq |S|$. Choosing S to be a maximum packing, we have that $\chi_d(G) \geq \rho(G)$. Further, if G does not have a perfect packing, then at least one additional color class is needed to contain the vertices that do not belong to the closed neighborhood of any vertex in S, and so $\chi_d(G) \geq \rho(G) + 1$. $\qquad\square$

The dominator chromatic number of a graph is related to its independence number as follows, where the independence number $\alpha(G)$ of a graph G is the maximum cardinality of an independent set in G.

Theorem 7 *([13, 15]) If G is a connected graph of order n, then $\chi_d(G) \leq n + 1 - \alpha(G)$.*

Proof If $n = 1$, then the result is trivial since in this case $\chi_d(G) = n = \alpha(G) = 1$. Hence, we may assume that $n \geq 2$. Let I be a maximum independent set in G, and consider the coloring $\mathcal{C}$ that colors all vertices in I with the same color, and colors all remaining $n - \alpha(G)$ vertices each with a different color. Each vertex in $V(G) \setminus I$ dominates the color class that contains it, noting that it is the unique vertex in that color class. By the connectivity of G and by the independence of the set I, every vertex in I has degree at least 1 and has all of its neighbor in $V(G) \setminus I$. Therefore,

(a) $\chi(G) = 2$ (a) $\chi_d(G) = 3$

Fig. 3 A double star $G = S(3, 3)$

by our choice of the coloring $\mathcal{C}$, every vertex in I dominates every color class that contains one of its neighbors. Hence, $\mathcal{C}$ is a dominator coloring of G, implying that $\chi_d(G) \leq |\mathcal{C}| = n + 1 - |I| = n + 1 - \alpha(G)$. $\qquad\square$

That the bound of Theorem 7 is sharp may be seen by taking, for example, a double star $G = S(\ell_1, \ell_2)$. We note that the (unique) proper 2-coloring of the double star is a dominator coloring of G, since no leaf dominates a color class. Hence, $\chi_d(G) \geq 3$. However, coloring all the leaves with one color, and coloring the two central vertices (the non-leaf vertices) with distinct colors, produces a proper 3-coloring that is a dominator coloring. Hence, $\chi_d(G) = 3$. In this example, G has order $n = \ell_1 + \ell_2 + 2$ and $\alpha(G) = \ell_1 + \ell_2 = n - 2$, and so $\chi_d(G) = n + 1 - \alpha(G)$. In the special case when $G = S(3, 3)$, we illustrate the χ-coloring and χ_d-coloring of G in Figure 3(a) and 3(b), respectively.

As observed earlier, the coloring of a graph G of order n that assigns a unique color to each vertex is a trivial dominator coloring of G, and so $\chi_d(G) \leq n$. By Observation 1, if G is a connected graph on at least two vertices, then $\chi_d(G) \geq \chi(G) \geq 2$. We state these observations formally as follows.

Observation 8 *If G is a connected graph of order $n \geq 2$, then $2 \leq \chi_d(G) \leq n$.*

A characterization of graphs achieving equality in the lower and upper bounds of Observation 8 is given by the following result.

Theorem 9 ([11, 15]) *If G is a connected graph of order $n \geq 2$, then the following holds.*

(a) *$\chi_d(G) = 2$ if and only if G is a complete bipartite graph.*
(b) *$\chi_d(G) = n$ if and only if G is a complete graph.*

Proof Suppose that $\chi_d(G) = 2$. By Observation 1, $\chi(G) = 2$, implying that the 2-coloring of G is a dominator coloring of G. Let V_1 and V_2 be the two color classes of G. If $|V_i| = 1$ for some $i \in [2]$, then $G = K_{1,n-1}$, and the desired result follows. Hence, we may assume that $|V_i| \geq 2$ for $i \in [2]$. Thus, no vertex can be a dominator of its own color class, implying that every vertex in V_i is a dominator of the color class V_{3-i} for $i \in [2]$, that is, $G = K_{n_1,n_2}$ where $n_i = |V_i|$. Hence if $\chi_d(G) = 2$, then G is a complete bipartite graph. The converse is immediate.

Suppose next that $\chi_d(G) = n$. By Theorem 7, $\chi_d(G) \leq n + 1 - \alpha(G)$. If G is not a complete graph, then $\alpha(G) \geq 2$, implying that $\chi_d(G) \leq n - 1$, a contradiction. Hence, G must be a complete graph. The converse is immediate. $\square$

The dominator chromatic number of a graph is related to its domination number. For a given graph G, let $\mathcal{A}(G)$ denote the set of all γ-sets in G. We next present an upper bound on the dominator chromatic number of a graph.

Theorem 10 *If G is a connected graph, then*

$$\chi_d(G) \leq \gamma(G) + \min_{S \in \mathcal{A}(G)} \{\chi(G - S)\},$$

and this bound is tight.

Proof Let S be an arbitrary γ-set of G, and let $\mathcal{C}$ be a proper coloring of the graph $G - S$ using $\chi(G - S)$ colors. We extend the coloring $\mathcal{C}$ to a coloring of the vertices of G by assigning to each vertex in S a new and distinct color. Let $\mathcal{C}'$ denote the resulting coloring of G, and note that $\mathcal{C}'$ uses $\gamma(G) + \chi(G - S)$ colors. Since S is a dominating set of G, every vertex in $V(G) \smallsetminus S$ is adjacent to at least one vertex of S. Since the color class of $\mathcal{C}'$ containing a given vertex of S consists only of that vertex, each vertex in $V(G) \smallsetminus S$ is adjacent to every vertex of some color class in the coloring $\mathcal{C}'$. Further, each vertex of S is a dominator of its own (singleton) color class. Hence, $\mathcal{C}'$ is a dominator coloring of G using $\gamma(G) + \chi(G - S)$ colors. This is true for every γ-set of G. The desired upper bound now follows by choosing S to be a γ-set of G that minimizes $\chi(G - S)$. The bound is achieved, for example, by taking G to be a complete graph. $\square$

The proof of Theorem 10 yields the following more general result.

Theorem 11 *If G is a connected graph, and $D(G)$ denotes the set of all dominating sets of G, then*

$$\chi_d(G) \leq \min_{S \in D(G)} \{ |S| + \chi(G - S) \}.$$

Gera [11, 12] established the following upper and lower bounds on the dominator chromatic number of an arbitrary graph in terms of its domination number and chromatic number.

Theorem 12 ([11, 12]) *Every graph G satisfies*

$$\max\{\gamma(G), \chi(G)\} \leq \chi_d(G) \leq \gamma(G) + \chi(G).$$

Proof By Observation 1, recall that $\chi(G) \leq \chi_d(G)$. To show that $\gamma(G) \leq \chi_d(G)$, consider a χ_d-coloring of G with color classes $V_1, \ldots, V_k$, where $k = \chi_d(G)$. Let v_i be an arbitrary vertex in the color class V_i for $i \in [k]$, and consider the set $D = \{v_1, \ldots, v_k\}$. Let v be an arbitrary vertex of G. By definition of a dominator coloring, the vertex v is a dominator of the color class V_i for at least one $i \in [k]$. In particular, the vertex $v = v_k$ or the vertex v is adjacent to the vertex v_k. This is true for every vertex v of G, implying that D is a dominating set of G. Hence, $\gamma(G) \leq |D| = \chi_d(G)$. This establishes the desired lower bound. The upper bound follows from Theorem 10, noting that $\chi(G - S) \leq \chi(G)$ for every proper subset $S \subset V(G)$. $\square$

Gera [12] established an intermediate value-type result for the dominator chromatic number and showed that for every triple (a, b, c) of integers where $1 \leq a \leq c$ and $2 \leq b \leq c$ is a *dominator realizable triple*, there exists a connected graph G such that $\gamma(G) = a$, $\chi(G) = b$, and $\chi_d(G) = c$.

That the lower bound of Theorem 12 is sharp may be seen by taking, for example, a complete bipartite graph G with both partite sets of cardinality at least 2. In this case, $\gamma(G) = \chi(G) = \chi_d(G) = 2$. To see that the upper bound is sharp, let G, for example, be a path P_n or a cycle C_n for some $n \geq 8$ even. In this case, $\gamma(G) = \lceil \frac{n}{3} \rceil$ and $\chi(G) = 2$, and so by Theorems 2 and 3, we have $\chi_d(G) = 2 + \lceil \frac{n}{3} \rceil = \chi(G) + \gamma(G)$.

3.2 Special Classes of Graphs

In this section, we consider the dominator chromatic number of certain classes of graphs.

3.2.1 Bipartite Graphs

As a special case of Theorem 12 when G is a bipartite graph, we have the following result.

Theorem 13 ([11, 12, 15]) *If G is a bipartite graph, then $\gamma(G) \leq \chi_d(G) \leq \gamma(G) + 2$.*

In order to characterize the graphs achieving equality in the lower bound of Theorem 13, we define a special subclass of bipartite graphs as follows.

Definition 1 A bipartite graph G is a *partially complete bipartite graph* if G can be obtained from the disjoint union of $k \geq 1$ complete bipartite graphs K_{x_i,y_i} with partite sets X_i and Y_i where $x_i = |X_i| \geq 2$ and $y_i = |Y_i| \geq 2$ for all $i \in [k]$ by adding edges between copies of these graphs so that the resulting graph is connected and the following conditions hold, where $X = \cup_{i=1}^{k} X_i$ and $Y = \cup_{i=1}^{k} Y_i$.

(a) For each set X_i where $i \in [k]$, there is no set $A \subseteq Y \setminus Y_i$ such that $|A \cap Y_j| = 1$ for all $j \in [k] \setminus \{i\}$ and the set A dominates the set X_i.
(b) For each set Y_i where $i \in [k]$, there is no set $A \subseteq X \setminus X_i$ such that $|A \cap X_j| = 1$ for all $j \in [k] \setminus \{i\}$ and the set A dominates the set Y_i.
(c) For each set X_i where $i \in [k]$, if $A \subseteq X_i$ dominates ℓ of the partite sets in Y, then $\ell \geq |A|$.

(d) For each set Y_i where $i \in [k]$, if $A \subseteq Y_i$ dominates ℓ of the partite sets in X, then $\ell \geq |A|$.

We note, for example, that every complete bipartite graph with both partite sets of cardinality at least 2 is a partially complete bipartite graph. In particular, $K_{2,n}$ is a partially complete bipartite graph for all $n \geq 2$.

Theorem 14 ([11]) *If G is a connected bipartite graph of order at least 2, then $\gamma(G) = \chi_d(G)$ if and only if G is a partially complete bipartite graph.*

Let Q_n be the n-dimensional hypercube, and so Q_n can be represented as the nth power of K_2 with respect to the Cartesian product operation $\square$, that is, $Q_1 = K_2$ and $Q_n = Q_{n-1} \,\square\, K_2$ for $n \geq 2$. Gera [11] established the following upper bound on the dominator chromatic number of an n-dimensional hypercube. The proof given in [11] is algorithmic in nature.

Theorem 15 ([11]) *For $n \geq 2$, $\chi_d(Q_{n+1}) \leq \chi_d(Q_n) + \gamma(Q_n)$.*

The following result established an upper bound on the dominator chromatic number of a connected bipartite graph in terms of its order.

Theorem 16 ([11]) *If G is a connected bipartite graph of order $n \geq 2$, then $\chi_d(G) \leq \frac{1}{2}(n + 2)$, and this bound is sharp.*

Proof Let X and Y be the partite sets of G, where $|X| \leq |Y|$. Coloring all vertices in Y with the same color and assigning a new color to each vertex of X produce a dominator coloring of G using $|X| + 1 \leq \frac{1}{2}n + 1$ colors. This establishes the desired upper bound.

That this bound is sharp may be seen by taking G to be the corona of an arbitrary connected bipartite graph F, and so $G = \mathrm{cor}(F)$. The graph G has order $n = 2|V(F)|$ and satisfies $\gamma(G) = |V(F)|$. Coloring all added vertices of degree 1 with the same color and assigning a new color to every vertex of F produce a dominator coloring of T using $|V(F)| + 1$ colors. Thus, $\chi_d(G) \leq |V(F)| + 1$. We note that each added vertex v of degree 1 either dominates its own class, in which case the vertex v is the only vertex of that color, or dominates the class of its unique neighbor, in which case its neighbor in F is the only vertex of that color. This implies that at least $|V(F)|$ vertices must receive a unique color. Since at least one additional color is needed for the remaining vertices of G, every dominator coloring of G uses at least $|V(F)| + 1$ colors. Thus, $\chi_d(G) \geq |V(F)| + 1$. As observed earlier, $\chi_d(G) \leq |V(F)| + 1$. Consequently, $\chi_d(G) = |V(F)| + 1 = \frac{1}{2}n + 1$. $\square$

3.2.2 Trees

Since no tree is a partially complete bipartite graph, we have the following consequence of Theorems 13 and 14.

Theorem 17 ([11]) *If T is a tree of order $n \geq 2$, then $\chi_d(T) = \gamma(T) + 1$ or $\chi_d(T) = \gamma(T) + 2$.*

By Theorem 2, if T is a path P_n where $n \geq 8$, then $\chi_d(T) = \lceil \frac{n}{3} \rceil + 2 = \gamma(T) + 2$. We note that if T is obtained from $k \geq 1$ vertex-disjoint copies of $K_{1,r}$ where $r \geq 2$ by adding a new vertex and joining it to the central vertex of each star, then $\chi_d(T) = k + 1 = \gamma(T) + 1$. More generally, if a tree T contains a γ-set D such that $V(T) \smallsetminus D$ is an independent set, then $\chi_d(T) = \gamma(T) + 1$, noting that we can color all vertices outside D with the same color and assign a new color to every vertex of D to produce a minimum dominator coloring of T using $\gamma(T) + 1$ colors. In particular, we note that both values for the dominator chromatic number in Theorem 17 are achievable for infinitely many trees. We say that a tree belongs to *dominator class i* if $\chi_d(T) = \gamma(T) + i$ for $i \in [2]$. It remains an open problem to characterize the dominator class 1 and dominator class 2 trees.

A sufficient condition for a tree to belong to dominator class 1 is the following.

Proposition 18 ([5, 15]) *If T is a nontrivial tree such that $\gamma(T) = \tau(T)$, then T belongs to dominator class* 1.

Proof Let D be a minimum vertex cover in T, and so $|D| = \tau(T) = \gamma(T)$. Since D is a vertex cover, the set $V \smallsetminus D$ is an independent set. Coloring all vertices in $V \smallsetminus D$ with the same color and assigning a new color to each vertex of D produce a dominator coloring of T using $|D| + 1 = \gamma(T) + 1$ colors. Thus, $\chi_d(T) \leq \gamma(T) + 1$. By Theorem 17, $\chi_d(T) \geq \gamma(T) + 1$. Consequently, $\chi_d(T) = \gamma(T) + 1$. $\qquad\square$

As observed in [5, 15], the converse of Proposition 18 is not true in general. For example, the tree T shown in Figure 4 belongs to dominator class 1, noting that $\chi_d(T) = 5 = \gamma(T) + 1$. However, $\gamma(T) = 4 < 5 = \tau(T)$. The 5-coloring shown in Figure 4 is a χ_d-coloring of the tree T.

In 2012, Boumediene Merouane and Chellali [4] provide a characterization of trees that belongs to dominator class 1.

Theorem 19 ([4]) *If T is a nontrivial tree, then $\chi_d(T) = \gamma(T) + 1$ if and only if there exists a γ-set, D, of T such that the set $V(T) \smallsetminus (D \cup N(A))$ is an independent set where $A = \{v \in D : pn(v, D) = \{v\}\}$, that is, A is the set of vertices in D, if any, that are isolated in $T[D]$ and have no D-external private neighbor.*

In practice, a tree may admit many minimum dominating sets, and it may not be easy to identify such a set satisfying the statement of Theorem 19. Therefore, in 2015, Boumediene Merouane and Chellali [5] provide a different characterization, which is more pleasing in the sense that it resulted in a polynomial time algorithm

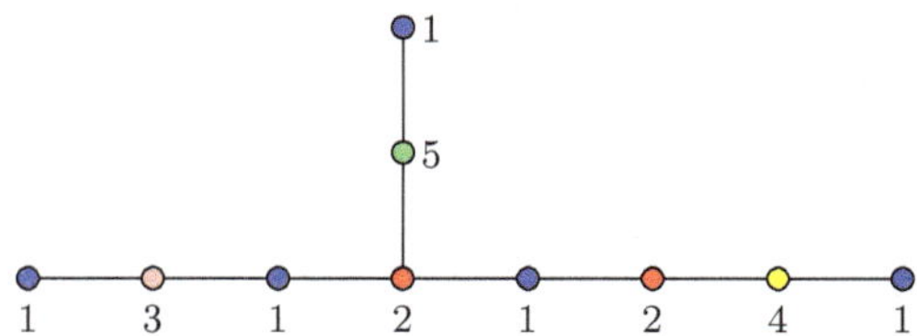

Fig. 4 A χ_d-coloring of a tree T

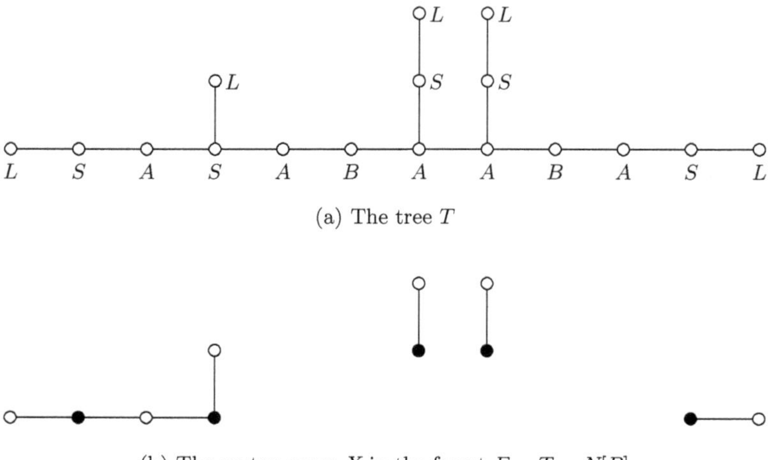

(a) The tree T

(b) The vertex cover X in the forest $F = T - N[B]$

Fig. 5 A tree T and its associated forest F

for computing the dominator chromatic number for every nontrivial tree. In order to state this result, we need some additional terminology.

Recall that a leaf of a tree is a vertex of degree 1 and a vertex with a leaf neighbor is a *support vertex*. Given a tree T, let L and S be the set of leaves and support vertices of T, respectively. Further, let A be the set of vertices of T that are neither leaves nor support vertices, but have a support vertex as a neighbor, that is, if $v \in A$, then $v \notin L \cup S$, but the vertex v is adjacent to a vertex in S. Further, let B be the set of vertices that have all their neighbors in A but do not belong to A, that is, $B = \{v \in V \setminus A : N(v) \subseteq A\}$. Let F be the forest obtained from T by deleting all vertices in $N[B]$, that is, $F = T - N[B]$, or, equivalently, F is the subgraph of T induced by the set $V(T) \setminus N[B]$. Let X be a minimum vertex cover of F containing all support vertices of T (if X contains a leaf of T, we simply replace this leaf by its neighbor in T). To illustrate these definitions, consider the tree T shown in Figure 5(a). The label of each vertex represents one of the sets, namely, L, S, A, or B, that it belongs to, as shown in Figure 5(a). The associated forest $F = T - N[B]$ is illustrated in Figure 5(b), where the vertices in the vertex cover X are given by the darkened vertices.

We are now in a position to state the characterization of trees that belong to dominator class 1 as given in [5].

Theorem 20 ([5]) *If T is a nontrivial tree, then $\chi_d(T) = \gamma(T) + 1$ if and only if the following three conditions hold.*

(a) *B is a packing.*
(b) *$B \cup X$ is a γ-set of T.*
(c) *$\gamma(F) = \tau(F)$.*

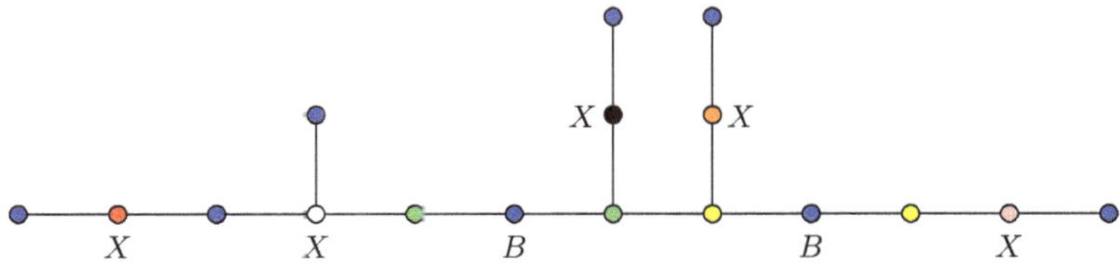

Fig. 6 A χ_d-coloring of the tree T

Suppose that T is a nontrivial tree satisfying the three conditions in the statement of Theorem 20. We color the vertices of T as follows.

- For each vertex v in B, we give a unique color to all vertices in $N(v)$.
- To each vertex in X, we give a unique (new) color.
- To all remaining vertices (including all vertices in $L \cup B$), we give the same, but new, color.

By condition (b), the set $X \cup B$ is a γ-set of T, and so $\gamma(T) = |B| + |X|$. Thus, the resulting coloring is a dominator coloring of T using $|B| + |X| + 1 = \gamma(T) + 1$ colors. Hence, $\gamma(T) + 1 \leq \chi_d(T) \leq |B| + |X| + 1 = \gamma(T) + 1$. Thus we must have equality throughout this inequality chain, implying that $\chi_d(T) = \gamma(T) + 1$.

To illustrate this coloring, consider the tree T shown earlier in Figure 5(a), where a vertex is labelled B or X if it belongs to the set B or X, respectively. To the one vertex in B, we color its two neighbors green, and to the other vertex in B, we color its two neighbors yellow. We color the five vertices in X with five new colors, namely, red, white, black, orange, and pink. Thereafter, we color all remaining vertices of T with a new color, namely, blue. The resulting coloring, illustrated in Figure 6, is a dominator coloring of T using $|B| + |X| + 1 = \gamma(T) + 1 = 8$ colors and is therefore a χ_d-coloring of T.

Based on Theorem 20, the authors in [5] give a quadratic time algorithm computing the dominator chromatic number of any nontrivial tree.

3.2.3 Chordal Graphs and Split Graphs

By Theorem 12, every graph G satisfies $\chi_d(G) \geq \gamma(G)$. In 2012, Chellali and Maffray [6] improved this bound by imposing certain structural restrictions on the graph.

Theorem 21 ([6]) *If G is a connected graph of order $n \geq 2$ that is C_4-free or is claw-free and different from C_4, then $\chi_d(G) \geq \gamma(G) + 1$.*

Since every chordal graph is C_4-free, as is every split graph, as an immediate consequence of Theorem 21, we have the following result.

Corollary 22 ([6]) *If G is a connected graph of order $n \geq 2$, then the following holds.*

(a) *If G is a chordal graph, then $\chi_d(G) \geq \gamma(G) + 1$.*
(b) *If G is a split graph, then $\chi_d(G) \geq \gamma(G) + 1$.*

We also remark that Theorem 17 follows immediately from Theorem 21, noting that every tree is, of course, C_4-free. Chellali and Maffray [6] characterized the split graphs that achieve equality in the bound of Corollary 22(b).

Theorem 23 ([6]) *If G is a connected split graph whose vertex set is partitioned into a clique Q and an independent set I such that Q is minimal, then $\chi_d(G) = \gamma(G) + 1$ if and only if every vertex of Q is a support vertex.*

3.2.4 Proper Interval Graphs and Block Graphs

In 2015, Panda and Pandey [32] study bounds on the dominator chromatic number for two important subclasses of chordal graphs, namely, proper interval graphs and block graphs.

A graph G is an *interval graph* if there exists a one-to-one correspondence between its vertex set and a family $\mathcal{F}$ of closed intervals in the real line, such that two vertices are adjacent if and only if their corresponding intervals intersect. Further, if no interval in $\mathcal{F}$ contains another interval in $\mathcal{F}$, then the graph G is called a *proper interval graph*. Panda and Pandey [32] establish the following lower and upper bounds for the dominator chromatic number of a proper interval graph in terms of its domination number and chromatic number. We note that the upper bound is a restatement of the result in Theorem 12.

Theorem 24 ([32]) *Every proper interval graph G satisfies*

$$\chi(G) + \gamma(G) - 2 \leq \chi_d(G) \leq \gamma(G) + \chi(G).$$

Moreover, all three values can be achieved by $\chi_d(G)$.

For a vertex v of G, the graph $G - v$ is the graph obtained from G by deleting v and deleting all edges of G incident with v. A vertex v is a *cut-vertex* of G if the number of components increases in $G - v$. A *block* of a graph G is a maximal connected subgraph of G that has no cut-vertex of its own. Thus, a block is a maximal 2-connected subgraph of G. The number of vertices in a block is called the *order of the block*. Any two blocks of a graph have at most one vertex in common, namely, a cut-vertex. If a connected graph contains a single block, we call the graph itself a *block*. A *block graph* is a connected graph in which every block is a clique. A block containing exactly one cut-vertex is called an *end block*. A *non-complete block graph* has at least two end blocks. Panda and Pandey [32] generalized the result of Theorem 17 to the class of block graphs. (We note that every tree is a block graph, in which every block is a copy of K_2.)

Theorem 25 ([32]) *If G is a block graph of order at least 2 with k blocks where each block has the same order, then*

$$\chi_d(G) = \gamma(G) + \chi(G) - 1 \quad \text{or} \quad \chi_d(G) = \gamma(G) + \chi(G).$$

Further, both values can be achieved by $\chi_d(G)$.

To illustrate the tightness of the bounds, for $k \geq 1$, let $G_{k,1}$ be a block graph with $2k$ blocks $B_1, B_2, \ldots, B_{2k}$, each having order $s \geq 3$, and $2k - 1$ cut-vertices $v_1, v_2, \ldots, v_{2k-1}$ such that $V(B_i) \cap V(B_{i+1}) = \{v_i\}$ for all $i \in [2k - 1]$. The resulting graph $G = G_{k,1}$ satisfies $\gamma(G) = k$, $\chi(G) = s$ and $\chi_d(G) = k + s - 1$. When $k = 3$ and $s = 3$, the block graph $G_{k,1}$ is illustrated in Figure 7. We color each vertex of the γ-set, $\{v_1, v_3, v_5\}$, of $G_{3,1}$ with a unique color (namely, green, yellow, and pink), and we 2-color the remaining vertices with two new colors (namely, blue and red). The resulting 5-coloring is a χ_d-coloring of G.

For $k \geq 3$, let $G = G_{k,2}$ be a block graph with $2k + 1$ blocks, $B_1, B_2, \ldots, B_{2k+1}$, each having order k, and $2k$ cut-vertices $v_1, v_2, \ldots, v_{2k}$. All vertices in block B_{2k+1} are cut-vertices, say $v_{k+1}, v_{k+2}, \ldots, v_{2k}$. For each $i \in [k]$, B_i is an end block, having exactly one cut-vertex v_i. For each j where $k + 1 \leq j \leq 2k$, block B_j has exactly two cut-vertices v_j and v_{j-k}. The resulting graph $G = G_{k,2}$ satisfies $\gamma(G) = k$, $\chi(G) = k$ and $\chi_d(G) = 2k$. When $k = 3$, the block graph $G_{k,2}$ is illustrated in Figure 8. We color each vertex of the γ-set, $\{v_1, v_2, v_3\}$, of $G_{3,1}$ with a unique color (namely, green, yellow, and pink), and we 3-color the remaining vertices with three new colors (namely, blue, red, and black). The resulting 6-coloring is a χ_d-coloring of G.

As a consequence of Theorem 25, we have the following result. We note that if G is a block graph, then the clique number $\omega(G)$ of G is the maximum order among all blocks in G.

Corollary 26 ([32]) *If G is a non-complete block graph that contains an end block of order $\omega(G)$, then $\chi_d(G) = \gamma(G) + \chi(G) - 1$ or $\chi_d(G) = \gamma(G) + \chi(G)$.*

Panda and Pandey [32] characterize the block graphs G for which one of the end blocks is of maximum size (namely, $\omega(G)$) and $\chi_d(G) = \gamma(G) + \chi(G) - 1$.

3.2.5 P_4-Free Graphs

Chellali and Maffray [6] determined the dominator chromatic number of the class of graphs that are P_4-free by exploiting the structure of these graphs, namely, that if

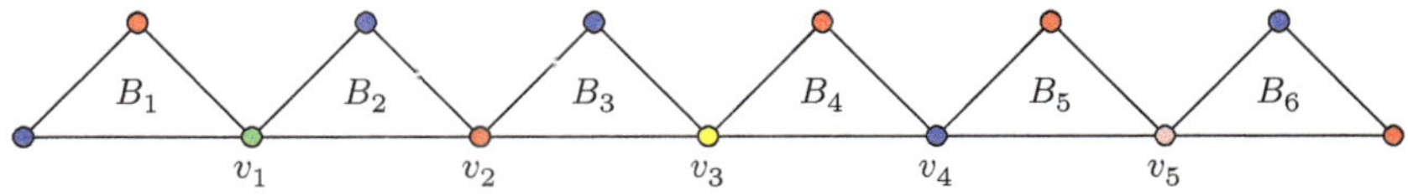

Fig. 7 A block graph $G_{3,1}$

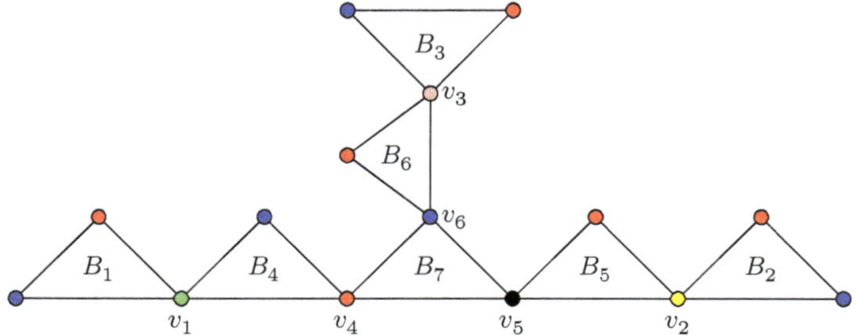

Fig. 8 A block graph $G_{3,2}$

G is a P_4-free graph of order at least 2, then G or its complement $\overline{G}$ is disconnected (see Seinsche [34]). We mention that P_4-free graphs are also known as *cographs*.

Theorem 27 ([6]) *If G is a P_4-free graph, then the following holds.*

(a) *If G is connected, then $\chi_d(G) = \chi(G)$.*
(b) *If G is disconnected with $k \geq 2$ components and h of these components have a universal vertex, then either G has a component H with a universal vertex and satisfies $\chi(G) = \chi(H)$, in which case $\chi_d(G) = \chi(G) + 2k - h - 1$, or G has no such component, in which case $\chi_d(G) = \chi(G) + 2k - h - 2$.*

3.2.6 Other Classes

We mention that the dominator chromatic number of other classes of graphs has also been studied, including degree splitting graph of some graphs [22], dragon and lollipop graphs [30], wheel related graphs [35], the generalized Petersen graph [31], and Mycielskian graphs [1]. However, we do not define these classes of graphs here.

3.3 Graph Products

In this section, we present some results on the dominator chromatic number in Cartesian products of graphs. The Cartesian product $G \square H$ of graphs G and H is the graph with vertex set $V(G) \times V(H) = \{(g, h) : g \in V(G) \text{ and } h \in V(H)\}$, where two vertices (g_1, h_1) and (g_2, h_2) in the *Cartesian product $G \square H$* of graphs G and H are adjacent if either $g_1 = g_2$ and $h_1 h_2$ is an edge in H or $h_1 = h_2$ and $g_1 g_2$ is an edge in G.

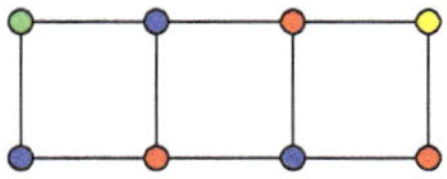

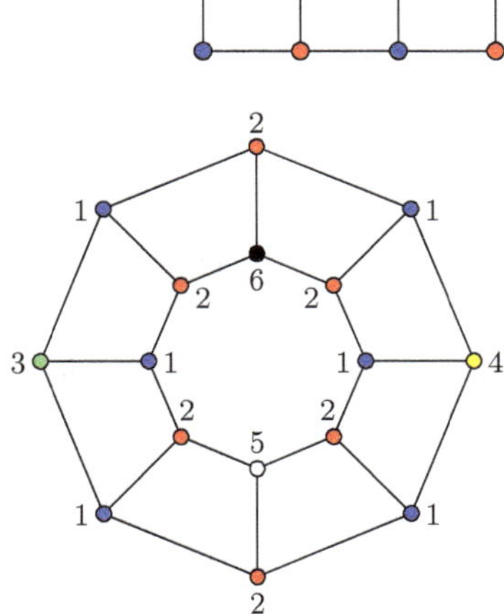

Fig. 9 A χ_d-coloring of $P_2 \,\square\, P_4$

Fig. 10 The circular ladder graph CL_8

In 2017, Chen, Zhao, and Zhao [8] determined the dominator chromatic number of Cartesian products of certain paths and cycles. The Cartesian product $P_m \,\square\, P_n$ of paths P_m and P_n is known as a $2 \times n$ grid graph.

Theorem 28 ([8]) $\chi_d(P_2 \,\square\, P_2) = 2$, $\chi_d(P_2 \,\square\, P_3) = \chi_d(P_2 \,\square\, P_3) = 4$, and for all $n \geq 5$,

$$\chi_d(P_2 \,\square\, P_n) = \left\lfloor \frac{n}{2} \right\rfloor + 3.$$

A dominator coloring of the 2×4 grid graph, for example, using four colors is shown in Figure 9.

The Cartesian product of a cycle C_n on $n \geq 3$ vertices and a path P_2 on two vertices is called a *circular ladder graph* CL_n of order $2n$; that is, $CL_n = C_n \,\square\, K_2$ (cf. [23]). A circular ladder graph is also called a *cycle prism* in the literature. We note that CL_n is bipartite if and only if n is even. The circular ladder graph CL_n is also called the *n-prism* in the literature. For example, let $G = C_8 \,\square\, K_2$ be the circular ladder graph CL_8 shown in Figure 10. We note that G is a bipartite graph and $\gamma(G) = 4$, and so, by Theorem 13, $\chi_d(G) \leq 6$. As shown in the proof of Theorem 12, we can find a dominator coloring of G using six colors as follows. We first 2-color the vertices of G with the colors 1 and 2 (depicted as the colors blue and red in Figure 10), and thereafter we recolor the vertices of a γ-set of G with the colors 3, 4, 5, *and* 6 (depicted as the colors green, yellow, white, and black, respectively, in Figure 10). The resulting 6-coloring is a dominator coloring of G. Thus, $\chi_d(G) \leq 6$. Moreover, as shown in the proof of Theorem 29, $\chi_d(G) \geq \gamma(G) + 2 = 6$. Consequently, $\chi_d(CL_8) = 6$.

In 2015, Manjula and Rajeswari [29] claimed to have proven that $\chi_d(CL_n) = n + 1$ for all $n \geq 9$. This result is incorrect. The correct value for the dominator chromatic

Fig. 11 A χ_d-coloring of
$CL_3 = P_2 \square C_3$

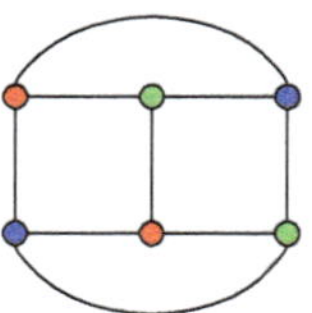

number of a circular ladder graph is given in Theorem 29 by Chen, Zhao, and Zhao
[8].

Theorem 29 *The dominator chromatic number of the circular ladder graph* $CL_n = C_n \square K_2$ *is given by* $\chi_d(CL_3) = 3$ *and for all* $n \geq 4$ *as follows.*

$$\chi_d(CL_n) = \begin{cases} \frac{1}{2}(n+4) & \text{when } n \equiv 0 \ (\mathrm{mod} \ 4) \\ \frac{1}{2}(n+5) & \text{when } n \ (\mathrm{mod} \ 4) \in \{1, 3\} \\ \frac{1}{2}(n+6) & \text{when } n \equiv 2 \ (\mathrm{mod} \ 4). \end{cases}$$

Proof Let $G = C_n \square K_2$ be the circular ladder graph CL_n where $n \geq 3$. A dominator
coloring of CL_3 using three colors is shown in Figure 11, showing that $\chi_d(CL_3) \leq 3$.
By Observation 1, $\chi_d(CL_3) \geq \chi(CL_3) = 3$. Consequently, $\chi_d(CL_3) = 3$. Hence in
what follows, we let $n \geq 4$.

Let $x_1 x_2 \ldots x_n x_1$ and $y_1 y_2 \ldots y_n y_1$ be the two disjoint copies of the cycle C_n used
to construct $CL_n = C_n \square K_2$ and where $x_i y_i \in E(CL_n)$ for $i \in [n]$. We note that
$\gamma(G) = \lfloor n/2 \rfloor + \lceil n/4 \rceil - \lfloor n/4 \rfloor$, that is, $\gamma(G) = n/2$ if $n \equiv 0 \ (\mathrm{mod} \ 4)$, $\gamma(G) = n/2 + 1$ if
$n \equiv 2 \ (\mathrm{mod} \ 4)$, and $\gamma(G) = (n+1)/2$ if $n \ (\mathrm{mod} \ 4) \in \{1, 3\}$.

We show firstly that $\chi_d(G) \leq \gamma(G) + 2$. If n is even, then we can apply Theorem
13 to yield $\chi_d(G) \leq \gamma(G) + 2$. Suppose that $n \equiv 3 \ (\mathrm{mod} \ 4)$. Thus, $n = 4k + 3$ for
some $k \geq 0$. In this case, the set

$$D = \bigcup_{i=0}^{k} \{x_{4i+1}, y_{4i+3}\}$$

is a γ-set of D, noting that D is a dominating set of G and $|D| = 2(k+1) = (n+1)/2$
$= \gamma(G)$. We note that removing the set D from G produced a graph $G - D \cong P_{6k+4}$.
We now 2-color the vertices of the path $G - D$, and thereafter we color each vertex
of D with a unique color. The resulting coloring of G is a dominator coloring of G
using $\gamma(G) + 2 = 2k + 4$ colors. Suppose next that $n \equiv 1 \ (\mathrm{mod} \ 4)$. Thus, $n = 4k + 1$
for some $k \geq 1$. In this case, we consider the set

$$S = \bigcup_{i=0}^{k-1} \{x_{4i+2}, y_{4i+4}\}.$$

We note that the set S is a packing in G. Further, the set S dominates all vertices
of G, except for the two vertices y_1 and x_n (note that here $x_n = x_{4k+1}$). Further, we

note that $S \cup \{x_1\}$ is a γ-set of G, implying that $|S| = \gamma(G) - 1$. We note that with the set S as defined above, the graph $G - S$ can be obtained from a path P_{6k} on $6k$ vertices given by

$$P: \; y_1 y_2 y_3 x_3 x_4 x_5 \ldots y_{4(k-1)+1} \, y_{4(k-1)+2} \, y_{4(k-1)+3} \, x_{4(k-1)+3} \, x_{4(k-1)+4} \, x_{4(k-1)+5}$$

that starts at the vertex y_1 and ends at the vertex x_{4k+1}, by adding the two vertices x_1 and y_{4k+1} and adding the four edges $x_1 y_1$, $x_1 x_{4k+1}$, $y_1 y_{4k+1}$, and $x_{4k+1} y_{4k+1}$. We now 2-color the vertices of the path P with the colors 1 and 2 and color x_1 and y_{4k+1} with the same new color 3 to produce a 3-coloring of $G - S$. Thereafter, we color the vertices of S with $|S| = \gamma(G) - 1$ new colors, one distinct color to each vertex. The resulting coloring is a dominator coloring of G using $\gamma(G) + 2$ colors. In particular, we note that y_1 and x_{4k+1} each are dominators of the color class $\{x_1, y_{4k+1}\}$ (whose vertices are colored 3), while every other vertex in $G - S$ is a dominator of the unique vertex in S that it is adjacent to. Further, each vertex v of S is a dominator of the color class that contains the vertex v (and is a singleton set consisting only of the vertex v).

To illustrate the above coloring, consider the case, for example, when $n = 9$. In this case, the set $S = \{x_2, y_4, x_6, y_8\}$ and is given by the set of darkened vertices in Figure 12(a). Further the 3-coloring of the graph $G - S$ is illustrated in Figure 12(b). We then extend this 3-coloring of $G - S$ to a 7-coloring of G by adding four new colors, one distinct color to each vertex in the set S. The resulting 7-coloring is a dominator coloring of G, implying that $\chi_d(CL_9) \le 7$.

In all the above cases, we have shown that $\chi_d(G) \le \gamma(G) + 2$. Further one can readily show that $\chi_d(G) \ge \gamma(G) + 2$. We present a proof of the simplest case when $n \equiv 0 \pmod 4$ as an illustration. In this case, $n = 4k$ for some $k \ge 1$. Further, $\gamma(G) = 2k$, and every γ-set of G is a packing. Each vertex $v \in D$ either dominates its own class, in which case the vertex v is the only vertex of that color, or dominates a color class that is a subset of its neighborhood, $N(v)$. Since the sets $N[v] = N(v) \cup \{v\}$ are vertex-disjoint sets for all $v \in D$, this implies that at least $|D| = \gamma(G)$ vertices must receive a unique color. Since at least two additional colors are needed for the remaining vertices of G, every dominator coloring of G uses at least $\gamma(G) + 2$ colors. Hence, $\chi_d(G) \ge \gamma(G) + 2$ in this case when $n \equiv 0 \pmod 4$. Analogous arguments show that $\chi_d(G) \ge \gamma(G) + 2$ in the three other cases when $n \pmod 4 \in \{1, 2, 3\}$. We

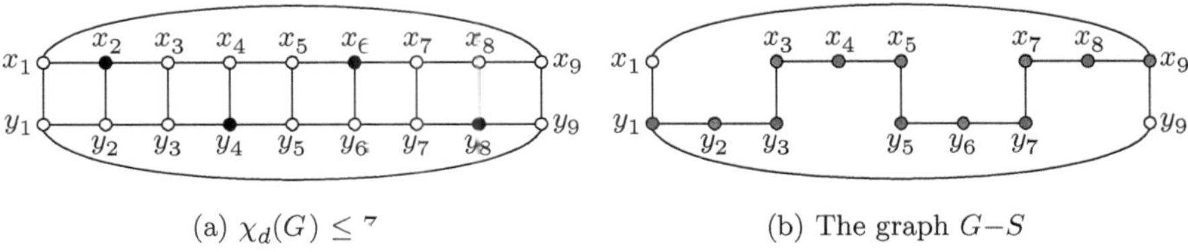

(a) $\chi_d(G) \le 7$ (b) The graph $G - S$

Fig. 12 A circular ladder graph $G = CL_9$

omit the details. Therefore, $\chi_d(G) = \gamma(G) + 2$. The desired result now follows from our earlier observation that $\gamma(G) = \lfloor n/2 \rfloor + \lceil n/4 \rceil - \lfloor n/4 \rfloor$. $\qquad\square$

We remark that Chen [7] continued the study of dominator chromatic number of Cartesian products of certain paths and cycles and considers the $3 \times n$ grid, $P_3 \,\square\, P_n$, the Cartesian product $P_3 \,\square\, C_n$. Two vertices (g_1, h_1) and (g_2, h_2) in the *direct product graph* $G \times H$ of graphs G and H are adjacent if $g_1 g_2 \in E(G)$ and $h_1 h_2 \in E(H)$. They also consider the dominator chromatic number of the Cartesian product $K_m \,\square\, K_n$ for $m, n \geq 2$.

Paulraja and Handrasekar [33] determined the dominator chromatic number for some classes of graphs, such as the direct product $K_m \times K_n$ for $m, n \geq 2$, and the direct product $(K_m \circ K_1) \times K_r$ for $m, n \geq 2$. They also present results on the Cartesian product $K_n \,\square\, Q_r$ for $r \geq 3$, where Q_r is the r-dimensional hypercube. We omit these results here.

3.4 Dominator Partition Number

We discuss briefly in this section the dominator partition number of a graph. In their introductory paper, Hedetniemi, Hedetniemi, Laskar, McRae, and Wallis [15] define a *dominator partition* of a graph G as a coloring (not necessarily proper) with the property that every vertex in G is adjacent to all other vertices in some color class (including possibly its own). The *dominator partition number* of G, which they denote by $\pi_d(G)$, is the minimum number of color classes in a dominator partition of G. We note that every dominator coloring is a dominator partition, but not conversely. Thus, $\pi_d(G) \leq \chi_d(G)$ for all graphs G. Hedetniemi et al. [15] provide the following lower and upper bounds on the dominator partition number of a graph in terms of the minimum and maximum degrees.

Theorem 30 ([15]) *If G is a graph of order n, then*

$$\frac{n}{1 + \Delta(G)} \leq \pi_d(G) \leq n - \delta(G).$$

Hedetniemi et al. [15] showed that the dominator partition number of a graph is surprisingly one of two possible values.

Theorem 31 ([15]) *If G is a graph of order n, then $\pi_d(G) = \gamma(G)$ or $\pi_d(T) = \gamma(G) + 1$.*

Proof The proof of the lower bound $\pi_d(G) \geq \gamma(G)$ is identical to the proof we presented earlier of Theorem 12. To prove the upper bound $\pi_d(G) \leq \gamma(G) + 1$, let $D = \{v_1, \ldots, v_k\}$ be a γ-set of G. Since the partition $\pi = \{V_1, \ldots, V_{k+1}\}$ of V, where $V_i = \{v_i\}$ for $i \in [k]$ and where $V_{k+1} = V \setminus D$, is a dominator partition of G, we have that $\pi_d(T) \leq \gamma(G) + 1$. $\qquad\square$

3.5 Algorithmic and Complexity Results

In this section, we consider the algorithmic complexity of the problem of computing the dominator chromatic number of an arbitrary graph. Formally, we consider the following decision problem:

GRAPH DOMINATOR k-COLORABILITY

Input: A graph G, and an integer $k \geq 1$.
Question: Does G have a dominator k-coloring?

Hedetniemi et al. [17] showed that to determine if a graph G has a dominator 3-coloring can be computed in polynomial time.

Theorem 32 ([17]) GRAPH DOMINATOR 3-COLORABILITY *is solvable in polynomial time.*

To show that the GRAPH DOMINATOR k-COLORABILITY is NP-complete for $k \geq 4$, we give a transformation from GRAPH k-COLORABILITY:

GRAPH k-COLORABILITY

Input: A graph G, and an integer $k \geq 4$.
Question: Does G have a k-coloring?

Theorem 33 ([13, 15]) *Graph Dominator k-Colorability is NP-complete for general graphs, for $k \geq 4$.*

Proof Let k be an integer greater than 3. GRAPH DOMINATOR k-COLORABILITY is clearly in the class NP since we can efficiently verify that an assignment of colors to the vertices of G is both a proper coloring and that every vertex dominates some color class.

Next we transform an instance of GRAPH $(k-1)$-COLORABILITY to an instance of GRAPH DOMINATOR k-COLORABILITY. Given an instance of GRAPH $(k-1)$-COLORABILITY, a graph G, and a $k-1$ coloring of G, construct an instance of GRAPH DOMINATOR k-COLORABILITY as follows. Let G' be the graph obtained from G by adding a new vertex v to G and adding all edges joining v to every vertex of G. We now consider the instance given by the graph G' and a dominator k-coloring of G.

Let $\mathcal{C}$ be a $(k-1)$-coloring of G, and let $\mathcal{C}'$ be the k-coloring of G' obtained from the coloring $\mathcal{C}$ by assigning a new color to the vertex v. Thus, the color class containing v consists only of the vertex v. Since $\{v\} \subseteq N_{G'}[u]$ for every vertex in G', every vertex in G' dominates some color class. Thus, C' is a dominator k-coloring of G'.

Conversely, suppose that G' has a dominator k-coloring C. Since v is adjacent to every other vertex in G', the vertex v is the only vertex in its color class. The removal of v produces a $(k-1)$-coloring of G.

It follows that G is $(k-1)$-colorable if and only if G' is dominator k-colorable. $\square$

By Theorem 33, it is NP-complete to decide if a graph admits a dominator coloring with at most four colors. Chellali and Maffray [6] characterized the graphs G such that $\chi_d(G) \leq 3$ and showed that their characterization leads to a polynomial time recognition algorithm for such graphs. A rough estimate of the complexity of their algorithm is $O(n^8)$. We note that this result that the problem "$\chi_d(G) \leq 3$" can be solved in polynomial time is in contrast the problem "$\chi(G) \leq 3$," which in NP-complete.

In 2009, Hedetniemi et al. [15] and in 2011 Arumugam, Raja Chandrasekar, Misra, Philip, and Saurabh [2] studied algorithmic aspects of dominator colorings in graphs. They established the following complexity result.

Theorem 34 ([2, 15]) *For $k \geq 4$ an integer,* GRAPH DOMINATOR k-COLORABILITY, *is NP-complete for bipartite, chordal, planar, or split graphs.*

Arumugam et al. [2] complemented the above hardness results by showing that the GRAPH DOMINATOR COLORABILITY is fixed-parameter tractable in certain classes. Informally, a *parameterization* of a problem assigns an integer k to each input instance, and a parameterized problem is *fixed-parameter tractable*, abbreviated FPT, if there is an algorithm that solves the problem in time $f(k) \cdot |I|^{O(1)}$, where $|I|$ is the size of the input and f is an arbitrary computable function that depends only on the parameter k. (For a discussion on parameterized complexity, we refer the reader to the 2013 book by Downey and Fellows [10].)

A graph is an *apex graph* if there exists a vertex in G whose removal from G yields a planar graph. A family $\mathcal{F}$ of graphs is *apex minor-free* if there is a specific apex graph H such that no graph in $\mathcal{F}$ has H as a minor. As an example, planar graphs are apex minor-free since no planar graph has K_5 as a minor. Apex graphs play an important role in aspects of graph minor theory and are closed under the operation of taking minors, that is, contracting an edge or removing an edge or vertex leads to another apex graph.

As remarked in [2], for $k \geq 4$ an integer, GRAPH DOMINATOR k-COLORABILITY, is not fixed-parameter tractable in general graphs unless P$=$NP. However, the problem is fixed-parameter tractable in apex minor-free graphs (which include planar graphs) and chordal graphs.

Theorem 35 ([2]) *For $k \geq 4$ an integer, Graph Dominator k-Colorability, is fixed-parameter tractable on apex minor-free graphs and on chordal graphs.*

Arumugam et al. [2] show that for $k \geq 4$ an integer, GRAPH DOMINATOR k-COLORABILITY, can be solved in "fast" fixed-parameter tractable time in split graphs.

Theorem 36 ([2]) *For $k \geq 4$ an integer, Graph Dominator k-Colorability, can be solved in $O(2^k \cdot n^2)$ time on a split graph on n vertices.*

Arumugam et al. [2] pose the problem of whether for $k \geq 4$ an integer, GRAPH DOMINATOR k-COLORABILITY, can be solved in polynomial time on interval graphs.

4 Total Dominator Colorings

The total version of dominator coloring in a graph was studied by several authors. The concept of total dominator colorings in graphs was first defined in the manuscript by Hedetniemi, Hedetniemi, McRae, Rall, and Hedetniemi [16] dated July 9, 2009. Subsequently, Hedetniemi, Hedetniemi, Hedetniemi, McRae, and Rall [17] continued the study of total dominator colorings in graphs in their manuscript dated February 18, 2011. The first published papers on the topic appears to be the 2012 paper by Vijayalekshmi [36] and the 2015 paper by Kazemi [26].

Formally, a *total dominator coloring*, abbreviated TD-*coloring*, of a graph G with no isolated vertex is a proper coloring of G in which each vertex of the graph is adjacent to every vertex of some other color class (different from its own color class). The *total dominator chromatic number* of G which we denote by χ_{td} (and denoted by $\chi_d^t(G)$ in [18, 26]) is the minimum integer k for which G has a TD-coloring with k colors. A χ_{td}-*coloring* of G is a coloring of G that uses $\chi_{td}(G)$ colors. Every total dominator coloring is a dominator coloring. Hence, we have the following observation.

Observation 37 *For every graph G without isolated vertices, we have $\chi_d(G) \leq \chi_{td}(G)$.*

Consider an arbitrary χ_{td}-coloring of G, and let S be a set consisting of one vertex from each of the resulting $\chi_{td}(G)$ color classes. Since every vertex in G is adjacent to every vertex of some color class (different from its own color class), the set S is a TD-set in G, implying that $\gamma_t(G) \leq |S| = \chi_{td}(G)$. Hence we have the following result, first observed by Vijayalekshmi [36] and Kazemi [26].

Observation 38 ([26, 36]) *For every graph G without isolated vertices, $\gamma_t(G) \leq \chi_{td}(G)$.*

Analogous results to Observation 8 and Theorem 9 hold for the total dominator chromatic number.

Theorem 39 ([26, 36]) *If G is a connected graph of order $n \geq 2$, then $2 \leq \chi_{td}(G) \leq n$. Moreover the following holds.*

Fig. 13 A χ_{td}-coloring of a path P_{14}

(a) $\chi_{\mathrm{td}}(G)=2$ *if and only if G is a complete bipartite graph.*
(b) $\chi_{\mathrm{td}}(G)=n$ *if and only if G is a complete graph.*

For disconnected graphs, we have the following upper and lower bounds on the total dominator chromatic number.

Theorem 40 ([36]) *If G is a disconnected graph with nontrivial components G_1, G_2, ..., G_k where $k \geq 2$, then*

$$2k - 2 + \max\{\chi_{\mathrm{td}}(G_i) \mid i \in [k]\} \leq \chi_{\mathrm{td}}(G) \leq \sum_{i=1}^{k} \chi_{\mathrm{td}}(G_i).$$

We remark that the total dominator chromatic number of a path and cycle is incorrectly determined in [26]. To state the total dominator chromatic number of a path P_n and a cycle C_n on n vertices, we shall need the following well-known result (see [19]).

Observation 41 *For $n \geq 3$, if $G \in \{P_n, C_n\}$, then we have*

$$\gamma_t(G) = \left\lfloor \frac{n}{2} \right\rfloor + \left\lceil \frac{n}{4} \right\rceil - \left\lfloor \frac{n}{4} \right\rfloor,$$

that is, $\gamma_t(G) = \frac{n}{2}$ if $n \equiv 0 \pmod 4$, $\gamma_t(G) = \frac{n}{2} + 1$ if $n \equiv 2 \pmod 4$, and $\gamma_t(G) = \frac{n+1}{2}$ for n odd.

Theorem 42 ([18]) *For $n \geq 2$, we have*

$$\chi_{\mathrm{td}}(P_n) = \begin{cases} \gamma_t(P_n) & \text{for } n \in \{2, 3, 6\} \\ \gamma_t(P_n) + 1 & \text{for } n \in \{4, 5, 7, 9, 10, 11, 14\} \\ \gamma_t(P_n) + 2 & \text{otherwise.} \end{cases}$$

For example, a χ_d-coloring of the path P_{14} (using $\gamma_t(P_{14})+1=8+1=9$ colors) is illustrated in Figure 13.

Thus, by Observation 41 and Theorem 42, we have the following closed formula for the total dominator chromatic number of a path of large order.

Theorem 43 ([18]) *For $n \geq 15$, $\chi_{\mathrm{td}}(P_n) = \left\lfloor \frac{n}{2} \right\rfloor + \left\lceil \frac{n}{4} \right\rceil - \left\lfloor \frac{n}{4} \right\rfloor + 2$.*

For $n \geq 16$, we define next a $\chi_{\mathrm{td}}(P_n)$-coloring, C_n^*, of a path P_n as follows. Let G be the path $v_1 v_2 \ldots v_n$, where $n \geq 16$. For each vertex v_i where $i \equiv 2, 3 \pmod 4$, assign a unique color. For each vertex v_i where $i \equiv 1 \pmod 4$, assign a new additional color, say 1. For each vertex v_i where $i \equiv 0 \pmod 4$, assign a further additional

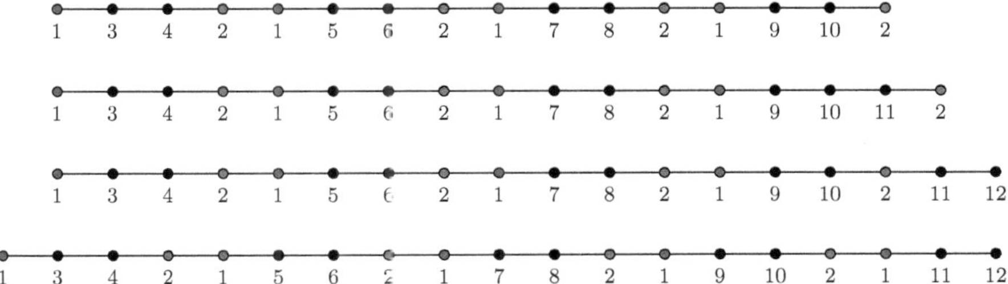

Fig. 14 A χ_{td}-coloring of a paths P_{16}, P_{17}, P_{18}, and P_{19}

color, say 2. Let C_n denote the resulting coloring. We now define a coloring C_n^* as follows. If $n \equiv 0$, 3 (mod 4), let $C_n^* = C_n$. If $n \equiv 1$ (mod 4), then recolor the vertex v_{n-1} (currently colored with color 2) with a new distinct color, and let C_n^* denote the resulting modified coloring. If $n \equiv 2$ (mod 4), then recolor the vertex v_{n-1} (currently colored with color 1) with a new distinct color, and let C_n^* denote the resulting modified coloring. The coloring C_n^* when $n \in \{16, 17, 18, 19\}$, for example, is illustrated in Figure 14. The darkened vertices in this coloring of C_n^* in Figure 14 form a γ_t-set of the path. A new color is assigned to each darkened vertex in the path.

Theorem 44 ([18]) $\chi_{td}(C_3) = 3$, $\chi_{td}(C_4) = 2$, and $\chi_{td}(C_{11}) = 8$. *For all other values of $n \geq 5$, we have $\chi_{td}(C_n) = \chi_{td}(P_n)$.*

4.1 Bounds on the Total Dominator Chromatic Number

By definition of a total dominator coloring, we have the following observation.

Observation 45 *If v is an arbitrary vertex in a graph G without isolated vertices, then in every dominator coloring of G, the open neighborhood $N(v)$ of v contains a color class.*

Theorem 46 *If G is a graph without isolated vertices, then $\chi_{td}(G) \geq \rho^o(G)$, with strict inequality if there is no perfect packing in G.*

Proof If S is an open packing in G, then by Observation 45, the open neighborhoods of vertices in S contain at least $|S|$ color classes, and so $\chi_{td}(G) \geq |S|$. Choosing S to be a maximum open packing, we have that $\chi_{td}(G) \geq \rho^o(G)$. Further, if G does not have a perfect open packing, then at least one additional color class is needed to contain the vertices that do not belong to the open neighborhood of any vertex in S, and so $\chi_{td}(G) \geq \rho^o(G) + 1$. $\qquad\square$

If H is any connected graph of order $k \geq 1$, then the 2-corona $G = H \circ P_2$ satisfies $\rho^o(G) = 2k = \chi_{td}(G)$, illustrating the existence of graphs G that contain a perfect

Fig. 15 The graph $C_4 \circ P_2$

open packing and satisfy $\rho^o(G) = \chi_{\mathrm{td}}(G)$. The graph $C_4 \circ P_2$, for example, is shown in Figure 15 (here, $H = C_4$).

If a graph G contains a perfect open packing, then it is not necessarily true that $\rho^o(G) = \chi_{\mathrm{td}}(G)$. The simplest example illustrating this is a path $G = P_4$, with $\rho^o(G) = 2$ and $\chi_{\mathrm{td}}(G) = 3$. More generally, if $G = P_n$ where $n \equiv 0 \pmod 4$ and $n \geq 8$, then G has a perfect open packing and $\rho^o(G) = \gamma_t(G)$. However in this case, by Theorem 42, we have $\chi_{\mathrm{td}}(G) = \gamma_t(G) + 2 = \rho^o(G) + 2$.

For a given graph G, let $\mathcal{A}_t(G)$ denote the set of all γ_t-sets in G. We next present an upper bound on the total dominator chromatic number of a graph.

Theorem 47 ([18, 26]) *If G is a connected graph without isolated vertices, then*

$$\chi_{\mathrm{td}}(G) \leq \gamma_t(G) + \min_{S \in \mathcal{A}_t(G)} \{\chi(G - S)\},$$

and this bound is tight.

Proof Let S be an arbitrary γ_t-set of G, and let $\mathcal{C}$ be a proper coloring of the graph $G - S$ using $\chi(G - S)$ colors. We now extend the coloring $\mathcal{C}$ to a coloring of the vertices of G by assigning to each vertex in S a new and distinct color. Let $\mathcal{C}'$ denote the resulting coloring of G, and note that $\mathcal{C}'$ uses $\gamma_t(G) + \chi(G - S)$ colors. Since S is a TD-set of G, every vertex in G is adjacent to at least one vertex of S. Since the color class of $\mathcal{C}'$ containing a given vertex of S consists only of that vertex, each vertex in G is adjacent to every vertex of some (other) color class in the coloring $\mathcal{C}'$. Hence, $\mathcal{C}'$ is a TD-coloring of G using $\gamma_t(G) + \chi(G - S)$ colors. This is true for every γ_t-set of G. The desired upper bound now follows by choosing S to be a γ_t-set of G that minimizes $\chi(G - S)$. The bound is achieved, for example, by taking G to be a complete graph. As shown in [18], the bound is also tight for infinitely many trees.   □

The proof of Theorem 47 yields the following more general result.

Theorem 48 *If G is a connected graph without isolated vertices, and* $\mathrm{TD}(G)$ *denotes the set of all total dominating sets of G, then*

$$\chi_{\mathrm{td}}(G) \leq \min_{S \in \mathrm{TD}(G)} \{ |S| + \chi(G - S) \}.$$

We observe that $\chi(G - S) \leq \chi(G)$ for every proper subset $S \subset V(G)$. This observation, together with the results of Observations 37 and 38, gives us the following analogous result to Theorem 12, thereby establishing upper and lower bounds on the total dominator chromatic number of an arbitrary graph in terms of its total domination number and chromatic number.

Theorem 49 ([26, 36]) *Every graph G without isolated vertices satisfies*

$$\max\{\gamma_t(G), \chi(G)\} \leq \chi_{\text{td}}(G) \leq \gamma_t(G) + \chi(G).$$

4.2 Special Classes of Graphs

In this section, we consider the total dominator chromatic number of certain classes of graphs.

4.2.1 Bipartite Graphs

As a special case of Theorem 49 when G is a bipartite graph, we have the following result.

Theorem 50 ([26, 36]) *If G is a bipartite graph, then* $\gamma_t(G) \leq \chi_{\text{td}}(G) \leq \gamma_t(G) + 2.$

For each $t \in \{0, 1, 2\}$, an infinite family $\mathcal{G}_t$ of bipartite graphs such that each graph $G \in \mathcal{G}_t$ satisfies $\chi_{\text{td}}(G) = \gamma_t(G) + t$ is constructed in [18] as follows.

Let $\mathcal{G}_0$ be the family of graphs G without isolated vertices that contain a TD-set S that is a perfect open packing in G and such that the neighborhood of each edge e in $G[S]$ induces a complete bipartite graph in G, that is, if $e = uv$ is an edge in $G[S]$, then the subgraph of G induced by the neighborhood, $N[e]$, of e is a complete bipartite graph K_{n_1,n_2} where $d(u) = n_1$ and $d(v) = n_2$. Let $G \in \mathcal{G}_0$. As an example, if H is an arbitrary graph, then the graph $G = H \circ P_2$ belongs to the family $\mathcal{G}_0$ since the set $S = V(G) \setminus V(H)$ is a TD-set that is a perfect open packing in G and the neighborhood of each edge e in $G[S]$ induces a complete bipartite graph $K_{1,2}$ in G.

Let $\mathcal{G}_1$ be the family of graphs that can be obtained from a graph H without isolated vertices by attaching any number of pendant edges, but at least one, to each vertex of H. For example, if H is an arbitrary isolate-free graph, then the corona $G = H \circ P_1$ of H belongs to the family $\mathcal{G}_1$.

Let $\mathcal{G}_2$ be the family of all paths P_n and cycles C_n, where $n \equiv 0 \pmod 4$ and $n \geq 8$.

Theorem 51 ([18]) *The following holds.*

(a) *If* $G \in \mathcal{G}_0$, *then* $\chi_{\text{td}}(G) = \gamma_t(G)$.
(b) *If* $G \in \mathcal{G}_1$, *then* $\chi_{\text{td}}(G) = \gamma_t(G) + 1$.
(c) *If* $G \in \mathcal{G}_2$, *then* $\chi_{\text{td}}(G) = \gamma_t(G) + 2$.

4.2.2　Trees

Recall that the dominator chromatic number of a tree is one of two values (see Theorem 17). However, the total dominator chromatic number of a tree is one of three values. By Theorem 50, if T is a tree, then $\gamma_t(T) \le \chi_{td}(T) \le \gamma_t(T) + 2$. Further, there are infinitely many trees T for which $\chi_{td}(T) = \gamma_t(T) + i$ for each $i \in [2]_0 = \{0, 1, 2\}$.

Theorem 52　([26, 36]) *If G is a tree, then $\gamma_t(T) \le \chi_{td}(T) \le \gamma_t(T) + 2$.*

The following properties of χ_{td}-colorings in a tree T are established in [18]. We say that a color class C in a given TD-coloring C of G is *free* if each vertex of G is adjacent to every vertex of some color class different from C.

Theorem 53　([18]) *If T is a nontrivial tree, then the following holds.*

(a) *If $\gamma_t(T) = \chi_{td}(T)$, then no $\chi_{td}(T)$-coloring contains a free color class.*
(b) *If $\chi_{td}(T) = \gamma_t(T) + 1$, then there exists a $\chi_{td}(T)$-coloring that contains a free color class.*
(c) *If $\chi_{td}(T) = \gamma_t(T) + 2$, then there exists a $\chi_{td}(T)$-coloring that contains two free color classes.*

The trees T satisfying $\gamma_t(T) = \chi_{td}(T)$ are characterized in [18]. Let $\mathcal{T}$ be the family of trees constructed as follows. Let $\mathcal{T}$ consist of the tree P_2 and all trees that can be obtained from a disjoint union of $k \ge 1$ stars each of order at least 3 by adding $k - 1$ edges joining leaf vertices in such a way that the resulting graph is connected and the center of each of the original k stars remains a support vertex.

Theorem 54　([18]) *If T is a nontrivial tree, then $\gamma_t(T) = \chi_{td}(T)$ if and only if $T \in \mathcal{T}$.*

In [18] a tight upper bound on the total dominator chromatic number of a tree in terms of its order is established, and the trees with maximum possible total dominator chromatic number are characterized. For this purpose, let $\mathcal{F}$ be the family of all trees T that can be obtained from a tree H of order at least 2 by selecting an arbitrary edge $e = uv$ in H and attaching a path of length 2 to each vertex of $V(H) \smallsetminus \{u, v\}$ so that the resulting paths are vertex-disjoint. We call H the *underlying tree* of T. A tree in the family $\mathcal{F}$ with underlying tree $H = P_5$, for example, is illustrated in Figure 16 (here the vertices of H are depicted by the darkened vertices).

Theorem 55　([18]) *If T is a tree or order $n \ge 2$, then $\chi_{td}(T) \le \frac{2}{3}(n + 1)$, with equality if and only if $T \in \mathcal{F}$.*

Fig. 16　A tree in the family $\mathcal{F}$

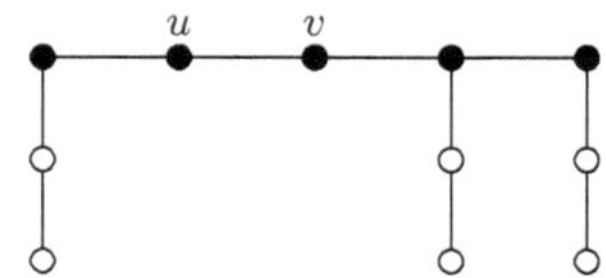

4.2.3 Mycielskian of a Graph

Let G be a graph without isolated vertices and with $V(G) = \{v_1, v_2, \ldots, v_n\}$. The Mycielskian $M(G)$ is the graph obtained from G by adding n new vertices $u_1, u_2, \ldots, u_n$ and an additional vertex v and then adding the edges vu_i for all $i \in [n]$. Further, for each edge $v_i v_j$ of G, we add the edges $u_i v_j$ and $v_i u_j$ to complete the construction of $M(G)$. For example, if $G = K_2$, then $M(G) = C_5$. If $G = C_5$, then $M(G)$ is the Grötzsch graph. Kazemi [24] proved that the dominator chromatic number of the Mycielskian of a graph is one of two values.

Theorem 56 ([24]) *If G is a graph without isolated vertices, then*

$$\chi_{td}(M(G)) = \chi_{td}(G) + 1 \quad \text{or} \quad \chi_{td}(M(G)) = \chi_{td}(G) + 2.$$

4.2.4 Circulants

Jalilolghadr, Kazemi, and Khodkar [20] studied total dominator colorings of circulant graphs $C_n(a, b)$ with two "jump sequences." For $n \geq 3$, let $1 \leq a_1 < \cdots a_k \leq \lfloor n/2 \rfloor$, and let $S = \{a_1, \ldots, a_k\}$. The graph G with vertex set $V(G) = [n]$ and edge set

$$E(G) = \{\{i, j\} : |i - j| \equiv a_i \ (\text{mod } n) \text{ for some } i \in [k]\}$$

is called a *circulant graph* with jump sequence S and denoted $C_n(S)$ or $C_n(a_1, \ldots, a_k)$. We note that $C_n(S)$ is a k-regular graph. Jalilolghadr et al. [20] prove the following result.

Theorem 57 ([20]) *If G is a circulant graph $C_n(a, b)$ where $n \geq 6$, $\gcd(a, n) = 1$ and $a^{-1}b \equiv 3 \ (\text{mod } n)$, then*

$$\chi_{td}(G) = \begin{cases} 2\lceil \frac{n}{8} \rceil & \text{for } n \in \{8, 9, 10\} \\ 2\lceil \frac{n}{8} \rceil + 1 & \text{for } n \equiv 1 \ (\text{mod } 8) \text{ or } n = 11 \\ 2\lceil \frac{n}{8} \rceil + 2 & \text{otherwise.} \end{cases}$$

4.2.5 Central Graphs

Kazemi and Kazemnejad [28] studied the total dominator chromatic number of central graphs, where they define the *central graph* $C(G)$ of a graph G as the graph obtained from G by subdividing every edge of G exactly once and adding all edges joining two vertices that were not adjacent in G. Among other results, they proved the following.

Theorem 58 ([28]) *If G is a connected graph of order $n \geq 4$, then the following holds.*

(a) $\chi_{td}(C(G)) \geq \frac{2}{3}n + 1$.

(b) $\chi_{td}(C(G)) \leq n + \lceil \frac{k}{2} \rceil$ *where k is the order of a longest path in G.*

(c) $\chi_{td}(C(G)) \leq n + 1$ *if* $\Delta(G) \leq n - 2$.

(d) $\chi_{td}(C(G)) \leq n + \lceil \frac{n}{2} \rceil$, *with equality if and only if* $G \cong K_n$.

4.3 Graph Products

In this section, we present some results due to Kazemi [25] on the total dominator chromatic number in Cartesian products ($\square$) and direct products ($\times$) of two graphs.

Theorem 59 ([25]) *If G and H are two graphs without isolated vertices, then*

$$\chi_{td}(G \times H) \leq \chi_{td}(G) \cdot \chi_{td}(H).$$

Theorem 60 ([25]) *For $q \geq p \geq 2$, if G is a complete p-partite graph and H is a complete q-partite graph, then* $\chi_{td}(G \times H) = p + 2$. *In particular,* $\chi_{td}(K_p \times K_q) = p + 2$.

Theorem 61 ([25]) *If G and H are two graphs without isolated vertices, then*

$$\max\{\chi_{td}(G), \chi_{td}(H)\} \leq \chi_{td}(G \,\square\, H) \leq \min\{\chi_{td}(G) \cdot n(H), \chi_{td}(H) \cdot n(G)\}.$$

Theorem 62 ([25]) *If G is a graph without isolated vertices, then* $\chi_{td}(G) \leq \chi_{td}(G \,\square\, K_2) \leq 2\chi_{td}(G)$.

4.4 Algorithmic and Complexity Results

We consider in this section the problem of finding the total dominator coloring number of an arbitrary graph. Formally, we consider the following decision problem:

GRAPH TOTAL DOMINATOR k-COLORABILITY

Input: A graph G, and an integer $k \geq 4$.
Question: Does G have a total dominator k-coloring?

An identical proof to that of Theorem 33 can be used to show that the GRAPH TOTAL k-DOMINATOR COLORABILITY is NP-complete for general graphs by transforming it from an instance of GRAPH DOMINATOR k-COLORABILITY.

Theorem 63 ([17, 26]) GRAPH TOTAL DOMINATOR k-COLORABILITY *is NP-complete for general graphs, for $k \geq 4$.*

5   Concluding Comments

In this chapter, we have surveyed selected results on the dominator chromatic number and total dominator chromatic number of a graph. Other results can be found, for example, in [3, 21, 27]. We close with a small list of open problems.

Problem 1 Find graphs, or classes of graphs, G satisfying the following.

(a) $\chi_d(G) = \gamma(G)$.
(b) $\chi_d(G) = \chi(G)$.
(c) $\chi_d(G) = \gamma(G) + \chi(G)$.
(d) $\chi_{td}(G) = \gamma_t(G)$.
(e) $\chi_{td}(G) = \chi(G)$.
(f) $\chi_{td}(G) = \gamma_t(G) + \chi(G)$.

Problem 2 Characterize the nontrivial trees T satisfying the following.

(a) $\gamma_t(T) = \chi_{td}(T) + 1$.
(b) $\gamma_t(T) = \chi_{td}(T) + 2$.

Problem 3 Characterize the graphs G satisfying $\chi_d(G) = \chi_{td}(G)$.

Problem 4 Determined the dominator chromatic number and the total dominator chromatic number of the $m \times n$ grid graph, $P_m \,\square\, P_n$, for all $m, n \geq 2$.

Problem 5 For any dominator (or total dominator) coloring, one can construct a so-called *dominator digraph* (*total dominator digraph*, respectively) which is an orientation of some of the edges of G such that for every vertex u, you orient the edge uv from u to v if u dominates the color class of vertex v. We note that for dominator colorings, this digraph will contain loops, if a vertex forms a singleton color class. However, the total dominator digraph will have no loops. We also note that these digraphs will have some unoriented edges which can be deleted. Study the resulting dominator digraphs and total dominator digraphs.

References

1. A.M. Abid, T.R. Ramesh Rao, Dominator coloring of Mycielskian graphs. Australas. J. Combin. **73**(2), 274–279 (2019)
2. S. Arumugam, K. Raja Chandrasekar, N. Misra, G. Philip, S. Saurabh, Algorithmic aspects of dominator colorings in graphs, in *Combinatorial Algorithms*. Lecture Notes in Comput. Sci., vol. 7056 (Springer, Heidelberg, 2011), pp. 19–30

3. S. Arumugam, J. Bagga, K.R. Chandrasekar, On dominator colorings in graphs. Proc. Indian Acad. Sci. Math. Sci. **122**(4), 561–571 (2012)

4. H. Boumediene Merouane, M. Chellali, On the dominator colorings in trees. Discuss. Math. Graph Theory **32**(4), 677–683 (2012)

5. H. Boumediene Merouane, M. Chellali, An algorithm for the dominator chromatic number of a tree. J. Comb. Optim. **30**(1), 27–33 (2015)

6. M. Chellali, F. Maffray, Dominator colorings in some classes of graphs. Graphs Combin. **28**, 97–107 (2012)

7. Q. Chen, Dominator colorings of Cartesian product of graphs. Util. Math. **109**, 155–172 (2018)

8. Q. Chen, C. Zhao, M. Zhao, Dominator colorings of certain Cartesian products of paths and cycles. Graphs Combin. **33**(1), 73–83 (2017)

9. E. Cockayne, S.M. Hedetniemi, S.T. Hedetniemi, Dominating partitions of graphs. Tech. Report., 1979. Unpublished manuscript

10. R.G. Downey, M.R. Fellows, *Fundamentals of Parameterized Complexity*. Texts in Computer Science (Springer, London, 2013), xxx+763 pp. ISBN: 978-1-4471-5558-4; 978-1-4471-5559-1

11. R. Gera, On the dominator colorings in bipartite graphs. Inform. Technol. New Gen. **ITNG'07**, 947–952 (2007)

12. R. Gera, On dominator colorings in graphs. Graph Theory Notes N. Y. **LII**, 25–30 (2007)

13. R. Gera, S. Horton, C. Rasmussen, Dominator colorings and safe clique partitions. Congress. Num. **181**, 19–32 (2006)

14. S.M. Hedetniemi, S.T. Hedetniemi, A.A. McRae, Dominator colorings of graphs (2006). Unpublished manuscript

15. S.M. Hedetniemi, S.T. Hedetniemi, R. Laskar, A.A. McRae, C.K. Wallis, Dominator partitions of graphs. J. Combin. Inform. Systems Sci. **34**, 183–192 (2009)

16. S.M. Hedetniemi, S.T. Hedetniemi, A.A. McRae, D.F. Rall, J.T. Hedetniemi, Dominator colorings of graphs, 9 July 2009. Unpublished 16-page manuscript

17. J.T. Hedetniemi, S.M. Hedetniemi, S.T. Hedetniemi, A.A. McRae, D.F. Rall, Total dominator partitions and colorings of graphs, 18 Feb 2011. Unpublished 18-page manuscript

18. M.A. Henning, Total dominator colorings and total domination in graphs. Graphs Combin. **31**(4), 953–974 (2015)

19. M.A. Henning, A. Yeo, *Total Domination in Graphs (Springer Monographs in Mathematics)* (2013). ISBN-13: 978-1461465249

20. P. Jalilolghadr, A.P. Kazemi, A. Khodkar, Total dominator coloring of circulant graphs $C_n(a, b)$. Manuscript, 1 May 2019. arXiv:1905.00211

21. M.I. Jinnah, A. Vijayalekshmi, Total dominator colorings in graphs. Ph.D Thesis, University of Kerala, 2010

22. R. Kalaivani, D. Vijayalakshimi, On dominator coloring of degree splitting graph of some graphs. J. Phys. Conf. Series **1139** 012081 (6 pp.) (2018)

23. G.Y. Katona, L.F. Papp, The optimal rubbling number of ladders, prisms and Möbius-ladders. Discrete Appl. Math. **209**, 227–246 (2016)

24. A.P. Kazemi, Total dominator chromatic number and Mycieleskian graphs. Manuscript, 29 July 2013. arXiv:1307.7706

25. A.P. Kazemi, Total dominator coloring in product graphs. Util. Math. **94**, 329–345 (2014)

26. A.P. Kazemi, Total dominator chromatic number of a graph. Trans. Comb. 4(2), 57–68 (2015)

27. A.P. Kazemi, Total dominator chromatic number of Mycieleskian graphs. Util. Math. **103**, 129–137 (2017)

28. F. Kazemnejad, A.P. Kazemi, Total dominator chromatic number of central graphs. Manuscript, 16 Jan 2018. arXiv:1801.05137

29. T. Manjula, R. Rajeswari, Dominator coloring of prism graph. Appl. Math. Sci. **9**(38), 1889–1894 (2015)

30. T. Manjula, R. Rajeswari, Dominator chromatic number of some graphs. Int. J. Pure. Appl. Math. **119**(7), 787–795 (2018)

31. J. Maria Jeyaseeli, N. Movarraei, S. Arumugam, Dominator coloring of generalized Petersen graphs, in *Theoretical Computer Science and Discrete Mathematics*. Lecture Notes in Comput. Sci., vol. 10398 (Springer, Cham, 2017), pp. 144–151
32. B.S. Panda, A. Pandey, On the dominator coloring in proper interval graphs and block graphs. Discrete Math. Algorithms Appl. **7**(4), 1550043 (17 pp.) (2015)
33. P. Paulraja, K.R. Chandrasekar, Dominator colorings of products of graphs, in *Theoretical Computer Science and Discrete Mathematics*. Lecture Notes in Comput. Sci., vol. 10398 (Springer, Cham, 2017), pp. 242–250
34. D. Seinsche, On a property of the class of n-colorable graphs. J. Comb. Theory Ser. B **16**, 191–193 (1974)
35. S.K. Vaidya, M.S. Shukla, Dominator coloring of some wheel related graphs. Int. J. Math. Sci. Comput. **6**(1), 4–6 (2016)
36. A. Vijayalekshmi, Total dominator colorings in graphs. Int. J. Adv. Res. Technol. **1**(4), 1–6 (2012)

Irredundance

C. M. Mynhardt and A. Roux

The concept of irredundance in graphs was introduced in 1978 by Cockayne, Hedetniemi and Miller [42] because of its relevance to dominating sets. Informally, a set X of vertices in a graph G is irredundant if each vertex in X dominates a vertex of G (perhaps itself) that is not dominated by any other vertex in X. More formally, in terms of private neighbours, X is *irredundant* if $\mathrm{pn}(x, X) = N[x] - N[X - \{x\}] \neq \varnothing$ for each $x \in X$, that is, if each $x \in X$ has an *X-private neighbour* (which could be x itself). If a set X has a vertex x without private neighbours, that is, if $N[x] \subseteq N[X - \{x\}]$, we say that x is *redundant in X* (in which case X is not an irredundant set).An irredundant set is *maximal irredundant* if it has no irredundant proper superset. The *lower* and *upper irredundant numbers* $\mathrm{ir}(G)$ and $\mathrm{IR}(G)$ are, respectively, the smallest and largest cardinalities of a maximal irredundant set of G. If X is a maximal irredundant set of cardinality $\mathrm{ir}(G)$, we call X an $\mathrm{ir}(G)$-*set* or simply an *ir-set*, depending on circumstances. An $\mathrm{IR}(G)$-*set* or *IR-set* is defined similarly; the same holds for any other domination-type parameter.

This chapter is organised as follows. To begin, we consider the partition of $V(G)$ associated with an irredundant set in Section 1. Here, we discuss the concepts of private neighbours, the private neighbour cube and generalised irredundance. The chain of lower and upper domination, independence and irredundance numbers is

The author "C. M. Mynhardt" was supported by the Natural Sciences and Engineering Research Council of Canada.

C. M. Mynhardt (✉)
Department of Mathematics and Statistics, University of Victoria, Victoria, BC, Canada
e-mail: kieka@uvic.ca

A. Roux
Department of Mathematical Sciences, Stellenbosch University, Stellenbosch, South Africa
e-mail: rianaroux@sun.ac.za

T. W. Haynes et al. (eds.), *Structures of Domination in Graphs*, Developments in Mathematics 66, https://doi.org/10.1007/978-3-030-58892-2_6

presented in Section 2. We discuss equality of parameters in the domination chain in Section 3, where we cover lower and upper irredundance perfect graphs, as well as other cases of equality, such as graphs with ir $=$ IR, ir $= \gamma$, $\alpha =$ IR or $\Gamma =$ IR. Bounds involving other graph parameters, including Nordhaus-Gaddum- and Gallai-type results, can be found in Section 4. Differences between and ratios of parameters in the domination chain are covered in Section 5 , criticality and stability concepts in Section 6, irredundance on chessboards in Section 7 and irredundant Ramsey numbers in Section 8. Finally, we discuss reconfiguration of irredundant sets in Section 9 and complexity in Section 10. We state open problems and conjectures throughout the text where appropriate.

1   Partition of $V(G)$ Associated with an Irredundant Set

Let us examine the properties of an irredundant set more closely. As noted by Cockayne, Grobler, Hedetniemi and McRae [41], we can associate a weak partition[1] of $V(G)$ with each irredundant set X, namely,

$$V(G) = X \cup Y \cup C \cup R, \text{ where}$$

Y consists of vertices in $V(G) - X$ that belong to private neighbourhoods of
  vertices in X,
C consists of vertices in $V(G) - X$ that have at least two neighbours in X, and
R is the set of vertices not dominated by X.

In Figure 1, the set X is indicated by coloured (red and yellow) discs, Y by blue squares, C by white discs and R by green triangles. The set X is further partitioned as

$$X = Z \cup I, \text{ where}$$

I is the set of vertices that are isolated in $G[X]$, indicated by yellow discs, and
$Z = X - I$, indicated by red discs.

The blue private neighbours in Y and the observation that $5 \in \mathrm{pn}(5, X)$ confirm that X is irredundant. Closer scrutiny however reveals that X is not maximal irredundant. For any $y \in R$, y is a private neighbour of itself in the set $X \cup \{y\}$, and for any z adjacent to y, y is a private neighbour of z in $X \cup \{z\}$. Hence we must

---

[1]In a weak partition, some of the parts could be empty.

also examine the private neighbourhood of each $x \in X$ in these supersets of X to determine whether or not X is maximal irredundant. Since $\mathrm{pn}(1, X) = \{6, 7\} \subseteq N(12), \mathrm{pn}(1, X \cup \{12\}) = \varnothing$, which means that $X \cup \{12\}$ is not irredundant. However, $\mathrm{pn}(1, X \cup \{6\}) = \{7\}$ and $\mathrm{pn}(6, X \cup \{6\}) = \{12\}$, and since all other vertices in X also have $X \cup \{6\}$-private neighbours, $X \cup \{6\}$ is irredundant. The set $X \cup \{14\}$ is likewise irredundant, as $4 \in \mathrm{pn}(4, X \cup \{14\})$ and $14 \in \mathrm{pn}(14, X \cup \{14\})$. The following result, which was used implicitly in earlier work and first formalised in [41], and which extends a result by Bollobás and Cockayne [10], provides a certificate for an irredundant set to be maximal irredundant (or not).

Theorem 1.1 ([41]) *An irredundant set X of a graph G is maximal irredundant if and only if, for each $u \in R = V(G) - N[X]$ and each $v \in N[u]$, there exists a vertex $x \in X$ such that v dominates $\mathrm{pn}(x, X)$.*

Proof. Assume that X is a maximal irredundant set of G for which the conclusion in the statement of the theorem does not hold. Then there exist vertices $u \in R$ and $v \in N[u]$ such that v does not dominate the private neighbourhood of any $x \in X$. Consider the set $X' = X \cup \{v\}$. The stated property of v implies that $\mathrm{pn}(x, X') \neq \varnothing$ for each $x \in X$. Moreover, since u is not dominated by any $x \in X$ but $u \in N[v]$, it follows that $u \in \mathrm{pn}(v, X')$. Therefore X' is irredundant, contrary to the maximality of X.

Conversely, assume that X is an irredundant set of G for which the conclusion of the statement holds. Consider any $v \in V(G) - X$. If v is undominated by X or adjacent to a vertex u that is undominated by X, then, by assumption, there exists a vertex $x \in X$ such that v dominates $\mathrm{pn}(x, X)$. This implies that x is redundant in $X \cup \{v\}$. On the other hand, if v and all its neighbours are dominated by X, then v is redundant in $X \cup \{v\}$. In either case, it follows that $X \cup \{v\}$ is not irredundant. Since v is arbitrary, we conclude that X is maximal irredundant. $\blacksquare$

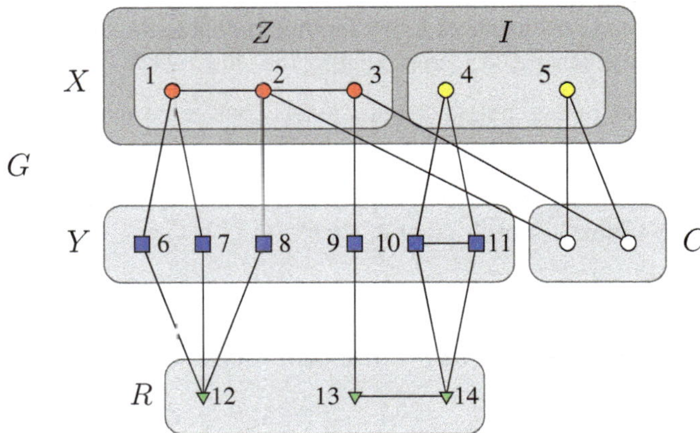

Fig. 1 The partition of the vertex set of a graph associated with an irredundant set X

If $v \in V(G) - X$ and $\mathrm{pn}(x, X) \subseteq N[v]$, we say that v *annihilates* x. Let G' be the graph obtained from G in Figure 1 by adding the edges 6 7 and 9 14. Then 12, 6 and 7 annihilate 1; 8 annihilates 2; 9, 13 and 14 annihilate 3; and 10 and 11 annihilate 4. Therefore X is maximal irredundant (but not dominating) in G'.

1.1 Private Neighbours

The elements of an irredundant set X could have one or both of two types of private neighbours: for $x \in X$, the vertex y is an

(i) *X-self-private neighbour* (*X-spn*) of x if $y = x$ and x is isolated in $G[X]$ (the vertices 4, 5 in Figure 1),
(ii) *X-external private neighbour* (*X-epn*) of x if $y \in V(G) - X$ and $N(y) \cap X = \{x\}$.

The set of X-external private neighbours of $x \in X$ is denoted by $\mathrm{epn}(x, X)$. In Figure 1, $\mathrm{pn}(4, X) = \{4\} \cup \mathrm{epn}(4, X) = \{4, 10, 11\}$. Bollobás and Cockayne [10] proved the following fundamental result.

Theorem 1.2 ([10]) *Every graph G without isolated vertices has a minimum dominating set X in which each vertex has an X-epn.*

Proof. Among all minimum dominating sets of G, let X be one that maximises the number of edges in $G[X]$. We show that X has the desired property. Assume to the contrary that $\mathrm{epn}(x, X) = \varnothing$ for some $x \in X$. By the minimality of X, $\mathrm{pn}(x, X) \neq \varnothing$, and the only possibility is that x is isolated in $G[X]$. Since G is isolate-free, x is adjacent to a vertex $u \in V(G) - X$. Since $\mathrm{epn}(x, X) = \varnothing$, u is adjacent to a vertex $y \in X - \{x\}$. Consider the set $X' = (X - \{x\}) \cup \{u\}$. Since x is adjacent to u, and each neighbour of x in $V(G) - X$ is adjacent to a vertex in $X - \{x\}$, X' dominates G. However, u is adjacent to $y \in X'$, whereas x is nonadjacent to all vertices in X. Hence X', having the same cardinality as X, is a minimum dominating of G such that $G[X']$ contains more edges than $G[X]$, contradicting the choice of X. ∎

A third type of private neighbour of $x \in X$ is considered in [76]: the vertex y is an

(iii) *X-internal private neighbour* (*X-ipn*) of x if $y \in X - \{x\}$ and $N(y) \cap X = \{x\}$. In Figure 1, vertex 3 is an X-ipn of 2.

1.2 The Private Neighbour Cube and Generalised Irredundance

Using all combinations of the three types of private neighbours, Fellows, Fricke, Hedetniemi and Jacobs [76] constructed the so-called private neighbour cube (illus-

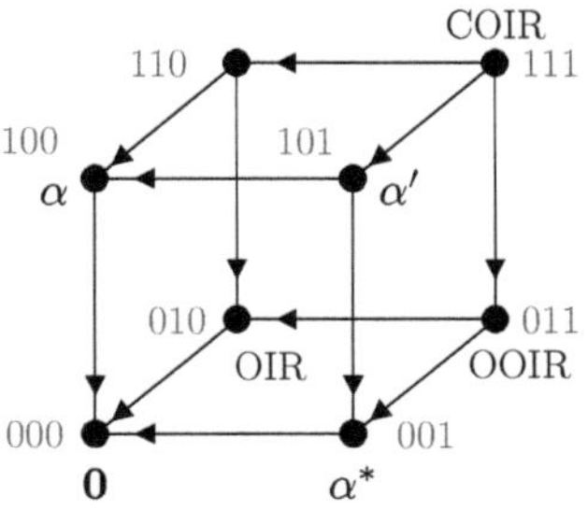

Fig. 2 The private neighbour cube of Fellows, Fricke, Hedetniemi and Jacobs [76]

trated in Figure 2) and obtained six additional types of irredundance. Combining the private neighbourhood types *and* their negations (e.g. each $x \in X$ has an X-epn but neither an X-spn nor an X-ipn), Cockayne [30] obtained further generalised irredundance concepts which were investigated in greater depth by Finbow [77]. Most of these generalisations, as well as others obtained by, for example, imposing structural requirements on X, are beyond the scope of this chapter, and we mainly consider irredundance in undirected graphs as defined in [42]. Apart from brief definitions to explain the private neighbour cube, we consider only two other concepts of irredundance, open or OC-irredundance, introduced by Farley and Shacham [62], and CO-irredundance, defined by Fellows et al. [76].

The vertices of the private neighbour cube are, simultaneously, types of sets defined according to which types of private neighbours their vertices possess and the maximum cardinality of such a set X in a graph G. The types of sets are represented by binary strings of length 3 as described below.

Type 000: Each vertex in X has a private neighbour which is not an X-ipn, an X-epn or an X-spn. Since we only consider these three types of private neighbours, the only set of this type is the empty set.

Type 001: Each vertex in X has an X-ipn. This is only possible if the subgraph $G[X]$ of G induced by X consists of disjoint copies of K_2. Following [76], we call such a set X a *strong matching set* and denote the maximum cardinality of a strong matching set in G by $\alpha*(G)$.

Type 010: Each vertex in X has an X-epn; that is, $\text{epn}(x, X) = N(x) - N[X - \{x\}] \neq \varnothing$ for each $x \in X$. Since this definition involves <u>o</u>pen and <u>c</u>losed neighbourhoods, we obtain the concept of *OC-irredundance*, usually called *open irredundance*. The *lower* and *upper open irredundant numbers* $\text{oir}(G)$ and $\text{OIR}(G)$ of G are, respectively, the smallest and largest cardinalities of a maximal open irredundant set of G. Open irredundant sets were studied by Farley and Shacham [62].

Type 011: Each vertex in X has an X-ipn or an X-epn; that is, $N(x) - N(X - \{x\}) \neq \varnothing$ for each $x \in X$. Since this definition involves two <u>o</u>pen neighbourhoods, we obtain the concept of *open-open* and *OO-irredundance*. The *lower* and *upper open irredundant numbers* $\text{ooir}(G)$ and $\text{OOIR}(G)$ are defined in the obvious manner. OO-irredundant sets were considered by Farley and Proskurowski [61] and Farley and Shacham [62].

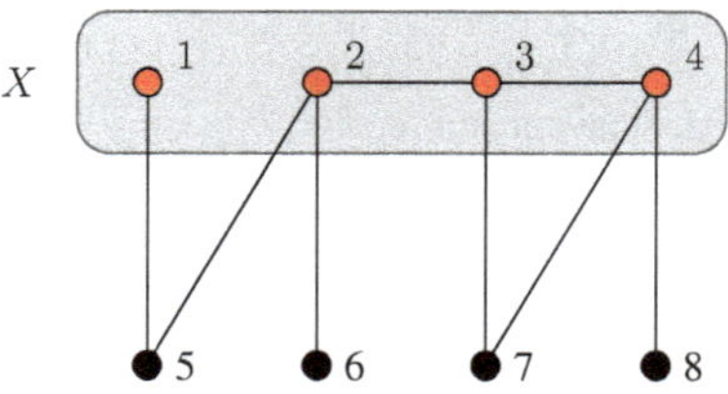

Fig. 3 A CO-irredundant set X

Type 100:   Each vertex in X is an X-spn, that is, X is an independent set. The associated parameters, namely, the independence number $\alpha(G)$ and the independent domination number $i(G)$, are well established in graph theory.

Type 101:   Each vertex in X is an X-spn or has an X-ipn. For such a set X, known as a 1-*dependent set*, $\Delta(G[X]) \leq 1$. The maximum cardinality of a 1-dependent set of G is the 1-*dependence number* $\alpha^1(G)$. These sets were studied by Fink and Jacobson [80].

Type 110:   Each vertex in X is an X-spn or has an X-epn, that is, X is an irredundant set. Since we require that $\mathrm{pn}(x, X) = N[x] - N[X - \{x\}] \neq \varnothing$ for each $x \in X$, the concept of <u>c</u>losed neighbourhood occurs twice in this definition. Thus, irredundance can also be called CC-irredundance.

Type 111:   Each vertex in X is an X-spn or has an X-epn or X-ipn, that is, $N[x] - N(X - \{x\}) \neq \varnothing$ for each $x \in X$. Since this definition involves <u>c</u>losed and <u>o</u>pen neighbourhoods, we get the concept of *CO-irredundance*. That is, a set X is CO-irredundant if each $x \in X$ has a private neighbour of at least one of the three types mentioned in Section 1.1. For example, the set X in Figure 3 is CO-irredundant: 1 is an X-spn, $\mathrm{epn}(2, X) = \{6\}$, $\mathrm{ipn}(3, X) = \{2, 4\}$ and $\mathrm{epn}(4, X) = \{8\}$. The *lower* and *upper CO-irredundant numbers* $\mathrm{coir}(G)$ and $\mathrm{COIR}(G)$ are defined in the obvious manner.

The partial order on the set of parameters (indicated by arrows, from large to small) in the private neighbourhood cube is defined by the lexicographic order of the binary strings that represent the respective sets.

2   The Domination Chain

The relationships between dominating, independent and irredundant sets of vertices of a graph are well-known [92, Chapter 3], and, following [41], we summarise them below.

$$
\begin{array}{ccccc}
\text{maximal} & & \text{minimal} & & \\
\text{independent} & & \text{dominating} & & \\
\updownarrow & \overset{(I)}{\Rightarrow} & \updownarrow & \overset{(II)}{\Rightarrow} & \text{maximal irredundant} \\
\text{independent and} & & \text{dominating and} & & \\
\text{dominating} & & \text{irredundant} & &
\end{array}
\qquad (1)
$$

It also follows from the definitions that any irredundant set is CO-irredundant, and any open irredundant set is irredundant. Simmons [122] showed that a total dominating set is minimal total dominating if and only if it is CO-irredundant. The implications (*I*) and (*II*) in (1) lead to the *domination chain* (2), first mentioned in [42]. For any graph G,

$$\mathrm{ir}(G) \leq \gamma(G) \leq i(G) \leq \alpha(G) \leq \Gamma(G) \leq \mathrm{IR}(G). \tag{2}$$

For open and CO-irredundance the inequalities,

$$\mathrm{oir}(G) \leq \gamma(G) \leq \mathrm{OIR}(G) \leq \mathrm{IR}(G) \text{ for any graph } G \text{ without isolated vertices,}$$

and

$$\mathrm{IR}(G) \leq \mathrm{COIR}(G) \text{ for any graph } G$$

hold (see [92, pp. 91–92]). Farley and Shacham [62] and Favaron [66] gave examples of graphs with OIR $< i$, $\Gamma <$ OIR, ir $<$ oir and oir $<$ ir.

For the parameters in the domination chain (2), Allan and Laskar [1] and Bollobás and Cockayne [10] further showed that $\gamma(G) \leq 2\,\mathrm{ir}(G) - 1$ (see Theorem 2.1 below), but all the other ratios of parameters in (2) are unbounded for general graphs: for i/γ, consider $K_{n,n}$; for α/i, consider $K_{1,n}$; for Γ/α, consider $K_2\square K_n$; and for IR $/\,\Gamma$, consider $K_2\square K_{n+1}$ and delete an edge between the two copies of K_{n+1}.

Theorem 2.1 ([1, 10]) *For any graph G, $\gamma(G) \leq 2\,\mathrm{ir}(G) - 1$.*

Proof. We use the notation defined in Section 1 and illustrated in Figure 1. Let X be an ir-set of G. If X dominates G, we are done; hence assume $R = V(G) - N[X] \neq \varnothing$. Let $u \in R$. By Theorem 1.1, there exists a vertex $x \in X$ such that u annihilates x. Since X does not dominate u, necessarily $\varnothing \neq \mathrm{epn}(x, X) \subseteq N(u)$. Define $X_R = \{x \in X : \mathrm{epn}(x, X) \neq \varnothing \text{ and } \mathrm{epn}(x, X) \subseteq N(u) \text{ for some } u \in R\}$. For each $x \in X_R$, choose a vertex $x' \in \mathrm{epn}(x, X)$ and define $Y_R = \{x' : x \in X_R\}$. Note that $Y_R \subseteq Y$ and $|Y_R| = |X_R| \leq |X|$. Moreover, $X \cup Y_R$ dominates G. Since $X \subsetneqq X \cup Y_R$, the maximality of X (as an irredundant set) implies that $X \cup Y_R$ is not irredundant. As stated in (1), this implies that $X \cup Y_R$ is not a minimal dominating set. Therefore $\gamma(G) \leq |X \cup Y_R| - 1 = |X| + |Y_R| - 1 \leq 2\,\mathrm{ir}(G) - 1$. ∎

Subject to $\gamma(G) \leq 2\,\mathrm{ir}(G) - 1$ and two other obvious restrictions, the differences between the parameters in (1) can be arbitrary in a connected graph, as proved in [47].

Theorem 2.2 ([47]) *For any positive integers $k_1 \leq \cdots \leq k_6$ such that (a) $k_1 = 1 \Rightarrow k_2 = k_3 = 1$, (b) $k_4 = 1 \Rightarrow k_1 = \cdots = k_6 = 1$ and (c) $k_2 \leq 2k_1 - 1$, there exists a connected graph G such that* $\mathrm{ir}(G) = k_1$, $\gamma(G) = k_2$, $i(G) = k_3$, $\alpha(G) = k_4$, $\Gamma(G) = k_5$ *and* $\mathrm{IR}(G) = k_6$.

Much work has been done to bound the ratios or prove equality of pairs of the parameters in the domination chain for special graph classes, or bound the parameters, their differences or their sums using other graph parameters. Except for a few results on open and CO-irredundance, we only mention results involving ir and IR.

3 Equality of Parameters in the Domination Chain

For graph parameters π, λ, we say that G is a (π, λ)-*graph* if $\pi(G) = \lambda(G)$. Thus, the (i, α)-graphs are precisely the well-covered graphs, and the (γ, Γ)-graphs are the well-dominated graphs, terms which work well(!) for independence and domination, but less well for irredundance and not at all if π and λ refer to different concepts. Many instances of (ir, γ)- and (α, IR)-graphs are mentioned in [77, pp. 23–24] and [92, pp. 77–84], and we do not repeat those results here. We write $H \trianglelefteq G$ ($H \lhd G$) if H is an induced (proper) subgraph of G.

More recent research focuses instead on (π, λ)-perfect graphs: a graph G is (π, λ)-*perfect* if $\pi(H) = \lambda(H)$ for every $H \trianglelefteq G$. For a trivial example, note that each component of a P_3-free graph is complete; hence P_3-free graphs are (ir, IR)-perfect graphs. Graphs with maximum degree $\Delta \leq 2$ are (ir, i)- and (α, IR)-perfect graphs (but not (i, α)-graphs). Also, (χ, ω)-perfect graphs, where χ and ω denote the chromatic and clique number, respectively, are the classical perfect graphs. We discuss (ir, γ)-perfect and (Γ, IR)-perfect graphs in Sections 3.1 and 3.2, respectively. In Sections 3.3–3.5, we consider equality results that do not involve (π, λ)-perfect graphs.

3.1 Lower Irredundance Perfect Graphs

A graph is *(lower) irredundance perfect* if it is (ir, γ)-perfect, and k-(ir, γ)-*perfect* if $\mathrm{ir}(H) = \gamma(H)$ for each induced subgraph H with $\mathrm{ir}(H) \leq k$. A graph G is *minimally* (ir, γ)-*imperfect* if G is not (ir, γ)-perfect and $\mathrm{ir}(H) = \gamma(H)$ for every $H \lhd G$. The (ir, i)-perfect graphs form a proper subset of the (ir, γ)-perfect graphs. If G does not contain any of the graphs $H_1, \ldots, H_k$ as induced subgraphs, we say that G is $(H_1, \ldots, H_k)$-*free*. Eight graphs $T_1, G_1, \ldots, G_7$ are depicted in Figure 4, and in the rest of this section, when we refer to T_1 or G_i, $i = 1, \ldots, 7$, we mean a graph in this figure. A *(generalised) spider* $\mathrm{Sp}(\ell_1, \ldots, \ell_k)$, $\ell_i \geq 1$, $k \geq 2$, is a tree obtained from the star $K_{1,k}$ with centre u by subdividing the edge uv_i $\ell_i - 1$ times, $i = 1, \ldots, k$.

An early result on (ir, γ)-perfect graphs by Faudree, Favaron and Li [63] concerns P_4-free graphs, also known as *cographs*, a subclass of chordal graphs.

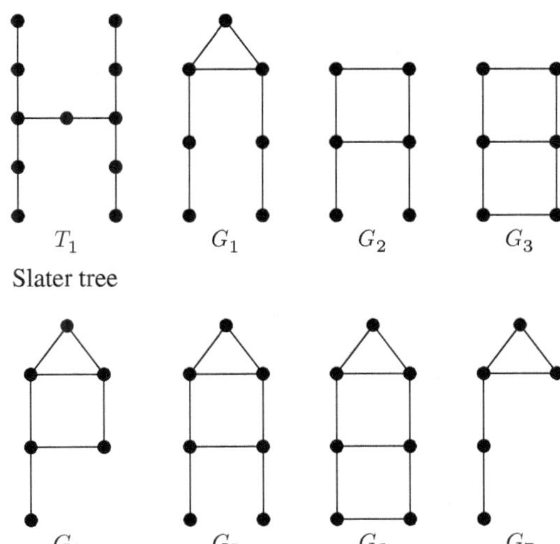

Fig. 4 Graphs used in Section 3.1

We prove the first part of the following theorem as an illustration of how results of this nature are obtained.

Theorem 3.1 ([63])

(i) *Any P_4-free graph, that is, any cograph, is* (ir, γ)*-perfect.*
(ii) *Any $(P_4, K_{3,3})$-free or $(K_{1,3}, G_1)$-free graph is* (ir, i)*-perfect.*

Proof of (i). Let G be a P_4-free graph and consider an ir-set X of G. We show that X is a dominating set of G. Suppose to the contrary that $R = V(G) - N[X] \neq \varnothing$ and consider $u \in R$. As in the proof of Theorem 2.1, there exists a vertex $x \in X$ such that u annihilates x; that is, $x \in Z$ (see Figure 1) and $\varnothing \neq \mathrm{epn}(x, X) \subseteq N(u)$. Let z be a vertex in Z adjacent to x, and let $w \in \mathrm{epn}(x, X)$. Then $G[\{y, x, w, u\}] \cong P_4$, contrary to our hypothesis. It follows that X dominates G; hence $\gamma(G) = \mathrm{ir}(G)$. Since each induced subgraph of a P_4-free graph is P_4-free, $\gamma(H) = \mathrm{ir}(H)$ for each $H \trianglelefteq G$, that is, G is (ir, γ)-perfect. ∎

Favaron [64] showed that graphs that contain none of six forbidden induced subgraphs are (ir, γ)-perfect and conjectured that (P_6, G_2, G_3)-free graphs are (ir, γ)-perfect. This conjecture was proved by Puech [115], who also proved a similar result which involves two forbidden subgraphs.

Theorem 3.2 ([115]) *Every (P_6, G_2, G_3)-free and every (P_6, G_4)-free graph is* (ir, γ)*-perfect.*

Since both P_6 and G_4 contain P_5 as an induced subgraph, the class of P_5-free graphs is included in the class of (P_6, G_4)-free graphs. Therefore we obtain the following corollary, first conjectured by Faudree et al. [63].

Corollary 3.3 ([115]) *Every P_5-free graph is* (ir, γ)*-perfect.*

Puech in turn conjectured that every (P_6, G_5, G_6)-free graph is (ir, γ)-perfect, a proof of which was given by Volkmann and Zverovich [127]. Note that Theorem 3.4 implies Theorem 3.2 and thus also the truth of the conjectures in [63, 64, 115].

Theorem 3.4 ([127]) *Every (P_6, G_5, G_6)-free graph is (ir, γ)-perfect.*

Puech [117] determined all pairs of connected graphs (X, Y) such that every sufficiently large graph containing neither X nor Y as induced subgraph is (ir, γ)-perfect.

Theorem 3.5 ([117]) *Let (X, Y) be a pair of connected graphs and let n_0 be a given positive integer. A graph G is (X, Y)-free implies that G is (ir, γ)-perfect for any connected graph of order at least n_0 if and only if one of the following statements holds:*

- $X \trianglelefteq P_5$ *and* Y *is arbitrary*
- $X \trianglelefteq P_6$ *and* $Y \trianglelefteq G_4$
- $X \trianglelefteq G_1$ *and* $Y \trianglelefteq \mathrm{Sp}(1, 1, 2)$
- $X \trianglelefteq G_7$ *and* $Y \trianglelefteq \mathrm{Sp}(1, 1, 3)$.

Building on work by Bollobás and Cockayne [10], Favaron [64], and Laskar and Pfaff [106], Henning [96] stated the following necessary and sufficient condition for a chordal graph to be (ir, γ)-perfect.

Theorem 3.6 ([96, 106])

(i) *A chordal graph is (ir, γ)-perfect if and only if it is (T_1, G_1)-free.*
(ii) *A tree is (ir, γ)-perfect if and only if it is T_1-free.*

Henning [96] characterised 2-(ir, γ)-perfect graphs in terms of 12 forbidden induced subgraphs and also conjectured that a graph G is (ir, γ)-perfect if and only if it is 4-(ir, γ)-perfect. However, Volkmann and Zverovich [128], in a paper that contains an excellent summary of results on (ir, γ)-perfect graphs up to that point, constructed a minimal irredundance imperfect counterexample F^* (shown in Figure 5) to Henning's conjecture; F^* is 4-(ir, γ)-perfect but $\mathrm{ir}(F^*) = 5$ and $\gamma(F^*) = 6$. The analytical proof that F^* is 4-(ir, γ)-perfect is quite long, but this fact can be verified by computer. The set $\{2, 4, 7, 9, 13, 15\}$ (red vertices in Figure 5) is a minimum dominating set of F^*, while $\{3, 4, 8, 13, 14\}$ (blue-circled vertices, with one private neighbour each indicated by brown squares) is an ir-set. They, in turn, formulated the following conjectures, which remain unresolved to date. (For what it is worth, the authors of this survey surmise that both conjectures are false.)

Conjecture 3.7 Volkmann and Zverovich [128, 2002]

(i) *A graph is (ir, γ)-perfect if and only if it is 5-(ir, γ)-perfect.*
(ii) *The number of minimally (ir, γ)-imperfect graphs is finite.*

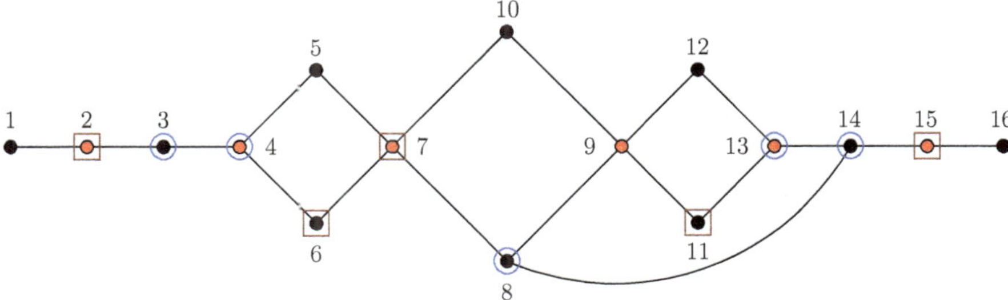

Fig. 5 A 4-(ir, γ)-perfect graph F^* with ir-set $\{3, 4, 8, 13, 14\}$ and γ-set $\{2, 4, 7, 9, 13, 15\}$ [128]

Cographs are (χ, ω)-perfect (i.e. perfect in the original sense) and (ir, γ)-perfect. Not all perfect graphs are (ir, γ)-perfect, e.g. the Slater tree is perfect but not (ir, γ)-perfect. On the other hand, C_5 is (ir, γ)-perfect but not perfect.

Problem 3.8 *Characterise the intersection of the two classes of (χ, ω)-perfect and (ir, γ)-perfect graphs.*

3.2 Upper Irredundance Perfect Graphs

Jacobson and Peters [102] defined a graph to be upper irredundance perfect if it is (α, IR)-perfect. On the other hand, Gutin and Zverovich [90] and Zverovich and Zverovich [133] defined a graph to be *upper domination perfect* if it is (α, Γ)-perfect and *upper irredundance perfect* if it is (Γ, IR)-perfect. To be consistent with the corresponding definition for the lower parameters, we prefer the latter definition. Fortunately, the definition of (π, λ)-perfect graphs is unambiguous, and we use it here. Moreover, Gutin and Zverovich [90] showed that any (α, Γ)-perfect graph is also (Γ, IR)-perfect, from which it follows that the classes of (α, Γ)-perfect and (α, IR)-perfect graphs are identical. We present the short proof here.

Theorem 3.9 ([90]) *If G is a (α, Γ)-perfect graph, then it is (Γ, IR)-perfect.*

Proof. Assume G is (α, Γ)-perfect and consider any $H \trianglelefteq G$. Then H is also (α, Γ)-perfect. Let X be an $\text{IR}(H)$-set and consider the subgraph F of H induced by $N[X]$. Then X is a dominating set of F. Since X is an IR-set of H, $\text{pn}_H(x, X) \neq \varnothing$ for each $x \in X$. But $\text{pn}_H(x, X) \subseteq N[X]$ for each $x \in X$; hence $\text{pn}_H(x, X) = \text{pn}_F(x, X) \neq \varnothing$ for each $x \in X$. Therefore X is irredundant in F, and (see (1)) it follows that X is a minimal dominating set of F. Hence $\Gamma(F) \geq |X| = \text{IR}(H)$. Since H is (α, Γ)-perfect, $\alpha(H) = \Gamma(H)$ and $\alpha(F) = \Gamma(F)$. Thus

$$\text{IR}(H) \leq \Gamma(F) = \alpha(F) \leq \alpha(H) = \Gamma(H) \leq \text{IR}(H),$$

that is, $\Gamma(H) = \text{IR}(H)$. Hence G is (Γ, IR)-perfect. ∎

To see that a (Γ, IR)-perfect graph need not be (α, IR)-perfect, note that $\alpha(K_2 \square K_3) = 2$ and $\Gamma(K_2 \square K_3) = \mathrm{IR}(K_2 \square K_3) = 3$. Therefore $K_2 \square K_3$ is not (α, IR)-perfect. For any vertex v, $\Gamma(K_2 \square K_3 - v) = \mathrm{IR}(K_2 \square K_3 - v) = 2$, and it now follows easily that for each $H \lhd K_2 \square K_3$, either $\Gamma(H) = \mathrm{IR}(H) = 2$ or $\Gamma(H) = \mathrm{IR}(H) = 1$. Thus $K_2 \square K_3$ is (Γ, IR)-perfect. Not much is known about (α, Γ)-imperfect (Γ, IR)-perfect graphs, but Cockayne, Favaron, Goddard, Grobler and Mynhardt [36] showed that if $\mathrm{IR}(G) > \alpha(G) = 2$, then $\mathrm{IR}(G) = \max\{r : K_r \square K_2 \trianglelefteq G\}$.

Problem 3.10 *Characterise* (α, Γ)-*imperfect* (Γ, IR)-*perfect graphs.*

We next define a few classes of graphs that occur in the discussion below. A *Meyniel graph* is a graph in which every odd cycle of length five or more has at least two chords. A *parity graph* is a graph in which every two induced paths between the same two vertices have the same parity; these graphs include distance-hereditary graphs and bipartite graphs and can be shown to be Meyniel graphs. A vertex v of a graph G is a *simplicial vertex* if $N[v]$ induces a clique, and a clique is a *simplex* if it contains a simplicial vertex. A vertex is v *peripheral* if v is simplicial either in G or in $\overline{G}$. A graph G is called a *peripheral graph* if every induced subgraph of G has a peripheral vertex. This family contains all chordal graphs and co-chordal graphs (graphs whose complements are chordal). A graph is *perfectly orderable* if its vertex set admits a linear order $<$ such that no induced $P_4 : (a, b, c, d)$ has $a < b$ and $d < c$. Let $\mathcal{C}$ be the collection of all maximal cliques of a graph G. A set $S \subseteq V(G)$ such that $|S \cap C| = 1$ for each $C \in \mathcal{C}$ is called a *stable transversal* of G. (Note that a stable transversal is a maximal independent set.) If each induced subgraph of G (including G itself) has a stable transversal, then G is *strongly perfect*. A graph G is *absorbently perfect* if every induced subgraph H of G contains a minimal dominating set that has a nonempty intersection with each maximal clique of H. Since each maximal independent set is minimal dominating, each strongly perfect graph is absorbently perfect (but the converse is false), and each absorbently perfect graph is perfect (i.e. (χ, ω)-perfect) [91].

Jacobson and Peters [102] characterised (α, IR)-perfect graphs, a result that encompasses those obtained in several other articles, e.g. [27, 38, 84, 101]. We provide a proof of this result because it illustrates the use of the partition of the vertex set of a graph associated with an irredundant set, as given in Section 1. Two sets of vertices $X = \{x_1, \ldots, x_k\}$ and $Y = \{y_1, \ldots, y_k\}$ of G are *independently matched* if the only edges between X and Y are $x_i y_i$, $i = 1, 2, \ldots, k$.

Property 1 A graph G has Property 1 if, for any pair of vertex subsets X and Y that are independently matched, $\alpha(G[X \cup Y]) \geq |X|$. A graph with Property 1 is called a *Property 1 graph*.

Theorem 3.11 ([102]) *A graph is* (α, IR)-*perfect if and only if it is a Property 1 graph.*

Proof. Suppose G is a Property 1 graph and consider any $H \trianglelefteq G$. Let X be an IR-set of H, I the set of vertices that are isolated in $H[X]$, and $Z = X - I$. If $Z = \varnothing$, then

$\alpha(H) \geq |I| = \mathrm{IR}(H)$ and thus $\alpha(H) = \mathrm{IR}(H)$. Assume $Z \neq \varnothing$. For each $z \in Z$, let $y_z \in \mathrm{EPN}(z, X)$ and $Y' = \{y_z : z \in Z\}$. By the private neighbour property, Z and Y' are independently matched in H and thus also in G. Since G is a Property 1 graph, $G[Z \cup Y']$ has an independent set, say A, such that $|A| = |Z|$. Since each vertex in I is nonadjacent to each vertex in Z (by the definition of I) and each vertex in Y' (by the private neighbour property), $A \cup I$ is the desired independent set of cardinality $|Z| + |I| = \mathrm{IR}(H)$. Thus $\alpha(H) = \mathrm{IR}(H)$, and it follows that G is (α, IR)-perfect.

Conversely, suppose that $\alpha(H) = \mathrm{IR}(H)$ for each $H \trianglelefteq G$. Suppose that for some integer $k \geq 1$, there exist disjoint sets $X, Y \subseteq V(G)$ that are independently matched. Let $H = G[X \cup Y]$. Then each vertex in X has an X-external private neighbour in Y; hence $\mathrm{IR}(H) \geq k$. By assumption $\alpha(H) \geq k$, hence G has Property 1. $\blacksquare$

As shown in [102] and [90], respectively, strongly perfect graphs and absorbently perfect graphs have Property 1 and therefore are (α, IR)-perfect. The following (not necessarily disjoint) classes of graphs have been shown to be strongly perfect and thus (α, IR)-perfect:

- comparability graphs, chordal graphs and complements of chordal graphs [6],
- perfectly orderable graphs [28],
- peripheral graphs [107],
- Meyniel graphs (and thus parity graphs) [119],
- graphs such that all odd cycles have a common vertex (Volkmann, as cited in [90]),
- permutation graphs, cographs, bipartite graphs, grids (easy to verify).

It is easy to see that not all perfect graphs are (Γ, IR)-perfect: let G be the graph obtained from $K_2 \square K_n$, $n \geq 5$, by deleting two nonadjacent vertices. Then G is perfect by the strong perfect graph theorem, but $\Gamma(G) = 2$ and $\mathrm{IR}(G) = n - 2$.

Other classes of (α, IR)-perfect graphs are P_4-free graphs and $(P_5, K_2 \square K_3)$-free graphs [27, as cited in [102]]; the latter result is an improvement of one in [38]. In addition, circular arc graphs have Property 1 and are therefore (α, IR)-perfect, as was proved in [84].

Gutin and Zverovich [90] also proved the previously mentioned result on $(P_5, K_2 \square K_3)$-free graphs and gave a forbidden subgraph characterisation of (α, IR)-perfect graphs (in terms of infinitely many forbidden subgraphs).

Dohmen, Rautenbach and Volkmann [56] generalised (α, Γ)- and (Γ, IR)-perfect graphs as follows: For $k \geq 0$, a graph G is k-(α, Γ)-*perfect* (k-(Γ, IR)-*perfect*, respectively) if $\Gamma(H) - \alpha(H) \leq k$ ($\mathrm{IR}(H) - \Gamma(H) \leq k$, respectively) for each $H \trianglelefteq G$. They showed that if G is k-(α, Γ)-perfect, it is also k-(Γ, IR)-perfect; hence we refer to these graphs as k-(α, IR)-*perfect*. They generalised Property 1 to Property A(k) and showed that G is k-(α, IR)-perfect if and only if it has Property A(k).

Property A(k) A graph G has Property A(k) if, for any pair of vertex subsets X and Y that are independently matched, $\alpha(G[X \cup Y]) \geq |X| - k$.

3.3   (ir, IR)-*Graphs*

We begin with the characterisations of the classes of bipartite and chordal (ir, IR)-graphs (a.k.a. well irredundant graphs) given by Topp and Vestergaard [125]. The corona of K_3 with an end-vertex deleted is known as the *bull*, usually denoted B.

Theorem 3.12 ([125]) *Let G be a nontrivial connected graph.*

  (i) *If G is bipartite, then G is an* (ir, IR)-*graph if and only if G is a* (γ, Γ)-*graph if and only if* $G = C_4$ *or* $G = H \circ K_1$ *for some connected bipartite graph H.*
 (ii) *If G is chordal, then G is an* (ir, IR)-*graph if and only if every vertex of G belongs to exactly one simplex, and if G has the bull B as induced subgraph, then the unique vertex of degree two in B is not a simplicial vertex of G.*
(iii) *(Corollary to (ii)) If G is a block graph, then G is an* (ir, IR)-*graph if and only if G is a generalised corona* $H \circ \{H_v : v \in V(H)\}$, *where H is a connected block graph and every graph of the family* $\{H_v : v \in V(H)\}$ *is complete.*

Topp and Vestergaard [125] also characterised the (ir, IR)-graphs belonging to a class of graphs containing 5-cycles, while Finbow and van Bommel [79] characterised the (ir, IR)-graphs belonging to a class of planar graphs containing many copies of K_4; these characterisations are quite lengthy, and we omit them here.

3.4   (ir, γ)-*Graphs*

The *inflated graph* G_I of a graph G is obtained by replacing every vertex x_i of degree d_i by a clique X_i of order d_i, and each edge $x_i x_j$ by an edge uv, where $u \in V(X_i)$, $v \in V(X_j)$, and different edges of G are replaced by nonadjacent edges in G_I. Dunbar and Haynes [58] conjectured that inflated graphs are (ir, γ)-graphs, but Favaron [68] (see Theorem 5.12) showed that in general the difference between the parameters can be arbitrarily large. However, Puech [116] proved the conjecture for inflated trees.

Theorem 3.13 ([116]) *If T is a tree, then* $\mathrm{ir}(T_I) = \gamma(T_I)$.

3.5   (α, IR)- *and* (Γ, IR)-*Graphs*

Let $(P, \leq)$ be a poset. The *upper bound graph* G of P has $V(G) = P$, and $xy \in E(G)$ if and only if $x \neq y$, and there exists $z \in P$ such that $x, y \leq z$. Cheston, Hare, Hedetniemi and Laskar [27, as cited in [102]] showed that if G is an upper bound graph, then $\alpha(G) = \mathrm{IR}(G)$. However, G is not necessarily (α, IR)-perfect; an illustrating example is given in [102].

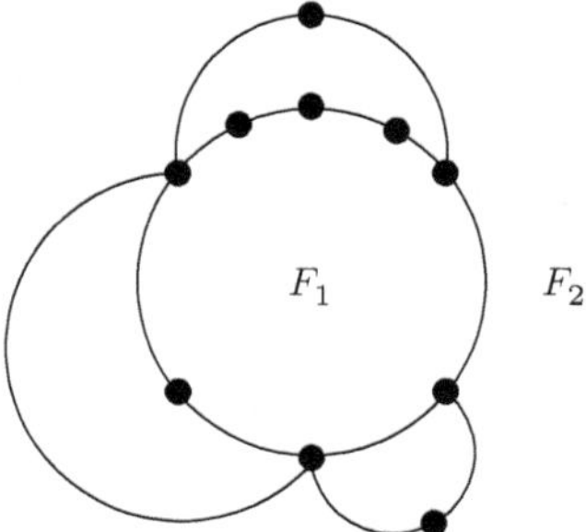

Fig. 6 A flower with large faces F_1 and F_2

Dunbar and Haynes [58] proved that inflated graphs of trees are (α, IR)-graphs, while Favaron [69] characterised 2-connected graphs whose inflated graphs are (α, IR)-graphs. A *flower* is a 2-connected planar graph with a plane representation in which all edges lie on the boundary of one or both of two faces, called the *large faces*. The $f - 2$ other faces are called the *petals*. See Figure 6 for an example of a flower.

Theorem 3.14 *Let G be a graph of order $n \geq 2$.*

(i) [58] *If T is a tree, then $\alpha(T_I) = \mathrm{IR}(T_I) = n - 1$.*

(ii) [69] *If G is 2-connected, then $\alpha(G_I) = \mathrm{IR}(G_I)$ if and only if G is a subdivision of K_4 in which the boundaries of all four faces have odd length or a flower in which the boundaries of all petals have odd length. (In either case $\alpha(G_I) = \mathrm{IR}(G_I) = n$.)*

Favaron [65] (see Proposition 4.4) showed that $\mathrm{IR}(G) \leq n - \delta(G)$ for any n-vertex graph G. Cockayne and Mynhardt [51] characterised graphs for which equality holds, and it turns out that these graphs are (Γ, IR)-graphs and sometimes (α, IR)-graphs.

Proposition 3.15 ([51]) *Let G be a connected graph of order n and minimum degree δ.*

(i) *If $\mathrm{IR}(G) = n - \delta$, then $\Gamma(G) = \mathrm{IR}(G)$, and if, in addition, $\delta < \frac{n}{2}$, then $\alpha(G) = \mathrm{IR}(G)$.*

(ii) *If $\mathrm{IR}(G) = n/2$, then $\Gamma(G) = \mathrm{IR}(G)$.*

4 Bounds Involving Other Graph Parameters

As can be expected, there are many bounds for ir and IR in terms of other graph parameters. We mention bounds for ir and IR separately and distinguish between bounds involving only one of these parameters among those in the domination chain, the sum of two parameters $\pi(G)$ and $\lambda(G)$ (called *Gallai-type results* after Gallai's Theorem [23, Theorem 12.11]) and the sum $\pi(G) + \pi(\overline{G})$ and product

$\pi(G) \cdot \pi(\overline{G})$ of a parameter π of a graph and its complement (called *Nordhaus-Gaddum-type results* after the Nordhaus-Gaddum Theorem [23, Theorem 14.23]). We first consider bounds that hold for general graphs (with perhaps some degree restrictions) and then bounds for specific classes of graphs.

4.1 General Graphs

4.1.1 Bounds for ir

The upper bounds $ir(G) \leq n/2$ and $ir(G) \leq n - \Delta$ for any graph G of order n follow directly from the classical bounds $\gamma(G) \leq n/2$ of Ore (see [92, Theorem 2.1]) and $\gamma(G) \leq n - \Delta$ of Berge (see [92, Theorem 2.11]).

Domke, Dunbar and Markus [57] showed that it is possible to find graphs for which $\gamma = n - \Delta$ and $ir < n - \Delta$. They construct an infinite class of such graphs, the smallest of which is the graph G (depicted in Figure 7) obtained as follows: begin with $P_7 : (v_1, \ldots, v_7)$, join v_3 and v_5, join a new vertex x to v_2, v_3, v_5, v_6 and another new vertex y to x. Then $\Delta(G) = \deg(x) = 5$, $\gamma(G) = 4$ (many γ-sets) and $ir(G) = 3 = |\{y, v_3, v_5\}|$.

Topp and Vestergaard [125] characterised graphs with $ir = n/2$, obtaining the same class of graphs as those for which $\gamma = n/2$, as determined in [113] and later in [81], namely, graphs for which each component is either C_4 or a corona $H \circ K_1$, where H is an arbitrary connected graph. Favaron and Mynhardt [74] gave a fairly complicated characterisation of graphs for which $ir = n - \Delta$.

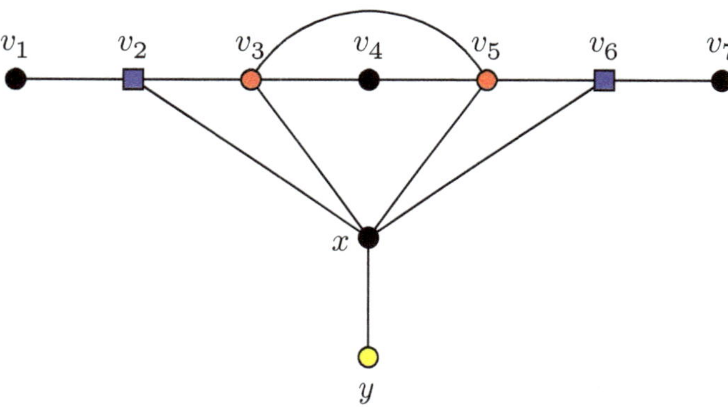

Fig. 7 A graph G satisfying $\gamma(G) = n - \Delta = 4$ (the black vertices form a γ-set) and $ir(G) = 3 = |\{y, v_3, v_5\}|$

Blidia, Chellali and Maffray [9] improved Berge's bound for $\mathrm{ir}(G)$. For any vertex v of a graph G, let $\alpha'_v(G)$ denote the cardinality of a maximum matching of $G - N[v]$ and $\alpha'_\Delta(G) = \max\{\alpha'_v(G) : v \in V(G) \text{ and } \deg(v) = \Delta\}$.

Theorem 4.1 ([9]) *For any graph G and $\mu \in \{\mathrm{ir}, \gamma, i\}$, $\mu(G) \leq n - \Delta - \alpha'_\Delta$.*

Bollobás and Cockayne [11] obtained the lower bound $\mathrm{ir}(G) \geq n/(2\Delta - 1)$, which is attained by paths P_n, $n \equiv 0 \pmod 3$. This bound was improved by Cockayne and Mynhardt [50].

Theorem 4.2 ([50]) *If G has order n and maximum degree $\Delta \geq 2$, then $\mathrm{ir}(G) \geq \frac{2n}{3\Delta}$. The bound is sharp for each value of $\Delta \geq 2$.*

Burger, Henning and van Vuuren [17] found a lower bound for ir in terms of the *lower packing number* $\rho_L(G)$, which is the minimum cardinality of a maximal 2-packing of G.

Theorem 4.3 ([17]) *If G is a connected graph such that $\mathrm{ir}(G) > 1$, then $\mathrm{ir}(G) \geq \frac{2}{3}(1 + \rho_L(G))$.*

This bound is sharp (e.g. for the graph G_1 in Figure 4). When Δ is large relative to n, it can also be a better lower bound for ir than the bound in Theorem 4.2. For example, the graph G_1 can be generalised by replacing its triangle with $F \cong K_r$, where $r \geq 4$, and appending a path of length 2 to each but one vertex of F. Denote this graph by H_r. See Figure 8 for the case $r = 6$, where a ρ_L-set is shown in green and an ir-set in blue. In general, $\rho_L(H_r) = \mathrm{ir}(H_r) = r - 1$. Since H_r has order $n = 3r - 2$ and maximum degree $\Delta = r$, Theorem 4.2 gives $\mathrm{ir}(H_r) \geq \left\lceil \frac{6r-4}{3r} \right\rceil = 2$, while Theorem 4.3 gives $\mathrm{ir}(H_r) \geq \frac{2}{3}(1 + \rho_L(H_r)) = \frac{2r}{3}$. On the other hand, the former bound is better for, e.g. paths P_n, since $\left\lceil \frac{2n}{3\Delta} \right\rceil = \mathrm{ir}(P_n) = \left\lceil \frac{n}{3} \right\rceil$ and $\rho_L(P_n) = \left\lceil \frac{n}{5} \right\rceil$.

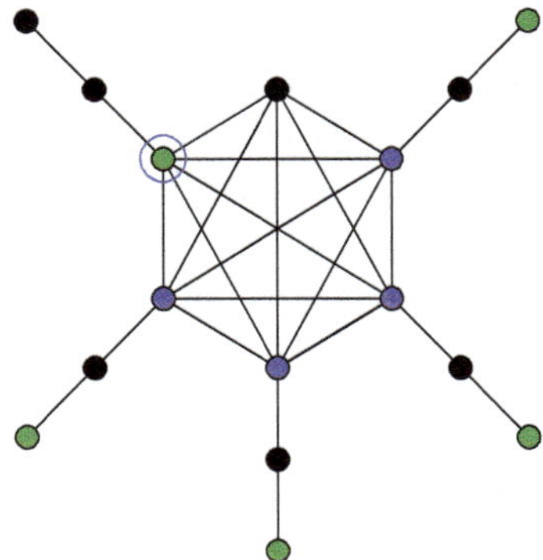

Fig. 8 The graph H_6 for which $\rho_L(H_6) = \mathrm{ir}(H_6) = 5$

4.1.2 Bounds for coir and oir

Finbow [78] proved lower bounds for the CO-irredundant number. For any graph G of order n and maximum degree Δ,

$$
\operatorname{coir}(G) \geq
\begin{cases}
\dfrac{n}{2} & \text{if } \Delta = 2 \\[2ex]
\dfrac{4n}{13} & \text{if } \Delta = 3 \\[2ex]
\dfrac{2n}{3\Delta - 3} & \text{if } \Delta \geq 4.
\end{cases}
$$

The bounds are best possible and the extremal graphs are characterised. Lower bounds for the open irredundance number were obtained in [35], and an extremal graph in each case was exhibited. For any graph G of order n, maximum degree Δ and without isolated vertices, if $\Delta = 1$, then $\operatorname{oir}(G) = \frac{n}{2}$; if $\Delta = 2$, then $\operatorname{oir}(G) \geq \frac{n}{3}$, otherwise

$$
\operatorname{oir}(G) \geq
\begin{cases}
\dfrac{2n}{11} & \text{if } \Delta = 3 \\[2ex]
\dfrac{n}{8} & \text{if } \Delta = 4 \\[2ex]
\dfrac{(3\Delta - 1)n}{2\Delta^3 - 5\Delta^2 + 8\Delta - 1)} & \text{if } \Delta \geq 5.
\end{cases}
$$

4.1.3 Bounds for IR

Since every vertex of a maximum independent set S of a graph G of order n has at least δ neighbours in $V(G) - S$, $\alpha(G) \leq n - \delta$. Favaron [65] showed that the same bound holds for $\operatorname{IR}(G)$ (which can also be deduced from results by [38] mentioned below), while Henning and Slater [100] and Cockayne and Mynhardt [51] bounded IR for regular graphs. Graphs for which equality holds in either case were characterised in [51].

Proposition 4.4

(i) [65] *For any graph G of order n, $\operatorname{IR}(G) \leq n - \delta$.*

(ii) [51] *$\operatorname{IR}(G) = n - \delta$ if and only if G is one of the following graphs:*
$G = \overline{K_{n-\delta}} \vee H$, *where H is any graph of order $\delta(G)$ and minimum degree at least $2\delta(G) - n$,*

or $\delta \geq n/2$ and $G = (K_2 \square K_{n-\delta}) \vee F$, where F is any graph of order $2\delta(G) - n$ and minimum degree at least $3\delta - 2n$.

Proposition 4.5 ([51, 100]) *For any $r \geq 1$, if G is an r-regular graph of order n, then $\mathrm{IR}(G) \leq n/2$. Equality holds if and only if each component of G is either an r-regular (hence balanced) bipartite graph or $K_2 \square H$, where H is an $r - 1$-regular graph.*

Bacsó and Favaron [4] generalised the bound in Proposition 4.5 to non-regular graphs. The bound is better than the bound $\mathrm{IR} \leq n - \delta$ in Proposition 4.4 for small δ, namely, when $\delta + \Delta < n$.

Proposition 4.6 ([4]) *For any graph G, $\mathrm{IR}(G) \leq \frac{n}{1+\delta/\Delta}$. Equality holds if and only if G is a bipartite graph such that all vertices in the same partite set have the same degree, or G is a regular graph described in Proposition 4.5.*

Aouchiche, Favaron and Hansen [3] obtained further upper bounds as well as a lower bound for IR in terms of order and maximum degree.

Proposition 4.7 ([3])

(i) *For any graph G of order n, $\mathrm{IR}(G) \leq n - \lceil 2\sqrt{n-1} \rceil + \Delta$, and there exists a graph that achieves equality in the bound.*
(ii) *For any connected graph G of order n, $\mathrm{IR}(G) \leq \frac{1}{2}\Delta \cdot \lceil \frac{n}{2} \rceil$, with equality if and only if G is a path, an even cycle or a claw.*
(iii) *Let G be a connected graph of order $n \in \{4, 6\}$ or $n \geq 8$. Then $\mathrm{IR}(G) \geq \lceil 2\sqrt{n} \rceil - \Delta$. The bound can be reached for any $n \in \{4, 6\}$ or $n \geq 8$.*

Finally, Hedetniemi, Jacobs and Laskar [94] showed that $\mathrm{IR}(G) \leq r(N(G))$ and $\mathrm{OIR}(G) \leq r(G)$, where $r(G)$ denotes the rank of the adjacency matrix $A(G)$ of G and $r(N(G))$ the rank of the closed neighborhood matrix $N(G) = A(G) + I$.

4.1.4 Nordhaus-Gaddum-Type Results

Cockayne and Mynhardt [45] showed that for every graph G of order n,

$$\mathrm{IR}(G) \cdot \mathrm{IR}(\overline{G}) \leq \left\lceil \frac{n(n+2)}{4} \right\rceil.$$

The graph G attains the bound if and only if G or $\overline{G}$ consists of (*i*) a set X of $\lfloor \frac{n}{2} \rfloor$ independent vertices, (*ii*) a set Y of $\lceil \frac{n}{2} \rceil$ vertices where $G[Y]$ is complete and $X \cap Y = \{x\}$ and (*iii*) an arbitrary set E of edges that join vertices in $X - \{x\}$ to vertices in $Y - \{x\}$.

For CO-irredundance, it was shown in [44] that for any graph G of order n,

$$\mathrm{COIR}(G) + \mathrm{COIR}(\overline{G}) \leq n + 2 \text{ and}$$

$$\mathrm{COIR}(G) \ \mathrm{COIR}(\overline{G}) \leq \left\lfloor (n+2)^2/4 \right\rfloor,$$

and that the bound can be attained for all even values of n. For open irredundance, Cockayne [32] showed that for any graph G of order $n \geq 16$,

$$\mathrm{OIR}(G) + \mathrm{OIR}(\overline{G}) \leq 3n/4 \text{ and}$$

$$\mathrm{OIR}(G) \cdot \mathrm{OIR}(\overline{G}) < 9n^2/64.$$

The bound for $\mathrm{OIR}(G) + \mathrm{OIR}(\overline{G})$ can be attained if $n \equiv 0 (\mathrm{mod}\, 4)$ and extremal graphs were exhibited.

4.1.5 Gallai-Type Results for ir and IR

Cockayne et al. [38] proved that if G has no isolated vertices and X is an irredundant set, then $V(G) - X$ is dominating. They deduced that $\gamma(G) + \mathrm{IR}(G) \leq n$ and hence $\mathrm{ir}(G) + \mathrm{IR}(G) \leq n$. If $\delta \geq 1$ and $\gamma(G) + \mathrm{IR}(G) = n$, then $\alpha(G) = \Gamma(G) = \mathrm{IR}(G)$ (the converse is false). If $\delta \geq 2$, then $\mathrm{ir}(G) + \mathrm{IR}(G) \leq \gamma(G) + \mathrm{IR}(G) \leq n - \delta + 2$. They conjectured that $i(G) + \mathrm{IR}(G) \leq 2(n + \delta - \sqrt{2n\delta})$. This conjecture was proved by Wang [129] and by Favaron [65], who also presented graphs that attain equality.

Proposition 4.8 *For any graph G of order n,*

 (i) *[65, 129]* $i(G) + \mathrm{IR}(G) \leq 2(n + \delta - \sqrt{2n\delta})$,
(ii) *[65]* $i(G) + 2\sqrt{\delta\,\mathrm{IR}(G)} \leq n + \delta$.

Chellali and Volkmann [24] used Brooks's Theorem on the chromatic number χ and the result in Theorem 4.1 to bound $\chi(G) + \mu(G)$ and $\mu(G) \cdot \chi(G)$, $\mu \in \{\mathrm{ir}, \gamma, i\}$. Recall that $\alpha'_v(G)$ denotes the cardinality of a maximum matching of $G - N[v]$ and $\alpha'_\Delta(G) = \max\{\alpha'_v(G) : v \in V(G) \text{ and } \deg(v) = \Delta\}$.

Proposition 4.9 ([24]) *For any graph G of order n and maximum degree Δ, and $\mu \in \{\mathrm{ir}, \gamma, i\}$,*

$$\chi(G) + \mu(G) \leq n + 1 - \alpha'_\Delta.$$

For $\mu \in \{\mathrm{ir}, \gamma\}$, equality holds if and only if

$$G = H \cup (t_1 C_4) \cup (t_2 K_1) \cup \bigcup_{i \in I}(F_i \circ K_1),$$

where $H \in \{K_{\Delta+1}, C_5, C_7\}$ and $\Delta(H) = \Delta$, t_i is a nonnegative integer for $i = 1, 2$, I is a (possibly empty) set of indices and F_i, $i \in I$, is a connected graph.

Theorem 4.10 ([24]) *Let $G \neq C_5$, C_7 be a connected graph of order $n \geq 4$ and $\mu \in \{\mathrm{ir}, \gamma, i\}$. Then*

$$\mu(G)\chi(G) \leq \left\lfloor \frac{(n - \alpha'_\Delta(G))^2}{4} \right\rfloor.$$

Equality holds if and only if

- $G \in \{K_4, C_9, C_{11}\}$, *or*
- $\chi(G) = \Delta$, $\mu(G) = n - \Delta - \alpha'_\Delta(G)$ *with either* $n - \alpha'_\Delta(G) - 2\Delta = 0$ *when* $n - \alpha'_\Delta(G)$ *is even, or* $n - \alpha'_\Delta(G) - 2\Delta = \pm 1$ *if* $n - \alpha'_\Delta(G)$ *is odd.*

If G is bipartite, then equality holds if and only if $G \in \{C_4, P_4, P_5, P_7\}$.

Corollary 4.11 ([24]) *If G is a connected graph of order $n \geq 4$, then* $\mathrm{ir}(G)\chi(G) \leq n^2/4$.

4.2 Specific Graph Classes

4.2.1 Trees

As mentioned above, $\mathrm{ir}(G) \leq n - \Delta$ for all graphs G. Domke et al. [57] showed that equality holds for a tree T if and only if $T = K_1, K_{1,r}, r \geq 1, P_4$ or a spider obtained from $K_{1,r}$ by subdividing at most $r - 1$ edges.

The lower bound $\mathrm{ir}(G) \geq \frac{2n}{3\Delta}$ given in Theorem 4.2 can be improved for trees, as shown by Cockayne [33] and Poschen and Volkmann [114].

Theorem 4.12

(i) [114] *For a tree T with order n and ℓ leaves,* $\mathrm{ir}(T) \geq \frac{n+2-\ell}{3}$. *Equality holds if and only if the distance between each pair of distinct leaves in T is congruent to 2 (mod 3).*

(ii) [33] *If $T \neq K_{1,n-1}$ is a tree of order n and maximum degree $\Delta \geq 3$, then* $\mathrm{ir}(T) \geq \frac{2(n+1)}{2\Delta+3}$.

The bound in Theorem 4.12(*i*) is better if there are relatively few leaves, while the bound in Theorem 4.12(*ii*) is better if many non-leaf vertices have degree Δ. The trees for which equality holds in (*ii*) in the case where ir is even were also characterised in [33].

4.2.2 Claw-Free Graphs

Favaron [70] investigated upper bounds for IR in claw-free graphs, as well as graphs which attain or nearly attain equality in the bounds.

Theorem 4.13 ([70]) *Every connected claw-free graph G of order n satisfies* $\mathrm{IR}(G) \leq \frac{n+1}{2}$. *If* $\mathrm{IR}(G) = \frac{n+1}{2}$, *then* $\alpha(G) = \mathrm{IR}(G)$, *and if* $\mathrm{IR}(G) = \frac{n}{2}$, *then* $\mathrm{IR}(G) = \Gamma(G)$.

Corollary 4.14 *If G is a claw-free graph of order n and $\delta(G) \geq 2$, then $\mathrm{IR}(G) \leq \lfloor \frac{n}{2} \rfloor$.*

Proof. As proved in [108], if G is a claw-free graph of order n and minimum degree δ, then $\alpha(G) \leq 2n/(\delta + 2)$. Since $\delta \geq 2$, $\alpha(G) \leq n/2$. By Theorem 4.13, if $\mathrm{IR}(G) = \frac{n+1}{2}$, then $\alpha(G) = \frac{n+1}{2}$, which is not the case.   ∎

Favaron also described an infinite class $\mathcal{F}$ of claw-free graphs with $\mathrm{IR}(G) = \frac{n+1}{2}$ and showed that this class characterises claw-free graphs with $\mathrm{IR}(G) = \frac{n+1}{2}$. She then investigated claw-free graphs G such that $\mathrm{IR}(G) = \lfloor \frac{n}{2} \rfloor$.

Theorem 4.15 ([70]) *Let G be a connected claw-free graph of order $n \geq 7$.*

(i) *If n is even and $\mathrm{IR}(G) = \frac{n}{2}$, then $\delta(G) = \frac{n}{2}$ or $1 \leq \delta(G) \leq \frac{n}{4}$. Moreover, for every integer r between 1 and n/4 or equal to n/2, there exists a connected claw-free graph G of order n such that $\mathrm{IR}(G) = n/2$ and $\delta(G) = r$.*

(ii) *If n is odd and $\mathrm{IR}(G) = \frac{n-1}{2}$, then $\frac{n-1}{2} \leq \delta(G) \leq \frac{n+1}{2}$ or $1 \leq \delta(G) \leq \frac{n+3}{4}$. Moreover, for every integer $r \in \{1, \ldots, \lfloor \frac{n+3}{4} \rfloor\} \cup \{\frac{n-1}{2}, \frac{n+1}{2}\}$, there exists a connected claw-free graph G of order n such that $\mathrm{IR}(G) = \frac{n-1}{2}$ and $\delta(G) = r$.*

4.2.3 Other Graphs

For inflated graphs, Dunbar and Haynes [58] stated it as an open problem to bound $\mathrm{IR}(G_I)$. Favaron [68] proves that (*i*) $\mathrm{IR}(G_I) \leq m(G)$ for every graph G of size m without isolated vertices, where equality holds if G is bipartite, and (*ii*) $\mathrm{IR}(G_I) \leq \lfloor n^2(G)/4 \rfloor$.

Favaron and Puech [75] showed that the lower irredundance number ir of a plane, cylindrical or toroidal grid of order $m \times n$ (i.e. $G \square H$, where $G \in \{P_m, C_m\}$ and $H \in \{P_n, C_n\}$) is at least $mn/5$ and is asymptotically equal to $mn/5$ when m and n tend to infinity.

5 Differences Between Parameters in the Domination Chain

Many authors have obtained results on the differences between irredundance numbers (usually IR) and other parameters in the domination chain, mostly in terms of order, or order and maximum degree, and occasionally also involving the chromatic number χ. Others have bounded the ratios of irredundance numbers to domination and independence numbers, mostly for special graph classes.

5.1 Differences Between Lower Parameters

Allan, Laskar and Hedetniemi [2] observed that the inequality $2\,\text{ir}(G) - \gamma(G) \geq 1$ obtained in [1, 10] can be improved to $2\,\text{ir}(G) - \gamma(G) \geq (k+1)$, where k is the maximum number of isolated vertices in an ir-set. Zverovich [137] bounded $\gamma - \text{ir}$ in terms of order, and $i - \text{ir}$ in terms of maximum degree and order.

Theorem 5.1 ([137])

(i) *For any graph G of order $n \geq 3$, $\gamma(G) - \text{ir}(G) \leq \left\lfloor \frac{n-3}{4} \right\rfloor$.*

(ii) *For any graph G of order n with maximum degree $\Delta \geq 3$,*

$$i(G) - \text{ir}(G) \leq \min\left\{ \left\lfloor \frac{2\Delta - 3}{2\Delta - 1} n \right\rfloor, \left\lfloor \frac{\Delta - 1}{\Delta} n - \frac{\Delta}{2} \right\rfloor \right\} - 1.$$

5.2 Differences Between Upper Parameters

Henning and Slater [100] conjectured that $\Gamma(G) = \text{IR}(G)$ if G is a cubic graph. The conjecture is false (see Theorem 5.14), but generated considerable interest in bounding the differences $\text{IR} - \alpha$ and $\text{IR} - \Gamma$ for general graphs and the ratio IR/Γ for special classes of graphs. Rautenbach [118] was the first to bound the abovementioned differences in terms of order and to characterise the extremal graphs.

Theorem 5.2 ([118])

(i) *For any graph G of order $n \geq 4$, $\text{IR}(G) - \alpha(G) \leq \left\lfloor \frac{n-4}{2} \right\rfloor$.*

(ii) *For any graph G of order $n \geq 6$, $\text{IR}(G) - \Gamma(G) \leq \left\lfloor \frac{n-4}{2} \right\rfloor$.*

For even values of $n \geq 6$, equality holds in (i) if and only if $G = K_{n/2}\square K_2$. If $n \geq 7$ and odd, equality holds in (i) if and only if $G = K_{(n-1)/2}\square K_2$ together with a vertex u which is either an isolated vertex, or u is adjacent to all vertices of one of the copies of $K_{(n-1)/2}$, or there is a pair of adjacent vertices x and y, one from each copy of $K_{(n-1)/2}$ such that u is adjacent to all vertices except x and y (and itself).

For even values of $n \geq 8$, equality holds in (ii) if and only if G is obtained from $K_{(n+2)/2}\square K_2$ by deleting two nonadjacent vertices. If $n \geq 7$ and odd, equality holds in (ii) if and only if G is one of the following types of graphs: let H be the graph obtained from $K_{(n+1)/2}\square K_2$ by deleting two nonadjacent vertices and denote the vertex sets of the two copies of $K_{(n-1)/2}$ in H by V_1 and V_2. For $i = 1, 2$, let $v_i \in V_i$ be the vertex of degree $\frac{n-3}{2}$, i.e. the vertex not adjacent to any vertex in V_j, $j \neq i$. Form the graph G by adding a new vertex w, where w is either (a) isolated, or (b) adjacent precisely to all vertices in (say) V_1, or (c) adjacent precisely to all, except possibly one, vertices in (say) $V_1 - \{v_1\}$, or (d) there is a pair of adjacent vertices $x \in V_1$ and $y \in V_2$ such that w is adjacent to all vertices except x and y (and itself).

Rautenbach also bounded $\mathrm{IR}(G) - \alpha(G)$ (and hence $\mathrm{IR}(G) - \Gamma(G)$ and $\Gamma(G) - \alpha(G)$) in terms of order and maximum degree and conjectured that if $\Delta(G) \leq 3$, then $\mathrm{IR}(G) - \alpha(G) \leq \lfloor \frac{n}{6} \rfloor$.

Theorem 5.3 ([118]) *For any graph G with maximum degree Δ, $\mathrm{IR}(G) - \alpha(G) \leq$* $\left\lfloor \frac{(\Delta-1)^2}{2\Delta^2} n \right\rfloor$.

Bacsó and Favaron [4] and Zverovich [137] obtained Rautenbach's conjecture as a corollary to the following result.

Theorem 5.4 *Let G be a connected graph of order n, chromatic number χ and maximum degree $\Delta \geq 2$. Then*

(i) [4, 137] $\mathrm{IR}(G) - \alpha(G) \leq \left\lfloor \frac{\Delta-2}{2\Delta} n \right\rfloor$.

(ii) [4] *Equality holds if and only if (a) G is a path or a cycle or (b) $\Delta \geq 3$ and $G = K_{n/2} \square K_2$, in which case $\mathrm{IR} = \Gamma = n/2$, $\alpha = n/\Delta$ and $\chi = \Delta$.*

We see from Theorem 5.4(*ii*) that for fixed $\Delta \geq 3$, there exist arbitrarily large connected Δ-regular graphs that satisfy equality: for arbitrary $r \geq 1$, join one copy of $rK_{n/2}$ to another one by a perfect matching that ensures the graph is connected. Favaron also mentioned that the stronger result $\mathrm{IR}(G) - \alpha(G) \leq \frac{\chi-2}{2\chi} n$ holds for $\chi \geq 2$. When $\chi < n/2$, this bound is better than Rautenbach's bound $\mathrm{IR}(G) - \alpha(G) \leq \frac{n-4}{2}$ (Theorem 5.2(*i*)).

The bound in Theorem 5.4(*i*), although asymptotically the same as Rautenbach's bound, is better for specific values of Δ and immediately proves Rautenbach's conjecture. (The conjecture was subsequently also proved in [109, 130].)

Corollary 5.5 ([4, 137]) *If $\Delta(G) = 3$, then $\mathrm{IR}(G) - \alpha(G) \leq \lfloor \frac{n}{6} \rfloor$.*

Zverovich also showed that this bound can be improved for triangle-free cubic graphs. A *generalised Petersen graph* can be described as a cubic graph obtained by joining the vertices of a regular polygon to the corresponding vertices of a star polygon.

Theorem 5.6 ([137])

(i) *If G is a triangle-free cubic graph of order n, then*

$$\mathrm{IR}(G) - \alpha(G) \leq \left\lfloor \frac{n}{7} \right\rfloor.$$

Equality is attained by any generalised Petersen graph of order 14.

(ii) *If G contains no K_q ($q \geq 3$) and $\Delta \geq 1$, then*

$$\mathrm{IR}(G) - \alpha(G) \leq \left\lfloor \frac{\Delta + q - 4}{2(\Delta + q)} n \right\rfloor.$$

Henning [97, p. 57] questioned whether there exists a cubic graph G such that $ir(G) < \gamma(G) < i(G) < \alpha(G) < \Gamma(G) < \mathrm{IR}(G)$. That this is indeed the case was proved by Zverovich and Zverovich [135]. The graph in their construction has connectivity 2, and they asked whether there exists a 3-connected cubic graph with the same property.

Theorem 5.7 ([135]) *For any nonnegative integers $k_1, \ldots, k_5$, there exists a cubic graph G with connectivity 2 such that*

$$\gamma(G) - \mathrm{ir}(G) \geq k_1, \ i(G) - \gamma(G) \geq k_2, \ \alpha(G) - i(G) \geq k_3,$$

$$\Gamma(G) - \alpha(G) \geq k_4 \text{ and } \mathrm{IR}(G) - \Gamma(G) \geq k_5.$$

5.3 *Ratios of Lower Parameters*

As shown in [1, 10], $\gamma(G)/\mathrm{ir}(G) < 2$ for any graph G. To see that this ratio can be arbitrarily close to 2, construct the graph G_k, $k \geq 2$, as follows. (See Figure 9 for $k = 3$.) Begin with the corona $K_k \circ K_1$ and subdivide each pendant edge. Let $X = \{x_1, \ldots, x_k\}$ be the vertex set of K_k, $Y = \{y_1, \ldots, y_k\}$ the vertices of degree 2 and $R = \{r_1, \ldots, r_k\}$ the end-vertices; assume that each y_i is adjacent to x_i and r_i. For each pair of distinct integers $i, j \in \{1, \ldots, k\}$, join a new vertex c_{ij} to x_i and x_j. This is the graph G_k. Now X is a maximal irredundant set in which pn $x_i, X) = \{y_i\}$ for each i, and it is not difficult to see that X is an ir-set and $\mathrm{ir}(G_k) = k$. Let D be a dominating set of G_k. To dominate the end-vertices, $\{y_i, r_i\} \cap D \neq \varnothing$ for each i; hence $|D \cap (Y \cup R)| \geq k$. To dominate c_{ij}, $\{x_i, x_j, c_{ij}\} \cap D \neq \varnothing$. After some thought, this implies that $|D \cap (X \cup \{c_{ij} : i, j \in \{1, \ldots, k\}\})| \geq k - 1$. Since $X \cup Y - \{x_k\}$ dominates G_k, $\gamma(G_k) = 2k - 1$. Therefore $\lim_{k \to \infty} \gamma(G_k)/\mathrm{ir}(G_k) = 2$.

For many graph classes, though, the bound on the ratio can be improved. We summarise the results below. The *cyclomatic number* $\mu(G)$ of G is given by $\mu(G) = |E(G)| - |V(G)| + k(G)$, where $k(G)$ denotes the number of components of

Fig. 9 The graph G_3 with $\mathrm{ir}(G_3) = 3$ and $\gamma(G_3) = 5$

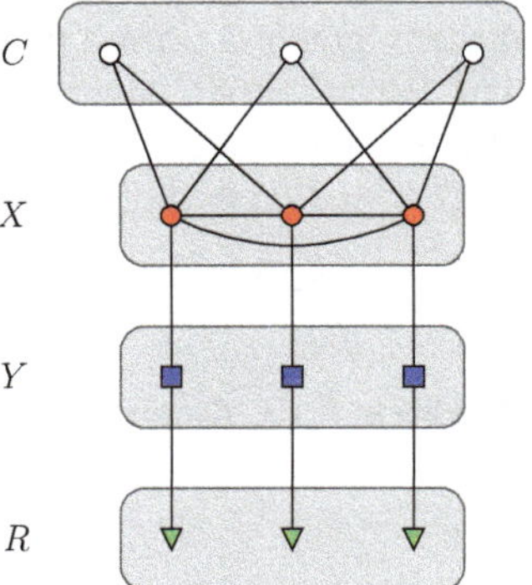

G. A *claw-free block graph* is a graph, all of whose blocks are claw-free. A *block-cactus graph* is a graph whose blocks are either complete or induced cycles.

Theorem 5.8

(i) [55] *For any tree T, $\gamma(T)/\operatorname{ir}(T) < \frac{3}{2}$.*
(ii) [126] *For any block graph G and for any graph G with cyclomatic number $\mu(G) \leq 2$, $\gamma(G)/\operatorname{ir}(G) \leq \frac{3}{2}$. The bound $\frac{3}{2}$ is best possible for block graphs and does not hold if $\mu(G) \geq 3$.*
(iii) [73] *If G is a claw-free graph, an inflated graph or the line graph of a triangle-free or a bipartite graph, then $\gamma(G)/\operatorname{ir}(G) \leq \frac{3}{2}$.*
(iv) [72] *If G is a claw-free block graph, then $\gamma(G)/\operatorname{ir}(G) \leq \frac{7}{4}$.*
(v) [136] *If G is a block-cactus graph having $\pi(G)$ induced cycles of length 2 (mod 4), then $\gamma(G)/\operatorname{ir}(G) \leq (8\pi(G)+6)/(5\pi(G)+4)$.*

Volkmann [126] conjectured that $\gamma(G)/\operatorname{ir}(G) < \frac{8}{5}$ for any cactus graph. Theorem 5.8(v) implies this inequality.

Corollary 5.9 ([136]) *If G is a block-cactus graph, then $\gamma(G)/\operatorname{ir}(G) < \frac{8}{5}$. The bound is asymptotically best possible.*

Henning and Slater [100] showed that the difference $\gamma - \operatorname{ir}$ can be arbitrary for cubic graphs. For the graphs they constructed, $\gamma/\operatorname{ir} \geq \frac{15}{13}$. This suggests the following problem. Denote the maximum ratio of two parameters π and λ by $\max\left\{\frac{\pi}{\lambda}\right\}$. Hence, by Corollary 5.9, $\max\left\{\frac{\gamma}{\operatorname{ir}}\right\} \to \frac{8}{5}$ for block-cactus graphs.

Problem 5.10 *Determine or bound $\max\left\{\frac{\gamma}{\operatorname{ir}}\right\}$ for cubic graphs. By the abovementioned construction, $\max\left\{\frac{\gamma}{\operatorname{ir}}\right\} \geq \frac{15}{13}$.*

Since a tree is (ir, γ)-perfect if and only if it is T_1-free, where T_1 is the Slater tree in Figure 4, and $\gamma(T_1)/\operatorname{ir}(T_1) = \frac{5}{4}$, we also have the following problem.

Problem 5.11 *Determine $\max\left\{\frac{\gamma}{\operatorname{ir}}\right\}$ for trees. Is it true that $\max\left\{\frac{\gamma}{\operatorname{ir}}\right\} = \frac{5}{4}$ for trees?*

For inflated graphs, Dunbar and Haynes [58] conjectured that $\operatorname{ir}(G_I) = \gamma(G_I)$ for any graph *G*. Puech proved this conjecture if *G* is a tree (see Theorem 3.13), but Favaron [68] gave a construction of 2-connected graphs *G* to show that the difference can be arbitrarily large. For the graphs G^k constructed, the ratio $\gamma(G_I^k)/\operatorname{ir}(G_I^k) \geq \lim_{k\to\infty} \frac{5k+2}{4k+2} = \frac{5}{4}$.

Theorem 5.12 ([68]) *For every positive integer k, there exists a 2-connected graph G^k such that $\gamma(G_I^k) - \operatorname{ir}(G_I^k) \geq k$.*

Problem 5.13 *Determine $\max\left\{\frac{\gamma}{\operatorname{ir}}\right\}$ for inflated graphs.*

5.4 Ratios of Upper Parameters

Henning and Slater [100] conjectured that $\Gamma(G) = \mathrm{IR}(G)$ if G is cubic, but this was proved to be false in [51] and [118].

Theorem 5.14

(i) [51] *For any positive integer k, there exists a 2-connected cubic graph H_k such that* $\mathrm{IR}(H_k) - \Gamma(H_k) \geq k$.

(ii) [118] *For any positive integer k and any integer $r \geq 3$, there exists a connected r-regular graph $H_{r,k}$ such that* $\mathrm{IR}(H_{r,k}) - \Gamma(H_{r,k}) \geq k$.

The ratio IR / Γ for the cubic graphs H_k in Theorem 5.14(i) is $\mathrm{IR} / \Gamma = \frac{7}{6}$, while the ratios for the graphs constructed by Rautenbach in Theorem 5.14(ii) are

$$
\mathrm{IR} / \Gamma \geq
\begin{cases}
\frac{18}{17} & \text{for cubic graphs,} \\[4pt]
\frac{16}{15} & \text{for 4-regular graphs} \\[4pt]
\frac{3r}{2r+4} & \text{for } r\text{-regular graphs where } r \geq 6 \text{ is even} \\[4pt]
\frac{4r}{3r+4} & \text{for } r\text{-regular graphs where } r \geq 5 \text{ is odd.}
\end{cases}
$$

We know the ratio IR / α is unbounded for regular graphs ($K_2 \square K_{\mathrm{IR}}$), as is the ratio IR / Γ for non-regular graphs (e.g. $K_2 \square K_{\mathrm{IR}+1} - e$, where e is an edge joining a vertex of one copy of $K_{\mathrm{IR}+1}$ to another).

Cockayne and Mynhardt [49] exhibited an infinite class of triangle-free graphs for which the difference between the upper irredundance and domination numbers is arbitrarily large, thus answering a question of [63]. The ratio for the given graphs G_k is

$$
\mathrm{IR}(G_k)/\Gamma(G_k) = \frac{4k}{3k+2}.
$$

Problem 5.15 *Determine* $\max\left\{\frac{\mathrm{IR}}{\Gamma}\right\}$ *for (i) r-regular graphs and (ii) triangle-free graphs.*

6 Criticality and Stability

We next consider how the upper and lower irredundance numbers change with the removal of a vertex or an edge or with the addition of an edge. For a graph parameter π, a graph G is:

- π-*critical* (π^+-*critical*) if $\pi(G-v) < \pi(G)$ ($\pi(G-v) > \pi(G)$) for every vertex $v \in V(G)$,

- π-*edge-critical* (π^+-*edge-critical*) if $\pi(G+e) < \pi(G)$ ($\pi(G+e) > \pi(G)$) for every edge $e \in E(\overline{G})$,
- π-*edge-removal critical* (π^--*edge-removal critical*), abbreviated to π-*ER-critical* (π^--*ER-critical*), if $\pi(G-e) > \pi(G)$ ($\pi(G-e) < \pi(G)$) for every edge $e \in E(G)$.

6.1 Criticality

All edgeless graphs with more than one vertex are π-critical and π-edge critical for $\pi \in \{ir, IR\}$. Furthermore, the complete graph K_n, $n \geq 2$, is IR-ER-critical, while the star $K_{1,n}$, $n \geq 1$, is ir-ER-critical. Topp [124] and Grobler [85] showed that there exist no π^+-critical graphs for $\pi \in \{ir, IR\}$, no ir$^+$-edge-critical graphs and no IR$^-$-ER-critical graphs.

Theorem 6.1 *For any nontrivial graph G,*

(i) [124] $IR(G - v) \leq IR(G)$ *for all* $v \in V(G)$;
(ii) [85] $ir(G - v) \leq ir(G)$ *for at least one* $v \in V(G)$;
(iii) [85] *if* $G \not\cong K_n$, *then* $ir(G + uv) \leq ir(G)$ *for at least one* $uv \in E(\overline{G})$;
(iv) [85] *if* $G \not\cong \overline{K_n}$, *then* $IR(G - uv) \leq IR(G)$ *for at least one* $uv \in E(G)$.

Proof.

(i) An IR-set S of $G - v$ is also irredundant in G and therefore $IR(G) \geq |S| \geq IR(G - v)$.
(ii) Let S be an ir-set of G. If S is also dominating, then for $v \in V(G) - S$, the set S is still dominating in $G - v$. From the domination chain, it follows that $ir(G - v) \leq \gamma(G - v) \leq |S| = ir(G)$.

   If S is not dominating, then S is an irredundant set of $G - v$ for $v \in R = V(G) - N[S]$. If S is not a maximal irredundant set of $G - v$, then there exists a vertex $x \in V(G - v) - S$ such that $S \cup \{x\}$ is an irredundant set of $G - v$. But then $S \cup \{x\}$ is also an irredundant set of G, contradicting the maximality of S in G. Hence S is a maximal irredundant set of $G - v$ and $ir(G - v) \leq |S| = ir(G)$.
(iii) If $ir(G) = \gamma(G)$, then let S be a dominating ir-set of G. Since S is also a dominating set of $G + uv$, it follows from the domination chain that $ir(G + uv) \leq \gamma(G + uv) \leq |S| = ir(G)$.

   If $ir < \gamma(G)$, let S be an ir-set of G. Then there exists an edge $uv \in E(\overline{G})$ such that $u, v \in R$. If not, all vertices in R are adjacent and $S' = S \cup \{x\}$, with $x \in R$, is a dominating set of G. Since S' is not irredundant, S' is not a minimal dominating set and $\gamma(G) < |S| + 1$. That is, $\gamma(G) \leq |S| = ir(G)$, a contradiction. From Theorem 1.1, it follows that S is a maximal irredundant set of $G + uv$ and so $ir(G + uv) \leq |S| = ir(G)$.

(iv) Let S be an IR-set of G. If S is independent, then S is also an independent irredundant set of $G - uv$ for every $uv \in E(G)$. Hence $\mathrm{IR}(G) = |S| \leq \mathrm{IR}(G - v)$.

If S is not independent, then there exists an edge $uv \in E(G)$ such that $u, v \in S$. Since S is an irredundant set of $G - uv$, $\mathrm{IR}(G) = |S| \leq \mathrm{IR}(G - uv)$. ∎

6.1.1 Criticality of ir

First consider how the deletion of a vertex influences the lower irredundance number. While the deletion of a vertex can decrease the domination number by at most one, Favaron [67] showed that this is not the case for the irredundance number of a graph.

Theorem 6.2 ([67]) *For every graph G and every vertex v of G such that* $\mathrm{ir}(G-v) \geq 2$, $\mathrm{ir}(G - v) \geq (\mathrm{ir}(G) + 1)/2$, *and this bound is sharp.*

To illustrate the sharpness of the bound, a graph G is constructed as follows. Let A and A' be two copies of $K_{1,n}$, with $V(A) = \{x_0, \ldots, x_n\}$ and $V(A') = \{x_0', \ldots, x_n'\}$ where x_0 and x_0' are the centres. Add the edges $x_i x_i'$ for $i = 0, \ldots, n$. Let $r \geq 2(n+1)$. For each each $i = 0, \ldots, n$, add an independent set $Y_i = \{y_{ik} : k = 1, \ldots, r\}$, and join each vertex in Y_i to x_i'. For each pair i, j with $0 \leq i \neq j \leq n$, add an independent set $Z_{ij} = \{z_{ijk} : k = 1, \ldots, r\}$, and join each vertex in Z_{ij} to x_i and x_j. Finally, add a pendant vertex v to x_0. Then $V(A)$ is an ir-set of $G - v$, so $\mathrm{ir}(G - v) = n + 1$, and the set $(V(A) - \{x_n\}) \cup V(A')$ is an ir-set of G (of cardinality $2n + 1$). The case where $n = 2$ is illustrated in Figure 10.

Since $\mathrm{ir} \leq \gamma$, it follows that if G is γ-critical and $\mathrm{ir} = \gamma$, then G is also ir-critical. However, the classes of γ-critical and ir-critical graphs do not coincide. Grobler and Roux [89] constructed two classes of graphs that are γ-critical but not ir-critical, while Roux [121] illustrated the existence of graphs that are ir-critical but not γ-critical.

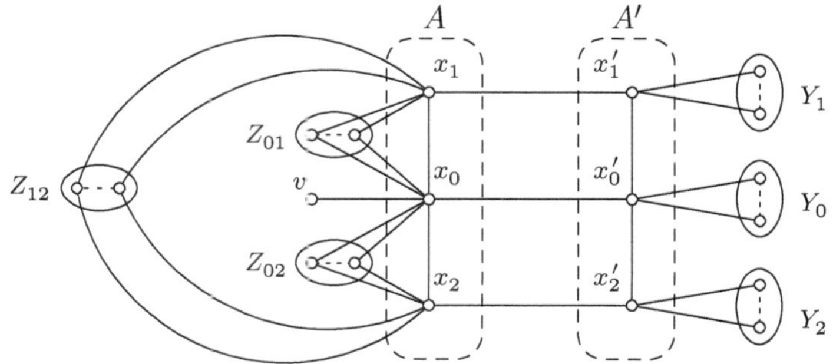

Fig. 10 A graph G having ir(G)-set $\{x_0, x_1, x_0', x_1', x_2'\}$ and ir$(G - v)$-set $\{x_0, x_1, x_2\}$

Similar to ir-critical graphs, if graph G is γ-edge-critical and $\gamma = $ ir, then G is also ir-edge-critical. It is however still unknown whether there exist graphs which are γ-edge-critical but not ir-edge-critical and vice versa.

Problem 6.3 *Investigate the intersection of the classes of γ-edge-critical and ir-edge-critical graphs.*

As for γ and i, the disjoint union of stars are also ir-ER-critical. That these are not the only ir-critical graphs was shown in [87] and [37] where connected ir-ER-critical graphs for ir $= 2$ and ir $= 3$, respectively, were characterised. Whether ir^--ER-critical graphs exist is still an open question.

Problem 6.4 *Does there exist a graph G for which $\text{ir}(G - e) < \text{ir}(G)$ for all $e \in E(G)$?*

6.1.2 Criticality of IR

Grobler and Mynhardt [86] showed that the classes of IR-critical and Γ-critical graphs coincide and characterised these graphs as follows:

Theorem 6.5 ([86]) *If G is a connected graph with n vertices, then the following statements are equivalent.*

  (i) *G is Γ-critical,*
 (ii) *$n > 2$, and for each Γ-set S and $T = V(G) - S$, S and T are independently matched,*
(iii) *$\Gamma(G) = n/2$ and no Γ-set has isolated vertices,*
(iv) *G is IR-critical.*

IR-edge-critical graphs were independently characterised by Grobler and Mynhardt [88] and Dunbar, Monroe and Whitehead [59] as precisely the graphs $K_a \vee bK_1$ or $K_a \vee (bK_1 \cup (K_c \Box K_2))$ for $a, b \geq 0$ and $c \geq 3$, where $V(K_0) = \varnothing$. Grobler and Mynhardt [88] also showed that the classes of IR-edge critical graphs and Γ-edge-critical graphs coincide.

Dunbar et al. [59] conjectured that there do not exist any IR^+-edge-critical graphs. Cockayne, Favaron and Mynhardt [37] disproved this conjecture by exhibiting an infinite class of IR^+-edge-critical graphs for which IR $= 2$.

Theorem 6.6 ([37]) *The graph $\overline{C_m \Box C_n}$ is IR^+-edge-critical if and only if $m, n \notin \{3, 4, 6\}$.*

Since $\overline{C_m \Box C_n}$ is also Γ^+-edge-critical, we ask the following questions:

Problem 6.7

  (i) *Do there exist graphs that are IR^+-edge-critical but not Γ^+-edge-critical?*
 (ii) *Find IR^+-edge-critical graphs with IR > 2.*

Let S be an irredundant set of G and $uv \in E(G)$. Then S is a *uv-irredundant set* if $u \in S$ and $v \in \mathrm{pn}(u, S)$, and either u is an isolated vertex of S and v does not annihilate any vertex in $S - \{u\}$, or there exists a vertex $s \in N[v] - S$ that does not annihilate any vertex in S. Graphs which are IR-ER-critical were characterised in [87] as graphs for which there exists a uv-irredundant IR-set in G for each edge $uv \in E(G)$.

6.2 Stability

When considering the effect on $\mathrm{IR}(G)$ of the addition of edges from $\overline{G}$, a graph G is defined to be IR-*insensitive* if $\mathrm{IR}(G + e) = \mathrm{IR}(G)$ for every $e \in E(\overline{G})$. Dunbar et al. [59] characterised IR-insensitive bipartite graphs without isolated vertices.

Theorem 6.8 ([59]) *A bipartite graph G containing no isolated vertex is IR-insensitive if and only if every independent set $X \subset V(G)$ satisfies the condition that $|X| \leq |N(X)|$.*

Wang and Hua [131] introduced the *stability number* of a graph G, denoted by $\mathrm{SN}(G)$, as the maximum cardinality among all sets of edges $E' \subseteq E(G)$ such that $\mathrm{IR}(G - E') = \mathrm{IR}(G)$. They showed that for any non-empty connected graph of order $n \geq 2$, $\mathrm{SN}(G) \leq n - 2$, with equality when G is a star. In a more general result, they showed that $\mathrm{SN}(G) \leq \Delta(\mathrm{IR}(G) - 1) - 1$ for any non-empty connected graph G with $\mathrm{IR}(G) \geq 2$.

7 Chessboards

Although we have not discussed the exact determination of ir and IR for specific classes of graphs (such as paths and cycles, for which this is easy), no survey on irredundance can be complete without mentioning chessboard problems. Hedetniemi, Hedetniemi and Reynolds [93] gave a complete survey of results on domination, independence and irredundance on the different types of chessboards up to 1998. Here we concentrate on the queens, kings and grid graphs, which, as far as we could ascertain, are the only chessboard graphs for which new irredundance results have appeared since then. We summarise the known exact values in Table 1 and present bounds in Section 7.2.

7.1 Exact Values

We only provide the provenance of results in Table 1 that do not appear in [93].

Table 1 Known exact values for Q_n, K_n and G_n

n		3	4	5	6	7	8	9	10	11	12
Queens	ir	1	2	3 [19]	3 [19]	4 [52]	5 [103]	5 [103]	5 [103]	5 [103]	6 [103]
Q_n	IR	2	4	5	7	9	11	13 [103]	15 [103]		
Kings	ir	1	3	4	4	8	9	9			16 [71]
K_n	IR	4	4	9	9	16	17 [103]	25 [103]	27 [103]	36 [103]	
Grids	ir	3	4 [75]								
G_n	IR	5	8	13	18	25	32	41	50	61	72

For all values of n for which $\mathrm{ir}(Q_n)$ is known, namely, $1 \leq n \leq 13$, $\mathrm{ir}(Q_n) = \gamma(Q_n)$. We do not surmise that equality holds for all n. The value $\mathrm{IR}(K_8) = 17$ is the only known case for which $\mathrm{IR}(K_n) > \alpha(K_n)$, as $\alpha(K_8) = 16$ and $\alpha(K_n) = \mathrm{IR}(K_n)$ for all $n \leq 7$.

Problem 7.1

(i) *Find more values of* $\mathrm{ir}(Q_n)$. *What is the smallest* n *such that* $\mathrm{ir}(Q_n) < \gamma(Q_n)$?
(ii) *Find more values of* $\Gamma(Q_r)$ *and* $\mathrm{IR}(Q_n)$. *What is the smallest* n *such that* $\Gamma(Q_n) < \mathrm{IR}(Q_n)$?
(iii) *What is the smallest* n *such that* $\Gamma(K_n) < \mathrm{IR}(K_n)$?

7.2 Bounds

7.2.1 Bounds for the Queens Graph

Apart from the exact values for $\mathrm{ir}(Q_n)$ in Table 1 and the value $\mathrm{ir}(Q_{13}) = \gamma(Q_{13}) = 7$ [103], not much is known about irredundance numbers of queens graphs. The best bounds known are given by

$$\left\lceil \frac{n+1}{4} \right\rceil \leq \mathrm{ir}(Q_n) \leq \gamma(Q_n) \leq \frac{200}{393}n + O(1),$$

where the lower bound follows from the bound $\mathrm{ir}(G) \geq (\gamma(G) + 1)/2$ for all graphs and Spencer's bound $\gamma(Q_n) \geq (n - 1)/2$ as cited in [29] and the upper bound (for $\gamma(Q_n)$) follows from a result in [20] and a suitable dominating configuration for Q_{129} in [112]. For all known values of $\mathrm{ir}(Q_n)$, $\mathrm{ir}(Q_n) = \gamma(Q_n)$, so the lower bound above appears to be weak.

Burger, Cockayne and Mynhardt [15] showed that $\mathrm{IR}(Q_n) \geq \Gamma(Q_n) \geq 2n - 5$ for odd $n \geq 5$, and $\mathrm{IR}(Q_n) \geq \Gamma(Q_n) \geq 2n - 6$ for even $n \geq 6$. If $n \geq 18$, then $\mathrm{IR}(Q_n) \geq \Gamma(Q_n) \geq \frac{5n}{2} - O(1)$. They also obtained the upper bound $\mathrm{IR}(Q_n) \leq \left\lfloor 6n + 6 - 8\sqrt{n + \sqrt{n} + 1} \right\rfloor$ for $n \geq 9$, improving the previous bound in [31].

Hedetniemi et al. [93, Open Problem 7] stated that it seems very likely that $\mathrm{IR}(Q_n) \leq 5n$ or possibly even $\mathrm{IR}(Q_n) \leq 4n$. This statement was disproved by Kearse and Gibbons [104], who obtained the bound $\mathrm{IR}(Q_{k^3}) \geq 6k^3 - 29k^2 - O(k)$ for every $k \geq 6$, which implies that $\mathrm{IR}(Q_n) \geq 6n - O(n^{2/3})$. They showed by computer that for $n = 17576 = 26^3$, $\mathrm{IR}(Q_n) > 5n$, and concluded with the remark that it *seems* likely that $6n - O(n^{2/3})$ is also an upper bound for $\mathrm{IR}(Q_n)$.

To summarise, the best known bounds for $\mathrm{IR}(Q_n)$, for n large enough, are

$$6n - O(n^{2/3}) \leq \mathrm{IR}(Q_n) \leq \left\lfloor 6n + 6 - 8\sqrt{n + \sqrt{n} + 1} \right\rfloor.$$

We end this section by mentioning that domination and irredundance numbers for queens on hexagonal boards were studied in [18].

7.2.2 Bounds for the Kings Graph

Favaron, Fricke, Pritikin and Puech [71] obtained the following lower and upper bounds for $\mathrm{ir}(K_n)$ and $\mathrm{IR}(K_n)$:

$$\left\lceil \frac{n^2}{9} \right\rceil \leq \mathrm{ir}(K_n) \leq \left\lfloor \frac{n+2}{3} \right\rfloor^2 \tag{3}$$

and

$$\left\lceil \frac{(n-1)^2}{3} \right\rceil \leq \mathrm{IR}(K_n) \leq \left\lfloor \frac{n^2}{3} \right\rfloor, \tag{4}$$

where upper bound in (4) holds for $n \geq 6$ and all the other bounds hold for all n. For $n \equiv 0 \pmod{3}$, the bounds in (3) are equal; hence $\mathrm{ir}(K_n) = n^2/9$ in this case. The upper bound in (3) can be improved to $\mathrm{ir}(K_n) \leq \left\lfloor \frac{n+2}{3} \right\rfloor^2 - 1$ if $n \equiv 4 \pmod{9}$.

7.2.3 Bounds for Grids

As mentioned in Section 4.2, Favaron and Puech [75] showed that $\mathrm{ir}(G_n) \geq n^2/5$ and $\mathrm{ir}(G_n)$ is asymptotically equal to $n^2/5$ when n tends to infinity. Since grids are (α, IR)-perfect, all their upper parameters are equal (as given in [38]), namely,

$$\alpha(G_n) = \Gamma(G_n) = \mathrm{IR}(G_n) = \begin{cases} \frac{n^2}{2} & \text{if } n \text{ is even} \\[2mm] \frac{n^2+1}{2} & \text{if } n \text{ is odd.} \end{cases}$$

8 Irredundant Ramsey Numbers

In order to solve a problem in formal logic, Frank Plumpton Ramsey (22 February 1903–19 January 1930) proved, in passing, his now-famous theorem as a "minor lemma". This result led to an area in extremal graph theory known as **Ramsey Theory**. A full version of the lemma, now known as **Ramsey's Theorem**, can be found in [23, Theorem 20.1]; we state the well-known special case for graphs here. (In almost all cases, the edge colourings referred to below are not proper edge colourings.)

Theorem 8.1 (Ramsey's Theorem for Graphs) *For any $k \geq 2$ positive integers $n_1, \ldots, n_k$, there exists a positive integer N such that any k-edge colouring of K_N produces a monochromatic K_{n_i} for some i $(1 \leq i \leq k)$.*

For fixed integers $n_1, \ldots, n_k$, the smallest integer N such that Ramsey's Theorem holds is called the *Ramsey number $r(n_1, \ldots, n_k)$*. Consider a k-edge colouring of K_N in the colours $1, \ldots, k$, and let H_i be the spanning subgraph of K_N whose edges are coloured i $(1 \leq i \leq k)$. We restate Ramsey's Theorem in terms of the clique numbers $\omega(H_i)$.

Ramsey's Theorem in Terms of Clique Numbers *For any $k \geq 2$ positive integers $n_1, \ldots, n_k$, there exists a positive integer N such that, for any edge-decomposition of K_N into spanning subgraphs $H_1, \ldots, H_k$, $\omega(H_i) \geq n_i$ for some i $(1 \leq i \leq k)$.*

Since the clique number of a graph equals the independence number of its complement, we can rephrase Ramsey's Theorem in terms of independence numbers.

Ramsey's Theorem in Terms of Independence Numbers *For any $k \geq 2$ positive integers $n_1, \ldots, n_k$, there exists a positive integer N such that, for any edge-decomposition of K_N into spanning subgraphs $H_1, \ldots, H_k$, $\alpha(\overline{H_i}) \geq n_i$ for some i $(1 \leq i \leq k)$.*

Since any independent set is irredundant, we obtain the following result as a corollary to Ramsey's Theorem.

Corollary 8.2 (Ramsey's Theorem for Irredundance) *For any $k \geq 2$ positive integers $n_1, \ldots, n_k$, there exists a positive integer N such that, for any edge-decomposition of K_N into spanning subgraphs $H_1, \ldots, H_k$, $\mathrm{IR}(\overline{H_i}) \geq n_i$ for some i $(1 \leq i \leq k)$.*

For fixed integers $n_1, \ldots, n_k$, the smallest integer N such that Corollary 8.2 holds is called the *irredundant Ramsey number $s(n_1, \ldots, n_k)$*. With the exception of $s(3, 3, 3)$, irredundant Ramsey numbers have only been determined for cases where $k = 2$. Note that $s(m, n)$ is the smallest integer N such that any graph G of order N satisfies $\mathrm{IR}(G) \geq m$ or $\mathrm{IR}(\overline{G}) \geq n$.

By "mixing" independence, upper domination and upper irredundance numbers in the definition of $s(m, n)$, we obtain four additional types of Ramsey numbers.

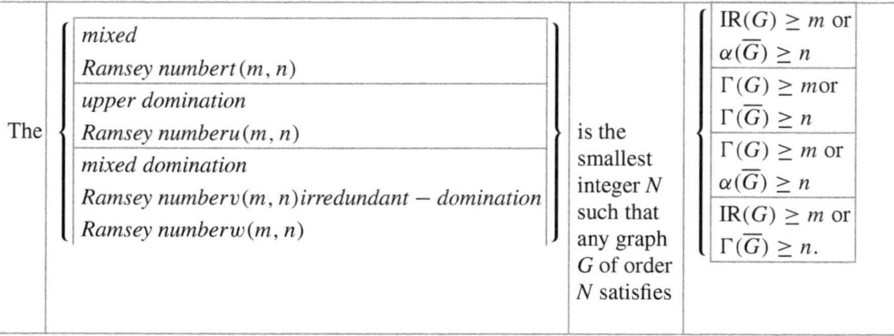

The
- *mixed Ramsey number* $t(m, n)$
- *upper domination Ramsey number* $u(m, n)$
- *mixed domination Ramsey number* $v(m, n)$
- *irredundant − domination Ramsey number* $w(m, n)$

is the smallest integer N such that any graph G of order N satisfies

- $IR(G) \geq m$ or $\alpha(\overline{G}) \geq n$
- $\Gamma(G) \geq m$ or $\Gamma(\overline{G}) \geq n$
- $\Gamma(G) \geq m$ or $\alpha(\overline{G}) \geq n$
- $IR(G) \geq m$ or $\Gamma(\overline{G}) \geq n$.

Since $\alpha(G) \leq \Gamma(G) \leq IR(G)$, the following inequalities are immediate. For all positive integers m and n,

$$s(m, n) \leq w(m, n) \leq \left\{ \begin{array}{c} t(m, n) \\ u(m, n) \leq v(m, n) \end{array} \right\} \leq r(m, n).$$

Hence all the usual upper bounds for $r(m, n)$ (see, e.g. [23, Chapter 20]) and lower bounds for $s(m, n)$ also hold for the other types of Ramsey numbers. The definitions imply that, like $r(m, n)$, $s(m, n) = s(n, m)$ and $u(m, n) = u(n, m)$ for all m and n, but this is not true for t, v and w.

E. J. Cockayne conceived the concept of irredundant Ramsey numbers and supervised R. C. Brewster's Masters thesis [12] on the topic. Mixed Ramsey numbers were introduced by Cockayne, Hattingh, Kok and Mynhardt [39]; upper domination Ramsey numbers by Oellermann and Shreve, as cited in [98]; mixed domination Ramsey numbers by Henning and Oellermann [99]; and irredundant-domination Ramsey numbers by Burger and van Vuuren [21].

8.1 Exact Values

Exactly like the classical Ramsey numbers, it is easy to see that $f(1, n) = f(n, 1) = 1$ and $f(2, n) = f(n, 2) = n$ for each n, where $f \in \{r, s, t, u, v, w\}$, hence we are only interested in $f(m, n)$ where $m, n \geq 3$. We list the known values of these Ramsey numbers in Table 2, where we only mention the provenance of the non-classical numbers, as those of $r(m, n)$ are freely available on the internet.

There is only one known Ramsey number $r(n_1, \ldots, n_k)$, $k \geq 3$, namely, $r(3, 3, 3) = 17$. There is similarly only one known irredundant Ramsey number $s(n_1, \ldots, n_k)$, $k \geq 3$, namely, $s(3, 3, 3) = 13$, determined by Cockayne and Mynhardt [46, 48]. The only other information we have on the numbers $f(3, 3, 3)$ is the bounds $13 \leq u(3, 3, 3) \leq 14$, as shown by Henning and Oellermann [98].

Table 2 Known Ramsey numbers $f(m, n), f \in \{r, s, t, u, v, w\}$

n	3	4	5	6	7	8
$s(3, n)$	6 [13]	8 [13]	12 [13]	15 [14]	18 [40][1]	21 [22]
$w(3, n)$	6 [21]	8 [21]	12 [21]	15 [21]	18 [16]	21 [22]
$w(n, 3)$		8 [21]	12 [21]	15 [21]		
$u(3, n)$	6 [99]	8 [99]	12 [99]	15 [99]		
$v(3, n)$	6 [99]	9 [99]	12 [99]	15 [99]		
$v(n, 3)$		8 [21]	13 [21]	17 [21]		
$t(3, n)$	6 [39]	8 [39]	13 [39]	15 [99]	18 [21]	22 [21]
$t(n, 3)$		9 [39]	12 [39]	17 [21]		
$r(3, n)$	6	9	14	18	23	28[2]
$s(4, n)$		13 [34]				
$w(4, n)$		13 [21]				
$u(4, n)$		13 [21]				
$v(4, n)$		14 [21]				
$t(4, n)$		14 [21]				
$r(4, n)$		18	25			

[1] Also [26], without a computer search (which was used in [40]).

[2] The Ramsey number $r(3, 9) = 36$ is also known.

Table 2 reveals obvious gaps in the literature on irredundant Ramsey numbers. With increasing computing power, some missing results, stated below, should be within reach.

Problem 8.3

(i) *Determine $w(7, 3)$, $u(3, 7)$, $v(3, 7)$, $v(7, 3)$ and $t(7, 3)$.*

(ii) *Determine or bound $w(8, 3)$, $u(3, 8)$, $v(3, 8)$, $v(8, 3)$ and $t(8, 3)$.*

(iii) *Burger and van Vuuren [22] showed that $24 \leq s(3, 9) \leq t(3, 9) \leq 27$. Improve these bounds or determine $s(3, 9)$ and $t(3, 9)$. Determine or bound $f(3, 9)$ and $f(9, 3)$ for $f \in \{t, u, v, w\}$.*

(iv) *Determine $f(3, 3, 3)$ for $f \in \{t, u, v, w\}$.*

The *CO-irredundant Ramsey number* $c(m, n)$ is the smallest integer N such that every graph G of order N satisfies $\mathrm{COIR}(G) \geq m$ or $\mathrm{COIR}(\overline{G}) \geq n$. They were defined in [122] and also studied in [43, 53]. Since $\mathrm{IR}(G) \leq \mathrm{COIR}(G)$ for all graphs, $c(m, n) \leq s(m, n)$ for all values of m and n. Hence $c(1, n) = 1$ for all n. Also, since $\mathrm{COIR}(K_2) = \mathrm{COIR}(\overline{K_2}) = 2$, $c(2, n) = 2$ for all n. The following non-trivial numbers have been determined:

$$c(3, n) = n \text{ for all } n \geq 3,$$

$$c(4, 4) = 6, \quad c(4, 5) = 8, \quad c(4, 6) = 11 \text{ and } c(4, 7) = 14.$$

As far as we know, there has been no hybridisation between CO-irredundant and the other types of Ramsey numbers mentioned above.

Problem 8.4 *Determine more CO-irredundant Ramsey numbers.*

8.2 Bounds

Analogies of bounds for $r(m, n)$ hold for the other types of Ramsey numbers as well. For any $f \in \{r, s, t, u, v, w\}$ and integers m, n, the growth property $f(m, n) \leq \min\{f(m+1, n), f(m, n+1)\}$ holds, as does the following recursive upper bound.

Proposition 8.5 *For any $f \in \{r, s, t, u, v, w\}$ and integers m, n, $f(m, n) \leq f(m-1, n) + f(m, n-1)$; this inequality is strict if $f(m-1, n)$ and $f(m, n-1)$ are both even.*

Chen, Hattingh and Rousseau [25] obtained an asymptotic lower bound for $t(m, n)$ that was soon improved to a bound for $s(m, n)$ by Erdös and Hattingh [60] and Krivelevich [105].

Theorem 8.6 ([60, 105]) *For every $m \geq 3$, there exists a positive constant c_m such that, for sufficiently large n,*

$$s(m, n) > c_m \left(\frac{n}{\log n} \right)^{\frac{(m^2 - m - 1)}{2(m-1)}}.$$

Chen et al. [25] also obtained an upper bound for $t(3, n)$, while Rousseau and Speed [120] bounded $t(3, n)$, $t(4, n)$ and $t(m, 3)$; their bound for $t(3, n)$ is an improvement of the bound in [25] for n large enough.

Theorem 8.7 *For every positive integer n,*

$$t(3, n) \leq \begin{cases} \sqrt{\frac{5}{2}} n^{\frac{3}{2}} & [25] \\[2ex] \dfrac{5 n^{\frac{3}{2}}}{\sqrt{\log n}} & [120] \end{cases}$$

$$t(4, n) \leq C \frac{5^{\frac{n}{2}}}{\sqrt{\log n}} \quad [120].$$

Theorem 8.8 ([120]) *For some positive constant c,*

$$c \left(\frac{m}{\log m} \right)^2 < t(m, 3) < (1 + o(1)) \frac{m^2}{\log m}.$$

Since there exist graphs G such that $\alpha(G) = 2$ and $IR(G) = k$ for arbitrary $k \geq 2$, the following interesting result is, perhaps, not entirely a surprise.

Theorem 8.9 ([120]) $\lim_{n \to \infty} \dfrac{t(3, n)}{r(3, n)} = 0.$

Problem 8.10 ([120]) *Is it true that, for every $m \geq 3$,* $\lim_{n \to \infty} \dfrac{t(m, n)}{r(m, n)} = 0$?

We can ask the same question for other types of Ramsey numbers. Let

$$(f, g) \in \{(c, s), (s, t), (s, w), (w, t), (w, u), (u, v), (w, r), (v, r)\}.$$

Problem 8.11 *Is it true that, for every $m \geq 3$,* $\lim_{n \to \infty} \dfrac{f(m, n)}{g(m, n)} = 0$?

9 Reconfiguration

Reconfiguration problems are concerned with determining conditions under which a feasible solution to a given problem can be transformed into another such solution via a sequence of feasible solutions in such a way that any two consecutive solutions are adjacent according to a specified adjacency relation. Reconfiguration problems model, for example, situations where we wish to implement a sequence of predefined elementary changes in order to transform a given configuration to a more desirable one while the intermediate steps are also feasible.

The reconfiguration of dominating sets was introduced by Subramanian and Sridharan [123]. They defined the γ-*graph* $\gamma \cdot G$ of the graph G to be the graph whose vertex set consists of all the $\gamma(G)$-sets, where two sets D and D' are adjacent if and only if $|D \cap D'| = |D| - 1$; that is, there exist vertices $v \in D$, $v' \in D'$ such that $D' = (D - \{v\}) \cup \{v'\}$. Here, v and v' need not be adjacent in G. This version of the γ-graph is also known as the "*single vertex replacement adjacency model*" or the *jump γ-graph*. Fricke, Hedetniemi, Hedetniemi and Hutson [82] studied the "*slide adjacency model*" or simply the *slide γ-graph $G(\gamma)$*, whose vertex set also consists of all the $\gamma(G)$-sets, but two sets D and D' are adjacent in $G(\gamma)$ if and only if there exist **adjacent** vertices $v \in D$, $v' \in D'$ such that $D' = (D - \{v\}) \cup \{v'\}$.

An initial question of Fricke et al. was to determine exactly which graphs are γ-graphs; they showed that every tree is the slide γ-graph of some graph and conjectured that every graph is such a graph. Connelly, Hedetniemi and Hutson [54] proved this conjecture. It is easy to see that if H is realisable as a γ-graph, then it is the γ-graph of infinitely many graphs.

Irredundance graphs were first considered by Mynhardt and Teshima [111]. For any parameter $\pi \in \{ir, i, \alpha, \Gamma, IR\}$ (and some other domination parameters), they defined the *slide π-graph $G(\pi)$* of G similar to the slide γ-graph $G(\gamma)$ and showed that every graph is the (slide) π-graph of some graph, where $\pi \in \{ir, \Gamma\}$, while not every graph is an i-graph. The graph G_H constructed to show that a given graph H is the Γ-graph of G_H satisfies $\Gamma(G_H) = \mathrm{IR}(G_H)$, but has more IR-sets than Γ-sets; hence H is not an IR-graph of G_H. They left the problem of whether all graphs are IR-graphs open. However, Mynhardt and Roux [110] showed that, although all disconnected graphs can be realised as IR-graphs, this does not hold for connected graphs.

Theorem 9.1 ([110]) *Every disconnected graph is the* IR-*graph of infinitely many graphs.*

To find connected graphs that are not IR-graphs, Mynhardt and Roux used the external private neighbours in a given IR-set to find more IR-sets. The result is of interest in its own right, and we include the short proof. For an irredundant set X, we weakly partition X into subsets Z and I (one of which may be empty), where each vertex in I is isolated in $G[X]$ and each vertex in Z has at least one external private neighbour. (This partition is not necessarily unique. Isolated vertices of $G[X]$ with external private neighbours can be allocated arbitrarily to Z or I.) For each $z \in Z$, let $z' \in \mathrm{epn}(z, X)$ and define $Z' = \{z' : z \in Z\}$. Let $X' = (X - Z) \cup \{Z'\}$; note that $|X| = |X'|$. The set X' is called a *flip-set* of X.

Proposition 9.2 *If X is an* IR-*set of G, then so is any flip-set X' of X.*

Proof. Consider any $x \in X'$. With notation as above, if $x \in I = X - Z = X' - Z'$, then x is isolated in $G[X]$. Since each vertex in Z' is an X-external private neighbour of some $z \in Z$, no vertex in Z' is adjacent to x. Therefore x is isolated in $G[X']$. Hence assume $x \in Z'$. Then $x = z'$ for some $z \in Z$, so z' is adjacent to $z \in V(G) - X'$. Now z is non-adjacent to all vertices in I because the latter vertices are isolated in $G[X]$, and z is nonadjacent to all vertices in $Z' - \{z'\}$, because each $v' \in Z' - \{z'\}$ is an X-external private neighbour of some $v \in Z - \{z\}$. Therefore $z \in \mathrm{epn}(z', X')$, that is, $z \in \mathrm{epn}(x, X')$. It follows that X' is irredundant. Since $|X'| = |X|$, X' is an IR-set of G. $\blacksquare$

Proposition 9.2 explains to some extent why a given connected graph H is not an IR-graph: for a possible source graph G, an IR-set X and its flip-set X' often belong to different components of $G(\mathrm{IR})$ because G lacks the necessary IR-sets to form an $X - X'$ path in $G(\mathrm{IR})$.

An IR-*tree* is a tree that is an IR-graph. All complete graphs are IR-graphs; hence K_1 and K_2 are IR-trees. To formulate results on IR-trees, we define some classes of trees. The *(generalised) spider* $\mathrm{Sp}(\ell_1, \ldots, \ell_k)$, $\ell_i \geq 1$, $k \geq 2$, is a tree obtained from the star $K_{1,k}$ with centre u by subdividing the edge uv_i $\ell_i - 1$ times, $i = 1, \ldots, k$. The *double star* $S(k, n)$ is the tree obtained by joining the centres of the stars $K_{1,k}$ and $K_{1,n}$. The *double spider* $\mathrm{Sp}(\ell_1, \ldots, \ell_k; m_1, \ldots, m_n)$ is obtained from $S(k, n)$ by subdividing the edges of the $K_{1,k}$-subgraph $\ell_i - 1$ times, $i = 1, \ldots, k$, and the edges of the $K_{1,n}$-subgraph $m_i - 1$ times, $i = 1, \ldots, n$.

Theorem 9.3 ([110])

(i) *Stars $K_{1,k}$, $k \geq 2$ (trees of diameter 2), are not* IR-*trees.*

(ii) *The double star $S(2, 2)$ is the unique smallest IR-tree with diameter three (and the unique smallest non-complete IR-tree).*

(iii) *The double spider $\mathrm{Sp}(1, 1;1, 2)$ is the unique smallest IR-tree with diameter four.*

(iv) *The cycles C_5, C_6, C_7 and the paths P_3, P_4, P_5 are not IR-graphs.*

(v) *The only connected IR-graphs of order four are K_4 and C_4.*

As mentioned in [110], a direct proof that P_5 is not an IR-graph is somewhat simpler than the proof of Theorem 9.3(*iii*), but not simple enough to easily generalise to a proof that P_n or C_n, $n \geq 5$, is not an IR-graph. The authors thus stated the following conjecture and open problems.

Conjecture 9.4 ([110]) *Paths P_n, $n \geq 3$, and cycles C_n, $n \geq 5$, are not IR-graphs.*

Problem 9.5 ([110])

(i) *Determine which spiders, double spiders and double stars are IR-trees.*

(ii) *Prove or disprove: Complete graphs and $K_m \square K_n$, where $m, n \geq 2$, are the only connected claw-free IR-graphs.*

10 Complexity

We conclude by briefly addressing complexity and algorithmic questions pertaining to irredundance. The decision problems that correspond to determining the lower and upper irredundance numbers of general graphs are NP-complete [76, 95]. Computing $\mathrm{ir}(G)$ remains NP-complete on the classes of bipartite graphs [95], split (hence also chordal) graphs [106], partial k-trees [132], planar cubic graphs [5] and irredundance perfect graphs [128]. It was however shown in [134] that graphs belonging to a subclass of irredundant perfect graphs, the class of locally well-dominated graphs, are polynomial-time solvable. A linear-time algorithm also exists for determining the lower irredundance number of a tree [7].

The upper irredundance number coincides with the independence number for bipartite and chordal graphs and can therefore be solved in polynomial time for these graph classes [38, 102]. For planar cubic graphs [5] and k-regular graphs with $k \geq 6$ [83], the problem of computing $\mathrm{IR}(G)$ remains NP-complete.

In [8] Binkele et al. considered both exact and parameterised algorithms and showed that $\mathrm{ir}(G)$ can be computed in $\mathcal{O}^*(1.99914^n)$ time and polynomial space. Furthermore, an algorithm for determining $\mathrm{IR}(G)$ runs in $\mathcal{O}^*(1.9369^n)$ time and polynomial space, where the running time can be improved to $\mathcal{O}^*(1.8475^n)$ for exponential space.

References

1. R.B. Allan, R.C. Laskar, On domination and some related concepts in graph theory, in *Proceedings of the Ninth Southeastern Conference on Combinatorics, Graph Theory, and Computing (Florida Atlantic Univ., Boca Raton, Fla., 1978).* Congr. Numer., vol. XXI (Utilitas Math., Winnipeg, 1978), pp. 43–56
2. R.B. Allan, R.C. Laskar, S.T. Hedetniemi, A note on total domination. Discrete Math. **49**(1), 7–13 (1984)
3. M. Aouchiche, O. Favaron, P. Hansen, Variable neighborhood search for extremal graphs. XXII. Extending bounds for independence to upper irredundance. Discrete Appl. Math. **157**, 3497–3510 (2009)
4. G. Bacsó, O. Favaron, Independence, irredundance, degrees and chromatic number in graphs. Discrete Math. **259**, 257–262 (2002)
5. C. Bazgan, L. Brankovic, K. Casel, H. Fernau, On the complexity landscape of the domination chain, in *Algorithms and Discrete Applied Mathematics.* Lecture Notes in Computer Science, vol. 9602 (Springer, Cham, 2016), pp. 61–72
6. C. Berge and P. Duchet, Strongly perfect graphs, in *Annals of Discrete Mathematics*, vol. 21 (North-Holland, Amsterdam, 1984), pp. 57–61
7. M. Bern, E.L. Lawler, A.L. Wong, Linear-time computation of optimal subgraphs of decomposable graphs. J. Algorithms **8**, 216–235 (1987)
8. D. Binkele-Raible, L. Brankovic, M. Cygan, H. Fernau, J. Kneis, D. Kratsch, A. Langer, M. Liedloff, M. Pilipczuk, P. Rossmanith, J.O. Wojtaszczyk, Breaking the 2^n-barrier for irredundance: two lines of attack. J. Discrete Algorithms **9**, 214–230 (2011)
9. M. Blidia, M. Chellali, F. Maffray, Extremal graphs for a new upper bound on domination parameters in graphs. Discrete Math. **306**, 2314–2326 (2006)
10. B. Bollobás, E.J. Cockayne, Graph-theoretic parameters concerning domination, independence, and irredundance. J. Graph Theory **3**, 241–249 (1979)
11. B. Bollobás, E.J. Cockayne, The irredundance number and maximum degree of a graph. Discrete Math. **49**(2), 197–199 (1984)
12. R.C. Brewster, Irredundant Ramsey numbers, Master's Thesis, University of Victoria (1988)
13. R.C. Brewster, E.J. Cockayne, C.M. Mynhardt, Irredundant Ramsey numbers for graphs. J. Graph Theory **13**, 283–290 (1989)
14. R.C. Brewster, E.J. Cockayne, C.M. Mynhardt, The irredundant Ramsey number $s(3, 6)$. Quaest. Math. **13**, 141–157 (1990)
15. A.P. Burger, E.J. Cockayne, C.M. Mynhardt, Domination and irredundance in the queens' graph. Discrete Math. **163**, 47–66 (1997)
16. A.P. Burger, J.H. Hattingh, J.H. van Vuuren, The mixed irredundant Ramsey numbers $t(3, 7) = 18$ and $t(3, 8) = 22$. Quaest. Math. **37**, 571–589 (2014)
17. A.P. Burger, M.A. Henning, J.H. van Vuuren, On the ratios between packing and domination parameters of a graph. Discrete Math. **309**, 2473–2478 (2009)
18. A.P. Burger, C.M. Mynhardt, Queens on hexagonal boards. Papers in honour of Stephen T. Hedetniemi. J. Combin. Math. Combin. Comput. **31**, 97–111 (1999)
19. A.P. Burger, C.M. Mynhardt, Small irredundance numbers for queens graphs. Papers in honour of Ernest J. Cockayne. J. Combin. Math. Combin. Comput. **33**, 33–43 (2000)
20. A.P. Burger, C.M. Mynhardt, An improved upper bound for queens domination numbers. Discrete Math. **266**, 119–131 (2003)
21. A.P. Burger, J.H. van Vuuren, Avoidance colourings for small nonclassical Ramsey numbers. Discrete Math. Theor. Comput. Sci. **13**, 81–96 (2011)
22. A.P. Burger, J.H. van Vuuren, The irredundance-related Ramsey numbers $s(3, 8) = 21$ and $w(3, 8) = 21$. Util. Math. (to appear)
23. G. Chartrand, L. Lesniak, P. Zhang, *Graphs & Digraphs*, 6th edn. (Chapman and Hall/CRC, Boca Raton, 2016)

24. M. Chellali, L. Volkmann, Lutz Relations between the lower domination parameters and the chromatic number of a graph. Discrete Math. **274**, 1–8 (2004)
25. G. Chen, J.H. Hattingh, C.C. Rousseau, Asymptotic bounds for irredundant and mixed Ramsey numbers. J. Graph Theory **17**, 193–206 (1993)
26. G. Chen, C.C. Rousseau, The irredundant Ramsey number $s(3, 7)$. J. Graph Theory **19**, 263–270 (1995)
27. G.A. Cheston, E.O. Hare, S.T. Hedetniemi, R.C. Laskar, Simplicial graphs. Congr. Numer. **67**, 105–113 (1988)
28. V. Chvátal, Perfectly ordered graphs, in *Annals of Discrete Mathematics*, vol. 21 (North-Holland, Amsterdam, 1984), pp. 63–65
29. E.J. Cockayne, Chessboard domination problems. Discrete Math. **86**, 13–20 (1990)
30. E.J. Cockayne, Generalized irredundance in graphs: hereditary properties and Ramsey numbers. J. Combin. Math. Combin. Comput. **31**, 15–31 (1999)
31. E.J. Cockayne, Irredundance in the queens' graph. Ars Combin. **55**, 227–232 (2000)
32. E.J. Cockayne, Nordhaus-Gaddum results for open irredundance. J. Combin. Math. Combin. Comput. **47**, 213–223 (2003)
33. E.J. Cockayne, Irredundance, secure domination and maximum degree in trees. Discrete Math. **307**, 12–17 (2007)
34. E.J. Cockayne, G. Exoo, J.H. Hattingh, C.M. Mynhardt, The irredundant Ramsey number $s(4, 4)$. Util. Math. **41**, 119–128 (1992)
35. E.J. Cockayne, O. Favaron, S. Finbow, C.M. Mynhardt, Open irredundance and maximum degree in graphs. Discrete Math. **308**, 5358–5375 (2008)
36. E.J. Cockayne, O. Favaron, W. Goddard, P.J.P. Grobler, C.M. Mynhardt, Changing upper irredundance by edge addition. Discrete Math. **266**, 185–193 (2003)
37. E.J. Cockayne, O. Favaron, C.M. Mynhardt, Irredundance-edge-removal-critical graphs. Util. Math. **60**, 219–228 (2001)
38. E.J. Cockayne, O Favaron, C. Payan, A.G. Thomason, Contributions to the theory of domination, independence and irredundance in graphs. Discrete Math. **33**, 249–258 (1981)
39. E.J. Cockayne, J.H. Hattingh, J. Kok, C.M. Mynhardt, Mixed Ramsey numbers and irredundant Turán numbers for graphs. Ars Combin. **29C**, 57–68 (1990)
40. E.J. Cockayne, J.H. Hattingh, C.M. Mynhardt, The irredundant Ramsey number $s(3, 7)$. Util. Math. **39**, 145–160 (1991)
41. E.J. Cockayne, P.J.P. Grobler, S.T. Hedetniemi, A.A. McRae, What makes an irredundant set maximal? J. Combin. Math. Combin. Comput. **25**, 213–223 (1997)
42. E.J. Cockayne, S.T. Hedetniemi, D.J. Miller, Properties of hereditary hypergraphs and middle graphs. Canad. Math. Bull. **21**, 461–468 (1978)
43. E. J. Cockayne, G. MacGillivray, J. Simmons, CO-irredundant Ramsey numbers for graphs. J. Graph Theory **34**, 258–268 (2000)
44. E.J. Cockayne, D. McCrea, C.M. Mynhardt, Nordhaus-Gaddum results for CO-irredundance in graphs. Discrete Math. **211**, 209–215 (2000)
45. E J Cockayne, C.M. Mynhardt, On the product of upper irredundance numbers of a graph and its complement. Discrete Math. **76**, 117–121 (1989)
46. E.J. Cockayne, C.M. Mynhardt, On the irredundant Ramsey number $s(3, 3, 3)$. Ars Combin. **29C**, 189–202 (1990)
47. E.J. Cockayne, C.M. Mynhardt, The sequence of upper and lower domination, independence and irredundance numbers of a graph. Discrete Math. **122**, 89–102 (1993)
48. E.J. Cockayne, C.M. Mynhardt, The irredundant Ramsey number $s(3, 3, 3) = 13$. J. Graph Theory **18**, 595–604 (1994)
49. E.J. Cockayne, C.M. Mynhardt, Triangle-free graphs with unequal upper domination and irredundance numbers. Bull. Inst. Combin. Appl. **21**, 100–103 (1997)
50. E.J. Cockayne, C.M. Mynhardt, Irredundance and maximum degree in graphs. Combin. Probab. Comput. **6**, 153–157 (1997)
51. E.J. Cockayne, C.M. Mynhardt, Domination and irredundance in cubic graphs. 15th British Combinatorial Conference (Stirling, 1995). Discrete Math. **167/168**, 205–214 (1997)

52. E.J. Cockayne, C.M. Mynhardt, Properties of queens graphs and the irredundance number of Q_7. Aust. J. Combin. **23**, 285–299 (2001)
53. E.J. Cockayne, C.M. Mynhardt, J. Simmons, The CO-irredundant Ramsey number $t(4, 7)$. Util. Math. **57**, 193–209 (2000)
54. E. Connelly, S.T. Hedetniemi, K.R. Hutson, A note on γ-Graphs. AKCE Int. J. Graphs Comb. **8**, 23–31 (2010)
55. P. Damaschke, Irredundance number versus domination number. Discrete Math. **89**, 101–104 (1991)
56. L. Dohmen, D. Rautenbach, L. Volkmann, A characterization of $\Gamma\alpha(k)$-perfect graphs. Discrete Math. **224**, 265–271 (2000)
57. G.S. Domke, J.E. Dunbar, L.R. Markus, Gallai-type theorems and domination parameters. Discrete Math. **167/168**, 237–248 (1997)
58. J.E. Dunbar, T.W. Haynes, Domination in inflated graphs. Congr. Numer. **118**, 143–154 (1996)
59. J.E. Dunbar, T.R. Monroe, C.A. Whitehead, Sensitivity of the upper irredundance number to edge addition. J. Combin. Math. Combin. Comput. **33**, 65–79 (2000)
60. P. Erdös, J. H. Hattingh, Asymptotic bounds for irredundant Ramsey numbers. Quaest. Math. **16**, 319–331 (1993)
61. A.M. Farley, A. Proskurowski, Computing the maximum order of an open irredundant set in a tree. Congr. Numer. **41**, 219–228 (1984)
62. A.M. Farley, N. Shacham, Senders in broadcast networks: open irredundancy in graphs. Congr. Numer. **38**, 47–57 (1983)
63. R. Faudree, O. Favaron, H. Li, Independence, domination, irredundance, and forbidden pairs. J. Combin. Math. Combin. Comput. **26**, 193–212 (1998)
64. O. Favaron, Stability, domination and irredundance in a graph. J. Graph Theory **10**(4), 429–438 (1986)
65. O. Favaron, Two relations between the parameters of independence and irredundance. Discrete Math. **70**(1), 17–20 (1988)
66. O. Favaron, A note on the open irredundance in a graph. Congr. Numer. **66**, 316–318 (1988)
67. O. Favaron, A note on the irredundance number after vertex deletion. Discrete Math. **121**, 51–54 (1993)
68. O. Favaron, Irredundance in inflated graphs. J. Graph Theory **28**, 97–104 (1998)
69. O. Favaron, Inflated graphs with equal independence number and upper irredundance number. Discrete Math. **236**, 81–94 (2001)
70. O. Favaron, Independence and upper irredundance in claw-free graphs. Discrete Appl. Math. **132**(1–3), 85–95 (2003)
71. O. Favaron, G.H. Fricke, D. Pritikin, J. Puech, Irredundance and domination in kings graphs. Discrete Math. **262**, 131–147 (2003)
72. O. Favaron, M.A. Henning, J. Puech, D. Rautenbach, On domination and annihilation in graphs with claw-free blocks. Discrete Math. **231**, 143–151 (2001)
73. O. Favaron, V. Kabanov, J. Puech, The ratio of three domination parameters in some classes of claw-free graphs. J. Combin. Math. Combin. Comput. **31**, 151–159 (1999)
74. O. Favaron, C.M. Mynhardt, On equality in an upper bound for domination parameters of graphs. J. Graph Theory **24**, 221–231 (1997)
75. O. Favaron, J. Puech, Irredundance in grids. Discrete Math. **179**, 257–265 (1998)
76. M. Fellows, G. Fricke, S.T. Hedetniemi, D.P. Jacobs, The private neighbor cube. SIAM J. Discrete Math. **7**, 41–47 (1994)
77. S. Finbow, Generalisations of irredundance in graphs. Doctoral Dissertation, University of Victoria, Canada (2003), p. 158, http://hdl.handle.net/1828/7913
78. S. Finbow, A lower bound for the CO-irredundance number of a graph. Discrete Math. **295**, 49–62 (2005)
79. S. Finbow, C.M. van Bommel, Triangulations and equality in the domination chain. Discrete Appl. Math. **194**, 81–92 (2015)

80. J.F. Fink, M.S. Jacobson, n-Domination in graphs, in *Graph Theory with Applications to Algorithms and Computer Science*, ed. by Y. Alavi et al. (Wiley, New York, 1985), pp. 283–300

81. J.F. Fink, M.S. Jacobson, L.F. Kinch, J. Roberts, On graphs having domination number half their order. Period. Math. Hung. **16**, 287–293 (1985)

82. G.H. Fricke, S.M. Hedetniemi, S.T. Hedetniemi, K.R. Hutson, γ-Graphs of graphs. Discuss. Math. Graph Theory **31**, 517–531 (2011)

83. G.H. Fricke, S.T Hedetniemi, D.P. Jacobs, Independence and irredundance in k-regular graphs. Ars Combin. **49**, 271–279 (1998)

84. M.C. Golumbic, R.C. Laskar, Irredundancy in circular arc graphs. Discrete Appl. Math. **44**, 79–89 (1993)

85. P.J.P. Grobler, Critical concepts in domination, independence and irredundance of graphs. Doctoral Dissertation, University of South Africa (South Africa), 1999

86. P.J.P. Grobler, C.M. Mynhardt, Vertex criticality for upper domination and irredundance. J. Graph Theory **37**, 205–212 (2001)

87. P.J.P. Grobler, C.M. Mynhardt, Domination parameters and edge-removal-critical graphs. Discrete Math. **231**, 221–239 (2001)

88. P.J.P. Grobler, C.M. Mynhardt, Upper domination parameters and edge-critical graphs. J. Combin. Math. Combin. Comput. **33**, 239–251 (2000)

89. P.J.P. Grobler, A. Roux, Coalescence and criticality of graphs. Discrete Math. **313**, 1087–1097 (2013)

90. G. Gutin, V.E. Zverovich, Upper domination and upper irredundance perfect graphs. Discrete Math. **190**, 95–105 (1998)

91. P.L. Hammer, F. Maffray, Preperfect graphs. Combinatorica **13**, 199–208 (1993)

92. T.W. Haynes, S.T. Hedetniemi, P.J. Slater, *Fundamentals of Domination in Graphs* (Marcel Dekker, New York, 1998)

93. S.T. Hedetniemi, S.M. Hedetniemi, R. Reynolds, Combinatorial problems on chessboards: II, in *Domination in Graphs: Advanced Topics*, ed. by T.W. Haynes, S.T. Hedetniemi, P.J. Slater (Marcel Dekker, New York, 1998), pp. 133–162

94. S.T. Hedetniemi, D.P. Jacobs, R.C. Laskar, Inequalities involving the rank of a graph. J. Combin. Math. Combin. Comput. **6**, 173–176 (1989)

95. S.T. Hedetniemi, R.C. Laskar, J. Pfaff, Irredundance in graphs: a survey. Congr. Numer. **48**, 183–193 (1985)

96. M.A. Henning, Irredundance perfect graphs. Discrete Math. **142**, 107–120 (1995)

97. M.A. Henning, Domination functions in graphs, in *Domination in Graphs: Advanced Topics*, ed. by T.W. Haynes, S.T. Hedetniemi, P.J. Slater (Marcel Dekker, New York, 1998), pp. 31–57

98. M.A. Henning, O.R. Oellermann, The upper domination Ramsey number $u(3, 3, 3)$. Discrete Math. **242**, 103–113 (2002)

99. M.A. Henning, O.R. Oellermann, On upper domination Ramsey numbers for graphs. Discrete Math. **274**, 125–135 (2004)

100. M.H. Henning, P.J. Slater, Inequalities relating domination parameters in cubic graphs. Discrete Math. **158**, 87–98 (1996)

101. M.S. Jacobson, K. Peters, Chordal graphs and upper irredundance, upper domination and independence. Discrete Math. **86**, 59–69 (1990)

102. M.S. Jacobson, K. Peters, A note on graphs which have upper irredundance equal to independence. Discrete Appl. Math. **44**, 91–97 (1993)

103. M.D. Kearse, P.B. Gibbons, Computational methods and new results for chessboard problems. Aust. J. Combin. **23**, 253–284 (2001)

104. M.D. Kearse, P.B. Gibbons, A new lower bound on upper irredundance in the queens' graph. Discrete Math. **256**, 225–242 (2002)

105. M. Krivelevich, A lower bound for irredundant Ramsey numbers. Discrete Math. **183**, 185–192 (1998)

106. R.C. Laskar, J. Pfaff, Domination and irredundance in split graphs, Technical Report No. 430, Dept. Math. Sci., Clemson Univ., Clemson, SC (1983)
107. J. Lehel, Peripheral graphs. Congr. Numer. **59**, 179–184 (1987)
108. H. Li, C. Virlouvet, Neighborhood conditions for claw-free Hamiltonian graphs. Ars Combin. **29A**, 109–116 (1990)
109. H. Liu, L. Sun, A note on the difference between the upper irredundance and independence numbers of a graph. Ars Combin. **72**, 199–202 (2004)
110. C.M. Mynhardt, A. Roux, Irredundance Graphs (November 2018, submitted)
111. C.M. Mynhardt, L.E. Teshima, A note on some variations of the γ-graph. J. Combin. Math. Combin. Comput. **104**, 217–230 (2018)
112. P.R.J. Östergrad, W.D. Weakley, Values of domination numbers of the Queen's graph. Electron. J. Combin. **8**, Research Paper 29, 19 (2001)
113. C. Payan, N.H. Xuong, Domination-balanced graphs. J. Graph Theory **6**, 23–32 (1982)
114. M. Poschen, L. Volkmann, A lower bound for the irredundance number of trees. Discuss. Math. Graph Theory **26**, 209–215 (2006)
115. J. Puech, Irredundance perfect and P_6-free graphs. J. Graph Theory **29**, 239–255 (1998)
116. J. Puech, The lower irredundance and domination parameters are equal for inflated trees. J. Combin. Math. Combin. Comput. **33**, 117–127 (2000)
117. J. Puech, Forbidden graphs and irredundant perfect graphs. J. Combin. Math. Combin. Comput. **36**, 215–228 (2001)
118. D. Rautenbach, On the differences between the upper irredundance, upper domination and independence numbers of a graph. Discrete Math. **203**, 239–252 (1999)
119. G. Ravindra, Meyniel's graphs are strongly perfect, in *Annals of Discrete Mathematics*, vol. 21 (North-Holland, Amsterdam, 1984), pp. 145–148
120. C.C. Rousseau, S.E. Speed, Mixed Ramsey numbers revisited. Combin. Probab. Comput. **12**, 653–660 (2003)
121. A. Roux, Vertex-criticality of the domination parameters of graphs. Master's Thesis, Stellenbosch University (2011)
122. J. Simmons, CO-irredundant Ramsey Numbers for Graphs. Masters Thesis, University of Victoria (1998)
123. K. Subramanian, N.S. Sridharan, γ-Graph of a graph. Bull. Kerala Math. Assoc. **5**, 17–34 (2008)
124. J. Topp, Domination, independence and irredundance in graphs. Diss. Math. **342**, 1–99 (1995)
125. J. Topp, P.D. Vestergaard, Well irredundant graphs. Discrete Appl. Math. **63**, 267–276 (1995)
126. L. Volkmann, The ratio of the irredundance and domination number of a graph. Discrete Math. **178**, 221–228 (1998)
127. L. Volkmann, V.E. Zverovich, Proof of a conjecture on irredundance perfect graphs. J. Graph Theory **41**, 292–306 (2002)
128. L. Volkmann, V.E. Zverovich, A disproof of Henning's conjecture on irredundance perfect graphs. Discrete Math. **254**, 539–554 (2002)
129. C.D. Wang, On the sum of two parameters concerning independence and irredundance in a graph. J. Math. (Wuhan) **5**, 201–204 (1985), and Discrete Math. **69**, 199–202 (1988)
130. C.X. Wang, J.Z. Mao, The difference between independence number and upper irredundance number for graphs with $\Delta \leq 3$. (Chinese). Acta Math. Sci. Ser. A (Chin. Ed.) **24**, 714–716 (2004)
131. D. Wang, H. Hua, On stability number of upper irredundance number in graphs. Int. J. Nonlinear Sci. **2**, 39–42 (2006)
132. T.V. Wimer, Linear algorithms on k-terminal graphs. Ph.D. Thesis, Dept. of Computer Science, Clemson University, Clemson, SC, August 1987
133. I.E. Zverovich, V.E. Zverovich, A semi-induced subgraph characterization of upper domination perfect graphs. J. Graph Theory **31**, 29–49 (1999)
134. I.E. Zverovich, V.E. Zverovich, Locally well-dominated and locally independent well-dominated graphs. Graphs Combin. **19**, 279–288 (2003)

135. I.E. Zverovich, V.E. Zverovich, The domination parameters of cubic graphs. Graphs Combin. **21**, 277–288 (2005)
136. V.E. Zverovich, The ratio of the irredundance number and the domination number for block-cactus graphs. J. Graph Theory **29**, 139–149 (1998)
137. V.E. Zverovich, On the differences of the independence, domination and irredundance parameters of a graph. Aust. J. Combin. **27**, 175–185 (2003)

The Private Neighbor Concept

Stephen T. Hedetniemi, Alice A. McRae, and Raghuveer Mohan

1 Introduction

Let $G = (V, E)$ be a graph of order $n = |V|$ and size $m = |E|$, and consider the family of all sets $S \subseteq V$ of vertices having some desired property $\mathcal{P}$. In graph theory, there are, of course, many types of sets that are studied, for example, as a function of the types of subgraphs $G[S]$ induced by S, the degrees of the vertices in S, and the relationships between the vertices in S and the vertices in $V \setminus S = \overline{S}$. But in particular, we would like to consider properties of sets which are called superhereditary. A property $\mathcal{P}$ is called *superhereditary* if whenever a set S has property $\mathcal{P}$, so does every superset of S.

In a similar way, one defines a property $\mathcal{P}$ to be *hereditary* if whenever a set S has property $\mathcal{P}$, so does every proper subset of S. It is easy to see, for example, that the property of being an independent set is hereditary.

Notice that if a set $S \subseteq V$ has a superhereditary property $\mathcal{P}$, then the entire vertex set V has property $\mathcal{P}$. Thus, the largest cardinality of a $\mathcal{P}$-set is $n = |V|$. What is interesting, in this case, is (i) the minimum cardinality of a $\mathcal{P}$-set, (ii) the maximum cardinality of a minimal $\mathcal{P}$-set, and (iii) the nature of minimal $\mathcal{P}$-sets. A $\mathcal{P}$-set S is called *minimal* if no proper subset $S' \subset S$ of S is a $\mathcal{P}$-set. A $\mathcal{P}$-set S is called 1-*minimal* if for every vertex $v \in S$, the set $S - \{v\}$ is not a $\mathcal{P}$-set. It is important to note that for superhereditary properties $\mathcal{P}$, a set S is a minimal $\mathcal{P}$-set if and only if S is a 1-minimal $\mathcal{P}$-set.

S. T. Hedetniemi
School of Computing, Clemson University, Clemson, SC, USA
e-mail: hedet@clemson.edu

A. A. McRae (✉) · R. Mohan
Computer Science Department, Appalachian State University, Boone, NC 28608, USA
e-mail: mcraeaa@appstate.edu; mohanr@appstate.edu

© The Author(s), under exclusive license to Springer Nature Switzerland AG 2021
T. W. Haynes et al. (eds.), *Structures of Domination in Graphs*, Developments in Mathematics 66, https://doi.org/10.1007/978-3-030-58892-2_7

In a 1-minimal $\mathcal{P}$-set, every vertex contributes in some way to the set S having property $\mathcal{P}$, and this contribution is absolutely necessary, else the set $S - \{v\}$ does not have property $\mathcal{P}$. It is the nature of this *private* contribution that leads to the general concept of a *private neighbor*.

2 Private Neighbors

To introduce this general concept, we present four examples of superhereditary properties $\mathcal{P}$ and for each, discuss the added property of being a minimal $\mathcal{P}$-set.

1. A *dominating set* in a graph $G = (V, E)$ is a set $S \subseteq V$ of vertices having the property that every vertex $v \in \overline{S}$ is adjacent to at least one vertex in S. This can also be stated as follows:

$$dominating\ set : (\forall v \in \overline{S})(\exists u \in S)[u \in (N(v) \cap S)].$$

In addition, a dominating set S is a *minimal* dominating set if every vertex $v \in S$ dominates at least one vertex, either itself or a vertex in $\overline{S}$, that no other vertex in S dominates. This can also be stated as follows:

$$minimal\ dominating\ set : (\forall u \in S)(\exists v \in V)[N[v] \cap S = \{u\}].$$

The vertex v in the minimal dominating set definition above is called a *private neighbor* of vertex u.

The *domination number* $\gamma(G)$ equals the minimum cardinality of a dominating set in G, while the *upper domination number* $\Gamma(G)$ equals the maximum cardinality of a minimal dominating set in G.

2. A *total dominating set* in a graph $G = (V, E)$ is a set $S \subseteq V$ of vertices having the property that every vertex $v \in V$ is dominated by a vertex in S, other than itself. This can also be stated as follows:

$$total\ dominating\ set : (\forall v \in V)(\exists u \in S - \{v\})[u \in (N(v) \cap S)].$$

In addition, a total dominating set S is a *minimal* total dominating set if every vertex $u \in S$ dominates at least one vertex in V, other than itself that no other vertex in S dominates and vertex u has at least one neighbor in S. This can also be stated as follows:

$$minimal\ total\ dominating\ set : (\forall u \in S)(\exists v \in V)[N(u) \cap S \neq \emptyset \ \wedge \ [N(v) \cap S = \{u\}].$$

The vertex v in the minimal total dominating set definition above is called a *private neighbor* of vertex u.

3. A *vertex cover* in a graph $G = (V, E)$ is a set $S \subseteq V$ of vertices having the property that for every edge $uv \in E$, either u or v (or both u and v) is a vertex in S. In

this case, we say that both vertices u and v *cover* edge uv. This can also be stated as follows:

$$\textit{vertex cover} : (\forall uv \in E)[\{u, v\} \cap S \neq \emptyset].$$

In addition, a vertex cover S is a *minimal* vertex cover if every vertex $v \in S$ covers at least one edge that is not covered by any other vertex in S. This can also be stated as follows:

$$\textit{minimal vertex cover} : (\forall u \in S)(\exists uv \in E)[\{u, v\} \cap S = \{u\}].$$

The edge uv v in the minimal vertex cover definition above is called a *private edge* of vertex u.

4. A *resolving set* in a graph $G = (V, E)$ is a set $S \subseteq V$ of vertices having the property that for every pair of vertices $u, v \in \overline{S}$, there is a vertex $w \in S$ such that $d(u, w) \neq d(v, w)$. In this case, we say that vertex w *resolves* the vertex pair u and v and denote this as $w \preccurlyeq u, v$. This can also be stated as follows:

$$\textit{resolving set} : (\forall u, v \in \overline{S})(\exists w \in S)[w \preceq u, v].$$

In addition, a resolving set S is a *minimal* resolving set if every vertex $u \in S$ resolves at least one vertex pair $v, w \in \overline{S}$ that is not resolved by any other vertex in S. This can also be stated as follows:

$$\textit{minimal resolving set} : (\forall w \in S)(\exists u, v \in \overline{S})[w \preceq u, v \wedge (\nexists x \in S)[x \neq w \wedge x \preceq u, v].$$

The pair of vertices u and v in the minimal resolving set definition above is called a *private pair* of vertex u.

3 Irredundant Sets

In 1978, Cockayne, Hedetniemi, and Miller [15] introduced the concept of private neighbors, as defined above in the definition of a minimal dominating set, and made the important distinction between a minimal dominating set and a set which isn't necessarily dominating, but nevertheless satisfies the added condition of being minimal dominating set. They therefore introduced the concept called *irredundance* in graphs as follows.

A set $S \subset V$ is called *irredundant* if $(\forall u \in S)(\exists v \in V)[N[v] \cap S = \{u\}]$. Stated in words, a set S is called irredundant if and only if every vertex $u \in S$ has at least one private neighbor, either itself, if $N[u] \cap S = \{u\}$, or a vertex $v \in \overline{S}$ such that $N(v) \cap S = \{u\}$; such a vertex $v \in \overline{S}$ is called an *external private neighbor* of u. An irredundant set S is called *maximal* if no proper superset of S is also an irredundant set.

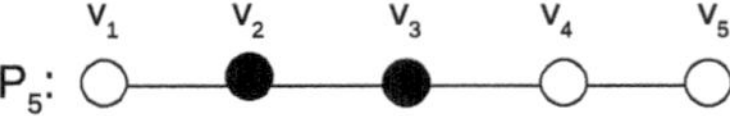

Fig. 1 The set $\{v_2, v_3\}$ is irredundant, but not dominating

The minimum cardinality of a maximal irredundant set in a graph is called the *lower irredundance number* and denoted $ir(G)$. The maximum cardinality of an irredundant set in a graph is called the *upper irredundance number* and denoted $IR(G)$.

It is important to note that even though the concept of irredundant sets arises from the condition that defines a minimal dominating set, an irredundant set, and indeed a maximal irredundant set, need not be a dominating set. Consider the example shown in Figure 1, the path P_5 with vertices labeled in order v_1, v_2, v_3, v_4, v_5. The set $S = \{v_2, v_3\}$ is both an irredundant set and a maximal irredundant set, but it is not a dominating set. In this set, v_2 is the only vertex in S that dominates v_1, and v_3 is the only vertex in S that dominates v_4. Thus, since both vertices in S have an external private neighbor, S is an irredundant set, but S is not a dominating set, since no vertex in S dominates v_5.

The idea of irredundance, and the associated concept of private neighbors, applies to the added conditions which hold if a set S having some property P is, in addition, minimal with respect to property P. In general, irredundance answers the question: why is a set minimal with respect to property P?

Suppose, for example, that a set has property P_2 if the subgraph $G[S]$ induced by a set S has minimum degree $\delta(G[S]) \geq 2$. If S is 1-minimal with respect to this property, then any vertex $v \in S$, if removed, would result in a subgraph $G[S - \{v\}]$ having a vertex of degree less than 2. And this means that every vertex in S must have a neighbor of degree 2 in $G[S]$. We could say, therefore, that a set S is P_2-irredundant if and only if every vertex $v \in S$ has a neighbor in S of degree 2 in $G[S]$. Notice that a set S, such that $G[S]$ is a 2-regular graph (every vertex of which has degree 2), for example, a cycle C_k, is both 1-minimal and minimal with respect to property P_2. However, a set S, such that $G[S] = K_3 \cup K_3$, is a P_2-set that is 1-minimal but is not minimal, since removing all three vertices in one K_3 creates another, smaller P_2-set.

Thus, we can observe that the reason an independent set is a maximal independent set is because it is also a dominating set. In fact, a set is maximal independent if and only if it is both an independent and a minimal dominating set.

And in the same way, the reason that a dominating set is a minimal dominating set is because it is also an irredundant set. In fact, a set is a minimal dominating set if and only if it is a dominating set and an irredundant set.

Because of the generality of the notion of private neighbors, the corresponding notion of irredundance is very general. Although we will discuss a variety of different types of irredundance in this chapter, a very comprehensive and in-depth study of the concept of irredundance in graphs can be found in the 2003 Ph.D. thesis

of Stephen Finbow [32]. In addition, see also the comprehensive chapter on the
irredundance numbers $ir(G)$ and $IR(G)$ by Mynhardt and Roux [49] in this volume.

4 The Basic Private Neighbors and Corresponding Irredundance Numbers

Given a vertex set $S \subseteq V$ in a graph $G = (V, E)$, we can define three kinds of *private
neighbors* of a vertex $v \in S$. If vertex v is not adjacent to any vertex in S, which is
equivalent to saying that v is an isolated vertex in the subgraph $G[S]$ induced by S,
or that $N(v) \cap S = \varnothing$, then we say that v is its *own private neighbor* with respect to
the set S, or that v is a *self-private neighbor* or an *spn*.

If vertex v is adjacent to a vertex $w \in V - S$ and w is not adjacent to any other
vertex in S, then we say that w is an *external private neighbor*, or an *epn*, of v. This
is equivalent to saying that $N(w) \cap S = \{v\}$. The *set of private neighbors* of a vertex
$v \in S$ is the set $pn[v] = N[v] - N[S - \{v\}]$.

If vertex v is adjacent to a vertex $w \in S$ and w is not adjacent to any other vertex
in S, then we say that w is an *internal private neighbor*, or an *ipn*, of v. This is
equivalent to saying that for some vertex $w \in S$, $N(w) \cap S = \{v\}$.

1. **independence numbers** $i(G)$ and $\alpha(G)$, the minimum and maximum cardinalities of a maximal independent set. Equivalently, the minimum and maximum
   cardinalities of maximal sets S such that every vertex $v \in S$ is its own private
   neighbor.
2. **irredundance numbers** $ir(G)$ and $IR(G)$, the minimum and maximum cardinalities of a maximal irredundant set. A set S is an *irredundant set* if for every vertex
   $v \in S$, $pn[v, S] = N[v] - N[S - \{v\}] \neq \varnothing$, that is, every vertex $v \in S$ either is an spn
   or has an epn with respect to the set S. Irredundant sets were first defined and
   studied by Cockayne, Hedetniemi, and Miller in 1978 [15]. See also an early
   survey paper by Hedetniemi, Laskar, and Pfaff [44] and a comprehensive paper
   showing the full generality of irredundance in graphs by Cockayne and Finbow
   [12].
3. **open irredundance numbers** $oir(G)$ and $OIR(G)$, the minimum and maximum
   cardinalities of a maximal open irredundant set in G. A set S is *open irredundant*
   if every vertex $u \in S$ has an external private neighbor. This is equivalent to saying
   that for every vertex $v \in S$, $N(v) - N[S - \{v\}] \neq \varnothing$. Open irredundance was first
   studied by Farley and Schacham [28] in 1983; see also Farley and Proskurowski
   [27] in 1984, Cockayne et al. [11] in 2003, and Cockayne et al. [22] in 2008.
4. **open-open irredundance numbers** $ooir(G)$ and $OOIR(G)$, the minimum and
   maximum cardinalities of a maximal open-open irredundant set in G. A set
   S is *open-open irredundant* if every vertex $u \in S$ has either an external or an
   internal private neighbor. This is equivalent to saying that for every vertex $v \in S$,
   $N(v) - N(S - \{v\}) \neq \varnothing$. Open-open irredundance was introduced by Cockayne,
   Finbow, and Swarts [23] in 2010.

5. **closed-open irredundance numbers** $coir(G)$ and $COIR(G)$, the minimum and maximum cardinalities of a maximal closed-open irredundant set in G. A set S is *closed-open irredundant* if every vertex $u \in S$ has either itself as a private neighbor, an external private neighbor, or an internal private neighbor. This is equivalent to saying that for every $v \in S$, $N[v] - N(S - \{v\}) \neq \emptyset$.

6. **strong matching number** $\alpha_*(G)$, the maximum cardinality of a set $S \subseteq V$ such that every vertex in S has an internal private neighbor. For such a set S, the induced subgraph $G[S]$ consists of a disjoint union of complete graphs K_2; the set of edges in the subgraph induced by such sets are called *strong* or *induced matchings*. These sets were introduced independently by Cameron [6] in 1989 and later by Golumbic and Laskar [41] in 1993.

7. **1-dependence number** $\alpha^1(G)$, the maximum cardinality of a set $S \subseteq V$ such that every vertex has either itself as a private neighbor or has an internal private neighbor. For such a set S, the induced subgraph $G[S]$ consists of a disjoint collection of K_1's or K_2's, or equivalently, the subgraph $G[S]$ has maximum degree $\Delta = 1$. These sets were introduced by Fink and Jacobson [33] in 1985, who called these 1-*dependent sets*. In general, a *k-dependent set* is a set S such that $\Delta(G[S]) \leq k$.

These seven types of irredundance numbers all fit into a natural cube of inequalities, as was shown by Fellows, Fricke, Hedetniemi, and Jacobs [31] in 1994, cf. Figure 2, where an arrow $u \to v$ indicates that the parameter associated with vertex u is greater than or equal to the parameter associated with vertex v. As the inequalities in Figure 2 also show, we have the following sequence, which is called the *Domination Chain*, which was first observed by Cockayne, Hedetniemi, and Miller in 1978 [15].

$$ir \leq \gamma \leq i \leq \alpha \leq \Gamma \leq IR$$

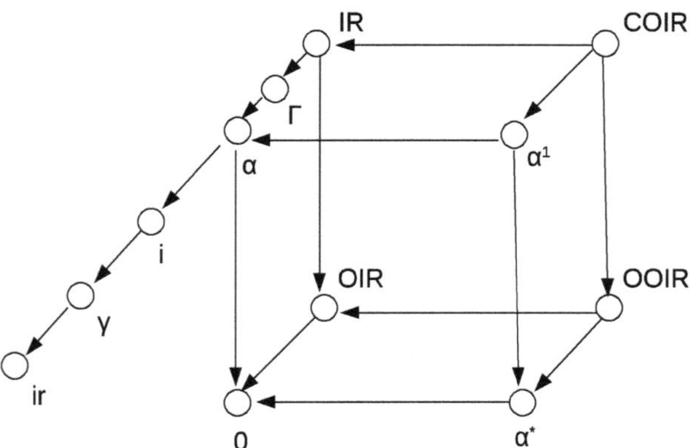

Fig. 2 The private neighbor cube

5 Generalized Irredundance by Cockayne

In 1999, Cockayne [10] considerably generalized the private neighbor cube of Fellows et al. [31]. He considered the 32 classes of sets S, which can be defined in terms of the types of private neighbors that vertices in a set S may or may not have.

Let p mean that a vertex v is its own private neighbor (an spn), meaning that v is not adjacent to any vertex in S.

Let q mean that v has a private neighbor inside S (an ipn), meaning that v is adjacent to a vertex u in S and no other vertex in S is adjacent to u.

Let r mean that v has an external private neighbor (an epn), meaning that it has a neighbor $u \in V - S$ and no other vertex in S is adjacent to u or, equivalently, $N(u) \cap S = \{v\}$.

Therefore, assuming that all vertices must have at least one type of private neighbor, there are five types of vertices, as follows. Note that it is not possible for a vertex to have both a self-private neighbor and an internal private neighbor; thus, there are no vertices of types 1,1,0 or 1,1,1.

	p	q	r	
Type 1:	0	0	1	vertex v must have an epn, but has no spn or ipn.
Type 2:	0	1	0	vertex v must have an ipn but has no spn or epn.
Type 3:	0	1	1	vertex v must have both an ipn and an epn, but has no spn.
Type 4:	1	0	0	vertex v must have an spn but has no ipn or epn.
Type 5:	1	0	1	vertex v must have an spn and an epn, but has no ipn.

There are, therefore, $2^5 = 32$ types of sets of vertices S (all possible subsets of these five types), with prescribed private neighbor requirements for the vertices in S, as given in the table below.

	p	q	r	0	1	2	3	4	5	6	7	8	9	10	11	12	13	14	15
1.	0	0	1	0	0	0	0	0	0	0	0	0	0	0	0	0	0	0	0
2.	0	1	0	0	0	0	0	0	0	0	0	1	1	1	1	1	1	1	1
3.	0	1	1	0	0	0	0	1	1	1	1	0	0	0	0	1	1	1	1
4.	1	0	0	0	0	1	1	0	0	1	1	0	0	1	1	0	0	1	1
5.	1	0	1	0	1	0	1	0	1	0	1	0	1	0	1	0	1	0	1

	p	q	r	16	17	13	19	20	21	22	23	24	25	26	27	28	29	30	31
1.	0	0	1	1	1	1	1	1	1	1	1	1	1	1	1	1	1	1	1
2.	0	1	0	0	0	0	0	0	0	0	0	1	1	1	1	1	1	1	1
3.	0	1	1	0	0	0	0	1	1	1	1	0	0	0	0	1	1	1	1
4.	1	0	0	0	0	1	1	0	0	1	1	0	0	1	1	0	0	1	1
5.	1	0	1	0	1	0	1	0	1	0	1	0	1	0	1	0	1	0	1

Consider the types of sets that are defined by each column. We will mention just a few.

Column 0 defines a set in which no vertex can have a private neighbor of any kind. So, for example, the set of all vertices in a cycle C_n constitutes such a set, since no vertex in a cycle has a private neighbor of any kind.

Column 2 defines a set in which every vertex must be an spn but can obviously have no ipn and can have no epn either. We can call this a {4}-set, since all vertices are of type 4. This is an independent set in which no vertex has an external private neighbor.

Column 12 is a {2,3}-set, which defines an induced matching in which every vertex must have an ipn and may or may not also have an epn.

Most of these sets have not been studied, but several are of particular interest. Cockayne [13] shows that of these 32 types of sets, only 12 define sets that are hereditary and therefore might warrant further study. They are the following:

Column 1 defines a set in which every vertex has both a self-private neighbor and an external private neighbor. This is called a {5}-set. This is an independent set in which every vertex has an external private neighbor.

Column 3 is a {4,5}-set, in which every vertex has an spn and may or may not have an epn. This is a standard independent set.

Column 5 is a {3,5}-set, in which every vertex has an epn and either an spn or an ipn.

Column 7 is a {3,4,5}-set, in which every vertex either has an spn, with or without an epn, or does not have an spn, but has both an ipn and an epn.

Column 9 is a {2,5}-set, in which every vertex either has an ipn, but no spn or epn, or has no ipn, but has both an spn and an epn.

Column 11 is a {2,4,5}-set, in which every vertex either has an spn, with or without an epn, or has an ipn with no spn or epn.

Column 13 is a {2,3,5}-set, in which every vertex either has an ipn, and may or may not have an epn, or has both an spn and an epn.

Column 15 is a {2,3,4,5}-set, which is a 1-dependent set, that is, a set in which every vertex has either an spn or an ipn.

Column 21 is a {1,3,5}-set, which is an open irredundant set, in which every vertex has an epn.

Column 23 is a {1,3,4,5}-set, which is an irredundant set, in which every vertex has either an spn or an epn.

Column 29 is a {1,2,3,5}-set, which is an open-open irredundant set, in which every vertex has either an ipn or an epn.

Column 31 is a {1,2,3,4,5}-set, which is a closed-open irredundant set, in which every vertex has at least one type of private neighbor.

6 Total Irredundance Numbers

The Domination Chain, which follows from the definitions of independent sets, dominating sets, and irredundant sets, raises the question of whether there is a type of irredundance related to total dominating sets. Total irredundance was introduced in 2002 by Favaron, Haynes, Hedetniemi, Henning, and Knisley [30]. See also Hedetniemi, Hedetniemi, and Jacobs [45] in 1993.

A set S is *total irredundant* if and only if for every vertex $v \in V$, $N[v] - N[S - \{v\}] \neq \varnothing$. The total irredundance numbers, $ir_t(G)$ and $IR_t(G)$, equal the minimum and maximum cardinalities of a maximal total irredundant set. Notice that the irredundance numbers $ir(G)$ and $IR(G)$ are defined in terms of two conditions, at least one of which must hold **for every vertex** $v \in S$. By contrast, the total irredundance numbers $ir_t(G)$ and $IR_t(G)$ are defined in terms of two conditions, at least one of which must hold **for every vertex** $v \in V$.

We note in passing that other types of total irredundance can be defined in terms of the conditions that must hold for every vertex $v \in V$ instead of every vertex $v \in S$. Thus, for example, one can define:

total open irredundance: for every vertex $v \in V$, $N(v) - N[S - \{v\}] \neq \varnothing$.

This means that every vertex v, either in S or in $V - S$, is adjacent to some vertex in $V - S$ that no other vertex in S is adjacent to.

To the best of our knowledge, none of these types of total irredundance have been defined and studied. Not only this, but one can define other types of irredundance in terms of private neighbor conditions that hold only for vertices in $V - S$. For example, one could define:

irredundance: **for every v in S**, $N[v] - N[S - \{v\}] \neq \varnothing$.
total irredundance: **for every v in V**, $N[v] - N[S - \{v\}] \neq \varnothing$.
external irredundance: **for every v in V − S**, $N[v] - N[S - \{v\}] \neq \varnothing$.

This means that for an external irredundant set S, every vertex in $V - S$ either is not adjacent to any vertex in S or is adjacent to some vertex in $V - S$ that no vertex in S is adjacent to. To the best of our knowledge, external irredundance has not been studied.

7 The Covering Chain, a Dual of the Domination Chain

In 2015, Arumugam, Hedetniemi, Hedetniemi, Sathikala, and Sudha [3] showed that there is an inequality chain that is complementary to the well-known Domination Chain,

$$ir \leq \gamma \leq i \leq \alpha \leq \Gamma \leq IR$$

This idea starts with the following well-known Theorem of Gallai [38], in which $\alpha(G)$ denotes the maximum cardinality of an independent set of vertices and $\beta(G)$ denotes the minimum cardinality of a vertex cover.

Theorem 1 (Gallai) *For any graph G of order n, $\alpha(G) + \beta(G) = n$.*

The idea introduced by Arumugam et al. is that if the Domination Chain begins with the concept of independence and $\alpha(G)$, then there might be an inequality chain that begins with the concept of a vertex cover and $\beta(G)$. Notice that independence is a hereditary property (every subset of an independent set is also an independent set), while the property of being a vertex cover is superhereditary (every superset of a vertex cover is also a vertex cover).

In order to develop this idea, we will need to quickly review a few definitions.

1. $\beta(G)$, the *vertex covering number*, equals the minimum number of vertices in a *vertex cover*, that is, a set $S \subseteq V$ having the property that for every edge $uv \in E$, either $u \in S$ or $v \in S$, or both.
2. $\beta^+(G)$, the *upper vertex covering number*, equals the maximum number of vertices in a minimal vertex cover of G.
3. $\beta'(G)$, the *edge covering number*, equals the minimum number of edges in an *edge cover*, that is, a set $F \subseteq E$ having the property that every vertex $v \in V$ is incident to at least one edge in F.
4. $\beta'^+(G)$, the *upper edge covering number*, equals the maximum number of edges in a minimal edge cover of G.
5. $\alpha'(G)$, the *matching number*, equals the maximum number of edges in a *matching*, that is, a set $F \subset E$, no two edges in F have a vertex in common.
6. $\Psi(G)$, the *upper enclaveless number*, equals the maximum number of vertices in a set S such that S has no *enclave*, that is, a vertex $v \in S$ such that $N[v] \subseteq S$.
7. $\psi(G)$, the *lower enclaveless number*, equals the minimum number of vertices in a maximal enclaveless set S.

It can be observed that since the property of being a vertex cover is superhereditary, it follows that a vertex cover S is minimal if and only if S is 1-minimal.

Before presenting the next several results, it is worthwhile pointing out that if a set S is a (minimal) vertex cover, then the complement $V - S$ must be a (maximal) independent set. Conversely, if S is a (maximal) independent set, then the complement $V - S$ must be a (minimal) vertex cover.

Proposition 1 (Arumugan et al.) *A vertex cover S of a graph G is a minimal vertex cover if and only if S is a vertex cover and is enclaveless.*

Proof. Let S be a minimal vertex cover of G. Then for every vertex $v \in S$, $S - \{v\}$ is not a vertex cover of G, and this must mean that v has at least one neighbor, say $w \in V - S$, so that the edge vw is not covered by $S - \{v\}$. Hence, $N[v] \nsubseteq S$. Thus, S is an enclaveless set.

Conversely, let S be an enclaveless vertex cover. If S is not a minimal vertex cover, then there exists a vertex $v \in S$ such that $S - \{v\}$ is a vertex cover. Hence, $N[v] \subseteq S$, so that v is an enclave in S, which is a contradiction. Thus, S is a minimal vertex cover. $\square$

Proposition 2 (Arumugam et al.) *Every minimal vertex cover S in a graph G is a maximal enclaveless set of G.*

Proof Let S be a minimal vertex cover in G. It follows from Proposition 1 that S is enclaveless. If S is not a maximal enclaveless set, then there exists a vertex $u \in V - S$ such that $S \cup \{u\}$ is enclaveless. Hence, there exists a vertex $w \in N(u) \cap (V - S)$, and the edge uw has both its ends in $V - S$, which is a contradiction. Thus, S is a maximal enclaveless set. $\qquad\square$

Corollary 1 *For any graph G,*

$$\psi \leq \beta \leq \beta^+ \leq \Psi.$$

Since the property of being an enclaveless set is hereditary, an enclaveless set S is maximal if and only if S is a 1-maximal enclaveless set.

Proposition 3 (Arumugam et al.) *An enclaveless set S is maximal enclaveless if and only if S is enclaveless and $V - S$ is irredundant.*

Proof Let S be a maximal enclaveless set. Then for any $u \in V - S$, $S \cup \{u\}$ contains an enclave, say v. Hence, $N[v] \subseteq S \cup \{u\}$, but $N[v] \not\subseteq S$. If $v \neq u$, then $N[v] \cap (V - S) = \{u\}$, and thus, v is an external private neighbor of u with respect to the set $V - S$. However, if $v = u$, then $N(u) \subseteq S$, and therefore u is not adjacent to any vertex of $V - S$. Thus, in either case, u has a private neighbor, either external or itself, with respect to the set $V - S$, and therefore, $V - S$ is an irredundant set.

Conversely, let S be enclaveless and $V - S$ be an irredundant set. Then for any $u \in V - S$, there exists $v \in N[u]$ such that $N[v] \cap (V - S) = \{u\}$. If $u = v$, then u is an enclave of $S \cup \{u\}$. If $v \neq u$, then $v \in S$ and v is an enclave of $S \cup \{u\}$. Therefore, S is a maximal enclaveless set. $\qquad\square$

This gives rise to a new definition.

Definition 1 *A subset S of V is called a co-irredundant set if $V - S$ is an irredundant set of G. The co-irredundance number $cir(G)$ equals the minimum cardinality of a co-irredundant set in G. The upper co-irredundance number $CIR(G)$ equals the maximum cardinality of a minimal co-irredundant set in G.*

Note that the property of being a co-irredundant set is superhereditary, and thus a set S is a minimal co-irredundant set if and only if it is a 1-minimal co-irredundant set.

Proposition 4 (Arumugam et al.) *Every maximal enclaveless set S in a graph G is a minimal co-irredundant set.*

Proof If S is a maximal enclaveness set, then it follows from Proposition 3 that $V - S$ is irredundant and hence, S is co-irredundant. If S is not a minimal co-irredundant set, then there exists $u \in S$ such that $S - \{u\}$ is co-irredundant. Hence,

$(V-S)\cup\{u\}$ is an irredundant set. Let v be a private neighbor of u with respect to $(V-S)\cup\{u\}$. Clearly, $v\neq u$, since then $N[u]\subseteq S$ is an enclave in S, a contradiction. However, if $v\in S$ and $N[v]\subseteq S$. Thus, v is an enclave in S, which is a contradiction. Thus, S is a minimal co-irredundant set. $\qquad\square$

Corollary 2   *For any graph G,*

$$cir \leq \psi \leq \beta \leq \beta^+ \leq \Psi \leq CIR.$$

This inequality chain above is called the *Covering Chain* of a graph G. But much more can be said about this, since it is closely related to the Domination Chain, as follows.

Theorem 2 (Arumugam et al.)   *For any graph G,*

(i) $ir(G) + CIR(G) = n,$
(ii) $IR(G) + cir(G) = n.$

Thus, the Covering Chain of a graph G can also be written as follows.

$$n - IR \leq n - \Gamma \leq n - \alpha \leq n - i \leq n - \gamma \leq n - ir$$

Hence, the Covering Chain is the dual of the Domination Chain.

Proposition 5 (Arumugam et al.)   *For any graph G without isolated vertices,* $\gamma(G) \leq cir(G).$

Proposition 6 (Arumugam et al.)   *For any graph G without isolated vertices,* $IR(G) \leq \Psi(G).$

Proof We know that

(i) $cir(G) + IR(G) = n,$
(ii) $\gamma(G) \leq cir(G),$
(iii) $\gamma(G) + \Psi(G) = n.$

From this it follows that $IR(G) \leq \Psi(G)$. $\qquad\square$

We can combine the inequalities given in Propositions 5 and 6 into the Domination Chain and Covering Chain to get two larger chains, which we can call the Extended Domination Chain and the Extended Covering Chain, respectively.

ir	$\leq$	γ	$\leq$	i	$\leq$	α	$\leq$	Γ	$\leq$	IR	$\leq$	Ψ	$\leq$	CIR
$+$		$+$		$+$		$+$		$+$		$+$		$+$		$+$
CIR	$\geq$	Ψ	$\geq$	β^+	$\geq$	β	$\geq$	ψ	$\geq$	cir	$\geq$	γ	$\geq$	ir
$=$		$=$		$=$		$=$		$=$		$=$		$=$		$=$
n		n		n		n		n		n		n		n

Arumugam et al. also add the following interesting comparison of two inequality chains:

$$\gamma \;\leq\; i \;\leq\; \alpha \;\leq\; \Gamma \;\leq\; IR \;\leq\; \Psi$$
$$\gamma \;\leq\; cir \;\leq\; \psi \;\leq\; \beta \;\leq\; \beta^{+} \;\leq\; \Psi$$

$$ir \;\leq\; \gamma \;\leq\; cir \;\leq\; \psi \;\leq\; \beta \;\leq\; \beta^{+} \;\leq\; \Psi \;\leq\; CIR$$
$$CIR \;\geq\; \Psi \;\geq\; IR \;\geq\; \Gamma \;\geq\; \alpha \;\geq\; i \;\geq\; \gamma \;\geq\; ir$$

8 Domination in Terms of Perfection in Graphs

To the definitions given above, of independent sets, dominating sets, and irredundant sets, we now introduce several concepts and parameters having to do with what is called *perfection* in graphs. These concepts were first introduced in 1999 by Fricke, Haynes, Hedetniemi, Hedetniemi, and Henning [35] and later developed in detail by J. T. Hedetniemi, S. M. Hedetniemi, and S. T. Hedetniemi in 2013 [47]. In the remainder of this section, we review the results given in this 2013 paper.

In the notation that follows, a subscript (α_S) refers to a parameter α having some condition on the vertices in S, for example, the vertex independence number is defined in terms of a condition on the vertices in S, that no two of them are adjacent.

Similarly, a superscript (α^{V-S}) refers to a parameter having some condition on the vertices in $V - S$, for example, the domination number is defined in terms of a condition on the vertices in $V - S$, that every vertex in $V - S$ has at least one neighbor in S.

If a parameter α requires some condition on all vertices in V, no subscript or superscript appears.

Definition 2 *Given a set $S \subseteq V$ in a graph $G = (V, E)$, a vertex $v \in V$ is said to be S-perfect if $|N[v] \cap S| = 1$, that is, the closed neighborhood $N[v]$ contains exactly one vertex in S.*

Notice that if a vertex $v \in S$ is S-perfect, then it has no neighbors in S, and if a vertex $v \in V - S$ is S-perfect, then it has exactly one neighbor in S.

Definition 3 *Given a set $S \subseteq V$ in a graph G, a vertex v is almost S-perfect if it is either S-perfect or is adjacent to an S-perfect vertex.*

Fig. 3 A perfect
neighborhood set

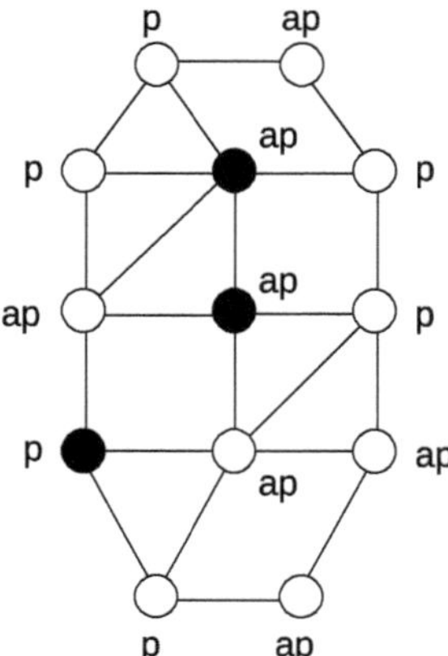

When a set *S* has been given and is assumed, we simply say that a vertex is *perfect* or *almost perfect*, without referring to the set *S*.

Definition 4 *A set $S \subset V$ is perfect if every vertex $v \in S$ is S-perfect and is almost perfect if every vertex $v \in S$ is almost S-perfect; for brevity we say that an almost perfect set is an ap-set. Let $\theta_{ap}(G)$ and $\Theta_{ap}(G)$ equal the minimum and maximum cardinalities of a maximal ap-set in G.*

Definition 5 *A set S is externally perfect if every vertex in $V - S$ is S-perfect and is externally almost perfect if every vertex in $V - S$ is either S-perfect or is adjacent to an S-perfect vertex; for brevity we say that an externally almost perfect set is an eap-set. Let $\theta^{ap}(G)$ and $\Theta^{ap}(G)$ equal the minimum and maximum cardinality of a minimal eap-set in G.*

In the graph in Figure 3, given in the paper by Hedetniemi et al., a vertex labeled "p" is perfect, while a vertex labeled "ap" is almost perfect. The three shaded vertices form a set *S* that is almost perfect (two vertices in *S* are almost perfect and the third is perfect) and is externally almost perfect (every vertex in $V - S$ is either perfect or is adjacent to a perfect vertex).

Definition 6 *A set S is a perfect neighborhood set if every vertex $v \in V$ is either perfect or is adjacent to a perfect vertex. Let $\theta(G)$ and $\Theta(G)$ equal the minimum and maximum cardinalities of a perfect neighborhood set in G, and let $\theta_p^{ap}(G)$ and $\Theta_p^{ap}(G)$ equal the minimum and maximum cardinalities of an independent perfect neighborhood set in G.*

Notice that the three shaded vertices in Figure 2 form a perfect neighborhood set.

Definition 7 *A set S is an eap irredundant, eap dominating, or eap independent set if it is a maximal irredundant, minimal dominating, or maximal independent set that is also eap. Thus, every vertex $v \in V - S$ is either perfect or is adjacent to a perfect vertex. Let $ir^{ap}(G)$, $\gamma^{ap}(G)$, $i^{ap}(G)$, $\alpha^{ap}(G)$, $\Gamma^{ap}(G)$, and $IR^{ap}(G)$ denote the minimum and maximum cardinalities of such sets.*

Given these definitions, we can relate them to independent, dominating, and irredundant sets, for example, the concept of a set S being perfect is equivalent to the concept of a set being independent.

Proposition 7 *A set S is perfect if and only if it is independent.*

Corollary 3 *For any graph G, $\theta_p^{ap}(G) \leq i(G) = i^{ap}(G)$.*

Corollary 4 *For any graph G, $\alpha(G) = \alpha^{ap}(G) = \Theta_p^{ap}(G)$.*

Recall that a dominating set S is called *perfect* if every vertex $v \in V - S$ has exactly one neighbor in S and is called an *efficient dominating set* if S is independent, dominating, and perfect.

Proposition 8 *A set S is externally perfect if and only if S is a perfect dominating set.*

Proposition 9 *A set S is completely perfect if and only if S is an efficient dominating set (or a perfect code).*

The concept of being almost perfect (ap) is equivalent to the concept of being irredundant.

Proposition 10 *A set S is almost perfect if and only if S is irredundant.*

Proof If a set S is almost perfect, then every vertex $u \in S$ is either perfect or adjacent to a perfect vertex. Either u is an isolated vertex in $G[S]$, in which case it is perfect and is its own private neighbor, or u is adjacent to a perfect vertex, say w. But w cannot be in S. Thus, $w \in V - S$ and w is perfect because $|N[w] \cap S| = |\{u\}| = 1$. This means that w is an external private neighbor of u. Thus, every vertex $u \in S$ has a private neighbor, and hence S is irredundant.

Conversely, if S is irredundant, then every vertex $u \in S$ either is its own private neighbor, in which case it is perfect, or has an external private neighbor, say w. But in this case w is perfect and therefore u is adjacent to a perfect vertex. Therefore, S is almost perfect. $\qquad\square$

Corollary 5 *For any graph G,*

$$ir = \theta_{ap} \leq \Theta_{ap} = IR(G)$$

Distance-2 dominating sets are closely related to externally almost perfect sets.

Proposition 11 *If a set S is eap, then it is a distance-2 dominating set.*

The Domination Chain can now be considerably expanded in terms of the concept of perfection.

Theorem 3 *For any graph G, the following system of inequalities holds.*

$$
\begin{array}{ccccccccccc}
& & \theta^{ap} & \leq & \theta & \leq & \theta^{ap}_p & \leq & \Theta^{ap}_p & \leq & \Theta & \leq & \Theta_{ap} \\
& & & & \mathrel{\rotatebox{90}{$\leq$}} & & \mathrel{\rotatebox{90}{$\leq$}} & & \| & & \| & & \| \\
\theta_{ap} & = & ir & \leq & \gamma & \leq & i & \leq & \alpha & \leq & \Gamma & \leq & IR \\
& & \mathrel{\rotatebox{90}{$\leq$}} & & \mathrel{\rotatebox{90}{$\leq$}} & & \| & & \| & & \mathrel{\rotatebox{-90}{$\leq$}} & & \mathrel{\rotatebox{-90}{$\leq$}} \\
\theta^{ap} & \leq & ir^{ap} & \leq & \gamma^{ap} & \leq & i^{ap} & \leq & \alpha^{ap} & \leq & \Gamma^{ap} & \leq & IR^{ap} \\
\mathrel{\rotatebox{-90}{$\leq$}} & & \mathrel{\rotatebox{-90}{$\leq$}} & & & & & & & & & & \\
\gamma_{\leq 2} & & \theta & & & & & & & & & &
\end{array}
$$

A similar pair of inequality chains holds when independent sets are considered.

Proposition 12 *For any graph G, the following inequalities hold.*

$$(i) \quad \gamma_{\leq 2} \leq \gamma^{ap}_{\leq 2} \leq i^{ap}_{\leq 2} \leq i^{ap} = i.$$

$$(ii) \quad \gamma_{\leq 2} \leq i_{\leq 2} \leq i^{ap}_{\leq 2} \leq i^{ap} = i.$$

If we define $\gamma^{ap}_d(G)$ to equal the minimum cardinality of a dominating set that is externally almost perfect, as distinct from $\gamma^{ap}(G)$, which equals the minimum cardinality of a *minimal* dominating set that is externally almost perfect, then we get the following refinement.

Proposition 13 *For any graph G,*

$$\gamma \leq \gamma^{ap}_d \leq \gamma^{ap} \leq i.$$

As given in Hedetniemi et al., the fact that each of these inequalities can be strict is illustrated by the unicyclic graph G in Figure 4. For this graph, the set $S_1 = \{1, 3, 4, 6\}$ shows that $\gamma(G) = 4$; for $\gamma^{ap}_d(G) = 6$, let $S_2 = \{1, 2, 3, 4, 5, 6\}$; for $\gamma^{ap}(G) = 8$, let $S_3 = \{1, 6, 7, 8, 9, 10, 11, 12\}$; and for $i(G) = 9$, let $S_4 = \{1, 7, 8, 9, 4, 13, 14, 15, 16\}$.

In summary, the concept of perfection in graphs provides a framework for unifying the concepts of independent sets, dominating sets, irredundant sets, perfect and efficient dominating sets, and perfect neighborhood sets. For example:

(i) A set is independent if and only if it is a perfect set.
(ii) The independence parameters $i(G)$ and $\alpha(G)$ can be expressed as maximal independent sets whose complements are almost perfect, that is, $i(G) = i^{ap}(G)$ and $\alpha(G) = \alpha^{ap}(G)$.
(iii) A set is an irredundant set if and only if it is an almost perfect set.
(iv) The parameters $ir(G)$ and $IR(G)$ can be expressed in terms of almost perfect sets, namely, $ir(G) = \theta_{ap}(G)$ and $IR(G) = \Theta_{ap}(G)$.

Fig. 4 $\gamma < \gamma_d^{ap} < \gamma^{ap} < i$

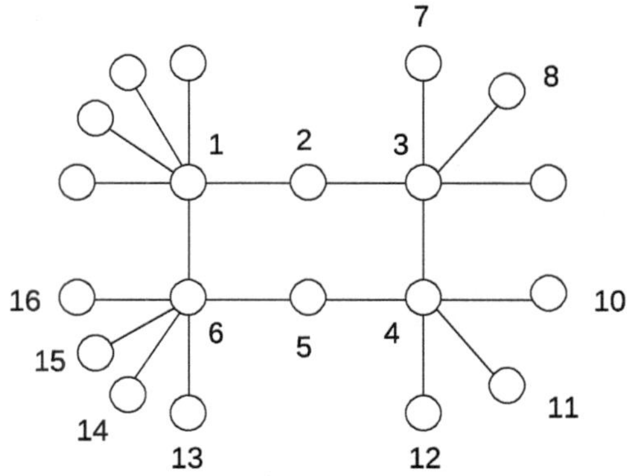

(v) The theorem in Fricke et al. [35], that $\Theta(G) = \Gamma(G)$, established an equality between two seemingly unrelated parameters. This result is now clearer. In particular, the inequality chain, $\alpha(G) \leq \Gamma(G) \leq IR(G)$, can now be stated equivalently as $\Theta_p^{cp}(G) \leq \Theta(G) \leq \Theta_{ap}(G)$, since $\Theta_p^{ap}(G) = \alpha(G)$, $\Theta(G) = \Gamma(G)$, and $\Theta_{ap}(G) = IR(G)$.

(vi) An expanded inequality chain exists between the domination and independence parameters:

$$\gamma \leq \gamma_d^{ap} \leq \gamma^{ap} \leq i \leq \alpha \leq \Gamma^{ap} \leq \Gamma.$$

Further papers on perfect neighborhood sets in graphs can be found by Cockayne, Hedetniemi, Hedetniemi, and Mynhardt [20], by Favaron and Puech [29], and by Hedetniemi, Hedetniemi, and Henning [46].

9 Partitions Involving Irredundant Sets

A *(proper) coloring* of a graph G is a vertex partition $V = \{V_1, V_2, \ldots, V_k\}$ such that for every $1 \leq i \leq k$, V_i is an independent set. Since the property of being independent is a hereditary property, one seeks the minimum order of a partition into independent sets. For example, the *chromatic number* $\chi(G)$ equals the minimum order of a partition of V into independent sets.

Continuing in the same manner, the property of being an irredundant set is hereditary. Therefore, it would be natural to consider the minimum order of a partition of V into irredundant sets.

As introduced by Haynes, Hedetniemi, Hedetniemi, McRae, and Slater in 2008 [43], the *irratic number* $\chi_{ir}(G)$ equals the minimum order of a partition of V into irredundant sets.

Clearly, since every independent set is irredundant, for any graph G, $\chi_{ir}(G) \leq \chi(G)$.

This inequality immediately raises the question: can you prove that for any planar graph G, $\chi_{ir}(G) \leq 4$, without appealing to the Four Color Theorem?

Haynes et al. show the following.

1. $\chi_{ir}(G) = 1$ if and only if $G = \overline{K_n}$.
2. $\chi_{ir}(G) = n$ if and only if $G = K_n$.
3. If $\chi(G) = 2$, then $\chi_{ir}(G) = 2$.
4. $\chi_{ir}(G \circ K_1) = 2$.
5. $\chi_{ir}(G \square K_2) = 2$.

The authors provide bounds for $\chi_{ir}(G)$ in terms of other well-known graphical parameters and take a closer look at graphs with $\chi_{ir}(G) = 2$, called *bi-irratic graphs*. Although every nonempty bipartite graph is bi-irratic, the problem of characterizing the class of bi-irratic graphs remains open. The authors also study complexity questions and establish the NP-completeness of the problem of determining if a given graph is bi-irratic.

In 2012, Arumugam and Chandrasekar [2] prove that the problem of deciding if the vertices of a graph can be partitioned into two open irredundant sets, that is, whether $\chi_{oir}(G) = 2$, is NP-complete.

A *complete coloring* of a graph G is a proper vertex coloring of G having the property that for every two distinct colors i and j, there exist adjacent vertices colored i and j. The maximum positive integer k for which G has a complete k-coloring is called the *achromatic number* $\Psi(G)$ of G. A *Grundy coloring* of a graph G is a proper vertex coloring of G having the property that for every two colors (positive integers) i and j with $i < j$, every vertex colored j has a neighbor colored i. The maximum positive integer k for which a graph G has a Grundy k-coloring is the *Grundy number* $\Gamma r(G)$ of G. For every graph G, these four coloring parameters satisfy the inequalities:

$$\chi_{ir} \leq \chi \leq \Gamma r \leq \Psi.$$

For these four coloring numbers, Chartrand, Hedetniemi, Okamoto, and Zhang [7] showed in 2011 that if a, b, c, *and* d are integers with $2 \leq a \leq b \leq c \leq d$, then there exists a nontrivial connected graph G with $\chi_{ir}(G) = a$, $\chi(G) = b$, $\Gamma r(G) = c$, and $\Psi(G) = d$ if and only if $d = 2$ or $c \neq 2$.

10 The Mystery of the Domination Chain: $?? \leq ir(G) \leq \gamma(G) \leq i(G) \leq \alpha(G) \leq \Gamma(G) \leq IR(G) \leq ??$

Notice that the parameters $i(G)$ and $\alpha(G)$ are the min and max parameters associated with the hereditary property, say P_1, of being an independent set.

Notice next that the parameters $\gamma(G)$ and $\Gamma(G)$ are the min and max parameters associated with the superhereditary property, P_2, of being a dominating set.

The third pair of parameters $ir(G)$ and $IR(G)$ are the min and max parameters associated with the hereditary property P_3 of being a maximal irredundant set.

This would lead one to think that there should exist a pair of parameters, call them $\psi(G)$ and $\Psi(G)$, that are associated with some superhereditary property, P_4, and we would have the following inequality chain.

$$\psi? \ \leq \ ir \ \leq \ \gamma \ \leq \ i \ \leq \ \alpha \ \leq \ \Gamma \ \leq \ IR \ \leq \ \Psi?$$

... and maybe still another similar pair of parameters, call them ϕ and Φ, that are associated with some hereditary property, P_5, and we would have the following inequality chain.

$$\phi? \ \leq \ \psi? \ \leq \ ir \ \leq \ \gamma \ \leq \ i \ \leq \ \alpha \ \leq \ \Gamma \ \leq \ IR \ \leq \ \Psi? \ \leq \ \Phi?$$

Does this chain of inequalities continue indefinitely, alternating between hereditary and superhereditary properties, or does it terminate?

Much of the discussion in this section, but not all, can be found in the 1997 paper by Cockayne, Hattingh, Hedetniemi, Hedetniemi, and McRae [17].

As discussed in this paper, the domination chain starts with the concept of an independent set, which is a hereditary property. Because of this, one can say that an independent set S is *maximal* if and only if it is 1-*maximal*, which means that for every vertex $v \in V - S$, the set $S \cup \{v\}$ is not an independent set. This, in turn, is equivalent to saying that for every vertex $v \in V - S$, there exists at least one vertex $u \in S$ such that v is adjacent to u. And this is the definition of a *dominating set*. So we can say that the maximality condition for an independent set is the definition of a dominating set.

In much the same way, the concept of a dominating set is a superhereditary property. Because of this, we can say that a dominating set S is *minimal* if and only if it is 1-*minimal*, which means that for every vertex $v \in S$, the set $S - \{v\}$ is not a dominating set. This, in turn, is equivalent to saying that either vertex v has no neighbor in S, and therefore is not dominated by any vertex in $S - \{v\}$, in which case we say that v is its own *self-private neighbor*, or *spn*, or v is the only vertex that dominates some vertex $w \in V - S$, in which case we say that w is an *external private neighbor* or *epn* of v. This condition is the definition of an *irredundant set*.

Because of these three definitions, the Domination Chain emerges.

But now we come to the property of being an irredundant set. It is easy to see that this, like the property of being an independent set, is a hereditary property, since every subset of an irredundant set is also an irredundant set. Thus, one can say that a set S is a maximal irredundant if and only if it is a 1-*maximal* irredundant set. And this means that a set S is a maximal irredundant set if and only if for every vertex $v \in V - S$, $S \cup \{v\}$ is not an irredundant set. This means that when you add v to the set S, some vertex in $S \cup \{v\}$ does not have a private neighbor.

This is equivalent to saying that a maximal irredundant set S is a set that is irredundant and has the added property that for every vertex $v \in V - S$, either

Condition (i): v does not have a private neighbor in the set $S \cup \{v\}$ or

Condition (ii): v has a private neighbor in $S \cup \{v\}$, but there exists some vertex $u \in S$ that has a private neighbor with respect to S but does not have a private neighbor with respect to $S \cup \{v\}$.

Conditions (i) and (ii) give rise to the following type of sets.

Definition 8 *A set S is called external redundant if for every vertex $v \in V - S$, either (i) v does not have a private neighbor in the set $S \cup \{v\}$ or (ii) v has a private neighbor in $S \cup \{v\}$, but there exists some vertex $u \in S$ that has a private neighbor in S but does not have a private neighbor in $S \cup \{v\}$. Let $er(G)$ and $ER(G)$ equal the minimum and maximum cardinality of a minimal external redundant set in G.*

Notice that an external redundant set need not be irredundant, but, by definition, every maximal irredundant set is external redundant.

Another way of thinking about the maximality condition of an irredundant set is the following. For any set $S \subseteq V$, let $pnc(S) = |\{v \in S : N[v] - N[S - \{v\}] \neq \varnothing\}|$; this is called the *private neighbor count* of set S, which equals the number of vertices in S having either an spn or an epn. Thus, a set S is irredundant if and only if $pnc(S) = |S|$.

We say that a set S is *pnc-maximal* if for every vertex $v \in V - S$, $pnc(S \cup \{v\}) \leq pnc(S)$. Thus, an irredundant set is a maximal irredundant set if and only if $pnc(S) = |S|$ and S is pnc-maximal or adding a vertex in $V - S$ to S cannot increase the private neighbor count.

A concept closely related to external redundance was introduced in 1997 by Cockayne, Grobler, Hedetniemi, and McRae [16] and later studied by Cockayne, Favaron, Puech, and Mynhardt in 1998 [18] and [19], and by Puech in 2000 [50].

Given a set $S \subset V$, let $R = V - N[S]$ be the set of vertices not dominated by any vertex in S. Let $u \in S$ be a vertex for which $pn[u, S] = N[u] - N[S - \{u\}] \neq \varnothing$, that is, vertex u has at least one private neighbor (spn or epn) with respect to S. A vertex $v \in V - S$ is said to *annihilate* u if v is adjacent to every vertex in $pn[u, S]$. This means that u has a private neighbor with respect to the set S, but does not have a private neighbor with respect to the set $S \cup \{v\}$.

A set S is said to be *R-annihilated*, or an *Ra-set*, if every vertex in R annihilates some vertex in S. This is a property satisfied by every maximal irredundant set. The *R-annihilated number $ra(G)$* equals the minimum cardinality of an Ra-set in G.

Let $A = \{v \in V - S | v$ annihilates some $u \in S\}$. For some subset $U \subset V - S$, we say that S is *U-annihilated* if $U \subseteq A$.

Theorem 4 (Cockayne, Grobler et al.) *A set $S \subset V$ in a graph $G = (V, E)$ is maximal irredundant if and only if S is irredundant and $N[R]$-annihilated.*

Notice that $N[R]$-annihilated sets are equivalent to eternal redundant sets. Cockayne, Grobler et al. then define the following parameters:

$\gamma_{\leq 2}(G)$, minimum cardinality of a distance-2 dominating set;

$ra(G)$, the minimum cardinality of an R-annihilated set;
$rai(G)$, the minimum cardinality of an irredundant R-annihilated set.
$\theta(G)$, the (lower) perfect neighborhood number.
$\theta_i(G)$, the (lower) independent perfect neighborhood number.
$\rho_L(G)$, the lower 2-packing number.
$\rho(G)$, the upper 2-packing number.

Theorem 5 (Cockayne, Grobler et al.) *For any connected graph G,*

$$\gamma_{\leq 2} \leq \left\{ \begin{array}{c} ra \leq \left\{ \begin{array}{c} rai \\ er \end{array} \right\} \leq ir \\ \theta \leq \theta_i \leq \rho_L \leq \rho \end{array} \right\} \leq \gamma \leq i.$$

It so happens that the property of being external redundant, pnc-maximal, or $N[R]$-annihilated is neither superhereditary nor hereditary.

Recall that the maximality condition of the hereditary property of being independent is the definition of the superhereditary property of being a dominating set.

And the minimality condition of the superhereditary property of being a dominating set is the definition of the hereditary property of being an irredundant set.

However, the maximality condition of the hereditary property of being an irredundant set seems to be the property of being an external redundant set. Yet, somewhat surprisingly, the property of being an external redundant set is not superhereditary. In Figure 5 below, the set $S = \{1, 3, 4, 5, 8, 11\}$ is a 1-minimal external redundant set, since removing any one of these vertices results in a set that is no longer external redundant. However, you can remove both vertices 1 and 3 and still have an external redundant set.

This is not expected. And we are puzzled. Why isn't the maximality condition of an irredundant set superhereditary? We don't know.

Fig. 5 A 1-minimal external redundant set

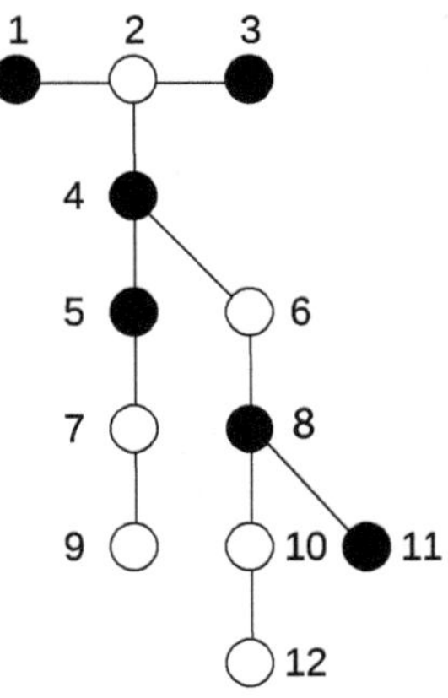

S = {1, 3, 4, 5, 8, 11} 1-minimal
er, but S' = S – {1, 3} is er.

Nevertheless, this has given rise to the definitions of $er(G)$ and $ER(G)$, and with these two definitions, we get the following extended inequality chain:

$$er \leq ir \leq \gamma \leq i \leq \alpha \leq \Gamma \leq IR \leq ER.$$

Once again, we are led to ask: is there still another pair of parameters which extend this inequality chain? As of now, none are known.

Here is still another thought about the sequence: independent set, dominating set, irredundant set, and external redundant set.

The property of being an independent set is what could be called 1-*local*. In order to decide if a set S is independent, all you have to do is to look at the neighbors $N(v)$ of every vertex $v \in S$ and see if another member of S appears.

In order to decide if a set S is a dominating set, all you have to do is to look at every vertex $w \in V - S$ and make sure that $N(w) \cap S \neq \varnothing$. This is also a 1-local check.

But now, in order to decide if a set S is an irredundant set, you have to work a bit harder. Once again, you have to look at every vertex in $v \in S$ and see if $N(v) \cap S = \varnothing$. If this is true, then vertex v is its own private neighbor. But if v is not its own private neighbor, then you must see if v has at least one neighbor, say $y \in N(v) \cap (V - S)$, for which $N(y) \cap S = \{v\}$. Such a vertex y is then an external private neighbor of v. This then becomes a 2-local, or distance-2, check, in that you have to look the neighbors of your neighbors.

But now, consider the problem of deciding if an irredundant set is a *maximal* irredundant set, or an external redundant set, you must look at every vertex in $V - S$. Here, it is possible that there could be a vertex $z \in V - S$, such that $d(z, v) \geq 3$ for all vertices $v \in S$, in which case the set $S \cup \{z\}$ would be an irredundant set. Thus, you could say that the property of being a maximal irredundant set, or, equivalently, an external redundant set, is 3-local.

And consider this. There is a very simple, greedy, linear algorithm for computing the vertex independence number $\alpha(T)$ for any tree T (cf. Mitchell, Hedetniemi, and Goodman in 1975 [48] and a more general algorithm for chordal graphs in 1972 by Gavril [39]). A second, slightly more complex, but still linear, algorithm exists for computing the independent domination number $i(T)$, for any tree T (cf. Beyer, Proskurowski, Hedetniemi, and Mitchell in 1977 [5]). A third, more complex, but still linear algorithm exists for computing the domination number $\gamma(T)$, for any tree T (cf. Cockayne, Goodman, and Hedetniemi in 1975 [14]). And finally, there exists a much more complex, but still linear, algorithm for computing the lower irredundance number $ir(T)$, for any tree T (cf. Bern, Lawler, and Wong in 1985 [4]). But this algorithm consists of a 20-by-20 table of some 400 possible combinations of the irredundance states of a vertex v and its parent in a rooted tree T. In [40], Goddard and Hedetniemi propose, without proof, an algorithm for computing $er(T)$, for any tree T, consisting of a 23-by-23 table of combinations of external redundant states of a vertex v and its parent in a rooted tree T. The authors state, "We believe that the table for external redundance is correct. For a proof of this, it would be sufficient to prove that none of the [23] classes needs to be divided. This is a lengthy and tedious argument and is omitted."

11   Broadcast Irredundance in Graphs

In 2015, Ahmadi, Fricke, Schroeder, Hedetniemi, and Laskar [1] introduced the concept of broadcast irredundance in graphs. In this section, we review the basic definitions and results of this model of irredundance in graphs.

The following concepts and definitions of broadcasts in graphs were introduced by Erwin in 2004 [26] and developed further by Dunbar, Erwin, Haynes, Hedetniemi, and Hedetniemi in 2006 [25].

A function $f : V \to \{0, 1, 2, \ldots\}$ defined on the vertex set V of a graph $G = (V, E)$ is called a *broadcast* if for every vertex $v \in V$, $f(v) \leq ecc(v)$. Intuitively this means, for example, that if a vertex v is assigned a broadcast power of $4, f(v) = 4$, then all vertices within distance 4 or less of v can hear a broadcast from vertex v. The *cost* $f(V)$ of a broadcast f is defined as $f(V) = \Sigma_{v \in V} f(v)$.

Given a broadcast function f, let $V_f^0 = \{v \mid f(v) = 0\}$ and $V_f^+ = V - V_f^0 = \{u \mid f(u) > 0\}$. The vertices in V_f^+ are called *broadcast* vertices.

Given a broadcast f and a broadcast vertex v, the *broadcast neighborhood* of v is the set $N_f[v] = \{u \mid d(u, v) \leq f(v)\}$. We say that every vertex in the broadcast neighborhood $N_f[v]$ can *hear* a broadcast from v or is *broadcast dominated by* v.

Thus, a vertex u with $f(u) = 0$ *hears* a broadcast if there exists a vertex v for which $d(u, v) \leq f(v)$. The set of vertices that a vertex u hears is the set $H(u) = \{v \in V_f^+ \mid d(u, v) \leq f(v)\}$. Define $H(f) \subseteq V$ to equal the set of vertices that hear a broadcast defined by f. Finally, we say that a broadcast g satisfies $g \leq f$, if for every vertex $v \in V$, $g(v) \leq f(v)$.

A broadcast f is called a *dominating broadcast* if for every vertex $u \in V$ with $f(u) = 0$, $H(u) \neq \varnothing$, or equivalently if $H(f) = V$. The *broadcast domination number* $\gamma_b(G)$ of a graph G equals the minimum weight $f(V)$ of a dominating broadcast f in G.

We say that a dominating broadcast f is *minimal* if there does not exist a dominating broadcast g for which $g \leq f$.

The following characterization of minimal dominating broadcasts is due to Erwin [26].

Theorem 6   (Erwin [26]) *A dominating broadcast f on a graph G is minimal if and only if the following two conditions are satisfied:*

1. *for every broadcast vertex v with $f(v) \geq 2$, there exists a vertex $u \in V_f^0$ such that $H(u) = \{v\}$ and $d(u, v) = f(v)$,*
2. *for every broadcast vertex v with $f(v) = 1$, there exists a vertex $u \in N[v]$ such that $H(u) = \{v\}$.*

Let $S \subseteq V$ be a minimal dominating set in a graph G. The *characteristic function* f_S of S is the broadcast function defined as follows: $f_S(v) = 0$ if $v \notin S$, and $f_S(v) = 1$ if $v \in S$.

Proposition 14 (Erwin [26]) *If $S \subseteq V$ is a minimal dominating set in a graph G, then the characteristic function f_S is a minimal dominating broadcast.*

Corollary 6 *For any graph G, $\gamma_b(G) \le \gamma(G)$.*

It is worth noting that the broadcast domination number of a graph can be considerably smaller than its domination number. For example, let $S(K_{1,n})$ denote a *subdivided star*, that is, a graph having one central vertex of degree n, to which are attached n paths of length 2. It is easy to see that $\gamma_b(S(K_{1,n})) = 2 < \gamma(S(K_{1,n})) = n$.

One can also speak of independent broadcasts.

A broadcast f is called *independent* if for every broadcast vertex $v \in V_f^+$, $|H(v)| = 1$, that is, $H(v) = \{v\}$. In other words, no broadcast vertex can hear a broadcast from any other vertex. The *broadcast independence number* $\alpha_b(G)$ of a graph G equals the maximum cost $f(V)$ of an independent broadcast in G, while the *lower broadcast independence number*, $i_b(G)$, equals the minimum cost of a maximal independent broadcast in G.

Proposition 15 *The characteristic function f_S of every maximal independent set S is (i) an independent broadcast, but not necessarily a maximal independent broadcast, and (ii) a minimal dominating broadcast.*

Note that for a path P_4 with vertices labeled in order v_1, v_2, v_3, v_4, the independent broadcast function f defined by $f(v_1) = f(v_4) = 1$ and $f(v_2) = f(v_3) = 0$ is the characteristic function of the maximum independent set $S = \{v_1, v_4\}$. But this is not a maximal independent broadcast, since the independent broadcast function $g(v_1) = g(v_4) = 2$ and $g(v_2) = g(v_3) = 0$ satisfies $g \ne f$ and $g \ge f$. In fact, $\alpha_b(P_4) = 4$.

Corollary 7 *For any graph G, $\alpha(G) \le \alpha_b(G)$.*

Corollary 8 *For any graph G,*

$$\gamma_b \le i_b \le \alpha \le \alpha_b \le \Gamma_b.$$

It is interesting to note that the two parameters $i(G)$ and $i_b(G)$ are not comparable.

Erwin's characterization of minimal dominating broadcasts in effect defines *irredundant* broadcasts, which were introduced in 2015 by Ahmadi, Fricke, Schroeder, Hedetniemi, and Laskar [1] as follows.

Definition 9 *A broadcast function $f : V \to \{0, 1, \ldots \ldots\}$ is called irredundant if it satisfies the following two conditions:*

(i) *for every broadcast vertex v with $f(v) \ge 2$, there exists a vertex $u \in V_f^0$ such that $H(u) = \{v\}$ and $d(u, v) = f(v)$,*

(ii) *for every broadcast vertex v with $f(v) = 1$, there exists a vertex $u \in N[v]$ such that $H(u) = \{v\}$.*

Stated equivalently, a broadcast function f is irredundant if reducing the broadcast value assigned to any broadcast vertex strictly decreases the number of vertices that hear a broadcast, that is, for any broadcast $g \leq f$, $|H(g)| < |H(f)|$. This is the analog of saying that every vertex in an irredundant set S has a private neighbor or that the number of vertices dominated by an irredundant set S is strictly greater than the number of vertices dominated by any proper subset $S' \subset S$.

Given an irredundant broadcast f, every vertex $w \in N_f[v]$ for which $H(w) = \{v\}$ is called a *private broadcast neighbor* of v or a *private f-neighbor* of v. Note that a broadcast vertex v can be its own private f-neighbor, while any other private f-neighbor of v must be a vertex $w \in V_f^0$.

An irredundant broadcast f is *maximal* if there does not exist an irredundant broadcast g such that (i) $g \neq f$ and (ii) for every vertex $v \in V$, $g(v) \geq f(v)$.

Definition 10 *The broadcast irredundance number $ir_b(G)$ equals the minimum cost of a maximal irredundant broadcast in G. Similarly, the upper broadcast irredundance number $IR_b(G)$ equals the maximum cost of an irredundant broadcast.*

To this definition we can add the following. An irredundant broadcast f is called 1-*maximal* if increasing the value of $f(v)$ of any one vertex $v \in V$ creates a function f' that is no longer an irredundant broadcast, either because it is no longer a broadcast function or because some broadcast vertex no longer meets Condition (i) or Condition (ii) of the definition of an irredundant broadcast.

To illustrate this definition, Ahmadi et al. consider the path P_8, with vertices labeled in order v_1, ..., v_8. Let f be the irredundant broadcast function defined by $f(v_4) = f(v_5) = 2$ and $f(v_1) = f(v_2) = f(v_3) = f(v_6) = f(v_7) = f(v_8) = 0$. This broadcast function f is irredundant because broadcast vertex v_4 has vertex v_2 as a private broadcast neighbor, while broadcast vertex v_5 has vertex v_7 as a private broadcast neighbor. It can be seen that if the value of any of vertex other than v_4 and v_5 is increased, then either vertex v_4 or v_5 will no longer have a private broadcast neighbor. Similarly, if the value of either v_4 or v_5 is increased, then vertex v_5 or v_4 will no longer have a private broadcast neighbor. Thus, f is a 1-maximal irredundant broadcast function. However, f is not a maximal irredundant broadcast, since the function g defined by $g(v_4) = g(v_5) = 3$ and $g(v_1) = g(v_2) = g(v_3) = g(v_6) = g(v_7) = g(v_8) = 0$ is an irredundant broadcast for which $g \geq f$.

Proposition 16 *An irredundant broadcast function can be 1-maximal but not maximal, but every maximal irredundant broadcast is 1-maximal.*

Consider the path P_6, with vertices labeled v_1, ..., v_6, the maximal irredundant set $S = \{v_3, v_4\}$, and the broadcast function f_S defined by $f_S(v_3) = f_S(v_4) = 1$ and $f_S(v_1) = f(v_2) = f(v_5) = f_S(v_6) = 0$. It is easy to see that f_S is an irredundant broadcast. However, it is not a maximal irredundant broadcast. The function g defined by

$f_S(v_3) = f_S(v_4) = 2$ and $f_S(v_1) = f(v_2) = f(v_5) = f_S(v_6) = 0$ is an irredundant broadcast function for which $g \geq f_S$.

Proposition 17 *The characteristic function of every maximal irredundant set is an irredundant broadcast, but is not necessarily a maximal irredundant broadcast function.*

Proposition 18 (Ahmadi et al.) *Every γ_b-broadcast function is a maximal irredundant broadcast.*

Proof Let f be a γ_b-broadcast for an arbitrary graph G of order n. It follows that any broadcast $g \neq f$ with $g \leq f$ cannot be a dominating broadcast, since f is a minimal dominating broadcast. This means that f is an irredundant broadcast since for any such a broadcast g, with $g \leq f$, $|H(g)| < |H(f)| = n$. It follows that f is a maximal irredundant broadcast since any broadcast h with $h \geq f$ must have $|H(h)| = n$, and therefore since $f \leq h$, and $|H(f)| = n = |H(h)|$, h cannot be an irredundant broadcast. $\square$

Corollary 9 *For any graph G,*

$$ir_b \leq \gamma_b \leq \Gamma_b \leq IR_b.$$

Finally, Ahmadi et al. establish the following Broadcast Domination Chain:

$$ir_b \leq \gamma_b \leq i_b \leq \alpha_b \leq \Gamma_b \leq IR_b.$$

In addition they show the following.

Proposition 19 (Ahmadi et al.) *For any graph G, $IR(G) \leq IR_b(G)$.*

At the end of their paper, Ahmadi et al. ask the following questions:

1. What is the relationship between $ir(G)$ and $ir_b(G)$?
2. Can $ir_b(T)$ and $IR_b(T)$ be computed in polynomial time for a tree T?
3. What is the complexity of the following decision problem?

   MAXIMAL IRREDUNDANT BROADCAST
   INSTANCE: Graph $G = (V, E)$, function $f : V \rightarrow \{0, 1, 2, \ldots \ldots\}$
   QUESTION: Is f a maximal irredundant broadcast function?

12   Roman Irredundance in Graphs

In 2016 Chellali, Haynes, S.M. Hedetniemi, S. T. Hedetniemi, and McRae [8] introduced a Roman Domination Chain, comparable in many respects to the Domination Chain and the Broadcast Domination Chain. In this section, we present the necessary definitions for this chain of inequalities.

A function $f : V \rightarrow \{0, 1, 2\}$ is a *Roman dominating function* on a graph $G = (V, E)$ if for every vertex $v \in V$ with $f(v) = 0$, there exists a neighbor $u \in N(v)$ with $f(u) = 2$. The *weight* of a Roman dominating function is $f(V) = \sum_{v \in V} f(v)$. The *Roman domination number* $\gamma_R(G)$ equals the minimum weight of a Roman dominating function on G, and the *upper Roman domination number* $\Gamma_R(G)$ equals the maximum weight of a minimal Roman dominating function on G. The graph theoretical introduction of Roman domination was introduced in 2004 by Cockayne, Dreyer, S. M. Hedetniemi, and S. T. Hedetniemi [21].

Given any function of the form $f : V \rightarrow \{0, 1, 2\}$, it is convenient to define the following three sets, for $i \in \{0, 1, 2\}$, $V_i = \{v \in V : f(v) = i\}$. Thus, we can denote such a function f by $f = \{V_0, V_1, V_2\}$. And the weight of such a function is $f(V) = |V_1| + 2|V_2|$.

A function $f = \{V_0, V_1, V_2\}$ is called *irredundant* if

1. V_1 is an independent set,
2. no vertex in V_1 is adjacent to a vertex in V_2,
3. every vertex $v \in V_2$ has a private neighbor in V_0 with respect to the set V_2, that is, there exists a vertex $w \in V_0$ such that $N(w) \cap V_2 = \{v\}$.

A Roman irredundant function is *maximal* if increasing the value assigned to any vertex results in a function that is no longer irredundant. The *(lower) Roman irredundance number* $ir_R(G)$ equals the minimum weight of a maximal Roman irredundant function on G. The *upper Roman irredundance number* $IR_R(G)$ equals the maximum weight of an irredundant function on G.

This leaves us to define the Roman independence numbers.

A Roman dominating function is called *independent* if $V_1 \cup V_2$ is an independent set. The *independent Roman domination number* $i_R(G)$ equals the minimum weight of an independent Roman dominating function on G.

The *Roman independence number* $\alpha_R(G)$ equals the maximum weight of an irredundant, independent Roman dominating function on G.

With these definitions, Chellali et al. were able to prove the following theorem.

Theorem 7 (Chellali et al.) *For any graph G,*

$$ir_R \leq \gamma_R \leq i_R \leq \alpha_R \leq \Gamma_R \leq IR_R.$$

Proof Sketch.

1. $i_R(G) \leq \alpha_R(G)$ follows from the definitions of $i_R(G)$ and $\alpha_R(G)$ and the fact that every i_R-function on a graph G is a Roman irredundant function.
2. $\gamma_R(G) \leq i_R(G)$ and $\alpha_R(G) \leq \Gamma_R(G)$ follow the fact that $i_R(G)$ and $\alpha_R(G)$ are both realized by Roman independent, irredundant dominating functions, while $\gamma_R(G)$ and $\Gamma_R(G)$ are realized by Roman irredundant dominating functions.
3. $\Gamma_R(G) \leq IR_R(G)$ follows from the definition of $\Gamma_R(G)$ that every Γ_R-function is an irredundant function.
4. $ir_R(G) \leq \gamma_R(G)$ follows, first of all, from the observation that every graph G has a γ_R-function that is irredundant. It only remains to show that every irredundant

γ_R-function is *maximal* irredundant. This follows from the observation that if any vertex is assigned a larger value, then the resulting function will no longer be an irredundant function: a. by definition no vertex in V_2 can have its value increased or it won't be a Roman dominating function, b. no vertex in V_0 can be increased to 1 since it would be adjacent to a vertex in V_2 and no longer be an irredundant function, and c. no vertex in V_0 or V_1 can be increased to 2 since it could not have a private neighbor in V_0. $\qquad\square$

Chellali et al. were also able to establish the following inequalities:

1. $ir(G) \leq ir_R(G)$.
2. $\alpha(G) \leq \alpha_R(G) \leq 2\alpha(G)$.
3. $\Gamma(G) \leq \Gamma_R(G)$.
4. $IR(G) \leq IR_R(G)$.

13 Fractional Irredundance

Many discrete, or integer-valued, graph theory concepts have fractional counterparts, and irredundance is no exception. This can be explained as follows.

Let $G = (V, E)$ be a graph and let Y be an arbitrary set of real numbers, finite or infinite, positive or negative. A function $f : V \to Y$ is called a Y-*dominating function* if for every $v \in V$, $f(N[v]) = \Sigma_{u \in N[v]} f(u) \geq 1$. In other words, the closed neighborhood sum $f(N[v])$ of every vertex $v \in V$ is at least one.

The *weight* of a Y-dominating function f is $w(f) = f(V) = \Sigma_{u \in V} f(u)$. The Y-*domination number* $\gamma_Y(G)$ equals the minimum weight of a Y-dominating function f on G.

A Y-dominating function f is called *minimal* if there does not exist another Y-dominating function g, $f \neq g$, with $g(v) \leq f(v)$ for every $v \in V$. The *upper Y-domination number* $\Gamma_Y(G)$ equals the maximum weight $w(f)$ of a minimal Y-dominating function f on G.

When $Y = \{0, 1\}$, $\gamma_{\{0,1\}}(G) = \gamma(G)$, the standard *domination number* of a graph G, and $\Gamma_{\{0,1\}}(G) = \Gamma(G)$, the *upper domination number* of G.

When $Y = [0, 1]$ is the closed unit interval, $\gamma_{[0,1]}(G) = \gamma_f(G)$, the *fractional domination number* of a graph G, and $\Gamma_{[0,1]}(G) = \Gamma_f(G)$, the *upper fractional domination number* of G.

In 2006 Fricke, Hedetniemi and Jacobs [36] introduced the following.

A function $g : V \to [0, 1]$ is called an *irredundant function* if for every vertex $v \in V$ with $g(v) > 0$, there exists a vertex $w \in N[v]$ such that $g(N[w]) = 1$.

An irredundant function f is called *maximal* if there does not exist an irredundant function h, $h \neq g$, with $h(v) \geq g(v)$ for every $v \in V$.

It is easy to see that the characteristic function χ_S of a (maximal) irredundant set S is a (maximal) irredundant function.

Definition 11 *The fractional irredundance number $ir_f(G)$ is the infimum, $ir_f(G) = \inf\{g(V) : g$ is a maximal irredundant function on $G\}$, and the fractional upper irredundance number $IR_f(G)$ is the supremum, $IR_f(G) = \sup\{g(V) : g$ is a maximal irredundant function on $G\}$.*

In 1988, Domke, Hecetniemi, and Laskar [24] point out that for the Hajós graph G, $\gamma_f(G) < \gamma(G)$ (cf. Figure 6), where $\gamma_f(G) = 3/2 < \gamma(G) = 2$.

In 1996, Fricke, Hedetniemi, and Jacobs [34] point out that for the path P_7, $ir_f(P_7) = 2 < ir(P_7) = 3$. They also note that while the infimum of $g(P_7) = 2$ over all maximal irredundant functions, no maximal irredundant function has this value (cf. Figure 7).

In 1990, Cheston, Fricke, Hedetniemi, and Jacobs [9] present a graph G with $\Gamma(G) < \Gamma_f(G)$ (cf. Figure 8).

Given that these strict inequalities can hold for some graphs G, $\gamma_f(G) < \gamma(G)$, $\Gamma(G) < \Gamma_f(G)$, and $ir_f(G) < ir(G)$, the following result due to Fricke, Hedetniemi, and Jacobs [36] is somewhat surprising.

Theorem 8 *For any graph G, $IR(G) = IR_f(G)$.*

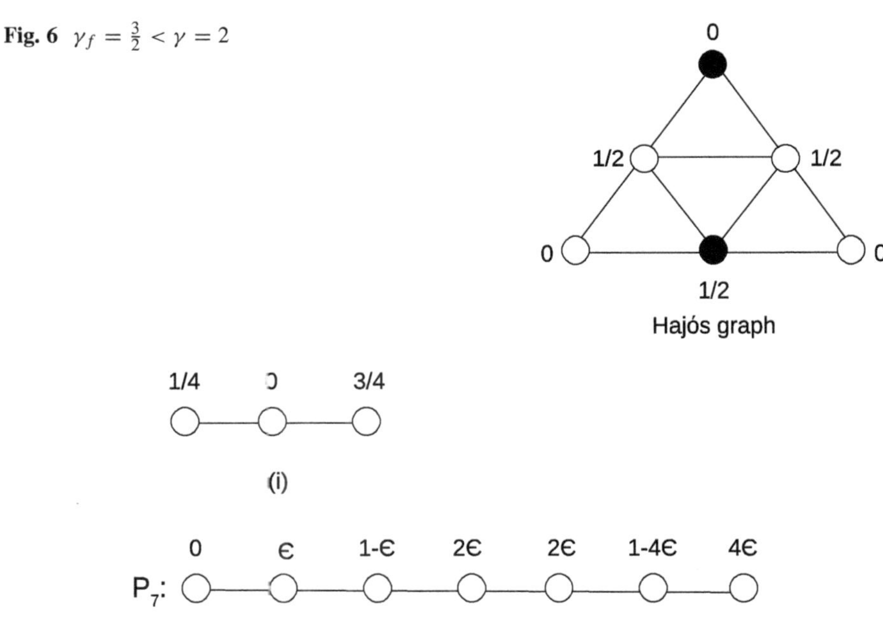

Fig. 6 $\gamma_f = \frac{3}{2} < \gamma = 2$

Fig. 7 (i) Irredundant but not maximal. (ii) Some maximal irredundant functions

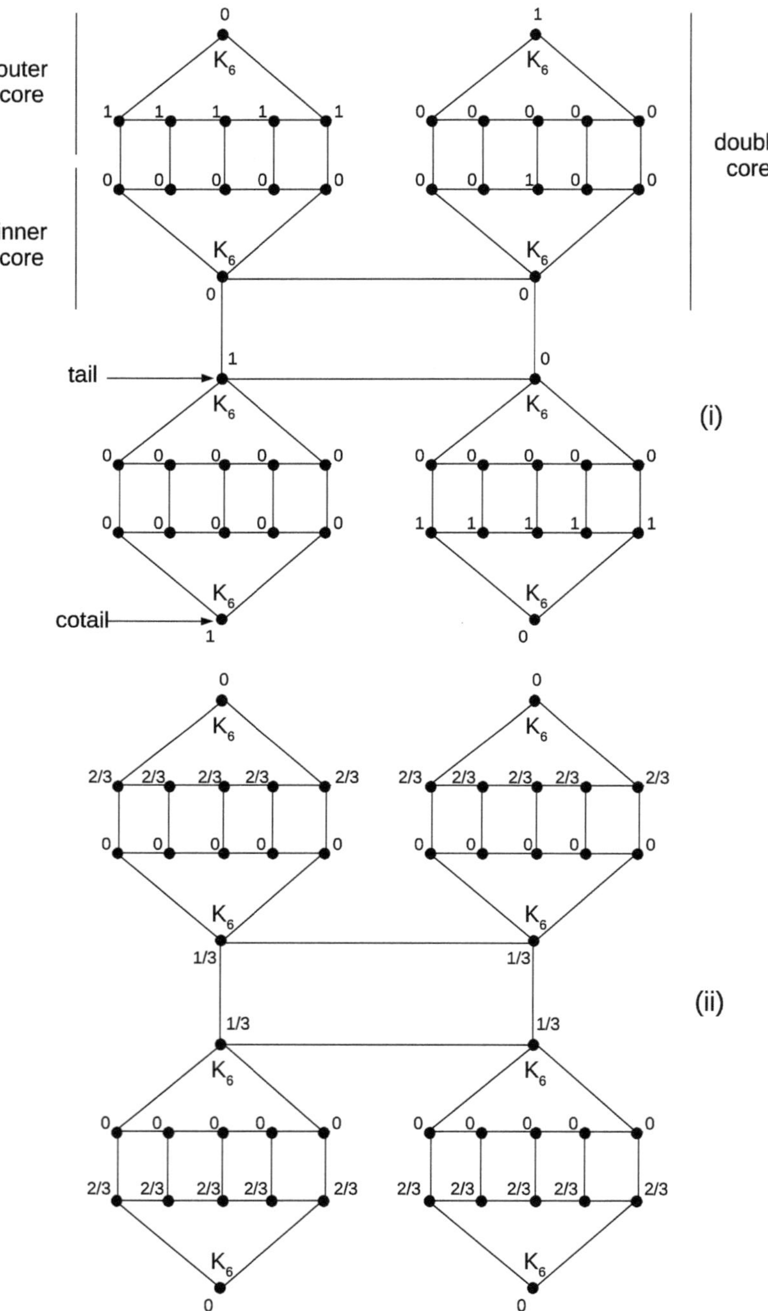

Fig. 8 An example where $\Gamma(G) = 14 < \Gamma_f(G) = 14\frac{2}{3}$

To give the reader some idea of what is involved in proving this result, but without providing the details, it depends on the following ten lemmas.

Let the real-valued functions defined on graphs G of order n be points in the Euclidean metric space $\mathcal{R}^n$, and recall that (i) a set X is *closed* if every convergent sequence in X has its limit in X, (ii) any finite intersection of closed sets is closed, and (iii) a set X is *compact* if every infinite sequence in X has a subsequence which converges in X.

Lemma 1 (Heine-Borel) *A set $X \subseteq \mathcal{R}^n$ is compact if and only if it is closed and bounded.*

Lemma 2 *A continuous real-valued function on a compact metric space achieves a maximum.*

Lemma 3 *In $\mathcal{R}^n$, the set $\mathcal{I}$ of all irredundant functions is a compact set.*

Lemma 4 *The quantity $f(V)$ achieves a maximum on $\mathcal{I}$.*

As defined in [9], a function $g : V \rightarrow [0, 1]$ is called *irreducible* if for every vertex $v \in V$ with $g(v) > 0$, there exists a vertex $w \in N[v]$ such that $g(N[w]) \leq 1$. Note that all irredundant functions are irreducible.

Lemma 5 *Every maximal irreducible function is maximal irredundant.*

Lemma 6 *For any irreducible function g, there exists a maximal irreducible function g' for which $g \leq g'$.*

Lemma 7 *If g is irreducible and $g(V) = IR_f(G)$, then g is irredundant.*

Lemma 8 *The set $\mathcal{M} = \{f \in \mathcal{I} : f(V) = IR_f(G)\}$ is closed.*

For a function f on V, let $z(f) = \{v \in V : f(v) = 0\}$, and let $\mathcal{Z}$ be the functions $f \in \mathcal{M}$ which maximize $z(f)$. The objective is to show that the functions in $\mathcal{Z}$ are 0–1 functions.

Lemma 9 *There exists a function $g \in \mathcal{Z}$ having the smallest positive value b over all functions in $\mathcal{Z}$.*

Let g and b be as in the previous lemma, and let v_b be a vertex for which $g(v_b) = b$.

Lemma 10 *For any vertex $w \neq v_b$ with $g(w) \neq 0$, there exists a vertex $x \in N[w] - N[v_b]$ with $g(N[x]) = 1$.*

The proof of Theorem 8 follows by showing that a maximum irredundant function g having smallest value $g(v_b) = b > 0$ among the functions in $\mathcal{Z}$ is a 0–1 function.

A similar theorem was proved in 2016 by Fricke, O'Brien, Schroeder, and Hedetniemi [37].

A real-valued function $g : V \rightarrow [0, 1]$ is called *open irredundant* if for every vertex $v \in V$ with $g(v) > 0$, there exists a vertex w adjacent to v such that $g(N[w]) = 1$.

An open irredundant function g is *maximal* if there does not exist an open irredundant function h such that $g \neq h$ and $g(v) \leq h(v)$, for every $v \in V$.

Definition 12 *The fractional open irredundance number $oir_f(G)$ is the infimum, $oir_f(G) = inf\{g(V) : g$ is a maximal open irredundant function on $G\}$, and the fractional upper open irredundance number $OIR_f(G)$ is the supremum, $OIR_f(G) = sup\{g(V) : g$ is a maximal open irredundant function on $G\}$.*

Notice the slight distinction between open irredundant functions and irredundant functions; for irredundant functions, there must be a vertex $w \in N[v]$ (closed neighborhood) with $g(N[w]) = 1$, while for open irredundant functions, there must be a such a vertex $w \in N(v)$ (open neighborhood) with $g(N[w]) = 1$, that is, $w \neq v$.

Theorem 9 (Fricke et al.) *For any graph G, $OIR(G) = OIR_f(G)$.*

This result is proved roughly as follows.

A function $g : V \to [0, 1]$ is *open irreducible* or *oiru* if for every vertex $v \in V$ with $g(v) > 0$, there exists a vertex $w \in N(v)$ (open neighborhood) such that $g(N[w]) \leq 1$.

In the special case that for every $v \in V$ with $g(v) > 0$, there exists a vertex $w \in N(v)$ such that $g(N[w]) = 1$, we say that g is *fractional open irredundant*.

Furthermore, if g is a fractional open irredundant function such that $g : V \to \{0, 1\}$, then g is *open irredundant*.

Examples of each type of function are shown by Fricke et al. [37] in Figure 9. Thus,

$$OIRU_f(G) = sup\{g(V) | g \text{ is a fractional oiru function}\},$$
$$OIR_f(G) = sup\{g(V) | g \text{ is a fractional open irredundant function}\}$$
$$OIR(G) = sup\{g(V) | g \text{ is an open irredundant function}\}.$$

Note that since all open irredundant functions are fractional open irredundant, and all fractional open irredundant functions are oiru, we immediately have that

$$OIR(G) \leq OIR_f(G) \leq OIRU_f(G).$$

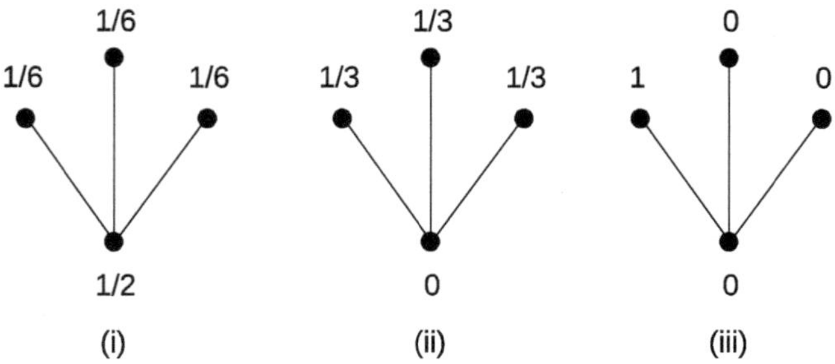

Fig. 9 (i) An oiru function. (ii) A fractional open irredundant function. (iii) An open irredundant function

Fricke et al. show that, in fact, equality holds, a key step of which is showing that the supremum $OIRU_f(G)$ is a maximum, that is, there exists an oiru function g such that $g(V) = OIRU_f(G)$.

14   Open Problems Involving Irredundance

The inequalities in the Domination Chain and Covering Chain suggest that these parameters might have relationships to others that have been studied in the literature. We mention just a few of these possibilities.

1. It has been shown that $IR(G) \leq ER(G)$, where $ER(G)$ equals the maximum cardinality of an external redundant set (cf. p. 97 of [42]). How does $ER(G)$ compare with $\Psi(G)$ and $CIR(G)$?
2. What can you say about the parameter $cir(G)$? We have observed that $\gamma(G) \leq cir(G) \leq \psi(G)$.
3. It has been observed that (i) $\gamma(G) \leq \alpha_1(G)$; (ii) $\gamma(G) \leq \beta_1(G)$; (iii) $\gamma(G) \leq \beta_2(G)$, where $\beta_2(G)$ is the 2-maximal matching number (cf. p. 59 of [42]); and (iv) $\gamma(G) \leq 2ir(G) - 1$. How do these bounds compare with either $i(G)$ or $cir(G)$?
4. It has been observed that for many classes of graphs, including bipartite, chordal, circular arc, cographs, and permutation graphs, just to name a few, the upper three parameters are all equal, that is, $\alpha(G) = \Gamma(G) = IR(G)$ (cf. p. 81 of [42]). Can these equalities be extended for these classes of graphs to: $\alpha(G) = \Gamma(G) = IR(G) = \Psi(G) = CIR(G)$?
5. The concept of irredundance is inherently about the concepts of private neighbors. In searching for the next concept after independence, domination, and irredundance, several authors proposed the study of external redundance, R-annihilated sets, private neighbor counts, and pnc-maximal sets. With this in mind, one can study the maximum number of private neighbors of a given type, or of given types, in sets S, not the number of vertices in S which have at least one private neighbor, but the total number of vertices in V which are a private neighbor of some vertex in S. One can make the following definitions:

   $IR^*(S)$, the number of vertices in V that are a private neighbor (self or external) of a vertex in S; $IR(G) \leq IR^*(G) = max\{IR^*(S) : S \subseteq V\}$.

   $OIR^*(S)$, the number of vertices in V that are an external private neighbor of a vertex in S; $OIR(G) \leq OIR^*(G) = max\{OIR^*(S) : S \subseteq V\}$.

   $OOIR^*(S)$, the number of vertices in V that are either an external or an internal private neighbor of a vertex in S; $OOIR(G) \leq OOIR^*(G) = max\{OOIR^*(S) : S \subseteq V\}$.

   $COIR^*(S)$, the number of vertices in V that are a private neighbor (self, external, or internal) of a vertex in S; $COIR(G) \leq COIR^*(G) = max\{COIR^*(S) : S \subseteq V\}$.

   One can extend this to other parameters, such as the following:

$IR_\alpha^*(S)$, the number of vertices in V that are a private neighbor (self or external) of a vertex in an independent set S; $IR_\alpha^*(G) = max\{IR_\alpha^*(S) : S$ an independent set in $G\} \geq \alpha(G)$.

$IR_\gamma^*(S)$, the number of vertices in V that are a private neighbor (self or external) of a vertex in a dominating set S; $IR_\gamma^*(G) = max\{IR_\gamma^*(S) : S$ a dominating set in $G\} \geq \Gamma(G)$.

$IR_{ir}^*(S)$, the number of vertices in V that are a private neighbor (self or external) of a vertex in an irredundant set S; $IR_{ir}^*(G) = max\{IR_{ir}^*(S) : S$ an irredundant set in $G\} \geq IR(G)$.

References

1. D. Ahmadi, G.H. Fricke, C. Schroeder, S.T. Hedetniemi, R.C. Laskar, Broadcast irredundance in graphs. Congr. Numer. **224**, 17–31 (2015)
2. S. Arumugam, K.R. Chandrasekar, A note on graphs with open irratic number two. Bull. Inst. Combin. Appl. **65**, 27–32 (2012)
3. S. Arumugam, S.T. Hedetniemi, S.M. Hedetniemi, L. Sathikala, S. Sudha, The covering chain of a graph. Util. Math. **98**, 183–196 (2015)
4. M.W. Bern, E.L. Lawler, A.L. Wong, Why certain subgraph computations require only linear time, in *Proc. 26th Ann. Symp. on the Foundations of Computer Science*, Portland, OR (1985), pp. 117–125
5. T. Beyer, A. Proskurowski, S. Hedetniemi, S. Mitchell, Independent domination in trees, in *Proc. Eighth Southeastern Conf. on Combinatorics, Graph Theory and Computing*. Congr. Numer., vol. 19 (1977), pp. 321–328
6. K. Cameron, Induced matchings. Discrete Appl. Math. **24**, 97–102 (1989)
7. G. Chartrand, S.T. Hedetniemi, F. Okamoto, P. Zhang, A four colorings theorem. J. Combin. Math. Combin. Comput. **77**, 75–87 (2011)
8. M. Chellali, T.W. Haynes, S.M. Hedetniemi, S.T. Hedetniemi, A.A. McRae, A Roman domination chain. Graphs Combin. **32**(1), 79–92 (2016)
9. G.A. Cheston, G. Fricke, S.T. Hedetniemi, D.P. Jacobs, On the computational complexity of upper fractional domination. Discrete Appl. Math. **27**, 195–207 (1990)
10. E.J. Cockayne, Generalized irredundance in graphs: hereditary properties and Ramsey numbers. J. Combin. Math. Combin. Comput. **31**, 15–31 (1999)
11. E.J. Cockayne, Nordhaus-Gaddum results for open irredundance. J. Combin. Math. Combin. Comput. **47**, 213–223 (2003)
12. E.J. Cockayne, S. Finbow, Generalised irredundance in graphs: Nordhaus-Gaddum bounds. Discuss. Math. Graph Theory **24**(1), 147–160 (2004)
13. E.J. Cockayne, C.M. Mynhardt, On a conjecture concerning irredundant and perfect neighbourhood sets in graphs. J. Combin. Math. Combin. Comput. **31**, 241–253 (1999)
14. E.J. Cockayne, S.E. Goodman, S.T. Hedetniemi, A linear algorithm for the domination number of a tree. Inform. Process. Lett. **4**, 41–44 (1975)
15. E.J. Cockayne, S.T. Hedetniemi, D.J. Miller, Properties of hereditary hypergraphs and middle graphs. Canad. Math. Bull. **21**, 461–468 (1978)
16. E.J. Cockayne, P.J.P. Grobler, S.T. Hedetniemi, A.A. McRae, What makes an irredundant set maximal?. J. Combin. Math. Combin. Comput. **25**, 213–223 (1997)
17. E.J. Cockayne, J.H. Hattingh, S.M. Hedetniemi, S.T. Hedetniemi, A.A. McRae, Using maximality and minimality conditions to construct inequality chains. Discrete Math. **176**(1–3), 43–61 (1997)
18. E.J. Cockayne, O. Favaron, C.M. Mynhardt, J. Puech, An inequality chain of domination parameters for trees. Discuss. Math. Graph Theory **18**, 127–142 (1998)

19. E.J. Cockayne, O. Favaron, J. Puech, C.M. Mynhardt, Packing, perfect neighborhood, irredundant and R-annihilated sets in graphs. Australas. J. Combin. **18**, 253–262 (1998)
20. E.J. Cockayne, S.M. Hedetniemi, S.T. Hedetniemi, C.M. Mynhardt, Irredundant and perfect neighbourhood sets in trees. Discrete Math. **188**, 253–260 (1998)
21. E J. Cockayne, P.M. Dreyer Sr., S.M. Hedetniemi, S.T. Hedetniemi, On Roman domination in graphs. Discrete Math. **278**, 11–22 (2004)
22. E.J. Cockayne, O. Favaron, S. Finbow, C.M. Mynhardt, Open irredundance and maximum degree in graphs. Discrete Math. **308**(23), 5358–5375 (2008)
23. E.J. Cockayne, S. Finbow, J.S. Swarts, OO-irredundance and maximum degree in paths and trees. J. Combin. Math. Combin. Comput. **73**, 223–236 (2010)
24. G.S. Domke, S.T. Hedetniemi, R.C. Laskar, Fractional packings, coverings and irredundance in graphs. Congr. Numer. **66**, 227–238 (1988)
25. J.E. Dunbar, D.J. Erwin, T.W. Haynes, S.M. Hedetniemi, S.T. Hedetniemi, Broadcasts in graphs. Discrete Appl. Math. **154**(1), 59–75 (2006)
26. D.J. Erwin, Dominating broadcasts in graphs. Bull. Inst. Combin. Appl. **42**, 89–105 (2004)
27. A.M. Farley, A. Proskurowski, Computing the maximum order of an open irredundant set in a tree. Congr. Numer. **41**, 219–228 (1984)
28. A.M. Farley, N. Schacham, Senders in broadcast networks: open irredundancy in graphs. Congr. Numer. **38**, 47–57 (1983)
29. O. Favaron, J. Puech, Irredundant and perfect neighborhood sets in graphs and claw-free graphs. Discrete Math. **197/198**, 269–284 (1999)
30. O. Favaron, T.W. Haynes, S.T. Hedetniemi, M.A. Henning, D.J. Knisley, Total irredundance in graphs. Discrete Math. **256**(1–2), 115–127 (2002)
31. M. Fellows, G. Fricke, S. Hedetniemi, D. Jacobs, The private neighbor cube. SIAM J. Discrete Math. **7**(1), 41–47 (1994)
32. S. Finbow, Generalisations of irredundance in graphs. Ph.D. Thesis, University of Victoria, Victoria, B.C., 2003, 158 pp.
33. J.F. Fink, M.S. Jacobson, *n*-domination in graphs, in *Graph Theory with Applications to Algorithms and Computer Science* (Wiley, New York, 1985), pp. 283–300
34. G.H. Fricke, S.T. Hedetniemi, D P. Jacobs, Maximal irredundant functions. Discrete Appl. Math. **68**, 267–277 (1996)
35. G.H. Fricke, T.W. Haynes, S.M. Hedetniemi, S.T. Hedetniemi, M.A. Henning, On perfect neighborhood sets in graphs. Discrete Math. **199**, 221–225 (1999)
36. G.H. Fricke, S.T. Hedetniemi, D.P. Jacobs, On the equivalence of the upper irredundance and fractional upper irredundance numbers of a graph. Bull. Inst. Combin. Appl. **48**, 99–106 (2006)
37. G.H. Fricke, T. O'Brien, C. Schroeder, S.T. Hedetniemi, On the equivalence of the upper open irredundance and fractional upper open irredundance numbers of a graph. J. Combin. Math. Combin. Comput. **99**, 187–198 (2016)
38. T. Gallai, Über extreme Punkt- und Kantenmengen. Ann. Univ. Sci. Budapest Eotvos Sect. Math. **2**, 133–138 (1959)
39. F. Gavril, Algorithms for minimum colorings, maximum clique, minimum coverings by cliques, and maximum independent set of a chordal graph. SIAM J. Comput. **1**, 180–187 (1972)
40. W. Goddard, S.T. Hedetniemi, A note on trees, tables, and algorithms. Networks **53**(2), 184–190 (2009)
41. M.C. Golumbic, R. Laskar, Irredundancy in circular arc graphs. Discrete Appl. Math. **44**, 79–90 (1993)
42. T.W. Haynes, S.T. Hedetniemi, P.J. Slater, *Fundamentals of Domination in Graphs* (Marcel Dekker, New York, 1998)
43. T.W. Haynes, S.M. Hedetniemi, S.T. Hedetniemi, A.A. McRae, P.J. Slater, Irredundant colorings of graphs. Bull. Inst. Combin. Appl. **54**, 103–121 (2008)
44. S.T. Hedetniemi, R. Laskar, J. Pfaff, Irredundance in graphs: a survey, in *Proc. Sixteenth Southeastern Conf. on Combinatorics, Graph Theory and Computing (Boca Raton, Fla., 1985)*. Congr. Numer., vol. 48 (1985), pp. 183–193

45. S.M. Hedetniemi, S.T. Hedetniemi, D.P. Jacobs, Total irredundance in graphs: theory and algorithms. Ars Combin. **35A**, 271–284 (1993)
46. S.M. Hedetniemi, S.T. Hedetniemi, M.A. Henning, The algorithmic complexity of perfect neighborhoods in graphs. J. Combin. Math. Combin. Comput. **25**,183–192 (1997)
47. J.T. Hedetniemi, S.M. Hedetniemi, S.T. Hedetniemi, Perfection in graphs, a new look at irredundance. J. Combin. Math. Combin. Comput. **85**, 129–139 (2013)
48. S. Mitchell, S. Hedetniemi, S. Goodman, Some linear algorithms on trees, in *Proc. Sixth Southeastern Conf. on Combinatorics, Graph Theory and Computing* (1975), pp. 467–483
49. C.M. Mynhardt, A. Roux, Irredundance, in *Structures of Domination in Graphs*, ed. by T.W. Haynes, S.T. Hedetniemi, M.A. Henning (Springer, Berlin, 2021), pp. 135–182
50. J. Puech, R-annihilated and independent perfect neighborhood sets in chordal graphs. Discrete Math. **215**(1–3), 181–199 (2000)

An Introduction to Game Domination in Graphs

Michael A. Henning

AMS Subject Classification: 05C65, 05C69

1 Introduction

Although competitive optimization graph games are well studied in the literature, the domination game which we discuss in this chapter is relatively new and was only formally birthed in 2010 by Brešar, Klavžar, and Rall [4]. We remark that this domination game introduced in [4], which we formally define in Section 2.1, is very different from the competition-enclaveless game introduced in 2001 by Philips and Slater [46, 47] which is played by two players who take turns in constructing a maximal enclaveless set in a graph. (Some of the significant differences between these two games are explained in [24, Chapter 5].) We also remark that the domination game introduced in [4] is very different from the domination game introduced in 2002 by Alon, Balogh, Bollobas, and Szabo [1]. In [1], Alon et al. define the oriented game domination number of a graph G for which two players alternately orient an edge of G until all of the edges are oriented, their goals being to minimize and maximize the domination number of the resulting oriented graph.

Since Brešar et al. [4] first studied the concept of the domination game in graphs, it has subsequently attracted considerable interest. Our aim in this chapter is to introduce and familiarize the reader with three domination-type games and to present selected results on these games with the hope to encourage and stimulate continued research of the topic.

M. A. Henning (✉)
Department of Mathematics and Applied Mathematics, University of Johannesburg, Johannesburg, South Africa
e-mail: mahenning@uj.ac.za

© The Author(s), under exclusive license to Springer Nature Switzerland AG 2021
T. W. Haynes et al. (eds.), *Structures of Domination in Graphs*, Developments in Mathematics 66, https://doi.org/10.1007/978-3-030-58892-2_8

For a more comprehensive and thorough treatment of the domination game, we refer the reader to the forthcoming book entitled "Domination Games Played in Graphs" by Brešar, Henning, Klavžar, and Rall [11].

2 Domination-Type Games

In this section, we define three domination-type games, namely the domination game, the total domination game, and the independent domination game. We remark that many other domination-type games are studied in the literature, including the connected domination game, the paired-domination game, the competition-enclaveless game, the oriented domination game, the maker-breaker domination game, the disjoint domination game, the fractional domination game, the Grundy domination game, the Grundy total domination game, the Z-Grundy domination game, the L-Grundy domination game, to name a few. A survey of these domination-type games in graphs can be found in [11].

In this chapter we model the "Domination in Graphs: Core Concepts" book by Haynes, Hedetniemi, and Henning [23] in the sense that we focus exclusively on the three main domination parameters, namely the domination number, the total domination number, and the independent domination number. That is, we restrict our attention to the domination game, the total domination game, and the independent domination game. We begin with a formal definition of the domination game played in graphs.

2.1 *The Domination Game*

We recall that a vertex *dominates* itself and its neighbors. A *dominating set* of a graph G is a set S of vertices of G such that every vertex in G is dominated by a vertex in S. The *domination number* of G, denoted $\gamma(G)$, is the minimum cardinality of a dominating set in G.

The domination game on a graph G consists of two players, *Dominator* and *Staller*, who take turns choosing a vertex from G. Each vertex chosen must dominate at least one vertex not dominated by the vertices previously chosen. We call such a vertex a *playable* vertex. A *move* in the game, sometimes referred to in the literature as a *legal move* for emphasis, is a vertex chosen by a player. The game ends when there are no more moves available. Upon completion of the game, the set of chosen (played) vertices is a dominating set in G, but is not necessarily a minimal dominating set. The goal of Dominator is to end the game with a minimum number of vertices chosen, while Staller has the opposite goal and wishes to end the game with as many vertices chosen as possible.

The Dominator-start domination game and the Staller-start domination game are the domination game when Dominator and Staller, respectively, choose the first

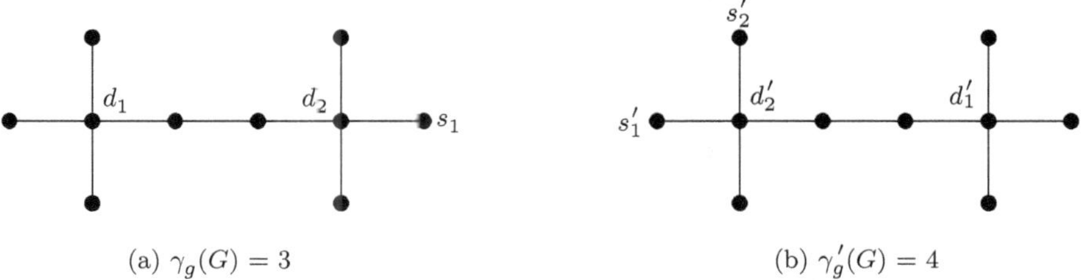

(a) $\gamma_g(G) = 3$

(b) $\gamma_g'(G) = 4$

Fig. 1 A graph G with $\gamma_g(G) = 3$ and $\gamma_g'(G) = 4$

vertex. These games are called the D-*game* and S-*game*, respectively. The *D-game domination number*, $\gamma_g(G)$, of G is the minimum possible number of moves in a D-game when both players play according to the rules, while the *S-game domination number*, $\gamma_g'(G)$, of G is defined analogously for the S-game. A sequence of moves by Dominator that achieves the minimum possible number of moves is called an *optimal sequence* for Dominator, while a sequence of moves by Staller that achieves the maximum possible number of moves is called an *optimal sequence* for Staller.

We denote the sequence of moves played in the D-game by $d_1, s_1, d_2, s_2, \ldots$, where d_i is the vertex chosen on Dominator's ith move, and s_i is the vertex chosen on Staller's ith move in response to Dominator's ith move. The sequence of moves played in the S-game is denoted by $s_1', d_1', s_2', d_2', \ldots$, where s_i' is the vertex played by Staller on her ith move and d_i' is the vertex played by Dominator on his ith move in response to Staller's ith move.

As an illustration, if G is the graph shown in Figure 1, then $\gamma_g(G) = 3$ and one optimal sequence in the D-game is given by d_1, s_1, d_2 as shown in Figure 1a. Moreover, $\gamma_g'(G) = 4$ and one optimal sequence in the S-game is given by s_1', d_1', s_2', d_2' as shown in Figure 1b.

As remarked earlier, the domination game in graphs which we discuss in the chapter was formally birthed in 2010 by Brešar, Klavžar, and Rall [4]. The domination game has subsequently been extensively studied in [5–10, 12, 13, 15, 19, 20, 24, 26, 28, 35, 36, 38–45. 49, 50, 53–56] and other papers.

2.2 The Total Domination Game

A vertex *totally dominates* another vertex if they are neighbors. A *total dominating set* of a graph G is a set S of vertices such that every vertex of G is totally dominated by a vertex in S. The *total domination number* of G, denoted $\gamma_t(G)$, is the minimum cardinality of a total dominating set in G.

The *total domination game* is defined analogously as the domination game, except that in the total version each vertex chosen must totally dominate at least one vertex not totally dominated by the set of vertices previously chosen. Such a

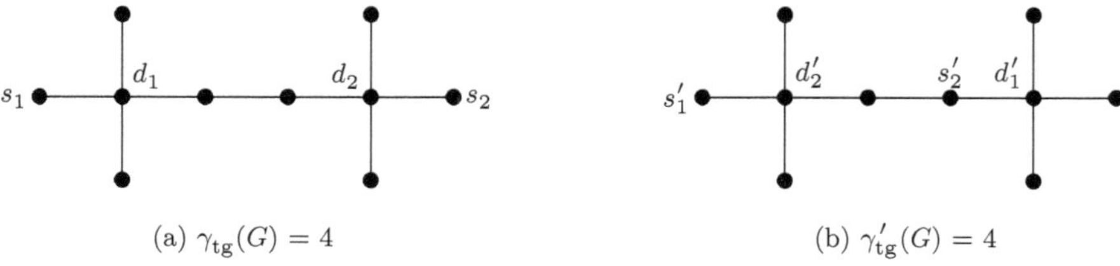

(a) $\gamma_{tg}(G) = 4$ (b) $\gamma'_{tg}(G) = 4$

Fig. 2 A graph G with $\gamma_{tg}(G) = 4$ and $\gamma'_{tg}(G) = 4$

chosen vertex is called a *move* (sometimes referred to as a *legal move* for emphasis) in the total domination game. The game ends when there is no legal move available. In this case, the set of vertices chosen is a total dominating set in G. Dominator's objective is to minimize the number of vertices chosen, while the goal of Staller is just the opposite, namely to end the game with as many vertices chosen as possible.

The Dominator-start (resp., Staller-start) total domination game is the total domination game when Dominator (resp., Staller) has the first move. As with the domination game, we refer to these simply as the *D-gama* and *S*-game, respectively. The *D-game total domination number*, $\gamma_{tg}(G)$, of G is the minimum number of moves in the D-game when both players follow a strategy to achieve their goals, while the *S-game total domination number*, $\gamma'_{tg}(G)$, is the maximum number of moves in a S-game when both players play optimally.

We adopt the same notation as in the domination game, and denote the sequence of moves played in the total version of the D-game by $d_1, s_1, d_2, s_2, \ldots$, and the sequence of moves played in the total version of the S-game by $s'_1, d'_1, s'_2, d'_2, \ldots$.

As an illustration, if G is the graph shown in Figure 2, then $\gamma_{tg}(G) = 4$ and one optimal sequence in the D-game is given by d_1, s_1, d_2, s_2 as shown in Figure 2a. Moreover, $\gamma'_{tg}(G) = 4$ and one optimal sequence in the S-game is given by s'_1, d'_1, s'_2, d'_2 as shown in Figure 2b.

The total version of the domination game was first investigated in 2015 by Henning, Klavžar, and Rall in [31], where it was demonstrated that these two versions differ significantly. The total domination game has subsequently been studied in [3, 14, 16–18, 27, 29–34, 37] and other papers.

2.3 The Independent Domination Game

An *independent dominating set* in G is a dominating set of G that is independent. The *independent domination number*, denoted $i(G)$, of G is the minimum cardinality of an independent dominating set in G. An independent set of vertices in G is a dominating set of G if and only if it is a maximal independent set. Thus, $i(G)$ is equivalently the minimum cardinality of a maximal independent set of vertices in G.

The *independent domination game*, called the *competition-independence game* by Philips and Slater [46, 47], is defined analogously as the domination game, except that in the independent version each vertex chosen must not be adjacent to any vertex previously chosen. More formally, adopting the notation coined by Goddard and Henning [22], the game is played by two players, Diminisher and Sweller, on some graph G. They take turns in constructing a maximal independent set I of G. That is, each turn a player chooses a vertex that is not adjacent to any of the vertices already chosen until there is no such vertex. Such a chosen vertex is called a *move* (or *legal move*, for emphasis) in the independent domination game. The game ends when there is no legal move available. In this case, the set I of vertices chosen is an independent dominating set in G. The goal of Diminisher is to make the final set I as small as possible and for Sweller to make the final set I as large as possible.

The Diminisher-start independent domination game and the Sweller-start independent domination game are the independent domination game when Diminisher and Sweller, respectively, choose the first vertex. As with the domination and total domination game, these games are called the D-*game* and S-*game*, respectively. The *D-game independent domination number*, $I_d(G)$, of G is the minimum possible number of moves in a D-game when both players follow a strategy to achieve their goals, while the *S-game independent domination number*, $I_s(G)$, is the number of moves in a S-game when both players play optimally.

As before, we denote the sequence of moves played in the independent domination version of the D-game by $d_1, s_1, d_2, s_2, \ldots$, and the sequence of moves played in the independent domination version of the S-game by $s_1', d_1', s_2', d_2', \ldots$.

As an illustration, if G is the graph shown in Figure 3, then $I_d(G) = 5$ and one optimal sequence in the D-game is given by d_1, s_1, d_2, s_2, d_3 as shown in Figure 3a. Moreover, $I_s(G) = 5$ and one optimal sequence in the S-game is given by $s_1', d_1', s_2', d_2', s_3'$ as shown in Figure 3b.

In 2001 Philips and Slater [46, 47] introduced the independent domination game, which they called the *competition-independence game*. The independent domination game has not attracted the same amount of interest as the domination and total domination games, and has been studied in [22, 52] and other papers. The most significant difference between the domination and total domination game compared with the independent domination game (and the competition-enclaveless game defined in [46, 47]) is that the so-called *Continuation Principle*, which we present in Section 5, holds for both the domination game and total domination game, but does not hold for the independent domination game. This makes it very difficult to obtain general results or the independent domination game (and the competition-enclaveless game).

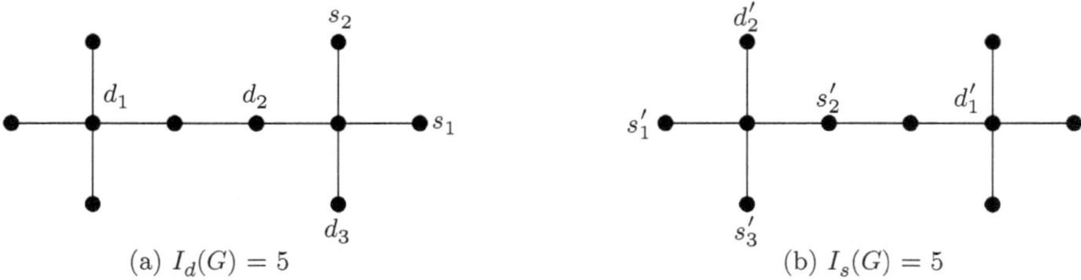

(a) $I_d(G) = 5$ (b) $I_s(G) = 5$

Fig. 3 A graph G with $I_d(G) = 5$ and $I_s(G) = 5$

3 Basic Properties

In their introductory paper on the domination game, Brešar, Klavžar, and Rall [4] established the following relationship between the domination number and the game domination number.

Theorem 1 ([4]) *For every graph G, we have $\gamma(G) \leq \gamma_g(G) \leq 2\gamma(G) - 1$.*

An analogous relationship between the total domination number and the game total domination number was established in the introductory paper by Henning, Klavžar, and Rall in [31] on the total domination game. The proof of Theorems 1 and 2 are along similar lines. We therefore present here only the proof of Theorem 2.

Theorem 2 ([31]) *If G is a graph with no isolated vertex, then $\gamma_t(G) \leq \gamma_{tg}(G) \leq 2\gamma_t(G) - 1$.*

Proof. Upon completion of the Dominator-start total domination game played on G, the vertices played by Dominator and Staller together form a total dominating set of G, implying that $\gamma_t(G) \leq \gamma_{tg}(G)$. To prove that $\gamma_{tg}(G) \leq 2\gamma_t(G) - 1$, Dominator adopts the following strategy. He selects an arbitrary minimum total dominating set D of G and orders the vertices of D. On each of his moves he plays a vertex from the set D sequentially according to this ordering that has not yet been played and is a legal move (and therefore totally dominates at least one vertex not totally by the set of vertices previously played by the two players). We note that when Dominator considers a vertex v in the ordered set D, either the vertex v is a legal move, in which case he plays the vertex v, or the vertex v is not a legal move, in which case he considers the next vertex in the ordering, if such a vertex exists. In both cases, after Dominator has played his move, the vertex v can never be a legal move in the remaining part of the game and Dominator therefore never considers the vertex v again. Once Dominator has considered all vertices according to his ordering of the set D, every vertex is totally dominated by the set of vertices previously played by Dominator and Staller, and hence no more moves are legal. Thus, Dominator plays at most $|D|$ moves and Staller at most $|D| - 1$ moves. In this way, Dominator can guarantee that the game finishes in at most $2|D| - 1 = 2\gamma_t(G) - 1$ moves, implying that $\gamma_{tg}(G) \le 2\gamma_t(G) - 1$. $\qquad\qquad\square$

A significant difference between the domination game and the independent domination game is that upon completion of the domination game, the set of played vertices is a dominating set although not necessarily a minimal dominating set, while upon completion of the independent domination game, the set of played vertices is always a maximal independent set. Thus, the independent domination game numbers of a graph G are always squeezed between the independent domination number $i(G)$ of G and the independence number $\alpha(G)$ of G, which is the maximum cardinality of an independent set in G. We state this formally as follows.

Theorem 3 *If G is a graph of order n, then*

$$i(G) \le I_d(G) \le \alpha(G) \quad \text{and} \quad i(G) \le I_s(G) \le \alpha(G).$$

A graph G is *well-covered* if all of the maximal independent set in G have the same cardinality. The problem of determining which graphs have the property that every maximal independent set of vertices is also a maximum independent set was proposed in 1970 by Plummer [48] and has subsequently been extensively studied in the literature. As observed earlier, upon completion of the independent domination game, the set of played vertices is always a maximal independent set. Hence, any sequence of legal moves by Diminisher and Sweller (regardless of strategy) in the independent domination game played in a well-covered graph of order n will always lead to the game ending in $\alpha(G)$ moves. Thus as a consequence of Theorem 3, we have the following interesting connection between the independent domination game and the class of well-covered graphs.

Theorem 4 *If G is a well-covered graph, then $I_d(G) = I_s(G) = \alpha(G)$.*

The game domination number and the game total domination number are related as follows.

Theorem 5 ([31]) *If G is a graph on at least two vertices, then $\gamma_g(G) \le 2\gamma_{tg}(G) - 1$.*

Proof. By Theorem 1, we have $\gamma_g(G) \le 2\gamma(G) - 1$. Since every total dominating set is by definition a dominating set of G, the inequality $\gamma(G) \le \gamma_t(G)$ holds. By Theorem 2, we have $\gamma_t(G) \le \gamma_{tg}(G)$. These observations imply that $\gamma_g(G) \le 2\gamma(G) - 1 \le 2\gamma_t(G) - 1 \le 2\gamma_{tg}(G) - 1$. $\square$

As observed in [31], to see that Theorem 5 is close to being optimal consider the following examples. For any integer $k \ge 2$, let G_k be the graph obtained from the complete graph on k vertices by attaching k leaves to each of its vertices. As shown in [31], $\gamma_{tg}(G_k) = k + 1$ and $\gamma_g(G_k) = 2k - 1$, and so $\gamma_g(G_k) = 2\gamma_{tg}(G_k) - 3$. Thus we have the following result.

Theorem 6 ([31]) *If $n \ge 2$ is an integer and $\mathcal{G}_n$ denotes the class of all isolate-free graphs G of order n, then*

$$\sup_{n} \frac{\gamma_g(G)}{\gamma_{tg}(G)} = 2,$$

where the supremum is taken over all graphs $G \in \mathcal{G}_n$.

The game total domination number can be bounded by the domination number as follows.

Theorem 7 ([31]) *If G is a graph such that $\gamma(G) \geq 2$, then $\gamma(G) \leq \gamma_{tg}(G) \leq 3\gamma(G) - 2$.*

Proof. The lower bound follows immediately from the inequality chain $\gamma(G) \leq \gamma_t(G) \leq \gamma_{tg}(G)$. To prove the upper bound, let D be an arbitrary γ-set of G. Dominator adopts the following simple strategy in the total domination game. He selects vertices in D sequentially whenever such a move is legal. Once Dominator has played all allowable vertices in D, we note that at most $2|D| - 1 = 2\gamma(G) - 1$ moves have been made. At this point of the game all vertices that have a neighbor in the set D are totally dominated. There are two possible cases to consider.

Case 1: No vertex in D is currently totally dominated. In this case, the set D is an independent set and both Dominator and Staller only played vertices from D. Thus, exactly $|D| = \gamma(G)$ moves have been made at this point in the game. The only remaining legal moves that can be played in the total domination game are those that totally dominate vertices in D. There are therefore at most $|D|$ additional moves that are played in order to complete the game, implying that the total number of moves played is at most $2|D| = 2\gamma(G) \leq 3\gamma(G) - 2$ noting that $\gamma(G) \geq 2$.

Case 2: At least one vertex in D is currently totally dominated. In this case, the only legal moves remaining in the total domination game are those that totally dominate vertices in D, if any, are not yet totally dominated. This implies that at most $|D| - 1$ additional moves are required to complete the game. Therefore, the total number of moves played is at most $(2|D| - 1) + (|D| - 1) = 3\gamma(G) - 2$. $\square$

As shown in [31], both the lower and upper bounds in Theorem 7 are tight. We present here only an example illustrating the lower bound. For $k \geq 3$ and $\ell \geq 1$, let $G = F_{k,\ell}$ be obtained from a complete bipartite graph $K_{2,k}$ by selecting an arbitrary vertex v of degree 2 in $K_{2,k}$, and appending to it ℓ vertex-disjoint paths of length 2. We note that $\gamma(G) = \ell + 2$, and hence $\gamma_{tg}(G) \geq \gamma(G) = \ell + 2$. Let u and w be the neighbors of v that belong to the complete bipartite graph $K_{2,k}$. Suppose now that Dominator plays as his first move in the total domination game the vertex v. The only possible legal moves in the remainder of the game are the $\ell + 2$ neighbors of the vertex v in G. However, exactly one of u and w can be played in the game, while every support vertex of v (of degree 2) in G must be played. Thus, exactly $\ell + 2$ vertices are played in the game, namely the vertex v, exactly one of u and w, and all ℓ support vertices of G. This strategy of Dominator implies that $\gamma_{tg}(G) \leq \ell + 2$. As observed earlier, $\gamma_{tg}(G) \geq \ell + 2$. Consequently $\gamma(G) = \gamma_{tg}(G) = \ell + 2$.

4 Paths and Cycles

Determining exact values of the domination game parameters for even relatively simple classes of graphs is relatively complex. The exact values of the game domination number, the game total domination number, and the game independent domination number for paths and cycles are known.

In 2017 Košmrlj [43] determined the formulas for the game domination number for paths and cycles, and also gave optimal strategies for both players.

Theorem 8 ([43]) *If $n \geq 3$, then*

$$
\gamma_g(C_n) =
\begin{cases}
\left\lceil \frac{n}{2} \right\rceil - 1; & n \equiv 3 \,(\mathrm{mod}\,4), \\[2ex]
\left\lceil \frac{n}{2} \right\rceil; & otherwise,
\end{cases}
$$

and

$$
\gamma'_g(C_n) =
\begin{cases}
\left\lceil \frac{n-1}{2} \right\rceil - 1; & n \equiv 2 \,(\mathrm{mod}\,4), \\[2ex]
\left\lceil \frac{n-1}{2} \right\rceil; & otherwise.
\end{cases}
$$

Theorem 9 ([43]) *If $n \geq 1$, then*

$$
\gamma_g(P_n) =
\begin{cases}
\left\lceil \frac{n}{2} \right\rceil - 1; & n \equiv 3 \,(\mathrm{mod}\,4), \\[2ex]
\left\lceil \frac{n}{2} \right\rceil; & otherwise,
\end{cases}
$$

and $\gamma'_g(P_n) = \left\lceil \frac{n}{2} \right\rceil$.

For small n, the values of the game domination numbers for a cycle C_n and a path P_n are shown in Table 1.

The game total domination numbers for cycles and paths were determined in 2016 by Dorbec and Henning [18].

Theorem 10 ([18]) *If $n \geq 3$, then*

$$
\gamma_{tg}(C_n) =
\begin{cases}
\left\lfloor \frac{2n+1}{3} \right\rfloor - 1; & n \equiv 4 \,(\mathrm{mod}\,6), \\[2ex]
\left\lfloor \frac{2n+1}{3} \right\rfloor; & otherwise,
\end{cases}
$$

Table 1 The game domination numbers for small cycles and paths

n	3	4	5	6	7	8	9	n	3	4	5	6	7	8	9
$\gamma_g(C_n)$	1	2	3	3	3	4	5	$\gamma_g(P_n)$	1	2	3	3	3	4	5
$\gamma'_g(C_n)$	1	2	2	2	3	4	4	$\gamma'_g(P_n)$	2	2	3	3	4	4	5

and

$$\gamma'_{\mathrm{tg}}(C_n) = \begin{cases} \lfloor \frac{2n}{3} \rfloor - 1; & n \equiv 2 \ (\mathrm{mod}\ 6), \\[2mm] \lfloor \frac{2n}{3} \rfloor; & otherwise. \end{cases}$$

Theorem 11 *If $n \geq 1$, then*

$$\gamma_{\mathrm{tg}}(P_n) = \begin{cases} \lfloor \frac{2n}{3} \rfloor; & n \equiv 5 \ (\mathrm{mod}\ 6), \\[2mm] \lceil \frac{2n}{3} \rceil; & otherwise, \end{cases}$$

and $\gamma'_{tg}(P_n) = \left\lceil \frac{2n}{3} \right\rceil$.

For small n, the values of the game total domination numbers for a cycle C_n and a path P_n are shown in Table 2.

We remark that these results for paths and cycles show that for some families of graphs Dominator has an advantage (paths) and for some families of graphs Staller has an advantage (cycles). And for still other families neither player has an advantage by going first.

In 2002 Philips and Slater [47] determined the game independent domination numbers of paths and cycles.

Theorem 12 ([47]) *The following holds.*

(a) *For $n \geq 1$, $I_d(P_n) = \lfloor \frac{3n+4}{7} \rfloor$ and $I_s(P_n) = \lfloor \frac{3n+6}{7} \rfloor$.*
(b) *For $n \geq 3$, $I_d(C_n) = \lfloor \frac{3n+3}{7} \rfloor$ and $I_s(C_n) = \lfloor \frac{3n+2}{7} \rfloor$.*

We note that the game independent domination numbers for a path immediately provide the value for a cycle, since the first move in a cycle C_n on $n \geq 3$ vertices produces a path P_{n-3} on $n - 3$ vertices. Thus, $I_d(C_n) = 1 + I_s(P_{n-3})$ and $I_s(C_n) = 1 + I_d(P_{n-3})$.

For small n, the values of the game independent domination numbers for a cycle C_n and a path P_n are shown in Table 3.

Table 2 The game total domination numbers for small cycles and paths

n	3	4	5	6	7	8	9	n	3	4	5	6	7	8	9
$\gamma_{tg}(C_n)$	2	2	3	4	5	5	6	$\gamma_{tg}(P_n)$	2	3	3	4	5	6	6
$\gamma'_{tg}(C_n)$	2	2	3	4	4	4	6	$\gamma'_{tg}(P_n)$	2	3	4	4	5	6	6

Table 3 The game independent domination numbers for small cycles and paths

n	3	4	5	6	7	8	9	n	3	4	5	6	7	8	9
$I_d(C_n)$	1	2	2	2	3	3	4	$I_d(P_n)$	1	2	2	3	3	4	4
$I_s(C_n)$	1	2	2	3	3	3	4	$I_s(P_n)$	2	2	3	3	3	4	4

5 Continuation and Total Continuation Principles

A *partially dominated graph* is a graph together with a declaration that some vertices are already dominated and need not be dominated in the rest of the game. More formally, if G is a graph and $S \subseteq V(G)$, then a partially dominated graph $G|S$ is a graph together with a declaration that the vertices from S are already dominated. We use $\gamma_g(G|S)$ (resp. $\gamma_g'(G|S)$) to denote the number of moves remaining in the game on $G|S$ under optimal play when Dominator (resp. Staller) has the next move. In 2013 Kinnersley, West, and Zamani in [38] presented the following key lemma, named the *Continuation Principle*.

Lemma 13 (Continuation Principle) *If G is a graph and $A, B \subseteq V(G)$ with $B \subseteq A$, then $\gamma_g(G|A) \leq \gamma_g(G|B)$ and $\gamma_g'(G|A) \leq \gamma_g'(G|B)$.*

As a consequence of the Continuation Principle whenever x and y are legal moves for Dominator in the domination game and $N[x] \subseteq N[y]$, then Dominator will play y instead of x, while Staller will play x instead of y. As a further consequence, we have the fundamental property of the domination game that the number of moves in the D-game and the S-game when played optimally can differ by at most 1.

Theorem 14 *If G is a graph, then $|\gamma_g(G) - \gamma_g'(G)| \leq 1$.*

There are graphs H_1, H_2, and H_3 such that $\gamma_g(H_1) = \gamma_g'(H_1)$, $\gamma_g(H_2) = \gamma_g'(H_2) + 1$, and $\gamma_g(H_3) = \gamma_g'(H_3) - 1$. For example, by Theorems 8 and 9 the following holds where $k \geq 1$ is an arbitrary integer. If $H_1 = C_n$ where $n = 4k$, then $\gamma_g(H_1) = \gamma_g'(H_1) = 2k$. If $H_2 = C_n$ where $n = 4k + 2$, then $\gamma_{tg}(H_2) = \gamma_{tg}'(H_2) + 1 = 2k + 1$. If $H_3 = P_n$ where $n = 4k + 3$, then $\gamma_{tg}(H_3) = \gamma_{tg}'(H_3) - 1 = 2k + 1$.

A *partially totally dominated graph* is a graph together with a declaration that some vertices are already totally dominated and need not be totally dominated in the rest of the game. If G is a graph and $S \subseteq V(G)$, then a partially dominated graph $G|S$ is a graph together with a declaration that the vertices from S are already totally dominated. We use $\gamma_{tg}(G|S)$ (resp. $\gamma_{tg}'(G|S)$) to denote the number of moves remaining in the total domination game on $G|S$ under optimal play when Dominator (resp. Staller) has the next move.

The proof of the Continuation Principle can be modified to work for several variants of the domination game. In their introductory paper on the total domination game, the authors in [31] showed that the Continuation Principle also holds for the total version of the game.

Lemma 15 (Total Continuation Principle) *If G is a graph and $A, B \subseteq V(G)$ with $B \subseteq A$, then $\gamma_{tg}(G|A) \leq \gamma_{tg}(G|B)$ and $\gamma_{tg}'(G|A) \leq \gamma_{tg}'(G|B)$.*

Proof. Two games will be played in parallel, Game 1 on the partially totally dominated graph $G|A$ and Game 2 on the partially totally dominated graph $G|B$. The first of these will be the real game, while Game 2 will only be imagined by Dominator. In Game 1, Staller will play optimally while in Game 2, Dominator will play optimally. In Game 2, Dominator will copy each move of Staller played in

Game 1, imagine that Staller played this move in Game 2, and then reply with an optimal move in Game 2. If this move is legal in Game 1, Dominator plays it in Game 1 as well. Otherwise, if the game is not yet over, Dominator plays any other legal move in Game 1. We prove next the following claim.

Claim 1 *In each stage of the games, the set of vertices that are totally dominated in Game 2 is a subset of the set of vertices that are totally dominated in Game 1.*

Proof. We proceed by induction. Since $B \subseteq A$, this is true at the start of the games. Suppose now that Staller has (optimally) selected a vertex u in Game 1. Applying the induction assumption, the vertex u is a legal move in Game 2 because a new vertex v that was totally dominated by u in Game 1 is not yet dominated in Game 2. According to his strategy, Dominator copies the move of Staller by playing the vertex u in Game 2, and then replies with an optimal move in the imagined Game 2. If this move is legal in Game 1, Dominator plays it in Game 1 as well. Otherwise, if the game is not yet over, Dominator plays any other legal move in Game 1. In either case the set of vertices that are dominated in Game 2 is a subset of the set of vertices that are dominated in Game 1. By induction, this proves the desired claim. ($\square$)

Claim 1 implies that Game 1 finishes no later than Game 2. Suppose that m_2 moves are played Game 2. Since Dominator was playing optimally in Game 2, we note that $m_2 \leq \gamma_{tg}(G|B)$. Since Staller was playing optimally in Game 1 and Dominator has a strategy to finish Game 1 in m_2 moves, we infer that $\gamma_{tg}(G|A) \leq m_2$. Therefore, $\gamma_{tg}(G|A) \leq m_2 \leq \gamma_{tg}(G|B)$. Hence if Dominator is the first to play, then the desired result follows. In our earlier argument we made no assumption who starts first. Thus in both cases, Game 1 will finish no later than Game 2. Hence the conclusion holds for γ'_{tg} as well; that is, $\gamma'_{tg}(G|A) \leq \gamma'_{tg}(G|B)$. $\square$

As a consequence of the Total Continuation Principle whenever x and y are legal moves for Dominator and $N(x) \subseteq N(y)$, then Dominator will play y instead of x, while Staller will play x instead of y. The following fundamental property of the total domination game that the number of moves in the D-game and the S-game when played optimally can differ by at most 1 follows readily from the Total Continuation Principle. We present a proof of Theorem 16 along analogous lines to a proof of the result in Theorem 14.

Theorem 16 *If G is a graph with no isolated vertex, then $|\gamma_{tg}(G) - \gamma'_{tg}(G)| \leq 1$.*

Proof. Consider the D-game and let v be the first move of Dominator. Let $A = N(v)$ and consider the partially totally dominated graph $G|A$. Further let $B = \varnothing$ and note that $G|B = G$. By our choice of the vertex v as an optimal first move of Dominator, we have $\gamma_{tg}(G) = 1 + \gamma'_{tg}(G|A)$. By the Total Continuation Principle, $\gamma'_{tg}(G|A) \leq \gamma'_{tg}(G|B) = \gamma'_{tg}(G)$. Therefore, $\gamma_{tg}(G) \leq \gamma'_{tg}(G|A) + 1 \leq \gamma'_{tg}(G) + 1$. Next we consider the S-game and let v be the first move of Staller. As before, let $A = N(v)$ and $B = \varnothing$, and consider the partially totally dominated graph $G|A$. By our choice of the vertex v as an optimal first move of Staller, we have $\gamma'_{tg}(G) = 1 + \gamma_{tg}(G|A)$. By the Total Continuation Principle, $\gamma_{tg}(G|A) \leq \gamma_{tg}(G|B) = \gamma_{tg}(G)$, implying that $\gamma'_{tg}(G) \leq \gamma_{tg}(G|A) + 1 \leq \gamma_{tg}(G) + 1$. $\square$

There are graphs G_1, G_2, and G_3 such that $\gamma_{\text{tg}}(G_1) = \gamma'_{\text{tg}}(G_1)$, $\gamma_{\text{tg}}(G_2) = \gamma'_{\text{tg}}(G_2) + 1$, and $\gamma_{\text{tg}}(G_3) = \gamma'_{\text{tg}}(G_3) - 1$. For example, by Theorems 10 and 11 the following holds where $k \geq 1$ is an arbitrary integer. If $G_1 = C_n$ where $n = 6k + 3$, then $\gamma_{\text{tg}}(G_1) = \gamma'_{\text{tg}}(G_1) = 4k + 2$. If $G_2 = C_n$ where $n = 6k + 1$, then $\gamma_{\text{tg}}(G_2) = \gamma'_{\text{tg}}(G_2) + 1 = 4k + 1$. If $G_3 = P_n$ where $n = 6k + 5$, then $\gamma_{\text{tg}}(G_3) = \gamma'_{\text{tg}}(G_3) - 1 = 4k + 3$.

If the Continuation Principle holds for some variant of the domination game, then the number of moves in the D-game and the S-game when played optimally on such a game can differ by at most 1. Conversely, if for some variant of the domination game the number of moves in the D-game and the S-game when played optimally differ by more than 1, then the Continuation Principle does not hold for such a game.

There are several variants of the domination games for which the Continuation Principle does not hold. One such variant is the independent domination game. Indeed in the independent domination game, the number of moves in the D-game and the S-game when played optimally can often differ by an arbitrarily large constant. As a simple example, let G be a star $K_{1,k}$ where k is arbitrary large. In the D-game, the first vertex played by Diminisher is the central vertex (of degree k) and the game immediately ends. However, in the S-game, the first vertex played by Sweller is a leaf, thereby forcing all k leaves to be played in the independent domination game. Thus in this example, $I_d(G) = 1$ and $I_s(G) = k$.

Using the Continuation Principle, in 2013 Kinnersley, West, and Zamani in [38] showed that the D-game in a partially dominated forest with no isolated vertex can never exceed its S-game.

Theorem 17 ([38]) *If F is a partially dominated forest with no isolated vertex, then $\gamma_g(F) \leq \gamma'_g(F)$.*

Using the Total Continuation Principle, in 2017 Henning and Rall [29] showed that the D-game in a partially total dominated forest with no isolated vertex can never exceed its S-game.

Theorem 18 ([29]) *If F is a partially totally dominated forest with no isolated vertex, then $\gamma_{\text{tg}}(F) \leq \gamma'_{\text{tg}}(F)$.*

6 Upper Bounds and Conjectured Upper Bounds

In this section, we present selected upper bounds and conjectured upper bounds on the game domination number and the game total domination number.

6.1 Domination Game Bounds

In 2013 Kinnersley, West, and Zamani in [38] were the first to prove a general upper bound on the game domination number of an isolate-free graph in terms of its order.

Theorem 19 ([38]) *If G is an isolate-free graph G of order n, then $\gamma_g(G) \le \lceil \frac{7}{10}n \rceil$.*

The upper bound of Theorem 19 was subsequently improved by Bujtás [12] and Henning and Kinnersley [26] using completely different proof techniques. The ingenious approach adopted by Bujtás [12] colors the vertices of the graph with three colors that reflect three different types of vertices and associates a weight with each vertex, and analyses the weight decrease resulting from each played vertex as the game unfolds. The proof method in [26] proves a strong inductive statement in a partially dominated graph with a set of vertices predominated, and from this they deduce the desired upper bound on the game domination number. As a consequence of these results we have the following improved upper bound on the game domination number of an isolate-free graph in terms of its order.

Theorem 20 ([12, 26]) *If G is an isolate-free graph of order n, then*

$$\gamma_g(G) \le \frac{2}{3}n \quad \text{and} \quad \gamma_g'(G) \le \frac{2}{3}n.$$

Much of the interest in the domination game was generated by the so-called $\frac{3}{5}$-Conjecture posed in 2013 by Kinnersley, West, and Zamani [38]. There are two $\frac{3}{5}$-Conjectures: one for isolate-free forests, and one for general isolate-free graphs. We state both conjectures.

Conjecture 1 ([38]) *If G is an isolate-free forest of order n, then $\gamma_g(G) \le \frac{3}{5}n$.*

Conjecture 2 ([38]) *If G is an isolate-free graph of order n, then $\gamma_g(G) \le \frac{3}{5}n$.*

Conjecture 1 for isolate-free forests is referred to as the $\frac{3}{5}$-Forest Conjecture, and Conjecture 2 for general isolate-free graphs as the $\frac{3}{5}$-Graph Conjecture. It is not known whether the $\frac{3}{5}$-Forest Conjecture implies the $\frac{3}{5}$-Graph Conjecture.

If the above two $\frac{3}{5}$-Conjectures are true, then the upper bound is tight. The simplest example is to take $G \cong kP_5$ where $k \ge 1$ is an arbitrary integer. The graph G has order $n = 5k$. By Theorem 9, we have $\gamma_g(P_5) = \gamma_g'(P_5) = 3$. The optimal strategy of Staller is whenever Dominator plays on a component of G, Staller plays on that component if at least one vertex in that component has not yet been dominated and adopts an optimal strategy on the component. If, however, Dominator previous move played on a component of G results in all vertices of that component dominated, then Staller plays in a component with at least one vertex not yet dominated and adopts an optimal strategy on the component. In this way, Staller can guarantee that three vertices are played from each component. This shows that

$\gamma_g(G) = 3k = \frac{3}{5}n$. In Section 7, we show that there exist forests G of arbitrarily large order n satisfying $\gamma_g(G) = \gamma_g'(G) = \frac{3}{5}n$.

In 2016, Henning and Kinnersley [26] proved the $\frac{3}{5}$-Graph Conjecture for the class of graphs of minimum degree at least 2.

Theorem 21 ([26]) *If G is a graph of order n with $\delta(G) \geq 2$, then*

$$\gamma_g(G) \leq \frac{3n}{5} \quad \text{and} \quad \gamma_g'(G) \leq \frac{3n-1}{5}.$$

Bujtás [12] established the following improved upper bound on the game domination number for the class of graphs of minimum degree at least 3.

Theorem 22 ([12]) *If G is a graph of order n with $\delta(G) \geq 3$, then*

$$\gamma_g(G) \leq \frac{34}{61}n \quad \text{and} \quad \gamma_g'(G) \leq \frac{34n-27}{61}.$$

More generally, Bujtás [12] proved the following remarkable result for graphs with large minimum degree.

Theorem 23 ([12]) *If G is a graph of order n with minimum degree $\delta(G) = \delta \geq 4$, then*

$$\gamma_g(G) \leq \left(\frac{15\delta^4 - 28\delta^3 - 129\delta^2 + 354\delta - 216}{45\delta^4 - 195\delta^3 + 174\delta^2 + 174\delta - 216} \right) n.$$

As an immediate consequence of Theorem 23, we have the following upper bound on the game domination number in terms of its order with given minimum degree.

Corollary 24 ([12]) *If G is a graph of order n with minimum degree $\delta(G)$, then the following holds.*

(a) *If $\delta(G) = 4$, then $\gamma_g(G) \leq \frac{37}{72}n < 0.5139n$.*
(b) *If $\delta(G) \geq 5$, then $\gamma_g(G) \leq \frac{2102}{4377}n < 0.4803n$.*

The $\frac{3}{5}$-Graph Conjecture has yet to be settled in general for graphs that contain vertices of degree 1.

6.2 Total Domination Game Bounds

We now shift our attention to upper bounds on the game total domination number. If G is a graph of order n that consists of a disjoint union of copies of K_2, then $\gamma_{tg}(G) = n$. Hence it is only of interest to consider upper bounds on the game total domination number of a graph in which every component has order at least 3. The first general upper bound on the game total domination number was given in 2017 by Henning, Klavžar, and Rall [32].

Theorem 25 ([32]) *If G is a graph of order n in which every component contains at least three vertices, then*

$$\gamma_{tg}(G) \le \frac{4}{5}n \quad \text{and} \quad \gamma'_{tg}(G) \le \frac{4n + 2}{5}.$$

In 2018 Bujtás [14] obtained a new improved upper bound on the game total domination number that improves the $\frac{4}{5}$-bound established in Theorem 25.

Theorem 26 ([14]) *If G is a graph of order n in which every component contains at least three vertices, then*

$$\gamma_{tg}(G) \le \frac{11}{14}n \quad \text{and} \quad \gamma'_{tg}(G) \le \frac{11n + 6}{14}.$$

Bujtás's bound in Theorem 26 is the best general upper bound on the game total domination number to date. In 2016, Henning, Klavžar, and Rall [32] posed the game total domination $\frac{3}{4}$-Conjecture.

Conjecture 3 ([32]) *If G is a graph of order n in which every component contains at least three vertices, then $\gamma_{tg}(G) \le \frac{3}{4}n$.*

As remarked in [32], if the game total domination $\frac{3}{4}$-Conjecture is true, then the upper bound is best possible. The simplest example is to take $G \cong kP_8$ where $k \ge 1$ is an arbitrary integer. The graph G has order $n = 8k$. By Theorem 11, we have $\gamma_{tg}(P_8) = \gamma'_{tg}(P_8) = 6$. The optimal strategy of Staller is whenever Dominator starts playing on a component of G, Staller plays on that component and adopts her optimal strategy on the component. Since $\gamma_{tg}(P_8) = 6$, which is even, Staller can continue this strategy until the completion of the game. This shows that $\gamma_{tg}(G) = 6k = \frac{3}{4}n$.

In 2016 Bujtás, Henning, and Tuza [16] studied upper bounds on the game total domination number over the class of graphs with minimum degree at least 2. For this purpose, they introduced a transversal game in hypergraphs, and establish a tight upper bound on the game transversal number of a hypergraph with all edges of size at least 2 in terms of its order and size. As an application of this result, they established the following result which proves the game total domination $\frac{3}{4}$-Conjecture for the class of graphs of minimum degree at least 2, noting that $\frac{8}{11} < \frac{3}{4}$.

Theorem 27 ([16]) *If G is a graph of order n with $\delta(G) \ge 2$, then $\gamma_{tg}(G) < \frac{8}{11}n$.*

In 2016 Henning and Rall [27] proved the game total domination $\frac{3}{4}$-Conjecture holds in a general graph G (with no isolated vertex) if we remove the minimum degree at least 2 condition, but impose the weaker condition that the degree sum of adjacent vertices in G is at least 4 and add the requirement that no two vertices of degree 1 are at distance 4 apart in G.

Theorem 28 ([27]) *The game total domination $\frac{3}{4}$-Conjecture is true over the class of graphs G that satisfy both conditions* (a) *and* (b) *below:*

(a) *The degree sum of adjacent vertices in G is at least* 4.
(b) *No two vertices of degree* 1 *are at distance exactly* 4 *apart in G.*

As a consequence of Theorem 28 and its proof, we have the game total domination $\frac{3}{4}$-Conjecture is true over the class of graphs with minimum degree at least 2 and, moreover, Dominator can complete the game total domination played in at most $3n/4$ moves by following a greedy strategy. Despite the pleasing progress made over the past few years, the game total domination $\frac{3}{4}$-Conjecture has yet to be settled in general for graphs that contain vertices of degree 1.

6.3 Independent Domination Game Bounds

We close this section with a discussion of bounds on the game independent domination number. In 1982 Berge [2] proved that the independence number $\alpha(G)$ of a well-covered graph G of order n without an isolated vertex is at most $\frac{1}{2}n$, and the extremal graphs for this result are known as the *very well-covered graphs* and were characterized in 1982 by Favaron [21]. In the case of well-covered graphs, as a consequence of Theorem 29 and Berge's result we have the following upper bound on the game independent domination number in the class of well-covered graphs.

Theorem 29 *If G is a well-covered graph of order n, then $I_d(G) = I_s(G) \le \frac{1}{2}n$.*

However, for general graphs the situation is more complex for the independent domination game compared with the domination game or the total domination game. The upper bound proofs established earlier for the game domination number and game total domination number rely heavily of the Continuation Principle and Total Continuation Principle, respectively. As eluded to earlier, the Continuation Principle does not hold for the independent domination game, and the known proof techniques and methods used to date to establish upper bounds for the game domination number and game total domination number cannot therefore be applied to the game independent domination number. Without the Continuation Principle at our disposal, general results on the independent domination game are difficult to obtain.

Although there are general upper bounds on the game domination number and game total domination number given by Theorems 20 and 26, respectively, to date no general upper bound on the game independent domination number is known, except for the trivial upper bound of the order of the graph.

For the class of trees, we have the following game independent domination $\frac{3}{4}$-Conjecture posed in 2018 by Goddard and Henning [22] for the Diminisher-start independent domination game.

Conjecture 4 ([22]) *If T is a tree of order $n \geq 2$, then $I_d(T) \leq \frac{3}{4}n$.*

If Conjecture 4 is true, then this conjecture is somewhat sharp, in the sense that there are trees T with $I_d(T) \geq 3n/4 - o(n)$. Such trees are constructed in [22] as follows. For k a large odd integer, let T_k be a tree with diameter 5 where the two central vertices and both have degree $k + 1$ and all their neighbors have degree $k + 1$. The resulting tree has order $n = 2(k^2 + k + 1)$. The first move of Diminisher is a support vertex, and the first move of Sweller is one of the two centers. Once a vertex is played in the independent domination game, that vertex and all its neighbors are deleted from the graph since they are no longer playable. Thus, after the first two moves, Sweller can ensure that the resulting reduced graph has k^2 isolated vertices, with the remaining non-isolated vertices belonging to $k - 1$ stars $K_{1,k}$. In these remaining stars, Sweller can then get to choose a leaf in half of these stars. The final independent set has size $\frac{3}{2}(k^2 + 1)$, implying that $I_d(T) = \frac{3}{2}(k^2 + 1) = \frac{3}{4}n - \frac{3}{2}k = \frac{3}{4}n - o(n)$, noting that $k = \frac{1}{2}(\sqrt{2n - 3} - 1)$.

For the Sweller-start independent domination game, we have the following game independent domination $\frac{3}{7}$-Conjecture posed in [24].

Conjecture 5 ([24]) *If T is a tree of order $n \geq 2$, then $I_s(T) \geq \frac{3}{7}n$.*

By Theorem 12, if $T = P_n$ where $n \equiv 0 \pmod 7$, then $I_s(P_n) = \frac{3}{7}n$. Hence, if Conjecture 4 is true, then the conjectured bound is sharp. Further, this would imply that with Sweller playing first on a tree, the path is extremal in the S-game.

7 Trees

In this section, we present results on the game domination number, the game total domination number, and the game independent domination number in trees.

7.1 The Domination Game in Trees

In 2015 Brešar, Klavžar, Košmrlj, and Rall [8] studied the domination game in trees and obtained the following lower bound on the game domination number of a tree in terms of its order and maximum degree.

Theorem 30 ([8]) *If T is a tree of order n and maximum degree Δ, then*

$$\gamma_g(T) \geq \left\lceil \frac{2n}{\Delta + 3} \right\rceil - 1,$$

and this bound is asymptotically best possible.

That the bound in Theorem 30 is asymptotically best possible may be seen by considering the family of caterpillars constructed in [8]. For positive integers s and t where $s \geq t + 1$, let $T = T(s, t)$ be the tree obtained from a path P_t by attaching $s - 1$ leaves to each vertex of the path. The resulting tree is a caterpillar of order $n = st$ and maximum degree $\Delta = s + 1$ satisfying $\gamma_g(T) = 2t - 1$. For a fixed t and for s sufficiently large, the term $\lceil 2n/(\Delta + 3) \rceil - 1 = \lceil 2st/(s+4) \rceil - 1$ can be made arbitrarily close to $2t - 1$.

Upper bounds on the game domination number of a tree have been studied extensively in the literature. In 2015, Bujtás [13] proved that if G is an isolate-free forest, then $\gamma_g(G) \leq \frac{5}{8}n(G)$. Moreover, if no two leaves in the forest are at distance 4 apart, then Bujtás [13] proved that $\frac{3}{5}$-Forest Conjecture holds. In 2016 Schmidt [50] extended her result and proved the 3/5-conjecture for weakly $S(K_{1,3})$-free forests, where $S(K_{1,3})$ is the graph obtained from a star $K_{1,3}$ by subdividing every edge once and where a weakly $S(K_{1,3})$-free forests is an isolate-free forest without induced $S(K_{1,3})$ whose leaves are leaves of the forest as well. On 3rd March 2016 Marcus and Peleg, announced they had proven the $\frac{3}{5}$-Forest Conjecture in an unpublished manuscript [44].

Theorem 31 ([44]) *The domination game $\frac{3}{5}$-Forest Conjecture is true; that is, if G is an isolate-free forest of order n, then $\gamma_g(G) \leq \frac{3}{5}n$.*

It remains an open problem to characterize the isolate-free forests that achieve equality in the $\frac{3}{5}$-Forest Conjecture. In 2013 Brešar, Klavžar, Košmrlj, and Rall [5] presented a construction that yields an infinite family of trees that attain the bound in the $\frac{3}{5}$-Forest Conjecture. Motivated by their construction, in 2017 by Henning and Löwenstein in [28] gave a larger construction of extremal trees. In order to explain this construction, they defined the notion of a 2-wing as follows.

Definition 1 ([28]) A tree T is a 2-*wing* if T has maximum degree at most 4 with no vertex of degree 3, and with the vertices of degree 2 in T precisely the support vertices of T, except for one vertex of degree 2 in T. This exceptional vertex of degree 2 in T that is not a support vertex, we call the gluing vertex of T.

The smallest 2-wing is a path on five vertices, with its central vertex as the gluing vertex. A 2-wing with gluing vertex v is illustrated in Figure 4.

Definition 2 ([28]) A tree T belongs to the family $\mathcal{T}$ if T is obtained from $k \geq 1$ vertex-disjoint 2-wings by adding $k - 1$ edges between the gluing vertices.

Theorem 32 ([28]) *If $T \in \mathcal{T}$ has order n, then $\gamma_g(T) = \gamma_g'(T) = \frac{3}{5}n$.*

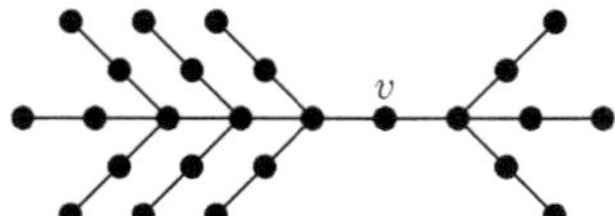

Fig. 4 A 2-wing with gluing vertex v

In 2017 the following conjecture was posed in [28] and has yet to be resolved.

Conjecture 6 ([28]) *If F is an isolate-free forest of order n satisfying $\gamma_g(F) = \frac{3}{5}n$, then every component of F belongs to the family $\mathcal{T}$.*

7.2 The Total Domination Game in Trees

Despite the pleasing progress made to date, the game total domination $\frac{3}{4}$-Conjecture has yet to be proven in the class of trees. Several partial results, however, have been obtained. For example, as a consequence of Theorem 28, we have the following result for the trees.

Theorem 33 ([27]) *The game total domination $\frac{3}{4}$-Conjecture is true over the class of trees T in which every support vertex of T has degree at least 3 and no two leaves are at distance exactly 4 apart in T.*

We close this section with the remark that a characterization of trees with equal total domination and game total domination numbers is given by Henning and Rall [29]. However, since this construction is relatively complex, we omit it here.

7.3 The Independent Domination Game in Trees

The independent domination game is very non-trivial even when played on trees with maximum degree at most 3. In this highly restricted class of graphs, tight bounds on the game independent domination number have yet to be determined. The following bounds for subcubic graphs are given by Goddard and Henning [22].

Proposition 34 ([22]) *If G is a connected graph of order $n \geq 2$ and maximum degree at most 3, then*

$$\frac{n}{4} \leq I_d(G) \quad \text{and} \quad I_s(G) \leq \frac{3n}{4}.$$

Proof. Let $\Delta = \Delta(G)$, and so $\Delta \leq 3$. The lower bound $I_d(G) \geq \frac{1}{4}n$ follows immediately from Theorem 3, noting that $\frac{n}{4} \leq \frac{n}{\Delta+1} \leq \gamma(G) \leq i(G) \leq I_d(G)$. To prove the upper bound $I_s(G) \leq \frac{3}{4}n$, let I be an α-set of G, and let $\overline{I} = V(G) \setminus I$. Let $[I, \overline{I}]$ be the set of edges of G between I and $\overline{I}$. Every vertex in $\overline{I}$ has at most three neighbors in I, while every vertex in I has at least one neighbor in $\overline{I}$, implying that $|I| \leq |[I, \overline{I}]| \leq 3|\overline{I}| = 3(n - |I|)$, and so $\alpha(T) = |I| \leq \frac{3}{4}n$. By Theorem 3, we therefore have that $I_s(G) \leq \alpha(G) \leq \frac{3}{4}n$. $\square$

The best known bounds for the game independent domination number in the class of trees with maximum degree at most 3 are given by Goddard and Henning [22]. We omit the proof.

Theorem 35 ([22]) *If T is a tree of order $n \geq 2$ and maximum degree at most 3, then the following holds.*

(a) $\frac{1}{4}n \leq I_d(T) \leq \frac{4}{7}n$.
(b) $\frac{3}{8}n \leq I_s(T) \leq \frac{3}{4}n$.

We note that if $T = K_{1,3}$, then $n = 4$, $I_d(T) = 1 = \frac{1}{4}n$ and $I_s(G) = 3 = \frac{3}{4}n$. Thus the lower bound $I_d(T) \geq \frac{1}{4}n$ and the upper bound $I_s(G) \leq \frac{3}{4}n$ cannot be improved. However, it remains an open problem to determine tight upper bounds (that hold for trees of sufficiently large order). Trees with maximum degree 3 and of large order n with $I_d(T) > \frac{1}{2}n$ and $I_s(T) > \frac{1}{2}n$ are constructed in [22].

Proposition 36 ([22]) *There exist trees T of maximum degree 3 and of arbitrarily large order n such that*

$$I_d(T) \geq \left(\frac{1}{2} + \varepsilon\right) n \quad \text{and} \quad I_s(T) \geq \left(\frac{1}{2} + \varepsilon\right) n$$

for some small ε.

8 Computational Complexity

The algorithmic complexity of determining the game domination number of a given graph was studied by Brešar, Dorbec, Klavžar, Košmrlj, and Renault [9]. For this purpose, they considered the following two game domination problems.

D-GAME DOMINATION PROBLEM

Input: A graph G, and an integer ℓ.
Question: Is $\gamma_g(G) \leq \ell$?

> **S-Game Domination Problem**
>
> *Input:* A graph G, and an integer ℓ.
> *Question:* Is $\gamma_g'(G) \le \ell$?

Brešar et al. [9] presented a reduction to the Game Domination Problem from the POS-CNF problem, which is known to be log-complete in PSPACE (see [51] for the complexity result on this problem). Using this reduction and careful analysis, they show that the complexity of both the D-Game Domination Problem and S-Game Domination Problem is in the class of PSPACE-complete problems.

Theorem 37 ([9]) *Both the D-Game Domination Problem and the S-Game Domination Problem are* log-*complete in PSPACE.*

Hence the decision version of the game domination problem is computationally harder than any NP-complete problem, unless NP=PSPACE. Klavžar, Košmrlj, and Schmidt [40] studied the D-Game Domination Problem and S-Game Domination Problem when the integer ℓ is fixed. In this case, when ℓ is not part of the input, they were able to solve the game domination problems in polynomial time.

Theorem 38 ([40]) *If G is a graph of order n with maximum degree Δ and ℓ is a fixed integer, then the D-Game Domination Problem and the S-Game Domination Problem can be solved in* $\mathrm{O}(\Delta \cdot n^\ell)$.

The algorithmic complexity of determining the game total domination number of a given graph was studied by Brešar and Henning [3] who considered the following game total domination problems.

> **D-Game Total Domination Problem**
>
> *Input:* A graph G, and an integer ℓ.
> *Question:* Is $\gamma_{tg}(G) \le \ell$?

> **S-Game Total Domination Problem**
>
> *Input:* A graph G, and an integer ℓ.
> *Question:* Is $\gamma_{tg}'(G) \le \ell$?

Analogously as in the Game Domination Problem, a reduction to the Game Total Domination Problem from the POS-CNF problem is presented, but using a different gadget graph. Using this reduction, they show that the complexity of both the D-GAME TOTAL DOMINATION PROBLEM and S-GAME TOTAL DOMINATION PROBLEM is in the class of PSPACE-complete problems.

Theorem 39 ([3]) *Both the D-Game Total Domination Problem and the S-Game Total Domination Problem are log-complete in PSPACE.*

References

1. N. Alon, J. Balogh, B. Bollobás, T. Szabó, Game domination number. Discrete Math **256**, 23–33 (2002)
2. C. Berge, Some common properties for regularizable graphs, edge-critical graphs and B-graphs, in *Theory and Practice of Combinatorics*, pp. 31–44. North-Holland Math. Stud., vol. **60**. *Ann. Discrete Math.*, vol. **12** (North-Holland, Amsterdam, 1982)
3. B. Brešar, M.A. Henning, The game total domination problem is log-complete in PSPACE. Inform. Process. Lett. **126**, 12–17 (2017)
4. B. Brešar, S. Klavžar, D.F. Rall, Domination game and an imagination strategy. SIAM J. Discrete Math. **24**, 979–991 (2010)
5. B. Brešar, S. Klavžar, G. Košmrlj, D.F. Rall, Domination game: extremal families of graphs for the 3/5-conjectures. Discrete Appl. Math. **161**, 1308–1316 (2013)
6. B. Brešar, S. Klavžar, D.F. Rall, Domination game played on trees and spanning subgraphs. Discrete Math. **313**, 915–923 (2013)
7. B. Brešar, P. Dorbec, S. Klavžar, G. Košmrlj, Domination game: effect of edge- and vertex-removal. Discrete Math. **330**, 1–10 (2014)
8. B. Brešar, S. Klavžar, G. Košmrlj, D.F. Rall, Guarded subgraphs and the domination game. Discrete Math. Theor. Comput. Sci. **17**, 161–168 (2015)
9. B. Brešar, P. Dorbec, S. Klavžar, G. Košmrlj, G. Renault, Complexity of the game domination problem. Theore. Comput. Sci. **648**, 1–7 (2016)
10. B. Brešar, P. Dorbec, S. Klavžar, G. Košmrlj, How long can one bluff in the domination game? Discuss. Math. Graph Theory **37**, 337–352 (2017)
11. B. Brešar, M.A. Henning, S. Klavžar, D.F. Rall, *Domination Games Played On Graphs* (in preparation)
12. Cs. Bujtás, On the game domination number of graphs with given minimum degree. Electr. J. Comb. **22**(3), #P3.29 (2015)
13. Cs. Bujtás, Domination game on forests. Discrete Math. **338**, 2220–2228 (2015)
14. Cs. Bujtás, On the game total domination number. Graphs Combin. **34**, 415–425 (2018)
15. C. Bujtás, S. Klavžar, Improved upper bounds on the domination number of graphs with minimum degree at least five. Graphs Combin. **32**, 511–519 (2016)
16. Cs. Bujtás, M.A. Henning, Zs. Tuza, Transversal game on hypergraphs and the $\frac{3}{4}$-conjecture on the total domination game. SIAM J. Discrete Math. **30**, 1830–1847 (2016)
17. Cs. Bujtás, M.A. Henning, Zs. Tuza, Bounds on the game transversal number in hypergraphs. Eur. J. Combin. **59**, 34–50 (2017)
18. P. Dorbec, M.A. Henning, Game total domination for cycles and paths. Discrete Appl. Math. **208**, 7–18 (2016)
19. P. Dorbec, G. Košmrlj, G. Renault, The domination game played on unions of graphs. Discrete Math. **338**, 71–79 (2015)
20. P. Dorbec, M.A. Henning, S. Klavžar, G. Košmrlj, Cutting lemma and union lemma for the domination game. Discrete Math. **342**, 1213–1222 (2019)
21. O. Favaron, Very well covered graphs. Discrete Math. **42**, 177–187 (1982)

22. W. Goddard, M.A. Henning, The competition-independence game in trees. J. Combin. Math. Combin. Comput. **104**, 161–170 (2018)
23. T.W. Haynes, S.T. Hedetniemi, M.A. Henning, *Domination in Graphs: Core Concepts.* Manuscript (Springer, New York, 2020)
24. M.A. Henning, My favorite domination game conjectures, in *Graph Theory-Favorite Conjectures and Open Problems*, vol 2. R. Gera. T.W. Haynes, S.T. Hedetniemi. Springer Monographs in Mathematics (Springer, New York, 2018), pp. 135–148. ISBN 978-3-319-97686-0
25. M.A. Henning, A. Yeo, *Total Domination in Graphs (Springer Monographs in Mathematics)* (2013). ISBN: 978-1-4614-6524-9 (Print) 978-1-4614-6525-6 (Online)
26. M.A. Henning, W.B. Kinnersley, Domination Game: A proof of the 3/5-Conjecture for graphs with minimum degree at least two. SIAM J. Discrete Math. **30**(1), 20–35 (2016)
27. M.A. Henning, D.F. Rall, Progress towards the total domination game $\frac{3}{4}$-Conjecture. Discrete Math. **339**, 2620–2627 (2016)
28. M.A. Henning, C. Löwenstein, Domination game: Extremal families for the 3/5-conjecture for forests. Discuss. Math. Graph Theory **37**(2), 369–381 (2017)
29. M.A. Henning, D.F. Rall, Trees with equal total domination and game total domination numbers. Discrete Appl. Math. **226**, 58–70 (2017)
30. M.A. Henning, S. Klavžar, Infinite families of circular and Möbius ladders that are total domination game critical. Bull. Malays. Math. Sci. Soc. **41**, 2141–2149 (2018)
31. M.A. Henning, S. Klavžar, D.F. Rall, Total version of the domination game. Graphs Combin. **31**, 1453–1462 (2015)
32. M.A. Henning, S. Klavžar, D. Rall, The 4/5 upper bound on the game total domination number. Combinatorica **37**, 223–251 (2017)
33. M.A. Henning, S. Klavžar, D. Rall, Game total domination critical graphs. Discrete Appl. Math. **250**, 28–37 (2018)
34. V. Iršič, Effect of predomination and vertex removal on the game total domination number of a graph. Discrete Appl. Math. **257**, 216–225 (2019)
35. T. James, P. Dorbec, A. Vijayakumar, Further progress on the heredity of the game domination number. *Theor. Comput. Sci. Discrete Math.* vol. 10398. Lecture Notes in Comput. Sci. (Springer, Cham, 2017), pp. 435–444
36. T. James, S. Klavžar, A. Vijayakumar, The domination game on split graphs. Bull. Malays. Math. Sci. Soc. **99**, 327–337 (2019)
37. Y. Jiang, M. Lu, Game total domination for cyclic bipartite graphs. Discrete Appl. Math. **265**, 120–127 (2019)
38. W.B. Kinnersley, D.B. West, R. Zamani, Extremal problems for game domination number. SIAM J. Discrete Math. **27**, 2090–2107 (2013)
39. S. Klavžar, D.F. Rall, Domination game and minimal edge cuts. Discrete Math. **342**, 951–958 (2019)
40. S. Klavžar, G. Košmrlj, S. Schmidt, On the computational complexity of the domination game. Iran. J. Math. Sci. Inform. **10**, 115–122 (2015)
41. S. Klavžar, G. Košmrlj, S. Schmidt, On graphs with small game domination number. Appl. Anal. Discrete Math. **10**, 30–45 (2016)
42. G. Košmrlj, Realizations of the game domination number. J. Comb. Optim. **28**, 447–461 (2014)
43. G. Košmrlj, Domination game on paths and cycles. Ars Math. Contemp. **13**, 125–136 (2017)
44. N. Marcus, D. Peleg, The Domination Game: Proving the 3/5 Conjecture on isolate-free forests. arXiv:1603.01181
45. M.J. Nadjafi-Arani, M. Siggers, H. Soltani, Characterisation of forests with trivial game domination numbers. J. Comb. Optim. **32**, 800–811 (2016)
46. J.B. Phillips, P.J. Slater, An introduction to graph competition independence and enclaveless parameters. Graph Theory Notes N. Y. **41**, 37–41 (2001)
47. J.B. Phillips, P.J. Slater, Graph competition independence and enclaveless parameters. Congr. Numer. **154**, 79–100 (2002)
48. M.D. Plummer, Some covering concepts in graphs. J. Comb. Theory **8**, 91–98 (1970)

49. W. Ruksasakchai, K. Onphaeng, C. Worawannotai, Game domination numbers of a disjoint union of paths and cycles. Quaest. Math. **42**, 1357–1372 (2019)
50. S. Schmidt, The 3/5-conjecture for weakly $S(K_{1,3})$-free forests. Discrete Math. **339**, 2767–2774 (2016)
51. T.J. Schaefer, On the complexity of some two-person perfect-information games. J. Comput. Syst. Sci. **16**(2), 185–225 (1978)
52. S.J. Seo, P.J. Slater, Competition parameters of a graph. AKCE Int. J. Graphs Comb. **4**(2), 183–190 (2007)
53. C. Weeranukujit, C. Lewchalermvongs, Domination game played on a graph constructed from 1-sum of paths. Thai J. Math. **17**, 34–45 (2019)
54. C. Worawannotai, N. Chantarachada, Game domination numbers of a disjoint union of chains and cycles of complete graphs. Chamchuri J. Math. **11**, 10–25 (2019)
55. K. Xu, X. Li, On domination game stable graphs and domination game edge-critical graphs. Discrete Appl. Math. **250**, 47–56 (2018)
56. K. Xu, X. Li, S. Klavžar, On graphs with largest possible game domination number. Discrete Math. **341**, 1768–1777 (2018)

Domination and Spectral Graph Theory

Carlos Hoppen, David P. Jacobs, and Vilmar Trevisan

1   Introduction

From the early days of graph theory within Mathematics and Computer Science, matrices have played an important role as data structures to store graphs. Let $G = (V, E)$ be a graph of order n where $V = \{v_1, \ldots, v_n\}$ and $E = \{e_1, \ldots, e_m\}$. A graph may be naturally stored by recording adjacencies between vertices or by recording the incidence structure between vertices and edges. Indeed, the *adjacency matrix* $A = A(G)$ of G is the $n \times n$ symmetric matrix where the entry $A_{ij} = 1$ if $\{v_i, v_j\} \in E$, and $A_{ij} = 0$ otherwise, while the *incidence matrix* $B = B(G)$ of G is the $n \times m$ matrix such that $B_{ij} = 1$ if $v_i \in e_j$, and $B_{ij} = 0$ otherwise.

Much more recently, an entire branch of graph theory was born of the interest in extracting properties of graphs from algebraic information about matrices associated with them. For square matrices, the spectrum is one such piece of information. Given a square matrix M of order n, a number λ is an *eigenvalue* of M if

$$M\mathbf{x} = \lambda\mathbf{x} \tag{1}$$

for some *nonzero* column vector $\mathbf{x}$, which is called an *eigenvector* for λ. Equation (1) is satisfied if and only if λ is a root of the characteristic polynomial $p(x) = \det(M - xI)$, which has degree n, so that any $n \times n$ real matrix has n complex eigenvalues, although some can be repeated. The multiset of eigenvalues is called the *spectrum* of the matrix and an eigenvalue's *multiplicity* is the number of times it occurs in

C. Hoppen · V. Trevisan (⊠)
UFRGS, 91509–900, Porto Alegre, Brazil
e-mail: choppen@ufrgs.br; trevisan@mat.ufrgs.br

D. P. Jacobs
Clemson University, 29634, Clemson, SC, USA
e-mail: dpj@clemson.edu

© The Author(s), under exclusive license to Springer Nature Switzerland AG 2021
T. W. Haynes et al. (eds.), *Structures of Domination in Graphs*, Developments in Mathematics 66, https://doi.org/10.1007/978-3-030-58892-2_9

the spectrum. For symmetric matrices, it is well-known that the eigenvalues are real numbers and that the eigenvectors associated with them produce an orthogonal basis of $\mathbb{R}^n$, that is, the inner product of two eigenvectors is zero.

Clearly, the adjacency matrix of a graph is such a symmetric matrix. Many other symmetric matrices have been introduced. Given a graph G with vertex set $V = \{v_1, \ldots, v_n\}$, let $D = D(G)$ be the diagonal matrix with entry $D_{ii} = d(v_i)$, the degree of v_i. Then the *Laplacian matrix* of G is the matrix $L(G) = D - A$ and the *signless Laplacian matrix* is the matrix $Q(G) = D + A$. We observe that $Q(G) = BB^T$, where B is the incidence matrix defined above, and that $L(G)$ may also be written in the form CC^T, where C is the incidence matrix of any orientation of G. This makes L and Q positive semidefinite, so that their eigenvalues are all non-negative. Other symmetric matrices associated with a graph G are its *normalized Laplacian matrix* $\mathscr{L}(G) = D^{-1/2}LD^{-1/2}$ and its *distance matrix*, $\mathscr{D}(G)$, for instance. The latter is the matrix indexed by the vertex set of G such that the entry ij is given by the distance between v_i and v_j.

The branch of graph theory that studies the properties of graphs through the eigenvalues of the matrices associated with them is known as *spectral graph theory*. Often these matrices are $A(G)$, $L(G)$, $Q(G)$, or $\mathscr{L}(G)$ defined above, but many other matrices have been considered in various contexts. Even if we restrict ourselves to the adjacency and the Laplacian matrices, eigenvalues and eigenvectors have been particularly useful for embedding graphs in the plane [4], for graph partitioning and clustering [44], in the study of random walks on graphs [13] and in the geometric description of data sets [15], just to mention a few examples. It has had a particularly relevant influence in Chemistry, where spectral parameters are widely used as molecular descriptors, to the point that some of the results are referred to as part of *chemical graph theory*. See [22, 30] for more information.

We should point out that, even though graph matrices are defined with respect to some labelling of the graph G, their spectrum does not depend on the labelling, so that isomorphic graphs share the same spectrum with respect to any fixed matrix. Given the important role of the *isomorphism problem* in graph theory, and the nice fact that eigenvalues and eigenvectors can be computed efficiently, it is quite natural that spectral approaches to the isomorphism problem have been a flourishing research theme in this area. In an ideal world, we would be able to test graph isomorphism by simply computing the eigenvalues of both graphs. However, it is not true that two non-isomorphic graphs must have distinct spectra, and a graph G may have a *cospectral mate* H, namely a graph that is not isomorphic to G, but has the same spectrum as G. Figure 1, extracted from [9], shows two graphs that are cospectral with respect to both the adjacency and the normalized Laplacian matrices.

More generally, the seminal work of Schwenk [49] showed that *almost every* tree T has a mate T' for which $A(T)$ and $A(T')$ share the same spectrum, in the sense that among all non-isomorphic trees on at most n vertices, the fraction that has a cospectral mate tends to 1 as n tends to infinity. This property of trees was shown to hold for other matrices, see, for instance, McKay [45] for the Laplacian matrix. In sharp contrast with this, Haemers [31] conjectures that most graphs *do*

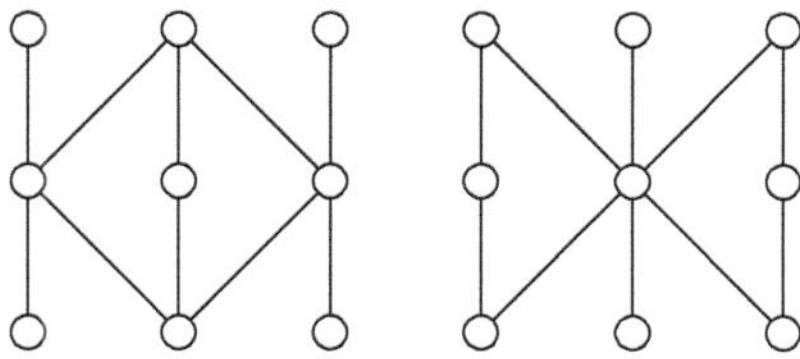

Fig. 1 Cospectral graphs G and H

not have cospectral mates with respect to the adjacency matrix. This is one of the main conjectures in this area. We refer the interested reader to [7, 53].

Why so many different matrices to study graphs? A natural question related to the previous paragraph is whether any particular matrix would *distinguish* more graphs than other matrices. This has been one of the driving forces to proposing new matrices. For instance, in 2009, Cvetković and Simić, in [16–18], established many properties of the signless Laplacian matrix, and argued that this matrix had less *spectral uncertainty* than other matrices, in the sense that more graphs are determined by their signless Laplacian spectrum than by their adjacency and Laplacian spectra. So far no definitive results have been obtained [54].

Moreover, as Butler and Chung point out in [10], each matrix has its advantages and disadvantages. As one might expect, since graph properties are often hard to compute, it would be unexpected that footprints in the spectrum of a matrix, which is computable in polynomial-time, could fully capture these properties. For example, a graph is *bipartite* if and only if the eigenvalues of $A(G)$ are symmetric about the origin. That is, for each eigenvalue λ in the spectrum, $-\lambda$ is an eigenvalue of the same multiplicity [7, Prop. 3.4.1]. On the other hand the multiplicity of the smallest eigenvalue of $L(G)$, which is 0, reveals how many *connected components* are in the graph, and the multiplicity of 0 in $Q(G)$ is the number of bipartite components in G [16]. The normalized Laplacian, introduced by Butler and Chung, will not be used in this chapter, but is closely connected to random walks in graphs. The focus of our chapter is the matrices $A(G)$, $L(G)$, $Q(G)$, and $\mathscr{D}(G)$.

Sometimes the relationship between the eigenvalue and graph parameter can be startling. To illustrate, consider the result by Delsarte and Hoffman [20], obtained in the 1970s, involving the *independence number* $\alpha(G)$, the cardinality of a maximum independent set of vertices in G, of a regular graph G. The theorem relates $\alpha(G)$ to the least eigenvalue of its adjacency matrix.

Theorem 1 *Let $n > d \geq 1$ be integers. Let $G = (V, E)$ be a d-regular graph on n vertices whose adjacency matrix has least eigenvalue λ_n. If $S \subseteq V$ is an independent set of G, then*

$$|S| \leq \frac{-\lambda_n}{d - \lambda_n} \cdot n. \qquad (2)$$

Moreover, given an independent set S, let $\mathbf{y}_S$ be the characteristic vector of S, that is, the entry corresponding to v_i is equal to 1 if $v_i \in S$, and is equal to 0 otherwise. Equality holds in (2) if and only if $\mathbf{y}_S - \frac{|S|}{n}\mathbf{1}$ is an eigenvector associated with λ_n.

Since the size $i(G)$ of a minimum independent dominating set in a graph G is the minimum size of a maximal independent set, Theorem 1 immediately implies

$$i(G) \leq \frac{-\lambda_n}{d - \lambda_n} \cdot n$$

for any d-regular graph G. To give an idea of the type of argument used for proving bounds in this area, we sketch the proof for Theorem 1.

Proof. Let G be a d-regular graph with vertex set $V = \{v_1, \ldots, v_n\}$ whose adjacency matrix has least eigenvalue λ_n. Since G is d-regular, $\lambda_1 = d$ is the largest eigenvalue of G, and it is associated with the eigenvector $\mathbf{1}$. The maximality of λ_1 relies on the Perron–Frobenius Theorem (see Theorem 8.4.4 in [37]). Let $\mathbf{x}_1 = \frac{1}{\sqrt{n}}\mathbf{1}$ and let $\mathbf{x}_2, \ldots, \mathbf{x}_n$ be eigenvectors associated with the remaining eigenvalues $\lambda_2 \geq \cdots \geq \lambda_n$, respectively, with the property that $\{\mathbf{x}_1, \ldots, \mathbf{x}_n\}$ is an orthonormal basis of $\mathbb{R}^n$, that is, the basis is orthogonal and $||\mathbf{x}_i|| = 1$, for each i.

Let $S \subset V$ be an independent set and let $\mathbf{x}_S$ be the vector whose entry corresponding to v_i is equal to $\frac{1}{\sqrt{n}}$ if $v_i \in S$, and is equal to 0 otherwise. Let $a_1, \ldots, a_n \in \mathbb{R}$ be such that $\mathbf{x}_S = a_1\mathbf{x}_1 + \cdots + a_n\mathbf{x}_n$. In particular, we have

$$||\mathbf{x}_S||^2 = \sum_{i=1}^{n} a_i^2 = \frac{|S|}{n}. \tag{3}$$

The inner product $\mathbf{x}_S \cdot \mathbf{x}_1$ satisfies $a_1 = \mathbf{x}_S \cdot \mathbf{x}_1 = \frac{|S|}{n}$. Since S is an independent set the quadratic form $\mathbf{x}_S \cdot A\mathbf{x}_S$ satisfies

$$\mathbf{x}_S \cdot A\mathbf{x}_S = \frac{1}{n} \sum_{v_i, v_j \in S} A_{i,j} = \frac{2}{n}|E(G[S])| = 0.$$

Now compute $\mathbf{x}_S \cdot A\mathbf{x}_S$ by expressing the rightmost $\mathbf{x}_S$ in terms of the basis, making use of the linearity of A, and noticing that inner products distribute over vector sums:

$$0 = \mathbf{x}_S \cdot A\mathbf{x}_S = \mathbf{x}_S \cdot \sum_{i=1}^{n} a_i A\mathbf{x}_i = \sum_{i=1}^{n} \mathbf{x}_S \cdot a_i A\mathbf{x}_i.$$

Next replace each $A\mathbf{x}_i$ with $\lambda_i\mathbf{x}_i$, and write the other $\mathbf{x}_S$ in terms of the basis. By orthogonality, each $\mathbf{x}_S \cdot \mathbf{x}_i = a_i$, and so the above equation becomes

$$0 = \sum_{i=1}^{n} \lambda_i a_i \mathbf{x}_S \cdot \mathbf{x}_i = \sum_{i=1}^{n} \lambda_i a_i^2.$$

Using the fact that λ_n is negative, writing $d = \lambda_1$, and applying (3) we get

$$0 \geq \lambda_1 a_1^2 + \lambda_n \sum_{i=2}^{n} a_i^2 = d a_1^2 + \left(||x_S||^2 - a_1^2 \right) \lambda_n = \frac{d|S|^2}{n^2} + \left(\frac{|S|}{n} - \frac{|S|^2}{n^2} \right) \lambda_n.$$

(4)

Inequality (4) leads to

$$|S| \leq \frac{-n\lambda_n}{d - \lambda_n},$$

as required.

This bound holds with equality if and only if the inequality in (4) holds with equality, that is, if and only if $a_i \neq 0$ implies that $i = 1$ or $\lambda_i = \lambda_n$. This means the vector $\mathbf{x}_S - a_1 \mathbf{x}_1 = \mathbf{x}_S - \frac{|S|}{n} \mathbf{x}_1$ is a linear combination

$$a_2 \mathbf{x}_2 + \ldots + a_n \mathbf{x}_n,$$

where, for each $i \geq 2$, either $a_i = 0$ or $\lambda_i = \lambda_n$. This implies that $\mathbf{x}_S - a_1 \mathbf{x}_1$ is a linear combination of eigenvectors associated with λ_n, and is therefore such an eigenvector. As a consequence,

$$\mathbf{y}_S - \frac{|S|}{n} \mathbf{1} = \left(\sqrt{n}\, \mathbf{x}_S - \frac{|S|}{n} \sqrt{n}\, \mathbf{x}_1 \right) = \sqrt{n} \left(\mathbf{x}_S - \frac{|S|}{n} \mathbf{x}_1 \right)$$

is an eigenvector associated with λ_n. This completes the proof. $\square$

In this chapter we will highlight some of the interesting and important results in spectral graph theory involving *domination* parameters that have appeared in the last 25 years. These results usually involve the well-known *domination number* $\gamma(G)$. However, we will also give some results involving the *total domination number* $\gamma_t(G)$, and the *signed domination number* $\gamma_s(G)$. Generally, there have been two kinds of spectral results involving domination in the literature: results that compare a *specific eigenvalue* to $\gamma(G)$, and results that compare the *number of eigenvalues* in an interval with $\gamma(G)$.

A meta-problem in this direction is the following. Let f_M be a spectral parameter associated with a graph matrix M and let $\mathscr{G}$ be a class of graphs. The problem is to determine the functions

$$\max\{f_M(G)\colon |V(G)| = n, G \in \mathscr{G}\} \text{ and } \min\{f_M\colon |V(G)| = n, G \in \mathscr{G}\}, \qquad (5)$$

and characterize the n-vertex graphs that attain the extremal values. For instance, if $f_A(G) = \lambda_k(G)$ is the k-th largest eigenvalue of the adjacency matrix of a graph G and $\mathscr{G}_\gamma$ is the set of all graphs with domination number γ, solving (5) would lead to upper and lower bounds on the value of $\lambda_k(G)$ in terms of its domination number. Bounds of this type may be often turned into upper and/or lower bounds on $\gamma(G)$ in terms of $\lambda_k(G)$.

The remainder of the chapter is organized as follows. Results involving domination and the adjacency matrix are in Section 2. This section includes a 1994 upper bound on $\gamma(G)$ by Rowlinson using the largest multiplicity in $A(G)$, and a beautiful inequality involving the index and $\gamma(G)$. In Section 3 we give several bounds for the largest and second smallest Laplacian eigenvalue using domination. We also exhibit an early eigenvector partition due to Brand and Seifter for constructing disjoint dominating sets, as well as results relating the number of Laplacian eigenvalues in an interval to $\gamma(G)$. In Section 4 we discuss the signless Laplacian matrix, and give bounds for the largest and second smallest eigenvalue using domination. Domination results involving the distance matrix are given in Section 5. We conclude this chapter with a few open problems

2 Adjacency Matrix

For a graph G, we say the *spectrum* of G is the multiset given by the eigenvalues of the adjacency matrix $A(G)$. Since $A(G)$ is a real symmetric matrix, its eigenvalues are real and we enumerate them as

$$\lambda_1 \geq \lambda_2 \geq \cdots \geq \lambda_n.$$

It is well-known that the spectrum of G determines some structure of G as, for example, the number of vertices, the number of edges, the number of triangles, whether G is bipartite and regular, among other properties.

The adjacency matrix is by far the most studied matrix in spectral graph theory. However, the literature relating the spectrum of the adjacency matrix of a graph and domination seems to be rare. Perhaps the first result about domination and matrices of a graph is the well-known fact that if G has no isolated vertices, then

$$\gamma(G) \leq r,$$

where r is the rank of the adjacency matrix of G. This result is from 1982 and is due to Van Nuffelen [55].

The first paper we found relating graph spectra to domination dates to 1994 by Rowlinson [48], and involves the notion of star partition of a graph G. A *star partition* of a graph G whose distinct eigenvalues are $\lambda_1, \ldots, \lambda_m$ is a partition $V(G) = X_1 \cup \cdots \cup X_m$ with the following two properties: the cardinality of each X_i is equal to the multiplicity of λ_i as an eigenvalue; λ_i is not an eigenvalue of the graph $G - X_i$ obtained from G by deleting all vertices in X_i, namely the subgraph of G induced by the complement $\overline{X_i}$ of G. It is known that every graph has a star partition. Rowlinson showed that if $X_1 \cup \cdots \cup X_m$ is a star partition of a graph G with no isolated vertices, then $\overline{X_i}$ is a dominating set of G. Moreover, for such a graph G, Rowlinson showed that

$$\gamma(G) \leq n - k,$$

where k is the largest multiplicity of an eigenvalue of $A(G)$. We remark that this improves on Van Nuffelen's bound, since the RHS of the latter may be viewed as $n - k_0$, where k_0 is the multiplicity of the eigenvalue 0.

2.1 Domination and Spectral Radius

The largest eigenvalue of the adjacency matrix $A(G)$ of G, namely λ_1, is called the *index* of G, while the *spectral radius* $\rho(G)$ of G is the maximum of the modulus of the eigenvalues of $A(G)$. By the Perron–Frobenius theory of matrices it is known that $\rho(G) = \lambda_1$. If G has at least two vertices and is connected, $\rho(G)$ is always positive, simple, and its associated eigenvector may be chosen with positive entries. By its many applications (see, for example, the book by Stevanović [51]), the spectral radius is likely to be the most studied spectral parameter of graphs.

Brualdi and Solheid [8] proposed the following general problem, which is a subproblem of (5) and became one of the classic problems of spectral graph theory:

> Given a set $\mathcal{G}$ of graphs, find $\min\{\rho(G) : G \in \mathcal{G}\}$ and $\max\{\rho(G) : G \in \mathcal{G}\}$, and characterize the graphs which achieve the minimum or maximum value.

In 2008, Stevanović, Aouchiche, and Hansen [52] studied this problem for the class of graphs having domination number γ. They characterize the graphs with n vertices having domination number γ with maximum spectral radius. The main result of the paper is the following. In the statements hereafter, given graphs G and H, and a positive integer m, $G \cup H$ denotes the disjoint union of G and H, mG denotes the disjoint union of m copies of G, and $\overline{G}$ is the complement of G.

Theorem 2 *If G is a graph on n vertices with domination number γ, then*

$$\rho(G) \leq n - \gamma.$$

Equality holds if and only if $G \cong K_{n-\gamma+1} \cup (\gamma - 1)K_1$ or, when $n - \gamma$ is even, $G \cong \overline{\frac{n-\gamma+2}{2} K_2} \cup (\gamma - 2)K_1$.

Of course we can restate the result above as an upper bound for domination number in terms of the spectral radius of G.

In order to explain their result for graphs with no isolated vertices, we need the following definition. The surjective split graph $SSG(n, k; a_1, \ldots, a_k)$, defined for positive integers n, k, a_1, $\ldots$, a_k, $3 \leq k \leq n$, satisfying $a_1 + \cdots + a_k = n - k$, $a_1 \geq \cdots \geq a_k$, is a split graph on n vertices formed from a clique K_{n-k} vertices and an independent set I with k vertices, in such a way that the ith vertex of I is adjacent to a_i vertices of K, and that no two vertices of I have a common neighbor in K. It

Fig. 2 $SSG(8, 4; 2, 2, 2, 2)$

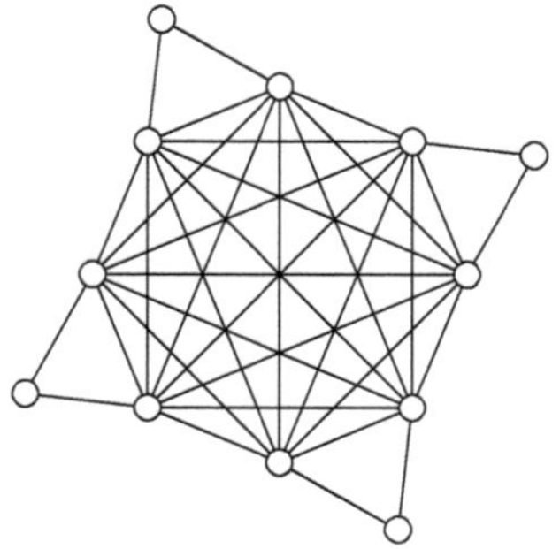

is easy to see that $\gamma(SSG(n, k; a_1, \ldots, a_k)) = k$. As an illustration see the graph of Figure 2.

Theorem 3 *If G is a graph on n vertices with no isolated vertices and domination number $\frac{n}{2} \geq \gamma \geq 3$, then*

$$\rho(G) \leq \rho(SSG(n, \gamma; n - 2\gamma + 1, 1, 1, \ldots, 1)),$$

with equality if and only if $G \cong SSG(n, \gamma; n - 2\gamma + 1, 1, 1, \ldots, 1)$.

Moreover the authors also characterize the graphs with no isolated vertices and maximum spectral radius having $\gamma \in \{1, 2\}$. Precisely, they show that if $\gamma = 1$, then $\rho(G) \leq \rho(K_n)$, if $\gamma = 2$ and n is even, then $\rho(G) \leq \rho(\frac{n}{2} K_2)$ and if $\gamma = 2$ and n is odd, then $\rho(G) \leq \rho((\frac{n-1}{2})K_2 \cup P_3)$.

In the paper [61], B-X. Zhu also deals with a Brualdi–Solheid problem, but restricting the set of candidates to bipartite graphs. Let $\mathscr{B}_\gamma^{(n)}$ be the set of bipartite graphs with n vertices and domination number γ. The author finds the graph of $\mathscr{B}_\gamma^{(n)}$ having maximum spectral radius:

Theorem 4 *Let $G \in \mathscr{B}_\gamma^{(n)}$. If G has the maximum spectral radius, then*

(i) $G \cong K_{1,n-1}$ *for $\gamma = 1$,*
(ii) $G \cong K_{\lfloor \frac{n-\gamma+2}{2} \rfloor, \lceil \frac{n-\gamma+2}{2} \rceil} \cup (\gamma - 2)K_1$ *for $\gamma \geq 2$.*

2.2 Domination and Energy

The *energy* $\mathscr{E}(G)$ of a graph G is defined as the sum of the absolute values of all eigenvalues of the adjacency matrix of the graph. This concept, introduced by Gutman in 1977 [29], has connections with theoretical Chemistry. Indeed, for the vast majority of conjugated hydrocarbons, the energy $\mathscr{E}(G)$ of a graph that models such a molecule is precisely the value of the total π-electron energy, calculated by the simple Hückel tight-binding molecular orbital (HMO). This allows one to apply

Fig. 3 Tree $T(n, \gamma)$

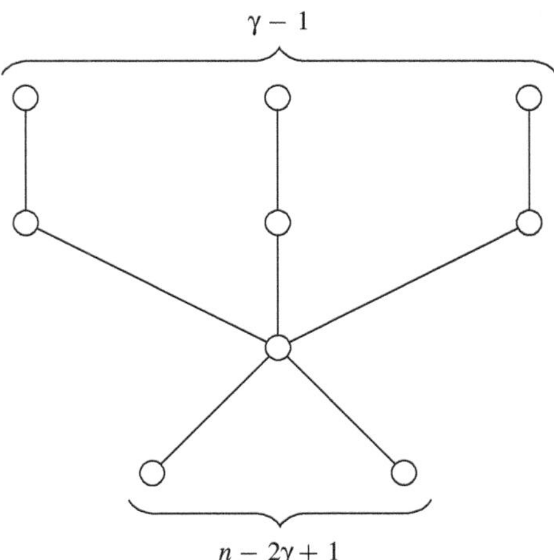

the energy $\mathscr{E}(G)$ to chemical and physical properties of organic molecules. Clearly, the graph theoretical definition is not restricted to molecular graphs.

In the paper [35], He, Wu, and Yu present sharp lower bounds for the energy of trees involving the domination number, determining also all extreme trees which attain these lower bounds. To state their result, we need the following definition.

For two given natural numbers $n > 2\gamma > 2$, the *wounded spider* is the tree $T(n, \gamma)$ obtained by subdividing exactly $\gamma - 1$ edges of the edges of the star $K_{1,n-\gamma}$. It is easy to see that $T(n, \gamma)$ has n vertices and domination number γ (Figure 3).

Consider the class $\mathscr{T}_{\gamma}^{(n)}$ of all trees having n vertices and domination γ. The authors show that $T(n, \gamma)$ is the unique tree with minimum energy among all elements of $\mathscr{T}_{\gamma}^{(n)}$.

By computing $\mathscr{E}(T(n, \gamma))$, they show that if a tree T has n vertices and domination number γ, then

$$\mathscr{E}(T) \geq 2\gamma - 4 + 2\sqrt{n - \gamma + 1} + 2\sqrt{n - 2\gamma + 1}.$$

In 2011, Xu and Feng [58] gave a shorter proof of the result about the minimum energy of $T(n, \gamma)$ over $\mathscr{T}_{\gamma}^{(n)}$. Moreover the paper characterizes the trees in $\mathscr{T}_{\gamma}^{(n)}$ where $n = k\gamma$ with maximal energy for $k = 2, 3, \frac{n}{4}, \frac{n}{3}, \frac{n}{2}$. In 2012, J. Zhu [62] shows that the tree with the second minimal energy is $B(n, \gamma)$ given in Figure 4.

2.3 Other Results

In the paper [61] that was mentioned above in relation with a Brualdi and Soldheid problem, B-X Zhu characterizes the unique graph whose least eigenvalue achieves

Fig. 4 Tree $B(n, \gamma)$

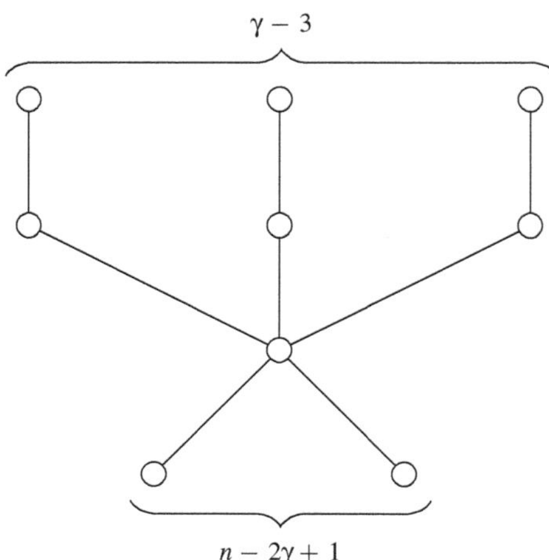

the minimum among all graphs with n vertices and domination number γ. The precise statement is the following.

Theorem 5 *Let G be a graph whose least eigenvalue λ_n is minimum among all graphs with n vertices and domination number γ. Then*

(i) $G \cong K_1 \vee K_{\lfloor \frac{n-1}{2} \rfloor, \lceil \frac{n-1}{2} \rceil}$ *for $\gamma = 1$ and $n \geq 6$.*

(ii) $G \cong K_{\lfloor \frac{n-\gamma+2}{2} \rfloor, \lceil \frac{n-\gamma+2}{2} \rceil} \cup (\gamma - 2)K_1$ *for $\gamma \geq 2$.*

A dominating set $S \subset V$ is an *efficient dominating set* if each vertex of G is dominated by precisely one vertex of S or, equivalently, if the minimum length of a path between any two vertices of S is at least three. Not every graph has an efficient dominating set, for example, the cycle C_4.

A subset $S \subset V(G)$ is a (k, τ)-regular set in G if it induces a k-regular subgraph in G and every vertex outside S has exactly τ neighbors in S. An efficient dominating set can also be defined as follows: a set S of vertices of a graph G is an efficient dominating set if $G[S]$ is a regular graph of degree 0 (i.e., S is an independent set) and every vertex of G outside S has precisely one neighbor in S. Thus an efficient dominating set can be viewed as a $(0, 1)$-regular set.

The efficient dominating set problem is the problem of determining whether a given graph has an efficient dominating set and finding such a set if it exists.

In the paper [11], Cardoso, Lozin, Luz, and Pacheco, using spectral results on (k, τ)-regular sets, as well as the theory of star complements, present a simplex-like algorithm for detecting a $(0, 1)$-regular set in an arbitrary graph. This particular algorithm can be used to find an efficient dominating set in any given graph or to conclude that such a set does not exist. The algorithm is not polynomial-time in general, however, the authors show that if -1 is not an eigenvalue of the adjacency matrix of the graph, it works in polynomial-time.

3 Laplacian Matrix

For a graph G of order n, its Laplacian eigenvalues always lie in the interval $[0, n]$. We number them

$$0 = \mu_n \leq \mu_{n-1} \leq \cdots \leq \mu_1.$$

The multiplicity of 0 is the number of connected components in G. There is a beautiful relationship between the Laplacian eigenvalues of G and of its complement $\overline{G}$. The eigenvalues of $\overline{G}$ are the following:

$$0 \leq n - \mu_1 \leq n - \mu_2 \leq \cdots \leq n - \mu_{n-1}.$$

In any graph, the average Laplacian eigenvalue is the average vertex degree. That is, $\sum_1^n \mu_i = \sum_1^n d_i(v_i)$. It is not surprising, then, that Laplacian eigenvalues reveal information about other properties of a graph.

3.1 Largest Eigenvalue

The largest Laplacian eigenvalue μ_1 is called the *Laplacian spectral radius*. We will now give several bounds for μ_1 using domination parameters. From what we can tell, *the earliest result relating Laplacian eigenvalues to the domination number* appears in the 1996 paper [6] by Brand and Seifter where they gave an upper bound for the Laplacian spectral radius.

Theorem 6 *Let G be a connected graph order n. If $\gamma(G) \geq 3$, then*

$$\mu_1 < n - \left\lceil \frac{\gamma(G) - 2}{2} \right\rceil. \tag{6}$$

If $\gamma(G) = 1$, then $\mu_1 = n$. If $\gamma(G) = 2$, then $\mu_1 \leq n$, and no better bound exists.

Obviously inequality (6) tells us something about μ_1 if we know $\gamma(G)$. But conversely, since Laplacian eigenvalues are always in $[0, n]$ their result implies that if μ_1 is close to n then $\gamma(G)$ is small. This is interesting because, unlike $\gamma(G)$, it is easy to compute μ_1.

In 2015 in [57], the upper bound for μ_1 in (6) was improved by Xing and Zhou who showed

$$\mu_1 \leq n - \gamma(G) + 2, \tag{7}$$

when $2 \leq \gamma(G) \leq n - 1$, and characterized the structure of extremal graphs for fixed n and $\gamma(G)$, without assuming connectivity. This is also a strict improvement over (6) when $\gamma(G) \geq 4$.

Let us now give some *lower bounds* on μ_1 based on domination parameters. In [46, Thr. 3] Nikiforov obtained the following lower bound on μ_1, and also characterized when equality occurs. A small domination number would imply a large μ_1.

Theorem 7 *If G is a graph containing an edge, then $\lceil \frac{n}{\gamma(G)} \rceil \leq \mu_1$.*

Proof. Let $H = (V, F)$ be a spanning subgraph consisting of a minimal set of edges such that $\gamma(H) = \gamma(G)$. It is easy to see that $H = \cup_{i=1}^{\gamma} S_i$, a disjoint union of $\gamma(G)$ stars, whose centers form a minimum dominating set of G. The largest star S_ℓ must have order at least $\lceil \frac{n}{\gamma(G)} \rceil$. Since H is a subgraph of G,

$$\mu_1(G) \geq \mu_1(H).$$

Since H is a disjoint union, $\mu_1(H)$ will be the maximum eigenvalue of the components. It is well-known that the largest Laplacian eigenvalue of a star S_ℓ is precisely its order for $\ell \geq 2$. This happens for the largest S_ℓ since G is not edgeless. Therefore

$$\mu_1(H) = \mu_1(S_\ell) = \ell \geq \left\lceil \frac{n}{\gamma(G)} \right\rceil.$$

Combining the last two lines completes the proof. $\qquad\qquad\square$

The vast majority of papers which relate domination to spectral properties of graphs contain results on the domination number $\gamma(G)$. The paper [50] by Shi, Kang, and Wu contains a lower bound of μ_1 using the *signed domination* number γ_s. A function $f : V \to \{-1, 1\}$ is called *signed dominating* if the sum of the values over any closed neighborhood is positive. Recall that the *closed neighborhood $N[v]$ of v* is the set of neighbors of v together with v. The signed domination number γ_s is the minimum weight over all signed dominating functions [21].

Theorem 8 ([50]) *Let G be a connected graph of order n. Then*

$$\frac{4n}{\gamma_s(G) + n} \leq \mu_1,$$

with equality holding if and only if $G = K_3$.

3.2 Second Smallest Eigenvalue

The second smallest Laplacian eigenvalue μ_{n-1} plays an important role in the structure of a graph. It is called the *algebraic connectivity* of the graph. The graph is connected if and only if $\mu_{n-1} > 0$. Much of our understanding of this eigenvalue is due to Fiedler [24]. It plays an important role in isoperimetric parameters through the so-called Cheeger inequalities (see [13]). As it turns out, the eigenvectors associated with the algebraic connectivity and with other small eigenvalues of the Laplacian matrix are widely used in graph partitioning [44].

We will now give some upper bounds on μ_{n-1} using γ and γ_t. An upper bound for the algebraic connectivity first appeared in 2005 in the paper by Lu, Liu, and Tian [43], who showed the following.

Theorem 9 *If G is a connected graph, then $\mu_{n-1} \leq \frac{n\,(n-2\gamma(G)+1)}{n-\gamma(G)}$.*

This is equivalent to

$$\mu_{n-1} \leq n - \frac{n}{n-\gamma(G)}(\gamma(G)-1). \tag{8}$$

Without assuming connectivity, in 2007 Nikiforov [46] showed that if $n \geq 2$, then

$$\mu_{n-1} \leq \begin{cases} n & \text{if}\,\gamma(G)=1 \\ n-\gamma(G) & \text{if}\,\gamma(G) \geq 2 \end{cases} \tag{9}$$

and also characterized when equality occurs.

As Har [33] showed in 2014 if one assumes no isolates, both results can be improved. Interestingly, Har cites the paper [43] but does not cite [46].

Theorem 10 *If G has no isolates then $\mu_{n-1} \leq n - 2(\gamma(G)-1)$.*

In the isolate-free case, clearly this improves upon (9). This theorem is also an improvement over inequality (8) since, when G has no isolates, one has $\gamma(G) \leq \frac{n}{2}$ and therefore $\frac{n}{n-\gamma(G)} \leq 2$.

Upper bounds for the second Laplacian eigenvalue are also given in the 2010 paper [3] by Aouchiche, Hansen, and Stevanović using the domination number. Here the authors assume the graph is connected.

The paper [50] also relates the *total domination* number γ_t to μ_{n-1}. A vertex set S is a total dominating set if every vertex in the graph is adjacent to some member of S, or if for every $v \in V$, $N(v) \cap S \neq \varnothing$. In a graph G without isolates, the total domination number $\gamma_t(G)$ is the minimum size of a total dominating set. The authors of [50] give two upper bounds on μ_{n-1} using γ_t. One of them is the following.

Theorem 11 *Let G be connected graph having $n \geq 3$ vertices, $G \neq K_n$. Then*

$$\mu_{n-1} \leq n - \gamma_t(G),$$

with equality holding if and only if $\overline{G}$ consists of a forest of K_2's or K_2's and isolates.

This bound is clearly better than (9) above since $\gamma(G) \le \gamma_t(G)$.

3.3 Disjoint Dominating Sets

A classic result of Ore [47, Thr. 13.1.5] states that in any graph without isolated vertices, if D is a minimal dominating set, then there exists another minimal dominating set *disjoint* from it.

Interestingly, Brand and Seifter's paper has a simple construction [6, Prop. 3.5] for also obtaining disjoint dominating sets. Their method uses a clever partition of an eigenvector of $L(G)$. However, it is not the usual partition formed by taking negative and non-negative entries. The method requires choosing an eigenvalue larger than $\Delta(G)$, the maximum degree. One might ask if this is always possible. The following result by Grone and Merris [27, Cor. 2] guarantees it.

Theorem 12 *If G has at least one edge, then $\mu_1 \ge \Delta(G) + 1$.*

Let $\mathbf{x}$ be an eigenvector for G. Then let D_-^0, D_+^0, and D_0^0 denote the vertices whose entries in $\mathbf{x}$ are negative, positive, and zero, respectively. For $i \ge 0$ recursively define

$$
\begin{aligned}
D_+^{i+1} &= D_+^i \cup \{v \in D_0^i \mid \{v, w\} \in E(G) \text{ for some } w \in D_-^i\} \\
D_0^{(i+1)'} &= D_0^i - D_+^{i+1} \\
D_-^{i+1} &= D_-^i \cup \{v \in D_0^{(i+1)'} \mid \{v, w\} \in E(G) \text{ for some } w \in D_+^{i+1}\} \\
D_0^{i+1} &= D_0^i - D_+^{i+1}.
\end{aligned}
$$

The construction stops when $D_0^m = \emptyset$. Now define $D_+ = D_+^m$ and $D_- = D_-^m$. By construction, D_- dominates the new vertices in D_+, and D_+ dominates the new vertices in D_-. Figure 5 shows the direction of the way vertices move from D_0 in the partition constructed above.

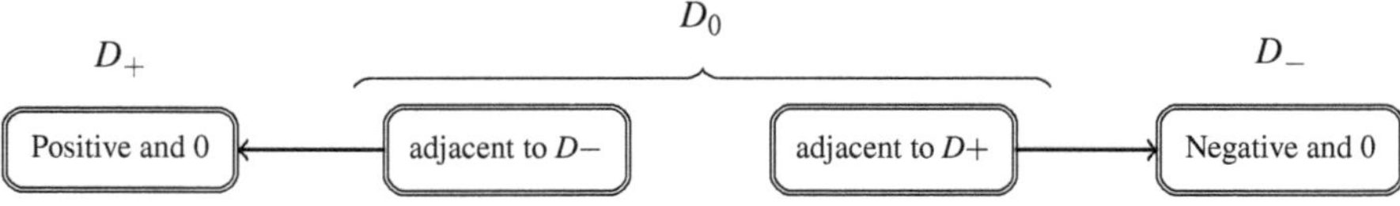

Fig. 5 Constructing disjoint domination sets

We are now ready to state the result in [6].

Theorem 13 *Let $G = (V, E)$ be a connected graph, and μ be an eigenvalue of $L(G)$, where $\mu > \Delta(G)$. Let $\mathbf{x}$ be an eigenvector for μ. Then D_+ and D_- are disjoint dominating sets.*

Proof. Clearly $D_+ \cup D_-$ forms a partition of V. Since $\mathbf{x}$ is an eigenvector we have $L(G)\mathbf{x} = \mu\mathbf{x}$, where $L(G) = D - A(G)$. Considering the row corresponding to v, one has

$$d(v)x_v - \sum_{\{v,w\}\in E} x_w = \mu x_v. \tag{10}$$

Suppose $v \in D_+^0$. Then the right side of (10) is positive. Since $\mu > \Delta(G) \geq d(v)$, there must be a negative term x_w in the summation, and v is dominated by $w \in D_-^0$. Hence D_+^0 is dominated by D_-^0.

On the other hand, suppose $v \in D_-^0$. A similar argument shows that the summation must contain a positive element. Therefore v is dominated by some $w \in D_+^0$. This shows that D_-^0 is dominated by D_+^0. If $D_0^0 = \emptyset$, then both D_-^0 and D_+^0 are dominating sets.

Now suppose there is a $v \in D_0^0$. By connectivity, at some point it will enter either D_+^i or D_-^i. Assume it enters D_+^i. This occurs because it is adjacent to some $w \in D_-^{i-1}$. Thus v is dominated by D_-^{i-1}. The other case is similar. $\square$

3.4 Laplacian Distribution

Here we are interested in results involving the number of Laplacian eigenvalues in an interval. If G is a graph and I is an interval, we let $m_G(I)$ denote the number of Laplacian eigenvalues of G in I, counting multiplicities.

In [28] the authors showed that in connected graphs G, $m_G[0, 1) \leq \alpha'(G)$, where $\alpha'(G)$ is the matching number of G. In other words, the number of Laplacian eigenvalues less than 1 is at most the matching number of G. In graphs without isolates it is known that $\alpha'(G) + \beta'(G) = n$, where $\beta'(G)$ denotes the graph's edge cover number. This implies that $m_G[1, n] \geq \beta'(G)$.

The following theorem in 2016 by Hedetniemi, Jacobs, and Trevisan [36] is an improvement on the result in [28] since $\gamma(G) \leq \alpha'(G)$ for any graph. Also the theorem holds for any graph regardless of its connectivity.

Theorem 14 *For any graph G, $m_G[0, 1) \leq \gamma(G)$.*

Corollary 1 *For any graph G, $m_G[1, n] \geq n - \gamma(G)$.*

Proof. Since there are n eigenvalues in $[0, n]$ one has $m_G[1, n] = n - m_G[0, 1)$. $\square$

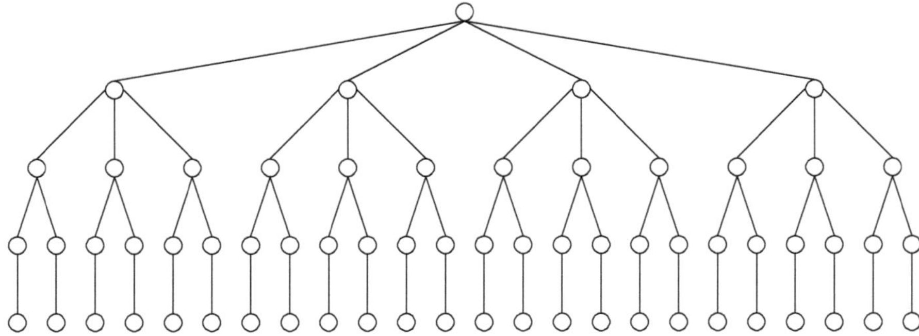

Fig. 6 Tree with $24 = m_T[0, 1) < \gamma(T) = 25$

For some simple classes of graphs one can have equality in Theorem 14.

Theorem 15 *If P_n is the path on n vertices then $m_{P_n}[0, 1) = \gamma(P_n) = \lceil \frac{n}{3} \rceil$.*

Proof. In ascending order, the eigenvalues P_n are $2 - 2\cos\left(\frac{i\pi}{n}\right)$, for $i = 0, \ldots,$ $n - 1$ (see [7, p. 9]). This implies that $0 \le \mu < 1$ if and only if $\frac{1}{2} < \cos\left(\frac{i\pi}{n}\right) \le 1$. Therefore, $i < \frac{n}{3}$. The largest such i is $\lceil \frac{n}{3} \rceil - 1$, so there are exactly $\lceil \frac{n}{3} \rceil$ such numbers, $m_{P_n}[0, 1) = \lceil \frac{n}{3} \rceil$. Since $\gamma(P_n) = \lceil \frac{n}{3} \rceil$, the theorem follows. $\qquad \square$

Using similar arguments one can show:

Theorem 16 *If C_n is the cycle on n vertices, then $m_{C_n}[0, 1) = 2\lceil \frac{n}{6} \rceil - 1$. Moreover,*

$$m_{C_n}[0, 1) = \begin{cases} \gamma(C_n) & \text{if } n \equiv 1, 2, 3 \bmod 6 \\ \gamma(C_n) - 1 & \text{if } n \equiv 0, 4, 5 \bmod 6. \end{cases}$$

There exist trees for which the inequality is strict, and Figure 6 depicts such a tree. To see this, we apply the algorithm in [5] which counts the number of Laplacian eigenvalues in any interval for trees and obtain $m_T[0, 1) = 24$. On the other hand, a minimum dominating set for the tree can be obtained by taking the 24 support vertices together with the root.

If we think of Theorem 14 as a *lower bound* of $\gamma(G)$, the following result by Cardoso, Jacobs, and Trevisan [12] gives an *upper bound*.

Theorem 17 *If G has minimum degree 1, then $\gamma(G) \le m_G[2, n]$.*

It should be noted that the result in Theorem 17 was obtained by Zhou, Zhou, and Du for trees in [60, Cor. 3.2]. As in the case of Theorem 14, the ratio can become arbitrarily large. However, for certain classes, the approximation ratio is small. The following results can be found in [12].

Theorem 18 *If T is a tree, $1 \le \frac{m_T[2,n]}{\gamma(T)} \le 2$.*

A connected graph of order n having $n-1+c$ edges is called *c-cyclic*. A generalization of Theorem 18 is the following.

Theorem 19 *If G is a c-cyclic graph where $c \geq 1$, then $1 \leq \frac{m_G[2,n]}{\gamma(G)} \leq c+1$.*

3.5   A Spectral Nordhaus–Gaddum Result

A *Nordhaus–Gaddum* inequality is a bound on the sum or product of a graph parameter for G and its complement $\overline{G}$. A result of Jaeger and Payan [41] states that

$$\gamma(G) + \gamma(\overline{G}) \leq n+1. \tag{11}$$

In [14] Cockayne and Hedetniemi proved the following.

Theorem 20 *For any graph G, $\gamma(G) + \gamma(\overline{G}) \leq n+1$ with equality if and only if $G = K_n$ or $G = \overline{K_n}$.*

Using the above results we can obtain a spectral Nordhaus–Gaddum inequality.

Theorem 21 *For any graph G, $m_G[0, 1) + m_{\overline{G}}[0, 1) \leq n+1$ with equality if and only if $G = K_n$ or $G = \overline{K_n}$.*

Proof. From Theorem 14 and (11) we get

$$m_G[0, 1) + m_{\overline{G}}[0, 1) \leq \gamma(G) + \gamma(\overline{G}) \leq n+1 \tag{12}$$

for any G, establishing the inequality. Now assume $G = K_n$ or $G = \overline{K_n}$. Since $m_{K_n}[0, 1) = 1$ and $m_{\overline{K_n}}[0, 1) = n$, we have equality. Conversely assume that $m_G[0, 1) + m_{\overline{G}}[0, 1) = n+1$. Then (12) implies that $\gamma(G) + \gamma(\overline{G}) = n+1$. Applying Theorem 20 it follows that $G = K_n$ or $G = \overline{K_n}$. This completes the proof. $\square$

To conclude the section, we mention that, more generally, computing the number of eigenvalues of a matrix M associated with a graph G that lie in a given real interval I is a problem that has been intensively studied in the last few years. An algorithm is said to *locate eigenvalues* for a graph class $\mathscr{C}$ if, for any graph $G \in \mathscr{C}$ and any real interval I, it finds the number of eigenvalues of G in the interval I. In recent years, efficient algorithms have been developed for the location of eigenvalues of the adjacency matrix in trees [38], threshold graphs [39] (also called *nested split graphs*), chain graphs [1] and cographs [40], for instance. Several of these algorithms have been adapted to the Laplacian matrix. There are also algorithms for general graphs that are very efficient when the graph admits a decomposition with "low complexity" (with respect to measures such as the clique-width, see for instance [25]).

4 Signless Laplacian Matrix

For a given graph $G = (V, E)$ of order $n = |V|$ and size $m = |E|$, the signless Laplacian spectrum of G is the multiset given by the eigenvalues of the matrix $Q(G) = D(G) + A(G)$. As Q is real symmetric and positive semidefinite, its eigenvalues are real and non-negative, and we order them as

$$q_1 \geq q_2 \geq \cdots \geq q_n.$$

In 2009, Cvetković and Simić, in the beautiful series of papers [16–18], introduced many properties of the signless Laplacian matrix for graphs and, in particular, it is known that the multiplicity of 0 as a Q-eigenvalue is the number bipartite components of G.

4.1 Bounds for the Index

The largest eigenvalue of the signless Laplacian matrix of G is called the signless Laplacian index of G. The first record we found in the literature relating the spectrum of the signless Laplacian matrix of a graph G and its domination number is from 2010 and due to Hansen and Lucas [32]. They considered relations between signless Laplacian eigenvalues and several graph parameters. In our context, the relevant result is the following.

Theorem 22 *Let G be a connected graph on $n \geq 4$ vertices with signless Laplacian index q_1 and domination number γ. Then*

1. $q_1 + \gamma \leq 2n - 1$,
2. $n \leq q_1 \cdot \gamma$.

Equality is attained in (1) *if and only G is the complete graph K_n.*

The case of general graphs has been addressed by Xing and Zhou [57], who were able to characterize all graphs for which the bound is tight.

Theorem 23 *Let G be a graph on $n \geq 4$ vertices with signless Laplacian index q_1 and domination number γ. Then*

$$q_1(G) \leq 2(n - \gamma),$$

with equality if and only if $G \cong K_{n-\gamma+1} \cup (\gamma - 1)K_1$ or when $\gamma \geq 2$ and $n - \gamma$ is even, $G \cong \frac{n-\gamma+2}{2} K_2 \cup (\gamma - 2)K_1$.

4.2 Bounds for the Smallest Signless Laplacian Eigenvalue

He and Zhou in 2014 [34] presented a sharp upper bound for the smallest signless Laplacian eigenvalue of a graph involving its domination number. They also determined extremal graphs which attain this bound. In order to state their result we need the following definition.

Let H be a graph on n vertices $v_1, v_2, \ldots, v_n$. The corona $G \circ K_1$ of a graph G is the graph obtained from G by attaching a leaf to every vertex of G.

Theorem 24 *Let G be a connected graph with even order $n = 2k \geq 6$, least signless Laplacian eigenvalue q_n and domination number $\gamma \geq 3$. Then*

$$q_n \leq 2k - 2\gamma + \frac{k - \sqrt{(k-2)^2 + 4}}{2},$$

with equality if and only if $G \simeq K_k \circ K_1$.

For n odd, we consider the family $\mathscr{F}$ of graphs given in Figure 7. The result of [34] is the following.

Theorem 25 *Let G be a connected graph with odd order $n = 2k + 1 \geq 7$, smallest signless Laplacian eigenvalue q_n, minimum degree δ and domination number $\gamma \geq 3$. If G satisfies one of the following conditions:*

1. *δ is even and $G \notin \mathscr{F}$ (see Figure 7),*
2. *$\delta = 1, n \geq 13$,*
3. *$\delta = 3, n \geq 17$,*
4. *$\delta = 5, n \geq 23$,*

   then

$$q_n \leq 2k - 2\gamma + \frac{k + 1 - \sqrt{(k-1)^2 + 4}}{2}.$$

We notice that this result does not tell anything about graphs having odd minimum degree $\delta > 5$. This case is not discussed in [34].

It is well-known that the smallest signless Laplacian eigenvalue $q_n(G) = 0$ if and only if G has a bipartite component. Hence it is natural to study lower bounds for

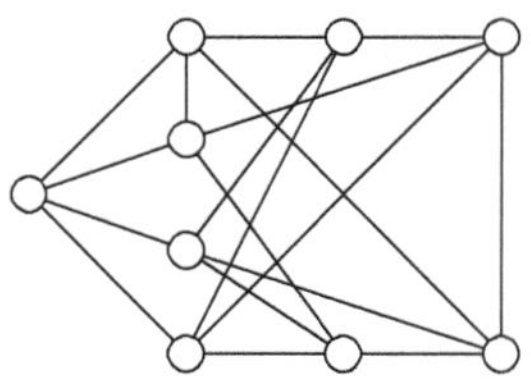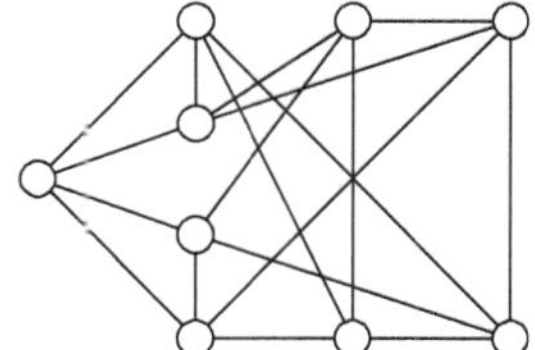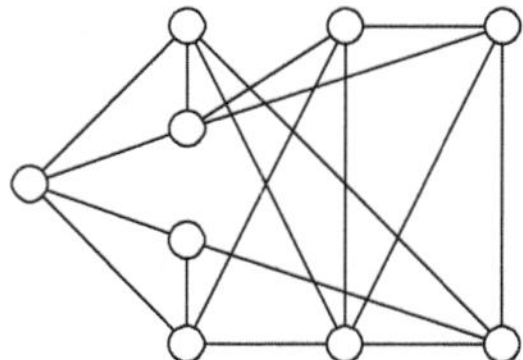

Fig. 7 Family $\mathscr{F}$ of graphs

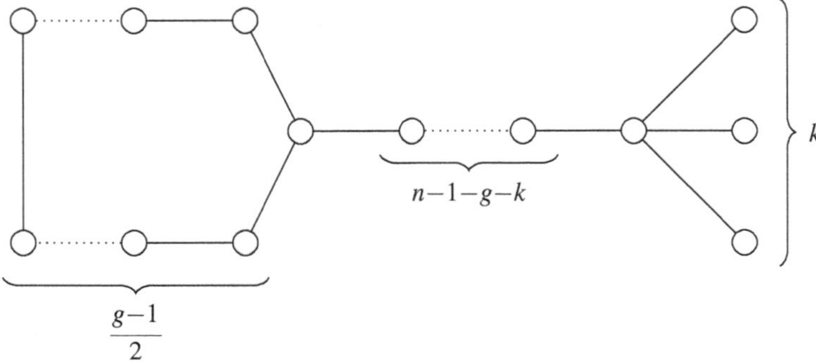

Fig. 8 Graph $U_n^k(g)$

the $q_n(G)$ when a connected graph G is not bipartite. In 2014, Fan and Tan [23] presented a lower bound for the least eigenvalue of the signless Laplacian of G in terms of the domination number. For a more precise statement of his result, we use the following notation.

Denote by $U_n^k(g)$ the unicyclic graph of order n, which is obtained from an odd cycle $C_g(g < n)$ and a star $S_{1,k}$ by identifying the end vertices of a path P_ℓ to one vertex of the cycle and the center of the star , where $\ell = n + 1 - g - k$ (see Figure 8).

It is easy to see that if $k \geq 2$, then

$$\gamma(U_n^k(g)) \leq \gamma(U_n^{k-1}(g)) \leq \cdots \leq \gamma(U_n^1(g)) := \gamma_{n,g}.$$

For fixed n and odd $g \in [3, n-1]$, for each $\gamma \in [\lceil \frac{g}{3} \rceil, \gamma_{n,g}]$, there exists one or more graphs $U_n^k(g)$ with domination number γ. The unique one with minimum k among those graphs is denoted by $W_n^\gamma(g)$.

Theorem 26 *Let G be a connected non-bipartite graph of order n with domination number $\gamma \leq \frac{n+1}{3}$. Then*

$$q_n(G) \leq q_n(W_n^\gamma(3)),$$

with equality if and only if $G = W_n^\gamma(3)$.

In an independent work in 2014 [59], Yu, Guo, Zhang, and Wu studied the same problem. With the above terminology, they determined exactly the structure of $W_n^\gamma(3)$. The precise statement is as follows.

Theorem 27 *Among all the non-bipartite graphs with both order $n \geq 4$ and domination number $\gamma \leq \frac{n+1}{3}$, we have*

(i) *If $n \in \{3\gamma - 1, 3\gamma, 3\gamma + 1\}$, then $W_n^1(3)$ is the unique graph with minimal q_n up to isomorphism;*

(ii) *If $n \geq 3\gamma + 2$, then $W_n^{n-\gamma}(3)$ is the unique graph with minimal q_n up to isomorphism.*

4.3 k-Domination and Bounds for Q-Eigenvalues

In the work by Liu and Lu [42], there are bounds for $q_2(G)$ and $q_n(G)$ based on the k-domination number. For an integer $k \geq 1$, a *k-dominating set* in G is a subset X of $V(G)$ such that each element of $V(G) \setminus X$ is adjacent to at least k vertices of X. (This is sometimes called a *k-fold-dominating set* to distinguish the situation where every vertex in the dominating set dominates all vertices at distance up to k.) The least cardinality of a k-dominating set is the *k-domination number* of G, denoted by $\gamma_k(G)$. The relevant results are as follows.

For a graph G, a partition $V(G) = V_1 \cup V_2$ is called (r, s)-local-regular if all vertices in V_1 have r neighbors in V_2 and all vertices in V_2 have s neighbors in V_1.

Theorem 28 *Let G be a graph of order $n \geq 2$ with maximum degree Δ, minimum degree δ, and average degree d. Let X be a minimum k-dominating set such that $|E(G[X])|$ is as large as possible. If $\delta \geq k$ and $|N(x) \cap X| \geq k - 1$ for each vertex $x \in X$, then*

$$q_2(G) \leq 2\delta - \frac{n(n - (1 + 1/k)|X| + 1)}{n - |X|},$$

with equality only if G is $(k, 1)$-local-regular graph. Moreover, if $k = 1$, then equality holds if and only if $G \cong K_{2,2}$.

Theorem 29 *Let G be a graph of order $n \geq 2$ with maximum degree Δ. Then*

$$q_n(G) \leq 2\Delta - \frac{nk}{\gamma_k(G)},$$

where the equality holds only if G is a $(k(n - \gamma_k(G))/\gamma_k(G), k)$-local-regular graph. Moreover, if G is connected and $k = 1$, then equality holds if and only if $G \cong K_n$.

5 Distance Matrices

Let G be a connected graph with vertex set $V(G)$ and edge set $E(G)$. For $u, v \in V(G)$, the distance between u and v in G is the length of a shortest path connecting them, denoted by $d(u, v)$. The distance matrix of G is defined as

$$\mathscr{D}(G) = (d(u, v))_{u,v \in V(G)}.$$

The eigenvalues of $\mathscr{D}(G)$ are called the distance eigenvalues of G. As $\mathscr{D}(G)$ is real and symmetric, the distance eigenvalues of G are real and ordered as

$$\partial_1 \geq \partial_2 \geq \cdots \geq \partial_n.$$

The distance spectral radius of G, ∂_1, is the largest distance eigenvalue of G. If $|V(G)| \geq 2$ and G is connected, then the matrix $\mathscr{D}(G)$ is irreducible and the Perron–Frobenius theorem implies that ∂_1 is positive, simple and there is a unique positive unit eigenvector $x(G)$ corresponding to ∂_1, which is called the *distance Perron vector* of G.

A remarkable property of the distance matrix given by Graham and Pollack [26] is a formula of the determinant of the distance matrix of a tree depending only on the order n. The determinant is given by

$$\det(\mathscr{D}) = (-1)^{n-1}(n - 1)2^{n-2},$$

implying that any nontrivial tree has a single, positive distance eigenvalue and $n - 1$ negative eigenvalues.

The work by Wang and Zhou [56] relates the spectral radius of a tree with its domination number. As we shall describe, the authors determine the unique tree of given domination number with minimum distance spectral radius and the unique tree of given domination number with maximum distance spectral radius.

Let $A(n, m)$ be the tree obtained from the star S_{n-m+1} by attaching a new leaf to each of $m - 1$ chosen leaves of S_{n-m+1}, where $1 \leq m \leq \lfloor \frac{n}{2} \rfloor$. It is easy to see that $\gamma(A(n, m)) = m$.

Theorem 30 *Let T be a tree on n vertices with domination number γ, where $1 \leq \gamma \leq \lfloor \frac{n}{2} \rfloor$. Then*

$$\partial_1(T) \geq \partial_1(A(n, \gamma)),$$

with equality if and only if $T \cong A(n, \gamma)$.

Let $D(n, a, b)$ be the tree obtained from the path P_{n-a-b} by attaching a and b leaves to the two end vertices, respectively, where $a \geq b \geq 1$ and $a + b \leq n - 1$.

Theorem 31 *Let T be a tree on n vertices with domination number γ, where $1 \leq \gamma < \lceil \frac{n}{3} \rceil$. Then*

$$\partial_1(T) \leq \partial_1\left(D\left(n, \left\lceil \frac{n - 3\gamma + 2}{2} \right\rceil, \left\lceil \frac{n - 3\gamma + 2}{2} \right\rceil\right)\right),$$

with equality if and only if $T \cong D\left(n, \lceil \frac{n-3\gamma+2}{2} \rceil, \lceil \frac{n-3\gamma+2}{2} \rceil\right)$.

The transmission D_u of a vertex u is the sum of the distances from u to all other vertices of in G, that is,

$$D_u = \sum_{v \in G} d(u, v).$$

If D^L is the diagonal matrix of the vertex transmissions, whose i-th entry is D_{u_i}, the *distance Laplacian matrix* of a graph G is the matrix

$$\mathscr{D}^L = D^L - \mathscr{D},$$

whose eigenvalues are going to be denoted by

$$\partial_1^L \geq \partial_2^L \geq \cdots \geq \partial_n^L.$$

Among the properties given in [2] of the distance Laplacian matrix $\mathscr{D}^L$ is that the spectrum of an n-vertex connected graph G with diameter at most 2 is given by

$$\partial_1^L(G) = 2n - \mu_{n-1} \geq \partial_2^L(G) = 2n - \mu_{n-2} \geq \cdots \geq \partial_{n-1}(G) = 2n - \mu_1$$
$$> \partial_n^L(G) = 0,$$

where the μ_i are the Laplacian eigenvalues.

In [19], the authors relate the distance Laplacian spectral radius ∂_1^L of a graph G and its domination number. The main results are summarized in the following theorem.

Theorem 32 *Let G be a connected graph of order $n \geq 2$ with domination number γ. Then the following hold:*

(i) *$\partial_1^L(G) \geq n + \gamma - 1$ with equality if and only if $G \cong K_n$.*
(ii) *If the diameter of G is d, then $\partial_1^L(G) \geq n + \gamma + d - 2$ with equality holding if and only if $G \cong K_n$ or, in the case $n = 2p$, $G \cong K_{\underbrace{2, 2, \ldots, 2}_{p}}$.*

(iii) *If $\partial_{n-1}^L(G) = n$, then $\gamma \leq 2$.*

6   Final Remarks and Open Problems

In writing this chapter, our aim was to collect some of the main results in spectral graph theory that involve domination parameters. As we have seen, many such results provide bounds on domination parameters in terms of eigenvalues or eigenvalue-based parameters. It also became clear that the connection between graph spectra and domination parameters is still not well understood. In this section, we propose directions for further investigation.

In the last decade, there have been considerable advances in generalized Brualdi–Solheid problems involving domination. Recall that these are problems where, for a set $\mathscr{G}$ of graphs and a spectral parameter f_M associated with a matrix M, one is asked to find $\min\{f_M(G) : G \in \mathscr{G}\}$ and $\max\{f_M(G) : G \in \mathscr{G}\}$, and to characterize the graphs which achieve the minimum or maximum value. Two prime examples are Theorems 2 and 3 in Section 2, which completely determine, respectively, the n-vertex graphs with domination number γ and the n-vertex graphs with domination number γ and no isolated vertices that maximize the spectral radius of the adjacency matrix. The maximization part has also been solved for the signless Laplacian matrix, as described in Section 4. However, the minimization part of the problem is often not studied, and we leave it here as an open question.

Problem 1 Let γ be a positive integer and $\mathscr{G}_\gamma^{(n)} = \{G : |V(G)| = n,\ \gamma(G) = \gamma\}$. Find the graphs $G \in \mathscr{G}_\gamma^{(n)}$ with minimum spectral radius.

The Laplacian matrix is an example where both the maximization and minimization part of the Brualdi–Solheid problem have been solved, but, for many other matrices, both parts are still open. This happens for the distance matrices of Section 5.

More generally, it is natural to consider the Brualdi–Solheid problem for other matrices and domination parameters.

Problem 2 Let γ be a positive integer, let $\gamma^*(G)$ be a domination parameter associated with a graph G and let M be a symmetric matrix associated with a graph G. Define $\mathscr{G}_\gamma^{*(n)} = \{G : |V(G)| = n,\ \gamma^*(G) = \gamma\}$. Find the graphs $G \in \mathscr{G}_\gamma^{*(n)}$ with maximum and minimum spectral radius (with respect to the matrix M).

Related to this problem, the result of Xing and Zhou in Section 3.1 describes the structure of extremal graphs with respect to the index of the Laplacian matrix. It turns out that these graphs contain isolated vertices for all $\gamma \geq 3$. It would be interesting to consider the case in which there are no isolates or where the graph is assumed to be connected.

Problem 3 Let γ be a positive integer and let $\mathscr{C}_\gamma^{(n)}$ be the set of n-vertex connected graphs. Find the graphs $G \in \mathscr{C}_\gamma^{(n)}$ with maximum and minimum index with respect to the Laplacian matrix.

This problem would also be interesting with respect to the signless Laplacian matrix, as the upper bound in Theorem 22 is tight if and only if $\gamma = 1$.

As we mentioned in Section 2.3, there has been interest in characterizing graphs with a given domination number such that the *least* eigenvalue is maximum or minimum. Another result in this direction, now in terms of the signless Laplacian matrix, is Theorem 27, which solves the problem of finding the smallest least Q-eigenvalue for graphs of order $n \geq 4$ and domination number $\gamma \leq \frac{n+1}{3}$.

Problem 4 Determine which graph(s) among all non-bipartite graphs on n vertices and domination number $\frac{n+1}{3} < \gamma \leq \frac{n}{2}$ has minimal least Q-eigenvalue.

In the same vein, there has been substantial interest in ordering the elements of a graph class according to the value of some parameter. So, in addition to finding the graph(s) that achieve the maximum or minimum value, one would be interested in graphs with the second largest value and so on. In more restricted settings, it may even be possible to completely order the elements in terms of this parameter. This may be quite hard if we consider all graphs with domination number γ. A particular question in this direction is stated in terms of the energy of trees with fixed domination number.

Problem 5 Let γ be a positive integer and $\mathscr{T}_\gamma^{(n)} = \{T : T \text{ tree}, |V(T)| = n, \gamma(G) = \gamma\}$. Order the elements of $\mathscr{T}_\gamma^{(n)}$ according to their energy.

As mentioned in Section 2.2, the trees with least energy and with second smallest energy in $\mathscr{T}_\gamma^{(n)}$ are already known, and the tree with maximum energy is known in some cases.

The distribution of the eigenvalues of a graph, or of the elements of a graph class, is a topic of great interest in spectral graph theory. In Section 3.4, we have described results relating the domination number with the number of eigenvalues in a given real interval. For instance, Theorem 14 determines that $m_G[0, 1) \leq \gamma(G)$, where $m_G[0, 1)$ is the number of Laplacian eigenvalues of G in $[0, 1)$. Even though we may have equality for some simple graph classes, the ratio between $m_G[0, 1)$ and $\gamma(G)$ can be arbitrarily large [12]. An open question is when Theorem 14 achieves equality.

Problem 6 Characterize graphs G for which $m_G[0, 1) = \gamma(G)$.

Another natural question is whether the ratio $\frac{\gamma(G)}{m_G[0,1)}$ is bounded for some particular graph class. The following is a particular problem in this direction.

Problem 7 Is the ratio $\frac{\gamma(T)}{m_T[0,1)}$ bounded for trees T?

To conclude this section, we also include a problem that is purely spectral, but has algorithmic consequences for efficient domination in light of the result of Cardoso, Lozin, Luz, and Pacheco [11] described in Section 2.3.

Problem 8 Describe the family of graphs G such that $\lambda = -1$ is not an eigenvalue of $A(G)$.

References

1. A. Alazemi, M. Andelić, S.K. Simić, Eigenvalue location for chain graphs. Linear Algebra Appl. **505**, 194–210 (2016). http://dx.doi.org/10.1016/j.laa.2016.04.030
2. M. Aouchiche, P. Hansen, Two Laplacians for the distance matrix of a graph. Linear Algebra Appl. **439**(1), 21–33 (2013). https://doi.org/10.1016/j.laa.2013.02.030. http://www.sciencedirect.com/science/article/pii/S0024379513001614
3. M. Aouchiche, P. Hansen, D. Stevanović, A sharp upper bound on algebraic connectivity using domination number. Linear Algebra Appl. **432**(11), 2879–2893 (2010). https://doi.org/10.1016/j.laa.2009.12.031
4. L. Babai, D.Y. Grigoryev, D.M. Mount, Isomorphism of graphs with bounded eigenvalue multiplicity. In: *Proceedings of the Fourteenth Annual ACM Symposium on Theory of Computing, STOC '82* (ACM, New York, NY, USA, 1982), pp. 310–324. http://doi.acm.org/10.1145/800070.802206
5. R.O. Braga, V.M. Rodrigues, V. Trevisan, On the distribution of Laplacian eigenvalues of trees. Discrete Math. **313**(21), 2382–2389 (2013). http://dx.doi.org/10.1016/j.disc.2013.06.017
6. C. Brand, N. Seifter, Eigenvalues and domination in graphs. Math. Slovaca **46**(1), 33–39 (1996)
7. A.E. Brouwer, W. Haemers, *Spectra of Graphs* (Springer, New York, 2012)
8. R. Brualdi, E. Solheid, On the spectral radius of complementary acyclic matrices of zeros and ones. SIAM J. Algebraic Discrete Methods **7**(2), 265–272 (1986). https://doi.org/10.1137/0607030
9. S. Butler, A note about cospectral graphs for the adjacency and normalized Laplacian matrices. Linear Multilinear Algebra **58**(3), 387–390 (2010). https://doi.org/10.1080/03081080902722741
10. S. Butler, F. Chung, Spectral graph theory, in *Handbook of Linear Algebra*, 2nd edn. (CRC Press, 2017)
11. D.M. Cardoso, V.V. Lozin, C.J. Luz, M.F. Pacheco, Efficient domination through eigenvalues. Discrete Appl. Math. **214**, 54–62 (2016). https://doi.org/10.1016/j.dam.2016.06.014. http://www.sciencedirect.com/science/article/pii/S0166218X16302931
12. D.M. Cardoso, D.P. Jacobs, V. Trevisan, Laplacian distribution and domination. Graphs Combin. **33**(5), 1283–1295 (2017). https://doi.org/10.1007/s00373-017-1844-x
13. F.R.K. Chung, *Spectral Graph Theory* (American Mathematical Society, Providence, 1997)
14. E.J. Cockayne, S.T. Hedetniemi, Toward a theory of domination in graphs. Networks **7**, 247–261 (1977)
15. R.R. Coifman, S. Lafon, Diffusion maps. Appl. Comput. Harmonic Anal. **21**(1), 5–30 (2006). http://dx.doi.org/10.1016/j.acha.2006.04.006. //www.sciencedirect.com/science/article/pii/S1063520306000546. Special Issue: Diffusion Maps and Wavelets
16. D. Cvetković, S.K. Simić, Towards a spectral theory of graphs based on the signless Laplacian. I. Publ. Inst. Math. (Beograd) (N.S.) **85**(99), 19–33 (2009). http://dx.doi.org/10.2298/PIM0999019C
17. D. Cvetković, S.K. Simić, Towards a spectral theory of graphs based on the signless Laplacian. II. Linear Algebra Appl. **432**(9), 2257–2272 (2010). http://dx.doi.org/10.1016/j.laa.2009.05.020
18. D. Cvetković, S.K. Simić, Towards a spectral theory of graphs based on the signless Laplacian. III. Appl. Anal. Discrete Math. **4**(1), 156–166 (2010). http://dx.doi.org/10.2298/AADM1000001C
19. K.C. Das, M. Aouchiche, P. Hansen, On distance Laplacian and distance signless Laplacian eigenvalues of graphs. Linear Multilinear Algebra **0**(0), 1–18 (2018). https://doi.org/10.1080/03081087.2018.1491522
20. P. Delsarte, An algebraic approach to the association schemes of coding theory. Philips Res. Repts. Suppl. **10**, 1–54 (1973)

21. J. Dunbar, S. Hedetniemi, M.A. Henning, P.J. Slater, Signed domination in graphs, in *Graph Theory, Combinatorics, and Algorithms*, vol. 1, 2 (Kalamazoo, MI, 1992) (Wiley, New York, 1995), pp. 311–321
22. S.N. Dusanka Janezic, Ante Milicevic, N. Trinajstic, *Graph-Theoretical Matrices in Chemistry* (CRC Press, Boca Raton, 2015)
23. Y.Z. Fan, Y.Y. Tan, The least eigenvalue of signless Laplacian of non-bipartite graphs with given domination number. Discrete Mathematics **334**, 20–25 (2014). https://doi.org/10.1016/j.disc.2014.06.021. http://www.sciencedirect.com/science/article/pii/S0012365X14002489
24. M. Fiedler, Algebraic connectivity of graphs. Czechoslovak Math. J. **23**(98), 298–305 (1973)
25. M. Fürer, C. Hoppen, D.P. Jacobs, V. Trevisan, Eigenvalue location in graphs of small clique-width. Linear Algebra Appl. **560**, 56–85 (2019). https://doi.org/10.1016/j.laa.2018.09.015. http://www.sciencedirect.com/science/article/pii/S0024379518304506
26. R.L. Graham, H.O. Pollak, On the addressing problem for loop switching. Bell Syst. Tech. J. **50**(8), 2495–2519 (1971). https://doi.org/10.1002/j.1538-7305.1971.tb02618.x
27. R. Grone, R. Merris, The Laplacian spectrum of a graph. II. SIAM J. Discrete Math. **7**(2), 221–229 (1994). https://doi.org/10.1137/S0895480191222653
28. J.M. Guo, X.L. Wu, J.M. Zhang, K.F. Fang, On the distribution of Laplacian eigenvalues of a graph. Acta Math. Sin. (Engl. Ser.) **27**(11), 2259–2268 (2011). http://dx.doi.org/10.1007/s10114-011-8624-y
29. I. Gutman, Acyclic systems with extremal Hückel π-electron energy. Theoretica Chimica Acta **45**, 79–87 (1977)
30. I. Gutman, X. Li, *Energies of Graphs - Theory and Applications* (University of Kragujevac and Faculty of Science Kragujevac, Kragujevac, 2016), p. 293
31. W.H. Haemers, Are almost all graphs determined by their spectrum? Not. S. Afr. Math. Soc. **47**(1), 42–45 (2016)
32. P. Hansen, C. Lucas, Bounds and conjectures for the signless Laplacian index of graphs. Linear Algebra Appl. **432**(12), 3319–3336 (2010). https://doi.org/10.1016/j.laa.2010.01.027. http://www.sciencedirect.com/science/article/pii/S0024379510000406
33. J. Har, A note on Laplacian eigenvalues and domination. Linear Algebra Appl. **449**, 115–118 (2014). http://dx.doi.org/10.1016/j.laa.2014.02.025
34. C.X. He, M. Zhou, A sharp upper bound on the least signless Laplacian eigenvalue using domination number. Graphs Comb. **30**(5), 1183–1192 (2014). https://doi.org/10.1007/s00373-013-1330-z
35. C.X. He, B.F. Wu, Z.S. Yu, Extremal energies of trees with a given domination number. MATCH Commun. Math. Comput. Chem. **64**(1), 169–180 (2010). http://match.pmf.kg.ac.rs/electronic_versions/Match64/n1/match64n1_169-180.pdf
36. S.T. Hedetniemi, D.P. Jacobs, V. Trevisan, Domination number and Laplacian eigenvalue distribution. Eur. J. Combin. **53**, 66–71 (2016). URL http://dx.doi.org/10.1016/j.ejc.2015.11.005
37. R.A. Horn, C.R. Johnson, *Matrix Analysis*, 2nd edn. (Cambridge University Press, New York, 2012)
38. D.P. Jacobs, V. Trevisan, Locating the eigenvalues of trees. Linear Algebra Appl. **434**(1), 81–88 (2011). http://dx.doi.org/10.1016/j.laa.2010.08.006
39. D.P. Jacobs, V. Trevisan, F. Tura, Eigenvalue location in threshold graphs. Linear Algebra Appl. **439**(10), 2762–2773 (2013). http://dx.doi.org/10.1016/j.laa.2013.07.030
40. D.P. Jacobs, V. Trevisan, F.C. Tura, Eigenvalue location in cographs. Discrete Appl. Math. **245**, 220–235 (2018). https://doi.org/10.1016/j.dam.2017.02.007. http://www.sciencedirect.com/science/article/pii/S0166218X17300926
41. F. Jaeger, C. Payan, Relations du type Nordhaus-Gaddum pour le nombre d'absorption d'un graphe simple. C. R. Acad. Sci. Paris Sér. A-B **274**, A728–A730 (1972)
42. H. Liu, M. Lu, Bounds of signless Laplacian spectrum of graphs based on the k-domination number. Linear Algebra Appl. **440**, 83–89 (2014). https://doi.org/10.1016/j.laa.2013.10.020. http://www.sciencedirect.com/science/article/pii/S002437951300640X
43. M. Lu, H. Liu, F. Tian, Bounds of Laplacian spectrum of graphs based on the domination number. Linear Algebra Appl. **402**, 390–396 (2005). https://doi.org/10.1016/j.laa.2005.01.006

44. U. Luxburg, A tutorial on spectral clustering. Stat. Comput. **17**(4), 395–416 (2007). http://dx.doi.org/10.1007/s11222-007-9033-z
45. B. McKay, On the spectral characterisation of trees. Ars Combin. **3**, 219–232 (1979)
46. V. Nikiforov, Bounds of graph eigenvalues. I. Linear Algebra Appl. **420**(2–3), 667–671 (2007). http://dx.doi.org/10.1016/j.laa.2006.08.020
47. O. Ore, *Theory of Graphs* (Amer. Math. Soc. Colloq. Publ., 1962)
48. P. Rowlinson, Dominating sets and eigenvalues of graphs. Bull. Lond. Math. Soc. **26**(3), 248–254 (1994). https://doi.org/10.1112/blms/26.3.248
49. A.J. Schwenk, Almost all trees are cospectral, in *New Directions in the Theory of Graphs* (Proc. Third Ann Arbor Conf., Univ. Michigan, Ann Arbor, MI, 1971), pp. 275–307 (Academic Press, New York, 1973)
50. W. Shi, L. Kang, S. Wu, Bounds on Laplacian eigenvalues related to total and signed domination of graphs. Czechoslov. Math. J. **60**(2), 315–325 (2010). https://doi.org/10.1007/s10587-010-0035-1
51. D. Stevanović, *Spectral Radius of Graphs* (Academic Press, New York, 2015)
52. D. Stevanović, M. Aouchiche, P. Hansen, On the spectral radius of graphs with a given domination number. Linear Algebra Appl. **428**(8), 1854–1864 (2008). https://doi.org/10.1016/j.laa.2007.10.024. http://www.sciencedirect.com/science/article/pii/S0024379507004831
53. E.R. van Dam, W.H. Haemers, Which graphs are determined by their spectrum? Linear Algebra Appl. **373**, 241–272 (2003)
54. E.R. van Dam, W.H. Haemers, Developments on spectral characterizations of graphs. Discrete Mathematics **309**(3), 576–586 (2009). https://doi.org/10.1016/j.disc.2008.08.019. http://www.sciencedirect.com/science/article/pii/S0012365X08005219. International Workshop on Design Theory, Graph Theory, and Computational Methods
55. C. Van Nuffelen, Rank and domination number, in *Graphs and Other Combinatorial Topics, Teubner-Texte Math.*, vol. 59, ed. by M. Fiedler. Proc. 3rd Czech. Symp. Graph Theory, Prague 1982, (1983), pp. 209–211
56. Y. Wang, B. Zhou, On distance spectral radius of graphs. Linear Algebra Appl. **438**(8), 3490–3503 (2013). https://doi.org/10.1016/j.laa.2012.12.024
57. R. Xing, B. Zhou, Laplacian and signless Laplacian spectral radii of graphs with fixed domination number. Math. Nachr. **288**(4), 476–480 (2015). http://dx.doi.org/10.1002/mana.201300331
58. K. Xu, L. Feng, Extremal energies of trees with a given domination number. Linear Algebra Appl. **435**(10), 2382–2393 (2011). https://doi.org/10.1016/j.laa.2010.09.008. http://www.sciencedirect.com/science/article/pii/S0024379510004714. Special Issue in Honor of Dragos Cvetkovic
59. G. Yu, S.G. Guo, R. Zhang, Y. Wu, The domination number and the least q-eigenvalue. Appl. Math. Comput. **244**, 274–282 (2014). https://doi.org/10.1016/j.amc.2014.06.076. http://www.sciencedirect.com/science/article/pii/S0096300314009205
60. L. Zhou, B. Zhou, Z. Du, On the number of Laplacian eigenvalues of trees smaller than two. Taiwanese J. Math. **19**(1), 65–75 (2015)
61. B.X. Zhu, The least eigenvalue of a graph with a given domination number. Linear Algebra Appl. **437**(11), 2713–2718 (2012). https://doi.org/10.1016/j.laa.2012.06.007
62. J. Zhu, Minimal energies of trees with given parameters. Linear Algebra Appl. **436**(9), 3120–3131 (2012). https://doi.org/10.1016/j.laa.2011.10.002. http://www.sciencedirect.com/science/article/pii/S0024379511006926

Varieties of Roman Domination

M. Chellali, N. Jafari Rad, S. M. Sheikholeslami, and L. Volkmann

1 Introduction

For a graph $G = (V, E)$, a function $f : V \rightarrow \{0, 1, 2\}$ is a *Roman dominating function*, or just an RDF, if every vertex u for which $f(u) = 0$ is adjacent to at least one vertex v for which $f(v) = 2$. The weight of a function f is $\omega(f) = \sum_{v \in V} f(v)$. The *Roman domination number* $\gamma_R(G)$ of a graph G is the minimum weight of an RDF of G. Roman domination was introduced in 2004 by Cockayne, Dreyer, and Hedetniemi [35] and is now well studied with over 200 papers published on it and its variations. For more on Roman domination, we refer the reader to the chapter written by the authors on Roman domination in [30]. In it, we covered the core results on Roman domination. In this chapter, we continue that survey to include variations of Roman domination. As of the time of this writing, there are at least twenty known Roman domination-related parameters that we review nine of them, each in a separate section. The remainder is surveyed in [31].

M. Chellali (✉)
LAMDA-RO Laboratory, Department of Mathematics, University of Blida, B.P. 270, Blida, Algeria

N. J. Rad
Department of Mathematics, Shahed University, Tehran, Iran

S. M. Sheikholeslami
Department of Mathematics, Azarbaijan Shahid Madani University, Tabriz, Iran
e-mail: s.m.sheikholeslami@azaruniv.ac.ir

L. Volkmann
Lehrstuhl II für Mathematik, RWTH Aachen, 52056, Aachen, Germany
e-mail: volkm@math2.rwth-aachen.de

© The Author(s), under exclusive license to Springer Nature Switzerland AG 2021
T. W. Haynes et al. (eds.), *Structures of Domination in Graphs*, Developments in Mathematics 66, https://doi.org/10.1007/978-3-030-58892-2_10

For a positive integer k and a function $f : V \to \{0, 1, \ldots, k\}$, the weight of f is $w(f) = \sum_{v \in V} f(v)$, and for $S \subseteq V$ we define $f(S) = \sum_{v \in S} f(v)$. So $w(f) = f(V)$. For every $i \in \{0, 1, \ldots, k\}$, let V_i be the set of vertices assigned the value i under a function f. Note that there is a 1-to-1 correspondence between the functions $f : V \to \{0, 1, \ldots, k\}$ and the ordered partitions $(V_0, V_1, \ldots, V_k)$ of V, so we will write $f = (V_0, V_1, \ldots, V_k)$.

For a graph G, we denote by $\gamma(G)$ the *domination number*, $\gamma_t(G)$ the *total domination number*, and $i(G)$ the *independent domination number*.

2 Weak Roman Domination

In 2003, Henning and Hedetniemi [47] considered a less restrictive version of Roman domination, which they called *weak Roman domination*, but still guaranteeing the defense of the Roman Empire from a single attack. Let $f = (V_0, V_1, V_2)$ be a function on a graph $G = (V, E)$. A vertex v with $f(v) = 0$ is said to be *undefended* with respect to f if it is not adjacent to a vertex u with $f(u) > 0$. A function f is called a *weak Roman dominating function* (WRDF) if each vertex v with $f(v) = 0$ is adjacent to a vertex u with $f(u) > 0$, such that the function $f' = (V_0', V_1', V_2')$ defined by $\acute{f}(v) = 1$, $\acute{f}(u) = f(u) - 1$, and $\acute{f}(w) = f(w)$ for all $w \in V \setminus \{v, u\}$, has no undefended vertex. The *weak Roman domination number* $\gamma_r(G)$ is the minimum weight of a WRDF in G. It should be noted that few papers have been published on weak Roman domination.

Few exact values on the weak Roman domination number have been established. For cycles and paths, Henning and Hedetniemi [47] obtained the following.

Proposition 2.1 ([47]) *For every* $n \geq 4$, $\gamma_r(C_n) = \gamma_r(P_n) = \left\lceil \frac{3n}{7} \right\rceil$.

Roushini Leely Pushpam and Malini Mai [76] extended the exact value on paths to 2-by-n grid graphs $G_{\{2,n\}}$ (Figure 1).

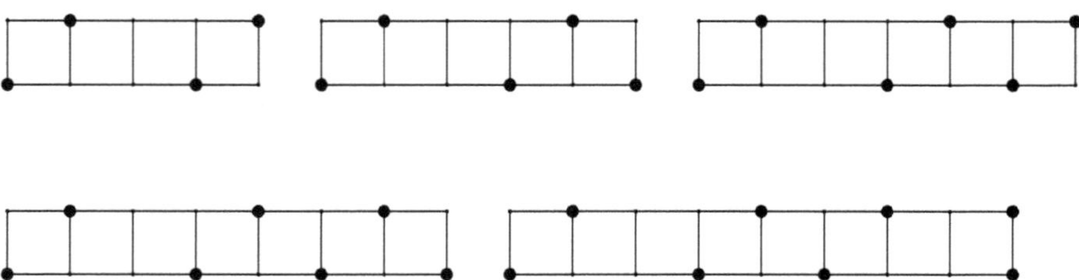

Fig. 1 The construction for $G_{\{2,n\}}$, where $n = 5k + i$, $0 \leq i \leq 4$. Filled-in circles denote vertices in V_1

Proposition 2.2 ([76]) *For any 2-by-n grid graph $G_{\{2,n\}}$,*

$$\gamma_r(G_{\{2,n\}}) = \begin{cases} \left\lfloor \left\lceil \frac{4n}{5} \right\rceil \right\rfloor & \text{if } n \equiv 0 \,(\mathrm{mod}\,5), \\ \left\lfloor \frac{4n}{5} \right\rfloor + 1 & \text{otherwise.} \end{cases}$$

Independently, Valveny and Rodríguez-Velázquez [81] and Zhu and Shao [84] showed that for any connected nontrivial graph G of order n, $\gamma_r(G) \leq \frac{2n}{3}$. Moreover, Zhu and Shao [84] characterized the connected graphs achieving equality.

Theorem 2.3 ([84]) *If G is a connected graph of order n, then $\gamma_r(G) = \frac{2n}{3}$ if and only if every vertex with degree at least 2 is adjacent to exactly two leaf neighbors.*

2.1 Relationships with γ_R and γ

Since for every WRDF $f = (V_0, V_1, V_2)$, $V \setminus V_0$ is a dominating set in G and every Roman dominating function on G is a WRDF, we have the following inequality chain which was observed in [47].

Theorem 2.4 ([47]) *For every graph G,*

$$\gamma(G) \leq \gamma_r(G) \leq \gamma_R(G) \leq 2\gamma(G). \tag{1}$$

Moreover, they provided a characterization of graphs G for which $\gamma(G) = \gamma_r(G)$ and the forests G for which $\gamma_r(G) = 2\gamma(G)$

Theorem 2.5 ([47]) *For any graph G, $\gamma(G) = \gamma_r(G)$ if and only if there exists a $\gamma(G)$-set S such that*

(i) *$pn(v, S)$ induces a clique for every $v \in S$;*
(ii) *for every vertex $u \in V(G) \setminus S$ that is not a private neighbor of any vertex of S, there exists a vertex $v \in S$ such that v dominates u and $pn(v, S) \cup \{u\}$ induces a clique.*

For the purpose of characterizing the class of trees T for which $\gamma(T) = \gamma_r(T)$, Roushini Leely Pushpam and Malini Mai [76] defined a family $\mathcal{F}$ of trees T satisfying the following four conditions (Figure 2).

- No vertex of T is a strong support.
- If $u \in V(T)$ is a non-support, which is adjacent to a support, then $N(u)$ contains exactly one vertex which is neither a support nor adjacent to a support, and all other members of $N(u)$ are either supports or adjacent to supports.
- For any vertex u of degree at least two, there exists at least one leaf v such that $d_T(u, v) \leq 3$.
- Two vertices which are neither supports nor adjacent to supports are not adjacent.

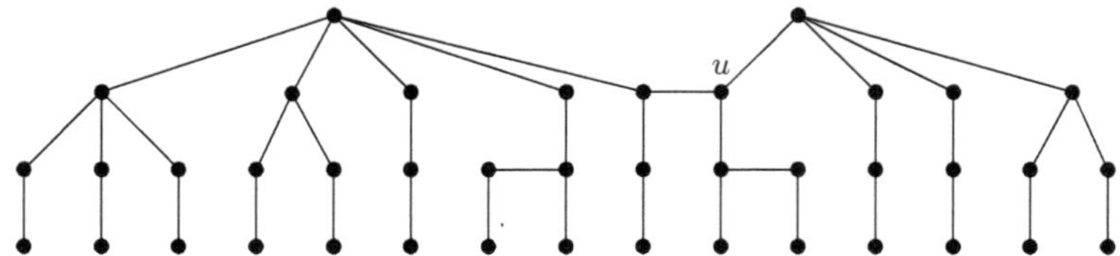

Fig. 2 A tree $T \in \mathcal{E}$

Theorem 2.6 ([76]) *For any tree T, $\gamma(T) = \gamma_r(T)$ if and only if $T \in \mathcal{F}$.*

Split graphs G such that $\gamma(G) = \gamma_r(G)$ were also characterized in [76] as follows.

Theorem 2.7 ([76]) *For any split graph G, $\gamma(G) = \gamma_r(G)$ if and only if G is the corona of $K_{\frac{n}{2}}$ for some even integer n.*

Clearly, if G is a graph with $\gamma_r(G) = \gamma_R(G)$, then every $\gamma_R(G)$-function is also a $\gamma_r(G)$-function. However, not every $\gamma_r(G)$-function is a $\gamma_R(G)$-function even when $\gamma_r(G) = \gamma_R(G)$. For example, the double star $S(2, 2)$ has three $\gamma_r(S(2, 2))$-functions but only two $\gamma_r(S(2, 2))$-functions are $\gamma_R(S(2, 2))$-functions. Hence, we say that $\gamma_r(G)$ and $\gamma_R(G)$ are *strongly equal*, denoted by $\gamma_r(G) \equiv \gamma_R(G)$, if every $\gamma_r(G)$-function is a $\gamma_R(G)$-function. It is worth mentioning that Haynes and Slater in [44] were the first to introduce strong equality between two parameters. Alvarado, Dantas, and Rautenbach [10] characterized the trees T with $\gamma_r(T) \equiv \gamma_R(T)$. Nevertheless, a characterization of trees T with $\gamma_r(T) = \gamma_R(T)$ was given in [84]. It was also shown in [10] that the problem of deciding whether $\gamma_r(G) = \gamma_R(G)$ for a given graph G is NP-hard.

Chellali et al. [27] gave an upper bound on the weak Roman domination number of connected claw-free graphs G in terms of their total domination number $\gamma_t(G)$.

Theorem 2.8 ([27]) *Let G be a nontrivial, connected, claw-free graph. Then,*

(i) $\gamma_r(G) \leq \frac{3}{2}\gamma_t(G)$;
(ii) *if further, G is $\{K_{1,3} + e\}$-free, then $\gamma_r(G) \leq \gamma_t(G)$.*

2.2 Nordhaus–Gaddum Type Bounds

Nordhaus–Gaddum type results on the weak Roman domination of a graph and its complement were provided by Valveny and Rodríguez-Velázquez in [81].

Theorem 2.9 ([81]) *The following statements hold for any graph G of order n.*

(i) $\gamma_r(G) + \gamma_r(\overline{G}) \leq n + 1$.
(ii) $\gamma_r(G)\gamma_r(\overline{G}) \leq \frac{(n+1)^2}{4}$.

Furthermore, if $G \not\cong C_5$ is a connected graph with $\delta(G) \geq 2$ and $\Delta(G) \leq n - 3$, then the following two statements hold.

(iii) $\gamma_r(G) + \gamma_r(\overline{G}) \le n - 1$ *if n is odd, and* $\gamma_r(G) + \gamma_r(\overline{G}) \le n$ *if n is even.*
(iv) $\gamma_r(G)\gamma_r(\overline{G}) \le \frac{(n-1)^2}{4}$ *if n is odd, and* $\gamma_r(G)\gamma_r(\overline{G}) \le \frac{n^2}{4}$ *if n is even.*

The reader can find in [81] two examples of graphs showing the sharpness of all inequalities in Theorem 2.9.

2.3 *Algorithmic and Complexity Results*

In [47], it is shown that the decision problem for the weak Roman domination is NP-complete, even when restricted to bipartite or chordal graphs by describing a polynomial transformation from the well-known NP-complete decision problem corresponding to the problem of computing the domination number $\gamma(G)$. Furthermore, Liu, Peng, and Tang [66] designed a linear-time algorithm for solving the weak Roman domination problem on block graphs. In [23], the authors have proven that the weak Roman domination problem can be solved in $O*(2^n)$ time needing exponential space, and have described an $O*(2.2279^n)$ algorithm using polynomial space. Moreover, they proved that the problem can be solved in linear time on interval graphs.

3   Independent Roman Domination

Independent Roman dominating functions were defined in [35] by Cockayne et al., but Adabi et al. [7] were the first to study these functions. A Roman dominating function $f = (V_0, V_1, V_2)$ is an *independent Roman dominating function* (IRDF) if the set $V_1 \cup V_2$ is independent. The *independent Roman domination number $i_R(G)$* is the minimum weight of an IRDF on G. Independent Roman domination has been studied in [7, 24, 38, 53, 57].

3.1   *Bounds on i_R*

It was observed in [7] that for every graph G of order n, $i_R(G) \le n$, with equality if and only if $G \simeq pK_2 \cup \overline{K_q}$, for any positive integers p and q with $n = 2p + q$. Other bounds on the independent Roman domination number have been established by Ebrahimi et al. [38], which are summarized by the following result.

Proposition 3.1 ([38]) *Let G be a graph of order n and girth $g(G)$. Then,*

(i) $i_R(G) \le n - \Delta(G) + 1$;
(ii) $i_R(G) \le n - \left\lceil \frac{\mathrm{diam}(G)-1}{3} \right\rceil$;
(iii) $\left\lfloor \frac{2g(G)+2}{3} \right\rfloor \le i_R(G) \le n - \left\lceil \frac{g(G)-2}{3} \right\rceil$.

We note that it had been previously known that $\gamma_R(G) \leq n - \Delta(G) + 1$. For the class of trees T, Ebrahimi et al. [38] gave an upper bound on the independent Roman domination number in terms of the order of T. Moreover, they provided a characterization of trees attaining this upper bound.

Theorem 3.2 ([38]) *For any tree T on $n \geq 3$ vertices, $i_R(T) \leq 4n/5$, and equality holds if and only if there exists a partition $V(T) = X_1 \cup \ldots \cup X_k$ of $V(T)$ such that each X_i induces a path P_5 and the subgraph induced by the central vertices of these paths is connected.*

The first Nordhaus–Gaddum inequality for the independent Roman domination number was given in [38].

Proposition 3.3 ([38]) *For any graph G of order $n \geq 3$,*

$$5 \leq i_R(G) + i_R(\overline{G}) \leq n + 3.$$

Equality holds in the lower bound if and only if G or $\overline{G}$ is K_3, or $(\delta(G), \Delta(G))$ or $(\delta(\overline{G}), \Delta(\overline{G})) = (1, n - 1)$, and in the upper bound if and only if G or $\overline{G}$ is C_5 or $\frac{n}{2}K_2$.

Proposition 3.4 ([38]) *If G is a connected graph of order n with diam $(G) \geq 3$, then*

$$6 \leq i_R(G) + i_R(\overline{G}) \leq n - \delta(G) + 4.$$

3.2   Relationships Between i_R and γ_R

The relationship between $i_R(G)$ and $\gamma_R(G)$ has been studied in [7, 24, 53, 57], where some interesting results bounding $i_R(G)$ in terms of $\gamma_R(G)$ are given. By definition, it is obvious that for any graph G, $\gamma_R(G) \leq i_R(G)$. Adabi et al. [7] showed that for a graph G, $\gamma_R(G) = i_R(G)$ if and only if there exists a $\gamma_R(G)$-function $f = (V_0, V_1, V_2)$ such that V_2 is independent. They also showed that the two parameters $\gamma_R(G)$ and $i_R(G)$ are equal for any graph G with maximum degree at most three. Jafari Rad and Volkmann [57] showed that $i_R(G) = \gamma_R(G)$ for a large class of graphs G, by proving the following result.

Theorem 3.5 ([57]) *Let $k \geq 2$ be an integer. If a graph G does not contain the star $K_{1,k+1}$ as an induced subgraph, then*

$$i_R(G) \leq (k - 1)\gamma_R(G) - 2(k - 2).$$

Corollary 3.6 ([57]) *If G is a claw-free graph, then $\gamma_R(G) = i_R(G)$.*

In [7], Adabi et al. presented for any graph G with $\Delta(G) \geq 3$ an upper bound for $i_R(G)$ in terms of $\gamma_R(G)$ and $\Delta(G)$, by showing that

$$i_R(G) \le \gamma_R(G) + \frac{(\gamma_R(G) - 2)}{2}(\Delta(G) - 3).$$

However, this bound has been improved by Jafari Rad [53] as follows. Let $k \ge 4$, and let H be a bipartite graph with partite sets A and B each of cardinality $k - 1$ such that $i_R(H) \ge k$ with the condition that if $i_R(H) = k$, then for every $i_R(H)$-function f, either $f(A) = 0$ or $f(B) = 0$. Let G_k be the graph obtained from H by adding two new vertices x and y, and adding edges xy, xu for all $u \in A$, and yv for all $v \in B$.

Theorem 3.7 ([53]) *For any connected graph G with $4 \le \Delta(G) \le 6$, $i_R(G) \le \frac{\Delta(G)+1}{4}\gamma_R(G)$, with equality if and only if G is a double star $S(\Delta, \Delta)$ or $G = G_\Delta$.*

Theorem 3.8 ([53]) *For any graph G with $\Delta(G) \ge 7$,*

$$i_R(G) \le \left\lceil (\Delta(G) - \frac{18}{5})\gamma_R(G) \right\rceil - 1.$$

The proof of Theorem 3.7 has allowed Jafari Rad to deduce also the following result.

Corollary 3.9 ([53]) *If G is a graph in which the vertices of degree at least 4 form an independent set, then $i_R(G) = \gamma_R(G)$.*

In [57], Jafari Rad and Volkmann defined Roman domination perfect graphs which is a concept closely related to domination perfect graphs introduced by Sumner [80] in 1990. A graph G is called *Roman domination perfect* if $\gamma_R(H) = i_R(H)$ for any induced subgraph H of G. They showed that a graph is Roman domination perfect if it does not contain eight forbidden induced subgraphs (the same as those given in [39]). In particular, chordal graphs that do not contain a double star $S(2, 2)$ as an induced subgraph are Roman domination perfect.

We close this subsection by mentioning that a constructive characterization of trees with strong equality between the Roman domination and independent Roman domination numbers was given by Chellali and Jafari Rad in [24].

3.3 Relationships Between i_R and i

It has been noticed in [7] that for any graph G, $i(G) \le i_R(G) \le 2i(G)$. Adabi et al. [7] were interested in the characterization of all graphs G with $i_R(G) = i(G) + k$, for $0 \le k \le i(G)$, and they obtained the following.

Theorem 3.10 ([7]) *Let G be any graph of order n. Then,*

(i) *$i_R(G) = i(G)$ if and only if $G = \overline{K}_n$;*
(ii) *$i_R(G) = i(G) + 1$ if and only if G has a vertex of degree $n - i(G)$;*
(iii) *for every integer $k \in \{2, \ldots, i(G)\}$, $i_R(G) = i(G) + k$ if and only if the following holds.*

> – *For any integer s with $1 \leq s \leq k - 1$, there is no independent set U_t of cardinality t such that $1 \leq t \leq s$ and $\left| \bigcup_{v \in U_t} N[v] \right| = n - i(G) - s + 2t$.*
> – *There is an independent set W_l of cardinality l for some integer $l \in \{1, \ldots, k\}$ such that $\left| \bigcup_{v \in W_l} N[v] \right| = n - i(G) - k + 2l$.*

The lower bound $i(G)$ has been improved by Chellali et al. [28] for all nontrivial connected graphs G. Let $\mathcal{G}$ be the family of connected graphs G of order n such that $\gamma(G) = n/(1 + \Delta(G))$. Recall that sets that are both dominating sets and packings are called *efficient dominating sets* and have been defined by Bange et al. [16]. Note that every graph of $\mathcal{G}$ admits an efficient dominating set in which every vertex has maximum degree. Let $\mathcal{F}$ be the family of graphs G such that G is the cycle C_4 or the corona of any connected $H \in \mathcal{G}$.

Theorem 3.11 ([28]) *Let G be a nontrivial connected graph with maximum degree Δ. Then, $i_R(G) \geq i(G) + \gamma(G)/\Delta$, with equality if and only if $G \in \mathcal{F}$.*

We note that for the class of trees, Chellali and Jafari Rad [25] gave a constructive characterization of trees T with $i_R(T) = 2i(T)$, answering an open question in [38].

It should be noted that to our knowledge, the relationship between $i_R(G)$ and the independence number $\alpha(G)$ has not been discussed before and could be an interesting subject for future work.

3.4 Algorithmic and Complexity Results

It was mentioned in [35] that Alice McRae has also shown that the decision problem corresponding to independent Roman domination is NP-complete, even when restricted to bipartite graphs. However, this result was never published. Liu and Chang [65] have studied the complexity and the algorithmic aspect of the independent Roman domination problem in a more general context. For real numbers $b \geq a > 0$, an *independent (a, b)-Roman dominating function* is an (a, b)-Roman dominating function f such that the set of vertices assigned a non-zero value is independent. Clearly, for $b = 2$ and $a = 1$, this is an independent Roman dominating function. Liu and Chang showed that for any fixed (a, b), the independent (a, b)-Roman domination problem is NP-complete for bipartite and chordal graphs. Moreover, using the framework of linear programming and the strong elimination ordering as a tool, they provided a linear-time algorithm for the weighted independent (a, b)-Roman domination problem with $2a \geq b \geq a > 0$ on strongly chordal graphs.

4 Roman k-Domination

Kämmerling and Volkmann [59] studied a generalization of the Roman dominating functions. A *Roman k-dominating function* (Rk-DF) on G is a function $f : V(G) \to \{0, 1, 2\}$ such that every vertex u for which $f(u) = 0$ is adjacent to at least k vertices $v_1, v_2, \ldots, v_k$ with $f(v_i) = 2$ for $i = 1, 2, \ldots, k$. The minimum weight of a Roman k-dominating function on a graph G is called the *Roman k-domination number* $\gamma_{kR}(G)$. An Rk-DF of minimum weight is called a $\gamma_{kR}(G)$-function. Clearly, the Roman 1-domination number γ_{1R} corresponds to the *Roman domination number* γ_R. Note that if $k \geq \Delta(G) + 1$, then $\gamma_{kR}(G) = |V|$. Hence, it is only interesting to consider graphs G when $k \leq \Delta(G)$. Roman k-domination has been further studied in [18, 19, 59, 64, 69] and elsewhere.

Let k be a positive integer. A subset $S \subseteq V(G)$ is a k-*dominating set* of G if every vertex of $V(G) - S$ is adjacent to at least k vertices of S. The k-*domination number* $\gamma_k(G)$ is the minimum cardinality of a k-dominating set of G. Note that the 1-domination number $\gamma_1(G)$ is the classical domination number $\gamma(G)$. Some properties of minimum Roman dominating functions given in [35] are generalized through the following result.

Proposition 4.1 ([59]) *Let $f = (V_0, V_1, V_2)$ be any $\gamma_{kR}(G)$-function of a graph G. Then,*

(i) *The complete bipartite graph $K_{k,k+1}$ is not a subgraph of $G[V_1]$.*

(ii) *If $w \in V_1$, then $|N_G(w) \cap V_2| \leq k - 1$.*

(iii) *If $A = \{u_1, \ldots, u_k\} \subseteq V_0$, then $|V_1 \cap N_G(u_1) \cap \ldots \cap N_G(u_k)| \leq 2k$.*

(iv) *V_2 is a minimum k-dominating set of $G[V_0 \cup V_2]$.*

(v) *Let $H = G[V_0 \cup V_2]$, and let $v \in V_2$. Then, there exists a vertex $u_1 \in N_H(v) \cap V_0$ such that u_1 has exactly $k - 1$ neighbors in $V_2 - \{v\}$. In addition, there exists either a second vertex $u_2 \in N_H(v) \cap V_0$ such that u_2 has exactly $k - 1$ neighbors in $V_2 - \{v\}$ or v has at most $k - 1$ neighbors in $V_2 - \{v\}$.*

(vi) *Let $v \in V_2$ such that $\deg_{G[V_2]}(v) = k - 1$ and v has precisely one neighbor in V_0, say w, with the property that w has exactly $k - 1$ neighbors in $V_2 - \{v\}$. If $S_1 \subseteq V_1$ is a set such that each vertex of S_1 has precisely $k - 1$ neighbors in $V_2 - \{v\}$, then $N_G(w) \cap S_1 = \varnothing$.*

(vii) *Let $S_2 \subseteq V_2$ be the set of vertices of degree at least k in $G[V_2]$, and let $C = \{x \in V_0 : |N_G(x) \cap V_2| \geq k + 1\}$. Then*

$$|V_0| \geq \max \left\{ |V_2| + \frac{|V_2| + |S_2|}{2} + |C| \right\}.$$

4.1 Bounds on γ_{kR} and Relationships with γ_k

Proposition 4.2 ([59]) *For any graph G,*

$$\gamma_k(G) \leq \gamma_{kR}(G) \leq 2\gamma_k(G).$$

Equality holds in the upper bound if and only if G has a $\gamma_{kR}(G)$-function $f = (V_0, V_1, V_2)$ with $V_1 = \varnothing$.

Proposition 4.3 ([59]) *If G is a graph of order n, then the following conditions are equivalent:*

(i) $\gamma_{kR}(G) = \gamma_k(G)$;
(ii) $\gamma_k(G) = n$;
(iii) $\Delta(G) \leq k - 1$.

An improvement of the upper bound in Proposition 4.2 was given by Bouchou et al. in [18] for the class of graphs with at most one cycle. Moreover, the authors characterized extremal graphs attaining the new upper bound.

Theorem 4.4 ([18]) *Let T be a tree of order $n \geq 3$ with $\Delta(T) \geq k \geq 2$. Then, $\gamma_{kR}(T) \leq 2\gamma_k(T) - k + 1$, with equality if and only if*

(i) $k = 2$ and T is the subdivision graph of another tree, or
(ii) $k = n - 1$ and T is a star.

Let $K_{1,p} + e$ denote the graph obtained from a star $K_{1,p}$ by adding an edge between two leaves of $K_{1,p}$. Let F be the graph obtained from a path P_5 whose vertices are labeled in order 1, 2, 3, 4, 5 by adding a new vertex x and edges $x2$ and $x4$.

Theorem 4.5 ([18]) *Let G be a unicyclic graph and $\Delta(G) \geq k \geq 3$. Then,*

$$\gamma_{kR}(G) \leq 2\gamma_k(G) - k + 1,$$

with equality if and only if either $k \in \{3, 4, n-1\}$ and $G = K_{1,k} + e$, or $k = 3$ and $G = F$.

Theorem 4.6 ([59]) *Let G be a graph of order n. Then,*

(i) $\gamma_{kR}(G) \geq \dfrac{2kn}{k + \Delta(G)}$;
(ii) $\gamma_{kR}(G) \geq \min\{n, \gamma_k(G) + k\}$;
(iii) *if $n \leq 2k$, then $\gamma_{kR}(G) = n$;*
(iv) *if $n \geq 2k + 1$, then $\gamma_{kR}(G) \geq 2k$;*
(v) *if $n \geq 2k + 1$ and $\gamma_k(G) = k$, then $\gamma_{kR}(G) = \gamma_k(G) + k = 2k$.*

Proposition 4.7 ([59]) *If G is a graph of order n with at most one cycle and $k \geq 2$ or a cactus graph with $k \geq 3$, then $\gamma_{kR}(G) = n$.*

Jafari Rad [55] established a probabilistic upper bound on the Roman k-domination number of a graph improving slightly the one obtained by Hansberg and Volkmann [43].

Theorem 4.8 ([55]) *Let G be a graph of order n, with minimum degree $\delta \geq 1$ and maximum degree Δ, and let k be a positive integer. If $\frac{\delta+1}{\ln(\delta+1)} \geq 2k$, then*

$$\gamma_{kR}(G) \leq \frac{2n}{\delta+1}\left(k\ln(\delta+1) - \ln(2) + \sum_{i=0}^{k-1} \frac{\delta^i}{i!(\delta+1)^{k-1}}\right)$$
$$- \frac{2n}{1+\Delta}\left(\frac{k\ln(\delta+1) - \ln(2)}{\delta+1}\right)^{1+\Delta}.$$

A slight improvement of Theorem 4.8 is given in [51] by using the well-known Brooks' Theorem for vertex coloring. Kämmerling and Volkmann proved a lower bound on $\gamma_{kR}(G) + \gamma_{kR}(\overline{G})$, and characterized the extremal graphs attaining this lower bound. However, a counterexample of the characterization was given by Mojdeh and Moghaddam [69] and the result becomes as follows.

Theorem 4.9 ([59, 69]) *If G is a graph of order n, then $\gamma_{kR}(G) + \gamma_{kR}(\overline{G}) \geq \min\{2n, 4k+1\}$ and the equality holds if and only if one of the following holds:*

(i) $n \leq 2k$;
(ii) $n = 2k+1$, and either $\gamma_k(G) = k$ or $\gamma_k(\overline{G}) = k$;
(iii) $k = 1$, $n \geq 4$ and G or $\overline{G}$ has a vertex of degree $n-1$ and its complement has a vertex of degree $n-2$.

4.2 Relationships Between γ_{kR} and γ_R

Lower bounds on the Roman k-domination in terms of the Roman domination number have been obtained by Bouchou, Blidia, and Chellali in [19].

Proposition 4.10 ([19]) *Let $k \geq 2$ be an integer and G a graph of order n such that $\gamma_{kR}(G) < n$. Then,*

(i) $\gamma_{kR}(G) \geq \gamma_R(G) + 2k - 4$;
(ii) *if further G is C_4-free, then $\gamma_{kR}(G) \geq \gamma_R(G) + 2k - 2$.*

Corollary 4.11 ([19]) *If G is a C_4-free graph of order n with $\gamma_{2R}(G) < n$, then $\gamma_{2R}(G) \geq \gamma_R(G) + 2$.*

The next result improves the lower bound of Proposition 4.10-(i) when $k = 2$.

Proposition 4.12 ([19]) *If G is a graph of order n with $\gamma_{2R}(G) < n$, then*

$$\gamma_{2R}(G) \geq \gamma_R(G) + 1.$$

284   M. Chellali et al.

For $k = 3$, a characterization of cubic graphs G attaining the lower bound of Proposition 4.10-(i) has been given in [19] as follows.

Theorem 4.13 ([19]) *Let G be a connected cubic graph of order n. Then $\gamma_{3R}(G) = \gamma_R(G) + 2$ if and only if $G = K_4$, $K_{3,3}$ or G is the complement graph of C_6.*

For $k = 2$, the following results have also been obtained in [19].

Theorem 4.14 ([19]) *Let G be a connected graph of order n with at most one cycle. Then $\gamma_{2R}(G) = \gamma_R(G) + 1$ if and only if $G \in \{P_3, C_3, P_4, C_4, P_5, C_5\}$.*

A *caterpillar* is a tree T with the property that the removal of its leaves results in a path $u_1, u_2, \ldots, u_s$ which is called the *spine* of T. A sequence of nonnegative integers $(t_1, t_2, \ldots, t_s)$, where t_i is the number of leaves adjacent to u_i for $s \geq 1$ is associated with T that can be denoted by $C(t_1, t_2, \ldots, t_s)$.

Let $\mathcal{H} = \{H_1, H_2, H_3, H_4, H_5\}$ as illustrated in Figure 3, and let $\mathcal{C}$ be the family of nine caterpillars illustrated in Figure 4.

Theorem 4.15 ([19]) *Let T be a tree of order n. Then, $\gamma_{2R}(T) = \gamma_R(T) + 2$ if and only if $T \in \{P_6, P_7, P_8\} \cup \mathcal{H} \cup \mathcal{C}$.*

For $\{K_{1,3}, K_{1,3} + e\}$-free graphs G, the authors of [19] provided a full characterization of graphs G such that $\gamma_{kR}(G) = \gamma_R(G) + t$, where $t \in \{2k - 3, 2k - 2, \lfloor \frac{n}{3} \rfloor\}$. Let R_n denote the complete graph of an even order n minus a perfect matching. Clearly, R_n is an $(n-2)$-regular graph of an even order n.

Theorem 4.16 ([19]) *Let G be a connected $\{K_{1,3}, K_{1,3} + e\}$-free graph of order n and k a positive integer with $2 \leq k \leq \min\left\{\Delta(G), \frac{n}{2}\right\}$. Then,*

(i) $\gamma_{kR}(G) = \gamma_R(G) + 2k - 3$ *if and only if $G \in \{P_3, C_3, P_4, C_4, P_5, C_5\}$ and $k = 2$ or $G = R_n$ and k is even or $n = 2k$;*

(ii) $\gamma_{kR}(G) = \gamma_R(G) + 2k - 2$ *if and only if $G \in \{P_6, C_6, P_7, C_7, P_8, C_8\}$ and $k = 2$, $G = K_n$, $G = R_n$ and k is odd with $n \geq 2k + 1$, or $G = K_p + R_q$ with $p \geq 1$, q is an even integer and $n \geq 2k$;*

(iii) $\gamma_{2R}(G) = \gamma_R(G) + \lfloor \frac{n}{3} \rfloor$ *if and only if $G = P_n$ or C_n for $n \geq 9$.*

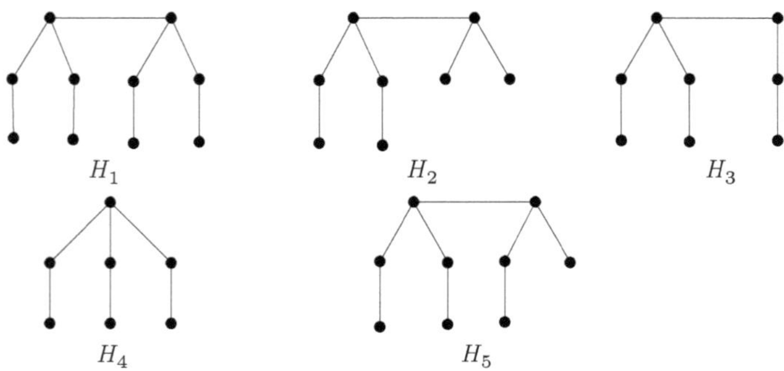

Fig. 3 Five trees H_i with $\gamma_{2R}(H_i) = \gamma_R(H_i) + 2$

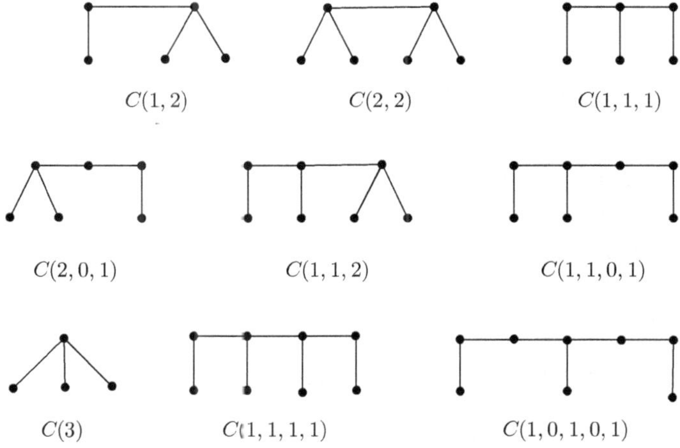

Fig. 4 Nine caterpillars C with $\gamma_{2R}(C) = \gamma_R(C) + 2$

4.3 k-Roman Graphs

Generalizing the definition of Roman graphs given in [35], Kämmerling and Volkmann defined a graph G to be a k-Roman graph if $\gamma_{kR}(G) = 2\gamma_k(G)$. k-Roman graphs have been studied mainly in [18], where the following necessary condition for graphs to be k-Roman is given.

Theorem 4.17 ([18]) *If G is a k-Roman graph with $k \geq 2$, then every vertex of G is adjacent to at most $k - 1$ leaves.*

For the case $k = \Delta$, Δ-Roman graphs were characterized as follows.

Theorem 4.18 ([18]) *A graph G is Δ-Roman if and only if G is a bipartite regular graph.*

For $k \geq 2$, it was shown in [18] that no tree is k-Roman, and for $k \geq 3$, no cactus is k-Roman. For 2-Roman unicyclic graphs, the following is obtained.

Theorem 4.19 ([18]) *A unicyclic graph G is a 2-Roman graph if and only if G is the subdivided graph of another unicyclic graph (possibly with a cycle on two vertices).*

4.4 Algorithmic and Complexity Results

It is shown in [64] that the decision problem corresponding to the problem of computing $\gamma_{kR}(G)$ is NP-complete even when restricted to bipartite graphs and chordal graphs. Moreover, for $k = 2$ the decision problem remains NP-complete for planar graphs. As of this writing, a linear algorithm for computing the Roman k-domination number for any tree has not yet been designed.

5   Roman {2}-Domination

In 2016, Chellali, Haynes, Hedetniemi, and McRae [29] defined a new variant of Roman dominating functions which they called Roman {2} -dominating functions. A *Roman {2}-dominating function* $f : V \rightarrow \{0, 1, 2\}$ has the property that for every vertex $v \in V$ with $f(v) = 0$, $f(N(v)) \geq 2$, that is, either there is a vertex $u \in N(v)$, with $f(u) = 2$, or at least two vertices $x, y \in N(v)$ with $f(x) = f(y) = 1$. In terms of the Roman Empire, this defense strategy requires that every location with no legion has a neighboring location with two legions, or at least two neighboring locations with one legion each. The minimum weight of a Roman {2}-dominating function of G is the *Roman {2}-domination number*, denoted $\gamma_{\{R2\}}(G)$. It should be noted that Roman {2}-dominating functions are closely related to {2} -dominating functions defined in [37] as functions $f : V \rightarrow \{0, 1, 2\}$ having the property that for every vertex $u \in V$, $f(N[u]) \geq 2$. Observe that a Roman {2}-dominating function f relaxes the restriction that for every vertex $u \in V$, $f(N[u]) = \sum_{v \in N[u]} f(v) \geq 2$ to only requiring that this property holds for every vertex assigned 0 under f. Also, for a Roman {2}-dominating function f, it is possible that $f(N[v]) = 1$ for some vertex with $f(v) = 1$. Moreover, it is worth noting that Roman {2}-domination was also studied in 2017 [48] and 2019 [62], where it was called *Italian domination*. The following property of $\gamma_{\{R2\}}(G)$-functions can be found in [29].

Proposition 5.1 ([29]) *For every graph G, there exists a* $\gamma_{\{R2\}}(G)$*-function* $f = (V_0, V_1, V_2)$ *such that either* $V_2 = \varnothing$ *or every vertex of* V_2 *has at least three private neighbors in* V_0 *with respect to the set* V_2.

In 2018, the independent version of Roman {2}-domination was initiated by Rahmouni and Chellali [74]. A Roman {2}-dominating function $f = (V_0, V_1, V_2)$ of G is an *independent Roman {2}-dominating function* (IR2DF) if the set $V_1 \cup V_2$ is independent. The *independent Roman {2}-domination number* $i_{\{R2\}}(G)$ is the minimum weight of an IR2DF on G. One of the main results of [74] relates i_R and $i_{\{R2\}}$.

Theorem 5.2 ([74]) *For every connected graph G of order n,* $i_R(G) - i_{\{R2\}}(G) \leq \frac{n}{4}$.

Note that the bound of Theorem 5.2 is sharp for cycles C_4 and C_8.

5.1   Bounds on $\gamma_{\{R2\}}$ and Relationships with γ, γ_2, γ_r, and γ_R

We begin by giving a lower bound and an upper bound on the Roman {2}-domination number established in [29] and [62], respectively.

Theorem 5.3 ([29]) *If G is a connected graph of order n and maximum degree* Δ, *then* $\gamma_{\{R2\}}(G) \geq 2n/(\Delta + 2)$.

Theorem 5.4 ([62]) *For all connected graphs G with* $n \geq 3$ *vertices,*

$$\gamma_{\{R2\}}(G) \leq \frac{3n}{4}.$$

Proof. It suffices to show that $\gamma_{\{R2\}}(T) \leq \frac{3}{4}n$ for an arbitrary spanning tree T of G. We use an induction on the order n of T. Clearly, the result holds for $n \in \{3, 4\}$, establishing the base case. Next, choose an edge e of T and let T_1 and T_2 be the components of $T - e$. If both T_1 and T_2 have at least three vertices, then the result follows from induction. So we must only consider the case when there is no such edge e, and thus T has diameter at most four. Clearly, if the diameter of T is two or three, then $\gamma_{\{R2\}}(T) \leq \frac{3}{4}n$. Hence, we assume that the diameter of T is four. If $T = P_5$, then $\gamma_{\{R2\}}(T) = 3 \leq \frac{3}{4}n$. Thus, let $T \neq P_5$. Then, T has a vertex of degree at least three and order $n \geq 6$. Let v be a vertex in T of minimum eccentricity. Then, v has degree at least three, and every neighbor of v has degree at most two (otherwise, the deletion of some edge incident with v provides two components each of order at least three). In this case, let $f(v) = 2, f(u) = 1$ for each leaf u of T that is not adjacent to v, and $f(w) = 0$ for each other vertex. This shows that $\gamma_{\{R2\}}(T) \leq \frac{3}{4}n$.

A characterization of connected extremal graphs attaining the upper bound in Theorem 5.4 was provided by Haynes, Henning, and Volkmann [46]. Let F be an arbitrary connected graph of order n_F, and let G be the graph of order $n = 4n_F$ obtained from F by adding to each vertex v of F three new vertices u, w and x and the edges uv, vw, and wx. It can be seen that in any minimum Roman $\{2\}$-dominating function f on such a graph G, $f(\{u, v, w, x\}) \geq 3$. Let $\mathcal{G}$ be the family of all such graphs G.

Theorem 5.5 ([46]) *Let G be a connected graph of order $n \geq 3$. Then, $\gamma_{\{R2\}}(G) = \frac{3}{4}n$ if and only if $G \in \mathcal{G}$.*

The parameters $\gamma_{\{R2\}}(G)$ and $\gamma_2(G)$ for arbitrary graphs G are related as follows.

Proposition 5.6 ([29]) *For every graph G, $\gamma(G) \leq \gamma_{\{R2\}}(G) \leq \gamma_2(G)$.*

It is shown in [62] that for any graph G, $\gamma_{\{R2\}}(G) = \gamma_2(G)$ if and only if there is a $\gamma_{\{R2\}}(G)$-function $f = (V_0, V_1, V_2)$ such that $V_2 = \varnothing$. In particular, Klostermeyer and MacGillivray showed that for cactus graphs G with $\delta(G) = 2$, $\gamma_{\{R2\}}(G) = \gamma_2(G)$. Hajibaba and Jafari Rad [41] were interested in graphs G such that $\gamma(G) = \gamma_{\{R2\}}(G)$ and they proved the following result.

Theorem 5.7 ([41]) *For any nontrivial graph G, $\gamma(G) = \gamma_{\{R2\}}(G)$ if and only if $\gamma(G) = \gamma_2(G)$.*

Moreover, Hajibaba and Jafari Rad provided a constructive characterization of all connected graphs G with $\gamma(G) = \gamma_{\{R2\}}(G)$ which solves the question of characterizing all connected graphs G with $\gamma(G) = \gamma_2(G)$ posed by Hansberg and Volkmann in [42].

On the other hand, Caro and Rodity [21] showed that $\gamma_2(G) \leq \frac{2}{3}n$ for every graph of order n with $\delta(G) \geq 2$, while Chen and Zhou [33] showed that $\gamma_2(G) \leq \frac{1}{2}n$ for every graph of order n with $\delta(G) \geq 3$. As a consequence, Proposition 5.6 leads

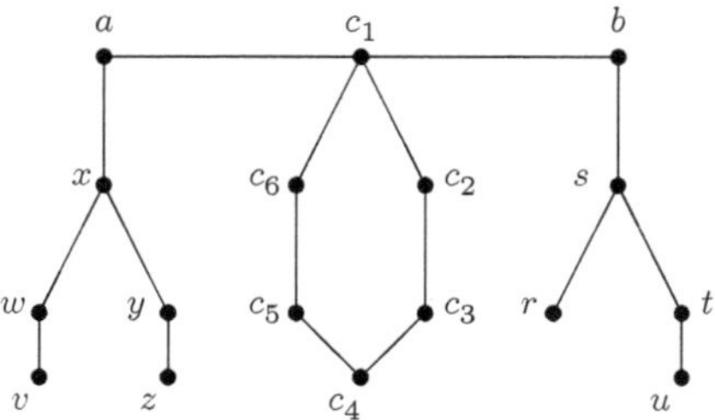

Fig. 5 Graph G with $\gamma(G) < \gamma_r(G) < \gamma_{\{R2\}}(G) < \gamma_R(G) < 2\gamma(G)$

to the following corollary that improves the upper bound of Theorem 5.4 for graphs G with $\delta(G) > 1$.

Corollary 5.8 ([62]) *Let G be a graph of order n. Then,*

(i) *if $\delta(G) \geq 2$, then $\gamma_{\{R2\}}(G) \leq \frac{2}{3}n$;*
(ii) *if $\delta(G) \geq 3$, then $\gamma_{\{R2\}}(G) \leq \frac{1}{2}n$.*

Extremal connected graphs attaining the upper bound in Corollary 5.8-(i) have been characterized by Haynes, Henning, and Volkmann [46]. For two graphs G and H, the corona graph $G \circ H$ is the graph obtained from one copy of G and $|V(G)|$ copies of H and joining the ith vertex of G to every vertex in ith copy of H. Let $\mathcal{G}_{\geq 2} = \{G \circ K_2 \mid G \text{ is a connected graph}\}$.

Theorem 5.9 ([46]) *Let G be a connected graph of order n with $\delta(G) \geq 2$. Then, $\gamma_{\{R2\}}(G) = \frac{2}{3}n$ if and only if $G \in \mathcal{G}_{\geq 2}$.*

For graphs with maximum degree $\Delta \leq 2$, any $\gamma_{\{R2\}}(G)$-function $f = (V_0, V_1, V_2)$ satisfying Proposition 5.1 must have $V_2 = \varnothing$, and so V_1 is a 2-dominating set of G. Because of this, $\gamma_{\{R2\}}(P_n) = \lceil (n+1)/2 \rceil$ and $\gamma_{\{R2\}}(C_n) = \lceil n/2 \rceil$.

An interesting string of inequalities is established in [29] relating the parameters γ, γ_r, $\gamma_{\{R2\}}$, and γ_R. This string extends the inequality chain (1) given in Subsection 2.1 as follows.

Theorem 5.10 ([29]) *For every graph G, $\gamma(G) \leq \gamma_r(G) \leq \gamma_{\{R2\}}(G) \leq \gamma_R(G) \leq 2\gamma(G)$.*

Furthermore, the authors of [29] provided the graph G illustrated in Figure 5 showing the strictness of all inequalities in Theorem 5.10. Indeed, for such a graph G, we have $\gamma(G) = 6$, $\gamma_r(G) = 8$, $\gamma_{\{R2\}}(G) = 9$, $\gamma_R(G) = 11$, and $2\gamma(G) = 12$.

In 2019, Martínez and Yero [68] gave a constructive characterization of trees T with $\gamma_{\{R2\}}(T) = \gamma_R(T)$. In [62], Klostermeyer and MacGillivray have shown that $\gamma_{\{R2\}}(T) \geq \gamma(T) + 1$ for any nontrivial tree T, and in [48], Henning and Klostermeyer characterized all trees T with $\gamma_{\{R2\}}(T) = \gamma(T) + 1$.

For positive integers r and s, let $F_{r,s}$ be the tree obtained from a double star $S(r, s)$ by subdividing every edge exactly once. Let $\mathcal{F}$ be the family of all such trees $F_{r,s}$;

that is, $\mathcal{F} = \{F_{r,s} \mid r, s \geq 1\}$. Also, let $\mathcal{T}$ be the family of trees $T_{k,j}$ of order $k \geq 2$, where $k \geq 2j + 1$ and $j \geq 0$, obtained from a star by subdividing j edges exactly once.

Theorem 5.11 ([48]) *Let T be a nontrivial tree. Then, $\gamma_{\{R2\}}(T) = \gamma(T) + 1$ if and only if $T \in \mathcal{F} \cup \mathcal{T}$.*

Henning and Klostermeyer [48] gave also a constructive characterization of all trees T with $\gamma_{\{R2\}}(T) = 2\gamma(T)$, which are called *Italian trees*.

5.2 Nordhaus–Gaddum Type Bounds

Nordhaus–Gaddum type results on the Roman $\{2\}$-domination of a graph and its complement have been established by Haynes, Henning, and Volkmann [46].

Theorem 5.12 ([46]) *If G is a graph of order $n \geq 3$, then $5 \leq \gamma_{\{R2\}}(G) + \gamma_{\{R2\}}(\overline{G}) \leq n + 2$. Further, if $\gamma_{\{R2\}}(G) \leq \gamma_{\{R2\}}(\overline{G})$, then $\gamma_{\{R2\}}(G) + \gamma_{\{R2\}}(\overline{G}) = 5$ if and only if there exists a vertex in G of degree $n - 1$ with a neighbor of degree 1 in G or with two adjacent neighbors of degree 2 in G.*

The upper bound in Theorem 5.12 has been slightly improved for graphs with no small components.

Theorem 5.13 ([46]) *If G is a graph of order $n \geq 16$ and having no component with fewer than three vertices, then $\gamma_{\{R2\}}(G) + \gamma_{\{R2\}}(\overline{G}) \leq n - 1$.*

5.3 Algorithmic and Complexity Results

In [29], it is shown that the decision problem corresponding to the problem of computing $\gamma_{\{R2\}}(G)$ is NP-complete even when restricted to bipartite graphs. Furthermore, Chen and Lu [34] showed that the problem remains NP-complete even for split graphs, and designed a linear-time algorithm for computing the value $i_{\{R2\}}(T)$ for any tree T, answering an open problem posed in [74]. Note that Rahmouni and Chellali showed that the decision problem corresponding to the problem of computing $i_{\{R2\}}(G)$ is NP-complete for bipartite graphs. Poureidi et al. [71] showed that the decision problem for computing the independent Roman $\{2\}$-domination number is NP-complete even when restricted to chordal graphs. Furthermore, aiming to answer a problem in [26] on a parameter, namely, *independent 2-Rainbow domination number*, they proposed a linear algorithm that in particular can compute the independent Roman $\{2\}$-domination number of a given tree, providing an answer to the open problem posed in [74].

6 Double Roman Domination

In 2016, Beeler, Haynes, and Hedetniemi [17] defined a stronger version of Roman domination which they called double Roman domination. This new strategy, where three legions can be deployed at a given location, offers a high level of defense ensuring that any attack can be defended by at least two legions.

A *double Roman dominating function* (DRDF) on a graph G is a function $f = (V_0, V_1, V_2, V_3)$ that satisfies the following conditions: (i) If $f(v) = 0$, then v must have one neighbor in V_3 or at least two neighbors in V_2; (ii) If $f(v) = 1$, then v must have at least one neighbor in $V_2 \cup V_3$. The *double Roman domination number* $\gamma_{dR}(G)$ equals the minimum weight of a double Roman dominating function on G, and a DRDF of G with weight $\gamma_{dR}(G)$ is called a γ_{dR}-function of G. Double Roman domination has been studied in [2, 3, 5, 13, 40, 56, 82, 83] and elsewhere. The following property of γ_{dR}-functions was given in [17].

Proposition 6.1 ([17]) *For any double Roman dominating function f', there exists a double Roman dominating function f of no greater weight than f' for which no vertex is assigned the value* 1.

According to Proposition 6.1, double Roman dominating functions can be assumed to be of the form $f : V(G) \rightarrow \{0, 2, 3\}$ such that if $f(v) = 0$, then either v has a neighbor w with $f(w) = 3$ or v has two neighbors x and y with $f(x) = f(y) = 2$. Therefore, in a very good sense, double Roman domination is equivalent to Roman $\{2\}$-domination as follows. If $S \subseteq V(G)$ is a solution to double Roman domination, then by subtracting one from each vertex assigned the value 2 or 3 we get a solution to Roman $\{2\}$-domination. Conversely, if S is a solution to Roman $\{2\}$-domination, then by adding one to every vertex assigned the value 1 or 2 we get a solution to double Roman domination.

The exact values on the double Roman domination number for paths and cycles have been established in [2].

Proposition 6.2 ([2]) *For* $n \geq 1$,

$$\gamma_{dR}(P_n) = \begin{cases} n & \text{if } n \equiv 0 \pmod 3 \\ n+1 & \text{if } n \equiv 1 \text{ or } 2 \pmod 3. \end{cases}$$

Proposition 6.3 ([2]) *For* $n \geq 3$,

$$\gamma_{dR}(C_n) = \begin{cases} n & \text{if } n \equiv 0, 2, 3, 4 \pmod 6 \\ n+1 & \text{if } n \equiv 1, 5 \pmod 6. \end{cases}$$

6.1 Bounds on γ_{dR} and Relationships with γ, γ_R, $\gamma_{\{R2\}}$, and γ_2

Various results relating $\gamma_{dR}(G)$ to $\gamma(G)$ and $\gamma_R(G)$ are presented by Beeler et al. [17]. We begin by the following result involving $\gamma_{dR}(G)$ and $\gamma_R(G)$ for connected graphs G.

Proposition 6.4 ([17]) *For any nontrivial connected graph G,*

$$1 + \gamma_R(G) \leq \gamma_{dR}(G) \leq 2\gamma_R(G) - 1.$$

Recall that a *wounded spider* is the graph obtained by subdividing at most $t - 1$ of the edges of a star $K_{1,t}$, for $t > 0$. The following result due to Zhang et al. [83] provides a characterization of trees T with $\gamma_{dR}(T) = 2\gamma_R(T) - 1$.

Theorem 6.5 ([83]) *Let T be a nontrivial tree. Then, $\gamma_{dR}(T) = 2\gamma_R(T) - 1$ if and only if T is a wounded spider.*

The domination and double Roman domination numbers are related as follows for arbitrary graphs.

Theorem 6.6 ([17]) *For any graph G, $2\gamma(G) \leq \gamma_{dR}(G) \leq 3\gamma(G)$.*

Both bounds of the previous theorem are sharp. Indeed, if G is a nontrivial star $K_{1,n-1}$, then $\gamma_{dR}(G) = 3\gamma(G) = 3$, and if $G = K_{2,k}$, for $k \geq 2$, then $\gamma_{dR}(G) = 2\gamma(G) = 4$. Moreover, Beeler et al. called a graph G having $\gamma_{dR}(G) = 3\gamma(G)$ a *double Roman graph*. They raised the problem of characterizing double Roman graphs, in particular the double Roman trees. These have been independently characterized constructively by Abdollahzadeh Ahangar et al. [3] and Henning and Jafari Rad [50]. For graphs G such that $\gamma_{dR}(G) = 2\gamma(G)$, the following necessary and sufficient was given in [17].

Theorem 6.7 ([17]) *For any graph G, $\gamma_{dR}(G) = 2\gamma(G)$ if and only if $\gamma(G) = \gamma_2(G)$.*

An improvement of the lower bound in Theorem 6.6 was given in [2] as follows. Recall that $\gamma_{\{R2\}}(G) \geq \gamma(G)$ holds for every graph G.

Proposition 6.8 ([2]) *For every connected graph G, $\gamma_{dR}(G) \geq \gamma_{\{R2\}}(G) + \gamma(G)$.*

Proof. Clearly, the result is valid for graphs of order $n \in \{1, 2\}$. Hence, let $n \geq 3$. According to Theorem 6.1, let $f = (V_0, V_1, V_2, V_3)$ be a $\gamma_{dR}(G)$-function with $V_1 = \varnothing$. Note that $V_2 \cup V_3$ dominates V_0 and so $\gamma(G) \leq |V_2| + |V_3|$. Let V_0' be the set of vertices of V_0 having at least one neighbor in V_3, and let $V_0'' = V_0 - V_0'$. Now define the function g on G as follows: $g(x) = f(x) - 1$ for all $x \in V_2 \cup V_3$ and $g(x) = f(x)$ for all $x \in V_0$. Clearly, g is a Roman $\{2\}$-dominating function on G, and so $\gamma_{\{R2\}}(G) \leq |V_2| + 2|V_3|$. Therefore, $\gamma_{dR}(G) = 2|V_2| + 3|V_3| = |V_2| + |V_3| + |V_2| + 2|V_3| \geq \gamma(G) + \gamma_{\{R2\}}(G)$.

As a consequence, the authors of [2] obtained the following result for the class of trees.

Proposition 6.9 ([2]) *Let T be a nontrivial tree. Then, $\gamma_{dR}(T) \geq 2\gamma(T) + 1$, with equality if and only if T is a wounded spider.*

We also note that a characterization of trees T with $\gamma_{dR}(T) = 2\gamma(T) + 2$ was given in [5]. On the other hand, since $\gamma_{\{R2\}}(G) \leq 2\gamma(G)$ holds for any graph G, Proposition 6.8 leads to $\gamma_{dR}(G) \geq \frac{3}{2}\gamma_{\{R2\}}(G)$, proved by Hajibaba and Jafari Rad in [40] who gave further a constructive characterization of trees T with $\gamma_{dR}(T) = \frac{3}{2}\gamma_{\{R2\}}(T)$. Moreover, it was also shown in [40] that for every graph G, $\gamma_{dR}(G) \leq 2\gamma_{\{R2\}}(G)$. This result improves an earlier upper bound from [2], where is proved that $\gamma_{dR}(G) \leq 2\gamma_2(G)$ for every graph G. Abdollahzadeh Ahangar et al. [2] also gave a lower bound on the double Roman domination number of a graph G in terms of the order, maximum degree, and the domination number.

Proposition 6.10 ([2]) *For any graph G of order n with maximum degree Δ,*

$$\gamma_{dR}(G) \geq \frac{2n}{\Delta} + \frac{\Delta - 2}{\Delta}\gamma(G).$$

This bound is sharp for even cycles and paths of order $3k$.

It is well known that $\gamma(G) \geq \frac{n}{\Delta+1}$ for every graph G of order n with maximum degree Δ. As an immediate consequence, Proposition 6.10 leads to the following corollary.

Corollary 6.11 ([82]) *If G is a graph of order n and maximum degree $\Delta \geq 1$, then* $\gamma_{dR}(G) \geq \left\lceil \frac{3n}{\Delta+1} \right\rceil$.

An upper bound on the double Roman domination number of connected graphs in terms of their order was obtained by Beeler et al. [17], who also characterized the graphs reaching this upper bound. Let $\mathcal{H}$ be the family of connected graphs G of order n that can be built from $n/4$ copies of P_4 by adding a connected subgraph on the set of centers of $\frac{n}{4}P_4$.

Theorem 6.12 ([17]) *If G is a connected graph of order $n \geq 3$, then $\gamma_{dR}(G) \leq \frac{5}{4}n$, with equality if and only if $G \in \mathcal{H}$.*

The authors [17] observed that every connected graph G having minimum degree at least two satisfies the inequality $\gamma_{dR}(G) \leq \frac{6n}{5}$ and posed the question whether this bound can be improved. This question has been settled by Amjadi et al. [13] by proving that $\gamma_{dR}(G) \leq \frac{8n}{7}$ except when G is a cycle C_5. However, this bound has been improved by Khoeilar et al. [60] as follows, where an infinite family of graphs attaining the new bound was also provided in [60].

Theorem 6.13 ([60]) *Let G be a graph of order n, $\delta(G) \geq 2$ and with no component isomorphic to C_5 or C_7. Then, $\gamma_{dR}(G) \leq \frac{11n}{10}$.*

Moreover, the authors in [17] also asked the question: which classes of graphs, or trees satisfy $\gamma_{dR}(G) \leq n$? This issue has been dealt in [2], where it is proved that for every graph G with minimum degree at least three, $\gamma_{dR}(G) \leq n$. Using the probabilistic method, Jafari Rad and Rahbani obtained the following upper bound.

Theorem 6.14 ([56]) *For a graph G of order n with minimum degree δ,*

$$\gamma_{dR}(G) \leq 3n \frac{\ln 2(1+\delta) - \ln 3 + 1}{1+\delta}.$$

6.2 Nordhaus–Gaddum Type Bounds

The first Nordhaus–Gaddum inequalities for the double Roman domination number were given in [56].

Theorem 6.15 ([56]) *If G is a graph of order $n \geq 2$, then $7 \leq \gamma_{dR}(G) + \gamma_{dR}(\overline{G}) \leq 2n + 3$. Equality holds for the lower bound if and only if G or $\overline{G}$ is K_2, and equality holds for the upper bound if and only if G or $\overline{G}$ is a complete graph.*

Theorem 6.16 ([56]) *If G is a graph of order $n \geq 3$, then $\gamma_{dR}(G) + \gamma_{dR}(\overline{G}) = 2n+2$ if and only if G or $\overline{G}$ is C_5, P_4, or a complete graph minus an edge.*

Theorem 6.17 ([56]) *Let G be a graph of order n. Then,*

  (i) *if $n \geq 240$ and $diam(G) = diam(\overline{G}) = 2$, then $\gamma_{dR}(G)\gamma_{dR}(\overline{G}) < \frac{15}{2}n$;*
  (ii) *if $n \geq 4$ and $diam(G) \geq 3$, then $\gamma_{dR}(G)\gamma_{dR}(\overline{G}) < \frac{15}{2}n$;*
 (iii) *if $n \geq 3$ and $\delta(G) = 1$, then $\gamma_{dR}(G)\gamma_{dR}(\overline{G}) \leq \frac{25}{4}n$.*

6.3 Algorithmic and Complexity Results

In [2], it is shown that the decision problem corresponding to the problem of computing $\gamma_{dR}(G)$ is NP-complete even when restricted to bipartite and chordal graphs. Furthermore, Zhang et al. [83] gave a linear-time algorithm to compute the value of $\gamma_{dR}(G)$ for any tree T, answering an open problem posed in [17]. Poureidi et al. [72] showed that the decision problem associated to double Roman domination is NP-complete even when restricted to planar graphs. Then, they showed that the problem of deciding whether a given graph is double Roman is NP-hard even when restricted to bipartite or chordal graphs. They also gave a linear algorithm that computes the double Roman domination number of a given unicyclic graph.

7 Total Roman Domination

A *total Roman dominating function* (TRDF) of a graph G with no isolated vertex is a Roman dominating function $f = (V_0, V_1, V_2)$ on G such that the subgraph induced by $V_1 \cup V_2$ under f has no isolated vertex. The *total Roman domination number* $\gamma_{tR}(G)$ is the minimum weight of a TRDF on G. A TRDF with minimum weight $\gamma_{tR}(G)$ is called a $\gamma_{tR}(G)$-function. The concept of total Roman dominating function in graphs was introduced in 2016 by Liu and Chang [65], namely total (a, b)-Roman domination for any given real numbers $b \geq a > 0$. Total Roman domination is studied in [1, 4, 12] and elsewhere. The following observation giving some properties of $\gamma_{tR}(G)$-functions can be found in [1].

Observation 7.1 ([1]) *Let G be a connected graph of order at least 3 and let $f = (V_0, V_1, V_2)$ be a $\gamma_{tR}(G)$-function. Then, the following hold.*

(i) $|V_2| \leq |V_0|$.
(ii) *If x is a leaf and y a support vertex in G, then $x \notin V_2$ and $y \notin V_0$.*
(iii) *If z has at least three leaf neighbors, then $f(z) = 2$ and at most one leaf neighbor of z belongs to V_1.*

The exact values of γ_{tR} for grids $G_{\{2,n\}}$ and $G_{\{3,n\}}$ are determined in [6].

Proposition 7.2 ([6]) *For* $n \geq 2$, $\gamma_{tR}(G_{\{2,n\}}) = \begin{cases} \frac{4n}{3} & \text{if } n \equiv 0 \pmod 3 \\ \frac{4n+2}{3} & \text{if } n \equiv 1 \pmod 3 \\ \frac{4n+4}{3} & \text{if } n \equiv 2 \pmod 3 \end{cases}$ *and*

$\gamma_{tR}(G_{\{3,n\}}) = 2n$.

Since the total Roman domination number of any nontrivial connected graph G of order n is at most n (simply assign a 1 to each vertex of G), it is interesting to characterize those graphs G for which $\gamma_{tR}(G) = n$. This problem has been considered in [1], where it was shown the following. Let $\mathcal{G}$ be the family of graphs that can be obtained from a $C_4 : v_1 v_2 v_3 v_4 v_1$ by adding $k_1 + k_2 \geq 1$ vertex-disjoint paths P_2 and joining v_1 to the end of k_1 such paths and joining v_2 to the end of k_2 such paths (possibly, $k_1 = 0$ or $k_2 = 0$). Let $\mathcal{H}$ be the family of graphs that can be obtained from a double star by subdividing each pendant edge once and subdividing the non-pendant edge $r \geq 0$ times.

Theorem 7.3 ([1]) *Let G be a connected graph of order $n \geq 2$. Then, $\gamma_{tR}(G) = n$ if and only if one of (i) G is a path or a cycle or (ii) G is a corona of some graph F or (iii) G is a subdivided star or (iv) $G \in \mathcal{G} \cup \mathcal{H}$.*

7.1   Bounds and Relations with Some Domination Parameters

In what follows, we present the bounds on γ_{tR} and some relationships with γ, γ_t, and γ_R. It is shown in [1] that for any graph G with no isolated vertex, $2\gamma(G) \leq \gamma_{tR}(G) \leq 3\gamma(G)$. Amjadi et al. [15] provided a constructive characterization of trees T with $\gamma_{tR}(T) = 2\gamma(T)$ or $\gamma_{tR}(T) = 3\gamma(T)$. Necessary conditions for graphs G attaining each bound are also established [1].

Theorem 7.4 ([1]) *Let G be a graph with no isolated vertex. If $\gamma_{tR}(G) = 2\gamma(G)$, then $\gamma(G) = \gamma_t(G)$ or there exists a set S of vertices of G such that the following hold.*

(i) *$G[S] = kK_2$ for some $k \geq 1$.*
(ii) *$\gamma(G - S) = \gamma_t(G - S)$.*
(iii) *No neighbor of a vertex of S in G belongs to a $\gamma_t(G - S)$-set.*

Theorem 7.5 ([1]) *Let G be a graph with no isolated vertex. If $\gamma_{tR}(G) = 3\gamma(G)$, then every $\gamma(G)$-set is a packing in G.*

As shown in [1], the converse of Theorem 7.5 is not true and can be seen by considering the graph G_k, for $k \geq 3$, obtained from a star $K_{1,k}$ by subdividing $k - 1$ edges twice and subdividing the remaining edge exactly once. Then, $\gamma(G_k) = k$, while $\gamma_{tR}(G_k) \leq 2(k + 1) < 3k$.

The total domination and total Roman domination numbers are related as follows for arbitrary graphs.

Theorem 7.6 ([1]) *If G is a graph with no isolated vertex, then $\gamma_t(G) \leq \gamma_{tR}(G) \leq 2\gamma_t(G)$. Further, the following holds.*

(i) *$\gamma_t(G) = \gamma_{tR}(G)$ if and only if G is the disjoint union of copies of K_2.*
(ii) *If $\gamma_{tR}(G) = 2\gamma_t(G)$ and S is an arbitrary $\gamma_t(G)$-set, then $epn(v, S) \neq \varnothing$ for all $v \in S$.*

According to Theorem 7.6, $\gamma_{tR}(G) \geq \gamma_t(G) + 1$ for every connected graph of order at least three. It is shown in [1] that connected graphs G of order $n \geq 3$ satisfy $\gamma_{tR}(G) = \gamma_t(G) + 1$ if and only if $\Delta(G) = n - 1$.

By analogy with Roman graphs defined in [35], Abdollahzadeh Ahangar et al. [1] called a *total Roman graph* any graph G satisfying $\gamma_{tR}(G) = 2\gamma_t(G)$. Moreover, they gave a necessary and sufficient condition for a graph to be a total Roman graph. It should be noted that a constructive characterization of total Roman trees was given by Amjadi et al. [12].

Proposition 7.7 ([1]) *Let G be a graph without isolated vertices. Then G is a total Roman graph if and only if there exists a $\gamma_{tR}(G)$-function $f = (V_0, V_1, V_2)$ such that $V_1 = \varnothing$.*

Obviously, for every graph G with no isolated vertex, $\gamma_R(G) \le \gamma_{tR}(G)$. An upper bound relating the total Roman domination number to the Roman domination number was given by Abdollahzadeh Ahangar et al. [1] as follows.

Theorem 7.8 ([1]) *If G is a graph of order n with no isolated vertex, then $\gamma_{tR}(G) \le 2\gamma_R(G) - 1$. Further, $\gamma_{tR}(G) = 2\gamma_R(G) - 1$ if and only if $\Delta(G) = n - 1$.*

An upper bound on the total Roman domination number of connected graphs different from stars in terms of their order, maximum degree, and matching number was obtained by Abdollahzadeh Ahangar et al. [6] who characterized in addition the graphs with girth at least 4 reaching this upper bound. Recall that the *matching number* $\alpha'(G)$ is the size of a largest matching in G.

Theorem 7.9 ([6]) *Let G be a graph of order $n \ge 4$, without isolated vertices, different from a star. Then, $\gamma_{tR}(G) \le n - \Delta(G) + \alpha'(G)$.*

They also gave a lower bound on the double Roman domination number of a graph G in terms of the order and maximum degree.

Theorem 7.10 ([6]) *For any graph G of order $n \ge 3$ with maximum degree Δ, $\gamma_{tR}(G) \ge \left\lceil \frac{2n}{\Delta} \right\rceil$. This bound is sharp for stars and double stars $S(p, p)$.*

The following Nordhaus–Gaddum inequalities for the total Roman domination are given in [14].

Theorem 7.11 ([14]) *If G and $\overline{G}$ are graphs of order n without isolated vertices, then*

$$\gamma_{tR}(G) + \gamma_{tR}(\overline{G}) \le n + 5.$$

Furthermore, this bound is sharp for a 5-cycle.

Theorem 7.12 ([14]) *If G and $\overline{G}$ are graphs of order n without isolated vertices, then*

$$\gamma_{tR}(G)\gamma_{tR}(\overline{G}) \le 6n - 5$$

with equality if and only if G is 5-cycle.

Campanelli and Kuziak [20] were interested in total Roman domination in the lexicographic product of two graphs. Recall that the *lexicographic product* of two graphs G and H is defined as the graph $G \cdot H$ with vertex set $V(G) \times V(H)$ and edge set $E(G \cdot H) = \{(u, v)(u', v') | uu' \in E(G) \text{ or } ((u = u' \text{ and } vv' \in E(H))\}$.

Theorem 7.13 ([20]) *If $G = (V, E)$ is a connected graph with a $\gamma_{tR}(G)$-function $h = (V_0, V_1, V_2)$, then for every graph H,*

$$\gamma_{tR}(G) \le \gamma_{tR}(G \cdot H) \le \gamma_{tR}(G) + |V_1|.$$

Clearly, if $V_1 = \varnothing$, then $\gamma_{tR}(G \cdot H) = \gamma_{tR}(G)$. In particular, if G is a total Roman graph, then $\gamma_{tR}(G \cdot H) = \gamma_{tR}(G)$ for any graph H. As an immediate consequence of Theorem 7.13, they obtained the following Vizing's-like result.

Corollary 7.14 ([20]) *If G and H are nontrivial connected graphs, then* $\gamma_{tR}(G \cdot H) \leq \gamma_{tR}(G)\gamma_{tR}(H)$.

The total Roman domination and total domination numbers are related for the lexicographic product of graphs as follows.

Proposition 7.15 ([20]) *If H is a graph and G is a graph without isolated vertices, then* $\gamma_{tR}(G \cdot H) \leq 2\gamma_t(G)$.

If one of the graphs G or H is a path or a cycle, Campanelli and Kuziak [20] established the following.

Theorem 7.16 ([20]) *If H is a graph and G is a path or a cycle of order n, then*

$$
n \leq \gamma_{tR}(G \cdot H) \leq \begin{cases} n & \text{if } n \equiv 0 \ (\mathrm{mod}\,4), \\ n+1 & \text{if } n \equiv 1, 3 \ (\mathrm{mod}\,4), \\ n+2 & \text{if } n \equiv 2 \ (\mathrm{mod}\,4). \end{cases}
$$

7.2 Algorithmic and Complexity Results

It is shown in [65] that the decision problem corresponding to the problem of computing $\gamma_{tR}(G)$ is NP-complete for bipartite graphs. Moreover, as of this writing, a linear algorithm for computing the total Roman domination number for any tree has not yet designed. Poureidi et al. [70] studied the complexity issue for problems of deciding whether for a graph G, $\gamma_{tR}(G) = 2\gamma(G)$, $\gamma_{tR}(G) = 2\gamma_t(G)$, or $\gamma_{tR}(G) = 3\gamma(G)$, and showed that the corresponding decision problems are NP-hard even when restricted to bipartite graphs.

8 Perfect Roman Domination

As defined in Livingston and Stout in [67], a *perfect dominating set* (PDS) is a dominating set S of G for which each vertex in $V - S$ is adjacent to exactly one vertex in S. The minimum cardinality of a PDS of a graph G is the *perfect domination number* $\gamma^p(G)$. Note that every graph G has a PDS since $V(G)$ is trivially such a set. Perfect domination has been studied by several authors, and for more details on this concept, the reader is referred to the survey in [61].

The study of the perfect Roman domination was initiated by Henning, Klostermeyer, and MacGillivray in [52]. A *perfect Roman dominating function* (PRDF) on a graph G is a Roman dominating function $f = (V_0, V_1, V_2)$ on G such that every vertex of V_0 is adjacent to exactly one vertex assigned in V_2. The minimum weight of a PRDF on G is the *perfect Roman domination number* $\gamma_R^p(G)$. A PRDF on G with weight $\gamma_R^p(G)$ is called a $\gamma_R^p(G)$-function.

8.1 Bounds

Note that every graph G has a PRDF since $f = (\varnothing, V(G), \varnothing)$ is trivially such a function, and thus $\gamma_R^p(G) \leq n$ for all G of order n. One can easily see that this bound is sharp if and only if $\Delta(G) \leq 1$.

Henning et al. [52] focused on the class of trees by giving an upper bound on the perfect Roman domination number in terms of the order. In addition, they characterized extremal trees attaining this upper bound. Let $\mathcal{T}$ be the family of all trees T whose vertex set can be partitioned into sets, each set inducing a path P_5 on five vertices, such that the subgraph induced by the central vertices of these P_5's is connected.

Theorem 8.1 ([52]) *If T is a tree of order $n \geq 3$, then $\gamma_R^p(T) \leq \frac{4}{5}n$, with equality if and only if $T \in \mathcal{T}$.*

The bound of Theorem 8.1 has been improved by Darkooti et al. [36] for trees T with $\ell(T) \geq 2s(T) - 2$, where $\ell(T)$ and $s(T)$ are the number of leaves and support vertices of T, respectively.

Theorem 8.2 ([36]) *For any tree T of order $n \geq 3$, $\gamma_R^p(T) \leq (4n - \ell(T) + 2s(T) - 2)/5$.*

Moreover, the same authors also showed that the decision problem associated with $\gamma_R^p(G)$ is NP-complete for bipartite graphs.

The question of whether the $\frac{4}{5}n$ upper bound on the perfect Roman domination number for trees remains valid for any connected graph G of order $n \geq 3$ was posed in [52]. This issue was addressed by Henning and Klostermeyer [49] for regular graphs, where a positive answer was given to some cases. Other than the case of cycles C_n for which $\gamma_R^p(C_n) \leq \frac{4}{5}n$, it is shown the following for $k \geq 3$.

Theorem 8.3 ([49]) *Let G be k-regular graph of order n. Then,*

(i) *if $k = 3$, then $\gamma_R^p(G) \leq \frac{3}{4}n$;*

(ii) *if $k \geq 4$, then $\gamma_R^p(G) \leq \left(\frac{k^2+k+3}{k^2+3k+1} \right) n$;*

(iii) *if $k \geq 3$ and G has girth at least 7, then $\gamma_R^p(G) \leq \left(\frac{k^2-k+4}{k^2+k+2} \right) n$.*

As an immediate consequence of Theorem 8.3, the following result is obtained.

Corollary 8.4 *If G is 4-regular or k-regular with $k \in \{5, 6, 7\}$ and with girth at least 7, then $\gamma_R^p(G) \leq \frac{4}{5}n$.*

The relationship between the perfect Roman domination and perfect domination numbers was raised in [32] and [79]. Obviously, if S is a minimum PDS of a graph G, then clearly $(V - S, \varnothing, S)$ is a PRDF of G, and thus $\gamma_R^p(G) \leq 2\gamma^p(G)$. We say that a graph G is a *perfect Roman graph* if $\gamma_R^p(G) = 2\gamma^p(G)$. An open problem is to characterize the perfect Roman graphs. A constructive characterization of perfect Roman trees was given in [79], where 7 operations have been defined to build such trees. Moreover, it is shown in [32] that $\gamma^p(G)$ may be larger or smaller than $\gamma_R^p(G)$. Clearly, for nontrivial stars G, we have $\gamma_R^p(G) > \gamma^p(G)$. To see the other situation, consider the following example of graphs given in [32]. Let H be the graph obtained from a double star $S(p, p)$, $(p \geq 3)$ with central vertices u, v by subdividing the edge uv with vertex w, and adding $2k$ $(k \geq 3)$ new vertices, where k vertices are attached to both u and w and the remaining k vertices are attached to both v and w. Then, $\gamma^p(H) = 2k + 3$ while $\gamma_R^p(H) = 5$ and so the difference $\gamma^p(H) - \gamma_R^p(H)$ can be even very large. In addition, the following bound relating the perfect Roman domination and perfect domination numbers for trees was proved in [32], and a constructive characterization of extremal trees was also given.

Theorem 8.5 ([32]) *For any tree T of order $n \geq 2$, $\gamma_R^p(T) \geq \gamma^p(T) + 1$.*

Haynes and Henning [45] extended the concept of perfect Roman domination to Italian domination (equivalently Roman {2}-domination) by defining a *perfect Italian dominating function*, abbreviated PIDF, on G as a function $f : V(G) \rightarrow \{0, 1, 2\}$ such that for every vertex $u \in V$ with $f(u) = 0$, the total weight assigned by f to the vertices of $N(u)$ is 2, that is, all the neighbors of u are assigned the weight 0 by f except for exactly one vertex v for which $f(v) = 2$ or for exactly two vertices v and w for which $f(v) = f(w) = 1$. The *perfect Italian domination number* of G, denoted $\gamma_I^p(G)$, is the minimum weight of a PIDF of G. It was shown in [45] that if G is a tree on $n \geq 3$ vertices, then $\gamma_I^p(G) \leq \frac{4}{5}n$. Haynes and Henning proposed in [45] the problem of determining the best possible constants c_G such that $\gamma_I^p(G) \leq c_G \times n$ for all graphs of order n when G is a planar or regular graph. This problem has been dealt by Lauri and Mitillos in [63], by proving that $c_G = 1$ when G is planar or split and $c_G = 2/3$ when G is cubic.

8.2 Algorithmic and Complexity Results

Lauri and Mitillos in [63] studied the complexity-theoretic questions for perfect Italian domination number. They proved that deciding whether a given graph G admits a perfect Italian dominating function of weight at most k is NP-complete, even when G is restricted to the class of bipartite planar graphs. They also strengthen the result of Chellali et al. [29] by showing that deciding whether G admits a Roman {2}-dominating function of weight at most k is NP-complete, even when G is both bipartite and planar.

9 Strong Roman Domination

In [11], Álvarez-Ruiz et al. defined a new version of Roman domination ensuring the defense of the Roman empire from multiple attacks, unlike with the strategy of Roman domination which only guarantees the defense of the empire from a single attack. Indeed, if several simultaneous attacks occur on weak places (places in which no army is stationed), then a strong place (in which two legions are deployed) may not be able to defend efficiently its neighbors. The new strategy suggested by Álvarez-Ruiz et al. [11] considers that a strong place should be able to defend itself and at least half of its weak neighbors.

For a given graph G of order n and maximum degree Δ, let $f : V(G) \rightarrow \{0, 1, \ldots, \lceil \frac{\Delta}{2} \rceil + 1\}$. Let $B_i = \{v \in V : f(v) = i\}$ for $i \in \{0, 1\}$, and $B_2 = V(G) - (B_0 \cup B_1)$. Then f is called a *strong Roman dominating function* (StRDF) for G, if every $v \in B_0$ has a neighbor w, such that $w \in B_2$ and $f(w) \geq 1 + \lceil \frac{1}{2} |N(w) \cap B_0| \rceil$. The *strong Roman domination number* $\gamma_{StR}(G)$ is the minimum weight of an StRDF on G.

Clearly, $\gamma_{StR}(G) = \gamma_R(G)$ for all connected graphs G with $\Delta(G) \leq 2$. In particular, if G is a path or a cycle of order n, then $\gamma_{StR}(G) = \lceil \frac{2n}{3} \rceil$. Various bounds on the strong Roman domination number obtained in [11] are gathered in the following proposition.

Proposition 9.1 ([11]) *Let G be a graph of order n with maximum degree Δ. Then,*

(i) $\gamma_R(G) \leq \gamma_{StR}(G) \leq \left(1 + \lceil \frac{\Delta}{2} \rceil\right) \gamma(G)$;

(ii) $\gamma_{StR}(G) \leq n - \lfloor \frac{\Delta}{2} \rfloor$;

(iii) $\gamma_{StR}(G) \geq \lceil \frac{n+1}{2} \rceil$;

(iv) $\gamma_{StR}(G) \leq n - \lfloor \frac{1 + \mathrm{diam}(G)}{3} \rfloor$;

(v) *if G has girth $g(G) \geq 3$, then* $\gamma_{StR}(G) \leq n - \lfloor \frac{g(G)}{3} \rfloor$.

The following result obtained by using a probabilistic approach is a generalization in some sense of the one presented in [35].

Proposition 9.2 ([11]) *Let G be a graph of order n, minimum degree δ, and maximum degree Δ, such that $\lceil \frac{\Delta}{2} \rceil < \delta$. Then,*

$$\gamma_{StR}(G) \leq \frac{(1 + \lceil \frac{\Delta}{2} \rceil)n}{\delta + 1} \left(\ln \left(\frac{\delta + 1}{1 + \lceil \frac{\Delta}{2} \rceil} \right) + 1 \right).$$

For the class of trees T, an upper bound on $\gamma_{StR}(T)$ in terms of the order is presented. Let H be a tree obtained from a star $K_{1,3}$ by subdividing each edge exactly once. Let $\mathcal{F}_p$ be the family of trees obtained from any tree T of order p by identifying each vertex of T with the central vertex of H so that the H's are vertex disjoint.

Theorem 9.3 ([11]) *If T is a tree of order n, then $\gamma_{StR}(T) \leq \frac{6}{7}n$, with equality if and only if $T \in \mathcal{F}_p$.*

Álvarez-Ruiz et al. have wondered if the $\frac{6}{7}n$ upper bound on the strong Roman domination number for trees remains valid for any connected graph G of order $n \geq 3$.

We close this section by mentioning some algorithmic and complexity results. It is shown in [11] that the decision problem corresponding to the problem of computing $\gamma_{StR}(G)$ is NP-complete for planar graphs. A problem of constructing a polynomial algorithm for computing the value of $\gamma_{StR}(T)$ for any tree T is proposed in [11]. This problem is answered by Poureidi et al. [73] who gave a linear algorithm that computes the strong Roman domination number of trees.

10 Edge Roman Domination

The study of the edge version of Roman domination was initiated by Roushini Leely Pushpam and Malini Mai in 2009 [75]. An *edge Roman dominating function* (ERDF) of a graph G is a function $f : E(G) \to \{0, 1, 2\}$ such that every edge e with $f(e) = 0$ is adjacent to some edge e' with $f(e') = 2$. The weight of an ERDF f is $w(f) = \sum_{e \in E(G)} f(e)$, and the *edge Roman domination number* of G, denoted by $\gamma_R'(G)$, is the minimum weight of an ERDF of G. An ERDF $f : E(G) \to \{0, 1, 2\}$ can be represented by the ordered partition (E_0, E_1, E_2) of $E(G)$, where $E_i = \{e \in E(G) : f(e) = i\}$ for $i \in \{0, 1, 2\}$. It is worth mentioning that the edge Roman domination number of G equals the Roman domination number of its line graph. Edge Roman domination has been studied in [8, 9, 22, 54, 75, 77] and elsewhere. It should be noted that as far as we know, no paper has dealt with either the complexity or the algorithmic aspect of the edge Roman domination problem.

Several properties of edge Roman dominating functions can be obtained analogously to Roman dominating functions given in [35]. Here are some summarized by the following result.

Proposition 10.1 ([75]) *Let $f = (E_0, E_1, E_2)$ be a minimum ERDF of an isolate-free graph G, such that $|E_2|$ is maximum. Then,*

(i) *E_1 is independent.*
(ii) *The edges of E_0 dominate the edges of E_1.*
(iii) *Each edge of E_0 is adjacent to at most one edge of E_1.*
(iv) *Let $e \in G[E_2]$ have exactly two private edges e_1 and e_2 in E_0 with respect to E_2. Then there do not exist edges $h_1, h_2 \in E_1$ such that (h_1, e_1, e, e_2, h_2) is the edge sequence of a path P_6.*

Also, it is obvious that for every graph G, $\gamma'(G) \leq \gamma_R'(G) \leq 2\gamma'(G)$, where $\gamma'(G)$ is the edge domination number of G. A characterization of trees with edge

Roman domination number twice the edge domination number was given by Jafari Rad in [54].

In [75], it is shown that $\gamma'_R(P_n) = \left\lfloor \frac{2n}{3} \right\rfloor$ and $\gamma'_R(C_n) = \left\lceil \frac{2n}{3} \right\rceil$. Akbari et al. [8] extended the exact value on paths to 2-by-n and 3-by-n grid graphs and proved that for $n \geq 2$, $\gamma'_R(G_{\{2,n\}}) = \left\lceil \frac{4n}{3} \right\rceil$ and $\gamma'_R(G_{\{3,n\}}) = 2n$. The problem of determining the edge Roman domination number for every m-by-n grid graph $G_{\{m,n\}}$ remains open.

For bounds on the edge Roman domination number, we begin with the class of trees T, where Akbari et al. [8] have bounded it in terms of the order and number of leaves of T.

Theorem 10.2 ([8]) *If T is a tree of order n with $\ell(T)$ leaves, then*

$$\left\lceil \frac{2(n - \ell(T) + 1)}{3} \right\rceil \leq \gamma'_R(T) \leq \left\lceil \frac{2(n-1)}{3} \right\rceil.$$

For arbitrary graphs, two upper bounds on the edge Roman domination number have been established by Akbari et al. [8].

Theorem 10.3 ([8]) *Let G be a graph of order n with maximum degree Δ. Then,*

(i) $\gamma'_R(G) \leq \frac{2\Delta}{2\Delta+1}n$;
(ii) *if G has a perfect matching, then* $\gamma'_R(G) \leq \frac{2\Delta-1}{2\Delta}n$.

Furthermore, the authors in [8] conjectured that $\gamma'_R(G) \leq \left\lceil \frac{\Delta}{\Delta+1}n \right\rceil$. This conjecture has been disproved by Chang, Chen, and Liu [22] who gave the following counterexample. Let $G(r, t)$ be the graph obtained from t copies of $K_{r,r+1}$ by adding edges $y^i_{r+1}y^{i+1}_1$ for $1 \leq i \leq t$ with $y^{t+1}_1 = y^1_1$, where the partite sets of the i-th $K_{r,r+1}$ are $X_i = \{x^i_1, \ldots, x^i_r\}$ and $Y_i = \{y^i_1, \ldots, y^i_{r+1}\}$. Note that $G(r, t)$ has order $n = (2r+1)t$ and maximum degree $\Delta = r+1$. It was shown then that $\gamma'_R(G(r,t)) = 2rt = \frac{2\Delta-2}{2\Delta-1}n > \left\lceil \frac{\Delta}{\Delta+1}n \right\rceil$ when $r \geq 2$ and t a multiple of $r+2$.

For the class of planar graphs, it is shown in [9] that if G is outerplanar, then $\gamma'_R(G) \leq \frac{4}{5}n$ and if G is planar claw-free, then $\gamma'_R(G) \leq \frac{6}{7}n$. Furthermore, the authors in [9] conjectured that $\gamma'_R(G) \leq \frac{6}{7}n$ for any planar graph G of order n. This conjecture was proved by Chang, Chen, and Liu in [22]. Also, they obtained several other bounds that we list below. Recall that a graph G is k-*degenerate* if for every subgraph H of G, $\delta(H) \leq k$.

Theorem 10.4 ([22]) *If G is a k-degenerate graph of n vertices, then $\gamma'_R(G) \leq \frac{2k}{2k+1}n$.*

Since any tree is 1-degenerate, the upper bound $\left\lfloor \frac{2n}{3} \right\rfloor$ in Theorem 10.2 becomes a simple consequence of Theorem 10.4. Moreover, since every graph is Δ-degenerate, the upper bound in Theorem 10.3 is also a straightforward consequence of Theorem 10.4.

Theorem 10.5 ([22]) *If G is a connected graph of order n with maximum degree Δ, then $\gamma_R'(G) \leq \frac{2\Delta-2}{2\Delta-1}n + \frac{2}{2\Delta-1}$.*

Theorem 10.6 ([22]) *If G is a subcubic graph of order n which does not contain $K_{3,3}$ as a component, then $\gamma_R'(G) \leq \frac{4}{5}n$.*

Theorem 10.7 ([22]) *If G is a graph of order n containing no subgraph isomorphic to a subdivision of $K_{2,3}$, then $\gamma_R'(G) \leq \frac{4}{5}n$.*

11 Open Problems

In this chapter, we have surveyed some results concerning nine Roman domination-related parameters. There is much scope for further research, here are some suggestions.

1. Determine the weak Roman domination number for every m-by-n grid graph $G_{\{m,n\}}$.
2. Since the problem of deciding whether $\gamma_r(G) = \gamma_R(G)$ for a given graph G is NP-hard, it is quite interesting to characterize other classes of graphs G, other than trees, such that $\gamma_r(G) = \gamma_R(G)$.
3. Design an algorithm for computing the Roman k-domination number for any tree T.
4. [Álvarez-Ruiz et al. [11]] Is it true that for every connected graph G of order $n \geq 3$, $\gamma_{StR}(T) \leq \frac{6}{7}n$?
5. What are the algorithmic, complexity, and approximation properties of edge Roman domination?
6. Determine the edge Roman domination number for every m-by-m grid graph $G_{\{m,m\}}$.

   In [78], it was shown that the generalized Petersen graph $P(n, 1)$ is a double Roman graph for any $n \equiv 2 \pmod 4$, while in [58], it is shown that the generalized Petersen graph $P(n, 2)$ is not double Roman for all $n \geq 3$.
7. [Jiang et al. [58]] Find other generalized Petersen graphs that are double Roman.

References

1. H. Abdollahzadeh Ahangar, M.A. Henning, V. Samodivkin, I.G. Yero, Total Roman domination in graphs. Appl. Anal. Discrete Math. **10**, 501–517 (2016)
2. H. Abdollahzadeh Ahangar, M. Chellali, S.M. Sheikholeslami, On the double Roman domination in graphs. Discrete Appl. Math. **232**, 1–7 (2017)
3. H. Abdollahzadeh Ahangar, J. Amjadi, M. Atapour, M. Chellali, S.M. Sheikholeslami, Double Roman trees. Ars Combin. **145**, 173–183 (2019)

4. H. Abdollahzadeh Ahangar, J. Amjadi, M. Chellali, S. Nazari-Moghaddam, S.M. Sheikholeslami, Total Roman reinforcement numbers in graphs. Discuss. Math. Graph Theory **39**, 787–803 (2019)
5. H. Abdollahzadeh Ahangar, J. Amjadi, M. Chellali, S. Nazari-Moghaddam, S.M. Sheikholeslami, Trees with double Roman domination number twice the domination number plus two. Iran. J. Sci. Technol. Trans. A Sci. **43**, 1081–1088 (2019)
6. H. Abdollahzadeh Ahangar, J. Amjadi, S.M. Sheikholeslami, M. Soroudi, On the total Roman domination number of graphs. Ars Combin. **150**, 225–240 (2020)
7. M. Adabi, E. Ebrahimi Targhi, N. Jafari Rad, M. Saied Moradi, Properties of independent Roman domination in graphs. Australas. J. Combin. **52**, 11–18 (2012)
8. S. Akbari, S. Ehsani, S. Ghajar, P. Jalaly Khalilabadi, S. Sadeghian Sadeghabad, On the edge Roman domination in graphs. (Preprint)
9. S. Akbari, S. Qajar, On the edge Roman domination number of planar graphs (Preprint)
10. J.D. Alvarado, S. Dantas, D. Rautenbach, Strong equality of Roman and weak Roman domination in trees. Discrete Appl. Math. **208**, 19–26 (2016)
11. M.P. Álvarez-Ruiz, T. Mediavilla-Gradolph, S.M. Sheikholeslami, J.C. Valenzuela-Tripodoro, I.G. Yero, On the strong Roman domination number of graphs. Discrete Appl. Math. **231**, 44–59 (2017)
12. J. Amjadi, S. Nazari-Moghaddam, S.M. Sheikholeslami, Total Roman domination number of trees. Australas. J. Combin. **69**, 271–285 (2017)
13. J. Amjadi, S. Nazari-Moghaddam, S.M. Sheikholeslami, L. Volkmann, An upper bound on the double Roman domination number. J. Comb. Optim. **36**, 81–89 (2018)
14. J. Amjadi, S.M. Sheikholeslami, M. Soroudi, Nordhaus-Gaddum bounds for total Roman domination. J. Comb. Optim. **35**, 126–133 (2018)
15. J. Amjadi, S.M. Sheikholeslami, M. Soroudi, On the total Roman domination in trees. Discuss. Math. Graph Theory **39**, 519–532 (2019)
16. D. Bange, A.E. Barkauskas, P.J. Slater, Efficient dominating sets in graphs, in *Applications of Discrete Math.*, ed. by R.D. Ringeisen, F.S. Roberts, (SIAM, Philadelphia, 1988), pp. 189–199
17. R.A. Beeler, T.W. Haynes, S.T. Hedetniemi, Double Roman domination. Discrete Appl. Math. **211**, 23–29 (2016)
18. M. Bouchou, M. Blidia, M. Chellali, A note on k-Roman graphs. Opuscula Math. **33**, 641–646 (2013)
19. M. Bouchou, M. Blidia, M. Chellali, Relations between the Roman k-domination and Roman domination numbers in graphs. Discrete Math. Algorithms Appl. **6**(3), 1450045, 11 pp. (2014)
20. N. Campanelli, D. Kuziak, Total Roman domination in the lexicographic product of graphs. Discrete Appl. Math. **263**, 88–95 (2019)
21. Y. Caro, Y. Roditty, A note on the k-domination number of a graph. Int. J. Math. Math. Sci. **13**, 205–206 (1990)
22. G.J. Chang, S.H. Chen, C.H. Liu, Edge Roman domination on graphs. Graphs Combin. **32**, 1731–1747 (2016)
23. M. Chapelle, M. Cochefert, J.F. Couturier, D. Kratsch, R. Letourneur, M. Liedloff, A. Perez, Exact algorithms for weak Roman domination. Discrete Appl. Math. **248**, 79–92 (2018)
24. M. Chellali, N. Jafari Rad, Strong equality between the Roman domination and independent Roman domination numbers in trees. Discuss. Math. Graph Theory **33**, 337–346 (2013)
25. M. Chellali, N. Jafari Rad, Trees with independent Roman domination number twice the independent domination number. Discrete Math. Algorithms Appl. **7**(4), 1550048, 8 pp. (2015)
26. M. Chellali, N. Jafari Rad, Independent 2-rainbow domination in graphs. J. Combin. Math. Combin. Comput. **94**, 133–148 (2015)
27. M. Chellali, T.W. Haynes, S.T. Hedetniemi, Bounds on weak Roman and 2-rainbow domination numbers. Discrete Appl. Math. **178**, 27–32 (2014)
28. M. Chellali, T.W. Haynes, S.T. Hedetniemi, Lower bounds on the Roman and independent Roman domination numbers. Appl. Anal. Discrete Math. **10**, 65–72 (2016)
29. M. Chellali, T.W. Haynes, S.T. Hedetniemi, A. MacRae, Roman {2}-domination. Discrete Appl. Math. **204**, 22–28 (2016)

30. M. Chellali, N. Jafari Rad, S.M. Sheikholeslami, L. Volkmann, Roman domination in graphs, in *Topics in Domination in Graphs* ed. by T.W. Haynes, S.T. Hedetniemi, M.A. Henning (Springer International Publishing, 2020)
31. M. Chellali, N. Jafari Rad, S.M. Sheikholeslami, L. Volkmann, Varieties of Roman domination II. AKCE Int. J. Graphs Combin. **17**, 966–984 (2020)
32. Z. Shao, S. Kosari, M. Chellali, S.M. Sheikholeslami, M. Soroudi, On a relation between the perfect Roman domination and perfect domination numbers of a tree. Math. **8**, ID 966, 13 pp. (2020)
33. B. Chen, S. Zhou, Upper bounds for f-domination number of graphs. Discrete Math. **185**, 239–243 (1998)
34. H. Chen, C. Lu, A note on Roman {2}-domination problem in graphs. arXiv:1804.09338v1 [math.CO] (2019)
35. E.J. Cockayne, P.A. Dreyer Jr., S.M. Hedetniemi, S.T. Hedetniemi. Roman domination in graphs. Discrete Math. **278**, 11–22 (2004)
36. M. Darkooti, A. Alhevaz, S. Rahimi, H. Rahbani, On perfect Roman domination number in trees: complexity and bounds. J. Comb. Optim. **38**, 712—720 (2019)
37. G.S. Domke, S.T. Hedetniemi, R.C. Laskar, G. Fricke, Relationships between integer and fractional parameters of graphs, in *Proc. Sixth Quadrennial Conf. on the Theory and Applications of Graphs, Western Michigan Univ*, ed. by Y. Alavi, G. Chartrand, O. Oellermann, A. Schwenk. Graph Theory, Combinatorics and Applications, vol. 1 (Wiley Interscience Publ., New York, 1991), pp. 371–387
38. E. Ebrahimi Targhi, N. Jafari Rad, C.M. Mynhardt, Y. Wu, Bounds for independent Roman domination in graphs. J. Combin. Math. Combin. Comput. **80**, 351–365 (2012)
39. J. Fulman, A note on the characterization of domination perfect graphs. J. Graph Theory **17**, 47–51 (1993)
40. M. Hajibaba, N. Jafari Rad, A note on the Italian domination number and double Roman domination number in graphs. J. Combin. Math. Combin. Comput. **109**, 169–183 (2019)
41. M. Hajibaba, N. Jafari Rad, On domination, 2-domination, and Italian domination numbers. Util. Math. **111**, 271–280 (2019)
42. A. Hansberg, L. Volkmann, On graphs with equal domination and 2-domination numbers. Discrete Math. **308**, 2277–2281 (2008)
43. A. Hansberg, L. Volkmann, Upper bounds on the k-domination number and the k-Roman domination number. Discrete Appl. Math. **157**, 1634–1639 (2009)
44. T.W. Haynes, P.J. Slater, Paired-domination in graphs. Networks **32**, 199–206 (1998)
45. T.W. Haynes, M.A. Henning, Perfect Italian domination in trees. Discrete Appl. Math. **260**, 164–177 (2019)
46. T.W. Haynes, M.A. Henning, L. Volkmann, Graphs with large Italian domination number. Bull. Malays. Math. Sci. Soc. **43**, 4273–4287 (2020)
47. M.A. Henning, S.T. Hedetniemi, Defending the Roman Empire - A new strategy. Discrete Math. **266**, 239–251 (2003)
48. M.A. Henning, W.F. Klostermeyer, Italian domination in trees. Discrete Appl. Math. **217**, 557–564 (2017)
49. M.A. Henning, W.F. Klostermeyer, Perfect Roman domination in regular graphs. Appl. Anal. Discrete Math. **12**, 143–152 (2018)
50. M.A. Henning, N. Jafari Rad, A characterization of double Roman trees. Discrete Appl. Math. **259**, 100–111 (2019)
51. M.A. Henning, N. Jafari Rad, Upper bounds on the k-tuple (Roman) domination number of a graph. Graphs Combin. (2020). https://doi.org/10.1007/s00373-020-02249-7
52. M.A. Henning, W.F. Klostermeyer, G. MacGillivray, Perfect Roman domination in trees. Discrete Appl. Math. **236**, 235–245 (2018)
53. N. Jafari Rad, Note on the independent Roman domination number of a graph. J. Combin. Math. Combin. Comput. **95**, 119–125 (2015)
54. N. Jafari Rad, A note on the edge Roman domination in trees. Elect. J. Graph Theory Appl. **5**, 1–6 (2017)

55. N. Jafari Rad, On the k-domination, k-tuple domination and Roman k-domination numbers in graphs. J. Combin. Math. Combin. Comput. to appear
56. N. Jafari Rad, H. Rahbani, Some progress on the double Roman domination in graphs. Discuss. Math. Graph Theory **39**, 41–53 (2019)
57. N. Jafari Rad, L. Volkmann, Roman domination perfect graphs. Annalele St. Univ. Ovid. Const. **19**, 167–174 (2011)
58. H. Jiang, P. Wu, Z. Shao, Y. Rao, J. Liu, The double Roman domination numbers of generalized Petersen graphs $P(n, \ 2)$. Mathematics **6**, 206 (2018)
59. K. Kämmerling, L. Volkmann, Roman k-domination in graphs. J. Korean Math. Soc. **46**, 1309–1318 (2009)
60. R. Khoeilar, H. Karami, M. Chellali, S.M. Sheikholeslami, An improved upper bound on the double Roman domination number of graphs with minimum degree at least two. Discrete Appl. Math. **270**, 159–167 (2019)
61. W. Klostermeyer, A taxonomy of perfect domination. J. Discrete Math. Sci. Cryptogr. 18, 105–116 (2015)
62. W.F. Klostermeyer, G. MacGillivray, Roman, Italian, and 2-domination. J. Combin. Math. Combin. Comput. **108**, 125–146 (2019)
63. J. Lauri, C. Mitillos, Perfect Italian domination on planar and regular graphs. Discrete Appl. Math. **285**, 676–687 (2020)
64. M. Li, On the k-Roman domination of graphs. J. Comput. Theor. Nanosci. **13**, 2705–2709 (2016)
65. C.-H. Liu, G.J. Chang, Roman domination on strongly chordal graphs. J. Comb. Optim. **26**, 608–619 (2013)
66. C.S. Liu, S.L. Peng and C.Y. Tang, Weak Roman domination on block graphs, in *Proceedings of the 27th Workshop on Combinatorial Mathematics and Computation Theory*, Providence University, Taichung, Taiwan, April 30-May 1, 2010
67. M. Livingston, Q.F. Stout, Perfect dominating sets. Congr. Numer. **79**, 187–203 (1990)
68. A.C. Martínez, I.G. Yero, A characterization of trees with equal Roman {2}-domination and Roman domination numbers. Commun. Comb. Optim. **4**, 95–107 (2019)
69. D.A. Mojdeh, S.M.H. Moghaddam, A correction to a paper on Roman k-domination in graphs. Bull. Korean Math. Soc. **50**, 469–473 (2013)
70. A. Poureidi, N. Jafari Rad, Algorithmic and complexity aspects of problems related to total Roman domination for graphs. J. Comb. Optim. **39**, 747–763 (2020)
71. A. Poureidi, N. Jafari Rad, Algorithmic aspects of the independent 2-rainbow domination number and independent Roman {2}-domination number. Discuss. Math. Graph Theory (to appear)
72. A. Poureidi, N. Jafari Rad, On algorithmic complexity of double Roman domination. Discrete Appl. Math. **285**, 539–551 (2020)
73. A. Poureidi, N. Jafari Rad, A linear algorithm for computing strong Roman domination number of trees (Submitted)
74. A. Rahmouni, M. Chellali, Independent Roman {2}-domination in graphs. Discrete Appl. Math. **236**, 408–414 (2018)
75. P. Roushini Leely Pushpam, T.N.M. Malini Mai, Edge Roman domination in graphs. J. Combin. Math. Combin. Comput. **69**, 175–182 (2009)
76. P. Roushini Leely Pushpam, T.N.M. Malini Mai, Weak Roman domination in graphs. Discuss. Math. Graph Theory **31**, 115–128 (2011)
77. P. Roushini Leely Pushpam, T.N.M. Malini Mai, Weak edge Roman domination in graphs. Australas. J. Combin. **51**, 125–138 (2011)
78. Z. Shao, H. Jiang, Z. Li, P. Wu, J. Zerovnik, X. Zhang, Discharging approach for double Roman domination in graphs. IEEE Access **6**, 63345–63351 (2018)
79. S.M. Sheikholeslami, M. Chellali, M. Soroudi, A characterization of perfect Roman trees. Discrete Appl. Math. **285**, 501–508 (2020)
80. D.P. Sumner, Critical concepts in domination. Discrete Math. **86**, 33–46 (1990)

81. M. Valveny, J.A. Rodríguez-Velázquez, Protection of graphs with emphasis on cartesian product graphs. Filomat **33**(1), 319–333 (2019)
82. L. Volkmann, Double Roman domination and domatic numbers of graphs. Commun. Comb. Optim. **3**, 71–77 (2018)
83. X. Zhang, Z. Li, H. Jiang, Z. Shao, Double Roman domination in trees. Inf. Proc. Lett. **134**, 31–34 (2018)
84. E. Zhu, Z. Shao, Extremal problems on weak Roman domination number. Inf. Proc. Lett. **138**, 12–18 (2018)

Part II
Domination in Selected Graph Families

Domination and Total Domination in Hypergraphs

Michael A. Henning and Anders Yeo

AMS Subject Classification: 05C65, 05C69

1 Introduction

Domination in hypergraphs has not been studied in as much depth as domination in graphs. Although the theory of domination in graphs is well developed in the literature, the theory of domination in hypergraphs is currently in its infant stages. In this chapter, we give a survey of selected results on domination and total domination in hypergraphs. We establish an essential connection in hypergraphs between dominating sets and transversals, and between total dominating sets and total transversals.

Hypergraphs are systems of sets, which are conceived as natural extensions of graphs. More formally, a *hypergraph* $H = (V(H), E(H))$ is a finite set $V(H)$ of elements, called *vertices*, together with a finite multiset $E(H)$ of subsets of $V(H)$, called *hyperedges*. If the hypergraph H is clear from the context, we simply write $V = V(H)$ and $E = E(H)$. We shall use the notation $n_H = |V|$ (or $n(H)$) and $m_H = |E|$ (or $m(H)$), and sometimes simply n and m without subscripts if

M. A. Henning
Department of Mathematics and Applied Mathematics, University of Johannesburg,
Johannesburg, South Africa
e-mail: mahenning@uj.ac.za

A. Yeo (✉)
Department of Mathematics and Applied Mathematics, University of Johannesburg,
Johannesburg, South Africa

Department of Mathematics and Computer Science, University of Southern Denmark, Odense,
Denmark

© The Author(s), under exclusive license to Springer Nature Switzerland AG 2021
T. W. Haynes et al. (eds.), *Structures of Domination in Graphs*, Developments
in Mathematics 66, https://doi.org/10.1007/978-3-030-58892-2_11

the hypergraph H is clear from the context, to denote the *order* and *size* of H, respectively. A vertex $v \in V$ is *incident* with a hyperedge $e \in E$ in H if $v \in e$, i.e., if the vertex v belongs to the edge e. A hypergraph H is *linear* if every two distinct edges of H intersect in at most one vertex.

The edge set E in a hypergraph $H = (V, E)$ is often allowed to be a multiset in the literature, but in this chapter we exclude multiple edges without loss of generality. We refer to the cardinality $|e|$ of an edge e in H as its *size*. In the problems studied here, we assume that every edge has size at least 2. Throughout this chapter, we simply refer to a *hyperedge* as an *edge*. An *isolated edge* in H is an edge in H that does not intersect any other edge in H.

A *k-edge* in H is an edge of size k. The *rank* of a hypergraph H is the maximum size of an edge on H. Thus, H is of rank k if $|e| \le k$ holds for each edge $e \in E$ in H. The hypergraph H is *k-uniform* if every edge of H is a k-edge. Every (simple) graph is a 2-uniform hypergraph. Thus, graphs are special hypergraphs. For $i \ge 2$, we denote the number of edges in H of size i by $e_i(H)$. The *degree* of a vertex v in H, denoted by $d_H(v)$ or $d(v)$ if H is clear from the context, is the number of edges of H, which contain v. A vertex of degree k is called a *degree-k vertex*. The minimum degree among the vertices of H is denoted by $\delta(H)$ and the maximum degree by $\Delta(H)$.

A *subhypergraph* of a hypergraph $H = (V, E)$ is a hypergraph $H' = (V', E')$ satisfying $V' \subseteq V$ and $E' \subseteq E$ (and where $E' \subseteq V'$ by definition of a hypergraph H'). Given a hypergraph $H = (V, E)$ and a nonempty set $X \subseteq V$, the *subhypergraph of H induced by X* is the hypergraph $H' = (V', E')$ where $E' = \{e \in E : e \subseteq X\}$ and where $V' \subseteq X$ is the set of vertices of H contained in at least one edge $e \subseteq X$.

A *2-section graph*, H_2, of a hypergraph H is defined as a graph H_2 with the same vertex set as H and in which two vertices are adjacent in the graph H_2 if and only if they belong to a common edge in H.

Two vertices x and y in a hypergraph H are *adjacent* if there is an edge e of H such that $\{x, y\} \subseteq e$. The *neighborhood* of a vertex v in H, denoted by $N_H(v)$, is the set of all vertices different from v that are adjacent to v, while the *closed neighborhood* of v is the set $N_H[v] = N_H(v) \cup \{v\}$. We call a vertex in $N_H(v)$ a *neighbor* of v in H. For a subset $X \subseteq V(H)$, the *neighborhood* of X is the set $N_H(X) = \cup_{x \in X} N_H(x)$, while the *closed neighborhood* of X is the set $N_H[X] = N_H(X) \cup X$.

Two vertices x and y are *connected* in the hypergraph H if there is a sequence $x = v_0, v_1, v_2 \ldots, v_k = y$ of vertices of H in which v_{i-1} is adjacent to v_i for $i \in [k]$. A *connected hypergraph* is a hypergraph in which every pair of vertices are connected. A *component* of H is a maximal connected subhypergraph of H. Thus, no edge in H contains vertices from different components.

A subset $T \subseteq V$ of vertices in a hypergraph H is a *transversal* (also called *vertex cover* or *hitting set* in many papers) if for every $e \in E$, $T \cap e \ne \varnothing$, that is, every edge has a vertex in T. A transversal of cardinality $\tau(H)$ is called a *τ-transversal of H*. A *total transversal* in H is a transversal T in H with the additional property that every vertex in T has at least one neighbor in T. The *total transversal number* $\tau_t(H)$ of H is the minimum size of a total transversal in H. A total transversal in H of cardinality $\tau_t(H)$ is called a *τ_t-total transversal of H*. Transversals in hypergraphs are well-studied in the literature (see, for example [6, 11, 15, 23, 27–30, 38, 42]).

For a subset $X \subset V$ of vertices in H, we define $H - X$ to be the hypergraph obtained from H by deleting the vertices in X and all edges incident with X, and deleting all isolated vertices, if any, from the resulting hypergraph. We note that if T' is a transversal in $H - X$, then $T' \cup X$ is a transversal in H. If $X = \{x\}$, then we write $H - X$ simply as $H - x$.

A *dominating set* in a hypergraph H is a subset of vertices $D \subseteq V$ such that for every vertex $v \in V \setminus D$, there exists an edge $e \in E$ for which $v \in e$ and $e \cap D \neq \varnothing$. Equivalently, a set D is a dominating set in H if every vertex outside the set D has a neighbor in D. The *domination number* $\gamma(H)$ is the minimum cardinality of a dominating set in H. A dominating set of H of cardinality $\gamma(H)$ we call a γ-*set of H*.

A *total dominating set*, abbreviated TD-set, in H is a subset of vertices $D \subseteq V(H)$ such that every vertex in H is adjacent with a vertex in D. Equivalently, a set D is a total dominating set in H if D is a dominating set in H with the additional property that for every vertex $v \in D$ there exists an edge $e \in E(H)$ for which $v \in e$ and $e \cap (D \setminus \{v\}) \neq \varnothing$. The *total domination number* $\gamma_t(H)$ is the minimum cardinality of a TD-set in H.

We remark that a set is a (total) dominating set in H if and only if it is a (total) dominating set in the 2-section graph H_2. Domination in hypergraphs was introduced in 2007 by Acharya [2] and studied in [4, 8, 22, 34, 35] and elsewhere, and also in the Ph.D. thesis of Bibin Jose [33]. Total domination in hypergraphs was introduced by Bujtás, Henning, Tuza, and Yeo in 2014 [9] and studied further by Henning and Yeo in 2015 [32].

We use the standard notation $[k] = \{1, \ldots, k\}$ and $[k]_0 = \{0, 1, \ldots, k\}$.

2 Domination in Hypergraphs

In this section, we focus on domination in hypergraphs. We proceed as follows. In Section 2.1, we consider disjoint dominating sets in hypergraphs. In Section 2.2 we discuss an interplay between domination and transversals in hypergraphs. As a consequence of the results in Section 2.2, we establish in Section 2.3 upper bounds on the domination numbers of uniform hypergraphs with minimum degree at least one. In Sections 2.4, 2.5, 2.6, and 2.7, we present results on hypergraphs with specified edge size having large domination numbers. In Section 2.8, we discuss a general setting of upper bounds on the domination number in terms of its order and size and present a general way to formulate the problem. In Section 2.9, we address the problem of finding the minimum order of a connected, k-uniform hypergraph with a given domination number. In Section 2.10, we present a Nordhaus–Gaddum-type result for the sum of domination parameters in hypergraphs and their complements. Hypergraphs with equality of the domination and transversal numbers are discussed in Section 2.11. In Section 2.12, we discuss a relationship between domination and matching in hypergraphs and present an upper bound on the domination number of a uniform hypergraph in terms of its matching number.

2.1 Disjoint Dominating Sets

The maximum number of vertex-disjoint dominating sets in a graph G is called the *domatic number* of G denoted by $\text{dom}(G)$. The domatic number of a graph was introduced in 1975 by Cockayne and Hedetniemi [12]. The *domatic number* of a hypergraph H, denoted by $\text{dom}(H)$, is defined analogously as the maximum number of vertex-disjoint dominating sets in H.

A fundamental result in domination theory in 1962 due to Ore [40] is the property that every graph without isolated vertices contains two disjoint dominating sets. Indeed as observed by Ore [40], if $G = (V, E)$ is a graph without isolated vertices, then the complement $V \smallsetminus D$ of every minimal dominating set D in G is a dominating set of G. Thus, every 2-uniform hypergraph H without isolated vertices contains two vertex-disjoint dominating sets; that is, $\text{dom}(H) \geq 2$. This property holds for k-uniform hypergraphs for all $k \geq 2$. To see this, recall that an *independent set* in a hypergraph $H = (V, E)$ is a set $S \subseteq V$ such that no two vertices of S are adjacent; that is, every edge of H intersects S in at most one vertex, and therefore no two vertices of S belong to the same edge in H.

Proposition 1 *For $k \geq 2$ an integer, every k-uniform hypergraph H without isolated vertices satisfies* $\text{dom}(H) \geq 2$.

Proof. Let S be a maximal independent set in the k-uniform hypergraph $H = (V, E)$. Thus, $|e \cap S| = 1$ for every edge $e \in E$. By the maximality of the set S, every vertex in $V \smallsetminus S$ is adjacent to at least one vertex in S; that is, if $v \in V \smallsetminus S$, then there exists at least one edge e_v such that $|e_v \cap S| = 1$. In particular, we note that the set S is a dominating set in H. Since H is without isolated vertices, every vertex in S is adjacent to at least one vertex outside S, implying that $V \smallsetminus S$ is a dominating set of H. Thus, H has at least two vertex-disjoint dominating sets, namely S and $V \smallsetminus S$. $\square$

We note that the result of Proposition 1 also holds for hypergraphs without isolated vertices and where every edge has size at least k. However it is not true that for any given $k \geq 3$, every k-uniform hypergraph without isolated vertices contains k vertex-disjoint dominating sets, as the following result shows.

Proposition 2 *For $k \geq 3$ an integer, there exist k-uniform hypergraphs H without isolated vertices satisfying* $\text{dom}(H) = 2$.

Proof. For a given integer $k \geq 3$, we construct a k-uniform hypergraph as follows. Let $H' = (V', E')$ be a complete $(k-1)$-uniform hypergraph on $3k-5$ vertices, that is, every $(k-1)$-element subset of V' forms an edge in H'. We now construct a k-uniform hypergraph H from H' by adding a new vertex v to each edge $e' \in E'$ in H' and replacing the edge e' with the edge $e' \cup \{v\}$ in such a way that all the new vertices are distinct and therefore have degree 1 in H. We show that $\text{dom}(H) = 2$. Suppose, to the contrary, that $\text{dom}(H) \geq 3$. Let D_1, D_2, D_3 be three vertex-disjoint dominating sets of H. We may assume that (D_1, D_2, D_3) is a partition of V, since vertices not in $D_1 \cup D_2 \cup D_3$ can be assigned randomly to some set D_i where $i \in [3]$. We note that

$$3k - 5 = |V'| = \sum_{i=1}^{3} |D_i \cap V(H')|.$$

Renaming the dominating sets D_1, D_2, D_3 if necessary, we may assume that $|D_1 \cap V(H')| \geq (3k - 5)/3$, which implies that $|D_1 \cap V(H')| \geq k - 1$. Thus, there exists an edge $e' \in E'$ of H' such that $e' \subseteq D_1 \cap V(H')$. Let v be the vertex added to the edge e' when constructing H from H' to produce the edge $e = e' \cup \{v\}$. At least one of the two dominating sets D_2 and D_3 does not contain the vertex v. Such a dominating set contains no vertex from the edge e and therefore does not dominate the vertex v, a contradiction. Hence, $\mathrm{dom}(H) \leq 2$. By Proposition 1, $\mathrm{dom}(H) \geq 2$. Consequently, $\mathrm{dom}(H) = 2$. $\square$

We remark that the hypergraph constructed in the proof of Proposition 2 is nonlinear. Even if we restrict our attention to linear hypergraphs, it is not true that for any given $k \geq 3$, every k-uniform linear hypergraph without isolated vertices contains k vertex-disjoint dominating sets.

Proposition 3 *For $k \geq 3$ an integer, there exist k-uniform linear hypergraphs H without isolated vertices satisfying* $\mathrm{dom}(H) < k$.

Proof. For a given integer $k \geq 3$, we construct a k-uniform, linear hypergraph as follows. Let $Y = \{y_1, y_2, \ldots, y_{k+1}\}$ be a set of $k + 1$ vertices. For each 2-element subset $\{y_i, y_j\}$ of Y where $1 \leq i < j \leq k + 1$, let X_{ij} be a set of $k - 2$ new vertices, and let

$$X = \bigcup_{1 \leq i < j \leq k+1} X_{ij}.$$

We note that

$$|Y| = k + 1 \quad \text{and} \quad |X| = (k - 2) \cdot \binom{k + 1}{2}.$$

Let $H = (V, E)$ be the k-uniform, linear hypergraph with vertex set $V = X \cup Y$ and with edge set $E = \{e_{ij} : 1 \leq i < j \leq k + 1\}$ where $e_{ij} = X_{ij} \cup \{y_i, y_j\}$. We note that each vertex in X has degree 1 in H, while each vertex in Y has degree k in H.

Suppose, to the contrary, that H contains k vertex-disjoint dominating sets, say $D_1, D_2, \ldots, D_k$. We may assume that $(D_1, D_2, \ldots, D_k)$ is a partition of V as vertices not in $D_1 \cup D_2 \cup \cdots \cup D_k$ can be assigned randomly to some set D_i where $i \in [k]$. Since $|Y| = k + 1$ and there are k dominating sets, two vertices of Y must belong to the same dominating set. Renaming vertices and dominating sets, if necessary, we may assume that $\{y_1, y_2\} \subseteq D_1$. We now consider the $k - 2$ vertices that belong to the set X_{12}. Each such vertex has degree 1 in H and is contained in the unique edge $e_{12} = X_{12} \cup \{y_1, y_2\}$ of H. Since $|X_{12}| = k - 2$ and there are $k - 1$ remaining dominating sets $D_2, \ldots, D_k$, at least one of these dominating sets contains no vertex from the edge e_{12}. However such a set dominates no vertex in X_{12}, a contradiction. Hence, H contains at most $k - 1$ vertex-disjoint dominating sets. $\square$

2.2 The Relationship Between Domination and Transversal

In this section, we discuss an interplay between domination in hypergraphs and transversals in hypergraphs established by Bujtás, Henning, and Tuza [8]. If H is a 2-uniform hypergraph, that is, when H is a graph, then it is well-known that every transversal in H is a dominating set of H, implying that $\gamma(H) \leq \tau(H)$. More generally, every transversal in an arbitrary hypergraph H without isolated vertices is a dominating set of H, implying that $\gamma(H) \leq \tau(H)$. We state this formally as follows.

Observation 4 *If H is a hypergraph without isolated vertices, then $\gamma(H) \leq \tau(H)$.*

We shall need the following definition and lemma from [8].

Definition 1 *For $k \geq 2$ an integer, let $\mathcal{H}_k$ denote the class of all k-uniform hypergraphs H with $\delta(H) \geq 1$. We note that if $H \in \mathcal{H}_k$, then $\displaystyle\sum_{v \in V} d_H(v) = k|E| = km_H$.*

Lemma 5 *([8]) For $k \in \mathbb{N}$ and $a, b \in \mathbb{R}$, the following hold:*

(a) *For every $H \in \mathcal{H}_k$, for $k \geq 2$, $an_H + bm_H > 0$ if and only if $b \geq 0$ and $a > -\frac{b}{k}$.*
(b) *For every $H \in \mathcal{H}_{k-1}$, for $k \geq 3$, $an_H + (a+b)m_H > 0$ if and only if $b \geq 0$ and $a > -\frac{b}{k}$.*

Proof. Suppose that $an_H + bm_H > 0$ for some $k \geq 2$ and for every $H \in \mathcal{H}_k$. The k-uniform complete hypergraph H of order $n_H = \ell$ and size $m_H = \binom{\ell}{k}$ gives $a\ell + b\binom{\ell}{k} > 0$, which implies that $b \geq 0$ as $\ell \to \infty$. The k-uniform hypergraph with exactly one edge of order $n_H = k$ and $m_H = 1$ gives $a \cdot k + b \cdot 1 > 0$, and so $a > -\frac{b}{k}$. Hence, both conditions $b \geq 0$ and $a > -\frac{b}{k}$ are necessary in part (a).

To prove the sufficiency part in (a), suppose that $b \geq 0$ and $-\frac{b}{k} < a$. We note that for every hypergraph $H \in \mathcal{H}_k$ we have by definition that $\delta(H) \geq 1$, and so $m_H = \frac{1}{k} \sum_{v \in V(H)} d_H(v) \geq \frac{1}{k} \cdot n_H$, or, equivalently, that $n_H \leq km_H$ holds. If $a \leq 0$, this implies that $an_H + bm_H \geq (ak + b)m_H > 0$. If $a > 0$, then trivially $an_H + bm_H > 0$ holds noting that $b \geq 0$ by supposition. This proves part (a).

To prove part (b), suppose that $k \geq 3$, $b \geq 0$, and $-\frac{b}{k} < a$. Similarly as before, $n_H \leq (k-1)m_H$ holds for every hypergraph $H \in \mathcal{H}_{k-1}$ with $k \geq 3$. If $a \leq 0$, this implies that $an_H + (a+b)m_H \geq (a(k-1) + (a+b))m_H = (ak+b)m_H > 0$. If $a > 0$, then trivially $an_H + (a+b)m_H > 0$ holds noting that $b \geq 0$ by supposition. This proves part (b). $\square$

We are now in a position to prove the following key relationship between the transversal number and the domination number of uniform hypergraphs.

Theorem 6 *([8]) For every integer $k \geq 3$ and for $a, b \in \mathbb{R}$ satisfying $b \geq 0$ and $a > -\frac{b}{k}$, the following equality holds:*

$$\sup_{H \in \mathcal{H}_k} \frac{\gamma(H)}{an_H + bm_H} = \sup_{H \in \mathcal{H}_{k-1}} \frac{\tau(H)}{an_H + (a+b)m_H}.$$

Proof. Let the parameters a, b, and k be fixed, where $k \geq 3$ is an integer and $a, b \in \mathbb{R}$ satisfy $b \geq 0$ and $a > -\frac{b}{k}$. We define

$$g_k = \sup_{H \in \mathcal{H}_k} \frac{\gamma(H)}{an_H + bm_H} \qquad \text{and} \qquad t_{k-1} = \sup_{H \in \mathcal{H}_{k-1}} \frac{\tau(H)}{an_H + (a+b)m_H}.$$

By Lemma 5, we have $an_H + bm_H > 0$ for every $H \in \mathcal{H}_k$ and $an_H + (a+b)m_H > 0$ for every $H \in \mathcal{H}_{k-1}$. We show first that $t_{k-1} \leq g_k$. Let H be an arbitrary hypergraph in the family $\mathcal{H}_{k-1}$, and so H is a $(k-1)$-uniform hypergraph and $\delta(H) \geq 1$. Let H have vertex set $V(H) = \{v_1, \ldots, v_n\}$ and edge set $E(H) = \{e_1, \ldots, e_m\}$. Let $H' \in \mathcal{H}_k$ be the k-uniform hypergraph constructed from H by adding to it m new vertices $u_1, \ldots, u_m$ and extending each edge e_i to the edge $e_i' = e_i \cup \{u_i\}$ for $i \in [m]$; that is, $V(H') = V(H) \cup \{u_1, \ldots, u_m\}$ and $E(H') = \{e_i' : i \in [m]\}$. We note that H' has $n' = n + m$ vertices and $m' = m$ edges.

Every dominating set of H' contains at least one vertex from every edge of H' in order to dominate the newly added vertices of degree 1, and so $\tau(H') \leq \gamma(H')$. By Observation 4, $\gamma(H') \leq \tau(H')$. Consequently, $\gamma(H') = \tau(H')$ holds. We also note that every transversal of H remains a transversal of H', and so $\tau(H') \leq \tau(H)$. Further we can always choose a τ-transversal T' of H' to contain no added vertices of degree 1 since if T' contains an added vertex u_i for some $i \in [m]$, then we can simply replace u_i in T' with a vertex in the edge e_i. Thus, T' is also a transversal of H, and so $\tau(H) \leq |T'| = \tau(H')$. Consequently, $\tau(H) = \tau(H')$. As observed earlier, $\tau(H') = \gamma(H')$, and so $\tau(H) = \gamma(H')$. By the definition of g_k, we therefore have

$$\tau(H) = \gamma(H') \leq g_k(an' + bm') = g_k(an + (a+b)m)$$

and hence

$$\frac{\tau(H)}{an + (a+b)m} \leq g_k$$

holds for every hypergraph $H \in \mathcal{H}_{k-1}$. Equivalently, $t_{k-1} \leq g_k$.

To prove the converse relation $t_{k-1} \geq g_k$, let F be an arbitrary hypergraph in the family $\mathcal{H}_k$, and so F is a k-uniform hypergraph and $\delta(F) \geq 1$. Let F have order n and size m. Let F' be a hypergraph obtained from F by successively deleting edges of F that do not contain any vertices of degree 1 in the resulting hypergraph at each stage. We note that F' is a k-uniform hypergraph with $n' = n$ vertices and $m' \leq m$ edges. When F is transformed into F', isolated vertices cannot arise, and so $F' \in \mathcal{H}_k$. Since removing edges cannot decrease the domination number, we note that $\gamma(F) \leq \gamma(F')$. Moreover, every edge of F' contains at least one vertex of degree 1 and hence $\tau(F') = \gamma(F')$. Consequently, $\tau(F') \geq \gamma(F)$. Deleting exactly one vertex of degree 1 from each edge of F', we obtain a $(k-1)$-uniform hypergraph F'' of order $n'' = n' - m'$ and of size $m'' = m'$ such that the transversal number remains unchanged. Thus,

$$\gamma(F) \leq \tau(F') = \tau(F'') \leq t_{k-1}(an'' + (a+b)m'') = t_{k-1}(an' + bm') \leq t_{k-1}(an + bm)$$

and hence

$$\frac{\gamma(F)}{an + (a+b)m} \le t_{k-1}$$

holds for every hypergraph $F \in \mathcal{H}_k$. Equivalently, $g_k \le t_{k-1}$. As shown earlier, $t_{k-1} \le g_k$. Consequently, $g_k = t_{k-1}$. $\square$

We show next that the uniformity condition in Theorem 6 can be relaxed. For this purpose, we shall need the following definition from [8].

Definition 2 *For an integer $k \ge 2$, let $\mathcal{H}_k^+$ ($\mathcal{H}_k^-$) denote the class of all hypergraphs H with $\delta(H) \ge 1$, in which every edge is of size at least k (at most k, respectively) .*

As shown in [8], every valid upper bound on the domination number of k-uniform hypergraphs of the form $\gamma(H) \le an_H + bm_H$ can be extended to hypergraphs with a less strict condition on edge sizes. The exact formulation will depend, however, on the sign of a. For $a \ge 0$, the result of Proposition 7 applies, while for $a \le 0$, the result of Proposition 8 applies. We state these results without proof.

Proposition 7 ([8]) *For any two nonnegative reals a and b (with $a + b > 0$) and for every integer $k \ge 2$, the following equality holds:*

$$\sup_{H \in \mathcal{H}_k^+} \frac{\gamma(H)}{an_H + bm_H} = \sup_{H \in \mathcal{H}_k} \frac{\gamma(H)}{an_H + bm_H}.$$

Proposition 8 ([8]) *For every integer $k \ge 2$ and for any two reals $a \le 0$ and $b > 0$, if $a > -\frac{b}{k}$, then the following equality holds:*

$$\sup_{H \in \mathcal{H}_k^-} \frac{\gamma(H)}{an_H + bm_H} = \sup_{H \in \mathcal{H}_k} \frac{\gamma(H)}{an_H + bm_H}.$$

2.3   Upper Bounds on the Domination Number

As a consequence of Theorem 6 and known results on the transversal number of a hypergraph, we establish in this section upper bounds on the domination number of a k-uniform hypergraph with minimum degree at least one. As remarked in [8], as a consequence of Theorem 6 the following two-way correspondence is obtained.

- For $k \ge 3$, if we have a general bound on the transversal number of the form $\tau(H) \le c_1 n_H + c_2 m_H$ with $-\frac{c_2}{k-1} < c_1 \le c_2$ for all $(k-1)$-uniform hypergraphs $H \in \mathcal{H}_{k-1}$, then the inequality $\gamma(H) \le c_1 n_H + (c_2 - c_1)m_H$ on the domination number necessarily holds for every k-uniform hypergraph $H \in \mathcal{H}_k$. Moreover, if the former bound is sharp, then the latter one is sharp, as well.

- For $k \geq 3$, similarly from every valid upper bound on the domination number of the form $\gamma(H) \leq an_H + bm_H$ for all k-uniform hypergraphs $H \in \mathcal{H}_k$ with real numbers $b \geq 0$, $a > -\frac{b}{k}$, we can derive the upper bound on the transversal number of the form $\tau(H) \leq an_H + (a+b)m_H$ for all $(k-1)$-uniform hypergraphs $H \in \mathcal{H}_{k-1}$. Moreover, if the former bound is sharp, then the latter one is sharp, as well.

We next present the results on the transversal number. For $k \geq 2$ an integer, the so-called *Tuza constant* c_k is defined by

$$c_k = \sup \frac{\tau(H)}{n_H + m_H},$$

where the supremum ranges over all k-uniform hypergraphs H. Erdős and Tuza [16, p. 1180] showed that $c_2 = \frac{1}{3}$. Chvátal and McDiarmid [11] and Tuza [43] independently established that $c_3 = \frac{1}{4}$, while Lai and Chang [38] showed that $c_4 = \frac{2}{9}$. We summarize these results below.

Theorem 9 *The following hold:*

(a) ([16, p. 1180]) $c_2 = \frac{1}{3}$.
(b) ([11, 43]) $c_3 = \frac{1}{4}$.
(c) ([38]) $c_4 = \frac{2}{9}$.

The precise value of c_k has yet to be determined for any values of k, with $k \geq 5$, some 30 years after the Tuza constants c_k were first introduced and studied. Applying probabilistic arguments, Alon [6] determined the asymptotic behavior of c_k as k grows.

Theorem 10 (Alon [6]) $c_k = (1 + o(1)) \left(\dfrac{\ln(k)}{k} \right)$ *as* $k \to \infty$.

Chvátal and McDiarmid [11] established the following upper bound on the transversal number of a uniform hypergraph in terms of its order and size.

Theorem 11 ([11]) *For $k \geq 2$, if H is a k-uniform hypergraph, then*

$$\tau(H) \leq \frac{n_H + \left\lfloor \frac{k}{2} \right\rfloor m_H}{\left\lfloor \frac{3k}{2} \right\rfloor}.$$

As a consequence of Alon's result in Theorem 10 and the relation established in Theorem 6 (with $b = 0$), we have the following asymptotic equality.

Theorem 12 ([8]) *As k tends to infinity,*

$$\sup_{H \in \mathcal{H}_k} \frac{\gamma(H)}{n_H} = (1 + o(1)) \left(\frac{\ln(k-1)}{k-1} \right) = (1 + o(1)) \left(\frac{\ln(k)}{k} \right).$$

Moreover, as an immediate consequence of the Chvátal–McDiarmid result in Theorem 11 and the relation given in Theorem 6, we obtain the following upper bound on the domination number of a uniform hypergraph without isolated edges. Recall that for $k \geq 2$, we denote by $\mathcal{H}_k$ the class of all k-uniform hypergraphs H with $\delta(H) \geq 1$.

Theorem 13 ([8]) *For $k \geq 3$, if $H \in \mathcal{H}_k$, then*

$$\gamma(H) \leq \frac{n_H + \left\lfloor \frac{k-3}{2} \right\rfloor m_H}{\left\lfloor \frac{3(k-1)}{2} \right\rfloor},$$

and this bound is sharp.

We remark that the k-uniformity condition in Theorem 13 can be relaxed to edge sizes at least k. We state this formally as follows.

Theorem 14 ([25]) *For $k \geq 3$, if H is a hypergraph with all edges of size at least k and with $\delta(H) \geq 1$, then*

$$\gamma(H) \leq \frac{n_H + \left\lfloor \frac{k-3}{2} \right\rfloor m_H}{\left\lfloor \frac{3(k-1)}{2} \right\rfloor}.$$

As an immediate consequence of Theorem 9 and the relation given in Theorem 6, we obtain the following upper bounds on the domination number of a hypergraph without isolated vertices and with edges sizes at least k where $k \in \{3, 4, 5\}$.

Theorem 15 *If H is a hypergraph with all edges of size at least k and with $\delta(H) \geq 1$, then the following hold:*

(a) *If $k = 3$, then $\gamma(H) \leq \frac{1}{3} n_H$.*

(b) *If $k = 4$, then $\gamma(H) \leq \frac{1}{4} n_H$.*

(c) *If $k = 5$, then $\gamma(H) \leq \frac{2}{9} n_H$.*

We remark that the result of Theorem 15(a) and 15(b) can also be deduced from a result in [26], which states that for $k \in \{3, 4\}$, if every edge in a graph G without isolated vertices and of order n is contained in a clique K_k, then $\gamma(G) \leq \frac{1}{k} n$.

2.4 Edge Size at Least Three

In this section, we present a characterization, due to Henning and Löwenstein [22], of the hypergraphs that achieve equality in the upper bound for the domination number given in Theorem 15(a). For this purpose, let $H_1, H_2, \ldots, H_{15}$ be the fifteen

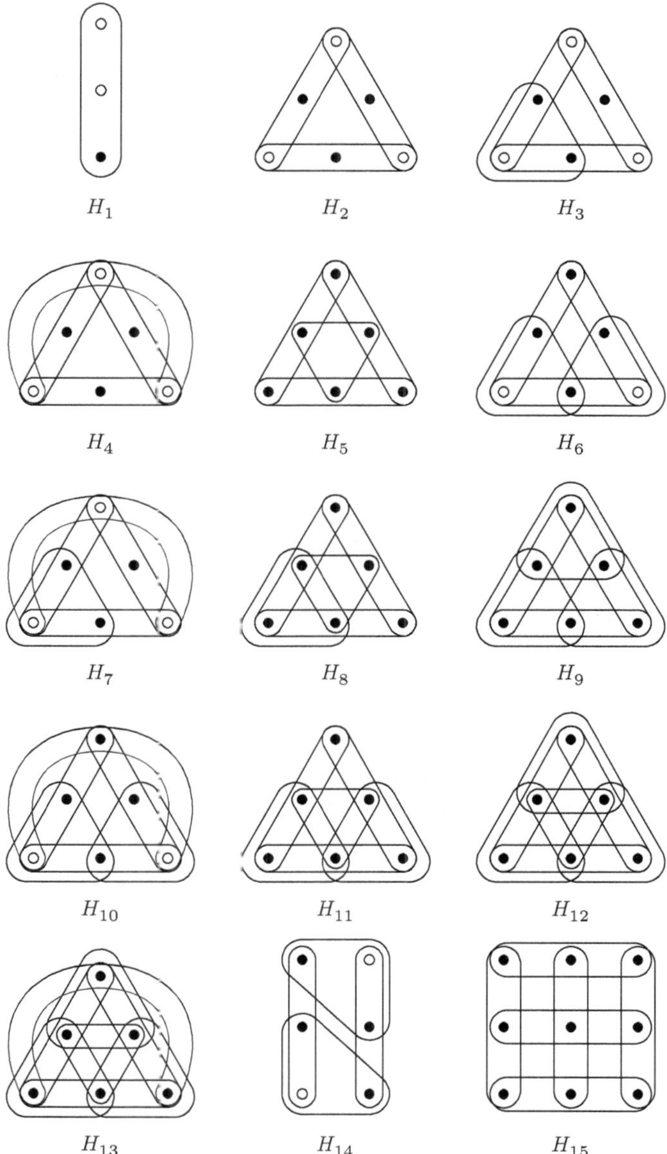

Fig. 1 The hypergraphs $H_1, H_2, \ldots, H_{15}$

hypergraphs shown in Figure 1. Let H_{under}, standing for *underlying hypergraph*, be a hypergraph every component of which is isomorphic to a hypergraph H_i for some $i \in [15]$. Each component of H_{under} we call a *unit* of H_{under}. In each unit, we 2-color the vertices with the colors black and white as indicated in Figure 1, and we call the white vertices the *link vertices* of the unit and the black vertices the *nonlink vertices*.

We now add edges between the units as follows. Let H be a hypergraph obtained from H_{under} by adding edges of size at least three, called *link edges*, in such a way that every added edge contains vertices from at least two units and contains only link vertices. Possibly, H is disconnected or $H = H_i$ for some $i \in [15]$. We call the hypergraph H_{under} an *underlying hypergraph* of H, and we let $\mathcal{U}(H_{under})$ denote the set of all units in H_{under}. Let $\mathcal{F}_{\geq 3}$ denote the family of all such hypergraphs H.

We are now in a position to present the characterization due to Henning and Löwenstein [22] of the hypergraphs, without isolated vertices and with all edges of size at least three, whose domination number is one-third their order.

Theorem 16 ([22]) *If H is a hypergraph with all edges of size at least three and with $\delta(H) \geq 1$, then $\gamma(H) \leq \frac{1}{3}n_H$ with equality if and only if $H \in \mathcal{F}_{\geq 3}$.*

2.5　Edge Size at Least Four

It remains an open problem (see Problem 7 in Chapter 4) to characterize the hypergraphs that achieve equality in the upper bound for the domination number given in Theorem 15(b). We remark that due to the interplay between domination and transversals in hypergraphs discussed in Section 2.2, a characterization of hypergraphs $H \in \mathcal{H}_4$ satisfying $\gamma(H) = \frac{1}{4}n_H$ builds on a characterization of the 3-uniform hypergraphs H satisfying $\tau(H) = \frac{1}{4}(n_H + m_H)$. A characterization of the extremal connected hypergraphs that achieve equality in the Chvátal–McDiarmid Theorem 11 for $k = 2$ and for all $k \geq 4$ is a relatively simple task. As shown in [24], there exist two such hypergraphs when k is even and one such hypergraph when k is odd. Surprisingly the case for $k = 3$ is much more challenging. The infinite extremal connected hypergraphs in this case were characterized by Henning and Yeo [27]. However, it remains an open problem to deduce the hypergraphs that achieve equality in Theorem 15(b) from the characterization in [27].

2.6　Edge Size at Least Five

In this section, we present a characterization due to Henning and Löwenstein [25] of the hypergraphs that achieve equality in the upper bound of Theorem 15(c). For this purpose, let H_9 be the hypergraph shown in Figure 2. Let H_{under} be a hypergraph

Fig. 2 The hypergraph H_9

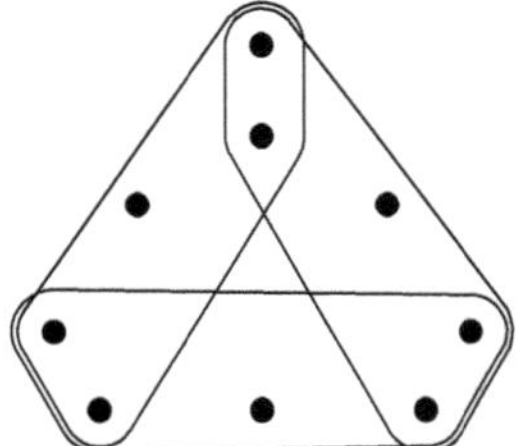

every component of which is isomorphic to H_9. Let H be a hypergraph obtained from H_{under} by adding edges of size at least five, called *link edges*, in such a way that every added edge contains only vertices of degree 2 in H_{under}. Possibly, H is disconnected or $H = H_9$. We call the hypergraph H_{under} an *underlying hypergraph* of H. Let $\mathcal{F}_{\geq 5}$ denote the family of all such hypergraphs H.

We are now in a position to present the characterization due to Henning and Löwenstein [25] of the hypergraphs with no isolated edge and with all edges of size at least five whose domination number is two-ninths their order.

Theorem 17 ([25]) *If H is a hypergraph with all edges of size at least five and with $\delta(H) \geq 1$, then $\gamma(H) \leq \frac{2}{9} n_H$ with equality if and only if $H \in \mathcal{F}_{\geq 5}$.*

2.7 A Characterization of Hypergraphs Achieving Equality in Theorem 14

In this section, we present a characterization due to Henning and Löwenstein [25] of the hypergraphs that achieve equality in the upper bound for the domination number given in Theorem 14. For this purpose, we first define some special hypergraphs.

For $k \geq 4$, let E_k denote the k-uniform hypergraph on k vertices with exactly one edge. The hypergraph E_4 is illustrated in Figure 3. For $k \geq 4$, the k-uniform hypergraph T_k is defined in [25] as follows. Let A, B, C, and D be vertex-disjoint sets of vertices with $|A| = \lceil k/2 \rceil$, $|B| = |C| = \lfloor k/2 \rfloor$, and $|D| = \lceil k/2 \rceil - \lfloor k/2 \rfloor$. In particular, if k is even, the set $D = \varnothing$, while if k is odd, the set D consist of a singleton vertex. Let T_k denote the k-uniform hypergraph with $V(T_k) = A \cup B \cup C \cup D$ and with $E(T_k) = \{e_1, e_2, e_3\}$, where $V(e_1) = A \cup B$, $V(e_2) = A \cup C$, and $V(e_3) = B \cup C \cup D$. The hypergraphs T_4 and T_5 are illustrated in Figure 3.

For odd $k \geq 5$, the hypergraph T_k^* is defined in [25] as follows. Let A, B, and C be vertex-disjoint sets of vertices with $|A| = |B| = (k+1)/2$ and $|C| = (k-1)/2$. Let T_k^* denote the hypergraph with $V(T_k^*) = A \cup B \cup C$ and with $E(T_k^*) = \{e_1, e_2, e_3\}$, where $V(e_1) = A \cup B$, $V(e_2) = A \cup C$, and $V(e_3) = B \cup C$. The hypergraph T_5^* is illustrated in Figure 3. We note that every edge in T_k^* has size at least k.

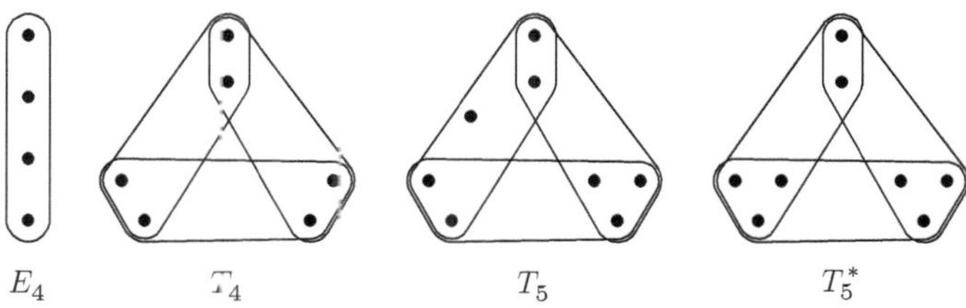

Fig. 3 The hypergraphs E_4, T_4, T_5, and T_5^*

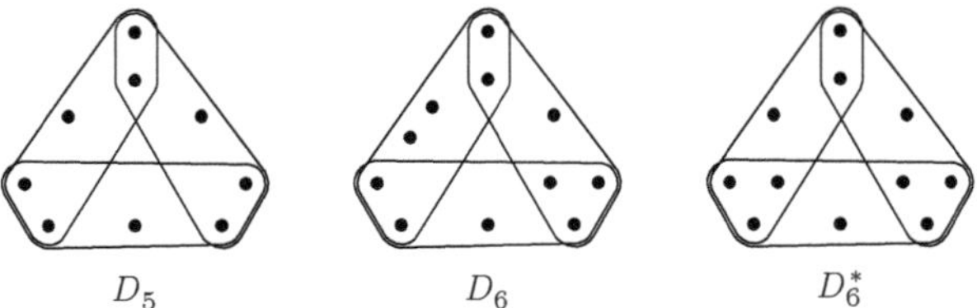

Fig. 4 The hypergraphs D_5, D_6, and D_6^*

The *expanded hypergraph*, abbreviated expa(H), of a hypergraph H is defined in [25] as the hypergraph obtained from H by expanding every edge in H by adding to it one new vertex, where all added vertices have degree 1 in expa(H). Thus for every edge $e \in E(H)$, if v_e denotes the new vertex added to e, where $v_e \neq v_f$ for edges $e \neq f$ in H, then expa(H) has edge set $\{e \cup \{v_e\} : e \in E(H)\}$ and vertex set

$$V(H) \cup \bigcup_{e \in E(H)} \{v_e\}.$$

For $k \geq 5$, let $D_k = \text{expa}(T_{k-1})$. The hypergraphs D_5 and D_6 are illustrated in Figure 4. We note that D_k is k-uniform. For even $k \geq 6$, let $D_k^* = \text{expa}(T_{k-1}^*)$. The hypergraph D_6^* is illustrated in Figure 4. We note that every edge in D_k^* is of size at least k.

The authors in [25] define a special family $\mathcal{D}_k$ of hypergraphs as follows. For odd $k \geq 5$, they define $\mathcal{D}_k = \{E_k, D_k\}$, and for even $k \geq 6$, they define $\mathcal{D}_k = \{E_k, D_k, D_k^*\}$.

We are now in a position to present the characterization given in [25] of the hypergraphs that achieve equality in the upper bound for the domination number given in Theorem 14.

Theorem 18 ([25]) *For $k \geq 3$, if H is a hypergraph with all edges of size at least k and with $\delta(H) \geq 1$, then*

$$\gamma(H) \leq \frac{n_H + \left\lfloor \frac{k-3}{2} \right\rfloor m_H}{\left\lfloor \frac{3(k-1)}{2} \right\rfloor},$$

with equality if and only if $H \in \mathcal{D}_k$.

2.8 General Setting

In the previous sections, we determined upper bounds of the form $an_H + bm_H$ on the domination number of k-uniform hypergraphs. In particular, we proved in Lemma 5 that $b \geq 0$ and $a > -\frac{b}{k}$ must hold in every valid upper bound. In this setting, we next present a general way to formulate the problem as described by Bujtás, Henning, and Tuza [8].

Problem 1 ([8]) *Given an integer $k \geq 2$, determine the shape of the surface $\Gamma_k(x, y, z)$, which is the subset of $D_k = \{(x, y, z) \mid y \geq 0 \ \wedge \ x > -y/k\} \subset \mathbb{R}^3$ defined by the rule*

$$z = \sup_{H \in \mathcal{H}_k} \frac{\gamma(H)}{x n_H + y m_H}.$$

In other words, for k given, determine $z = z(x, y)$ as a function of x and y.

The hypergraph H with $n_H = k$ vertices and $m_H = 1$ edge of size k has $\gamma(H) = 1$. Hence, we have the following simple general lower bound first observed in [8].

Observation 19 ([8]) *For every integer $k \geq 2$ and reals $y \geq 0$ and $x > -y/k$,*

$$z(x, y) \geq \frac{1}{kx + y}.$$

Bujtás, Henning, and Tuza [8] gave a complete solution for Problem 1 for $k \in \{2, 3\}$, showing that Observation 19 holds with equality in this case.

Theorem 20 ([8]) *For $k \in \{2, 3\}$, the surface $\Gamma_k(x, y, z)$ is determined by*

$$z(x, y) = \frac{1}{kx + y}.$$

An equivalent formulation of Theorem 20 gives us the following general upper bound on the domination number of a k-uniform hypergraph for $k \in \{2, 3\}$.

Theorem 21 ([8]) *For $k = 2$ and $k = 3$, the bound*

$$\gamma(H) \leq a n_H + b m_H$$

is valid for every k-uniform hypergraph H, which does not contain isolated vertices, if and only if both $ka + b \geq 1$ and $b \geq 0$ hold.

For example, to illustrate Theorem 21, let H be a k-uniform hypergraph without isolated vertices. For $k \in \{2, 3\}$ and taking $(a, b) = (\frac{1}{k}, 0)$, we have that $\gamma(H) \leq \frac{1}{k} m_H$. For $k = 2$, this yields the classical result due to Ore [40], while for $k = 3$, this yields the result of Theorem 15(a).

As remarked in [8], the general description of $\Gamma_k(x, y, z)$ in Problem 1 appears to be a rather hard problem, already for $k = 4$. Indeed it remains an open problem to give a complete solution for Problem 1 for any value of $k \geq 4$. In particular, it remains an open problem to determine whether the result of Theorem 21 also holds for $k = 4$. We know that the result of Theorem 21 does not hold for $k = 5$. We pose two open questions in these cases when $k = 4$ and $k = 5$ in Section 4.

2.9 *Hypergraphs with Given Domination Number*

In this section, we address the problem of finding the minimum number of vertices
that a connected k-uniform hypergraph with high domination number must contain.
Since all isolated vertices always are contained in every dominating set, we can
simply delete them and restrict our attention to hypergraphs without isolates. Let
$n(k, \gamma)$ be the minimum number of vertices that a k-uniform hypergraph with no
isolated vertices must contain if its domination number is at least γ.

2.9.1 The Case $\gamma = 1$

We observe that the k-uniform hypergraph with exactly one edge, which we denoted
by E_k in Section 2.7, has order k and is the smallest such hypergraph with $\gamma = 1$.
We state this trivial case formally as follows.

Observation 22 *For $k \geq 2$, we have $n(k, 1) = k$.*

2.9.2 The Case $\gamma = 2$

By a classical result due to Ore [40], if H is 2-uniform hypergraph without isolated
vertices, then $\gamma(H) \leq \frac{1}{2}n_H$. By Theorem 15 for $k \in \{3, 4\}$, if H is a k-uniform
hypergraph without isolated vertices, then $\gamma(H) \leq \frac{1}{k}n_H$. In particular, if $k \in \{2, 3,$
$4\}$ and $\gamma(H) = 2$, then $n_H \geq 2k$. If $k = 2$, then let F_2 be the 4-cycle, and let F_3 and
F_4 be the hypergraph shown in Figure 5(a) and 5(b), respectively. For $k \in \{2, 3, 4\}$,
if $H = F_k$, then $\gamma(H) = 2$ and $n_H = 2k$. Hence, we have the following result for
small k.

Observation 23 *For $k \in \{2, 3, 4\}$, we have $n(k, 2) = 2k$.*

The general case when $\gamma = 2$ was studied by Erdős, Henning, and Swart [17],
albeit in a graph theory setting. In order to state the result in [17], for each integer
$r \geq 2$, let I_r be the set of integers in the interval $I_r = [r^2 - r + 2, r^2 + r + 1]$. We
note further that if $(r - 1)^2 \leq k \leq (r + 1)^2$, then either $(r - 1)^2 \leq k \leq r^2 - r + 1$,

Fig. 5 The hypergraphs F_3
and F_4

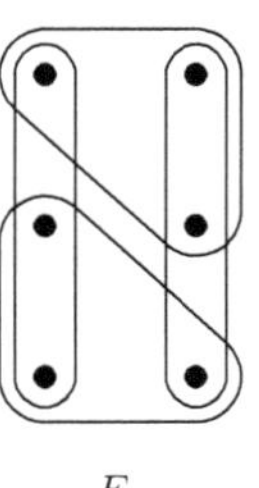

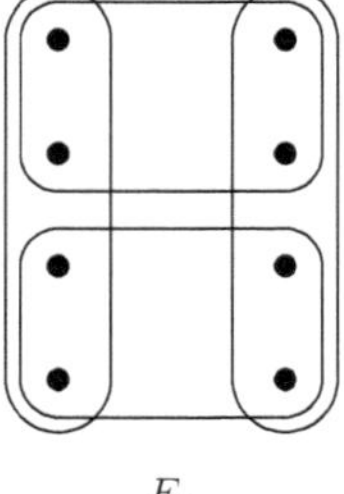

$$F_3 \qquad\qquad\qquad\qquad F_4$$

Table 1 Values of $n(k, 2)$ for small k

k	4	5	6	7	8	9	10	11	12	13
n(k,2)	8	9	11	12	14	15	16	18	19	20

in which case $k \in I_{r-1}$, or $r^2 - r + 2 \le k \le r^2 + r + 1$, in which case $k \in I_r$, or $r^2 + r + 2 \le k \le (r + 1)^2$, in which case $k \in I_{r+1}$. Hence, each integer $k \ge 4$ belongs to a unique interval I_r for some $r \ge 2$. Recall that $[r] = \{1, \ldots, r\}$ and $[r]_0 = \{0, 1, \ldots, r\}$. We note that if $k \in I_r$, then either $k = r^2 + 1 - i$ where $i \in [r - 1]_0$ or $k = r^2 + 1 + i$ where $i \in [r]$. We are now in a position to state the result in [17].

Theorem 24 ([17]) *If $k \ge 4$ is an arbitrary integer, then k belongs to a unique interval I_r for some $r \ge 2$, and the following holds:*

$$n(k, 2) = \begin{cases} (r + 1)^2 - i & \text{if} \quad k = r^2 + 1 - i \text{ and } i \in [r - 1]_0 \\ (r + 1)^2 + 1 + i & \text{if} \quad k = r^2 + 1 + i \text{ and } i \in [r]. \end{cases}$$

For example, in the special cases when $k \in I_2 = \{4, 5, 6, 7\}$ and $k \in I_3 = \{8, 9, \ldots, 13\}$ we summarize the results of Theorem 24 in Table 1 above.

For each integer $k \ge 4$, we next present an example of a k-uniform hypergraph H with $\gamma(H) = 2$ and of order $n_H = n(k, 2)$ given by the expression in Theorem 24. Let $k \ge 4$ be fixed. By our earlier observations, the integer k belongs to a unique interval I_r for some $r \ge 2$, and either $k = r^2 + 1 - i$ where $i \in [r - 1]_0$ or $k = r^2 + 1 + i$ where $i \in [r]$. We consider the two cases in turn.

Case 1. $k = r^2 + 1 - i$ *where* $i \in [r - 1]_0$. In this case, for $j \in [r + 1]$, let A_j be a set of vertices defined as follows. If $i = 0$, let $|A_j| = r$ for $j \in [r + 1]$. If $i \in [r - 1]$, let

$$|A_j| = \begin{cases} r - 1 & \text{if} \quad j \in [i] \\ r & \text{if} \quad j \in [r + 1] \setminus [i]. \end{cases}$$

Further, let the sets A_j be pairwise disjoint sets, and let

$$A = \bigcup_{j=1}^{r+1} A_j \qquad \text{and} \qquad B = \{v_1, v_2, \ldots, v_{r+1}\},$$

where the vertices in B are $r + 1$ additional vertices that do not belong to the set A. Let v be an arbitrary vertex in A_{r+1}. Let H be the hypergraph with vertex set $V(H) = A \cup B$ of order $n_H = |A| + |B| = (r + 1)^2 - i$ and with edge set $E(H) = \{e_1, e_2, \ldots, e_{r+1}\}$ where the edge e_j for $j \in [r + 1]$ is defined as follows:

$$e_j = \begin{cases} (A \cup \{v_j\}) \setminus (A_j \cup \{v\}) & \text{if} \quad i \in [r - 1] \text{ and } j \in [i] \\ (A \cup \{v_j\}) \setminus A_j & \text{if} \quad i = 0 \text{ or } i \in [r - 1] \text{ and } j \in [r + 1] \setminus [i]. \end{cases}$$

By construction, we note that $|e_j| = r^2 + 1 - i = k$ for all $j \in [r+1]$, and so H is a k-uniform hypergraph. Since no vertex of H dominates the set B, we note that $\gamma(H) \geq 2$. However, if $u_r \in A_r$ and $u_{r+1} \in A_{r+1} \setminus \{v\}$, then the set $\{u_r, u_{r+1}\}$ is an example of a dominating set of H, and so $\gamma(H) \leq 2$. Consequently, H is a k-uniform hypergraph of order $n_H = (r+1)^2 - i$ without isolated vertices satisfying $\gamma(H) = 2$.

Case 2. $k = r^2 + 1 + i$ and $i \in [r]$. In this case, for $j \in [r+1]$, let A_j be a set of vertices defined as follows:

$$|A_j| = \begin{cases} r+1 & \text{if} \quad j \in [i+1] \\ r & \text{if} \quad i < r \text{ and } j \in [r+1] \setminus [i+1]. \end{cases}$$

Further, let the sets A_j be pairwise disjoint sets, and let

$$A = \bigcup_{j=1}^{r+1} A_j \quad \text{and} \quad B = \{v_1, v_2, \ldots, v_{r+1}\},$$

where the vertices in B are $r+1$ additional vertices that do not belong to the set A. Let v be an arbitrary vertex in A_1. Let H be the hypergraph with vertex set $V(H) = A \cup B$ of order $n_H = |A| + |B| = (r+1)^2 + 1 + i$ and with edge set $E(H) = \{e_1, e_2, \ldots, e_{r+1}\}$ where the edge e_j for $j \in [r+1]$ is defined as follows:

$$e_j = \begin{cases} (A \cup \{v_j\}) \setminus A_j & \text{if} \quad j \in [i+1] \\ (A \cup \{v_j\}) \setminus (A_j \cup \{v\}) & \text{if} \quad i < r \text{ and } j \in [r+1] \setminus [i+1]. \end{cases}$$

By construction, we note that $|e_j| = r^2 + 1 + i = k$ for all $j \in [r+1]$, and so H is a k-uniform hypergraph. Since no vertex of H dominates the set B, we note that $\gamma(H) \geq 2$. However, if $u_1 \in A_1 \setminus \{v\}$ and $u_2 \in A_2$, then the set $\{u_1, u_2\}$ is an example of a dominating set of H, and so $\gamma(H) \leq 2$. Consequently, H is a k-uniform hypergraph of order $n_H = (r+1)^2 + 1 + i$ without isolated vertices satisfying $\gamma(H) = 2$.

2.9.3 The Case $\gamma \geq 3$

The case when $\gamma \geq 3$ was addressed by Bujtás, Patkós, Tuza, and Vizer [10] in a more general setting. For an integer $s \geq 1$, they define an *s-dominating set* of a hypergraph H as a dominating set D of H with the property that every vertex outside the set D has at least s neighbors inside the set D; that is, $|N_H(v) \cap D| \geq s$ for all vertices $v \in V(H) \setminus D$, where $N_H(v)$ denotes the open neighborhood of v. The authors in [10] also define an *s-tuple dominating set* of H as a dominating set D of H with the property that every vertex of H either belongs to D and has at least $s - 1$ neighbors inside the set D or belongs outside the set D and has at least s neighbors inside the set D; that is, $|N_H[v] \cap D| \geq s$ for all vertices $v \in V(H)$, where $N_H[v]$ denotes the closed neighborhood of v. We note that dominating sets are precisely the 1-dominating sets and 1-tuple dominating sets.

The *s-domination number* $\gamma(H, s)$ of a hypergraph H is the minimum size of an s-dominating set in H, and the *s-tuple domination number* $\gamma_\times(H, s)$ of H is the minimum size of an s-tuple dominating set in H. By definition, we have $\gamma(H, s) \leq \gamma_\times(H, s)$. For every pair γ and s of integers with $\gamma \geq s$, let $n(k, \gamma, s)$ denote the minimum number of vertices that a k-uniform hypergraph H with no isolated vertices must have if $\gamma(H, s) \geq \gamma$ holds, and let $n_\times(k, \gamma, s)$ denote the minimum number of vertices that a k-uniform hypergraph H with no isolated vertices must have if $\gamma_\times(H, s) \geq \gamma$ holds. As observed in [10], we have $n_\times(k, \gamma, s) \leq n(k, \gamma, s)$ and $n(k, \gamma) = n_\times(k, \gamma, 1) = n(k, \gamma, 1)$.

We are now in a position to state the result due to Bujtás, Patkós, Tuza, and Vizer [10].

Theorem 25 ([10]) *For integers* $\gamma \geq 2$ *and* $s \geq 1$ *with* $\gamma > s$, *we have*

$$k + k^{1-1/(\gamma-s+1)} \leq n_\times(k, \gamma, s) \leq n(k, \gamma, s) \leq k + (4 + o(1))k^{1-1/(\gamma-s+1)}.$$

In the special case in Theorem 25 when $s = 1$, we have the following result.

Theorem 26 ([10]) *For integers* $k \geq 2$ *and* $\gamma \geq 2$, *we have*

$$k + k^{1-\frac{1}{\gamma}} \leq n(k, \gamma) \leq k + (4 + o(1))k^{1-\frac{1}{\gamma}}.$$

2.10 Nordhaus–Gaddum-Type Results

In this section, we present a Nordhaus–Gaddum-type result for the sum of domination parameters in hypergraphs and their complements. Given a hypergraph $H = (V, E)$, the *complement* $\overline{H}$ of H is the hypergraph $\overline{H} = (V, \overline{E})$ where $\overline{E} = \{V \setminus e : e \in E\}$. In 2008, Hedetniemi et al. [20] defined the disjoint domination number $\gamma\gamma(G)$ of a graph G as the minimum sum of the cardinalities of two disjoint dominating sets in G. The *disjoint domination number* $\gamma\gamma(H)$ of a hypergraph H is defined analogously as the minimum sum of the cardinalities of two disjoint dominating set in H; that is,

$$\gamma\gamma(H) = \min\{|D_1| + |D_2| : D_1, D_2 \text{ are minimal dominating sets in } H \text{ with } D_1 \cap D_2 = \emptyset\}.$$

Recall that a γ-set of a hypergraph H is a dominating set of H of cardinality $\gamma(H)$. An *inverse dominating set* with respect to a given γ-set D of H is a dominating set D' of H such that $D' \subseteq V(H) \setminus D$. The *inverse domination number* $\gamma^{-1}(H)$ is defined as

$$\gamma^{-1}(H) = \min\{|D'| : D' \text{ is an inverse dominating set with respect to some } \gamma\text{-set } D \text{ of } H\}.$$

We remark that the inverse domination number of a graph was first defined in 1991 by Kulli and Sigarkanti [37]. Acharya [4] posed the problem of finding best possible lower and upper bounds for $\gamma\gamma(H) + \gamma\gamma(\overline{H})$. This problem was subsequently solved by Jose and Tuza [34], who proved an even stronger statement for the upper bound.

Theorem 27 ([34]) *For every integer $n \geq 4$, if H is a hypergraph of order n, then*

$$4 \leq \gamma\gamma(H) + \gamma\gamma(\overline{H}) \leq \gamma(H) + \gamma^{-1}(H) + \gamma(\overline{H}) + \gamma^{-1}(\overline{H}) \leq \max\{8, n+2\},$$

and the bounds are tight.

That the bounds of Theorem 27 are tight may be seen as follows. For the lower bound, Jose and Tuza [34] constructed a hypergraph $H = (V, E)$ with n vertices as follows. Let $V = (V_1, V_2, V_3, V_4)$ be a partition of the set V into four nonempty sets, and let $E = \{V_i \cup V_j : 1 \leq i < j \leq 4\}$. We note that H and $\overline{H}$ are isomorphic hypergraphs, and all the vertices are adjacent to each other in both H and $\overline{H}$. Hence, $\gamma\gamma(H) = \gamma\gamma(\overline{H}) = 2$. Therefore the lower bound is attainable for all $n \geq 4$.

To prove tightness in the upper bound, suppose first that $4 \leq n \leq 6$. In this case, Jose and Tuza [34] constructed a hypergraph $H = (V, E)$ with n vertices by once again taking a partition of the set $V = (V_1, V_2, V_3, V_4)$ into four nonempty sets, but now letting $E = \{V_1 \cup V_2, V_2 \cup V_3, V_3 \cup V_4\}$. The edge set of $\overline{H} = (V, \overline{E})$ is $\overline{E} = \{V_1 \cup V_2, V_1 \cup V_4, V_3 \cup V_4\}$. Thus, $\gamma\gamma(H) = \gamma\gamma(\overline{H}) = 4$.

Suppose next that $n \geq 7$. In this case, Jose and Tuza [34] let H be a double star $S(\lfloor (n-2)/2 \rfloor, \lceil (n-2)/2 \rceil)$, which is a tree with two (adjacent) nonleaf vertices one of which has $\lfloor (n-2)/2 \rfloor$ leaf neighbors and the other $\lceil (n-2)/2 \rceil$ leaf neighbors. Let x and y be the two nonleaf vertices, and let X and Y be the set of leaf neighbors of x and y, respectively. We note that there are exactly four minimal dominating sets, namely $D_1 = \{x\} \cup Y$, $D_2 = \{y\} \cup X$, $D_3 = \{x, y\}$, and $D_4 = X \cup Y$, forming two disjoint pairs, namely (D_1, D_2) and (D_3, D_4). Since both pairs partition $V(H)$, we note that $\gamma\gamma(H) = n$. Further, we note that $\gamma\gamma(\overline{H}) = 2$. Thus, $\gamma\gamma(H) + \gamma\gamma(\overline{H}) = n + 2$.

2.11 *Equality of Domination and Transversal Numbers*

Every transversal in a hypergraph H without isolated vertices is a dominating set in H, implying that $\gamma(H) \leq \tau(H)$ is valid for every such hypergraph H. Arumugam, Jose, Bujtás, and Tuza [7] investigated the hypergraphs H without isolated vertices satisfying $\gamma(H) = \tau(H)$.

We first consider the special case when H is a 2-uniform hypergraph, that is, when H is a graph. A *stem*, also called a *support vertex* in the literature, is a vertex in a graph G that is adjacent to a vertex of degree 1. Let Stem(G) denote the set of stems in G. Let $\mathcal{S}(G)$ denote the graph obtained from the graph G by deleting all edges contained entirely in $\mathrm{Stem}(G)$. We note that the transformation $\mathcal{S}$ does not create isolated vertices, unless G contains a component isomorphic to K_2. Arumugam et al. [7] provided the following characterization of connected graphs satisfying $\gamma(G) = \tau(G)$.

Theorem 28 ([7]) *For a connected graph G of order at least 3, $\gamma(G) = \tau(G)$ holds if and only if there exists a bipartition (A, B) of $\mathcal{S}(G)$ such that $\mathrm{Stem}(G) \subseteq A$, and moreover, for every pair $u, v \in A \setminus \mathrm{Stem}(G)$, if u and v have some common neighbor, then they have at least two common neighbors of degree two.*

As a consequence of the characterization given in Theorem 28, Arumugam et al. [7] showed that graphs G without isolated vertices and with $\gamma(G) = \tau(G)$ can be recognized in polynomial time.

Theorem 29 ([7]) *It can be decided in time*

$$O\left(\sum_{v \in V(G)} (d_G(v))^2 \right)$$

whether an arbitrary graph G without isolated vertices satisfies $\gamma(G) = \tau(G)$.

In contrast, Arumugam et al. [7] proved that the corresponding problem is NP-hard for hypergraphs, even if edges of size greater than 3 are excluded.

Theorem 30 ([7]) *It is NP-hard to decide whether a generic input linear hypergraph H of rank 3 and of minimum degree 1 satisfies $\gamma(H) = \tau(H)$.*

As remarked in [7], "it is very likely that the problem of testing $\gamma(H) = \tau(H)$ is harder than any problem in NP."

Arumugam et al. [7] also present several structural results on hypergraphs in which each subhypergraph H' without isolated vertices fulfills the equality $\gamma(H') = \tau(H')$. They also investigate hypergraphs for which the equality $\gamma(H) = \tau(H)$ holds hereditarily. That is, the property is required not only for the hypergraph H itself but also for all of its subhypergraphs or for all of its induced subhypergraphs. For this purpose, they define the following notions:

(P1) For a hypergraph H, the equality $\gamma = \tau$ *hereditarily* holds if every subhypergraph H' of H satisfies $\gamma(H') = \tau(H')$.

(P2) For a hypergraph H, the equality $\gamma = \tau$ *induced-hereditarily* holds if every induced subhypergraph H' of H satisfies $\gamma(H') = \tau(H')$.

By definition, property (P1) implies property (P2). Arumugam et al. [7] prove that in fact properties (P1) and (P2) are equivalent.

Theorem 31 ([7]) *For a hypergraph H, the equality $\gamma = \tau$ holds hereditarily if and only if it holds induced-hereditarily.*

The following notions are also defined in [7]:

(P3) A hypergraph H is *minimal for $\gamma < \tau$* if $\gamma(H) < \tau(H)$, but for every proper subhypergraph H' of H (without isolated vertices) the equality $\gamma(H') = \tau(H')$ holds.

(P4) A hypergraph H is *induced-minimal for $\gamma < \tau$* if $\gamma(H) < \tau(H)$, but for every proper induced subhypergraph H' of H (without isolated vertices) the equality $\gamma(H') = \tau(H')$ holds.

By definition, property (P3) implies property (P4). However, properties (P3) and (P4) are not equivalent since there exist hypergraphs, which are induced-minimal but not minimal for $\gamma < \tau$.

Arumugam et al. [7] prove that for any given integer $k \geq 2$, the number of hypergraphs of rank k that are minimal or induced-minimal for $\gamma < \tau$ is bounded.

Theorem 32 ([7]) *For every fixed $k \geq 2$, the following holds:*

(a) *There exist only finitely many hypergraphs of rank k, which are minimal for $\gamma < \tau$.*

(b) *There exist only finitely many hypergraphs of rank k, which are induced-minimal for $\gamma < \tau$.*

2.12   The Relationship Between Domination and Matching

In this section, we present an upper bound on the domination number of a uniform hypergraph in terms of its matching number. A matching in a hypergraph is a set of disjoint edges. Thus, if M is a matching in a hypergraph H, then $e \cap f = \varnothing$ for every pair of edges e and f in M. The matching number, $\alpha'(H)$, of a hypergraph H is the maximum size of a matching in H.

In order to cover every edge of a hypergraph H, we note that $\tau(H) \geq \alpha'(H)$ since a transversal of H must contain at least one vertex from every edge in a maximum matching in H. If H is a k-uniform hypergraph, then the union of the edges of a maximum matching of H forms a transversal of H, implying that $\tau(H) \leq k\alpha'(H)$. We state this formally as follows.

Observation 33 *If H is a k-uniform hypergraph, then $\tau(H) \leq k\alpha'(H)$.*

A hypergraph is *k-partite* if its vertex set can be partitioned into k sets such that every edge contains exactly one vertex from each of these partite sets. In particular, a k-partite hypergraph is a k-uniform hypergraph. A long-standing open problem, known as *Ryser's conjecture*, states that if H is a k-partite hypergraph, then $\tau(H) \leq (k-1)\alpha'(H)$. When $k = 2$, this is the classical theorem of König. When $k = 3$, Ryser's conjecture was proven by Aharoni [5]. However, the conjecture remains open for $k \geq 4$.

Motivated by Ryser's conjecture, Kang, Li, Dong, and Shan [36] gave the following Ryser-like relation between the domination number and matching number of a uniform hypergraph. The proof is based on the approach presented in the proof of Theorem 6 by Bujtás et al. [9].

Theorem 34 ([36]) *For $k \geq 2$, if H is a k-uniform hypergraph without isolated vertices, then $\gamma(H) \leq (k-1)\alpha'(H)$.*

Proof. Let H' be the hypergraph obtained from H by successively deleting edges of H that do not contain any vertices of degree 1 in the resulting hypergraph at each stage. We note that H' is a k-uniform hypergraph with $n_{H'} = n_H$ vertices and $m_{H'} \leq m_H$ edges. When H is transformed into H', isolated vertices cannot arise. Since removing edges cannot decrease the domination number, we note that $\gamma(H) \leq \gamma(H')$. We also note that removing edges cannot increase the matching number, and so $\alpha'(H') \leq \alpha'(H)$. Moreover, every edge of H' contains at least one vertex of degree 1, and so $\tau(H') = \gamma(H')$. Consequently, $\tau(H') \geq \gamma(H)$.

Deleting exactly one vertex of degree 1 from each edge of H', we obtain a $(k-1)$-uniform hypergraph H'' of order $n_{H'} - m_{H'}$ and of size $m_{H'}$. When constructing H'' from H', the transversal number and the matching number of H'' remains unchanged, that is, $\tau(H'') = \tau(H')$ and $\alpha'(H'') = \alpha'(H')$. These observations, together with the result of Observation 33, imply that

$$\gamma(H) \leq \tau(H') = \tau(H'') \leq (k-1)\alpha'(H'') = (k-1)\alpha'(H') \leq (k-1)\alpha'(H).$$

This establishes the desired upper bound, namely $\gamma(H) \leq (k-1)\alpha'(H)$. $\square$

To show that the upper bound is achievable, we recall the definition of a finite projective plane. For $q = p^n \geq 2$, where some prime p and some integer $n \geq 1$, a *finite projective plane* of order q, denoted by $PG(q)$, consists of a set of $q^2 + q + 1$ points and the same number of lines, having the following properties:

- Every line contains $q + 1$ points.
- Every point lies in $q + 1$ lines.
- Any two distinct points lie on a unique line.
- Any two distinct lines intersect in exactly one point.
- There are four points such that no line is incident with more than two of them.

A finite projective plane may be considered as a hypergraph, whose vertex set is the set of points and whose edge set is the set of lines of the plane. The hypergraph associated with a finite projective plane $PG(2, q)$ is a $(q + 1)$-uniform hypergraph on $q^2 + q + 1$ vertices in which every two edges intersect in exactly one vertex. Kang, Li, Dong, and Shan [36] showed that when $k - 2$ is a prime power, the upper bound in Theorem 34 on the domination number is tight.

Suppose that $k - 2$ is a prime power and consider the hypergraph H associated with a finite projective plane $PG(k - 2)$. By our earlier observations, H is a $(k - 1)$-uniform hypergraph in which every two edges intersect in exactly one vertex. We note that $\alpha'(H) = 1$. Let H' be the k-uniform hypergraph obtained from H by adding to it m_H new vertices of degree 1, one vertex for each edge in H. Thus, each edge of H (of size $k - 1$) is extended to an edge of size k by adding to it one new vertex of degree 1. We note that H' has $n_{H'} = n_H + m_H$ vertices and $m_{H'} = m_H$ edges. Further we note that $\tau(H') = \tau(H)$ and $\alpha'(H') = \alpha'(H)$. Since every edge of H' contains a vertex of degree 1, we have $\gamma(H') = \tau(H') = \tau(H)$. Since H is the hypergraph associated with a finite projective plane, the set of minimum transversals in H consists precisely of the set of edges of H. Thus, since H is a $(k-1)$-uniform hypergraph, each τ-transversal in H is an edge of H, which has size $k-1$, and so $\tau(H) = k-1$. Hence, $\gamma(H') = \tau(H) = k-1 = (k-1)\cdot 1 = (k-1)\alpha'(H')$. Thus, when $k-2$ is a prime power, the upper bound in Theorem 34 on the domination number is tight.

Shan, Dong, Kang, and Li [41] observed that the inequality in Theorem 34 still holds for arbitrary hypergraphs of rank k. We state this formally as follows.

Corollary 35 ([36]) *For $k \geq 2$, if H is a hypergraph of rank k without isolated vertices, then $\gamma(H) \leq (k-1)\alpha'(H)$.*

Shan, Dong, Kang, and Li [41] give a complete characterization of the extremal hypergraphs H of rank 3 without isolated vertices satisfying $\gamma(H) = 2\alpha'(H)$.

As remarked in [36], for $k \geq 4$ a constructive characterization of hypergraphs of rank k without isolated vertices satisfying $\gamma(H) = (k-1)\alpha'(H)$ seems difficult to obtain, even in the special case when H is an *intersecting hypergraph*, that is, hypergraphs in which every two edges have a nonempty intersection. We note that if H is an intersecting hypergraph, then $\alpha'(H) = 1$.

Dong, Shan, Kang, and Li [14] present structural properties of intersecting hypergraphs H with no isolated vertices and with rank k satisfying the equality $\gamma(H) = k-1$. Their main result is that all linear intersecting hypergraphs H with no isolated vertices and with rank 4 such that $\gamma(H) = 3$ can be constructed from the Fano plane, where a linear hypergraph is one in which every two edges intersect in at most one vertex. It remains, however, an open problem to characterize the nonlinear intersecting hypergraphs H with no isolated vertices and with rank 4 satisfying $\gamma(H) = 3$.

Li, Kang, Shan, and Dong [39] show that all the 5-uniform linear intersecting hypergraphs H with no isolated vertices satisfying $\gamma(H) = 4$ are generated by the finite projective plane of order 3.

3 Total Domination in Hypergraphs

In this section, we focus on total domination in hypergraphs. We proceed as follows. We first establish a relationship between the total transversal number and the total domination number of uniform hypergraphs due to Bujtás, Henning, Tuza, and Yeo [9]. Using this interplay between total transversals and total domination in hypergraphs, we prove tight asymptotic upper bounds on the total transversal number in terms of the number of vertices, the number of edges, and the edge size. We shall need the following definitions given in [9].

Definition 3 ([9]) *For an integer $k \geq 2$, let $\mathcal{H}_k$ be the class of all k-uniform hypergraphs containing no isolated vertices or isolated edges or multiple edges. Further, for $k \geq 3$ let $\mathcal{H}_k^*$ consist of all hypergraphs in $\mathcal{H}_k$ that have no two edges intersecting in $k-1$ vertices. We note that $\mathcal{H}_k^*$ is a proper subclass of $\mathcal{H}_k$.*

Definition 4 ([9]) *For an integer $k \geq 2$, let*

$$b_k = \sup_{H \in \mathcal{H}_k} \frac{\tau_t(H)}{n_H + m_H}.$$

We remark that it is not known if the supremum in Definition 4 is a maximum. Bujtás, Henning, Tuza, and Yeo [9] proved the following upper bounds on the total domination number of a uniform hypergraph in terms of its total transversal number, order, and size.

Theorem 36 ([9]) *For $k \geq 3$, if $H \in \mathcal{H}_k$, then*

$$\gamma_t(H) \leq \left(\max \left\{ \frac{2}{k+1}, b_{k-1} \right\} \right) n_H.$$

In view of Theorem 36, it is of interest to determine the values of b_k for $k \geq 2$. The value of b_k for small k was determined in [9].

Theorem 37 ([9]) *The following hold:*

(a) $b_2 = \frac{2}{5}$ and $b_3 = \frac{1}{3}$.

(b) $b_4 \leq \frac{1}{3}$.

(c) $b_k \leq \frac{2}{7}$ for all $k \geq 5$.

Henning and Yeo [32] continued the study of total transversals in hypergraphs and proved the following result.

Theorem 38 ([32]) *The following hold:*

(a) $b_4 = \frac{2}{7}$.

(b) $b_6 \leq \frac{1}{4}$.

(c) $b_k \leq \frac{2}{9}$ for all $k \geq 7$.

By Theorems 37 and 38, we observe that

$$b_{k-1} \leq \frac{2}{k+1} \qquad \text{for } k \in \{3, 4, 5, 6, 7, 8\}.$$

Hence as a consequence of Theorem 36 and Theorem 37, and the well-known fact (see, [13]) that if $H \in \mathcal{H}_2$, then $\gamma_t(H) \leq \frac{2}{3} n_H$, we have the following result.

Theorem 39 ([9, 32]) *For $k \in [8] \setminus [1]$, if $H \in \mathcal{H}_k$, then*

$$\gamma_t(H) \le \left(\frac{2}{k+1} \right) n_H,$$

and this bound is sharp.

That the upper bound in Theorem 39 is sharp may be seen as follows. Let $\mathcal{F}_k \subset \mathcal{H}_k$ be the subfamily of hypergraphs in $\mathcal{H}_k$ that can be obtained as follows. Let F be an arbitrary hypergraph in the family $\mathcal{H}_k$. For each vertex v in F, add k new vertices $v_1, v_2, \ldots, v_k$ and two new k-edges $e_v = \{v, v_1, \ldots, v_{k-1}\}$ and $f_v = \{v_1, v_2, \ldots, v_k\}$. Let $H \in \mathcal{H}_k$ denote the resulting k-uniform hypergraph. As observed in [9], every total dominating set of H must contain at least two vertices from $e_v \cup f_v$ for every vertex $v \in V(F)$, implying that $\gamma_t(H) \ge 2n_F$. The set $\bigcup_{v \in V(F)} \{v, v_1\}$, where the union is taken over all vertices $v \in V(F)$, forms a total dominating set of H, and so $\gamma_t(H) \le 2n_F$. Consequently, $\gamma_t(H) = 2n_F$. Since $n_H = (k+1)n_F$, this yield the following result from [9], implying that the upper bound in Theorem 39 is tight.

Observation 40 ([9]) *For $k \ge 2$, if $H \in \mathcal{F}_k$, then $\gamma_t(H) = \left(\frac{2}{k+1} \right) n_H$.*

Bujtás et al. [9] proved the following result, which is a strengthening of the upper bound of Theorem 36 if we restrict the edges to intersect in at most $k-2$ vertices.

Theorem 41 ([9]) *For $k \ge 4$, if $H \in \mathcal{H}_k^*$, then*

$$\gamma_t(H) \le \left(\max \left\{ \frac{2}{k+2}, b_{k-1} \right\} \right) n_H.$$

Corollary 42 ([9]) *For $k \ge 4$, if $H \in \mathcal{H}_k^*$, then $\gamma_t(H) \le \frac{1}{3} n_H$.*

Bujtás et al. [9] established the following tight asymptotic bound on b_k for sufficiently large k.

Theorem 43 ([9]) *For all $k \ge 2$, $b_k = (1 + o(1)) \dfrac{\ln(k)}{k}$.*

Theorem 43 implies that the inequality $b_{k-1} \le \frac{2}{k+1}$ is not true when k is large enough.

By definition, $\gamma_t(H) \ge \gamma(H)$ for every hypergraph H without isolated vertices. Hence as a consequence of Theorem 36, Theorem 43, and Corollary 12 (established in Section 2.3), we have the following result.

Theorem 44 ([9]) *For all $k \ge 3$,*

$$\sup_{H \in \mathcal{H}_k} \frac{\gamma_t(H)}{n_H} = (1 + o(1)) \left(\frac{\ln(k)}{k} \right).$$

We remark that Theorem 44 implies that Theorem 39 is not true for large k.

In view of Theorem 36, it is of interest to determine the value of b_k for $k \geq 2$. In Theorems 37 and 38, we have that $b_2 = \frac{2}{5}$, $b_3 = \frac{1}{3}$, and $b_4 = \frac{2}{7}$. Further, $b_5 \leq \frac{2}{7}$, $b_6 \leq \frac{1}{4}$, and $b_k \leq \frac{2}{9}$ for all $k \geq 7$. Thus, $b_{k-1} \leq \frac{2}{k+1}$ for $k \in \{3, 4, 5, 6, 7, 8\}$.

4 Conjectures and Open Problems

This chapter presents an overview of research on domination and total domination in hypergraphs. Results not presented in this chapter can be found, for example, in [1, 2, 18, 19, 21, 31]. We close this chapter with a conjecture and a list of open problems for future research.

Problem 2 *Characterize the hypergraphs H with all edges of size at least 4 and with $\delta(H) \geq 1$ that achieve equality in the upper bound $\gamma(H) \leq \frac{1}{4}n_H$ given in Theorem 15.*

Problem 3 *([8]) Given an integer $k \geq 4$, determine the shape of the surface $\Gamma_k(x, y, z)$, which is the subset of $D_k = \{(x, y, z) \mid y \geq 0 \ \wedge \ x > -y/k\} \subset \mathbb{R}^3$ defined by the rule*

$$z = \sup_{H \in \mathcal{H}_k} \frac{\gamma(H)}{xn_H + ym_H}.$$

In other words, for $k \geq 4$ given, determine $z = z(x, y)$ as a function of x and y.

Problem 4 *Prove or disprove: The bound $\gamma(H) \leq an_H + bm_H$ is valid for every 4-uniform hypergraph H without isolated vertices, if and only if both $4a + b \geq 1$ and $b \geq 0$ hold.*

Problem 5 *Prove or disprove: The bound $\gamma(H) \leq an_H + bm_H$ is valid for every 5-uniform hypergraph H without isolated vertices, if and only if both $\frac{9}{2}a + b \geq 1$ and $b \geq 0$ hold.*

Problem 6 *Determine the exact value of $n(k, 3)$ for all $k \geq 5$. From our earlier results, we note that for $k \in \{2, 3, 4\}$, we have $n(k, 3) = 3k$.*

Conjecture 1 *Prove Ryser's conjecture when $k \geq 4$; that is, prove that if $k \geq 4$ and H is a k-partite hypergraph, then $\tau(H) \leq (k-1)\alpha'(H)$.*

Problem 7 *Close the gap between the upper and lower bounds of Theorem 26.*

Problem 8 *([9]) Determine the exact value of b_k for $k \geq 5$.*

Problem 9 *([9]) Determine the smallest value of k for which $b_{k-1} > \frac{2}{k+1}$.*

References

1. B.D. Acharya, Interrelations among the notions of independence, domination and full sets in a hypergraph. Nat. Acad. Sci. Lett. **13**(11), 421–422 (1990)
2. B.D. Acharya, Domination in hypergraphs. AKCE Int. J. Graphs Combin. **4**, 117–126 (2007)
3. B.D. Acharya, "Domination in hypergraphs": corrigendum [see B.D. Acharya, Domination in hypergraphs. AKCE Int. J. Graphs Combin. **4**, 117–126 (2007)]. AKCE Int. J. Graphs Combin. **5**, 99–101 (2008)
4. B.D. Acharya, Domination in hypergraphs II. New directions, in *Proceedings of the International Conference - ICDM 2008*, Mysore (2008), pp. 1–16
5. R. Aharoni, Ryser's conjecture for tripartite 3-graphs. Combinatorica **21**(1), 1–4 (2001)
6. N. Alon, Transversal numbers of uniform hypergraphs. Graphs Combin. **6**, 1–4 (1990)
7. S. Arumugam, B.K. Jose, C. Bujtás, Zs. Tuza, Equality of domination and transversal numbers in hypergraphs. Discrete Appl. Math. **161**, 1859–1867 (2013)
8. C. Bujtás, M.A. Henning, Zs. Tuza, Transversals and domination in uniform hypergraphs. Eur. J. Combin. **33**, 62–71 (2012)
9. C. Bujtás, M.A. Henning, Zs. Tuza, A. Yeo, Total transversals and total domination in uniform hypergraphs. Electron. J. Combin. **21**(2), Paper 2.24, 22 pp. (2014)
10. C. Bujtás, B. Patkós, Zs. Tuza, M. Vizer, The minimum number of vertices in uniform hypergraphs with given domination number. Discrete Math. **340**, 2704–2713 (2017)
11. V. Chvátal, C. McDiarmid, Small transversals in hypergraphs. Combinatorica **12**, 19–26 (1992)
12. E.J. Cockayne, S.T. Hedetniemi, Optimal domination in graphs. IEEE Trans. Circ. Syst. CAS-2 **22**, 855–857 (1975)
13. E.J. Cockayne, R.M. Dawes, S.T. Hedetniemi, Total domination in graphs. Networks **10**, 211–219 (1980)
14. Y. Dong, E. Shan, L. Kang, S. Li, Domination in intersecting hypergraphs. Discrete Appl. Math. **251**, 155–159 (2018)
15. M. Dorfling, M.A. Henning, Linear hypergraphs with large transversal number and maximum degree two. Eur. J. Combin. **36**, 231–236 (2014)
16. P. Erdős, Zs. Tuza, Vertex coverings of the edge set in a connected graph, in *Graph Theory, Combinatorics, and Applications*, ed. by Y. Alavi, A.J. Schwenk (Wiley, New York, 1995), pp. 1179–1187
17. P. Erdős, M.A. Henning, H.C. Swart, The smallest order of a graph with domination number equal to two and with every vertex contained in a K_n. Ars Combin. **35A**, 217–223 (1993)
18. T.W. Haynes, S.T. Hedetniemi, P.J. Slater (eds.), *Fundamentals of Domination in Graphs* (Marcel Dekker, New York, 1998)
19. T.W. Haynes, S.T. Hedetniemi, P.J. Slater (eds.), *Domination in Graphs: Advanced Topics* (Marcel Dekker, New York, 1998)
20. S.M. Hedetniemi, S.T. Hedetniemi, R.C. Laskar, L. Markus, P.J. Slater, Disjoint dominating sets in graphs, in *Discrete Mathematics*. Ramanujan Mathematical Society Lecture Notes Series, vol. 7 (Ramanujan Mathematical Society, Mysore, 2008), pp. 87–100
21. M.A. Henning, Recent results on total domination in graphs: a survey. Discrete Math. **309**, 32–63 (2009)
22. M.A. Henning, C. Löwenstein, Hypergraphs with large domination number and edge sizes at least 3. Discrete Appl. Math. **160**, 1757–1765 (2012)
23. M.A. Henning, C. Löwenstein, Hypergraphs with large transversal number and with edge sizes at least four. Central Eur. J. Math. **10**(3), 1133–1140 (2012)
24. M.A. Henning, C. Löwenstein, A characterization of the hypergraphs that achieve equality in the Chvátal-McDiarmid Theorem. Discrete Math. **323**, 69–75 (2014)
25. M.A. Henning, C. Löwenstein, A characterization of hypergraphs with large domination number. Discuss. Math. Graph Theory **36**, 427–438 (2016)

26. M.A. Henning, H.C. Swart, Bounds on a generalized domination parameter. Quaest. Math. **13**(2), 237–253 (1990)
27. M.A. Henning, A. Yeo, Hypergraphs with large transversal number and with edge sizes at least three. J. Graph Theory **59**, 326–348 (2008)
28. M.A. Henning, A. Yeo, Total domination in 2-connected graphs and in graphs with no induced 6-cycles. J. Graph Theory **60**, 55–79 (2009)
29. M.A. Henning, A. Yeo, Strong transversals in hypergraphs and double total domination in graphs. SIAM J. Discrete Math. **24**(4), 1336–1355 (2010)
30. M.A. Henning, A. Yeo, Transversals and matchings in 3-uniform hypergraphs. Eur. J. Combin. **34**, 217–228 (2013)
31. M.A. Henning, A. Yeo, *Total Domination in Graphs*. Springer Monographs in Mathematics (2013). ISBN-13: 978-1461465249
32. M.A. Henning, A. Yeo, Total transversals in hypergraphs and their applications. SIAM J. Discrete Math. **29**(1), 309–320 (2015)
33. B.K. Jose, Domination in hypergraphs. Ph.D. thesis. Mahatma Gandhi University, 2011. http://hdl.handle.net/10603/25947
34. B.K. Jose, Zs. Tuza, Hypergraph domination and strong independence. Appl. Anal. Discrete Math. **3**, 237–358 (2009)
35. B.K. Jose, K.A. Germina. K. Abhishek, On some open problems of stable sets and domination in hypergraphs, unpublished manuscript (2009)
36. L. Kang, S. Li, Y. Dong, E. Shan, Matching and domination numbers in r-uniform hypergraphs. J. Comb. Optim. **34**(2), 656–659 (2017)
37. V.R. Kulli, S.C. Sigarkarti, Inverse domination in graphs. Nat. Acad. Sci. Lett. **14**(12), 473–475 (1991)
38. F.C. Lai, G.J. Chang, An upper bound for the transversal numbers of 4-uniform hypergraphs. J. Combin. Theory Ser. B **50**, 129–133 (1990)
39. S. Li, L. Kang, E. Shan, Y. Dong, The finite projective plane and the 5-uniform linear intersecting hypergraphs with domination number four. Graphs Combin. **34**(5), 931–945 (2018)
40. O. Ore, *Theory of Graphs*, vol. 38 (American Mathematical Society, Providence, 1962), pp. 206–212
41. E. Shan, Y. Dong, L. Kang, S. Li, Extremal hypergraphs for matching number and domination number. Discrete Appl. Math. **236** 415–421 (2018)
42. S. Thomassé, A. Yeo, Total domination of graphs and small transversals of hypergraphs. Combinatorica **27**, 473–487 (2007)
43. Zs. Tuza, Covering all cliques of a graph. Discrete Math. **86**, 117–126 (1990)

Domination in Chessboards

Jason T. Hedetniemi and Stephen T. Hedetniemi

1 Introduction

In this chapter we consider *chessboards*, which are finite, uniform (or regular) tessellations of the plane into identical *cells* or *squares*.

With square cells, we consider either the more common $n \times n$ chessboards, with n rows and n columns, or the *rectangular $m \times n$* chessboards. But we briefly consider these variations: (i) *triangular, or diamond*, shaped boards, (ii) *sawtooth* square boards, (iii) torus (or circular) boards (where the leftmost squares are adjacent to the corresponding rightmost squares and the topmost squares are adjacent to the corresponding bottommost squares), and (iv) *three-dimensional* square boards.

For each type of board, we consider the following chess pieces: (i) queens, (ii) kings, (iii) rooks, (iv) bishops, and (v) knights, each of which defines a graph, whose vertices correspond one-to-one with the squares of the board, and two squares are considered to be *adjacent* if and only if a piece of a given type can move from one square to the other square in one (legal) move. Thus, we define:

(i) the *queens graph Q_n*, where two squares are adjacent if and only if they lie on a common row, column, or diagonal;

(ii) the *kings graph K_n*, where two squares are adjacent if and only if they are next to each other on a common row, column, or diagonal;

(iii) the *rooks graph R_n*, where two squares are adjacent if and only if they lie on a common row or column;

J. T. Hedetniemi (⊠)
Wilkes Honors College, Florida Atlantic University, Jupiter, FL 33458, USA

S. T. Hedetniemi
School of Computing, Clemson University, Clemson, SC, USA
e-mail: hedet@clemson.edu

© The Author(s), under exclusive license to Springer Nature Switzerland AG 2021
T. W. Haynes et al. (eds.), *Structures of Domination in Graphs*, Developments
in Mathematics 66, https://doi.org/10.1007/978-3-030-58892-2_12

(iv) the *bishops graph B_n*, where two squares are adjacent if and only if they lie on a common diagonal; and

(v) the *knights graph N_n*, where two squares are adjacent if and only if you can move from one square to the other by a two-step process of moving either one square in one direction followed by two squares in a perpendicular direction, or similarly, by moving two squares in one direction followed by one square in a perpendicular direction.

Before continuing we should comment that the notation K_n is used throughout graph theory to denote the complete graph of order n. Similarly, the notation Q_n is used throughout graph theory to denote the n-dimensional cube graph. However, in the context of chessboards, K_n denotes the $n \times n$ kings graph and Q_n denotes the $n \times n$ queens graph.

For each chessboard graph, Q_n, K_n, R_n, B_n, and N_n, we consider what is known about the values of the following seven graph theory parameters, the first six of which form what is known as the *Domination Chain* of inequalities:

$$ir(G) \le \gamma(G) \le i(G) \le \alpha(G) \le \Gamma(G) \le IR(G),$$

and the seventh parameter is the total domination number, $\gamma_t(G)$. These parameters are defined in the Glossary in this volume. But with respect to chessboard problems these can be defined as follows:

(i) the *lower irredundance number $ir(G)$* equals the minimum number of pieces of one type that can be placed on a set S of squares so that every piece attacks or occupies a square that is not attacked by any other piece in S and no additional piece can be added to S which preserves this property.

(ii) the *domination number $\gamma(G)$* equals the minimum number of pieces of one type that can be placed on a set S of squares so that all other squares are attacked by a piece in S.

(iii) the *independent domination number $i(G)$* equals the minimum number of pieces of one type that can be placed on a set S of squares so that all other squares are attacked by a square in S, and no two pieces in S attack each other.

(iv) the *independence number $\alpha(G)$* equals the maximum number of pieces of one type that can be placed on a set S of squares, so that no two pieces in S attack each other.

(v) the *upper domination number $\Gamma(G)$* equals the maximum number of pieces of one type that can be placed on a set S of squares so that all squares are attacked by a piece in S, and every piece attacks or occupies a square not attacked by another piece in S.

(vi) the *upper irredundance number $IR(G)$* equals the maximum number of pieces of one type that can be placed on a set S of squares so that every piece attacks or occupies a square that is not attacked by any other piece in S.

(vii) the *total domination number $\gamma_t(G)$* equals the minimum number of pieces that can be placed on a set S of squares so that all other squares are attacked by a piece in S and every piece in S is attacked by another piece in S.

The focus on these seven domination parameters is not to suggest that they are the only four types of domination worth studying with respect to chessboards. Indeed, more than 50 types of domination have been defined and studied. But when applied to chessboards, these are the major types of domination that have been studied.

We add in closing this section that several papers have studied queens domination, when queens can only be placed on (i) the major diagonal, (ii) the column nearest the center column, or (iii) the border, or outermost, squares of the chessboard.

2 Historical Origins

The independence number $\alpha(G)$, the independent domination number $i(G)$, the domination number $\gamma(G)$, and the total domination number $\gamma_t(G)$, although not defined at the time as formal graph theory parameters, are all considered in the early mathematical studies of chessboard problems. This is documented by W. W. Rouse Ball in his book *Mathematical Recreations and Problems of Past and Present Times* [81], published in 1892, and by P. J. Campbell in his 1977 paper entitled "Gauss and the eight queens problem: a study in miniature of the propagation of historical error" [30].

In the book by Rouse Ball and Coxeter, *Mathematical Recreations and Essays* [79], we find the following passage on page 166:

"One of the classical problems connected with the chessboard is the determination of the number of ways in which eight queens can be placed on a chessboard (or, more generally, in which n queens can be placed on a board of n^2 cells) so that no queen can take any other. This was proposed originally by Franz Nauck in 1850."

Actually, Ball was in error here, as the problem was originally stated by a chess player, Max Bezzel [5] in 1848 (this is discussed in Campbell's paper, mentioned above). But Dr. Franz Nauck [73] is given credit, by Campbell, for being the first person to show that one can always place n non-attacking queens on an order n board. Thus, Nauck can be given credit for showing that the vertex independence number of the queens graph Q_n is n, that is, $\alpha(Q_n) = n$.

On page 119 [79], we find the following:

"MAXIMUM PIECES PROBLEM. The Eight Queens Problem suggests the somewhat analogous question of finding the maximum number of kings - or more generally of pieces of one type - which can be put on a board so that no one can take any other, and the number of solutions possible in each case."

It is clear that Ball had in mind the general idea of independent sets in graphs (non-attacking chess pieces of one type) and particularly of finding maximum independent sets, hence $\alpha(G)$.

Ball notes the following values for 8×8 chessboards:

$\alpha(Q_8) = 8$
$\alpha(R_8) = 8$

$$\alpha(B_8) = 14$$
$$\alpha(K_8) = 16$$
$$\alpha(N_8) = 32.$$

On page 119 [79], Ball continues:

"MINIMUM PIECES PROBLEM. Another problem of a somewhat similar character is the determination of the minimum number of kings - or more generally pieces of one type - which can be put on a board so as to command or occupy all the cells."

It is clear that Ball had in mind the general idea of dominating sets in graphs, and particularly of finding minimum dominating sets, hence $\gamma(G)$.

Ball notes the following values (for example, cf. Figure 1 which shows a minimum dominating set of 3 queens on Q_6):

$$\gamma(Q_8) = 5, \ \gamma(Q_7) = 4, \ \gamma(Q_6) = 3, \ \gamma(Q_5) = 3, \ \gamma(Q_4) = 2$$
$$\gamma(B_8) = 8$$
$$\gamma(R_8) = 8$$
$$\gamma(K_8) = 9$$
$$\gamma(N_8) = 12.$$

On page 120 [79], Ball continues:

"Jaenisch [43] proposed also the problem of the determination of the minimum number of queens which can be placed on a board of n^2 cells so as to command all the unoccupied cells, subject to the restriction that no queen shall attack the cell occupied by any other queen."

Thus, de Jaenisch should be given credit for the idea of the independent domination number $i(G)$ in graphs.

Ball gives the following values of $i(Q_n)$ (for example, cf. Figure 1, which shows a minimum independent dominating set of 4 queens on Q_7):

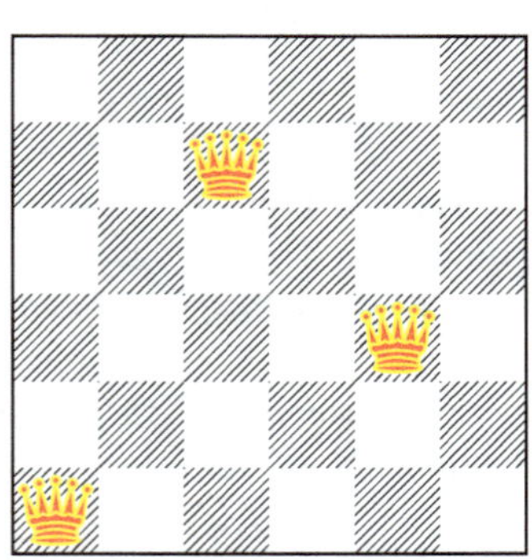

Fig. 1 $\gamma(Q_6) = 3$ and $i(Q_7) = 4$

Fig. 2 A minimum total dominating set of 12 kings on K_8

$$i(Q_4) = 3, i(Q_5) = 3, i(Q_6) = 4, i(Q_7) = 4, i(Q_8) = 5.$$

On p.120 [79], Ball continues:

"A problem of the same nature would be the determination of the minimum number of queens (or other pieces) which can be placed on a board so as to protect one another and command all the unoccupied cells."

Thus, Ball effectively defines the concept of the total domination number of a graph, something that would not be formally defined graph theoretically until 1980 by Cockayne, Dawes, and Hedetniemi [38].

Ball notes the following values for 8×8 chessboards, where (i, j) indicates a piece placed on the cell in row i and column j:

$\gamma_t(Q_8) = 5$: (2,4),(3,4),(4,4),(5,4),(8,4)

$\gamma_t(R_8) = 8$

$\gamma_t(B_8) = 10$: (2,4),(2,5),(3,4),(3,5),(4,4),(4,5),(6,4),(6,5),(7,4),(7,5)

$\gamma_t(N_8) = 14$: (3,2),(3,3),(3,6),(3,7),(4,3),(4,4),(4,5),(4,6),(6,3),(6,4),(6,5),(6,6),
   (7,3),(7,6)

$\gamma_t(K_8) = 16$.

It is interesting to note that Ball did not discuss the total domination numbers of kings graphs. The kings total domination number of the 8×8 chessboard was given as 12 by Garnick and Nieuwejaar in 1995 [56]. Although the authors did not present a solution, Figure 2 shows a simple solution.

3 Early Chessboard Domination

The literature on chessboard domination problems is far greater than can be reported in this chapter. We therefore make no claims of being comprehensive in covering

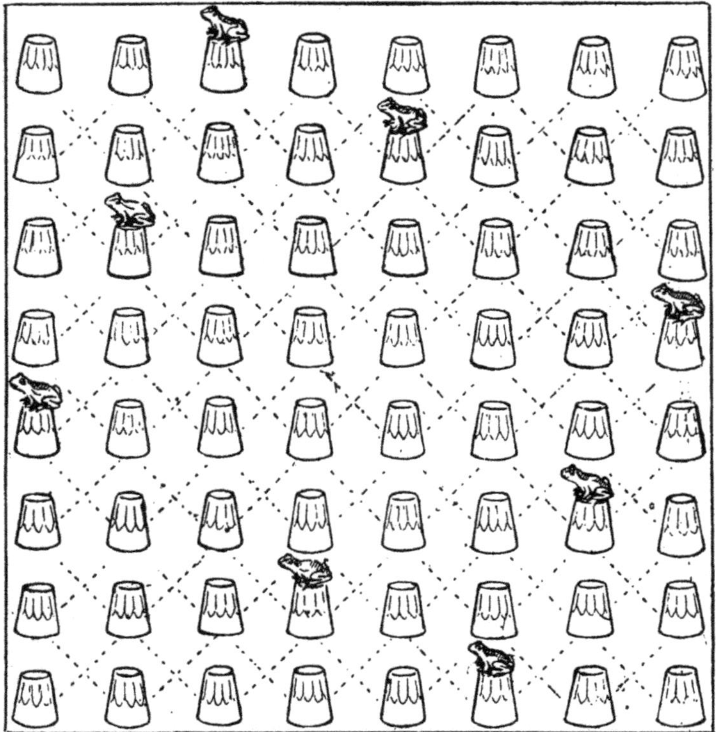

Fig. 3 Dudeney [46] p. 84

this literature. But certain publications stand out as being significant sources of information, either historically or in terms of key results, beginning with the book originally published in 1892 by W. W. Rouse Ball [81], which is cited in the previous section.

It should also be pointed out that the well-known English author and mathematician, Henry Ernest Dudeney, created many puzzles around the turn of the 20th century, which either directly or indirectly involve dominating sets of chess pieces on varying sizes of chessboards. Some of these puzzles can be seen in his book entitled *The Canterbury Puzzles and Other Curious Problems* [46], which is freely available on the web.

On page 84 of this book [46] is Puzzle 69. The Frogs and Tumblers. It shows eight frogs each sitting on a different tumbler in an 8×8 array of tumblers (cf. Figure 3). When viewed as queens, this set forms a maximum independent set on the queens graph Q_8 (cf. Figure 4). The puzzle is to move three queens to different squares in order to form another maximum independent set of 8 queens. It also suggests that there is only one such solution, up to isomorphism.

On page 108 of this book [46] is Puzzle 92. The Four Porkers. It shows four pigs placed on the squares of a 6×6 chessboard, which when viewed as queens,

Fig. 4 A maximum
independent set of 8 queens
on Q_8

Fig. 5 A minimum
independent dominating set
of 4 queens on Q_6

form a set of 4 independent dominating queens of Q_6 (cf. Figure 5). The puzzle is
to determine the number of different sets of 4 queens which form an independent
dominating set of Q_6 (cf. Figure 5).

In 1910 Pauls proves the following well-known result.

Theorem 1 (Pauls [76, 77]) *For queens graphs Q_n,*

  (i) $\alpha(Q_1) = \alpha(Q_2) = 1$
 (ii) $\alpha(Q_3) = 2$
(iii) *for $n \geq 4$, $\alpha(Q_n) = n$.*

The basic idea for proving this theorem for $n \geq 4$ is indicated in Figure 6.

Historically, the first true graph theory book, published by Dénes König in 1936
[67], is noteworthy in that essentially the (independent) domination number was
first formally defined as a graph theory concept, although not by this name; it was
called a *punktbasis*, or *point basis*. In his book [67], König presents the following
illustration that $i(Q_8) = \gamma(Q_8) = 5$ (cf. Figure 7).

In 1964 [96] the Yaglom brothers published the book *Challenging Mathematical
Problems with Elementary Solutions. Vol. I: Combinatorial Analysis and Probability
Theory*, which contains 9 chessboard domination problems. Notable among their

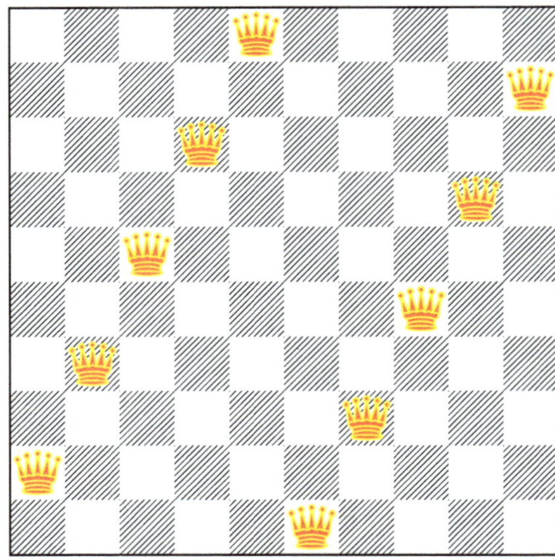

Fig. 6 Ahrens' maximum independent set of 10 queens on Q_{10}

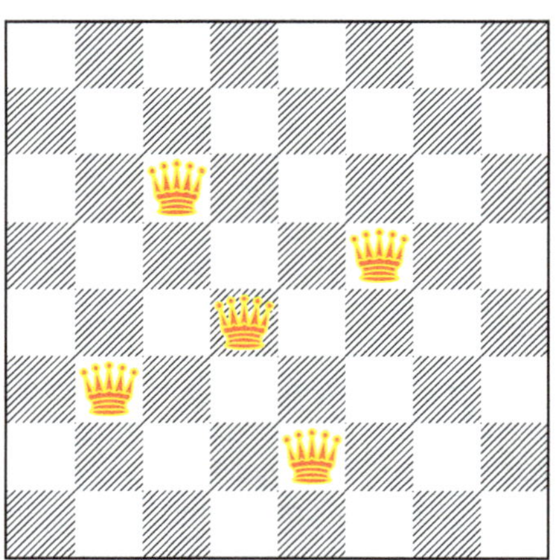

Fig. 7 A minimum (independent) dominating set of 5 queens on Q_8

contributions were the following three theorems about the kings domination number of square and rectangular chessboards, and the kings independent domination and independence numbers of square chessboards.

Theorem 2 (Yaglom and Yaglom [96]) *For kings graphs,* $\gamma(K_n) = \lfloor \frac{n+2}{3} \rfloor^2$.

Theorem 3 (Yaglom and Yaglom [96]) *For rectangular kings graphs,*

$$\gamma(K_{m,n}) = \lfloor \frac{m+2}{3} \rfloor \lfloor \frac{n+2}{3} \rfloor.$$

Theorem 4 (Yaglom and Yaglom [96]) *For kings graphs,* $i(K_n) = \alpha(K_n) = \lfloor \frac{n+1}{2} \rfloor^2$.

Chessboard domination problems often appeared in the many columns written by Martin Gardner in *Scientific American* in the 1970s, cf. [53–55].

In 1995 [52] Fricke, Hedetniemi, Hedetniemi, McRae, Wallis, Jacobon, Martin, and Weakley present the first comprehensive survey of chessboard domination results. This was followed in 1998 by a survey by Hedetniemi, Hedetniemi, and Reynolds [63] containing even more chessboard domination results.

In 2004 [90] Watkins, publishes the book *Across the Board: The Mathematics of Chessboard Problems*, in which Chapter Seven Domination, Chapter Eight Queens Domination, Chapter Nine Domination on Other Surfaces, and Chapter Ten Independence discuss some of the results reviewed in this chapter.

In 2008 [3] Bell and Stevens present a comprehensive 32-page survey of everything that is known about the n-queens problem, including the placement of non-attacking queens on a wide variety of chessboards, including n-dimensional boards, Möbius boards, and modular boards.

Finally, in 2018 [95] Weakley presents an excellent survey of research on the queens domination number in the last 25 years.

We should also add that there is a huge amount of chess information at http://www.kotesovec.cz.

4 Queens

In this section we focus on chessboard domination using only queens. In subsequent sections we will focus on chessboard domination using the other pieces of bishops, knights, kings, and rooks.

In 1977 [68] Larson shows that for primes of the form $n = 4k + 1$, elegant solutions can be constructed for the n-queens problem using the following simple rule: place a queen on the center square and then place other queens by making successive (2, 3) movements—two squares to the right and three squares upward, where the top and bottom edges of the board are identified, as well as the right and left edges of the board. The resulting queen placement for $n = 13$ is shown in Figure 8.

Larson shows that whenever u and v are positive integers and $u^2 + v^2$ is an odd prime p, then queens located at successive (u, v) movements from a queen on the center square of the $p \times p$ chessboard give a solution to the p-queens problem, and such solutions exist whenever p is a prime of the form $4k + 1$.

In 1984 [88] Wagner and Geist discuss the results of a programming assignment given to students in a graduate computer science class: write a program to solve the following variant of the 8-queens problem.

A *crippled queen CQ* is a chess queen that can move at most two squares at a time in any direction (vertical, horizontal, or diagonal). Find the maximum number $\alpha(CQ_n)$ of CQs that can be placed on an 8×8 chessboard so that no two CQs can attack one another. Find also the number of ways that this number of CQs can be so placed. If possible, generalize the program to compute $\alpha(CQ_{8,n})$ or $\alpha(CQ_{m,n})$.

Fig. 8 Thirteen non-attacking queens in a (2,3)-pattern on a 13 × 13 chessboard

Fig. 9 Thirteen independent crippled queens on an 8 × 8 chessboard

Figure 9 illustrates a crippled queens solution for the 8 × 8 board. Notice that in this solution every CQ is a knight's move away from at least one other CQ. Notice also that a crippled queen is very much like a super king, which can move two squares: one square in any direction, followed by a second square in any direction.

In 1986 [35] Cockayne and Hedetniemi introduce a variation of the standard queens domination problem, in which you seek to find the minimum number of queens which, when placed only on the main diagonal of an $n \times n$ chessboard, dominate all squares. Let $diag(n)$ denote this diagonal queens domination number; this is also denoted by $\gamma_{diag}(Q_n)$.

The authors show that the *diagonal queens domination problem* is equivalent to the problem of finding a midpoint-free, even-sum set of integers up to n, which, as well, is equivalent to that of finding a midpoint-free subset of $[n/2]$; this is a collection of integers up to $n/2$ not containing a three-term arithmetic progression.

A subset $K \subset N$ is called *midpoint-free* if for all $\{i, j\} \subseteq K$, $(i+j)/2 \notin K$, and K is called an *even-sum subset* if the sum of each pair of elements of K is even, i.e., its elements are either all odd or all even.

Table 1 Values of *diag(n)*

n	$diag(n)$	Minimum diagonal dominating set
7	4	$\{2, 4, 5, 6\}$
8	5	$\{2, 4, 5, 6, 8\}$
11	7	$\{1, 3, 5, 6, 7, 9, 11\}$
15	11	$N - \{2, 4, 8, 10\}$
20	15	$N - \{2, 4, 8, 10, 20\}$
24	18	$N - \{2, 4, 8, 10, 20, 22\}$
25	18	$N - \{1, 3, 7, 9, 19, 21, 25\}$
30	22	$N - \{1, 3, 7, 9, 19, 21, 25, 27\}$

The authors show that the diagonal queens domination number is related to the number-theoretic function $r_3(n)$, which equals the smallest number of integers in a subset of $\{1, 2, \ldots, n\}$ that must contain three terms in arithmetic progression.

Suppose that the squares of a chessboard are labeled (i, j), so that black and red squares have $(i + j)$ even or odd, respectively. A subset $K \subset N = \{1, 2, \ldots, n\}$ is called a *diagonal dominating set* if queens placed in positions $\{(k, k) : k \in K\}$ on the black major diagonal dominate the entire board.

Theorem 5 (Cockayne, Hedetniemi [35]) *A subset K is a diagonal dominating set if and only if $N - K$ is a midpoint-free, even-sum set.*

Proof. Let K be a diagonal dominating set and let $\{i, j\} \subseteq N - K$. Then square (i, j) is not covered by a queen along a row or column. Since only black squares are covered diagonally, square (i, j) must be black, which implies that $(i + j)$ is even, i.e., $N - K$ is an even-sum set. Since square (i, j) is covered, by a queen at square (k, k), for some $k \in K$, we have $i + j = 2k$. Hence, $(i + j)/2 \notin N - K$ and $N - K$ are midpoint-free.

Conversely, suppose $N - K$ is a midpoint-free, even-sum set. Place queens at $\{(k, k) | k \in K\}$. If (i, j) is a red square, i.e., $i + j$ is odd, then by the even-sum property, either i or j is in K and (i, j) is covered by a queen along a row or column. If (i, j) is a black square and is not covered by a row or a column, then $(i, j) \subseteq N - K$ and $i + 2j = 2l$, for some $l \in N$. Since $N - K$ is midpoint-free, $l \notin N - K$. Therefore, $l \in K$ and (i, j) are dominated by the queen at position (l, l). This completes the proof. $\square$

Corollary 6 $diag(n) = r - max \cdot |K| |K$ *is a midpoint-free, even-sum subset of N*}.

In Table 1, notice that the complements of the indicated minimum diagonal dominating sets are midpoint-free; for example, for $n = 25$, $\{1, 3, 7, 9, 19, 21, 25\}$ is a midpoint-free set.

See Figure 10 for a minimum diagonal queens dominating set on Q_8.

In 1985 [40] Cockayne, Gamble, and Shepherd consider another variation of the standard queens domination problem.

Denote by $col(n)$ the minimum number of queens on any single column that is required to dominate the $n \times n$ chessboard. It is easy to see that a column nearest the center is as good as any other. The authors show that like $diag(n)$, $col(n)$ is also related to the number-theoretic function $r_3(n)$, as follows.

Fig. 10 Diagonal minimum dominating set for Q_8

Let $A(n) = n - r_3(\lceil n/3 \rceil)$.

Let $B(n) = n - max_{k+l=\lceil n/2 \rceil}\{r_3(\lceil k/2 \rceil) + r_3(\lceil l/2 \rceil)\}$.

Theorem 7 (Cockayne, Gamble, Shepherd [40]) *For $n \geq 2$, $col(n) = min\{A(n), B(n)\}$.*

Corollary 8 *For any n, $diag(n) \leq col(n)$.*

They also raised the following question.

Question 9 *For all n, is $A(n) \geq B(n)$?*

In 1987 [78] Raghavan and Venkatesan prove the following bounds on the queens domination number. The proof of this upper bound is essentially the same as the proof attributed to Welch below.

Theorem 10 (Raghavan, Venkatesan [78]) *For any $n \geq 1$, $\lfloor \frac{1}{2}n \rfloor \leq \gamma(Q_n) \leq \lfloor \frac{2}{3}n \rfloor + 2$.*

This upper bound shows up in several subsequent papers, and can be proved in several different ways.

In 1988 [37] Cockayne and Spencer provide the following upper bound for the independent queens domination number.

Theorem 11 (Cockayne, Spencer [37]) *For any $n \geq 1$, $i(Q_n) \leq 0.705n + 2.305$.*

In 1990 [34] Cockayne surveys results known at the time on domination and independent domination numbers of the queens graph, the diagonal queens domination problem, domination by queens in a single column and domination, independent domination and total domination of the bishops graph. In this paper he presents the following basic result, attributed to L. Welch in an undated private communication to Cockayne. This construction is very similar to the same upper bound presented earlier by Raghavan and Venkatesan [78].

Theorem 12 (Welch [34]) *For $n = 3q + r$, where $0 \leq r \leq 2$, $\gamma(Q_n) \leq 2q + r$.*

Proof Sketch. For the case $n = 3q$, divide the $n \times n$ chessboard into nine $q \times q$ boards, where the top three boards are numbered B_1, B_2, and B_3, the middle three boards are numbered B_4, B_5, and B_6, and the bottom three boards are numbered B_7, B_8, and B_9. Place q queens on the main (northwest down to southeast) diagonal of board B_3. Place $q - 1$ queens on the diagonal immediately above the main diagonal of board B_7, and place the last queen in the bottom left corner of board B_7. It is easy to see that the q queens on the main diagonal of board B_3 dominate all squares in boards B_1, B_2, B_3, B_6, and B_9. Similarly, the q queens in board B_7 dominate all squares in boards B_4, B_7, and B_8. It only remains to show that all squares in the middle board B_5 are dominated by these $2q$ queens.

If $n = 3q + 1$ or $n = 3q + 2$, it is easy to add one or two extra queens in squares $(3q + 1, 3q + 1)$ and $(3q + 2, 3q + 2)$. $\qquad\square$

It is worth noting that one can always add just one queen to the pattern suggested in Welch's proof to obtain an upper bound for the connected domination number of the queens graph.

Cockayne also presents one of the most often quoted results in queens domination theory, attributed to P. H. Spencer, when he was an undergraduate research student at the University of Victoria in the summer of 1984 (C.M. Mynhardt, Private communication, 2020).

Theorem 13 (Spencer, 1984) *For any $n \geq 1$, $\gamma(Q_n) \geq (n - 1)/2$.*

Proof. Since $\gamma(Q_1) = \gamma(Q_2) = \gamma(Q_3) = 1$, we can assume that $n \geq 4$. It is easy to see that the set of $n - 2$ queens placed on the main diagonal on every square (i, i) except for $(1, 1)$ and $(3, 3)$ is a dominating set. Thus, for all $n \geq 4$, $\gamma(Q_n) \leq n - 2$, and therefore any minimum dominating set of queens on Q_n will have at least two rows and two columns with no queen on them.

Assume that the columns of Q_n are numbered $1, 2, \ldots, n$ from left to right, and that the rows are similarly numbered from top to bottom. Let S be a minimum dominating set of queens on Q_n. Let a be the leftmost column, b the rightmost column, c the lowest row, and d the highest row not containing a queen. By symmetry, we may assume that $\varepsilon_2 = d - c \leq \delta_1 = b - a$.

Consider the sets of squares S_a and S_b in columns a and b, respectively, which lie between rows c and $c + \delta_1 - 1$ inclusively, and let $S = S_a \cup S_b$. Thus, $|S_a| = |S_b| = \delta_1 - 1$.

Since $\delta_2 \leq \delta_1$, no diagonal intersects both S_a and S_b. Therefore, every queen diagonally dominates at most two squares of S, one in S_a and one in S_b. Furthermore, all queens situated above row c or below row $c + \delta_1 - 1$ do not dominate any squares of S by row or column.

By the definition of c, there are at least $c - 1$ queens above row c and each row below d is occupied by at least one queen, where $d = c + \delta_2 \leq c + \delta_1$. Therefore, since all of the $n - (c + \delta_1)$ rows below row $c + \delta_1$ are occupied, there are at least $n - (c + \delta_1)$ queens below row $c + \delta_1 - 1$.

It follows that at least $(c - 1) + (n - c - \delta_1) = n - \delta_1 - 1$ queens dominate at most 2 squares of S. The remaining queens, at most $\gamma(Q_n) - (n - \delta_1 - 1)$, may cover at most 4 squares of S. Since all of the $2\delta_1$ squares of S must be dominated, we must have

$$2(n - \delta_1 - 1) + 4(\gamma(Q_n) - (n - \delta_1 - 1)) \geq 2\delta_1,$$

which gives $\gamma(Q_n) \geq (n - 1)/2$, as required. $\square$

Note, by the way, that the bound $\gamma(Q_n) \geq (n - 1)/2$ means that for even values of n, $\gamma(Q_n) \geq n/2$; this is noted in many papers dealing with queens domination numbers. Thus, whenever for even n you can find a dominating set of $n/2$ queens, you know it is best possible, and this happens quite often.

In 1991 Grinstead, Hahne, and Van Stone (see also Eisenstein, Grinstead, Hahne and Van Stone [47]) prove the following two theorems, which were the best known bounds at the time, the second of which is very much like the theorem of Welch and the theorem of Raghavan and Venkatesen above.

Theorem 14 (Grinstead, Hahne, Van Stone [60]) *For any* $n \geq 1$,

$$\gamma(Q_n) \leq \frac{14}{23}n + O(1).$$

Theorem 15 (Grinstead, Hahne, Van Stone [60]) *For any* $n \geq 1$, $i(Q_n) \leq \frac{2}{3}n + O(1)$.

One basic pattern of independent dominating queens which achieves the upper bound in Theorem 15 is shown in Figure 11.

In 1991 Weakley studies $\Gamma(Q_n)$ and $\Gamma_t(Q_n)$ and proves the following two lower bounds.

Theorem 16 (W.D. Weakley, Private communication, July 26, 1991) *For* $n \geq 5$,

(i) $\Gamma(Q_n) \geq 2n - 5$.
(ii) $\Gamma_t(Q_n) \geq 2n - 5$.

He also shows that $n = 6$ is the smallest value for which $\Gamma(Q_n) > n$.

Weakley then studies the value of $\Gamma(Q_{m,n})$ and $\Gamma_t(Q_{m,n})$ for rectangular $m \times n$ chessboards, and shows the following.

Theorem 17 (W.D. Weakley, Private communication, July 26, 1991) *For any* $n \geq 1$,

(i) $\Gamma(Q_{2,n}) = IR(Q_{2,n}) = \lceil n/2 \rceil$.
(ii) $\Gamma_t(Q_{2,n}) = IR_t(Q_{2,n}) = \lceil n/2 \rceil$ *except for* $n = 1$, 2, 5, 6, *when* $\Gamma_t(Q_{2,n}) = IR_t(Q_{2,n}) = \lceil n/2 \rceil + 1$.

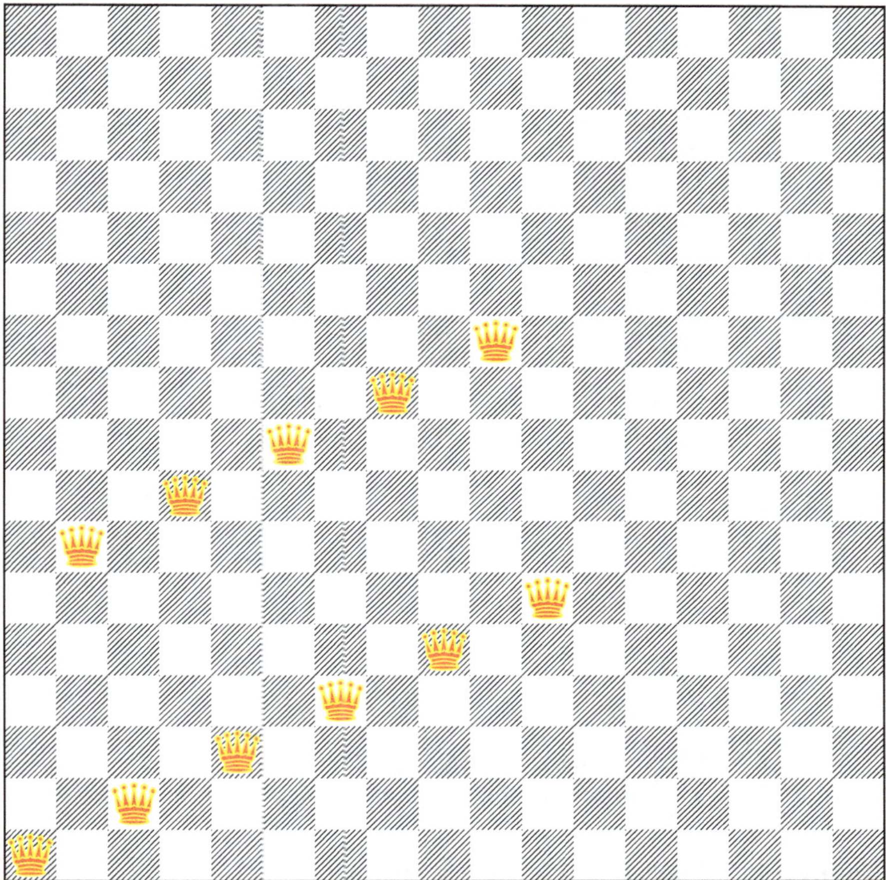

Fig. 11 A set of 11 independent queens dominating Q_{17}

(iii) $\Gamma(Q_{3,n}) \geq \lceil (n+1)/2 \rceil$.

(iv) $\Gamma(Q_{4,n}) \geq n$ and $\Gamma_t(Q_{4,n}) \geq n$.

The tables below illustrate the bounds for $\Gamma(Q_{3,n})$ and $\Gamma_t(Q_{4,n})$ (cf. Table 2), where a private neighbor of queen Q_k is the square numbered k. Notice that in Table 2 queen Q_3 is its own private neighbor. Notice also that the set of queens in Table 2 is not a total dominating set of queens because of queen Q_3.

In 1994 [19] and [20] Burger, in his master's thesis and PhD dissertation, gives a complete listing of all minimum queens dominating sets for Q_n, for $5 \leq n \leq 8$.

In 1994 Burger, Mynhardt, and Cockayne provide the following four exact values of $\gamma(Q_n)$ by exhibiting symmetric solutions.

Theorem 18 (Burger, Mynhardt, Cockayne [26]) *For* $k = 9$, 12, 13, 15, $\gamma(Q_{4k+1}) = 2k + 1$.

Table 2 $\Gamma(Q_{3,n}) \geq \lceil (n+1)/2 \rceil$ and $\Gamma(Q_{4,n}) \geq n$

Q_1	Q_2				Q_4	Q_5
			Q_3			
1	2				4	5

1		3		5
Q_1		Q_3		Q_5
	Q_2		Q_4	
	2		4	

Table 3 Best known results as of 2020

Chess pieces	ir	γ	i	α	Γ	IR
Queens Q_n				n	$\geq 2n - 5$	
Kings K_n		$(\lfloor (n+2)/3 \rfloor)^2$	$(\lfloor (n+2)/3 \rfloor)^2$	$(\lfloor (n+1)/2 \rfloor)^2$		
Rooks R_n	n	n	n	n	n	$2n - 4$
Bishops B_n	n	n	n	$2n - 2$	$2n - 2$	$4n - 14$
Knights N_n				[39]	[39]	[39]
Grid G_n		Known		[39]	[39]	[39]

Table 4 $\Gamma(Q_6) = 7$

		2	3	4	
Q_1	Q_2				
			5	6	7
Q_3	Q_4				
			1		7
Q_5	Q_6				Q_7

As we will see below, more results for $\gamma(Q_{4k+1})$ were to follow.

In 1995 Weakley proves the following two results, which he presented at a conference in 1992.

Theorem 19 (Weakley [92]) *For all k, $\gamma(Q_{4k+1}) \geq 2k + 1$*

Theorem 20 (Weakley [92]) *For $k \leq 6$, and $k = 8$, $\gamma(Q_{4k+1}) = i(Q_{4k+1}) = 2k + 1$.*

He also proves that $\gamma(Q_7) = i(Q_7) = 4$, which was stated, but not proved, by W. W Rouse Ball in 1892 [81].

In 1995 [52] Fricke et al. and in 1998 [63] Hedetniemi et al. publish two comprehensive surveys of the following 36 chessboard domination-related problems. (cf. Table 3).

Space limitations do not permit us to discuss the state of knowledge of all 36 problems. Thus, we only highlight a few.

The result that $\alpha(Q_n) = n$ is frequently attributed to Ahrens in 1910 [1], but was first shown by Pauls in 1874 [76]. The inequality $\Gamma(Q_n) \geq 2n - 5$ is due to Weakley (private communication dated July 26, 1991). An illustration of Weakley's construction of a maximum cardinality, minimal dominating set of seven queens on Q_6 is given in Table 4, with seven numbered queens and squares with an integer k indicating a private neighbor of queen Q_k.

Notice in Table 3 that for the rooks graph, all formulas are known, since these graphs have a simple clique structure. The results that $\gamma(R_n) = i(R_n) = \alpha(R_n) = n$ are

Fig. 12 $IR(R_n) = 2n - 4$

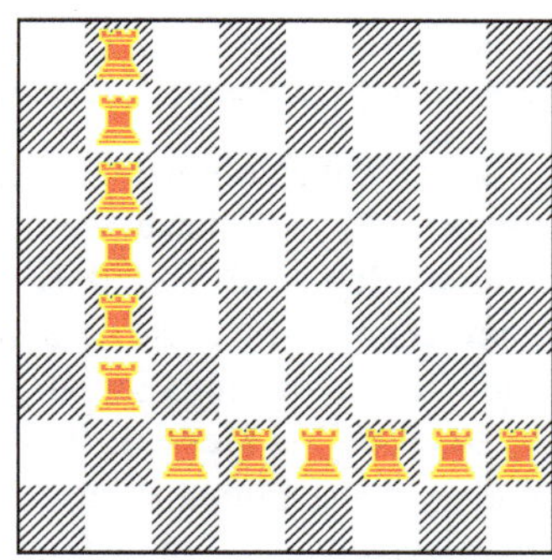

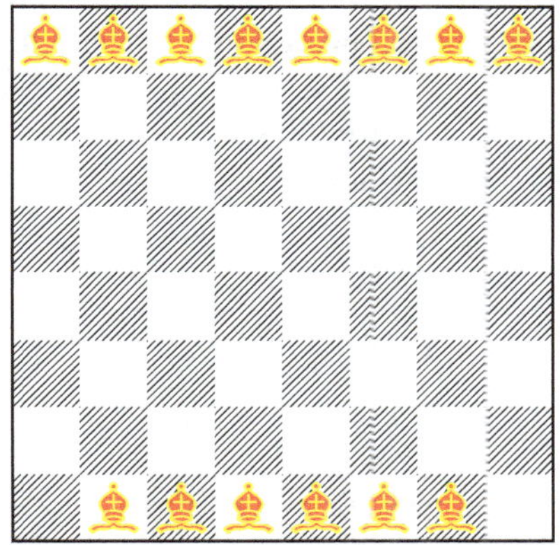

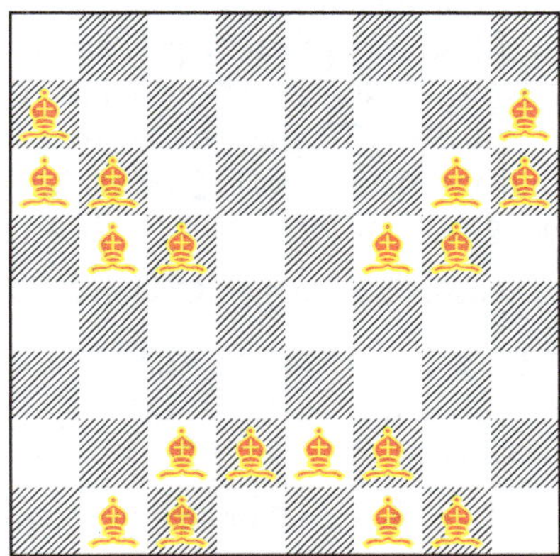

Fig. 13 $\alpha(B_n) = \Gamma(B_n) = 2n - 2$ and $IR(B_n) = 4n - 14$

due to Yaglom and Yaglom [96], as are the results that $\gamma(B_n) = i(B_n) = n$, and the result that $\alpha(B_n) = 2n - 2$.

The results, that $ir(R_n) = n$, $\Gamma(R_n) = n$ and $IR(R_n) = 2n - 4$, are attributed to Hedetniemi, Hedetniemi and Wallis, but are stated as unpublished in [52] (cf. Figure 12, where the rooks in column 2 all have a private neighbor in column 1, and the rooks in row 2 all have a private neighbor in row 1).

The result that $ir(B_n) = n$ is attributed to Wallis, but is stated as unpublished in [52], and the results that $\Gamma(B_n) = 2n - 2$, and $IR(B_n) = 4n - 14$ are attributed to Fricke, but are also stated as unpublished in [52]; cf. Figure 13 and for illustrations of Fricke's results.

Because of the following theorem, proved in 1981 by Cockayne, Favaron, Payan, and Thomason, it becomes easy to establish the values of α, Γ, and IR for knights graphs and grid graphs, since both of these are bipartite families of graphs.

Theorem 21 (Cockayne, Favaron, Payan, Thomason [39]) *If G is a bipartite graph, then $\alpha(G) = \Gamma(G) = IR(G)$.*

Corollary 22 (Cockayne, Favaron, Payan, Thomason [39]) *For all $n \geq 1$,*

$\alpha(N_n) = \Gamma(N_n) = IR(N_n) = \{n^2/2 \text{ for even } n; (n^2+1)/2, \text{ for odd } n\}.$
$\alpha(G_n) = \Gamma(G_n) = IR(G_n) = \{n^2/2 \text{ for even } n; (n^2+1)/2, \text{ for odd } n\}.$

Finally, the domination numbers of all grid graphs, that is, Cartesian products of the form $P_m \square P_n$, have been completely determined in a 2011 paper by Goncalves, Pinlou, Rao, and Thomasse [59], who present 16 formulas for the domination numbers of m-by-n grid graphs; 15 different formulas for $\gamma(G_{m,n})$ for $1 \leq m \leq 15$, and one final formula for $\gamma(G_{m,n})$, for all $m \geq 16$.

In 1995 [71] Messick, in his MS research paper, develops a genetic algorithm for finding near-optimal solutions for $IR(Q_n)$. One of his maximal irredundant sets of 11 queens on Q_8 is shown in Table 5, in which queen Q_k has as a private neighbor the square numbered k.

Messick's genetic algorithm also established the following lower bounds for $IR(Q_n)$, for $6 \leq n \leq 18$ (cf. Table 6).

In 1997 [27] Burger, Cockayne, and Mynhardt introduce the study of the upper domination number $\Gamma(Q_n)$ and the upper irredundance number $IR(Q_n)$ of the queens graph Q_n and present the following three results.

Theorem 23 (Burger, Cockayne, Mynhardt) *For all $n \geq 1$,*

   (i) $\gamma(Q_n) \leq 31n/54 + O(1)$;
  (ii) $\Gamma(Q_n) \geq 5n/2 - O(1)$;
 (iii) $IR(Q_n) \leq \lfloor 6n + 6 - 8\sqrt{n + \sqrt{n} + 1} \rfloor.$

They also mention that they have determined all 638 non-isomorphic independent dominating sets of size 5 of Q_8.

In 1997 [58] Gibbons and Webb, using simulated annealing and exhaustive search techniques, extend the known values of $\gamma(Q_n)$ and $i(Q_n)$ as shown in Tables 7 and 8:

Table 5 $IR(Q_8) \geq 11$

		Q_1		Q_2		Q_3	Q_4
		Q_5		Q_6			Q_7
1	5						
						Q_8	Q_9
2	6						
						Q_{10}	Q_{11}
3			8		10		
4	7		9		11		

Table 6 Lower bounds for $IR(Q_n)$

n	6	7	8	9	10	11	12	13	14	15	16	17	18
$IR(Q_n) \geq$	7	9	11	13	15	17	19	21	24	26	28	29	31

Table 7 New values of $\gamma(Q_n)$

n	29	41	45	57
$\gamma(Q_n) \geq$	15	21	23	29

Table 8 New values of $i(Q_n)$

n	14	15	16
$i(Q_n) \geq$	8	9	9

Fig. 14 $\gamma(Q_{13}) = 7$

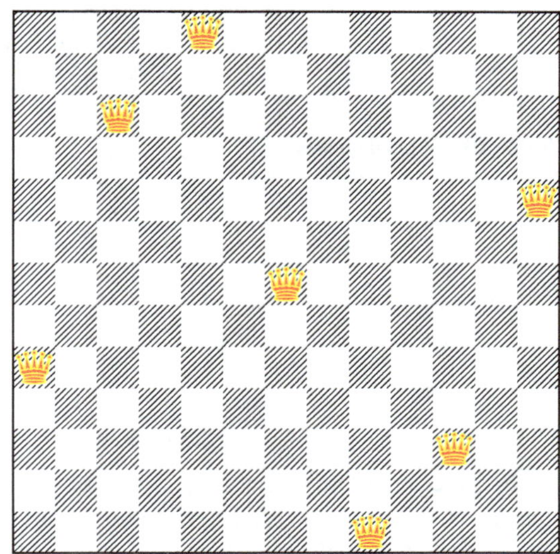

As a by-product, the number of non-equivalent ways of covering Q_n with k independent queens, for $1 \leq n \leq 15$ and $1 \leq k \leq 8$, as well as the case $n = 16$ and $k = 8$, are determined. As an illustration, they present the following minimum dominating set of seven queens on Q_{13} (cf. Figure 14); note that three queens lie symmetrically on the main diagonal, the four queens on border are symmetrically placed, and every other row and column contains exactly one queen. This is another case when $\gamma(Q_{4k+1}) = 2k + 1$.

In 2000 [23] Burger and Mynhardt provide the following two queens domination numbers:

$$\gamma(Q_{19}) = 10, \gamma(Q_{31}) = 16.$$

In 2000 [22] Burger and Mynhardt provide the following four queens domination numbers:

$$\gamma(Q_{30}) = 15, \gamma(Q_{59}) = 35, \gamma(Q_{77}) = 39, i(Q_{45}) = 23.$$

They also provide the following tabulation of known values of $\gamma(Q_{4k+1})$ (cf. Table 9).

In 2000 [22] Burger and Mynhardt add the following two values for the lower irredundance number of the queens graph:

$$ir(Q_5) = \gamma(Q_5) = 3, ir(Q_6) = \gamma(Q_6) = 3 \text{ (cf. Figure 1)}.$$

Table 9 $\gamma(Q_{4k+1})$

k	2	3	4	5	6	7	8	9	10
n	9	13	17	21	25	29	33	37	41
γ	5	7	9	11	13	15	17	19	21
k	11	12	13	14	15	16	17	18	19
n	45	49	53	57	61	65	69	73	77
γ	23	25	27	29	31	≤ 35	35		39

Table 10 Toroidal queens chessboard

•	•	Q	•	•	•	••	
	a	•	1				
b		•		2			
		•			3		c
		•				■	
		•			e		5
6		•		f			
	7	•	g				

In effect, they prove that there does not exist a maximal irredundant set of two queens on either Q_5 or Q_6. Thus, the minimum dominating sets of cardinality three for Q_5 and Q_6 are also minimum cardinality maximal irredundant sets on these two chessboards.

They also state, but without proof, that

$$ir(Q_7) = \gamma(Q_7) = 4 \text{ (cf. Figure 1).}$$

At the close of their paper, the authors offer the following interesting comment:

"The existence of a maximal irredundant set X of queens on Q_n with $|X| < \gamma(Q_n)$ for some n seems unlikely, as the (average) number of pns [private neighbors] per queen seems to increase rapidly as n increases, as does the cardinality of R, the set of open [undominated] squares, and hence the cardinality of $N[R]$. For every square in $N[R]$ to annihilate a queen in X (see Theorem 1) is a tall order.

"(Note: Harborth [private communication, January 2000] recently reported that $ir(Q_n) = \gamma(Q_n)$ for $n \leq 10$.)"

In 2001 [28] Burger, Cockayne, and Mynhardt introduce the study of domination in queens graphs on the torus, denoted Q_n^t, where the torus is the Cartesian product $C_n \square C_n$. In Q_n^t a diagonal is no longer a path with a beginning and an end; instead, it is a cycle, cf. Table 10, where the queen Q on the top row dominates toroidally every square in its top row, every square in its column, and all of the labeled squares, which form to diagonal cycles of length 8, where the black square ■ can be labeled both 4 and d. Thus, in the toroidal queens graph, square b is adjacent to square c, square 5 is adjacent to square 6, and squares 7 and g are both adjacent to the square labeled Q, and to each other, since they are in the same row.

An example of four toroidal queens which dominate Q_8^t is shown in Figure 15; note that it takes five queens to dominate Q_8; this is the smallest value of n, for

Fig. 15 Four minimum toroidal queens for Q_8

which $\gamma(Q_n^t) < \gamma(Q_n)$. It is interesting to note that the authors have determined that $\gamma^t(Q_{15}) = 5 < \gamma(Q_{15}) = 9$.

The situation for $i(Q_n)$ when compared with $i(Q_n^t)$ is also interesting, since they are not comparable. For example, the authors have shown that $i(Q_6) = 4 = i(Q_6^t)$, $i(Q_7) = 4 < i(Q_7^t) = 5$, yet $i(Q_8) = 5 < i(Q_7^t) = 4$.

The authors consider the independence, domination, and independent domination numbers of graphs obtained from the moves of queens on chessboards drawn on the torus, and determine exact values for each of these parameters in infinitely many cases.

In 2001 Cockayne and Mynhardt study the lower irredundance number of the queens graph, $ir(Q_n)$. The determination of $ir(Q_n)$ is no easy task. After some 15 pages of preliminary results and careful analysis, the authors prove the following theorem.

Theorem 24 (Cockayne, Mynhardt [36]) *For any $n \geq 8$, the queens graph Q_n does not have a maximal irredundant set of size three.*

Theorem 25 (Cockayne, Mynhardt [36]) *The queens graph Q_7 does not have a maximal irredundant set of size three.*

Since it is known that $\gamma(Q_7) = 4$ and $ir(Q_n) \leq \gamma(Q_n)$, we can conclude the following.

Corollary 26 (Cockayne, Mynhardt [36]) *For the queens graph Q_7, $ir(Q_7) = 4$.*

In 2001 [64] Kearse and Gibbons, using probabilistic and exhaustive search techniques, such as backtracking with refinements and enhancements, reduction methods, and local search techniques, establish the following queens domination numbers:

(i) $\gamma(Q_{15}) = \gamma(Q_{16}) = 9,$
(ii) $\gamma(Q_{19}) = 10,$
(iii) $\gamma(Q_{4k+1}) = 2k + 1,$ for $k = 16, 18, 20$ and $21,$
(iv) $i(Q_{18}) = 10,$
(v) $10 \le i(Q_{19}) \le 11,$ and
(vi) $i(Q_{22}) \le 12.$

Parameters closely related to γ and i are the irredundance numbers, ir and IR, and the upper domination number Γ. Kearse and Gibbons also show that:

(vii) $ir(Q_n) = \gamma(Q_n),$ for $n \le 13,$
(viii) $IR(Q_9) = \Gamma(Q_9) = 13,$
(ix) $IR(Q_{10}) = \Gamma(Q_{10}) = 15.$

For the kings graphs K_n, to be discussed further below, the authors establish the following results:

(x) $IR(K_8) = 17, IR(K_9) = 25, IR(K_{10}) = 27,$ and $IR(K_{11}) = 36.$

Kearse and Gibbons also calculate the numbers of non-isomorphic, minimum dominating sets and independent dominating sets in Q_n, for $n \le 15$ and $n \le 18$, respectively.

In 2001 [74] Östergard and Weakley publish what is arguably the most definitive paper on queens domination to date. Using a combination of known theoretical bounds for both $\gamma(Q_n)$ and $i(Q_n)$, along with advanced computer search algorithms, the authors determine quite a few new values and bounds for these two queens parameters, which we list here.

1. $\gamma(Q_n) = \lceil n/2 \rceil,$ for 17 values of n.
2. $i(Q_n) = \lceil n/2 \rceil,$ for 11 values of n.
3. One or both of $\gamma(Q_n)$ and $i(Q_n)$ is equal to one of $\{\lceil n/2 \rceil, \lceil n/2 \rceil + 1\}$, for 85 additional values of n.
4. $\gamma(Q_{4k+1}) = 2k + 1,$ for $k \le 32$.
5. For $n \le 120$, each of $\gamma(Q_n)$ and $i(Q_n)$ is either known, or known to have one of only two consecutive values.
6. $\gamma(Q_n) \le 69n/133 + (1).$
7. $i(Q_n) \le 61n/111 + \mathcal{O}(1).$
8. For all n, $(n-1)/2 \le \gamma(Q_n) \le i(Q_n).$

9. **Conjecture 27** *For all n, $i(Q_n) \le \lceil n/2 \rceil + 1$.*

10. If $n < 143$ and $n \ne 3, 11$, then $\gamma(Q_n) \ge n/2.$

The authors raise the question of whether $\gamma(Q_n) = (n-1)/2$ holds for any value of n other than $n = 3$ and $n = 11$. This question was finally answered in 2007 by Finozhenok and Weakley.

Theorem 28 (Finozhenok, Weakley [50]) *The only integers n for which*

$$\gamma(Q_n) = (n-1)/2 \text{ are } n = 3, 11.$$

In conclusion, Östergard and Weakley [74], together with the theorem of Finozhenok and Weakley, provide the following summary results:

$\gamma(Q_n) = i(Q_n) = (n-1)/2$, only for $n = 3, 11$.

$\gamma(Q_n) = \lceil n/2 \rceil$, for $n = 1, 2, 4$–$7, 9, 10, 12, 13, 17-19, 21, 23, 25, 27, 29$–$31$, $33, 37, 39, 41, 45, 49, 53, 57, 61, 65, 69, 71, 73, 77, 81, 85, 89, 91, 93, 97, 101, 105$, $109, 113, 115, 117, 121, 125, 129$–$131$.

$\gamma(Q_n) = \lceil n/2 \rceil + 1$, for $n = 8, 14, 15, 16$.

$\gamma(Q_n) \in \{\lceil n/2 \rceil, \lceil n/2 \rceil + 1\}$, for $n = 20, 22, 24$–$26, 28, 32, 34, 35, 36, 38, 40, 42$, $43, 44, 46, 47, 48, 50, 51, 52, 54, 55, 56, 58, 59, 60, 62, 63, 64, 66, 67, 68, 70$, $72, 74, 75, 76, 78, 79, 80, 82, 83, 84, 86, 87, 88, 90, 92, 94$–$96, 98$–$100, 102$–$104$, 106–$108, 110$–$112, 114, 116, 118$–$120, 122, 126, 132$.

$i(Q_n) = \lceil n/2 \rceil$, for $n = 1, 2, 5, 7, 9, 10, 13, 17, 21, 25, 33, 45, 57, 61, 69, 73, 77$, $81, 85, 89, 93, 97, 105, 109$.

$i(Q_n) = \lceil n/2 \rceil + 1$, for $n = 4, 6, 8, 12, 14-16, 18$.

$i(Q_n) \in \{\lceil n/2 \rceil, \lceil n/2 \rceil + 1\}$, for $n = 19, 20, 22, 23, 24, 26$–$30, 31, 32, 34$–$44, 46$–$56, 58$–$60, 62$–$68, 70$–$72, 74$–$76, 78$–$80, 82$–$84, 86$–$88, 90$–$92, 94$–$96, 98$–$104$, 106–$108, 110$–120.

In 2002 Burger and Mynhardt prove the following.

Theorem 29 (Burger, Mynhardt [24]) *For any $n \geq 1$, $\gamma(Q_n) \leq \frac{8}{15}n + \mathcal{O}(1)$.*

In 2002 Kearse and Gibbons provide the best known lower bounds for $IR(Q_n)$.

Theorem 30 (Kearse, Gibbons [65]) *For Q_n, $6n - O(n^{2/3}) \leq IR(Q_n)$.*

Theorem 31 (Kearse, Gibbons [65]) *For even $k \geq 6$, $6k^3 - 29k^2 - O(k) \leq IR(Q_{k^3})$.*

The authors conclude their paper with the following comment: "Finally, it seems likely, although not proven, that $6n - O(n^{2/3})$ is also an upper bound for $IR(Q_n)$."

In 2002 Weakley establishes improved upper bounds for the queens domination number and queens independent domination number.

Theorem 32 (Weakley [93], [94]) *For all $n \geq 1$, $\gamma(Q_n) \leq 34n/63 + \mathcal{O}(1) < 0.54n + \mathcal{O}(1)$.*

Theorem 33 (Weakley [93], [94]) *For all $n \geq 1$, $i(Q_n) \leq 19n/33 + \mathcal{O}(1) < 0.57n + \mathcal{O}(1)$.*

In 2003 Burger and Mynhardt provide improved upper bounds, in special cases, for both $\gamma(Q_n)$ and $\gamma(Q_n^t)$, where Q_n^t is the $n \times n$ queens graph on a torus. They present a 10-page proof of the following theorem.

Theorem 34 (Burger, Mynhardt [25]) *For all n large enough, $\gamma(Q_n) \leq \frac{101}{195}n + \mathcal{O}(1)$.*

For queens on a torus, the authors provide the following summary of known results:

$$\gamma(Q_{3k}^t) = \begin{cases} k & \text{if } k \equiv 1, 5, 7, 11 \ (\text{mod } 12) \\ k + 1 & \text{if } k \equiv 2, 10 \ (\text{mod } 12) \\ k + 2 & \text{if } k \equiv 0, 3, 4, 6, 8, 9 \ (\text{mod } 12) \end{cases}$$

They also show that if $n \equiv 2, 4 \ (\text{mod } 6)$, then $\lceil n/3 \rceil \le \gamma(Q_n^t) \le \frac{n}{2}$, and if $n \equiv 1,$ 5 (mod 6), then $\lceil n/3 \rceil \le \gamma(Q_n^t) \le \gamma(Q_n)$.

The authors show that if for some fixed k there is a dominating set of Q_{4k+1} of a certain type with cardinality $2k+1$, then for any n large enough, $\gamma(Q_n) \le [(3k+5)/(6k+3)]n + O(1)$. The same construction shows that for any $m \ge 1$ and $n = 2(6m - 1)(2k+1) - 1$, $\gamma(Q_n^t) \le [(2k+3)/(4k+2)]n + \mathcal{O}(1)$.

In 2003 Burger, Mynhardt, and Weakley prove the following for the queens domination number on a torus.

Theorem 35 (Burger, Mynhardt, Weakley [29]) *For all* $n \ge 1$, $\gamma(Q_{3n}^t) = 2n - \alpha(Q_n^t)$.

In 2003 [72] Mynhardt establishes improved upper bounds for $\gamma(Q_n^t)$ and $i(Q_n^t)$, for queens on a torus.

In 2005 [2] Amirabadi, in his MS research paper, develops search algorithms for approximating the total domination and connected domination numbers of queens graphs. His search algorithm produces results that are within three of proven lower bounds. It is known that for the first 130 values of n, $\gamma(Q_n)$ is either known or known to be one of two consecutive numbers. As a result of the author's computations, for the first 30 values of n, $\gamma_t(Q_n)$ and $\gamma_c(Q_n)$ are either known or known to be one of three consecutive numbers (cf. Table 11).

In 2006 [11] and later in 2016 [15] Burchett initiates the study of the paired domination number $\gamma_{pr}(Q_n)$, the total domination number $\gamma_t(Q_n)$, and the connected domination number $\gamma_c(Q_n)$ of queens graphs. Exact values for $\gamma_{pr}(Q_n)$, $\gamma_t(Q_n)$, and $\gamma_c(Q_n)$ are provided for the following values of n:

$\gamma_{pr}(Q_n)$: $2 \le n \le 10$, $n = 12, 13$, and $15 \le n \le 20$.
$\gamma_t(Q_n)$: $2 \le n \le 10$, $n = 12, 15, 17, 18, 19$.

Table 11 Values found for $\gamma_t(Q_n)$ and $\gamma_c(Q_n)$

n	1	2	3	4	5	6	7	8	9	10	11	12	13	14	15
$\gamma_t \ge$	x	2	2	2	3	3	4	4	5	6	6	7	7	8	8
Found	x	2	2	2	3	4	4	5	5	6	7	7	8	9	9
$\gamma_c \ge$	1	1	1	2	3	3	4	5	5	6	7	7	8	9	9
Found	1	1	1	2	3	4	4	5	5	6	7	8	8	9	10
n	16	17	18	19	20	21	22	23	24	25	26	27	28	29	30
$\gamma_t \ge$	9	9	10	10	11	12	12	13	13	14	15	15	16	16	17
Found	10	10	11	12	12	13	14	14	15	15	16	17	17	18	19
$\gamma_c \ge$	10	10	11	11	12	13	13	14	15	15	16	16	17	18	18
Found	11	12	12	12	13	14	15	16	16	17	18	19	19	19	20

$\gamma_c(Q_n)$: $2 \leq n \leq 11$, $n = 13, 14, 16, 17, 19, 20, 22, 23$.

The following bounds are also provided:

$$\tfrac{4}{7}(n-1) \leq \gamma_t(Q_n) \leq \gamma_{pr}(Q_n).$$
$$\gamma_t(Q_n) \leq \gamma_{pr}(Q_n) \leq 2n/3 + \mathcal{O}(1).$$
$$2n/3 - 1 \leq \gamma_c(Q_n) \leq 2n/3 + \mathcal{O}(1).$$

In 2008 [84] Sinko and Slater initiate the study of the *border queens domination problem*, that is, determining how few queens are needed to cover all of the squares of an $n \times n$ chessboard when the queens are restricted to squares on the border. We denote this number by $\gamma_{bor}(Q_n)$. In this paper the authors give the values of $\gamma_{bor}(Q_n)$ for $1 \leq n \leq 13$ shown in Table 12.

What is particularly interesting about this is the observation that $\gamma_{bor}(Q_{12}) = 10 > \gamma_{bor}(Q_{13}) = 9$. Thus, the border queens domination number is not monotonically non-decreasing. It has long been conjectured that the queens domination number is monotonically non-decreasing, that is, for all $n \geq 1$, $\gamma(Q_n) \leq \gamma(Q_{n+1})$.

The authors present solutions to the border queens domination number for $4 \leq n \leq 10$ as follows:

$$\gamma_{bor}(Q_n) = \begin{cases} 2 & \text{if } n = 4 : \{(a,2),(d,2)\} \\ 3 & \text{if } n = 5 : \{(b,5),(c,1),(d,5)\} \\ 4 & \text{if } n = 6 : \{(b,6),(c,1),(d,1),(e,6)\} \\ 5 & \text{if } n = 7 : \{(b,7),(c,7),(d,1),(e,7),(f,7)\} \\ 6 & \text{if } n = 8 : \{(b,8),(c,8),(d,1),(e,1),(f,8),(g,8)\} \\ 6 & \text{if } n = 9 : \{(a,7),(c,1),(e,1),(g,9),(i,3),(i,5)\} \\ 6 & \text{if } n = 10 : \{(a,1),(a,7),(d,1),(g,10),(j,1),(j,4)\} \end{cases}$$

They also establish by a computer search that $11 \leq \gamma_{bor}(Q_{14}) \leq 12$ and $9 \leq \gamma_{bor}(Q_{15}) \leq 13$.

We illustrate a solution for $\gamma_{bor}(Q_{10})$ in Figure 16. Note that the solutions given for $4 \leq n \leq 8$ above, all have a symmetry about the center column, while the solutions given for $n = 9$ and $n = 10$, although asymmetric, have a type of rotational symmetry.

For the general case, they establish the following bounds.

Theorem 36 (Sinko, Slater [84]) *For all $n \geq 4$,*

Table 12 Values of $\gamma_{bor}(Q_n)$

k	1	2	3	4	5	6	7	8	9	10	11	12	13
$\gamma(Q_n)$	1	1	1	2	3	3	4	5	5	5	5	6	7
$\gamma_{bor}(Q_n)$	1	1	2	2	3	4	5	6	6	6	9	10	9

Fig. 16 $\gamma_{bor}(Q_{10}) = 6$

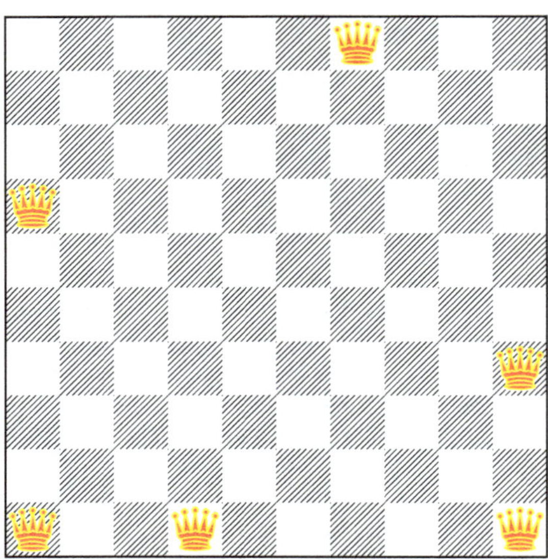

$$n(2 - 9/2n - \sqrt{8n^2 - 49n + 49}/2n) \le \gamma_{bor}(Q_n) \le n - 2.$$

The authors note that $lim_{n \to \infty}(2 - 9/2n - \sqrt{8n^2 - 49n + 49}/2n) = 2 - \sqrt{2}$.

For $n = 3t + 1$ they improve this upper bound to $2t + 1$ if $3t + 1$ is odd and $2t$ if $3t + 1$ is even.

In 2011 [12, 13] Burchett studies k-tuple domination in the rooks graph, in which it is required that every square not in S be attacked at least k times with a minimum number of rooks. He also continues the study of the border queens domination problem.

In 2014 [10] Brown considers a variation on the queens domination problem posed by Bell and Stevens [3] in their survey of the *n queens problem*. Bell and Stevens asked: given an $n \times n$ chessboard on which one queen has been arbitrarily placed, when is it possible to place $n - 1$ remaining queens to create an arrangement of n non-attacking queens? In [10] Brown considers the possibility that a solution to this *Initial Placement Problem* is always possible for $n > 6$, and proceeds to provide solutions for $n \equiv 0 \pmod 6$ and $n \equiv 2 \pmod 6$. He then conjectures that the Initial Placement Problem is solvable for all initial placements of two non-attacking queens when $n > 9$.

In 2016 Burchett provides new upper bounds for paired, total, and connected domination for the queens graph.

Theorem 37 (Burchett [14]) *For all $n \ge 1$,*

(i) $\lceil \frac{2n}{3} \rceil - 1 \le \gamma_c(Q_n) \le \lceil \frac{2n}{3} \rceil$.

(ii) *For $n \ge 21$ and $n \equiv 3, 4, 5 \pmod 6$, $\gamma_t(Q_n) \le \lceil \frac{2n}{3} \rceil - 1$.*

(iii) *For $n \ge 22$ and $n \equiv 4 \pmod 6$, $\gamma_{pr}(Q_n) \le \lceil \frac{2n}{3} \rceil - 1$.*

Fig. 17 Minimum queens connected dominating set for Q_{21}

For an illustration of this theorem, cf. Figure 17.

In 2017 [6] William Bird, in his PhD thesis, adds considerably to what is known about a variety of queens domination problems. Noteworthy are three new values of $\gamma(Q_n)$:

$$\gamma(Q_{20}) = 11,\ \gamma(Q_{22}) = 12,\ \text{and}\ \gamma(Q_{24}) = 13.$$

In each of these three cases the value was known to be one of two values, this value shown or one less. Thus, in each of these three cases, the value is $\gamma(Q_n) = \lceil n/2 \rceil + 1$. In addition Bird's sophisticated computer program was able to establish the following five new values: $i(Q_{19}) = 11$, $i(Q_{20}) = 11$, $i(Q_{22}) = 12$, $i(Q_{23}) = 13$, and $i(Q_{24}) = 13$; again, in all of these five cases, $i(Q_n) = \lceil n/2 \rceil + 1$.

Bird also adds many new values of the border queens domination number for $14 \leq n \leq 24$, given in boldface in the following Table 13.

Table 13 Values of $\gamma_{bor}(Q_n)$

n	10	11	12	13	14	1 5	16	17	18	19	20	121	22	23	24
$\gamma_{bor}(Q_n)$	6	9	10	9	**12**	**13**	**10**	**14**	**16**	**13**	**18**	**19**	**14**	**21**	**22**

Table 14 Number of solutions to $\gamma(Q_n)$ and $i(Q_n)$

n	3	4	5	6	7	8	9	10	11	12	13	14	15	16	17	18
$\gamma(Q_n)$	1	2	3	3	4	5	5	5	5	6	7	8	9	9	9	9
Solutions	1	3	37	1	13	638	21	1	1	1	41	588	25,872	43	22	2
$i(Q_n)$	1	3	3	4	4	5	5	5	5	7	7	8	9	9	9	10
Solutions	1	2	2	17	1	91	16	1	1	105	4	55	1314	16	2	28

It is interesting to note from Bird's data how frequently $\gamma_{bor}(Q_n) > \gamma_{bor}(Q_{n+1})$, that is, that this function fails to be monotonic non-decreasing. But these numbers suggest the following conjecture.

Conjecture 38 *For all $n \geq 1$, $\gamma_{bor}(Q_n) \leq \gamma_{bor}(Q_{n+2})$.*

Of equal interest, Bird determines the number of minimum dominating sets and minimum independent dominating sets for Q_n, for $3 \leq n \leq 18$, as shown in Table 14.

We close this section on queens domination by referring to the 2010 paper [49] by Fernau, in which he discusses three computational approaches to solving the queens domination problem: (i) backtracking, (ii) dynamic programming on subsets, and (iii) dynamic programming using treewidth, or path decompositions. He points out that the determination of the $\gamma(Q_n)$ sequence of integers is listed as Problem C18 in Richard Guy's book entitled "Unsolved Problems in Number Theory." [61].

At the end of this paper, Fernau discusses the perplexing problem that so little is known about the complexity of the queens domination problem.

QUEENS DOMINATING SET
Instance: Positive integer n, positive integer k.
Question: Does Q_n have a dominating set of cardinality at most k?

It is unknown if this problem is NP-hard. This would seem unlikely, since for the first 130 values of n, $\gamma(Q_n)$ is either known exactly or is known to be one of two consecutive integers. And yet no polynomial-time algorithm is known for computing the value of $\gamma(Q_n)$. Fernau also points out that since all known upper bounds for $\gamma(Q_n)$ are algorithmic in nature, except for additive $\mathcal{O}(1)$ constants, $\gamma(Q_n)$ can be approximated up to a factor of $\frac{138}{133}$. For a related complexity question involving n-queens, see Gent, Jefferson and Nightingale [57].

5 Bishops

In this section we review results about domination in *bishops graphs*. The bishops graph B_n is the graph whose vertices are the n^2 squares of the $n \times n$ chessboard, and two vertices are adjacent if and only if their corresponding squares lie on a common diagonal, which corresponds to a move of a bishop.

In 1986 Cockayne, Gamble, and Shepherd prove the following two basic theorems which determine the domination and total domination numbers of all bishops graphs. The fact that $\gamma(B_n) = n$ had previously been proved by Yaglom and Yaglom [96].

Theorem 39 (Cockayne, Gamble, Shepherd [41]) *For any n, $\gamma(B_n) = i(B_n) = n$.*

Proof Sketch. The set of squares of a nearest column to the center is an independent dominating set of the bishops graph; hence, $\gamma(B_n) \leq i(B_n) \leq n$. It remains to show that $\gamma(B_n) \geq n$. Assume that $\gamma(B_n) < n$. Then there must be a diagonal not having any bishop on it.

Assume that the northwest to southeast running diagonals are labeled sequentially 1, 2, …, $2n - 1$ starting in the southwest corner and proceeding to the northeast corner. Notice that for $1 \leq d \leq n$ diagonal d has d squares, and for $n + 1 \leq d \leq 2n - 1$, diagonal d has $2n - d$ squares.

Let r (and b) be the labels of the red (black) diagonal closest to the main diagonal which has no bishop. Without loss of generality, we may assume that $\{r, b\} \subset \{1, 2, \ldots, n\}$.

Diagonal r has r squares and these must be dominated. By the definition of r, there are bishops on each diagonal strictly between r and $2n - r$, else there is a row closer to the main diagonal which has no bishop. Hence, the number of red bishops in any dominating set satisfies

$$n_r \geq max\{r, n - r - 1\}.$$

Similarly,

$$n_b \geq max\{b, n - b - 1\}.$$

From these two inequalities we can deduce that $\gamma(B_n) \geq n$. $\qquad\square$

Theorem 40 (Cockayne, Gamble, Shepherd [41]) *For any $n \geq 3$,*

$$\gamma_t(B_n) = 2\lceil \frac{2}{3}(n - 1)\rceil.$$

Proof Sketch. The bishops graph B_n is the disjoint union of the red bishops graph R_n and the black bishops graph B_n. We summarize only the proof that $\gamma_t(B_n) = 2\lceil \frac{2}{3}(n - 1)\rceil$ for n even.

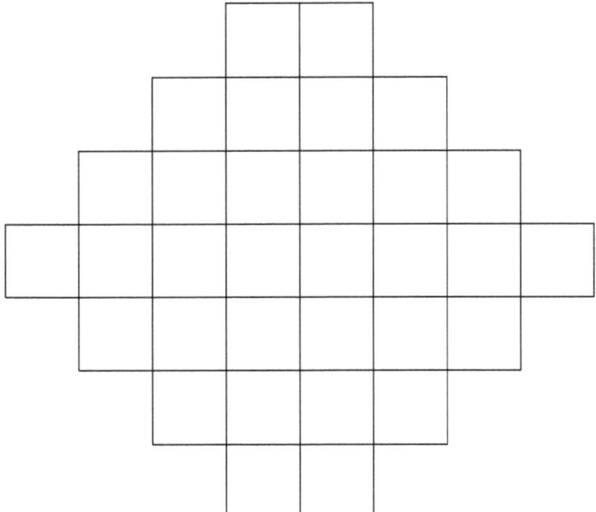

Fig. 18 Diamond-shaped chessboard

Notice that a bishops total dominating set of B_n is precisely a rook total dominating set of the diamond-shaped chessboard S_n, which has n rows and $n-1$ columns (cf. Figure 18). For ease of presentation, we use rooks, rows, and columns, rather than bishops and diagonals.

Lemma 41 ([41]) *For any n, S_n has a minimum rooks total dominating set with rooks on consecutive rows and columns.*

It follows from this lemma that some minimum rooks total dominating set of S_n may be used to construct a rooks total dominating set of an $m \times p$ rectangular board with property REL, i.e., a rook on every line (row or column). It is shown that such a board satisfies $m + p \geq n - 1$, and hence, if $s(m, p)$ equals the minimum number of rooks in an REL total dominating set of an $m \times p$ board, then $\gamma_t(B_n) = min_{m+p \geq n-1}\{s(m, p)\}$.

Lemma 42 ([41]) *For $p \leq m \leq 2p + 2$, $s(m, p) = \lceil \frac{2}{3}(m + p)\rceil$, and for $m > 2p + 2$, $s(m, p) = m$.*

Proof.

By establishing and solving a recurrence for $s(m, p)$. 						□

One may deduce from this that $\gamma_t(B_n) \geq \lceil \frac{2}{3}(n - 1)\rceil$. The final part of the proof exhibits a rooks total dominating set of S_n with $\lceil \frac{2}{3}(n - 1)\rceil$ rooks. 						□

Figure 19 illustrates a minimum independent dominating set of bishops on B_8 and a minimum total dominating set of bishops on B_8.

In 1994 [66] Koehler, in his MS research paper, initiates the study of chessboard domination problems in three-dimensional chessboards. For three-dimensional bishops graphs B_n^3, Koehler obtains the following results, cf. Table 15.

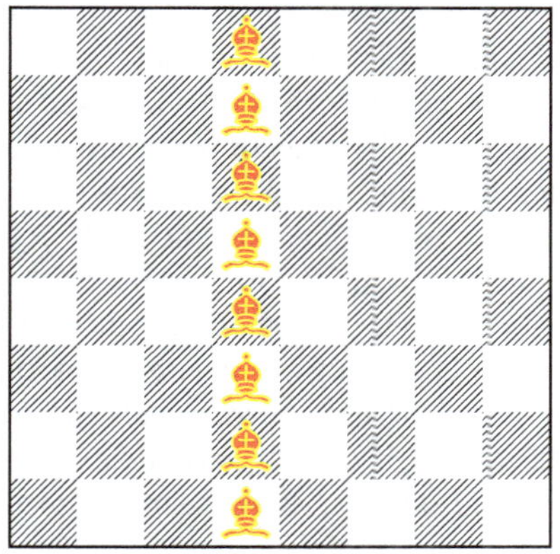 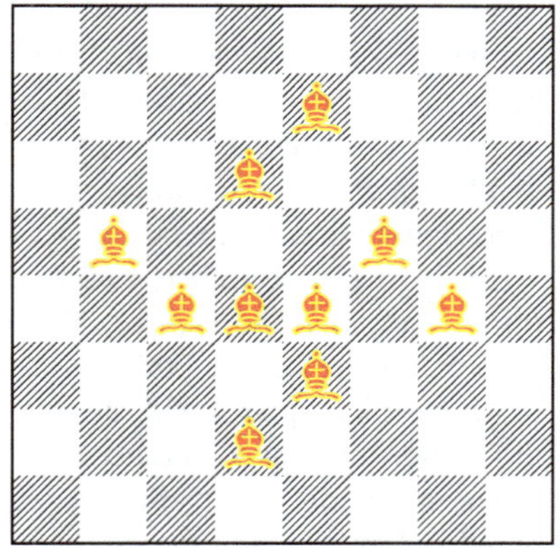

Fig. 19 Minimum bishops independent and total dominating set for B_8

Table 15 Values of $\gamma(B_n^3)$

n	1	2	3	4	5	6	7
$\gamma(B_n^3)$	1	2	3	8	≤ 13	≤ 18	≤ 27

Table 16 Minimal dominating set of 8 bishops on B_4^3

 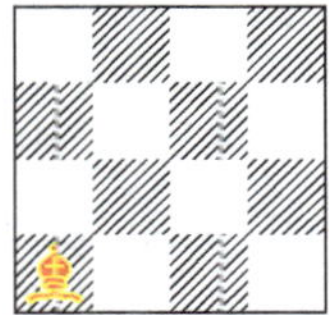  

Table 16 illustrates a minimum dominating set of 8 bishops on the three-dimensional bishops graph B_4^3. Each level is a 4×4 bishops graph, and there are four levels from left to right. A bishop dominates every square in its two diagonals on its level, and all squares above and below it on ascending or descending diagonals.

In 2002 [51] Fisher and Thalos consider bishops graphs on rectangular $k \times n$ boards, which we denote by $B_{k,n}$. They extend the result by Yaglom and Yaglom [96], and independently by Cockayne, Gamble, and Shepherd [41], that $\gamma(B_{n,n}) = n$, as follows.

Theorem 43 (Fisher, Thalos [51]) *For $B_{k,n}$,*

(i) *If $k < n$, then $\gamma(B_{k,n}) = 2\lfloor n/2 \rfloor$.*
(ii) *For $2 < 2k < n$, $\gamma(B_{k,n}) \leq 2\lfloor (k+n)/3 \rfloor$.*

The authors then make the following conjecture.

Conjecture 44 (Fisher, Thalos) *For $2 < 2k < n$, $\gamma(B_{k,n}) = 2\lfloor (k+n)/3 \rfloor$.*

They show that this conjecture is true when $k \leq 3$ or $n \leq 2k + 5$.

In 2016 [14] Burchett introduces the study of the *k-tuple domination number*, denoted $\gamma_{\times k}(B_n)$, which equals the minimum number of bishops in a set S so that every square not in S is attacked by at least k bishops, and every bishop is attacked by at least $k-1$ bishops. In this paper, for odd n, and $k \leq \lfloor \frac{n}{2} \rfloor$, the k-tuple domination number of B_n is shown to equal one of two possible values, and for even n, the k-tuple domination number is shown to be bounded between $nk - k$ and nk for $k \leq \frac{n}{2}$.

In 2016 [16] Burchett and Buckley introduce the concept of the *kth border bishop's domination number*. When $k = 1$, the kth border is the set of outermost squares on the board. For $k = 2$, the kth border consists of all squares adjacent to first border squares and, in general, the set of kth border square equals the set of square adjacent to the set of $k-1$ border squares. Let $\gamma_{bor,k}(B_n)$ denote the minimum number of bishops which can be placed only on kth border square in order to dominate all squares not containing a bishop. The authors point out that as k grows large with respect to n, kth border dominating sets might not exist. In fact, they show that for $k > \lfloor n/4 \rfloor + 1$, no kth border dominating sets exist.

In this paper, they prove the following result.

Theorem 45 (Burchett, Buckley) *If a kth border dominating sets exist for B_n, then*

(i) $\gamma_{bor,k}(B_n) = 2n - 4k + 2$, *for $n \equiv 2, 3 \pmod 4$, and*
(ii) $\gamma_{bor,k}(B_n) = 2n - 4k + 2$, *for $n \equiv 0, 1 \pmod 4$, unless $k = \lfloor n/4 \rfloor + 1$, in which case $\gamma_{bor,k}(B_n) = 2n - 4k + 4$, and $\gamma_{bor,k}(B_n) = 2n - 4k + 3$, respectively.*

In 2017 [70] Low and Kapbasov introduce the study of the vertex independence number of bishops $\alpha(B_{m,n})$ and kings $\alpha(K_{m,n})$ on $m \times n$ rectangular and *cylindrical* chessboards, where on a cylindrical chessboard the left and right edges of the board are identified. The authors only consider narrow boards, $1 \times n$, $2 \times n$, and $3 \times n$, and for each of these they determine the number of non-attacking bishops or kings positions.

6 Knights

The study of knights domination dates at least back to 1896, in *L'Intermédiaire des Mathématiciens*, Gauthier-Villars, Paris, Tome III (1896), p. 58, Tome IV (1897), p. 15, and Tome V (1898), p. 87 (cf. Ball [80]).

In 1910 Ahrens [1] presents known results for this problem, giving a covering for the 11×11 chessboard using 22 knights, which was known since 1896. The value $\gamma(N_{11}) \leq 21$ was provided by Lemaire in 1973 [69] (cf. Figure 20).

In 1967, in his *Scientific American* column *Mathematical Games*, Martin Gardner [53] discusses the knights covering problem for the $n \times n$ chessboard, and gives the best known solutions for various values of n (cf. Table 17).

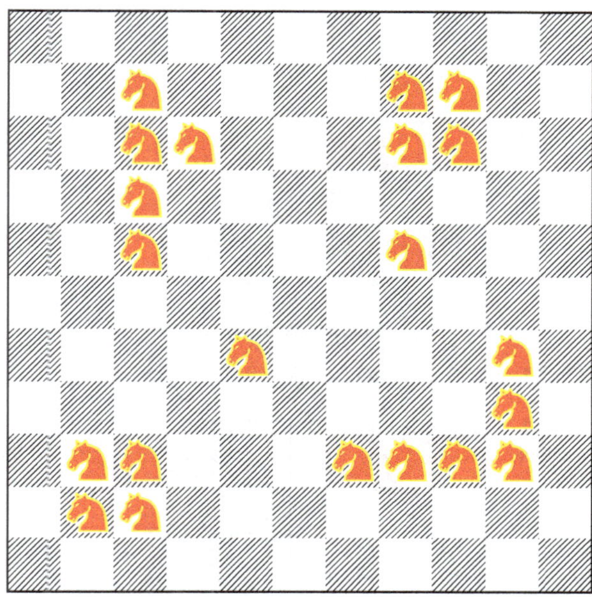

Fig. 20 Lemaire's dominating set of 21 knights on N_{11}

Table 17 Values of $\gamma(N_n)$ and number of solutions

n	3	4	5	6	7	8	9	10
$\gamma(N_n)$	4	4	5	8	10	12	14	16
# solutions	2	3	8	22	3	1	1?	1?

Gardner gives solutions for $\gamma(N_n)$, for $3 \leq n \leq 8$, suggesting that the readers find solutions for $n = 9, 10$, stating that both of these solutions were thought to be unique. Answers appeared in the following issue [54]. In the January, 1968 issue [55], Gardner presents a second solution for $\gamma(N_{10})$ that had been found by his readers. Also in that issue, Gardner gives the best known solutions for $\gamma(N_n)$ for $n = 11, 12, 13, 14, 15$, which use 22, 24, 28, 34, and 37 knights, respectively.

Proofs of the optimality of the values of $\gamma(N_n)$ given in Table 17 by Gardner, along with figures showing optimal solutions for $3 \leq n \leq 10$, due to Frank Rubin, can be found at the website: http://www.contestcen.com/knight.htm. Figure 21 illustrates two solutions.

In 1987 [62] Hare and Hedetniemi present a dynamic programming algorithm for computing the knights domination number $\gamma(N_n)$ on rectangular $m \times n$ chessboards, which is linear in n but exponential in m. Figure 22 illustrates a minimum dominating set of knights on $N_{8,10}$ that is the only minimum dominating set for this knights graph.

The authors present the following values of $\gamma(N_{m,n})$ in Table 18.

The authors make the following conjectures.

Conjecture 46 (Hare, Hedetniemi [62]) *For $k = 3$ and $n > 8$, $\gamma(N_{k,n})$ is given by the following:*

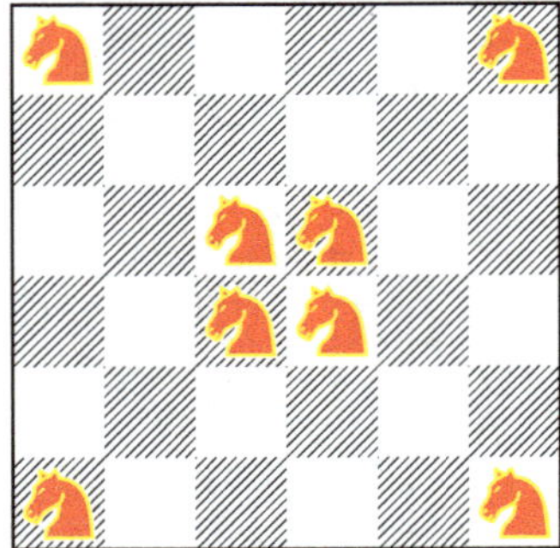 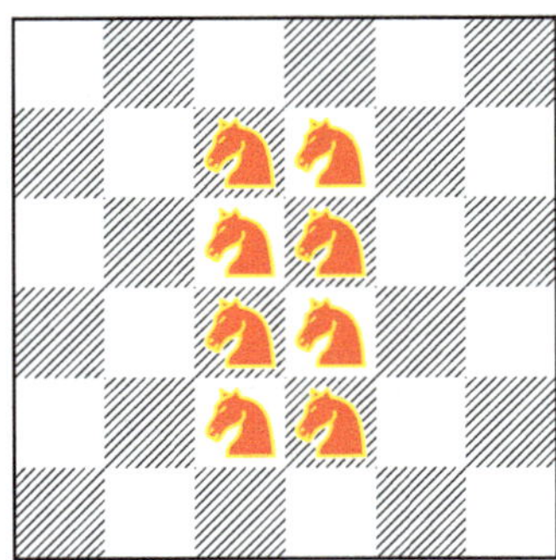

Fig. 21 Independent and total dominating knights for N_6

Fig. 22 Unique knights dominating set for $N_{8,10}$

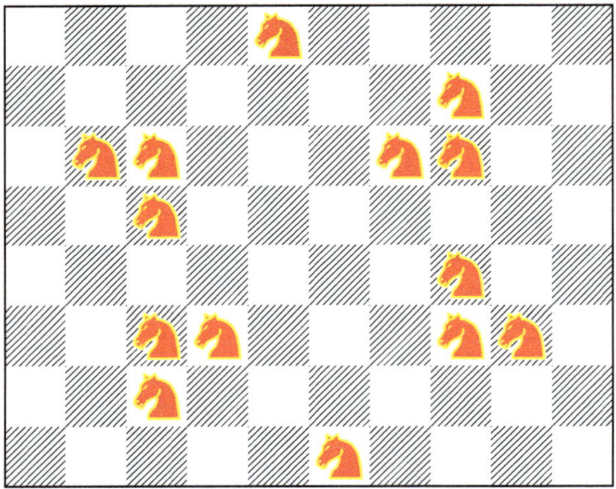

Table 18 Values of $\gamma(N_{m,n})$

$\gamma(N_{m,n})$	3	4	5	6	7	8	9	10	11	12
3	4	4	4	4	6	8	8	8	8	8
4		4	4	4	6	8	8	8	8	8
5			5	6	7	8	8	9	10	10
6				8	8	8	10	12	12	12
7					10	11	12	14	15	16
8						12	13	14	16	16
9							14	15	17	18
10								16		

$$\gamma(N_{k,n}) = \begin{cases} (2n+4)/3 & \text{for } n \equiv 1 \pmod 6 \\ (2n+5)/3 & \text{for } n \equiv 2 \pmod 6 \\ 4\lceil n/6 \rceil & \text{otherwise} \end{cases}$$

Conjecture 47 (Hare, Hedetniemi [62]) *For $k=4$ and $n>7$, $\gamma(N_{k,n})$ is given by the following:*

$$\gamma(N_{k,n}) = \begin{cases} (2n+4)/3 & \text{for } n \equiv 1 \pmod 6 \\ 4^-n/6\rceil & \text{otherwise} \end{cases}$$

Conjecture 48 (Hare, Hedetniemi [62]) *For $k=6$ and $n>5$, $\gamma(N_{k,n})$ is given by the following:*

$$\gamma(N_{k,n}) = \begin{cases} n+1 & \text{for } n \equiv 1 \pmod 4 \\ (2n+5)/3 & \text{for } n \equiv 2 \pmod 6 \\ 4^-n/4\rceil & \text{otherwise} \end{cases}$$

In 1994 Wallis, in his PhD thesis, introduces the study of domination in k-dimensional chessboards and gives the following theorem.

Theorem 49 (Wallis [89]) *For any n, $\alpha(N_n^k) = \Gamma(N_n^k) = IR(N_n^k) = \lceil \frac{n^k}{2} \rceil$.*

Proof Sketch. It is well known that the knights graph N_n is bipartite. Wallis shows that for any k, the k-dimensional knights graph is still bipartite. The theorem then follows from this well-known result.

Theorem 50 (Cockayne, Favaron, Payan, Thomason [39]) *For any bipartite graph G, $\alpha(G) = \Gamma(G) = IR(G)$.*

In 1995 [56] Garnick and Nieuwejaar initiate the study of total domination on rectangular chessboards by considering knights graphs and kings graphs. They observe that Rouse Ball had the idea of total domination in 1892. It is easy to see that for all n, $\gamma_t(R_n) = n$ for rooks graphs. And it is immediate that queens diagonal domination provides an upper bound, that is, $\gamma_t(Q_n) \le \gamma_{diag}(Q_n)$.

For knights total domination they provide the following results.

Theorem 51 (Garnick, Nieuwejaar [56]) *For all $m, n>4$,*

(i) $mn/8 < \gamma_t(N_{m,n})$,
(ii) $\gamma_t(N_{m,n}) \le (mn+5m+6n+56)/8$, *for $m \equiv n \pmod 2$,*
(iii) $\gamma_t(N_{m,n}) \le (mn+5m+5n+43)/8$, *for m odd and n even.*

Using a backtracking search algorithm, the authors were able to determine the following values of $\gamma_t(N_n)$ for square chessboards:

$$\gamma_t(N_n) = \begin{cases} 6 & \text{if } n = 4 \\ 7 & \text{if } n = 5 \\ 8 & \text{if } n = 6 \\ 10 & \text{if } n = 7 \\ 14 & \text{if } n = 8 \\ 18 & \text{if } n = 9 \end{cases}$$

They also provide improved upper bounds for $\gamma_t(N_n)$ as shown in Table 19.

7 Kings

As was mentioned in Section 3 Early Chessboard Domination, the following formulas are known for the kings graph.

Theorem 52 (Yaglom and Yaglom [96]) *For kings graphs,* $\gamma(K_n) = \lfloor \frac{n+2}{3} \rfloor^2$.

Theorem 53 (Yaglom and Yaglom [96]) *For rectangular kings graphs,*

$$\gamma(K_{m,n}) = \lfloor \frac{m+2}{3} \rfloor \lfloor \frac{n+2}{3} \rfloor.$$

Theorem 54 (Yaglom and Yaglom [96]) *For kings graphs,* $i(K_n) = \alpha(K_n) = \lfloor \frac{n+1}{2} \rfloor^2$.

Notice that while for any graph G, we have $\gamma(G) \le \gamma_t(G) \le 2\gamma(G)$, for the kings graph K_n, both of these bounds can be achieved, since $\gamma(K_4) = \gamma_t(K_4) = 4$ and $\gamma(K_7) = \gamma_t(K_7) = 9$, but $\gamma_t(K_6) = 2\gamma(K_6) = 8$ (cf. Figure 23).

In 1995 Garnick and Nieuwejaar initiate the study of total domination on rectangular chessboards for kings graphs.

For narrow boards $1 \le m \le 4$, it is easy to determine the kings total domination number.

Theorem 55 (Garnick, Nieuwejaar [56]) *For $n > 1$ and $m \le 3$,*

$$\gamma_t(K_{m,n}) = \begin{cases} n/2 & \text{for } n \equiv 0 \ (\text{mod } 4) \\ \lfloor n/2 \rfloor + 1 & \text{for } n \equiv 1, 2, 3 \ (\text{mod } 4) \\ 2\lceil n/3 \rceil & \text{for } m = 4 \end{cases}$$

Table 19 Upper bounds for $\gamma(N_n)$

n	13	14	15	16	17	18	19	20	21	22	23	24	25	30
$\gamma_t(N_n) \le$	32	39	44	48	57	61	66	75	80	86	94	101	109	152

Fig. 23 Four dominating kings (left) and eight total dominating kings (right) on a 6×6 chessboard

Table 20 Values of $\gamma_t(K_n)$

n	2	3	4	5	6	7	8	9	10	11	12
$\gamma_t(K_n)$	2	2	4	5	8	9	12	15	18	21	24

Table 21 Upper bounds for $\gamma_t(K_n)$

n	13	14	15	16	17	18	19	20	21	22	23	24	25	30
$\gamma(K_n) \leq$	29	33	38	43	48	54	60	63	72	80	87	95	102	146

They provide the following general lower and upper bounds for $\gamma_t(K_n)$.

Theorem 56 (Garnick, Nieuwejaar [56]) *For all* m, $n \geq 5$, $mn/7 \leq \gamma_t(K_{m,n}) \leq (mn + 2n + 89)/7$.

They provide both exact values and improved upper bounds for $\gamma_t(K_n)$, as shown in Tables 20 and 21.

In 2002 [91] Watkins and Ricci initiate the study of kings domination on a torus.

In 2003 Favaron, Fricke, Pritikin, and Puech establish the following results involving irredundant sets of kings.

Theorem 57 (Favaron, Fricke, Pritikin, Puech [48]) *For* $n \geq 6$, $(n - 1)^2/3 \leq IR(K_n) \leq n^2/3$.

Theorem 58 (Favaron, Fricke, Pritikin, Puech [48]) *For* $n \geq 6$, $\lceil(n - 2)^2/3\rceil + 3 \leq \Gamma(K_n) \leq n^2/3$.

Theorem 59 (Favaron, Fricke, Pritikin, Puech [48]) *For* $n \geq 1$, $\lceil n^2/9\rceil \leq ir(K_n) \leq \lfloor(n + 2)/3\rfloor^2$, *and* $ir(K_n) = n^2/9$ *when* $n \equiv 0 \pmod 3$.

In Table 22 the authors prove the first few values of $ir(K_n)$, $\Gamma(K_n)$, and $IR(K_n)$, cf. Figure 24 for a minimaximal irredundant set of kings on K_7.

Table 22 Values of $ir(K_n)$, $\gamma(K_n)$, $IR(K_n)$

n	1	2	3	4	5	6	7	8	9
$ir(K_n)$	1	1	1	3	4	4	8	9	9
$\Gamma(K_n)$	1	1	4	4	9	9	16		
$IR(K_n)$	1	1	4	4	9	9	16		

Fig. 24 $ir(K_7) = 8$

The authors also offer an intermediate value, or interpolation, theorem for the cardinalities of maximal independent sets in kings graphs K_n.

Theorem 60 (Favaron, Fricke, Pritikin, Puech [48]) *For any $n \geq 1$ and positive integer t such that $i(K_n) \leq t \leq \alpha(K_n)$, there exists a maximal independent set of t kings on K_n.*

8 Rooks

The structure of rooks graphs R_n is the simplest of all chessboard graphs. Therefore, the values of all seven domination parameters are fairly easy to establish.

Theorem 61 (Yaglom and Yaglom Yaglom and Yaglom [96]) *For $n \geq 1$, $\gamma(R_n) = i(R_n) = \alpha(R_n) = n$.*

Corollary 62 *For $n \geq 2$, $\gamma(R_n) = \gamma_t(R_n)$.*

The following three results are given, but stated as unpublished, in [52].

Theorem 63 (Hedetniemi, Hedetniemi, Wallis) *For $n \geq 1$, $ir(R_n) = n$.*

Theorem 64 (Hedetniemi, Hedetniemi, Wallis) *For $n \geq 1$, $\Gamma(R_n) = n$.*

Theorem 65 (Hedetniemi, Jacobson, Wallis) *For $n \geq 4$, $IR(R_n) = 2n - 4$, $IR(R_1) = 1$, $IR(R_2) = 2$, and $IR(R_3) = 3$.*

Table 23 Lower bound of $31 \leq \Gamma(R_5^3)$

Top band (three levels side by side):

	R	R	R						R					R
	R	R	R						R					R
	R	R	R	R					R					R
			R			R	R	R	R		R	R	R	

Bottom band (two levels side by side):

					R				
				R					
				R					
				R					
	R	R	R						

In 1994 Koehler, in his MS research paper, initiates the study of chessboard domination problems in three-dimensional rooks graphs.

Theorem 66 (Koehler [66]) *For $n \geq 1$, $\gamma(R_n^3) = i(R_n^3) = \frac{n^2}{2}$.*

Theorem 67 (Koehler [66]) *For $n \geq 1$, $\alpha(R_n^3) = n^2$.*

Theorem 68 (Koehler [66]) *For $n \geq 3$, $\Gamma(R_n^3) \geq 3(n-2)^2 + 4$.*

Table 23 illustrates a minimal dominating set of 31 rooks on the three-dimensional rooks graph R_5^3. Each level is a 5×5 rooks graph, and there are five levels. A rook dominates every vertex in its row and column on its level, and all rooks above and below it.

In 2008 [33] Chen and Ho initiate the study of rooks domination on what are called *sawtoothed chessboards*, or *STC* for short. These are chessboards whose boundary forms two staircases from left down to right without any holes inside. A rook at square (i, j) still dominates all squares in row i and column j. In this paper, the authors represent an STC by two particular graphs: a rooks graph and a *board graph*. They show that for an STC, the rooks graph is the line graph of the board graph, and the board graph is a bipartite permutation graph. Thus, the rooks domination problem on STCs can be solved by any algorithm for solving the edge domination problem on bipartite permutation graphs.

9 Other Varieties of Chessboard Domination Problems

In this concluding section we briefly mention a variety of other types of chessboard domination problems that have been considered in the literature; they are mixed and quite varied.

1. In 2009 [32] Chatham, Doyle, Fricke, Reitmann, Skaggs, and Wolff consider the general problem of placing a fixed number k of pawns on a chessboard in such a way as to influence either the maximum number of independent pieces

of one kind that can be placed on the board, or the minimum number of pieces necessary to cover all squares. The main result in this paper is that for each positive integer k and each $n > max\{87 + k, 25k\}$, it is possible to place k pawns and $n + k$ independent queens on Q_n. The authors consider the same problem for bishops and rooks.

2. In 1998 [87] Theron and Geldenhuys consider queens domination in *beehive* or *hexagonal* chessboards, in which each square is a hexagon. For example, in square $n \times n$ hexagonal chessboards they show that the diagonal queens domination number equals $n - 1$.

3. In 1999 [21] Burger and Mynhardt study queens domination on hexagonal boards and show that on hexagonal boards with $n \geq 1$ rows and diagonals, for $n \equiv 3 \pmod 4$, there are only two types of minimum dominating sets of queens. The authors also study the queens irredundance numbers on 5×7 hexagonal boards.

4. In 2000 [7] Bode and Harborth study the independence numbers of chess-like pieces on boards whose cells are either triangles or hexagons, and for many of these pieces they determine the independence numbers.

5. In 2000 [86] Theron and Burger study queens domination on hexagonal boards.

6. In 2003 [9] Bode, Harborth, and Harborth study the kings independence numbers on triangle-cell chessboards.

7. In 2003 [8] Bode and Harborth study three types of knights on triangle-cell and hexagonal boards, and determine the independence numbers for two of these types of knights, and for one residue class mod 4 for the third type.

8. In 2005 [82] Sinko and Slater introduce the study of several domination-related parameters in chessboards, called *influence parameters*. The *influence of a vertex* v in a graph G equals $I(v) = |N[v]|$, the number of vertices it dominates. The *influence of a set* S equals the sum of the influences of its vertices, that is, $I(S) = \Sigma_{v \in S} I(v) = \Sigma_{v \in S} |N[v]|$.

   A vertex set S is called an *efficient dominating set*, or a *perfect code* if for every vertex $v \in V$, $|N[v] \cap S| = 1$. Since not every graph G has an efficient dominating set, one can instead consider the maximum number of vertices that can be dominated by a set S subject to the restriction that no vertex is dominated more than once; this is called the *efficient domination number*, denoted $F(G)$. This restriction means that the set S must be a packing, that is, for every $u, v \in S$, $d(u, v) \geq 3$. Thus, $F(G) = max\{I(S) : S \text{ is a packing }\}$.

   Similarly one might seek to minimize the total amount of domination, given that every vertex must be dominated at least once. This gives rise to a parameter called the *total redundance* $R(G) = min\{I(S) : S \text{ dominates } V(G)\}$.

   In this paper, the authors consider the values of these and several other related parameters on rectangular rooks, kings, and knights graphs.

   In 2006 [83] Sinko and Slater study the efficient domination number $F(N_{m,n})$ on rectangular knights graphs $N_{m,n}$. They provide the following initial values in Table 24.

   In Table 25 we illustrate a set of three knights, labeled **N**, dominating a set of 19 squares at most once, labeled **X**.

Table 24 Small values of $F(N_{m,n})$

$N_{m,n}$	$N_{1,n}$	$N_{2,n}$	$N_{3,8t}$	$N_{3,3}$	$N_{3,4}$	$N_{4,4}$	$N_{5,5}$
$F(N_{m,n})$	n	$2n$	$20t$	7	12	12	19

Table 25 $F(N_{5,5}) = 19$

X	X		X	X
X		N		X
X	X	N		X
X	X	X	X	X
N	X		X	

Table 26 Values of knights $\gamma(N_{m,n})$ and number of solutions

m/n	3	4	5	6	7	8
3	4/8	4/15	4/6	4/2	6/10	8/1192
4		4/9	4/3	4/1	6/1	8/579
5			5/47	6/46	7/47	7/1
6				8/127	8/4	8/1
7					10/10	11/2
8						12/2

9. In 2009 [18] Burchett, Lane, and Lachniet consider the problem of the minimum number of rooks in set S such that every unoccupied square is covered by at least k rooks in S (k-domination), or in such a way that every square, including squares occupied by a rook, are covered at least k times (k-tuple domination).

10. In 2010 [85] Steinbach and Posthoff develop a computation methodology based on Boolean models to compute the domination, independent domination, and vertex independence numbers of rectangular $m \times n$ bishops graphs.

11. In 2011 [12] Burchett continues the study of k-tuple domination in the rooks graph, as well as the border queens domination problem.

12. In 2012 [4] Berghammer initiated the study of domination, independent domination, and total domination in rectangular $m \times n$ chessboards, by describing a simple computing technique, based on relational modeling, which is applicable to a variety of other chessboard problems. He presents tables for the domination and independence numbers, and the number of solutions, for rooks, kings, knights, and bishops, for $3 \leq m \leq 8$ and $3 \leq n \leq 8$, a sample of which is given in Table 26.

13. In 2013 [45] DeMaio and Tran study the domination number and vertex independence number of triangular-shaped hexagonal boards, having n hexagons on each of three exterior sides. We denote such boards by TR_n for rooks, TR_n for bishops, TN_n for knights, and TK_n for rooks. They show the following: (i) $\gamma(TR_n) = n$, (ii) $\alpha(TR_n) = n$, (iii) $\alpha(TB_n) = n$, and (iv) $\gamma(T N_n) \leq \Sigma_{i=1}^{\lceil n/5 \rceil}(4i - 1)$.

14. In 2014 [42] Cooper, Pikhurko, Schmitt, and Warrington solve the following problem posed by Martin Gardner: What is the smallest number of queens you can put on Q_n so that no additional queen can be added without creating three

in a row, column, or diagonal? The authors prove that this number is at least n, unless $n \equiv 3 \pmod 4$, in which case $n - 1$ may suffice.

15. In 2014 [44] DeMaio and Lightcap study kings total domination numbers on square $n \times n$ hexagonal boards.

16. In 2018 [17] Burchett and Chatham study more chessboard separation problems, such as the maximum number of independent rooks and bishops that can be placed on an $n \times n$ board containing k pawns, and all the values of k for which there is a placement of k pawns that allows the placement of $n + k$ independent rooks on an $n \times n$ board. They also study the same problem for bishops. For queens, they find lower bounds on the queens domination-, total domination-, paired domination-, and connected domination-separation numbers.

17. In 2018 [31] Chatham considers the domination number, the independence number, and the independent domination number of *dragon king* boards and *dragon horse* boards. A dragon king moves like a rook and a king, while a dragon horse moves like a bishop and a king. These are pieces from the chess-like game called *shogi*.

18. In 2018 [75] Pahlavsay, Palezzato, and Torielli consider 3-tuple total domination in rectangular rooks graphs. The authors give a formula for the 3-tuple total domination number of an $m \times n$ rooks graph.

References

1. W. Ahrens, *Mathematische Unterhaltungen und Spiele* (Druck und Verlag von B. G. Teubner, Leipzig, 1910), pp. 311–312
2. A. Amirabadi, Values of total domination numbers and connected domination numbers in queen graph. M.S. in Computer Science, Clemson University (2005)
3. J. Bell, B. Stevens, A survey of known results and research areas for n queens. Discrete Math. **309**, 1–31 (2008)
4. R. Berghammer, Relation-algebraic modeling and solution of chessboard independence and domination problems. J. Logic Algebraic Program. **81**, 625–642 (2012)
5. M. Bezzel, Schachfreund. Berliner Schachzeitung **3**, 363 (1848)
6. W. Bird, Computational methods for domination problems. Ph.D. Thesis, Dept. Computer Science, University of Victoria (2017)
7. J.-P. Bode, H. Harborth, Independent chess pieces on Euclidean boards. J. Combin. Math. Combin. Comput. **33**, 209–223 (2000)
8. J.-P. Bode, H. Harborth, Independence for knights on hexagon and triangle boards. Discrete Math. **272**, 27–35 (2003)
9. J.-P. Bode, H. Harborth, M. Harborth, King independence on triangle boards. Discrete Math. **266**(1–3), 101–107 (2003)
10. T.M. Brown, Non-attacking arrangements of n queens with initial placements. J. Recreat. Math. **38**(1), 23–37 (2014)
11. P.A. Burchett, Paired, total, and connected domination on the queen's graph. Congr. Numer. **178**, 207–222 (2006)
12. P.A. Burchett, On the border queens problem and k-tuple domination on the rook's graph. Congr. Numer. **209**, 179–187 (2011)

13. P.A. Burchett, An addendum regarding double domination on the queen's graph. Congr. Numer. **209**, 197 (2011)
14. P.A. Burchett, k-tuple domination on the bishop's graph. Util. Math. **101**, 351–358 (2016)
15. P.A. Burchett, Paired, total and connected domination on the queen's graph revisited. Open J. Discrete Math. **6**, 1–6 (2016)
16. P.A. Burchett, J. Buckley, The kth-border bishop domination problem. Util. Math. **100**, 3–21 (2016)
17. P.A. Burchett, R.D. Chatham, Some results for chessboard separation problems. AKCE Int. J. Graphs Comb. **15**(3), 251–260 (2018)
18. P.A. Burchett, D. Lane, J.A. Lachniet, k-tuple and k-domination on the rook's graph and other results. Congr. Numer. **199**, 187–204 (2009)
19. A.P. Burger, Domination in the Queens graph. Master's Thesis, University of South Africa (1994)
20. A.P. Burger, The queens domination problem. Doctoral Thesis, University of South Africa (1998)
21. A.P. Burger, C.M. Mynhardt, Queens on hexagonal boards. Papers in honour of Stephen T. Hedetniemi. J. Combin. Math. Combin. Comput. **31**, 97–111 (1999)
22. A.P. Burger, C.M. Mynhardt, Small irredundance numbers for queens graphs. Papers in honour of Ernest J. Cockayne. J. Combin. Math. Combin. Comput. **33**, 33–43 (2000)
23. A.P. Burger, C.M. Mynhardt, Properties of dominating sets of the queens graph Q_{4k+1}. Util. Math. **57**, 237–253 (2000)
24. A.P. Burger, C.M. Mynhardt, An upper bound for the minimum number of queens covering the $n \times n$ chessboard. Discrete Appl. Math. **121**(1–3), 51–60 (2002)
25. A.P. Burger, C.M. Mynhardt, An improved upper bound for queens domination numbers. Discrete Math. **266**, 119–131 (2003)
26. A.P. Burger, C.M. Mynhardt, E.J. Cockayne, Domination numbers for the queen's graph. Bull. Inst. Combin. Appl. **10**, 73–82 (1994)
27. A.P. Burger, E.J. Cockayne, C.M. Mynhardt, Domination and irredundance in the queen's graph. Discrete Math. **163**, 47–66 (1997)
28. A.P. Burger, C.M. Mynhardt, E.J. Cockayne, Queens graphs for chessboards on the torus. Australas. J. Combin. **24**, 231–246 (2001)
29. A.P. Burger, C.M. Mynhardt, W.D. Weakley, The domination number of the toroidal queens graph of size $3k \times 3k$. Australas. J. Combin. **28**, 137–148 (2003)
30. P.J. Campbell, Gauss and the eight queens problem: a study in miniature of the propagation of historical error. Hist. Math. **4**, 397–404 (1977)
31. D. Chatham, Independence and domination on shogiboard graphs. Recreat. Math. Mag. (8), 25–37 (2018)
32. R.D. Chatham, M. Doyle, G.H. Fricke, J. Reitmann, R.D. Skaggs, M. Wolff, Independence and domination separation on chessboard graphs. J. Combin. Math. Combin. Comput. **68**, 3–17 (2009)
33. H.-C. Chen, T.-Y. Ho, The rook problem on saw-toothed chessboards. Appl. Math. Lett. **21**, 1234–1237 (2008)
34. E.J. Cockayne, Chessboard domination problems. Discrete Math. **86**(1–3), 13–20 (1990)
35. E. Cockayne, S. Hedetniemi, On the diagonal queens domination problem. J. Combin. Theory Ser. B **42**, 137–139 (1986)
36. E.J. Cockayne, C.M. Mynhardt, Properties of queens graphs and the irredundance number of Q_7. Aust. J. Commun. **23**, 285–299 (2001)
37. E.J. Cockayne, P.H. Spencer, On the independent queens covering problem. Graphs Combin. **4**(2), 101–110 (1988)
38. E.J. Cockayne, R.M. Dawes, S.T. Hedetniemi, Total domination in graphs. Networks **10**, 211–219 (1980)
39. E.J. Cockayne, O. Favaron, C. Payan, A.G. Thomason, Contributions to the theory of domination, independence and irredundance in graphs. Discrete Math. **33**, 249–258 (1981)

40. E. Cockayne, B. Gamble, B. Shepherd, Domination of chessboards by queens in a column. Ars Combin. **19**, 105–118 (1985)
41. E.J. Cockayne, B. Gamble, B. Shepherd, Domination parameters for the bishop's graph. Discrete Math. **58**, 221–227 (1986)
42. A.S. Cooper, O. Pikhurko, J.R. Schmitt, G.S. Warrington, Martin Gardner's minimum no-3-in-a-line problem. Amer. Math. Monthly **121**(3), 213–221 (2014)
43. C.F. de Jaenisch, *Applications de l'Analyse Mathematique au Jeu des Echecs*. Petrograd (1862)
44. J. DeMaio, A. Lightcap, King's total domination number on the square of side n. Ars Combin. **113**, 139–150 (2014)
45. J. DeMaio, H.L. Tran, Domination and independence on a triangular honeycomb chessboard. College Math. J. **44**, 307–314 (2013)
46. H.E. Dudeney, *The Canterbury Puzzles and Other Curious Problems* (E. P. Dutton and Company, New York, 1908)
47. M. Eisenstein, C. Grinstead, B. Hahne, D. Van Stone, The queen domination problem. Congr. Numer. **91**, 189–193 (1992)
48. O. Favaron, G.H. Fricke, D. Pritikin, J. Puech, Irredundance and domination in kings graphs. Discrete Math. **262**(1–3), 131–147 (2003)
49. H. Fernau, Minimum Dominating Set of Queens: a trivial programming exercise? Discrete Appl. Math. **158**(4), 308–318 (2010)
50. D. Finozhenok, W.D. Weakley, An improved lower bound for domination numbers of the queen's graph. Australas. J. Combin. **37**, 295–300 (2007)
51. D.C. Fisher, P. Thalos, Bishop covers of rectangular boards. Ars Combin. **65**, 199–208 (2002)
52. G.H. Fricke, S.M. Hedetniemi, S.T. Hedetniemi, A.A. McRae, C.K. Wallis, M.S. Jacobson, H.W. Martin, W.D. Weakley, Combinatorial problems on chessboards: a brief survey, in *Graph Theory, Combinatorics, and Algorithms* (Kalamazoo,1992), vols. 1, 2 (Wiley-Intersci. Publ., Wiley, New York, 1995), pp. 507–528
53. M. Gardner, Mathematical games – problems that are built on the knight's move in chess. Sci. Am. 128–132 (1967)
54. M. Gardner, Mathematical games. Sci. Am. 125–129 (1967)
55. M. Gardner, Mathematical games. Sci. Am. 124–127 (1968)
56. D.K. Garnick, N.A. Nieuwejaar, Total domination of the $m \times n$ chessboard by kings, crosses, and knights. Ars Combin. **41**, 65–75 (1995)
57. I.P. Gent, C. Jefferson, P. Nightingale, Complexity of n-queens completion. J. Artif. Intell. Res. **59**, 815–848 (2017)
58. P.B. Gibbons, J.A. Webb, Some new results for the queens domination problem. Aust. J. Commun. **15**, 145–160 (1997)
59. D. Goncalves, A. Pinlou, M. Rao, S. Thomasse, The domination number of grids. SIAM J. Disc. Math. **25**(3), 1443–1453 (2011)
60. C.M. Grinstead, B. Hahne, D. Van Stone, On the queen domination problem. Discrete Math. **86**(1–3), 21–26 (1991)
61. R.K. Guy, *Unsolved Problems in Number Theory*, vol. I (Springer, Berlin, 1981)
62. E.O. Hare, S.T. Hedetniemi, A linear algorithm for computing the knight's domination number of a $k \times n$ chessboard. Congr. Numer. **59**, 115–130 (1987)
63. S. Hedetniemi, S. Hedetniemi, R. Reynolds, Combinatorial problems on chessboards: II, in *Domination in Graphs: Advanced Topics*, ed. by T. Haynes, S. Hedetniemi, P. Slater (Marcel Dekker, New York, 1998), pp. 507–528
64. M.D. Kearse, P.B. Gibbons, Computational methods and new results for chessboard problems. Aust. J. Commun. **23**, 253–284 (2001)
65. M.D. Kearse, P.B. Gibbons, A new lower bound on upper irredundance in the queen's graph. Discrete Math. **256**(1–2), 225–242 (2002)
66. E.G. Koehler, Domination problems in three-dimensional chessboard graphs. M.S. Computer Science, Clemson University (1994)
67. D. König, *Theorie der Endlichen und Unendlichen Graphen* (Akademische VerlagsgesellschaftM. B. H., Leipzig, 1936), later Chelsea, New York, 1950

68. L. Larson, A theorem about primes proved on a chessboard. Math. Mag. **50**(2), 69–74 (1977)
69. B. Lemaire, Covering the 11×11 chessboard with knights. J. Recreat. Math. **6**(4), 292 (1973)
70. R.M. Low, A. Kapbasov, Non-attacking bishop and king positions on regular and cylindrical chessboards. J. Integer. Seq. **20**(6), 26 (2017). Art. 17.6.1
71. B. Messick, Finding irredundant sets in queen's graphs using parallel genetic algorithms. M.S. in Computer Science, Clemson University (1995)
72. C.M. Mynhardt, Upper bounds for the domination numbers of toroidal queens graphs. Discuss. Math. Graph Theory **23**(1), 163–175 (2003)
73. F. Nauck, Briefwechsel mit allen für alle. Illustrirte Zeitung **15**, 182 (1850)
74. P.R. Östergard, W.D. Weakley, Values of domination numbers of the queen's graph. Electron. J. Combin. **8**(1), 19 (2001). Research Paper 29
75. B. Pahlavsay, E. Palezzato, M. Torielli, 3-tuple total domination number of rook's graphs. Preprint (2018)
76. E. Pauls, Das Maximalproblem der Damen auf dem Schachbrett. Dtsch. Schwesternztg. Organ für das Gesammte Schachleben **29**(5), 129–134 (1874)
77. E. Pauls, Das Maximalproblem der Damen auf dem Schachbrete, II. Dtsch. Schwesternztg. Organ für das Gesammte Schachleben **29**(9), 257–267 (1874)
78. V. Raghavan, S.M. Venkatesan, On bounds for a covering problem. Inform. Process. Lett. **25**, 281–284 (1987)
79. W.W. Rouse Ball, H.S.M. Coxeter, *Mathematical Recreations and Essays*, 13th edn. (Dover, Illinois, 1987)
80. W.W. Rouse Ball, *Mathematical Recreations and Essays*, 4th edn. (Macmillan, New York, 1905)
81. W.W. Rouse Ball, *Mathematical Recreations and Essays*, revision by H.S.M. Coxeter of the original 1892 Edition, Chapter 6, Minimum Pieces Problem, 3rd edn. (Macmillan, London, 1939)
82. A. Sinko, P. Slater, An introduction to influence parameters for chessboard graphs. Congr. Numer. **172**, 15–27 (2005)
83. A. Sinko, P.J. Slater, Efficient domination in knights graphs. AKCE Int. J. Graphs Comb. **3**(2), 193–204 (2006)
84. A. Sinko, P.J. Slater, Queen's domination using border squares and (A, B)-restricted domination. Discrete Math. **308**(20), 4822–4828 (2008)
85. B. Steinbach, C. Posthoff, New results based on Boolean models, in *Proceedings of the 9th International Workshop on Boolean Problems*, ed. by B. Steinbach (TU Bergakademie, Freiberg, 2010), pp. 29–36
86. W.F.D. Theron, A.P. Burger, Queen domination of hexagonal hives. J. Combin. Math. Combin. Comput. **32**, 161–172 (2000)
87. W.F.D. Theron, G. Geldenhuys, Domination by queens on a square beehive. Discrete Math. **178**(1–3), 213–220 (1998)
88. R. Wagner, R. Geist, The crippled queen placement problem. Sci. Comput. Program. **4**, 221–248 (1984)
89. C.K. Wallis, Domination parameters of line graphs, of designs and variations of chessboard graphs. Ph.D. Thesis, Dept. Mathematical Sciences, Clemson University (1994)
90. J.J. Watkins, *Across the Board: The Mathematics of Chessboard Problems* (Princeton University Press, Princeton, 2004)
91. J.J. Watkins, C. Ricci, Kings domination on a torus, in *The Ninth Quadrennial International Conference on Graph Theory, Combinatorics, Algorithms and Applications*. Electronic Notes in Discrete Mathematics, vol. 11 (Elsevier Sci. B. V., Amsterdam, 2002), 6 pp.
92. W.D. Weakley, Domination in the queen's graph, in *Graph Theory, Combinatorics, and Algorithms* (Kalamazoo, 1992), vols. 1, 2 (Wiley-Intersci. Publ., Wiley, New York, 1995), pp. 1223–1232
93. W.D. Weakley, Upper bounds for domination numbers of the queen's graph. Discrete Math. **242**(1–3), 229–243 (2002)

94. W.D. Weakley, Erratum to: "Upper bounds for domination numbers of the queen's graph" [Discrete Math. **242**(1–3), 229–243 (2002)]. Discrete Math. **277**(1–3), 321 (2004)
95. W.D. Weakley, Queens around the world in twenty-five years, in *Graph Theory, Favorite Conjectures and Open Problems*, ed. by R. Gera, T.W. Haynes, S.T. Hedetniemi (Springer, Berlin, 2018)
96. A.M. Yaglom, I.M. Yaglom, *Challenging Mathematical Problems with Elementary Solutions, Combinatorial Analysis and Probability Theory*, vol. I (Holden-Day Inc., San Francisco, 1964)

Domination in Digraphs

Teresa W. Haynes, Stephen T. Hedetniemi, and Michael A. Henning

AMS Subject Classification: 05C69

1 Introduction

Domination in digraphs is relatively unexplored if compared to its counterpart in graphs. In this chapter, we present selected results on domination in digraphs and give some background on the related topics of bases and kernels. The first two Ph.D. dissertations devoted to the study of domination in digraphs were written by Changwoo Lee [62] in 1994 and by Lisa Hansen [46] in 1997. A survey of results prior to 1998 on domination in directed graphs by Ghoshal, Laskar, and Pillone [43] is given in Chapter 15 of [54]. For completeness, many of these results are repeated

T. W. Haynes (⊠)
Department of Mathematics and Statistics, East Tennessee State University, Johnson City, TN, USA

Department of Mathematics and Applied Mathematics, University of Johannesburg, Johannesburg, South Africa
e-mail: haynes@etsu.edu

S. T. Hedetniemi
School of Computing, Clemson University, Clemson, SC, USA
e-mail: hedet@clemson.edu

M. A. Henning
Department of Mathematics and Applied Mathematics, University of Johannesburg, Johannesburg, South Africa
e-mail: mahenning@uj.ac.za

© The Author(s), under exclusive license to Springer Nature Switzerland AG 2021
T. W. Haynes et al. (eds.), *Structures of Domination in Graphs*, Developments in Mathematics 66, https://doi.org/10.1007/978-3-030-58892-2_13

here. We first present some terminology. For terminology and notation not found here, we refer the reader to the glossary in chapter "Glossary of Common Terms" of this volume.

1.1 Basic Terminology and Notation

Throughout this chapter, we let $D = (V, A)$ be a finite directed graph, or *digraph*, with a finite *vertex set* $V = V(D)$ and an *arc set* $A = A(D) \subseteq V \times V$, which is a subset of the Cartesian product $V \times V$, consisting of all ordered pairs of vertices in V, where neither loops (u, u) nor multiple arcs (u, v) and (u, v) are allowed, although pairs of opposite arcs, such as (u, v) and (v, u), are allowed. Also, $G = (V, E)$ stands for a simple, finite, *undirected* graph with vertex set $V(G)$ and *edge* set $E(G)$, which consists of a subset of the set of all unordered pairs $uv = vu$ of distinct vertices in V.

For two vertices $u, v \in V$ and an arc $(u, v) \in A$, we say that:

   (i) (u, v) is an arc *from u to v*,
  (ii) *u is adjacent to v*,
 (iii) *v is adjacent from u*,
 (iv) *v* is an *out-neighbor* of *u*,
  (v) *u* is an *in-neighbor* of *v*,
 (vi) *v* is a *successor* of *u* or the *terminal vertex* of the arc,
(vii) *u* is a *predecessor* of *v* or the *initial vertex* of the arc,
(viii) *u* and *v* are incident to arc (u, v), and
 (ix) arc (v, u) is the *reverse* of arc (u, v).

We also denote an arc (u, v) by $u \to v$. If both arcs (u, v) and (v, u) are in A, we denote this by $u \leftrightarrow v$; and this is called a *bidirected* or *symmetric* arc. A digraph $D = (V, A)$ is called *oriented* or *anti-symmetric* if for every $(u, v) \in A$, we have $(v, u) \notin A$, that is, D has no symmetric arcs. Equivalently, an *oriented digraph* can be obtained from a graph G by assigning a direction, either $u \to v$ or $v \to u$, to each edge uv of G.

The *outset* or *out-neighborhood* of a vertex $u \in V$ is the set of vertices $N_D^+(u) = \{v \mid u \to v \in A\}$, while the *inset* or *in-neighborhood* of vertex u is the set $N_D^-(u) = \{v \mid u \leftarrow v \in A\}$. The *outdegree* of vertex u, denoted $\mathrm{od}_D(u)$ or $d_D^+(u)$ in the literature, equals $|N_D^+(u)|$, while the *indegree* of u, denoted $\mathrm{id}_D(u)$ or $d_D^-(u)$ in the literature, equals $|N_D^-(u)|$. The maximum indegree of a digraph D, denoted $\Delta^-(D)$, is the maximum indegree among the vertices in D. The maximum outdegree of D is defined as expected and is denoted $\Delta^+(D)$. Similarly, the minimum indegree and minimum outdegree of D are denoted $\delta^-(D)$ and $\delta^+(D)$, respectively. The *degree* of a vertex v in D is $d_D(v) = \mathrm{od}_D(v) + \mathrm{id}_D(v)$. We note that

$$\sum_{v \in V(D)} \mathrm{od}_D(v) = \sum_{v \in V(D)} \mathrm{id}_D(v).$$

A digraph is *r-regular* if $\mathrm{od}_D(v) = \mathrm{id}_D(v) = r$ for every vertex v of D. We also define the *closed out-neighborhood* of a vertex v to equal $N_D^+[v] = N_D^+(v) \cup \{v\}$ and similarly the *closed in*-neighborhood to equal $N_D^-[v] = N_D^-(v) \cup \{v\}$. The *out-neighborhood* of a set S of vertices is $N_D^+(S) = \cup_{v \in S} N_D^+(v)$, and the *closed out-neighborhood* of S is $N_D^+[S] = \cup_{v \in S} N_D^+[v]$. And finally, the *in-neighborhood* of S is $N_D^-(S) = \cup_{v \in S} N_D^-(v)$, and the *closed in-neighborhood* of S is $N_D^-[S] = \cup_{v \in S} N_D^-[v]$.

Let $S \subseteq V$ and $u \in S$. A vertex $v \in V \setminus S$ is called a *private out-neighbor* of u with respect to S if $N_D^-(v) \cap S = \{u\}$, that is, v is an out-neighbor of u, $u \to v$, but is not an out-neighbor of any other vertex in S. The set of all private out-neighbors of u with respect to S is denoted by $\mathrm{pn}_D^+(u, S)$. Similarly, a vertex $v \in V \setminus S$ is called a *private in-neighbor* of u with respect to S if $N_D^+(v) \cap S = \{u\}$, that is, v is an in-neighbor of u, $u \leftarrow v$, but is not an in-neighbor of any other vertex in S. The set of all private in-neighbors of u with respect to S is denoted by $\mathrm{pn}_D^-(u, S)$.

If the digraph D is clear from context, we omit the subscript D from the above notational definitions. For example, we simply write $\mathrm{id}(u)$, $\mathrm{od}(u)$, $N^-(u)$, $N^+(u)$, $\mathrm{pn}^+(u, S)$, and $\mathrm{pn}^-(u, S)$, rather than $\mathrm{id}_D(u)$, $\mathrm{od}_D(u)$, $N_D^-(u)$, $N_D^+(u)$, $\mathrm{pn}_D^+(u, S)$, and $\mathrm{pn}_D^-(u, S)$, respectively. A vertex u is called:

(i) an *isolated vertex* if $\mathrm{od}(u) = \mathrm{id}(u) = 0$,
(ii) a *source* or *transmitter* if $\mathrm{id}(u) = 0$ and $\mathrm{od}(u) > 0$, and
(iii) a *sink* or *receiver* if $\mathrm{od}(u) = 0$ and $\mathrm{id}(u) > 0$.

Given two sets $R, S \subseteq V$, we let (R, S) denote the set of all arcs in A from R to S, that is, $(R, S) = \{(u, v) \in A \mid u \in R, v \in S\}$.

For any integer $k \geq 1$, we use the standard notation $[k] = \{1, \ldots, k\}$ and $[k]_0 = [k] \cup \{0\} = \{0, 1, \ldots, k\}$. A *directed walk* in a digraph $D = (V, A)$ from a vertex u to a vertex w, called a *(u, w)-walk*, is a sequence of vertices of the form $u = v_0, v_1, \ldots, v_k = w$ such that for every $i \in [k]$, we have $(v_{i-1}, v_i) \in A$. Such a (u, w)-walk has *length k*. A directed walk having no repeated edges is called a *directed trail*. A directed walk having no repeated vertices is called a *directed path*. A directed walk in which $v_0 = v_k$ is called a *closed directed walk*, and a closed walk in which all vertices, except v_0 and v_k, are distinct is called a *directed cycle* or a *circuit*. Let $\vec{C}_n$ denote the directed cycle on n vertices.

The *distance $d_D(u, v)$* from a vertex u to a vertex v in a digraph D is the minimum length of a directed (u, v)-path. If the digraph D is clear from the context, we write $d(u, v)$ rather than $d_D(u, v)$.

Given a digraph $D = (V, A)$, the *underlying graph of D* is the undirected graph $G(D) = (V, E)$, where $uv \in E$ if and only if $u \to v \in A$, $u \leftarrow v \in A$, or $u \leftrightarrow v \in A$. A digraph D is *connected* or *weakly connected* if its *underlying graph* $G(D)$ is connected.

A digraph D is said to be *strongly connected* if for every $u, w \in V$, there exist a directed (u, w)-path and a directed (w, u)-path. We note that one could consider the class of digraphs having the property that for every $u, w \in V$ either there is a directed walk from u to w or there is a directed walk from w to u.

A digraph $D = (V, A)$ is said to be *transitive* if (u, v), $(v, w) \in A$ implies that the arc $(u, w) \in A$. In other applications, a digraph D of order n is said to have a *transitive orientation* if there is an ordering of the vertices $v_1, v_2, \ldots, v_n$ such that for every

$i \in [n-1]$, we have $(v_i, v_{i+1}) \in A$. A digraph is *complete* if for every $u, v \in V$, either (u, v), (v, u), or both arcs are in A. A *tournament* is an oriented complete graph.

We denote the *degree* of a vertex v in an undirected graph G by $d_G(v)$, or simply by $d(v)$ if the graph G is clear from context. The average degree in G is denoted by $d_{av}(G)$. The minimum degree among the vertices of G is denoted by $\delta(G)$ and the maximum degree by $\Delta(G)$.

1.2   Domination and Independence

In this section we define independence and the types of domination in digraphs that will be discussed in this chapter. Let $D = (V, A)$ be a digraph with vertex set V and arc set A.

Definition 1   A set S of vertices in a digraph D is *independent* if no two vertices u, $v \in S$ are joined by an arc, that is, $(u, v) \notin A$ and $(v, u) \notin A$. The maximum cardinality of an independent set in a digraph D is called the *vertex independence number* of D and is denoted $\alpha(D)$, while the minimum cardinality of a maximal independent set of vertices in a digraph is the *lower vertex independence number*, denoted $\alpha_{\min}(D)$.

Definition 2   A set S of vertices in a digraph D is an *out-dominating set*, or just a *dominating set*, if for every vertex $v \in V \setminus S$, there exists a vertex $u \in S$ such that $u \to v \in A$, that is, every vertex in $V \setminus S$ is adjacent from a vertex in S. In other words, S is a dominating set of D if $V \setminus S \subseteq N^+[S]$. The minimum cardinality of dominating set in D is called the *out-domination number*, or simply the *domination number*, of D and is denoted $\gamma^+(D)$, or just $\gamma(D)$.

In general, we adopt the simplified terminology for out-dominating sets by omitting "out" and simply referring to dominating sets, domination number, and $\gamma(D)$.

Definition 3   A set S of vertices in a digraph D is an *in-dominating set* (also called a *converse dominating set* in the literature) if for every vertex $v \in V \setminus S$, there exists a vertex $u \in S$ such that $v \to u \in A$, that is, every vertex in $V \setminus S$ is adjacent to a vertex in S. In other words, S is an in-dominating set of D if $N^+(v) \cap S \neq \varnothing$. The minimum cardinality of an in-dominating set in a directed graph D is called the *in-domination number* of D and is denoted $\gamma^-(D)$.

Definition 4   A set S of vertices in a digraph D is a *twin dominating set* of D if it is both an in-dominating set and out-dominating set of D. The minimum cardinality of a twin dominating set is the *twin domination number* $\gamma^\pm(D)$ of D (also denoted $\gamma^*(D)$ in the literature).

To illustrate the above definitions, consider the digraph D shown in Figure 1. The darkened vertices in Figure 1(a) and 1(b) form a dominating set and an in-dominating set, respectively, of D, while the darkened vertices in Figure 1(c) form a twin dominating set of D.

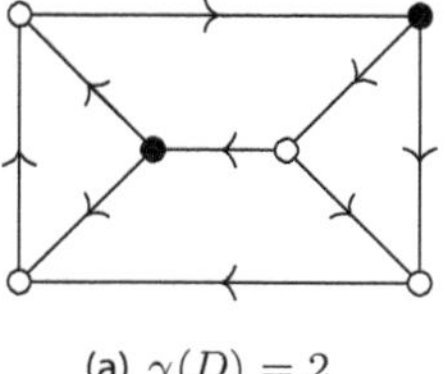 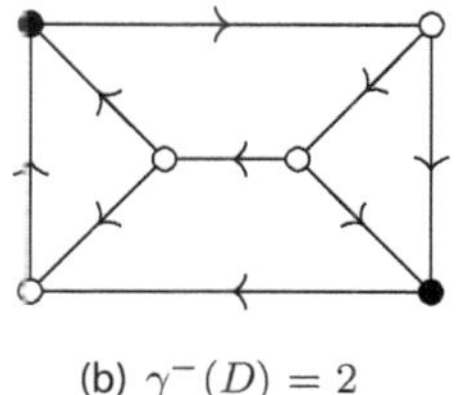 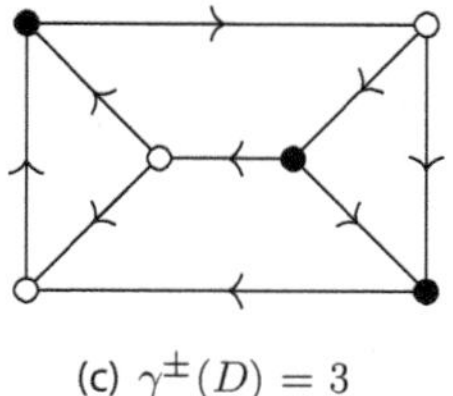

(a) $\gamma(D) = 2$ (b) $\gamma^-(D) = 2$ (c) $\gamma^{\pm}(D) = 3$

Fig. 1 A digraph D with $\gamma(D) = \gamma^-(D) = 2$ and $\gamma^{\pm}(D) = 3$

2 Background and History

In this section, we recognize and honor Dénes König for his pioneering work on domination in digraphs. His work on the *basis* of a digraph, which we shall see is an independent dominating set, comes some 30 years before any other mention of domination in the literature. Since König was the originator of domination in digraphs, we give several of his theorems along with their proofs. In the second part of this section, we present a brief overview of *kernels* in digraphs, which we shall see are independent in-dominating sets. We include some of Berge's early results on kernels with a sampling of proofs. We also give some results on the existence of kernels in digraphs. A survey of the expansive literature on kernels is beyond the scope of this chapter, so our brief overview is not meant to be complete. For more information we refer the reader to surveys by Boros and Gurvich [12] and Frankel [37], respectively.

2.1 *Basis of the Second Kind*

The concept of domination in digraphs was introduced as early as 1936 by König [61]. We present his original ideas in what follows, as they form a foundation on which many ideas for domination in digraphs can be built.

For any vertex $a \in V$ in a digraph $D = (V, A)$, let V_a equal the set consisting of a together with all vertices x for which there exists a directed path from a to x. If there is no vertex $b \in V$ such that $V_a \subset V_b$, then V_a is called a *basic set* with *source a*.

Theorem 1 ([61]) *Every vertex $a \in V$ of a finite directed graph $D = (V, A)$ is a member of some basic set of D.*

Proof Let $a \in V$. If V_a is a basic set, then clearly a is a member of a basic set. By definition, if V_a is not a basic set, then there exists a vertex $b \in V$ such that $V_a \subset V_b$, which implies that there must exist a directed path from b to a. Thus, if V_b is a basic set, then a is a member of the basic set V_b. Again, if V_b is not a basic set, then by definition, there exists a vertex $c \in V$ such that $V_a \subset V_b \subset V_c$. If V_c is a basic set, then a is a member of the basic set V_c. Since V is a finite set, this process must end with a vertex $x \in V$, such that V_x is a basic set containing a. $\qquad \square$

König pointed out that this theorem does not hold for infinite directed graphs, using the example of an infinite directed path $v_1, v_2, v_3, \ldots$, in which every arc has the form (v_{i+1}, v_i). It is easy to see that this infinite directed path has no basic set.

Theorem 2 ([61]) *No proper subset of a basic set is a basic set.*

Proof Suppose, to the contrary, that a basic set V_b contains a basic set V_a as a proper subset. Since there is a directed path from b to a, and since V_a is a basic set, V_a cannot be properly contained in another basic set. Thus, it follows that there must be a directed path from a to b. From this it follows that V_b must be a subset of V_a and thus that $V_a = V_b$. But this means that V_a is not a proper subset of V_b, a contradiction. $\qquad\square$

We can now define a basis of a directed graph.

Definition 5 A *basis* of a directed graph $D = (V, A)$ is a set $B \subset V$ having the following two properties:

(i) for every vertex $v \in V \setminus B$, there exist a vertex $u \in B$ and a directed path from u to v.
(ii) for every pair of vertices $u, v \in B$, there is no directed path from u to v.

Theorem 3 ([61]) *Every finite directed graph $D = (V, A)$ has a basis.*

Proof Let $\mathcal{V} = \{V_a, V_b, \ldots, V_k\}$ be the set of all basic sets of a finite directed graph $D = (V, A)$, and let $B = \{a, b, \ldots, k\}$ be sources for each of these basic sets. We claim that the set B is a basis of D.

Note that Theorem 1 says that every vertex $v \in V$ is a member of some basic set, say $v \in V' \in \mathcal{V}$. Assume that $v \in V \setminus B$. But $V' = V_w$ for some $V_w \in \mathcal{V}$ and $w \in B$, since $\mathcal{V}$ contains all basic sets. Thus, by definition there must be a directed path from w to v, and property (i) in Definition 5 is satisfied.

In order to show that B satisfies property (ii) in Definition 5, suppose, to the contrary, that for two sources a and b in B, where $V_a \neq V_b$, there is a directed path from a to b. But in this case, it follows that $V_b \subseteq V_a$. However, if $V_b \subset V_a$, then V_b cannot be a basic set, a contradiction. On the other hand, if $V_b = V_a$, then we contradict the supposition that $V_a \neq V_b$. $\qquad\square$

Theorem 4 ([61]) *If a vertex $a \in V$ is contained in a basis B in a directed graph $D = (V, A)$, then V_a is a basic set.*

Proof Assume that a vertex $a \in V$ is contained in a basis. Suppose, to the contrary, that V_a is not a basic set. Then there must exist a vertex $b \in V$ not contained in V_a such that V_a is a proper subset of V_b. Therefore, there must be a directed path from b to a. But if this is the case, then b does not belong to the basis B, since by property (ii) there can be no directed path between two vertices in a basis. Therefore, there must be a directed path from a vertex c of B to b, where $c \neq a$, for otherwise b would belong to V_a. The directed paths from c to b and from b to a imply, by Theorem 1, that there exists a directed path from c to a, contradicting property (ii) in the definition of a basis. $\qquad\square$

Theorem 5 ([61]) *Every basis B in a digraph $D = (V, A)$ consists of one source from each basic set.*

Proof By Theorem 4, every vertex of a basis B is a source of a basic set. In addition, two distinct vertices in B are never sources of the same basic set, since by property (ii) there can be no directed path between two vertices in B. It only remains to show that every basic set has a source in B. Suppose there exists a basic set V_a with source a such that $a \notin B$. By the definition of basis, there is a vertex $b \in B$ such that there is a directed path from b to a. But b is the source of a basic set V_b, and so the basic set V_a is a proper subset of the basic set V_b, contradicting Theorem 2. $\qquad\square$

Corollary 6 ([61]) *Every basis of a digraph D has the same cardinality, which equals the number of source vertices in D.*

Proof By Theorem 5, since every basis has one source from each basic set, every basis has a cardinality equal to the number of basic sets in D. $\qquad\square$

In his book, König pointed out that if every edge of a digraph D is symmetric, and the digraph D is basically an undirected graph, then the number of basic sets equals the number of components. König then defined a basis of the second kind as follows.

Definition 6 A *basis of the second kind* in a directed graph $D = (V, A)$ is a set $B \subset V$ satisfying the following two conditions:

(i) if v is a vertex in $V \setminus B$, then there is an arc (u, v) from a vertex $u \in B$ to v, and
(ii) there is no arc between two vertices in B.

Notice that by property (i) a basis of the second kind is a dominating set of D and by (ii) a basis of the second kind is an independent set of D. König noted that Corollary 6 is no longer true for bases of the second kind, i.e., for independent dominating sets.

In the case where a digraph D is symmetric, König's basis of the second kind appears to be the first time in the literature where an *independent dominating* set is defined in an undirected graph. It also, of course, defines an independent dominating set in a digraph for the first time. To illustrate a minimum independent dominating set in an undirected graph, König used as an example the classical problem of covering an 8×8 chessboard with the minimum number of queens. The *Queen's graph* consists of 64 vertices (one for each square on the chessboard), where two vertices/squares are adjacent if and only if a queen placed on one square can occupy the second square in 1 move. Thus, two vertices are adjacent if and only if they are in the same row, column, or diagonal. The minimum number of queens needed to cover the chessboard (the domination number of the Queen's graph) is 5. König's example of five queens, placed at the locations shown in Figure 2, covers the board with the added constraint that no two queens can attack each other, that is, this placement of these five queens represents a minimum independent dominating set of the Queen's graph.

Fig. 2 Minimum independent dominating set of queens

An independent dominating set of a digraph is also called a *solution* in the literature. In the context of games, a solution is defined by Von Neumann and Morgenstern in their now classic book [92]. We formally state the definition of a solution in terms of digraphs and give notation for a minimum independent dominating set.

Definition 7 A *solution* in a digraph D is an independent dominating set of D. The *solution number* of D, denoted $i^+(D)$, equals the minimum cardinality of a solution in D, that is, $i^+(D) = \alpha_{\min}(D)$.

Richardson [79] showed that every digraph with no odd cycles has at least one solution.

2.2 Kernels in Digraphs

In 1958, Berge [6] defined an in-dominating set, which he called an *absorbant set*. Although he called the in-domination number the absorption number and denoted it by $\beta(D)$, we shall continue with the terminology in-domination and denote the in-domination number as $\gamma^-(D)$, as defined in Section 1.2.

Definition 8 A *kernel* in a digraph D is an independent, in-dominating (absorbant) set of D. The *kernel number* of D equals the minimum cardinality of a kernel in D and is denoted $i^-(D)$.

The topic of kernels in digraphs has its roots in game theory and was introduced by Von Neumann and Morgenstern in 1944 [92]. Kernel applications have grown from n-person games and Nim-type games to more recent applications in artificial intelligence, combinatorics, and coding theory.

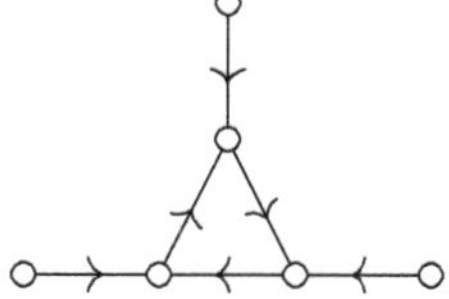

Fig. 3 A graph with a solution but no kernel

We note that not every digraph has a kernel; for example, a directed cycle $\vec{C}_5$ does not. Neither does $\vec{C}_5$ have a solution. The graph in Figure 3 has a solution, consisting of the three vertices of indegree zero, but it has no kernel.

For digraphs with kernels, Berge [6] proved the following.

Theorem 7 ([6]) *If S is a kernel, then S is both a maximal independent set and a minimal in-dominating set.*

Proof Let $S \subseteq V$ be a kernel in a digraph $D = (V, A)$. Since S is an in-dominating set, for each vertex $u \in V \setminus S$, there is an arc $(u, v) \in A$ where $v \in S$. Hence, $S \cup \{u\}$ is not an independent set, and so, S is a maximal independent set. Similarly, if $u \in S$, then $S \setminus \{u\}$ is not an in-dominating set since S is an independent set, and therefore there is no arc (u, v) for any $v \in S \setminus \{u\}$. Thus, S is a minimal in-dominating set. $\square$

Since not all digraphs have kernels, a natural question to ask is: What structural properties of digraphs imply the existence of a kernel? The existence of a kernel in a given digraph has been studied in many papers, including [5, 25, 26, 41, 79]. Berge [7] gave a necessary and sufficient condition for a vertex set to be a kernel in terms of its characteristic function. Recall that the characteristic function $\phi_S: V \to \{0, 1\}$ of a set S is defined as: $\phi_S(x) = 1$ if $x \in S$ and $\phi_S(x) = 0$ if $x \notin S$. We will assume that if a vertex x has no out-neighbors, then $\max\{\phi_S(y) \mid y \in N^+(x)\} = 0$.

Theorem 8 ([7]) *A set $S \subseteq V$ is a kernel of a digraph $D = (V, A)$ if and only if for every $x \in V$, $\phi_S(x) = 1 - \max\{\phi_S(y) \mid y \in N^+(x)\}$.*

Proof Let S be a kernel in a digraph D, and assume that ϕ_S is the characteristic function defined on it. If $x \in S$, then $\phi_S(x) = 1$. Since S is an independent set, no out-neighbor of x is in S. Thus, $\max\{\phi_S(y) \mid y \in N^+(x)\} = 0$, and therefore, $\phi_S(x) = 1 = 1 - \max\{\phi_S(y) \mid y \in N^+(x)\}$.

If $x \notin S$, then $\phi_S(x) = 0$. Since S is an in-dominating set, it follows that there must be a vertex $v \in S$ and an arc $(x, v) \in A$. Thus, $\max\{\phi_S(y) \mid y \in N^+(x)\} = 1$, and therefore, $\phi_S(x) = 0 = 1 - \max\{\phi_S(y) \mid y \in N^+(x)\}$.

Conversely, let S be a set for which, for every $x \in V$, $\phi_S(x) = 1 - \max\{\phi_S(y) \mid y \in N^+(x)\}$. If $x \in S$, then $\phi_S(x) = 1$. Thus, since $\phi_S(x) = 1 - \max\{\phi_S(y) \mid y \in N^+(x)\}$, it must follow that $\max\{\phi_S(y) \mid y \in N^+(x)\} = 0$, but this means that no out-neighbor of x is in S. If an in-neighbor of x, say y, is in S, then x is an out-neighbor of y, and therefore, $\phi_S(y) = 1$. But $\max\{\phi_S(x) \mid x \in N^+(y)\} = 1$, and so, $1 - \max\{\phi_S(x) \mid x \in N^-(y)\} = 0$, a contradiction. Therefore, S is an independent set.

Similarly, if $x \notin S$, then $\phi_S(x) = 0$. But since, for every $x \in V$, $\phi_S(x) = 1 - \max\{\phi_S(y) \mid y \in N^+(x)\}$, this must mean that $\max\{\phi_S(y) \mid y \in N^+(x)\} = 1$. Hence, at least one neighbor of x, say y, is in S. Therefore, S is an in-dominating set. $\square$

As early as 1936, König [61] proved the following result. A digraph $D = (V, A)$ is called *transitive* if whenever $(u, v) \in A$ and $(v, w) \in A$, then $(u, w) \in A$.

Theorem 9 ([61]) *If $D = (V, A)$ is a transitive digraph, then every minimal in-dominating set has the same cardinality. Furthermore, a set $S \subseteq V$ is a kernel if and only if S is a minimal in-dominating set.*

Corollary 10 *Every transitive digraph has a kernel, and all of its kernels have the same cardinality.*

In 1990 De la Vega [29] showed that although not all digraphs have kernels, probabilistically speaking, almost all digraphs do. Let $D(n, p) = (V, A)$ denote a *random digraph* of order n where for every $u, v \in V$, the arc (u, v) is chosen with probability p.

Theorem 11 ([29]) *For any probability p, where $0 \le p \le 1$, the probability that the random digraph $D(n, p)$ has a kernel goes to 1 as $n \to \infty$.*

Algorithms for determining all the kernels of a digraph D have been presented by Rudeanu [81] in 1966 and Roy [80] in 1970.

Many of the existence results for kernels are proved under an even stronger condition that the digraph is kernel-perfect. A digraph D is said to be *kernel-perfect* if D has a kernel and every induced subdigraph of D has a kernel. Meyniel conjectured that if every circuit of a digraph D has at least two chords, then D is kernel-perfect. Although Galeana-Sánchez [39] proved this conjecture to be false, the searching for a proof motivated results on sufficient conditions for the existence of a kernel in a digraph. The proof we present of the following result of Von Neumann and Morgenstern [92] is due to Berge [7].

Theorem 12 ([92]) *Every digraph D without directed cycles is kernel-perfect and has a unique kernel.*

Proof Given a digraph D having no directed cycles, define the set S_0 as the collection of sinks of D, and for each $k \ge 1$, define S_k as the set of all vertices u such that a longest (directed) path from u to a vertex in S_0 has length k. Thus,

$$S_0 = \{v \in V \mid N^+(v) = \emptyset\}.$$
$$S_1 = \{v \in V \mid N^+(v) \subseteq S_0\}.$$
$$S_2 = \{v \in V \mid N^+(v) \subseteq (S_0 \cup S_1)\}.$$

And in general, $S_k = \{v \in V \mid N^+(v) \subseteq (S_0 \cup S_1 \cup \ldots \cup S_{k-1})\}$.

Since D contains no directed cycles, the sets S_k form a partition of $V(D)$. One can then define a characteristic function $\phi_S(x) = 1 - \max\{\phi_S(y) \mid y \in N^+(x)\}$

iteratively, starting with the vertices $u \in S_0$ for each of which $\phi_S(u) = 1$, and then each vertex in S_1 receives the value 0. After this, a vertex x can be assigned a value $\phi_S(x)$ only after all of the vertices in $N^+(x)$ have been assigned a value, at which point the value of $\max\{\phi_S(y) \mid y \in N^+(x)\}$ can be determined. By Theorem 8, $S = \{v \in V \mid \phi_S(v) = 1\}$ is a kernel of D.

Since every subdigraph of D is acyclic, it follows that D is kernel-perfect. Moreover, the set S_0 of sinks is nonempty and unique, and so by definition, S_k is unique for each $k \geq 1$. The uniqueness of S follows from the fact that any kernel of D must contain S_0, and hence the vertices of $S_k \cap S$. $\qquad \square$

We next mention classical results due to Richardson [79] and Duchet [25].

Theorem 13 ([79]) *Every digraph D without odd directed cycles is kernel-perfect.*

Theorem 14 ([25]) *If every circuit in a digraph D has at least one symmetric arc, then D is kernel-perfect.*

Recall that a kernel in a digraph D is an independent set S such that every vertex not in S dominates some vertex in S, where as usual by "dominates" we mean "out-dominates," that is, a vertex u dominates a vertex v if there is an arc (u, v) from u to v. We next define a semi-kernel in a digraph. Recall that the distance $d_D(u, v)$ from a vertex u to a vertex v in a digraph D is the shortest directed path from u to v. We note that $d_D(u, v)$ may be very different from $d_D(v, u)$.

Definition 9 A set S of vertices in a digraph D is a *semi-kernel* if S is an independent set and every vertex not in S either dominates some vertex in S or dominates a vertex which in turn dominates some vertex in S. Thus, S is a semi-kernel in D if S is an independent set and for every vertex $v \in V(D) \setminus S$, there is a vertex $u \in S$ such that $d_D(v, u) \leq 2$.

As observed earlier, not all digraphs have kernels. However, every digraph has a semi-kernel. This result is attributed to Chvátal and Lovász [24]. However, in this paper they proved Theorem 16, which we state shortly. It is not clear if Theorem 16 immediately implies Theorem 15. The proof of the following result is due to Bondy [11].

Theorem 15 ([11]) *Every digraph has a semi-kernel.*

Proof Let D be a digraph and let H be a maximal induced acyclic subdigraph of D. By Theorem 12, the acyclic digraph H has a (unique) kernel. Let S be the kernel of H. We claim that S is a semi-kernel of D. Since S is a kernel of H, every vertex of $H - S$ dominates some vertex of S. Let v be an arbitrary vertex outside H, and so $v \in V(D) \setminus V(H)$. By our choice of H, there is a directed cycle C in the subdigraph of D induced by $V(H) \cup \{v\}$. The vertex v therefore dominates its successor v^+ on C. Since $v^+ \in V(H)$, either $v^+ \in S$, in which case v dominates a vertex of S, or $v^+ \notin S$, in which v^+ dominates a vertex of S and therefore v dominates a vertex which in turn dominates some vertex of S. Thus, S is a semi-kernel of D. $\qquad \square$

Definition 10 For any integer $k \geq 2$, a set S of vertices in a digraph D is a *k-dominating set* if S is an independent set and every vertex not in S can be reached from a vertex of S by a directed path of length at most k, that is, for every vertex $v \in V(D) \setminus S$, there is a vertex $u \in S$ such that $d(u, v) \leq k$.

We note that a 1-dominating set of a digraph D is an independent, (out-) dominating set of D. For $k \geq 1$, every k-dominating set is a $(k + 1)$-dominating set. In particular, every 1-dominating set is a 2-dominating set. Not every digraph has a 1-dominating set; for example, $\vec{C_5}$ does not. In 1974 Chvátal and Lovász [24] proved that every digraph has a 2-dominating set.

Theorem 16 ([24]) *Every digraph has a 2-dominating set.*

Proof We proceed by induction on the order n of a digraph D. For $n = 1$ or $n = 2$, the result is immediate. Let $n \geq 3$ and assume that every digraph of order less than n has a 2-dominating set. Let w be an arbitrary vertex of D. If $V(D) = N_D^+(w)$, then the set $\{w\}$ is a 1-dominating set and therefore also a 2-dominating set. Hence, we may assume that $V(D) \neq N_D^+(w)$. Let D' be the subdigraph of D induced by the set of vertices at distance at least 2 from w in D. Thus, $V(D') = \{v \in V(D) \mid d_D(w, v) \geq 2\}$. Further, $(x, y) \in A(D')$ if and only if $x, y \in V(D')$ and $(x, y) \in A(D)$. Applying the inductive hypothesis, the digraph D' contains a 2-dominating set S'. Suppose firstly that there is an arc from u to w for some vertex $u \in S'$. Therefore, $d_D(u, w) = 1$, and every vertex in $N_D^+(w)$ is reachable from u by a directed path of length at most 2, that is, $d_D(u, x) \leq 2$ for every vertex $x \in N_D^+[w]$. In this case, let $S = S'$. Suppose secondly that there is no arc from a vertex in S' to the vertex w, and so $d_D(u, w) \geq 2$ for all vertices $u \in S'$. In this case, we let $S = S' \cup \{w\}$. In both cases, the set S is a 2-dominating set of D. $\qquad\square$

As observed earlier, not every digraph has a 1-dominating set. In 1996 Jacob and Meyniel [59] proved that a digraph with no 1-dominating set contains at least three 2-dominating sets.

Theorem 17 ([59]) *Every digraph with no 1-dominating set contains at least three 2-dominating sets.*

Kernels have relations to Grundy functions in digraphs. We conclude this subsection with some results relating the two.

Definition 11 A non-negative function $g : V \to [n]_0$ from the vertex set V of a digraph D to the integers $[n]_0$ is called a *Grundy function* if for every vertex $u \in V$, $g(u)$ is the smallest non-negative integer not belonging to $\{g(v) \mid v \in N^+(u)\}$. It follows, therefore, that if g is a Grundy function, then the following hold.

(1) $g(u) = k$ implies that for each $0 \leq j < k$, there is a vertex $v \in N^+(u)$ with $g(v) = j$.
(2) $g(u) = k$ implies that for every $v \in N^+(u)$, $g(v) \neq g(u)$.

Proposition 18 ([7]) *If a digraph D has a Grundy function, then D has a kernel.*

Proof Let $g: V \to [n]_0$ be a Grundy function on a digraph $D = (V, A)$, and let $S = \{u \in V \mid g(u) = 0\}$. From condition (2) in Definition 11, we know that $g(u) = 0$ implies that for every $v \in N^+(u)$, $g(v) \neq g(u) = 0$, and therefore, S is an independent set.

If a vertex $v \notin S$, then $g(v) = k > 0$. From condition (1) in Definition 11, we know that $g(u) = k > 0$ implies that for each $j < k$, there is a vertex $u \in N^+(u)$ with $g(u) = j$, and in particular there is a vertex $w \in N^+(u)$ with $g(w) = 0$. Thus, S is an in-dominating set. Therefore, S is a kernel. $\qquad\square$

While it can be verified that if a graph has a kernel, it need not have a Grundy function, the following interesting connection to kernel-perfect digraphs was shown by Berge [7].

Theorem 19 ([7]) *Every kernel-perfect digraph has a Grundy function.*

Proof Let $D = D_0$ be a kernel-perfect digraph, and let S_0 be a kernel of D_0. It follows from the definition of a kernel-perfect digraph that the digraph $D_1 = D_0 - S_0$ is a kernel-perfect digraph. Therefore, let S_1 be a kernel of D_1. Let $D_2 = D_1 - S_1$ and let S_2 be a kernel in D_2. In general for $k \geq 1$, let S_k be a kernel of the subdigraph D_k. The resulting sets $S_0, S_1, \ldots, S_k$ form a partition of $V(D)$. Define a function $g: V \to [k]_0$ by $g(u) = j$ if and only if $u \in S_j$. It follows that g is a Grundy function of D.

If $g(u) = j$, then vertex u is a vertex in every digraph $D_0, D_1, \ldots, D_{j-1}$. And $S_0, S_1, \ldots, S_{j-1}$ are in-dominating sets of these digraphs, respectively. Therefore, for each $i < j$, there is a vertex $w \in S_i$ where $w \in N^+(u)$. Thus, condition (1) of a Grundy function (see Definition 11) is satisfied. If $g(u) = j$, then $u \in S_j$, which is an in-dominating set of the digraph D_j. This means that the set S_j is an independent set. Therefore, if $g(u) = j$, then each $v \in N^+(u)$ satisfies $g(v) \neq j$. Therefore, every kernel-perfect digraph D has a Grundy function g. $\qquad\square$

Fraenkel [36] has determined that deciding whether a finite digraph D has a kernel or a Grundy function is NP-complete, even when restricted to cyclic planar digraphs with $od(x) \leq 2$, $id(x) \leq 2$, and $od(x) + id(x) \leq 3$, and these bounds are best possible, since decreasing any of them results in a decision problem that can be solved in polynomial time. The proof of this theorem uses a simple transformation from 3-Satisfiability.

3 Bounds on In, Out, and Twin Domination Numbers

In this section, we present bounds on the domination, in-domination, and twin domination numbers of digraphs.

3.1 (Out)-Domination

We begin with some well-known results of Ore [75] on dominating sets of graphs.

Theorem 20 ([75]) *If G is a graph having no isolated vertices, then the complement $V \setminus S$ of any minimal dominating set S is a dominating set of G.*

Corollary 21 *The vertices of any graph G having no isolated vertices can be partitioned into two dominating sets.*

Corollary 22 *For any graph G of order n having no isolated vertices, $\gamma(G) \leq \frac{1}{2}n$.*

Fu was interested in possible analogs of these results of Ore for digraphs. For example, can the vertices of a digraph D without isolated vertices be partitioned into two (directed) dominating sets? Fu [38] obtained the following results on dominating sets of directed graphs.

Theorem 23 ([38]) *A dominating set S in a digraph D is a minimal dominating set if for each $u \in S$, there is no arc (u, v) for any vertex $v \in S$.*

Proof Assume that S is a dominating set of a digraph D having the property that for no two vertices $u, v \in S$, $(u, v) \in A$, that is, S is an independent set. Then it follows that for every $u \in S$, $S \setminus \{u\}$ is a not a dominating set since there is no vertex in $S \setminus \{u\}$ that dominates vertex u. Thus, S is a minimal dominating set of D. $\qquad\qquad\square$

As observed by Fu [38], in order that a digraph D has a dominating set S such that its complement $V \setminus S$ is also a dominating set, it is necessary and sufficient that each vertex $u \in S$ is dominated by a vertex in $V \setminus S$ and each vertex in $V \setminus S$ is dominated by a vertex in S. Moreover, in order that a digraph D has a dominating set S whose complement $V \setminus S$ is an in-dominating set, it is necessary and sufficient that each vertex in S dominates at least one vertex in $V \setminus S$.

Fu defined a digraph D to be *cyclic* or *strongly connected* if every pair of vertices are contained in a directed cycle.

Theorem 24 ([38]) *A strongly connected digraph D has a dominating set S whose complement $\overline{S} = V \setminus S$ is also a dominating set if and only if D contains a directed cycle of even length.*

Proof For the necessity part, assume that D has a dominating set S whose complement $\overline{S} = V \setminus S$ is also a dominating set. Assume that no vertices are colored. Select an arbitrary vertex $u \in S$. Color it blue. Since the complement $\overline{S}$ is a dominating set, there must be a vertex $v \in \overline{S}$ and an arc (v, u). Color vertex v red. Since S is a dominating set, there are a vertex $w \in S$ and an arc (w, v). If $w = u$, then we have found a directed cycle of length 2. If $w \neq u$, color vertex w blue. There must be a vertex $z \in \overline{S}$ which dominates w. If z has been previously colored, we have found a directed cycle beginning and ending in $\overline{S}$ and therefore having even length. If z has not been colored, color it red. Continuing in this way, all vertices encountered will either be in S and colored blue or in $\overline{S}$ and colored red.

Sooner or later we will have to encounter a previously colored vertex and hence have constructed a directed cycle of even length.

To prove the sufficiency, we assume that there is a directed cycle of even length, and we need only show that there is a way to assign the vertices of D either to S or $\overline{S}$, in such a way that both sets are dominating sets. We begin with any directed cycle C_0 of even length and alternately assign its vertices to S and $\overline{S}$. Thus, all of the vertices on C_0 are assigned to a dominating set of C_0. If this includes all vertices of D, then the theorem is proved. Thus, we may assume that there is an unassigned vertex, say w. Since D is strongly connected, w and u are on a directed cycle for any vertex u on C_0. We may then find a directed path from u to w and continue until a vertex is encountered which has already been assigned. The vertices on this directed path can be alternately assigned to either S or $\overline{S}$. This directed path may end with two consecutive vertices assigned to the same set, but each vertex thus encountered is always dominated by the vertex which precedes it on the directed path. Since w is an arbitrary unassigned vertex, every vertex of D can be assigned to one of S and $\overline{S}$.
$\square$

Corollary 25 ([38]) *A strongly connected digraph D has a dominating set S whose complement $\overline{S}$ is also a dominating set, and furthermore both S and $\overline{S}$ are in-dominating sets if and only if every vertex of V is in some directed cycle of even length.*

Corollary 26 ([38]) *In order that a strongly connected digraph D has a dominating set S whose complement $\overline{S}$ is an in-dominating set, it is sufficient that D contains a directed cycle of even length.*

Corollary 27 ([38]) *If D is a strongly connected digraph of order n having a cycle of even length, then $\gamma(D) \le \frac{1}{2}n$.*

We observe that if D is a Hamiltonian digraph of order n, then $\gamma(D) \le \left\lceil \frac{n}{2} \right\rceil$. In 1998 Lee [63] improved the result of Corollary 27 as follows.

Theorem 28 ([63]) *If D is a strongly connected digraph of order n, then $1 \le \gamma(D) \le \left\lceil \frac{n}{2} \right\rceil$.*

In order to prove Theorem 28, Lee [63] proved that if D is a directed tree of order n that contains a vertex u such that every vertex in D is reachable from u, that is, for every v in D different from u, there is a directed path from u to v, then $1 \le \gamma(D) \le \left\lceil \frac{n}{2} \right\rceil$. The proof of this result given in [63] is algorithmic in nature and finds a dominating set S in such a directed tree D satisfying $1 \le |S| \le \left\lceil \frac{n}{2} \right\rceil$. From this result, we can readily deduce Theorem 28, noting that a strongly connected digraph has as a subdigraph a directed spanning tree with the desired property.

Lee [62] proved the following upper bound on the domination number of a digraph D in terms of its order and the minimum indegree $\delta^-(D)$.

Theorem 29 ([62]) *If D is a digraph of order n with $\delta^-(D) = \delta^- \ge 1$, then*

$$\gamma(D) \le \left(\frac{\delta^- + 1}{2\delta^- + 1} \right) n.$$

As a consequence of Theorem 29, we have the following upper bound on the domination number of a digraph in which every vertex has indegree at least 1.

Corollary 30 ([62]) *If D is a digraph of order n with $\delta^-(D) \ge 1$, then $\gamma(D) \le \frac{2}{3}n$.*

Using standard probabilistic arguments, Lee [62] established the following upper bound on the domination of a digraph.

Theorem 31 ([62]) *If D is a digraph of order n with $\delta^-(D) = \delta^- \ge 1$, then*

$$\gamma(D) \le \left(1 - \left(\frac{1}{1+\delta^-} \right)^{\frac{1}{\delta^-}} + \left(\frac{1}{1+\delta^-} \right)^{\frac{1+\delta^-}{\delta^-}} \right) n.$$

We remark that when the minimum indegree $\delta^-(D)$ is small, namely, $\delta^-(D) \in \{1, 2\}$, then the upper bound given by Theorem 29 is better than that given by Theorem 31.

As before, let $D(n, p) = (V, A)$ denote a *random digraph* of order n where for every $u, v \in V$, the arc (u, v) is chosen with probability p. Let Q be a property of digraphs. If $\mathcal{A}$ is the set of digraphs of order n with property Q and the probability $\Pr(\mathcal{A})$ of $\mathcal{A}$ has limit 1 as $n \to \infty$, then we say *almost all digraphs have property Q* or *a random digraph has property Q almost surely*. Lee [62] established the following result for random digraphs.

Theorem 32 ([62]) *For a fixed p with $0 < p < 1$, a random digraph $D \in D(n, p)$ satisfies*

$$\gamma(D) = \lfloor k^* \rfloor + 1 \quad \text{or} \quad \gamma(D) = \lfloor k^* \rfloor + 2$$

almost surely, where $k^* = \log n - 2 \log \log n + \log \log e$ and where $\log$ denotes the logarithm with base $1/(1 - p)$.

Ghoshal, Laskar, and Pillone [43] determined lower and upper bounds on the domination number of a digraph in terms of its order and maximum outdegree.

Theorem 33 ([43]) *If D is a digraph of order n, then*

$$\frac{n}{1 + \Delta^+(D)} \le \gamma(D) \le n - \Delta^+(D).$$

Proof Let $x \in V$ be any vertex having maximum outdegree in D, that is, $od(x) = \Delta^+(D)$. Let $S = V \setminus N^+(x)$. It follows that S is an out-dominating set. Thus, $\gamma(D) \le |S| = n - \Delta^+(D)$. This establishes the upper bound. To prove the lower bound, let $S \subseteq V$ be a minimum dominating set of D, that is, $\gamma^+(D) = |S|$.

Every vertex in S dominates at most $\Delta^+(D)$ vertices outside S, implying that $n - |S| = |V \setminus S| \leq |S| \cdot \Delta^+(D)$, and so $\gamma^+(D) = |S| \geq 1/(1 + \Delta^+(D))$. $\qquad\square$

Hao and Qian [52] strengthened the lower bound of Theorem 33. The *Slater number* $\mathrm{sl}(D)$ of a digraph D is the smallest integer t such that adding t to the sum of the first t terms of the non-increasing outdegree sequence of D is at least as large as the order of D.

Theorem 34 ([52]) *If D is a digraph of order n, then*

$$\frac{n}{1 + \Delta^+(D)} \leq \mathrm{sl}(D) \leq \gamma(D).$$

Moreover, the authors [52] showed that the difference between $\mathrm{sl}(D)$ and $\left\lceil \frac{n}{1+\Delta^+(D)} \right\rceil$ can be arbitrarily large.

3.2 In-Domination

We turn our attention to bounds on the in-domination number of a digraph and give the following classical 1973 results due to Berge [7].

Proposition 35 ([7]) *If D is a digraph of order n and size m, then $\gamma^-(D) \geq n - m$.*

Proof Let $S \subseteq V$ be a minimum in-dominating set, that is, $\gamma^-(D) = |S|$. Since for every vertex $w \in V \setminus S$, there exist a vertex $v \in S$ and an arc (w, v), it follows that $n - |S| = |V \setminus S| \leq m$, and so $\gamma^-(D) = |S| \geq n - m$. $\qquad\square$

Proposition 36 ([7]) *For any digraph D of order n having maximum indegree $\Delta^-(D)$,*

$$\left\lceil \frac{n}{1 + \Delta^-(D)} \right\rceil \leq \gamma^-(D) \leq n - \Delta^-(D).$$

Proof Let $x \in V$ be any vertex having maximum indegree in D, that is, $\mathrm{id}(x) = \Delta^-(D)$. Let $S = V \setminus N^-(x)$. It follows that S is an in-dominating set. Thus, $\gamma^-(D) \leq |S| = n - \Delta^-(D)$. This establishes the upper bound. To prove the lower bound, let $S \subseteq V$ be a minimum in-dominating set of D, that is, $\gamma^-(D) = |S|$. Every vertex in S is dominated by at most $\Delta^-(D)$ vertices outside S, implying that $n - |S| = |V \setminus S| \leq |S| \cdot \Delta^-(D)$, and so $\gamma^-(D) = |S| \geq 1/(1 + \Delta^-(D))$. $\qquad\square$

We note that both bounds of Proposition 36 are sharp for a digraph of order n having $\Delta^-(D) = n - 1$.

3.3 Domination and In-Domination

In 1999 Chartrand, Harary, and Yue [19] proved the following upper bound on the sum of the domination number and the in-domination number of a digraph. Recall that $\vec{C}_3$ denotes the directed cycle on three vertices and an endvertex is a vertex of degree 1.

Theorem 37 ([19]) *If D is a digraph of order n with $\delta^-(D) \geq 1$ and $\delta^+(D) \geq 1$, then*

$$\gamma(D) + \gamma^-(D) \leq \frac{4}{3}n.$$

Further, equality holds if and only if $D = \vec{C}_3$, or if every vertex of D is an endvertex or is adjacent to exactly one endvertex and adjacent from exactly one endvertex.

In 2015 Hao and Qian [51] improved the upper bound of Theorem 37 as follows.

Theorem 38 ([51]) *Let D be a digraph of order n with $\delta^-(D) \geq 1$ and $\delta^+(D) \geq 1$. If $2k + 1$ is the length of a shortest odd circuit of D, then*

$$\gamma(D) + \gamma^-(D) \leq \left(\frac{2k + 2}{2k - 1} \right) n.$$

As a consequence of Theorem 38, we have the following result.

Corollary 39 ([51]) *If D is a digraph of order n with $\delta^-(D) \geq 1$ and $\delta^+(D) \geq 1$ with no odd directed cycle, then $\gamma(D) + \gamma^-(D) \leq n$.*

3.4 Twin Domination

In this section, we present results on the twin domination number of a digraph. We first present the following key lemma. Recall that for $r \geq 1$ an integer, a graph G is *r-degenerate* if every induced subgraph of G has minimum degree at most r. When we say that digraph D is *minimal* with respect to some property $\mathcal{P}$, we mean arc-minimal, that is, removing any arc from D destroys property $\mathcal{P}$.

Lemma 40 *If a digraph D is minimal with respect to the property of every vertex of D having indegree and outdegree at least k, then the underlying graph is 2k-degenerate.*

Proof Let D be a digraph that is minimal with respect to the property $\mathcal{P}$ that every vertex of D has indegree and outdegree at least k. Let G be the underlying (undirected) graph of D. We show that G is $2k$-degenerate. Suppose, to the contrary, that there is a set V' of vertices such that the subgraph, say G', of G induced by the set V' has minimum degree at least $2k + 1$. Let D' be the subdigraph of D

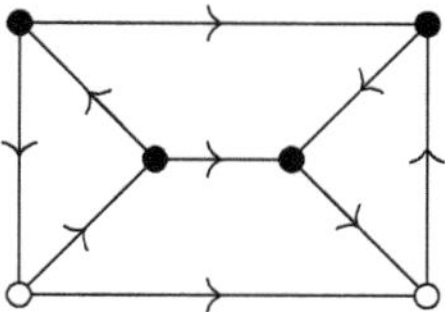

Fig. 4 A digraph D with $\gamma^{\pm}(D) = \frac{2}{3}n$

induced by the set V', and so G' is the underlying graph of the digraph D'. Each vertex $v \in V'$ has an excess of in- or out-arcs in D', noting that $d_{G'}(v) \geq 2k + 1$. Suppose there is an arc a_v whose removal from D' destroys the property of v having indegree and outdegree at least k. If $\mathrm{od}_{D'}(v) \geq k + 1$, then a_v is an arc into v and in this case $\mathrm{id}_{D'}(v) = k$. If $\mathrm{id}_{D'}(v) \geq k + 1$, then a_v is an arc out of v and in this case $\mathrm{od}_{D'}(v) = k$. Thus, the number of arcs incident to v whose removal from D' destroys property $\mathcal{P}$ is either zero or k. Hence, there are at most $k|V'|$ arcs in D whose removal destroys property $\mathcal{P}$. But every arc removal from D' destroys property $\mathcal{P}$ for some vertex of D', implying that there are at most $k|V'|$ arcs in D'. This in turn implies that every vertex has indegree and outdegree exactly k in D', and therefore G' is a $(2k)$-regular graph, contradicting the supposition that $\delta(G') \geq 2k + 1$. $\qquad\square$

In 2003 Chartrand, Dankelmann, Schultz, and Swart [20] established the following upper bound on the twin domination number of a digraph. We present here a simple proof of this result, using the key Lemma 40. Our proof is based on the fact that a k-degenerate graph has chromatic number at most $k + 1$, as shown by Szekeres and Wilf [88] in 1968. Recall that a vertex and an edge *cover* each other in a graph G if they are incident in G. A *vertex cover* in G is a set of vertices that covers all the edges of G. The *vertex cover number* $\beta(G)$ (also denoted by $\tau(G)$ or $\mathrm{vc}(G)$ in the literature) is the minimum cardinality of a vertex cover in G.

Theorem 41 ([20]) *If D is a digraph of order n with $\delta^-(D) \geq 1$ and $\delta^+(D) \geq 1$, then $\gamma^{\pm}(D) \leq \frac{2}{3}n$.*

Proof We may assume the digraph D is minimal with respect to this property of $\delta^-(D) \geq 1$ and $\delta^+(D) \geq 1$, since adding arcs cannot increase the twin domination number. With this assumption, the underlying graph G of D is 2-degenerate by Lemma 40 and hence 3-colorable. Thus, the independence number of G is at least $n/3$, which means that the vertex cover number of G is at most $2n/3$. But a vertex cover of G is a twin dominating set in D since $\delta^-(D) \geq 1$ and $\delta^+(D) \geq 1$. Thus, $\gamma^{\pm}(D) \leq \frac{2}{3}n$. $\qquad\square$

The simplest example of a digraph achieving equality in the upper bound of Theorem 41 is $\vec{C}_3$. As a further small example, the digraph D shown in Figure 4 has order $n = 6$ and satisfies $\gamma^{\pm}(D) = 4 = \frac{2}{3}n$, where the darkened vertices form a twin dominating set of D of cardinality 4.

In 2013 Arumugam, Ebadi, and Sathikala [4] gave the following upper bound on the twin domination number.

Theorem 42 ([4]) *If D is a digraph of order n and $\ell(D)$ is the length of a longest directed path in D, then $\gamma^{\pm}(D) \leq n - \left\lfloor \frac{\ell(D)}{2} \right\rfloor$.*

The bound of Theorem 42 is attained, for example, by directed paths and also by any digraph D obtained from a directed path $P_k: u_1, u_2, \ldots, u_k$ by adding a new vertex u_i' and arc (u_i', u_i) for each u_i for $i \in [k]$.

3.5 Reverse Domination

The digraph obtained from a digraph D by reversing all the arcs of D is called the *reverse digraph* (also called the *converse* in the literature) of D, denoted D^-. We note that $\gamma(D) = \gamma^-(D^-)$ for every digraph D. Thus by Theorem 37, if D is a digraph of order n with $\delta^-(D) \geq 1$ and $\delta^+(D) \geq 1$, then $\gamma(D) + \gamma(D^-) \leq \frac{4}{3}n$.

For $r \geq 1$, let $\mathcal{D}_r$ be the class of r-regular strongly connected digraphs. We note that the only 1-regular strongly connected digraphs are the directed cycles, and so $\mathcal{D}_1 = \{\vec{C}_n \mid n \geq 3\}$. Since a directed cycle is isomorphic to its reverse, if $D \in \mathcal{D}_1$, then $\gamma(D^-) - \gamma(D) = 0$. For $r \geq 2$, the difference $\gamma(D^-) - \gamma(D)$ can be arbitrarily large in the class $\mathcal{D}_r$, as shown by Gyürki [45] in the case when $r = 2$ and by Niepel and Knor [73] for all $r \geq 3$. However, for a fixed $r \geq 2$, it remains an open problem to determine the greatest ratio $\gamma(D^-)/\gamma(D)$ of an r-regular strongly connected digraph. The best known results to date are the following.

Theorem 43 ([45]) *For digraph $D \in \mathcal{D}_2$,* $\displaystyle \sup_{D \in \mathcal{D}_2} \frac{\gamma(D^-)}{\gamma(D)} \geq \frac{4}{3}$.

Theorem 44 ([45, 73]) *For $r \geq 3$, we have* $\displaystyle \sup_{D \in \mathcal{D}_r} \frac{\gamma(D^-)}{\gamma(D)} \geq \frac{7}{6}$.

4 Domination in Digraph Products

Vizing's conjecture [90] asserts that the domination number of the Cartesian product of two graphs is at least as large as the Cartesian product of their domination numbers. This conjecture was first stated in 1963 as a problem in [89] and later in 1968 formally posed as a conjecture in [90]. It is considered by many to be the main open problem in the area of domination in graphs. It is natural then that the study of domination in digraphs considers results for Cartesian products of digraphs.

The *Cartesian product* of two digraphs $G = (V(G), A(G))$ and $H = (V(H), A(H))$, denoted by $G \square H$, is the digraph with vertex set $V(G) \times V(H)$, and there exists an arc $((u_1, v_1), (u_2, v_2)) \in A(G \square H)$ if and only if either $(u_1, u_2) \in A(G)$ and $v_1 = v_2$

or $(v_1, v_2) \in A(H)$ and $u_1 = u_2$. Much of the work on Cartesian products in digraphs considers directed cycles.

In 2009 Shaheen [84] and in 2010 Liu, Zhang, Chen, and Meng [64, 93] independently determined the domination number of $\vec{C}_m \,\square\, \vec{C}_n$ for $m \le 6$ and arbitrary $n \ge 2$.

Theorem 45 ([64, 84, 93]) *For $n \ge 2$, the following hold.*

(a) $\gamma(\vec{C}_2 \,\square\, \vec{C}_n) = n$.
(b) $\gamma(\vec{C}_3 \,\square\, \vec{C}_n) = n$ *if* $n \equiv 0 \pmod 3$; *otherwise,* $\gamma(\vec{C}_3 \,\square\, \vec{C}_n) = n + 1$.
(c) $\gamma(\vec{C}_4 \,\square\, \vec{C}_n) = \frac{3}{2}n$ *if* $n \equiv 0 \pmod 8$; *otherwise,* $\gamma(\vec{C}_4 \,\square\, \vec{C}_n) = n + \left\lceil \frac{n+1}{2} \right\rceil$.
(d) $\gamma(\vec{C}_5 \,\square\, \vec{C}_n) = 2n$.
(e) $\gamma(\vec{C}_6 \,\square\, \vec{C}_n) = 2n + 2$.

Zhang et al. [93] also determined $\gamma(\vec{C}_m \,\square\, \vec{C}_n)$ when both m and n are divisible by 3.

Theorem 46 ([93]) *If* $m \equiv 0 \pmod 3$ *and* $n \equiv 0 \pmod 3$, *then* $\gamma(\vec{C}_m \,\square\, \vec{C}_n) = \frac{1}{3}mn$.

In 2013, Mollard [71] determined the exact values of $\gamma(C_m \square C_n)$ for m congruent to 2 modulo 3, with the exception of one subcase.

Theorem 47 ([71]) *If* $m, n \ge 2$, $m \equiv 2 \pmod 3$, $k = \lfloor m/3 \rfloor$, *and* $\ell = \lfloor n/3 \rfloor$, *then*

$$
\gamma(\vec{C}_m \,\square\, \vec{C}_n) =
\begin{cases}
n(k + 1) & \text{if } n = 3\ell \\
n(k + 1) & \text{if } n = 3\ell + 1 \text{ and } 2\ell \ge k \\
n(k + 1) & \text{if } n = 3\ell + 2 \text{ and } n \ge m \\
m(\ell + 1) & \text{if } n = 3\ell + 2 \text{ and } n \le m.
\end{cases}
$$

Furthermore, $\gamma(\vec{C}_m \,\square\, \vec{C}_n)$ *if* $n = 3\ell + 1$ *and* $2\ell < k$.

Zhang et al. [93] conjectured that if $k \ge 2$ where $k = \lfloor \frac{m}{3} \rfloor$, then $\gamma(\vec{C}_m \,\square\, \vec{C}_n) = k(n+1)$ for $n \not\equiv 0 \pmod 3$, but Mollard [71] disproved this conjecture by showing that it doesn't always hold when $n \equiv 1 \pmod 3$. For example, they noted that $\gamma(\vec{C}_{3k} \,\square\, \vec{C}_4) = \gamma(\vec{C}_4 \square \vec{C}_{3k}) = 3k + \left\lceil \frac{3k+1}{2} \right\rceil$ when $k \not\equiv 0 \pmod 8$, while the conjecture claims that $\gamma(\vec{C}_4 \,\square\, \vec{C}_{3k}) = 5k$. These values are different for $k \ge 3$.

Mollard [71] also established the following bounds.

Theorem 48 ([71]) *If* $m, n \ge 2$ *and* $k = \lfloor \frac{m}{3} \rfloor$, *then*

$$
\gamma(\vec{C}_m \,\square\, \vec{C}_n) \ge
\begin{cases}
nk & \text{if } m \equiv 0 \pmod 3 \\
nk + \frac{n}{2} & \text{if } m \equiv 1 \pmod 3 \\
nk + n & \text{if } m \equiv 2 \pmod 3.
\end{cases}
$$

In 2013 Shao, Zhu, and Lang [86] determined upper and lower bounds on $\gamma(\vec{C}_m \square \vec{C}_n)$ for the case when m is congruent to 1 modulo 3.

Theorem 49 ([86]) *If $k \geq 1$ and $n \geq 3$ are integers, then*

$$\left\lceil \frac{(2k+1)n}{2} \right\rceil \leq \gamma(\vec{C}_{3k+1} \square \vec{C}_n) \leq \left\lceil \frac{(2k+1)n}{2} \right\rceil + k.$$

Based on the bounds of Theorem 49, Shao et al. [86] determined the exact values of $\gamma(\vec{C}_m \square \vec{C}_n)$ for $m \in \{7, 10\}$.

We conclude this section by noting that Liu, Zhang, and Meng [65] investigated domination numbers of Cartesian products of directed paths in 2011, and Ma and Liu [67] studied the twin domination number of the Cartesian products of directed cycles in 2016. For domination and twin domination in other types of digraph products, see [15, 66, 68, 69, 74].

5   Domination in Oriented Graphs

Recall that an *oriented graph D* is a digraph that can be obtained from a graph G by assigning a direction to (i.e., orienting) each edge of G. The resulting digraph D is called an *orientation* of G. Thus, if D is an oriented graph, then for every pair u and v of distinct vertices of D, at most one of (u, v) and (v, u) is an arc of D. For example, a *tournament* is an oriented complete graph. Recall also that the independence number of a directed graph D is denoted by $\alpha(D)$. As before, unless otherwise stated, we refer to an out-dominating set in a digraph simply as a dominating set.

5.1   Oriented Graphs

In 1996 Chartrand, Vanderjagt, and Yue [18] studied domination in oriented graphs. They defined the *lower orientable domination number* of a graph G, which they denoted as $\mathrm{dom}(G)$ (denoted by $\gamma_d(G)$ in [17]), to equal the minimum domination number over all orientations of G. Further, they defined the *upper orientable domination number*, or simply the *orientable domination number*, of a graph G, which they denoted as $\mathrm{DOM}(G)$ (denoted by $\Gamma_d(G)$ in [17]), as the maximum domination number over all orientations of G. Thus,

$$\mathrm{dom}(G) \;= \min\{\gamma(D) \mid \text{ over all orientations } D \text{ of } G\}$$
$$\mathrm{DOM}(G) = \max\{\gamma(D) \mid \text{ over all orientations } D \text{ of } G\}.$$

The orientable domination number of a complete graph was first studied by Erdős in 1963 [28], albeit in disguised form. In 1962, Schütte [28] raised the question of given any positive integer $k > 0$, does there exist a tournament $T_{n(k)}$ on $n(k)$ vertices in which for any set S of k vertices, there is a vertex u that dominates all vertices in S. Erdős [28] showed, by probabilistic arguments, that such a tournament $T_{n(k)}$ does exist, for every positive integer k. The proof of the following bounds on the orientable domination number of a complete graph is along identical lines to that presented by Erdős [28]. This result can also be found in [78]. Here, log is to the base 2.

Theorem 50 ([28]) *For $n \geq 2$, $\log n - 2\log(\log n) \leq \mathrm{DOM}(K_n) \leq \log(n+1)$.*

This notion of orientable domination in a complete graph was subsequently extended to orientable domination of all graphs by Chartrand et al. [18]. They proved the following result.

Theorem 51 ([18]) *For every graph G, $\mathrm{dom}(G) = \gamma(G)$.*

In view of Theorem 51, it is not interesting to ask about the lower orientable domination number, $\mathrm{dom}(G)$, of a graph G since this is precisely its domination number, which is very well studied. We therefore focus our attention on the (upper) orientable domination number of a graph. Chartrand et al. [18] determined $DOM(G)$ for special classes of graphs, including paths, cycles, complete bipartite graphs, and regular complete tripartite graphs. They also proved the following result.

Theorem 52 ([18]) *For every graph G and for every integer c with $\mathrm{dom}(G) \leq c \leq DOM(G)$, there exists an orientation D of G such that $\gamma(D) = c$.*

In 2010 Blidia and Ould-Rabah [8] continued the study of domination in oriented graphs. For an oriented graph D, let $\alpha'(D)$ denote the matching number of D and let $s(D)$ denote the number of support vertices in the underlying graph of D. The authors in [8] proved the following result. In fact, they proved a slightly stronger result involving the irredundance number of an oriented graph (which we do not define here).

Theorem 53 ([8]) *If D is an oriented graph of order n, then $s(D) \leq \gamma(D) \leq n - \alpha'(D)$.*

Blidia and Ould-Rabah [8] characterized the oriented trees T satisfying $\gamma(T) - \alpha'(T)$ and the oriented graphs D satisfying $\gamma(D) = s(D)$ and $s(D) = n - \alpha'(D)$.

In 2011 Caro and Henning [16] also studied domination in oriented graphs. In this paper, they proved a Greedy Partition Lemma, which they used to present an upper bound on the orientable domination number of a graph in terms of its independence number. To state their result, let $\alpha \geq 1$ be an integer and let $\mathcal{G}_\alpha$ be the class of all graphs G with $\alpha \geq \alpha(G)$.

Theorem 54 ([16]) *For $\alpha \geq 1$ an integer, if $G \in \mathcal{G}_\alpha$ has order $n \geq \alpha$, then*

$$\mathrm{DOM}(G) \leq \alpha \left(1 + 2\ln\left(\frac{n}{\alpha}\right)\right).$$

The next result follows as a consequence of Theorem 54, where $\chi(G)$ denotes the chromatic number of G and $d_{av}(G)$ denotes the average degree in G.

Corollary 55 ([16]) *If G is a graph of order n, then the following hold.*

(a) $\text{DOM}(G) \le \alpha(G)\,(1 + 2\ln(\chi(G)))$.
(b) $\text{DOM}(G) \le \alpha(G)\,(1 + 2\ln(d_{av}(G) + 1))$.

For any integer $d \ge 1$, let $\mathcal{F}_d$ be the class of all graphs G whose complement is a d-degenerate graph. The property of being d-degenerate is a hereditary property that is closed under induced subgraphs, as is the property of the complement of a graph being d-degenerate. Applying their Greedy Partition Lemma for domination in oriented graphs, the authors in [16] proved the following result.

Theorem 56 ([16]) *For any integer $d \ge 1$, if $G \in \mathcal{F}_d$ has order n, then*

$$\text{DOM}(G) \le 2d + 1 + 2\ln\left(\frac{n - 2d + 1}{2}\right).$$

The following upper bound on the orientable domination number of a $K_{1,m}$-free graph is established in [16], where a graph is F-free if it does not contain F as an induced subgraph.

Theorem 57 ([16]) *For $m \ge 3$, if G is a $K_{1,m}$-free graph of order n with $\delta(G) = \delta$, then*

$$\text{DOM}(G) < 2(m - 1)n \ln\left(\frac{\delta + m - 1}{\delta + m - 1}\right).$$

Let $\mathcal{G}_n$ denote the family of all graphs of order n. We define

$$\text{NG}_{\min}(n) = \min\{\text{DOM}(G) + \text{DOM}(\overline{G})\}$$
$$\text{NG}_{\max}(n) = \max\{\text{DOM}(G) + \text{DOM}(\overline{G})\}$$

where the minimum and maximum are taken over all graphs $G \in \mathcal{G}_n$. The following Nordhaus-Gaddum-type bounds for the orientable domination of a graph were established in [16].

Theorem 58 ([16]) *The following hold.*

(a) $c_1 \log n \le \text{NG}_{\min}(n) \le c_2 (\log n)^2$ *for some constants c_1 and c_2.*
(b) $n + \log n - 2\log(\log n) \le \text{NG}_{\max}(n) \le n + \lceil \frac{n}{2} \rceil$.

Caro and Henning continued their study of the orientable domination number in [17]. They defined the *maximum average degree* in a graph G, denoted by $\text{mad}(G)$, as the maximum of the average degrees taken over all subgraphs H of G, that is,

$$\text{mad}(G) = \max_{H \subset G}\left\{\frac{2|E(H)|}{|V(H)|}\right\}.$$

Theorem 59 ([17]) *If G is a graph of order n, then the following hold.*

(a) $\mathrm{DOM}(G) \geq \alpha(G) \geq \gamma(G)$.
(b) $\mathrm{DOM}(G) \geq n/\chi(G)$.
(c) $\mathrm{DOM}(G) \geq \lceil(\mathrm{diam}(G)+1)/2)\rceil$.
(d) $\mathrm{DOM}(G) \geq n/(\lceil \mathrm{mad}(G)/2\rceil + 1)$.

Proof We present here only a proof of part (a). Let I be a maximum independent set in G, and let D be the digraph obtained from G by orienting all arcs from I to $V \setminus I$ and orienting all arcs in $G[V \setminus I]$, if any, arbitrarily. Every dominating set of D contains the set I, and so $\gamma(D) \geq |I|$. However, the set I itself is a dominating set of D, and so $\gamma(D) \leq |I|$. Consequently, $DOM(G) \geq \gamma(D) = |I| = \alpha(G) \geq \gamma(G)$. $\square$

As remarked in [17], since $\mathrm{mad}(G) \leq \Delta(G)$ for every graph G, as an immediate consequence of Theorem 59(d), we have that $\mathrm{DOM}(G) \geq n/(\lceil \Delta(G)/2\rceil + 1)$. The following lemma is useful when establishing upper bounds on the orientable domination number of a graph.

Lemma 60 ([17]) *Let $G = (V, E)$ be a graph and let $V_1, V_2, \ldots, V_k$ be subsets of V, not necessarily disjoint, such that $\cup_{i=1}^{k} V_i = V$. If $G_i = G[V_i]$ for $i \in [k]$, then*

$$\mathrm{DOM}(G) \leq \sum_{i=1}^{k} \mathrm{DOM}(G_i).$$

Proof Consider an arbitrary orientation D of G. Let D_i be the orientation of the edges of G_i induced by D, and let S_i be a γ-set of D_i for each $i \in [k]$. By Theorem 59(a), $\mathrm{DOM}(G_i) \geq \gamma(D_i) = |S_i|$ for each $i \in [k]$. Since the set $S = \cup_{i=1}^{k} S_i$ is a dominating set of D, we have that

$$\gamma(D) \leq |S| \leq \sum_{i=1}^{k} |S_i| \leq \sum_{i=1}^{k} \mathrm{DOM}(G_i).$$

Since this is true for every orientation D of G, the desired upper bound of $\mathrm{DOM}(G)$ follows. $\square$

As a consequence of Lemma 60, the authors in [17] proved the following upper bounds on the orientable domination number of a graph.

Theorem 61 ([17]) *If G is a graph of order n, then the following hold.*

(a) $\mathrm{DOM}(G) \leq n - \alpha'(G)$.
(b) *If G has a perfect matching, then* $\mathrm{DOM}(G) \leq n/2$.
(c) $\mathrm{DOM}(G) \leq n$ *with equality if and only if $G = \overline{K}_n$.*
(d) *If G has minimum degree δ and $n \geq 2\delta$, then* $\mathrm{DOM}(G) \leq n - \delta$.
(e) $\mathrm{DOM}(G) = n - 1$ *if and only if every component of G is a K_1-component, except for one component which is either a star or a complete graph K_3.*

Proof We present here only a proof of part (a). Let $M = \{u_1v_1, u_2v_2, \ldots, u_tv_t\}$ be a maximum matching in G, and so $t = \alpha'(G)$. Let $V_i = \{u_i, v_i\}$ for $i \in [t]$. If $n > 2t$, let $(V_{t+1}, \ldots, V_{n-2t})$ be a partition of the remaining vertices of G into $n - 2t$ subsets each consisting of a single vertex. By Lemma 60,

$$\mathrm{DOM}(G) \leq \sum_{i=1}^{n-t} \mathrm{DOM}(G_i) = t + (n - 2t) = n - t = n - \alpha'(G). \qquad \square$$

Applying results on the size of a maximum matching in a regular graph established in [57], we have the following consequence of Theorem 61(a).

Theorem 62 ([17]) *For $r \geq 2$, if G is a connected r-regular graph of order n, then*

$$\mathrm{DOM}(G) \leq \begin{cases} \max\left\{\left(\dfrac{r^2 + 2r}{r^2 + r + 2}\right) \times \dfrac{n}{2}, \dfrac{n+1}{2}\right\} & \text{if } r \text{ is even} \\[4mm] \dfrac{(r^3 + r^2 - 6r + 2)\,n + 2r - 2}{2(r^3 - 3r)} & \text{if } r \text{ is odd.} \end{cases}$$

The orientable domination number of a bipartite graph is precisely its independence number. Recall that König [60] and Egerváry [27] showed that if G is a bipartite graph, then $\alpha'(G) = \beta(G)$. Hence by Gallai's Theorem [42], if G is a bipartite graph of order n, then $\alpha(G) + \alpha'(G) = n$.

Theorem 63 ([17]) *If G is a bipartite graph, then* $\mathrm{DOM}(G) = \alpha(G)$.

Proof Since G is a bipartite graph, we have that $n - \alpha'(G) = \alpha(G)$. Thus, by Theorem 59(a) and Theorem 61(a), we have that $\alpha(G) \leq \mathrm{DOM}(G) \leq n - \alpha'(G) = \alpha(G)$. Consequently, we must have equality throughout this inequality chain. In particular, $\mathrm{DOM}(G) = \alpha(G)$. $\qquad \square$

In 2018 Harutyunyan, Le, Newman, and Thomassé [53] observed that in general there is no upper bound on the orientable domination number of a graph solely in terms of its independence number. Nevertheless, they showed that these two quantities can be related.

Theorem 64 ([53]) *If G is a graph of order n, then* $\mathrm{DOM}(G) \leq \alpha(G) \cdot \log n$.

Theorem 64 implies that when the independence number of an oriented graph is sufficiently large, it is possible to bound the orientable domination number of the graph purely in terms of its independence number.

Theorem 65 ([53]) *If D is a graph of order n and $\alpha(G) \geq \log n$, then* $\mathrm{DOM}(D) \leq (\alpha(D))^2$.

Harutyunyan et al. [53] concluded their paper with the following conjecture.

Conjecture 1 *There exists an integer k such that for any $\vec{C}_3$-free oriented graph D with $\alpha(D) = \alpha$, we have $\gamma(D) \leq \alpha^k$.*

The following result establishes an upper bound on the orientable domination number of a graph in terms of its independence number and chromatic number.

Theorem 66 ([17]) *If G is a graph of order n, then the following hold.*

(a) $\text{DOM}(G) \leq \alpha(G) \cdot \lceil \chi(G)/2 \rceil$.
(b) $\text{DOM}(G) \leq n - \lfloor \chi(G)/2 \rfloor$.
(c) $\text{DOM}(G) \leq (n + \alpha(G))/2$.

The following result establishes an upper bound on the orientable domination of a graph in terms of the chromatic number of its complement.

Theorem 67 ([17]) *If G is a graph of order n, then*

$$\text{DOM}(G) \leq \chi(\overline{G}) \cdot \log\left(\left\lceil \frac{n}{\chi(\overline{G})} \right\rceil + 1\right).$$

As a consequence of Theorem 67, we have the following result on the orientable domination number of a graph with sufficiently large minimum degree.

Theorem 68 ([17]) *If G is a graph of order n with minimum degree $\delta(G) \geq (k-1)n/k$ where k divides n, then $\text{DOM}(G) \leq \frac{n}{k}\log(k+1)$.*

Let $\text{Mop}(n) = \max\{\text{DOM}(G)\}$, where the maximum is taken over all maximal outerplanar graphs of order n.

Theorem 69 ([17]) *For maximal outerplanar graphs of order n, $\text{Mop}(n) = \lceil \frac{n}{2} \rceil$.*

5.2 Tournaments

Since a tournament is an oriented complete graph, many applications interpret a tournament as a competition graph. That is, a tournament on n vertices represents a competition between n teams (each represented by a vertex) in which the teams play each other once. No ties are allowed, and there is an arc from a vertex u to a vertex v if and only if u defeats v. The score of a vertex v is its outdegree (the number of teams it defeats). Hence, a dominating set S of a tournament represents a collection of teams such that every team not in S is defeated by at least one team in S. Tournaments are popular, in part, because of this pairwise comparison and ranking of competitors.

The following result is attributed by Moon to Erdős (cf. Moon [72] p. 28). As before, unless otherwise stated, log is to the base 2.

Theorem 70 (Erdős) *If T is a tournament with $n \geq 2$ vertices, then $\gamma(T) \leq \lceil \log n \rceil$.*

Proof The sum of the outdegrees of the vertices in a tournament $T = (V, A)$ of order n is the number of arcs in T, that is,

$$\sum_{u \in V} \mathrm{od}_T(u) = \frac{1}{2}n(n-1).$$

Thus, there must be a vertex $x \in V$ with $\mathrm{od}_T(x) \geq \lceil \frac{1}{2}(n-1) \rceil$. We remove this vertex x and all out-neighbors of x, thereby removing at least half the vertices. We now repeat this process on the remaining tournament, which has at most $\lceil \frac{1}{2}(n-1) \rceil$ vertices, by again selecting a vertex which dominates at least half of the remaining vertices and then deleting this second vertex and all of its out-neighbors. Repeating this process will produce a dominating set with no more than $\lceil \log n \rceil$ vertices. $\qquad\square$

A *random tournament* is obtained by orienting the edges of a complete graph randomly, independently, with equal probabilities. Let T_n be the probability space consisting of the random tournaments on n vertices. In 1997 Bollobás and Szabó [9] showed that the domination number of a random tournament is one of two values, where log is to the base 2. We remark that this result was obtained by Lee [62] in 1994.

Theorem 71 ([9, 62]) *A random tournament $T \in T_n$ has domination number $\lfloor k \rfloor + 1$ or $\lfloor k \rfloor + 2$, where $k = \log(n) - 2\log(\log(n)) + \log(\log(e))$.*

By Theorem 71, there are tournaments having arbitrarily large domination numbers. This leads to the question: Which tournaments have bounded domination number (not dependent on the order n of the tournament)? To partially answer this question, we first define a k-majority tournament.

Definition 12 As usual, by a linear order in a tournament, we mean with respect to the transitive orientation of the tournament. A tournament T is a *k-majority tournament* if there are $2k-1$ linear orders of $V(T)$ such that for all distinct vertices u and v in T, if u is adjacent to v, then u is before v in at least k of the $2k-1$ orders. Let $F(k)$ be the supremum of the size of a minimum dominating set in a k-majority tournament, where the supremum is taken over all k-majority tournaments, with no restriction on their size.

Trivially, $F(1) = 1$. In 2006 Alon, Brightwell, Kierstead, Kostochka, and Winkler [2] proved that $F(2) = 3$. To do this, they first showed that every 2-majority tournament has a dominating set of size at most 3, that is, $F(2) \leq 3$. We omit their proof.

To show that $F(2) \geq 3$, Alon et al. [2] provided the following example. Recall that if there is an integer x with $0 < x < p$ such that $x^2 \equiv q \pmod{p}$, then q is a *quadratic residue modulo p*. In practice, it suffices to restrict the range of x to $0 < x \leq \lfloor p/2 \rfloor$ because of the symmetry $(p - x)^2 \equiv x^2 \pmod{p}$. For example, the quadratic residues modulo 7 are given by 1, 2, 4 since $1^1 \equiv 1 \pmod{7}$, $2^2 \equiv 4 \pmod{7}$, and $3^2 \equiv 2 \pmod{7}$. Let T be the quadratic residue tournament whose vertices are the elements of the finite field $GF(7)$ in which $i \to j$ if and only if $i - j$ is a quadratic residue modulo 7,

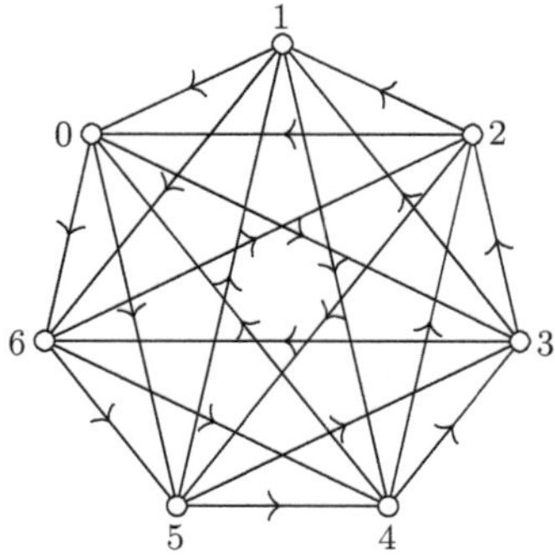

Fig. 5 A 2-majority tournament T' with $\gamma(T')=3$

i.e., $(i-j) \bmod 7 \in \{1, 2, 4\}$. Since the edges of T are preserved under translation, it suffices for us to consider the subtournament T' of T with vertex set $\{0, 1, \ldots, 6\}$ as illustrated in Figure 5.

No two vertices dominate T', while the set $\{0, 1, 2\}$, for example, is a dominating set of T', and so $\gamma(T')=3$. Further, T' is a 2-majority tournament realized by the orders P_1, P_2, and P_3, where

$$P_1 : 0 < 1 < 2 < 3 < 4 < 5 < 6,$$
$$P_2 : 4 < 6 < 1 < 3 < 5 < 0 < 2,$$
$$P_3 : 5 < 2 < 6 < 3 < 0 < 4 < 1.$$

Thus, T' is a 2-majority tournament satisfying $\gamma(T')=3$. As observed earlier, the edges of T are preserved under translation, implying that T is a 2-majority tournament satisfying $\gamma(T)=3$. This example shows that $F(2) \geq 3$. As observed earlier, $F(2) \leq 3$. Consequently, $F(2) = 3$. We state this result formally as follows.

Theorem 72 ([2]) *For 2-majority tournaments, $F(2)=3$.*

The value of $F(k)$ has yet to be determined for any value of $k \geq 3$. The following nontrivial result shows that $F(3) \geq 4$.

Theorem 73 ([2]) *There exists a 3-majority tournament T with $\gamma(T)=4$, that is, $F(3) \geq 4$.*

As observed earlier, there are tournaments having arbitrarily large domination numbers. Kierstead and Trotter (see [2] for a discussion) conjectured that this is not the case for k-majority tournaments for some fixed k. Alon et al. [2] proved this conjecture and showed that $F(k)$ is finite for each fixed k.

Theorem 74 ([2]) *For an arbitrary fixed integer $k \geq 1$, if T is a k-majority tournament, then*

$$\gamma(T) \leq 20(2 + o(1))k \log(k(2 \log 2)) \leq (80 + o(1))k \log(k).$$

We remark that their paper was the first to introduce the idea of using the VC dimension to study domination in tournaments, where the VC dimension (Vapnik-Chervonenkis dimension) of a hypergraph H is the largest cardinality of a vertex subset X *shattered by H*, that is, for any $Y \subseteq X$, the hypergraph H has an edge A such that $A \cap X = Y$. The upper bound in the following theorem follows as a consequence of Theorem 74.

Theorem 75 ([2]) *For an arbitrary fixed integer $k \geq 1$,*

$$\left(\frac{1}{5} + \mathrm{o}(1) \right) \frac{k}{\log k} \leq F(k) \leq (80 + \mathrm{o}(1)) k \log(k).$$

A tournament is *k-transitive* if its edge set can be partitioned into k sets each of which is transitively oriented. András Gyárfás made the conjecture that k-transitive tournaments have bounded domination number, and this was explored in 2014 by Pálvölgyi and Gyárfás [76].

Conjecture 2 (Gyárfás) *For each positive integer k, there exists a (least) $p(k)$ such that every k-transitive tournament has a dominating set of at most $p(k)$ vertices.*

We proceed further with the following definitions.

Definition 13 A class $\mathcal{C}$ of tournaments has *bounded domination* if there exists a constant c such that every tournament in $\mathcal{C}$ has domination number at most c. If S and T are tournaments, then T is called *S-free* if no subtournament of T is isomorphic to S. A tournament S is a *rebel* if the class of all S-free tournaments has bounded domination.

In 2018 Chudnovsky, Ringi, Chun-Hung, Seymour, and Thomassé [23] investigated the following conjecture posed by HeHui Wu.

Conjecture 3 (HeHui Wu) *Every tournament is a rebel.*

Chudnovsky et al. [23] disproved Conjecture 3. For this purpose, they defined the notion of a poset tournament.

Definition 14 A tournament T is a *poset tournament* if its vertex set can be ordered $\{v_1, \ldots, v_n\}$ such that for all $1 \leq i < j < k \leq n$, if v_j is adjacent from v_i and adjacent to v_k, then v_i is adjacent to v_k; that is, the "forward" edges under this linear order form the comparability graph of a partial order.

Chudnovsky et al. [23] observed that not every tournament is a poset tournament. Thereafter, they proved the following result, hence disproving Conjecture 3.

Theorem 76 ([23]) *Every rebel is a poset tournament.*

However, it remains an open problem to determine if every poset tournament is a rebel. Since Wu's Conjecture, that every tournament is a rebel, is false, it naturally raises the question: Which tournaments are rebels? Theorem 76 provides a partial

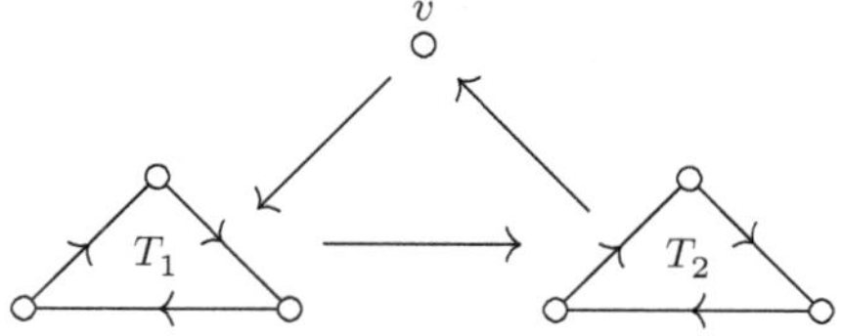

Fig. 6 The non-2-colorable tournament T_*

answer to this question. To further answer this question, we need the definition of a coloring of a tournament.

Definition 15 A k-coloring of a tournament T is a partition of $V(T)$ into k transitive sets, or, equivalently, into k acyclic sets. A tournament T with a k-coloring is called k-colorable.

Chudnovsky et al. [23] proved that Conjecture 3 is true for 2-colorable tournaments. Their proof followed from a direct application of VC dimension.

Theorem 77 ([23]) *All 2-colorable tournaments are rebels.*

A breakthrough in their paper [23] is that Chudnovsky et al. overcame the unboundedness of the VC dimension by showing that large shattered sets in a hypergraph are sparse, which turns out to be enough to carry over the proof of Theorem 76. This enabled them to give a non-2-colorable tournament T_* on seven vertices that satisfies Conjecture 3. Such a tournament T_* is constructed from a cyclic triangle by substituting a copy of a cyclic triangle for two of the three vertices of an original cyclic triangle. A sketch of the tournament T_* is given in Figure 6, where the arrow from v to the cyclic triangle T_1 indicates that all three arcs from v to T_1 are arcs out of v while the arrow from the cyclic triangle T_2 to v indicates that all three arcs from T_2 to v are arcs into v. Further, the arc from T_1 to T_2 indicates that every vertex in T_1 is adjacent to every vertex in T_2.

Theorem 78 ([23]) *The non-2-colorable tournament T_* is a rebel.*

Thus, Theorem 78 gives a counterexample to the converse of Theorem 77, that all rebels are 2-colorable. As a consequence of Theorem 78, the following result is proven, where the *odd girth* of a tournament T is the smallest k for which there exists a subtournament of T with k vertices that is not 2-colorable (and is undefined if T is 2-colorable).

Theorem 79 ([23]) *For $k \geq 8$, the class of tournaments with odd girth at least k has bounded domination.*

We close this section on domination in tournaments, with a brief discussion on what we define next as a domination graph of a digraph.

Definition 16 Two vertices x and y dominate an oriented graph $D = (V, A)$ if the set $\{x, y\}$ is a dominating set of D, that is, every vertex z different from x and y is adjacent from at least one of x and y, and so $(x, z) \in A$ or $(y, z) \in A$. The *domination graph* of an oriented graph D is the graph G with $V(G) = V(D)$ and with an edge between two vertices x and y if x and y dominate T, that is, if every other vertex loses to at least one of x and y.

Domination graphs were introduced and studied by Fisher et al. [30–35] and [21, 22], who largely considered the domination graphs of tournaments. In particular, Fisher et al. showed that the domination graph of a tournament is either an odd cycle with or without isolated and/or pendant vertices or a forest of caterpillars. They also showed that any graph consisting of an odd cycle with or without isolated and/or pendant vertices is the domination graph of some tournament.

6 Total Domination in Digraphs

There are several possibilities for defining the counterpart of a total dominating set in a digraph D. We consider four such versions in the following subsections.

6.1 Total Domination: Version 1

In this version of total domination, we define a set S in a digraph D to be a *total in-dominating set* if S is an in-dominating set in D with the added property that the subdigraph induced by S has no isolated vertices. Here we define the *total in-domination number* $\gamma_{ti}^-(D)$ of a digraph D to equal the minimum cardinality of such a set S according to Version 1. We note that if the underlying graph of D has no isolated vertices, then $V(D)$ is vacuously a total in-domination set of D, and so $\gamma_{ti}^-(D)$ is well-defined and $\gamma_{ti}^-(D) \le |V(D)|$.

6.2 Total Domination: Version 2

In this version of total domination, a set S in a digraph D is a *total dominating set* if S is a dominating set in D with the added property that the subdigraph induced by S has no isolated vertices. This is a version defined by Arumugam, Jacob, and Volkmann [3] in 2007 and Hao [49] in 2017. We define the *total domination number* $\gamma_t(D)$ of a digraph D with no isolated vertices to equal the minimum cardinality of such a set S according to Version 2. As with version 1 above, we note that $\gamma_t(D)$ is well-defined and $\gamma_t(D) \le |V(D)|$. Arumugam et al. [3] established the following lower bound on the total domination number of a digraph.

Theorem 80 ([3]) *If D is a digraph of order n, with maximum outdegree Δ^+ and without isolated vertices, then*

$$\gamma_t(D) \geq \left\lceil \frac{2n}{2\Delta^+ + 1} \right\rceil.$$

Hao and Chen [50] improved the lower bound in Theorem 80. For this purpose, they define the *out-Slater number* of a digraph D of order n as

$$\mathrm{sl}^+(D) = \min\{k : \lfloor k/2 \rfloor + (d_1^+ + d_2^+ + \cdots + d_k^+) \geq n\},$$

where $d_1^+, d_2^+, \ldots, d_k^+$ are the first k largest outdegrees of D.

Theorem 81 ([50]) *If D is a digraph of order n, with maximum outdegree Δ^+ and without isolated vertices, then*

$$\gamma_t(D) \geq \mathrm{sl}^+(D) \geq \left\lceil \frac{2n}{2\Delta^+ + 1} \right\rceil.$$

Further, the gap between the rightmost two numbers can be arbitrarily large.

The authors in [50] also determined the following lower bound on the total domination number of an oriented tree in terms of its order and number of vertices of outdegree 0.

Theorem 82 ([50]) *If T is an oriented tree of order $n \geq 2$, with n_0 vertices of outdegree 0 and with non-increasing outdegree sequence $d_1^+, d_2^+, \ldots, d_n^+$, then*

$$\gamma_t(T) \geq \mathrm{sl}^+(D) \geq \frac{2}{3}(n - n_0 + 1),$$

with equality if and only if $n - n_C \equiv 2 \pmod 3$ and $d_{k+1}^+ \leq 1$, where $k = \frac{2}{3}(n - n_0 + 1)$.

6.3 Total Domination: Version 3

In this version of total domination, a set S in a digraph $D = (V, E)$ is a *total in-dominating set* if every vertex in V is adjacent *to* a vertex in S, that is, $N^-(S) = V$. This is equivalent to saying that S is an in-dominating set and the subdigraph induced by S has no isolated vertices and no sources. The minimum cardinality of such a set could be called the *total absorption number*, denoted $\gamma_t^-(D)$. We note that every digraph D with $\delta^-(D) \geq 1$ has a total dominating set according to this definition since $V(D)$ is such a set. For example, the digraph D shown in Figure 7 satisfies $\gamma_t^-(D) = 3$, where the darkened vertices form a total dominating set of D of cardinality 3.

For a digraph $D = (V, E)$ and for a real-valued function $f : V \to \mathbb{R}$, the *weight* of f is $w(f) = \sum_{v \in V} f(v)$. Further, for $S \subseteq V$, we define $f(S) = \sum_{v \in S} f(v)$; in particular,

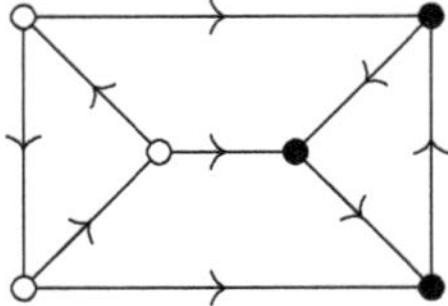

Fig. 7 A digraph D with $\gamma_t^-(D) = 3$

this means that $w(f) = f(V)$. Let $f : V \to \{0, 1\}$ be a function which assigns to each vertex of a graph an element of the set $\{0, 1\}$. We say f is a *total dominating function* if for every $v \in V$, the sum of the function values under f in every out-neighborhood of a vertex is at least 1, that is, for every vertex $v \in V$, we have

$$\sum_{u \in N^+(v)} f(u) \geq 1.$$

The total absorption number of D can be defined as

$$\gamma_t^-(D) = \min\{w(f) \mid f \text{ is a total dominating function on } D\}.$$

In order to present a lower bound on the total absorption number of a digraph, St-Louis, Gendron, and Hertz [87] in 2012 considered the fractional version of a total dominating set where vertices have fractional weights in the range $[0, 1]$. A real-valued function $f : V \to [0, 1]$ is called a *fractional total dominating function* of a digraph D if $\sum_{u \in N^+(v)} f(u) \geq 1$ for each $v \in V$. The minimum weight of a fractional total dominating function of D is the *fractional total domination number*, which we denote here by $\gamma_{tf}^-(D)$. Thus,

$$\gamma_{tf}^-(D) = \min\{w(f) \mid f \text{ is a fractional total dominating function for } D\}.$$

We remark that the fractional total domination number is readily viewed as a linear program. Thus we can talk of minimum, rather than infimum in the above definition. By definition, $\gamma_t^-(D) \geq \gamma_{tf}^-(D)$, and so the fractional version provides a lower bound on the total absorption number of D. The *girth* $g(D)$ of a digraph D is the number of vertices of the smallest directed cycle in D. St-Louis et al. [87] posed two conjectures, one of which is the following.

Conjecture 4 ([87]) *If D is a digraph with $\delta^+(D) \geq 1$, then $\gamma_{tf}^-(D) > g(D) - 1$.*

St-Louis et al. [87] proved that Conjecture 4 is equivalent to the 1978 Caccetta-Häggkvist Conjecture which we state below.

Conjecture 5 ([14]) *If D is a digraph of order n with $\delta^+(D) \geq r \geq 1$, then $g(D) \leq \lceil \frac{n}{r} \rceil$.*

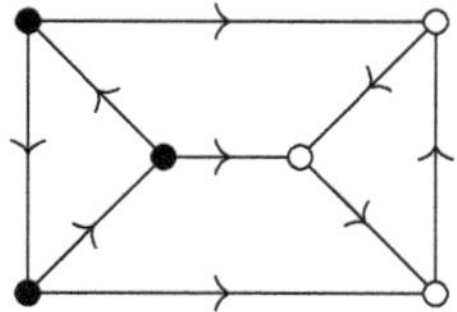

Fig. 8 A digraph D with $\gamma_{to}^{+}(D) = 3$

6.4 Total Domination: Version 4

In this version of total domination, a set S in a digraph $D = (V, E)$ is a *total dominating set* if every vertex in V is adjacent *from* a vertex in S, that is, $N^+(S) = V$. This is equivalent to saying that S is a dominating set and the subdigraph induced by S has no isolated vertices and no sinks. This is a version defined by Hansen, Lai, and Yue [47] in 1999 and by Schaudt [83] in 2012. We shall call this type of total dominating set a *total open dominating set* and let $\gamma_{to}^{+}(D)$ equal the minimum cardinality of a total open dominating set in a digraph D. For example, the digraph D shown in Figure 8 satisfies $\gamma_{to}^{-}(D) = 3$, where the darkened vertices form a total open dominating set of D of cardinality 3.

In 1999 Hansen et al. [47] defined the *lower orientable open domination number* $\mathrm{dom}_1(G)$ of a graph G as the minimum total open domination number among all orientations of G. The *upper orientable total open domination number* $\mathrm{DOM}_1(G)$ equals the maximum such total open domination number.

Theorem 83 ([47]) *For a connected graph G, $\mathrm{dom}_1(G)$ and $\mathrm{DOM}_1(G)$ exist if and only if G is not a tree.*

Hansen et al. [47] also investigated the function $\mathrm{DOM}_1(K_n)$. They showed this to be a non-decreasing function and unbounded and determined specific values. Analogous to Theorem 52, they proved the following result.

Theorem 84 ([47]) *For every integer c with $\mathrm{dom}_1(K_n) \leq c \leq \mathrm{DOM}_1(K_n)$, there exists an orientation D of K_n such that $\gamma_{to}^{+}(D) = c$.*

In 2012 Schaudt [83] studied efficient total domination in digraphs, where an *efficient total dominating set* of a digraph D is a total open dominating set S with the property that for each vertex v of D, there is a unique vertex $u \in S$ that is adjacent to v. Graphs that permit an orientation having such a set were studied in [83]. Further, complexity results and characterizations were given.

6.5 Fractional Domination in Digraphs

In Section 6.3, we considered the fractional version of total domination in digraphs. In this section, we present results on the fractional version of domination in digraphs. Adopting our earlier notation, a real-valued function $f : V \rightarrow \mathbb{R}$ in a digraph D is a *dominating function* if for every $v \in V$, the sum of the function values under f in every closed out-neighborhood of a vertex is at least 1, that is, for every vertex $v \in V$, we have

$$\sum_{u \in N^+[v]} f(u) \geq 1.$$

The domination number of D can be defined as

$$\gamma(D) = \min\{w(f) \mid f \text{ is a dominating function on } D\}.$$

A real-valued function $f : V \to [0, 1]$ is called a *fractional dominating function* of a digraph D if $\sum_{u \in N^+[v]} f(u) \geq 1$ for each $v \in V$. The minimum weight of a fractional dominating function of D is the *fractional domination number*, which we denote here by $\gamma_f(D)$. Thus,

$$\gamma_f(D) = \min\{w(f) \mid f \text{ is a fractional dominating function for } D\}.$$

In 1982 Sands, Sauer, and Woodrow [82] (also due to Erdős) posed the following conjecture.

Conjecture 6 ([82]) *For each n, there is a (least) positive integer $f(n)$ so that every finite tournament whose edges are colored with n colors contains a set S of $f(n)$ vertices with the property that for every vertex u not in S, there is a monochromatic directed path from u to a vertex of S.*

A *complete multidigraph* is a directed graph in which multiple arcs and circuits of length 2 are allowed and such that there always exists an arc between two distinct vertices. A tournament, for example, is a complete multidigraph in the special case when the directed graph is simple (and contains no multiple arcs or circuits of length 2). As remarked in [13], the transitive closure of each color class is a quasi-order (i.e., a transitive digraph); hence, the Erdős-Sands-Sauer-Woodrow conjecture can be restated as follows.

Conjecture 7 ([82]) *For every k, there exists an integer $f(k)$ such that if T is a complete multidigraph whose arcs are the union of k quasi-orders, then $\gamma(T) \leq f(k)$.*

In 2019 Bousquet, Lochet, and Thomassé [13] succeeded in proving this long-standing 1982 Erdős-Sands-Sauer-Woodrow conjecture. The main ingredient in their proof is that the fractional domination number of complete multidigraphs (and therefore of tournaments) is bounded.

Theorem 85 ([13]) *For every k, if T is a complete multidigraph whose arcs are the union of k quasi-orders, then*

$$\gamma(T) = O(\ln(2k) \cdot k^{k+2}).$$

Harutyunyan, Le, Newman, and Thomassé [53] continued the study of fractional domination in digraphs Recall that in general there is no upper bound on the domination number of an oriented graph solely in terms of its independence number. However, by Theorem 64, if G is a graph of order n, then $\mathrm{DOM}(G) \le \alpha(G) \cdot \log n$. In contrast to this result, Harutyunyan et al. [53] showed that for any digraph, its fractional domination number is at most twice its independence number.

Theorem 86 ([53]) *For every digraph D, we have $\gamma_f(D) \le 2\alpha(D)$, and this bound is sharp.*

The authors in [53] presented two proofs of Theorem 86. The first proof uses the duality of linear programming, while the second proof is by induction. To show sharpness of the bound, given an arbitrary small real number $\epsilon > 0$, for any integer $k \ge 1$, they constructed a digraph D such that $\alpha(G) = k$ and $\gamma_f(D) > 2k - \epsilon$. Further, they showed that almost surely a random tournament has fractional domination number close to the upper bound of 2.

7 The Oriented Version of the Domination Game

In 2002 Alon, Balogh, Bollobás, and Szabó [1] introduced and first studied the oriented domination game, which belongs to the growing family of competitive optimization graph games. The oriented domination game describes a process in which two players with conflicting goals alternately orient an edge of a graph G until all of the edges are oriented. One player's goal is to minimize the domination number of the resulting oriented graph, while the other player wants to maximize it.

Formally, the oriented domination game on a graph G consists of two players, *Minimizer* and *Maximizer* (called *Dominator* and *Avoider* in [1]), who take turns orienting an unoriented edge of a graph G, until all edges are oriented. The goal of Minimizer is to minimize the domination number of the resulting digraph, while the goal of Maximizer is to maximize the domination number. The Minimizer-start oriented domination game is the oriented domination game when Minimizer plays first. The *oriented game domination number* $\gamma_{\mathrm{og}}(G)$ of G is the minimum possible domination number of the resulting digraph when both players play according to the rule that on each move a player may only orient an unoriented edge. To illustrate the game, Alon et al. [1] determined the oriented game domination number of a complete graph.

Proposition 87 ([1]) *For a complete graph K_n of order $n \ge 4$, we have $\gamma_{\mathrm{og}}(K_n) = 2$.*

Proof Minimizer's strategy is to pick two arbitrary vertices, say u and v. On each of his turns, Minimizer orients an edge from u or v to a vertex w different from u and v. His strategy is to orient these edges in such a way that at least one of u and v is oriented towards w. He can always achieve his goal as follows. Whenever Maximizer orients the edge uw from w to u, then Minimizer immediately replies by orienting the edge vw from v to w, if it is not already oriented. Analogously, whenever Maximizer orients the edge vw from w to v, then Minimizer immediately replies by orienting the edge uw from u to w, if it is not already oriented. In this way, he ensures that the set $\{u, v\}$ is a dominating set in the resulting oriented graph. Thus, $\gamma_{\mathrm{og}}(K_n) \leq 2$.

To show that $\gamma_{\mathrm{og}}(K_n) \geq 2$, Maximizer adopts the following strategy. Maximizer can clearly prevent a source in the oriented graph resulting when $n = 4$. In the case when $n \geq 5$, there exists a collection of n edge-disjoint paths of length 2, one centered at each of the n vertices of K_n (see [10]). Maximizer's strategy is whenever Minimizer orients one of these edges from a central vertex on one of these paths, Maximizer responds by orienting the other edge of the corresponding path towards the central vertex. In this way, Maximizer guarantees that the indegree of each vertex in the resulting oriented graph becomes at least 1, implying that $\gamma_{\mathrm{og}}(K_n) \geq 2$. Consequently, $\gamma_{\mathrm{og}}(K_n) \geq 2$. $\square$

In [1], the authors obtained a sharp lower bound for the oriented game domination number of trees.

Theorem 88 ([1]) *If G is a tree of order n, then $\frac{1}{2}n \leq \gamma_{\mathrm{og}}(G) \leq \frac{2}{3}n$.*

The proof of Theorem 88 implies that the upper bound holds for any connected graph G, as Minimizer can concentrate his attention on a spanning tree T of G and play according to his strategy in the tree T. Whenever Maximizer orients an edge not in T, Minimizer continues to orient edges according to his strategy in the tree. As shown in [1], both bounds in Theorem 88 are sharp. For graphs with minimum degree at least 2, the following improved upper bound was given in [1].

Theorem 89 ([1]) *If G is a graph of order n with $\delta(G) \geq 2$, then $\gamma_{\mathrm{og}}(G) \leq \frac{1}{2}n$.*

If G is a graph of order n with maximum degree Δ, then a trivial lower bound on the domination number is $\gamma(G) \geq n/\Delta$. In the oriented domination game, Maximizer orients half of the edges. As observed by Alon et al. [1], Maximizer might succeed in decreasing the outdegree of each vertex to about $\Delta/2$, in which case the resulting domination number is at least $2n/\Delta$. This prompted them to pose the following conjecture.

Conjecture 8 ([1]) *If G is a graph of order n with maximum degree Δ, then*

$$\gamma_{\mathrm{og}}(G) \geq \left(\frac{2}{(1 + \mathrm{o}(1))\Delta} \right) n.$$

Conjecture 8 has yet to be settled. The best general lower bound to date on the oriented game domination number in terms of the maximum degree and order of the graph is the following result in [1].

Theorem 90 ([1]) *If G is a graph of order n with maximum degree Δ, then*

$$\gamma_{\mathrm{og}}(G) \geq \left(\frac{4}{3\Delta + 7} \right) n.$$

Nordhaus-Gaddum-type inequalities for the oriented domination game are given in [1]. Here, $\overline{G}$ denotes the complement of a graph G.

Theorem 91 ([1]) *If G is a graph of order n, then $\gamma_{\mathrm{og}}(G) + \gamma_{\mathrm{og}}(\overline{G}) \leq n + 2$, and this bound is sharp.*

We note that if G is the complete graph K_n where $n \geq 4$, then $\gamma_{\mathrm{og}}(\overline{G}) = n$ and, by Proposition 87, $\gamma_{\mathrm{og}}(G) = 2$. Thus, if $G = K_n$, then $\gamma_{\mathrm{og}}(G) + \gamma_{\mathrm{og}}(\overline{G}) = n+2$, showing sharpness of the bound in Theorem 91. We close this section with the following conjecture posed in [1], that the inequality in Theorem 91 can be strengthened for connected graphs.

Conjecture 9 ([1]) *If both G and its complement $\overline{G}$ are connected graphs of order n, then*

$$\gamma_{\mathrm{og}}(G) + \gamma_{\mathrm{og}}(\overline{G}) \leq \frac{2}{3}n + 3.$$

8 Concluding Comments

In this chapter, we have surveyed selected results on domination in digraphs. Many results have been omitted to prevent the chapter from growing too large. For example, topics such as signed domination in digraphs, efficient domination in digraphs, packing in digraphs, reinforcement numbers of digraphs, rainbow domination in digraphs, and Roman domination in digraphs, to name a few, are omitted. Additional references on domination in digraphs can be found in [40, 44, 48, 55, 56, 58, 70, 77, 85, 91]. Due to space limitations, we have also omitted proofs of many important results on domination in digraphs presented in this chapter, including the proofs of results due to Alon, Brightwell, Kierstead, Kostochka, and Winkler [2]; Chudnovsky, Ringi, Chun-Hung, Seymour, and Thomassé [23]; Harutyunyan, Le, Newman, and Thomassé [53]; and Bousquet, Lochet, and Thomassé [13] which have significantly impacted the latest developments in the field of domination in digraphs and tournaments. We apologize for these omissions.

References

1. N. Alon, J. Balogh, B. Bollobás, T. Szabó, Game domination number. Discrete Math. **256**, 23–33 (2002)
2. N. Alon, G. Brightwell, H.A. Kierstead, A.V. Kostochka, P. Winkler, Dominating sets in k-majority tournaments. J. Combin. Theory Ser. B **96**(3), 374–387 (2006)
3. S. Arumugam, K. Jacob, L. Volkmann, Total and connected domination in digraphs. Australas. J. Combin. **39**, 283–292 (2007)
4. S. Arumugam, K. Ebadi, L. Sathikala, Twin domination and twin irredundance in digraphs. Appl. Anal. Discrete Math. **7**, 275–284 (2013)

5. C. Balbuena, H. Galeana-Sánchez, M. Guevara, A sufficient condition for kernel-perfectness of a digraph in terms of semikernel modulo F. Acta Math. Appl. Sin. **28**, 340–356 (2012). English Series

6. C. Berge, Théorie des graphes et ses applications (French), in *Collection Universitaire de Mathematiques*, vol. II (Dunod, Paris, 1958), viii+277 pp.

7. C. Berge, *Graphs and Hypergraphs* (North Holland, New York, 1973)

8. M. Blidia, L. Ould-Rabah, Bounds on the domination number in oriented graphs. Australas. J. Combin. **48**, 231–241 (2010)

9. B. Bollobás, T. Szabó, Domination in oriented graphs. Proceedings of the Twenty-eighth Southeastern International Conference on Combinatorics, Graph Theory and Computing (Boca Raton, FL, 1997). Congr. Numer. **123**, 55–64 (1997)

10. B. Bollobás, T. Szabó, The oriented cycle game. Discrete Math. **186**, 55–67 (1998)

11. J.A. Bondy, Short proofs of classical theorems. J. Graph Theory **44**, 159–165 (2003)

12. E. Boros, V. Gurvich, Perfect graphs, kernels, and cores of cooperative games. Discrete Math. **306**, 2336–2354 (2006)

13. N. Bousquet, W. Lochet, S. Thomassé, A proof of the Erdős-Sands-Sauer-Woodrow conjecture. J. Combin. Theory Ser. B **137**, 316–319 (2019)

14. L. Caccetta, R. Häggkvist, On minimal digraphs with given girth. Congr. Numer. **21**, 181–187 (1978)

15. H. Cai, J. Liu, L. Qian, The domination number of strong product of directed cycles. Discrete Math. Algorithms Appl. **6**(2), 1450021, 10 pp. (2014)

16. Y. Caro, M.A. Henning, A greedy partition lemma for directed domination. Discrete Optim. **8**, 452–458 (2011)

17. Y. Caro, M.A. Henning, Directed domination in oriented graphs. Discrete Appl. Math. **160**, 1053–1063 (2012)

18. G. Chartrand, D. Vanderjagt, B.Q. Yue, Orientable domination in graphs. Congr. Numer. **119**, 51–63 (1996)

19. G. Chartrand, F. Harary, B.Q. Yue, On the out-domination and in-domination numbers of a digraph. Discrete Math. **197/198**, 179–183 (1999)

20. G. Chartrand, P. Dankelmann, M. Schultz, H.C. Swart, Twin domination in digraphs. Ars Combin. **67**, 105–114 (2003)

21. H.H. Cho, F. Doherty, S.R. Kim, J.R. Lundgren, Domination graphs of regular tournaments II. Congr. Numer. **130**, 95–111 (1998)

22. H.H. Cho, S.R. Kim, J.R. Lundgren, Domination graphs of regular tournaments. Discrete Math. **252**, 57–71 (2002)

23. M. Chudnovsky, R. Kim, C.-H. Liu, P. Seymour, S. Thomassé, Domination in tournaments. J. Combin. Theory Ser. B **130**, 98–113 (2018)

24. V. Chvátal, L. Lovász, Every directed graph has a semi-kernel, in *Hypergraph Seminar*, ed. by C. Berge, D.K. Ray-Chaudhuri (Springer, New York, 1974), p. 175

25. P. Duchet, Graphes Noyau-parfaits. Ann. Discrete Math. **9**, 93–101 (1980)

26. P. Duchet, A sufficient condition for a digraph to be kernel-perfect. J. Graph Theory **11**, 81–85 (1987)

27. E. Egerváry, On combinatorial properties of matrices. Mat. Lapok **38**, 16–28 (1931)

28. P. Erdős, On Schütte problem. Math. Gaz. **47**, 220–222 (1963)

29. W.F. De la Vega, Kernels in random graphs. Discrete Math. **82**, 213–217 (1990)

30. D.C. Fisher, J.R. Lundgren, S.K. Merz, K.B. Reid, Domination graphs of tournaments and digraphs. Congr. Numer. **108**, 97–107 (1995)

31. D.C. Fisher, J.R. Lundgren, S.K. Merz, K.B. Reid, The domination and competition graphs of a tournament. J. Graph Theory **29**, 103–110 (1998)

32. D.C. Fisher, J.R. Lundgren, S.K. Merz, K.B. Reid, Connected domination graphs of tournaments. J. Combin. Math. Combin. Comput. **31**, 169–176 (1999)

33. D.C. Fisher, D. Guichard, J.R. Lundgren, S.K. Merz, Domination graphs with nontrivial components. Graphs Combin. **17**(2), 227–228 (2001)

34. D.C. Fisher, D. Guichard, J.R. Lundgren, S.K. Merz, K.B. Reid, Domination graphs of tournaments with isolated vertices. Ars Combin. **66**, 299–311 (2003)
35. D.C. Fisher, D. Guichard, J.R. Lundgren, S.K. Merz, K.B. Reid, Domination graphs with 2 or 3 nontrivial components. Bull. ICA **40**, 67–76 (2004)
36. A. Fraenkel, Planar kernel and Grundy with $d \leq 3$, $d_{out} \leq 2$, $d_{in} \leq 2$ are NP-complete. Discrete Appl. Math. **3**, 257–262 (1981)
37. A.S. Fraenkel, Combinatorial games: selected bibliography with a Succinct Gourmet introduction. Electron. J. Combin. **42**, #DS2 (2007)
38. Y. Fu, Dominating set and converse dominating set of a directed graph. Amer. Math. Mon. **75**, 861–863 (1968)
39. H. Galeana-Sánchez, A counterexample to a conjecture of Meyniel on kernel-perfect graphs. Discrete Math. **41**, 105–107 (1982)
40. H. Galeana-Sánchez, C. Hernandez-Cruz, On the existence of (k, l)-kernels in digraphs with a given circumference. AKCE Int. J. Graphs Comb. **10**(1), 15–28 (2013)
41. H. Galeana-Sánchez, V. Neumann-Lara, On kernels and semikernels of digraphs. Discrete Math. **48**, 67–76 (1984)
42. T. Gallai, Über extreme Punkt- und Kantenmengen. Ann. Univ. Sci. Budapest Eötvös Sect. Math. **2**, 133–138 (1959)
43. J. Ghoshal, R. Laskar, D. Pillone, Topics on domination in directed graphs, in *Domination in Graphs: Advanced Topics*, ed. by T.W. Haynes, S.T. Hedetniemi, P.J. Slater (Marcel Dekker, New York, 1998)
44. S. Gonzalez Hermosillo de la Maza, C. Hernandez-Cruz, On the existence of 3- and 4-kernels in digraphs. Discrete Math. **341**(3), 638–651 (2018)
45. Š. Gyürki, On the difference of the domination number of a digraph and of its reverse. Discrete Appl. Math. **160**, 1270–1276 (2012)
46. L. Hansen, Domination in digraphs. Ph.D. Dissertation, Western Michigan University, 1997, 109 pp.
47. L. Hansen, Y.-L. Lai, B.Q. Yue, Orientable open domination of graphs. J. Combin. Math. Combin. Comput. **29**, 159–181 (1999)
48. G. Hao, On the domination number of digraphs. Ars Combin. **134**, 51–60 (2017)
49. G. Hao, Total domination in digraphs. Quaest. Math. **40**, 333–346 (2017)
50. G. Hao, X. Chen, A note on lower bounds for the total domination number of digraphs. Quaest. Math. **40**, 553–562 (2017)
51. G. Hao, J. Qian, On the sum of out-domination number and in-domination number of digraphs. Ars Combin. **119**, 331–337 (2015)
52. G. Hao, J. Qian, Bounds on the domination number of a digraph. J. Combin. Optim. **35**(1), 64–74 (2018)
53. A. Harutyunyan, T. Le, A. Newman, S. Thomassé, Domination and fractional domination in digraphs. Electron. J. Combin. **25**(3), Paper 3.32, 9 pp. (2018)
54. T.W. Haynes, S.T. Hedetniemi, P.J. Slater (eds.) *Domination in Graphs: Advanced Topics* (Marcel Dekker, New York, 1998)
55. P. Hell, C. Hernandez-Cruz, On the complexity of the 3-kernel problem in some classes of digraphs. Discuss. Math. Graph Theory **34**(1), 167–185 (2014)
56. M.A. Henning, A survey of selected recent results on total domination in graphs. Discrete Math. **309**, 32–63 (2009)
57. M.A. Henning, A. Yeo, Tight lower bounds on the size of a matching in a regular graph. Graphs Combin. **23**, 647–657 (2007)
58. M.A. Henning, A. Yeo, *Total Domination in Graphs*. Springer Monographs in Mathematics (2013). ISBN: 978-1-4614-6524-9 (Print) 978-1-4614-6525-6 (Online)
59. H. Jacob, H. Meyniel, About quasi-kernels in a digraph. Discrete Math. **154**, 279–280 (1996)
60. D. König, Graphen und Matrizen. Math. Riz. Lapok **38**, 116–119 (1931)
61. D. König, *Theorie der endlichen und unendlichen graphen* (Akademische Verlagsgesellschaft M. B. H., Leipzig, 1936), and Chelsea Publ. Co., New York, 1950
62. C. Lee, Domination in digraphs. Ph.D. Dissertation, Michigan State University, 1994, 59 pp.

63. C. Lee, Domination in digraphs. J. Korean Math. Soc. **35**, 843–853 (1998)
64. J. Liu, X. Zhang, X. Chen, J. Meng, On domination number of Cartesian product of directed cycles. Inform. Process. Lett. **110**(5), 171–173 (2010)
65. J. Liu, X. Zhang, J. Meng, On domination number of Cartesian product of directed paths. J. Comb. Optim. **22**(4), 651–662 (2011)
66. J. Liu, X. Zhang, J. Meng, Domination in lexicographic product digraphs. Ars Combin. **120**, 23–32 (2015)
67. H. Ma, J. Liu, The twin domination number of Cartesian product of directed cycles. J. Math. Res. Appl. **36**(2), 171–176 (2016)
68. H. Ma, J. Liu, The twin domination number of lexicographic product of digraphs (English, Chinese summary). J. Nat. Sci. Hunan Norm. Univ. **39**(6), 80–84, 93 (2016)
69. H. Ma, J. Liu, The twin domination number of strong product of digraphs (English summary). Commun. Math. Res. **32**(4), 332–338 (2016)
70. A. Meir, J.W. Moon, Relations between packing and covering numbers of a tree. Pacific J. Math. **61**, 225–233 (1975)
71. M. Mollard, On the domination of Cartesian product of directed cycles: results for certain equivalence classes of lengths. Discuss. Math. Graph Theory **33**(2), 387–394 (2013)
72. J.W. Moon, *Topics on Tournaments* (Holt, Rinehartand Winston, New York, 1968)
73. L. Niepel, M. Knor, Domination in a digraph and in its reverse. Discrete Appl. Math. **157**(13), 2973–2977 (2009)
74. L. Niepel, M. Knor, Domination in the cross product of digraphs. Ars Combin. **97**, 271–279 (2010)
75. O. Ore, *Theory of Graphs* (American Mathematical Society, Providence, 1962)
76. D. Pálvölgyi, A. Gyárfás, Domination in transitive colorings of tournaments. J. Combin. Theory Ser. B **107**, 1–11 (2014)
77. A. Ramoul, M. Blidia, A new generalization of kernels in digraphs. Discrete Appl. Math. **217**, 673–684 (2017)
78. K.B. Reid, A.A. McRae, S.M. Hedetniemi, S.T. Hedetniemi, Domination and irredundance in tournaments. Australas. J. Combin. **29**, 157–172 (2004)
79. M. Richardson, On weakly ordered systems. Bull. Amer. Math. Soc. **52**, 113–116 (1946)
80. B. Roy, *Algebre moderne et theorie des Graphes*, vol. 2 (Dunod, Paris, 1970), Chapter VI
81. S. Rudeanu, Notes ur l' existence et l' unicite du noyau d' un graphe. Revue Francaise Rech. Operat. **33**, 20–26 (1964)
82. B. Sands, N. Sauer, R. Woodrow, On monochromatic paths in edge-coloured digraphs. J. Comb. Theory Ser. B **33**, 271–275 (1982)
83. O. Schaudt, Efficient total domination in digraphs. J. Discrete Algorithms **15**, 32–42 (2012)
84. R.S. Shaheen, Domination number of toroidal grid digraphs. Util. Math. **78**, 175–184 (2009)
85. R.S. Shaheen, Total domination number of products of two directed cycles. Util. Math. **92**, 235–250 (2013)
86. Z. Shao, E. Zhu, F. Lang, On the domination number of Cartesian product of two directed cycles. J. Appl. Math., Art. ID 619695, 7 pp. (2013)
87. P. St-Louis, B. Gendron, A. Hertz, Total domination and the Caccetta-Häggkvist conjecture. Discrete Optim. **9**(4), 236–240 (2012)
88. G. Szekeres, H.S. Wilf, An inequality for chromatic number of a graph. J. Combin. Theory Ser. B **4**, 1–3 (1968)
89. V.G. Vizing, The Cartesian product of graphs. Vyčisl Sistemy **9**, 30–43 (1963)
90. V.G. Vizing, Some unsolved problems in graph theory. Uspehi Mat. Nauk **23**(6(144)), 117–134 (1968)
91. L. Volkmann, Signed total Roman domination in digraphs. Discuss. Math. Graph Theory **37**, 261–272 (2017)
92. J. Von Neumann, O. Morgenstern, *Theory of Games and Economic Behavior* (Princeton University Press, Princeton, 1944)
93. X. Zhang, J. Liu, X. Chen, J. Meng, Domination number of Cartesian products of directed cycles. Inform. Process. Lett. **111**(1), 36–39 (2010)

Part III
Algorithms and Complexity

Algorithms and Complexity of Signed, Minus, and Majority Domination

Stephen T. Hedetniemi, Alice A. McRae, and Raghuveer Mohan

1 Introduction to Y-Dominating Functions

In this chapter, we discuss the algorithmic and complexity results for several types of domination, which can be formulated as follows.

Let $G = (V, E)$ be a graph and let Y be an arbitrary set of real numbers, finite or infinite, positive or negative. A function $f : V \to Y$ is called a *Y-dominating function* if for every $v \in V$, $f(N[v]) = \Sigma_{u \in N[v]} f(u) \geq 1$. In other words, the closed neighborhood sum $f(N[v])$ of every vertex $v \in V$ is at least one.

The *weight* of a dominating function f is $w(f) = f(V) = \Sigma_{u \in V} f(u)$. The *Y-domination number* $\gamma_Y(G)$ equals the minimum weight of a Y-dominating function f on G.

A Y-dominating function f is called *minimal* if there does not exist another Y-dominating function g, $f \neq g$, with $g(u) \leq f(u)$ for every $v \in V$. The *upper Y-domination number* $\Gamma_Y(G)$ equals the maximum weight $w(f)$ of a minimal Y-dominating function f on G.

We say that a Y-dominating function is *efficient* if $f(N[v]) = 1$ for every vertex $v \in V$. As we will see below, depending on the set Y of real numbers, a graph G may or may not have an efficient Y-dominating function.

By changing closed neighborhoods to open neighborhoods, we can define the following. A function $g : V \to Y$ is called a *Y-total dominating function* if for every $v \in V$, $g(N(v)) = \Sigma_{u \in N(v)} g(u) \geq 1$. In other words, the open neighborhood sum $g(N(v))$ of every vertex $v \in V$ is at least one. The *weight* of a total dominating

S. T. Hedetniemi
School of Computing, Clemson University, Clemson, SC, USA
e-mail: hedet@clemson.edu

A. A. McRae (✉) · R. Mohan
Computer Science Department, Appalachian State University, Boone, NC 28608, USA
e-mail: mcraeaa@appstate.edu; mohanr@appstate.edu

T. W. Haynes et al. (eds.), *Structures of Domination in Graphs*, Developments in Mathematics 66, https://doi.org/10.1007/978-3-030-58892-2_14

function g is $w(g) = g(V) = \Sigma_{u \in V} g(u)$. The *Y-total domination number* $\gamma_{Yt}(G)$ equals the minimum weight of a Y-total dominating function g on G.

A Y-total dominating function g is called *minimal* if there does not exist another Y-total dominating function h, $h \neq g$, with $h(v) \leq g(v)$ for every $v \in V$. The *upper Y-total domination number* $\Gamma_{Yt}(G)$ equals the maximum weight $w(g)$ of a minimal Y-dominating function g on G.

We say that a Y-total dominating function is *total, or open, efficient* if $f(N(v)) = 1$ for every vertex $v \in V$.

By varying the set Y of real numbers, we produce different types of dominating functions, those involving closed neighborhood sums, $f(N[v]) = \Sigma_{u \in N[v]} f(u) \geq 1$, and those involving open neighborhood sums, $g(N(v)) = \Sigma_{u \in N(v)} g(u) \geq 1$.

The first four below use closed neighborhood sums.

1.1 $Y = \{0, 1\}$ with $f(N[v]) \geq 1$

When $Y = \{0, 1\}$, $\gamma_{\{0,1\}}(G) = \gamma(G)$, the standard *domination number* of a graph G, and $\Gamma_{\{0,1\}}(G) = \Gamma(G)$, the *upper domination number* of G.

1.2 $Y = [0, 1]$ with $f(N[v]) \geq 1$

When $Y = [0, 1]$, $\gamma_{[0,1]}(G) = \gamma_f(G)$, the *fractional domination number* of a graph G, and $\Gamma_{[0,1]}(G) = \Gamma_f(G)$, the *upper fractional domination number* of G.

1.3 $Y = \{-1, 1\}$ with $f(N[v]) \geq 1$

When $Y = \{-1, 1\}$, $\gamma_{\{-1,1\}}(G) = \gamma_s(G)$, the *signed domination number* of a graph G, and $\Gamma_{\{-1,1\}}(G) = \Gamma_s(G)$, the *upper signed domination number* of G.

1.4 $Y = \{-1, 0, 1\}$ with $f(N[v]) \geq 1$

When $Y = \{-1, 0, 1\}$, $\gamma_{\{-1,0,1\}}(G) = \gamma^-(G)$, the *minus domination number* of a graph G, and $\Gamma_{\{-1,0,1\}}(G) = \Gamma^-(G)$, the *upper minus domination number* of G.

The next four types of domination use open neighborhood sums.

1.5 $Y = \{0, 1\}$ with $f(N(v)) \geq 1$

When $Y = \{0, 1\}$, $\gamma_{\{0,1\}t}(G) = \gamma_t(G)$, the standard *total domination number* of a graph G, and $\Gamma_{\{0,1\}t}(G) = \Gamma_t(G)$, the *upper total domination number* of G.

1.6 $Y = [0, 1]$ with $f(N(v)) \geq 1$

When $Y = [0, 1]$, $\gamma_{[0,1]t}(G) = \gamma_{ft}(G)$, the *fractional total domination number* of a graph G, and $\Gamma_{[0,1]t}(G) = \Gamma_{ft}(G)$, the *upper fractional total domination number* of G.

1.7 $Y = \{-1, 1\}$ with $f(N(v)) \geq 1$

When $Y = \{-1, 1\}$, $\gamma_{\{-1,1\}t}(G) = \gamma_{st}(G)$, the *signed total domination number* of a graph G, and $\Gamma_{\{-1,1\}t}(G) = \Gamma_{st}(G)$, the *upper signed total domination number* of G.

1.8 $Y = \{-1, 0, 1\}$ with $f(N(v)) \geq 1$

When $Y = \{-1, 0, 1\}$, $\gamma_{\{-1,0,1\}t}(G) = \gamma_t^-(G)$, the *minus total domination number* of a graph G, and $\Gamma_{\{-1,0,1\}t}(G) = \Gamma_t^-(G)$, the *upper minus total domination number* of G.

1.9 $Y = \{-1, 1\}$ with $f(N[v]) \geq 1$ for at least half of the vertices $v \in V$

When $Y = \{-1, 1\}$, $\gamma_{maj}(G)$, the *majority domination number* of a graph G, and $\Gamma_{maj}(G) = \Gamma_s(G)$, the *upper majority domination number* of G.

An excellent discussion of properties of Y-dominating functions, as of 1998, can be found in a chapter by Henning, entitled "Dominating Functions in Graphs" in [19]. This same volume contains a chapter entitled "Majority Domination and Its Generalizations," by Hattingh [17]. A comprehensive presentation of signed and minus domination appears in a chapter by Kang and Shan, entitled "Signed and Minus Dominating Functions in Graphs in [22]. In this chapter, we will focus primarily on algorithms and complexity of signed domination and minus domination. Later in the chapter, we will focus on majority domination in graphs.

434 S. T. Hedetniemi et al.

Figures 1, 2, 3, 4, 5, 6, 7, 8, and 9 illustrate the above dominating functions. To avoid clutter in the figures, we adopt the following colors – red for an assigned value of -1, yellow for 0 or absence, green for $+1$ or presence, and shades of blue for fractional values. We indicate within the respective diagrams what fractional values are used.

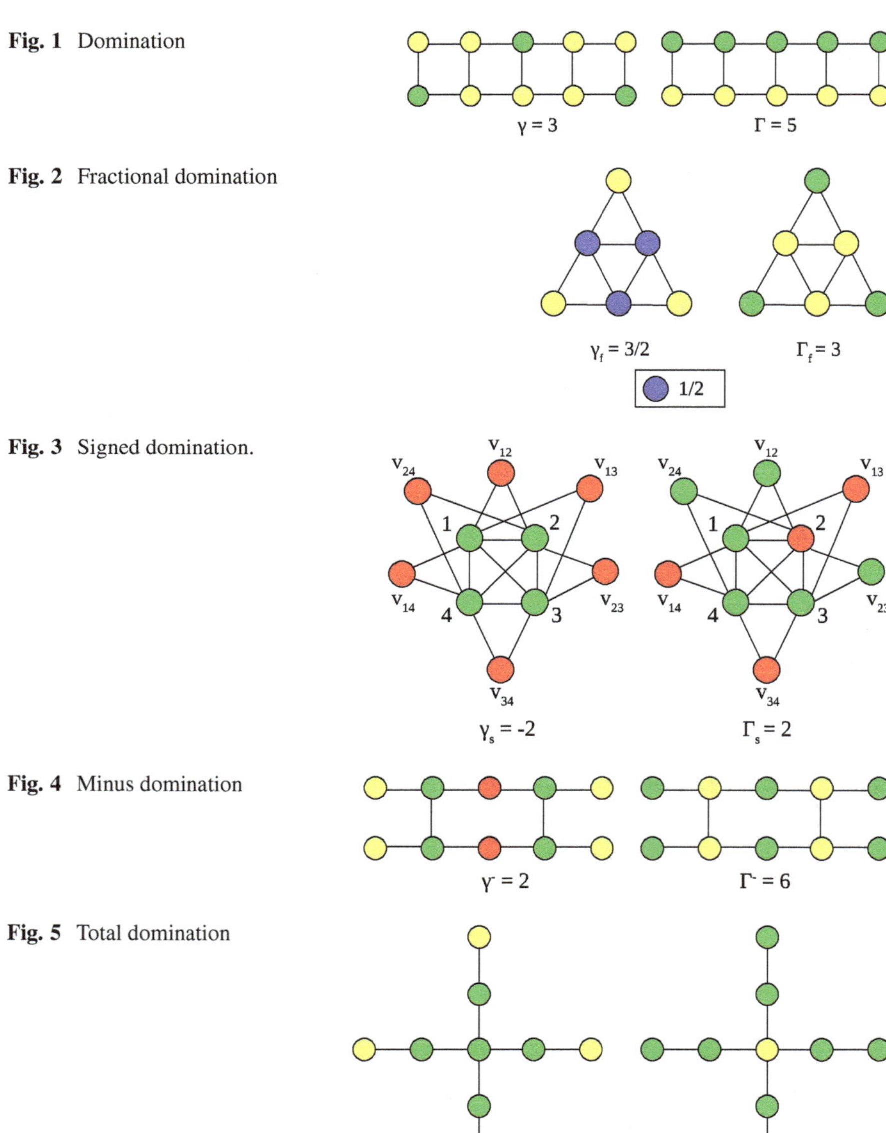

Fig. 6 Fractional total domination

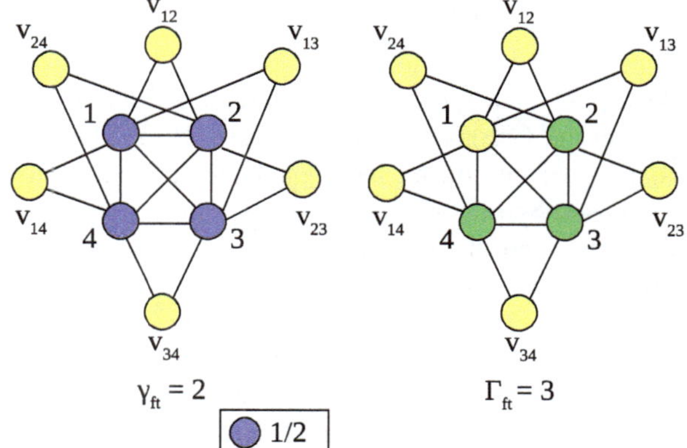

Fig. 7 Signed total domination

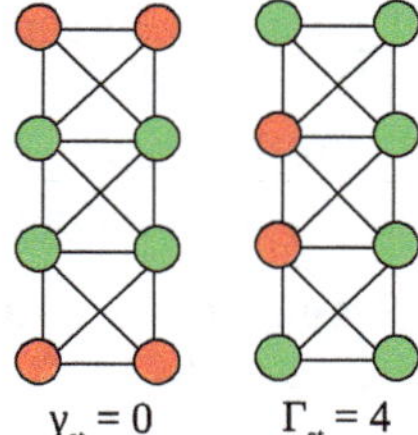

Fig. 8 Minus total domination

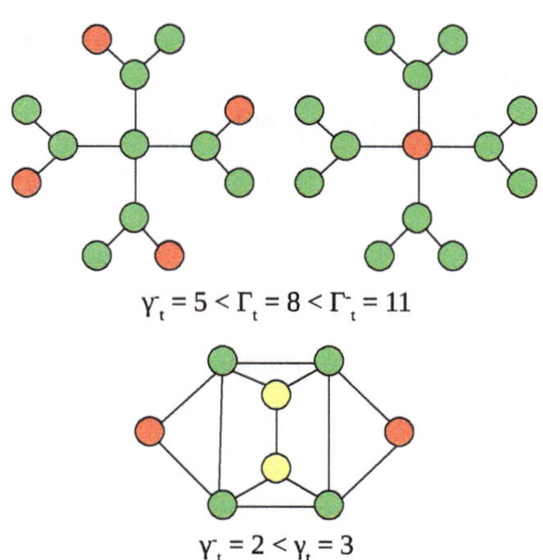

Fig. 9 Signed majority domination

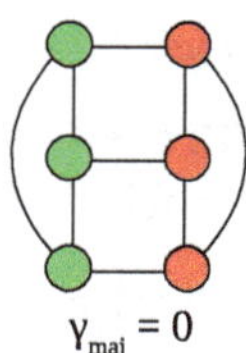

2 Signed Domination

At the Seventh Quadrennial International Conference on the Theory and Applications of Graphs (Kalamazoo, MI, 1992), Dunbar, Hedetniemi, Henning, and Slater [10] introduced the signed domination number $\gamma_s(G)$, where $Y = \{-1, 1\}$, motivated by preferences and voting applications in social networks. In this paper, they observed that even though the closed neighborhoods of all vertices can have positive sums, meaning that a majority of vertices in every neighborhood have the value $+1$, the total sum of all values $f(V)$ can be considerably negative. An excellent example of this is the left graph G in Figure 3, obtained from a complete graph K_n with vertices $V(K_n) = \{v_1, v_2, \ldots, v_n\}$ to which are added $\binom{n}{2}$ vertices as follows: for every v_i and v_j, add a vertex v_{ij} of degree 2 that is adjacent only to vertices v_i and v_j. Now construct the signed dominating function f in which $f(v) = 1$ (color green) for every $v \in V(K_n)$ and $f(v_{ij}) = -1$ (color red), for every $1 \le i < j \le n$. It can be seen that for every vertex $w \in V(G)$, $f(N[w]) = 1$, and thus, this is an efficient signed dominating function, yet $f(V(G)) = n - \binom{n}{2}$.

In this paper, the authors present the following characterization of a minimal signed dominating function.

Proposition 1 (Dunbar et al. [10]) *A signed dominating function $f: V \to \{-1, 1\}$ on a graph $G = (V, E)$ is minimal if and only if for every vertex $v \in V$ with $f(v) = 1$, there exists a vertex $u \in N[v]$ with $1 \le f(N[u]) \le 2$.*

The authors point out that while for any negative integer k, there exist bipartite graphs, chordal graphs, and outerplanar graphs G with $\gamma_s(G) \le k$; for other classes of graphs, the signed domination number cannot be negative.

The decision problem for signed domination is the following.

SIGNED DOMINATING FUNCTION (SDF)
Instance: Graph $G = (V, E)$, positive integer k.
Question: Does G have a signed dominating function f of weight $f(V) \le k$?

UPPER SIGNED DOMINATING FUNCTION (USDF)
Instance: Graph $G = (V, E)$, positive integer k.
Question: Does G have a minimal signed dominating function f of weight $f(V) \ge k$?

The first paper to address the complexity of signed domination and upper signed domination was by Hattingh, Henning, and Slater in 1995 [18] who showed the following.

Theorem 2 (Hatting, Henning, Slater [18]) *SDF is NP-complete for bipartite and chordal graphs.*

Proof Sketch. Use a transformation from the following, well-known NP-complete problem.

DOMINATING SET (DOMSET)
Instance: Graph $G = (V, E)$, integer k.

Question: Does G have a dominating set of cardinality at most k?

Given an instance of DOMSET, a graph $G = (V, E)$ and a positive integer k, construct a graph H by attaching to each vertex $v \in V(G)$, $deg(v) + 1$ paths of length two. Thus, if $n = |V(G)|$ and $m = |E(G)|$, then $|V(H)| = n + 2 \Sigma_{v \in V}(deg(v) + 1) = 3n + 4m$, and $|E(H)| = m + 2 \Sigma_{v \in V}(deg(v) + 1) = 2n + 5m$. Thus, H can be constructed in polynomial time.

It is easy to see that any signed dominating function defined on the constructed graph H must assign the value $+ 1$ to the two vertices on all attached paths of length two and can assign the value $- 1$ to all vertices not in a dominating set of G and $+ 1$ to all vertices in a dominating set of G. Based on this observation, one can then prove that $\gamma(G) \le k$ if and only if $\gamma_s(H) \le 3n + 4m - 2(n - k) = 4m + n + 2k$.

Notice that if G is either bipartite or chordal, so is the constructed graph H. $\qquad \square$

If an instance G of DOMSET is planar, then the constructed graph H in the above proof is also planar. Since DOMSET is known to be NP-complete for planar graphs [15], we can conclude the following as well.

Corollary 3 (Hatting, Henning, Slater [18]) *SDF is NP-complete for planar graphs.*

Theorem 4 (Hatting, Henning, Slater [18]) *USDF is NP-complete for bipartite and chordal graphs.*

Proof Sketch. Use a transformation from One-in-Three 3SAT.

ONE-IN-THREE 3SAT

Instance: Set U of variables, a collection C of 3-variable clauses over U, no clause containing a negated variable.

Question: Is there a truth assignment for U such that each clause $c \in C$ has exactly one true variable?

Given an instance $U = \{u_1, u_2, \ldots, u_n\}$ and $C = \{c_1, c_2, \ldots, c_m\}$ of ONE-IN-THREE 3SAT, construct the following bipartite graph H. Let $H_1, H_2, \ldots, H_n$ be n copies of the path P_5 with vertices labeled u, v_1, v_2, v_3, v_4, where the vertices in copy H_i are labeled $u_i, v_{i,1}, v_{i,2}, v_{i,3}, v_{i,4}$.

Corresponding to each 3-variable clause c_j, say $u_2 \vee u_3 \vee u_5$, associate a path P_4, one end vertex of which is labeled c_j, and add edges between c_j and each of the vertices labeled $u_2, u_3,$ and u_5.

It can be shown that the instance (U, C) of ONE-IN-THREE 3SAT has a one-in-three satisfying truth assignment if and only if G has a minimal signed dominating function of weight at least k, where $k = 3n + 4m$. For example, given a ONE-IN-THREE 3SAT satisfying truth assignment, assign the value $f(u_i) = 1$ and $f(v_{i,1}) = -1$ if u_i is assigned the value TRUE, and assign the value $f(u_i) = -1$ and $f(v_{i,1}) = +1$ if u_i is assigned the value FALSE; then assign the value $+ 1$ to all remaining vertices. This can be seen to be a minimal signed dominating function of weight $k = 3n + 4m$. $\qquad \square$

Fig. 10 The graph L
(reproduced with from [18])

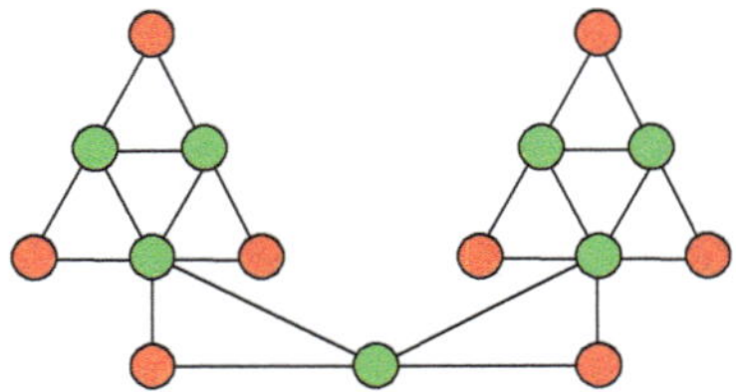

Although DOMSET can be solved in polynomial time for the fixed parameter k, simply by checking all $\binom{n}{k}$ sets of size k, SDF does not have this property. Consider the following decision problem.

ZERO SIGNED DOMINATING FUNCTION (ZSDF)
Instance: Graph $G = (V, E)$
Question: Does G have a signed dominating function of weight at most $k = 0$?

Theorem 5 (Hatting, Henning, Slater [18]) *ZSDF is NP-complete for bipartite or chordal graphs.*

Proof Sketch. Show that a polynomial algorithm for solving ZSDF can be used to solve SDF. Given an instance $G = (V, E)$ of SDF, construct the graph $H = G \cup \bigcup_{i=1}^{j} L_i$, where the graph L_i is a copy of the graph shown in Figure 10.
    Note that if G is chordal, then so is H, since the graphs L_i are chordal. Note also that $\gamma_s(L) = -1$, as shown. One can then show that $\gamma_s(G) \leq j$ if and only if $\gamma_s(H) \leq 0$.   □

Hattingh, Henning, and Slater present a linear algorithm for computing the value of $\gamma_s(T)$ for any tree T; since it is quite simple, we present it here.

Algorithm Tree SDF (Hattingh et al. [18]).
Input:
array *parent*$[1, \ldots, n]$ representing the vertices of a tree T rooted at vertex v_n, with vertices labeled $v_1, v_2, \ldots, v_n$ in such a way that the parent of vertex v_j is a vertex v_k with $j < k$, that is, if *parent*$[j] = k$ then $j < k$
    array *deg*$[1, \ldots, n]$, where *deg*$[i] = deg(v_i)$ in T
    array *sum*$[1, \ldots, n]$
    array $f[1, \ldots, n]$
Output: a signed dominating function $f : V \rightarrow \{-1, 1\}$ of minimum weight.
    begin
    for $i = 1$ **to** n **do**
        1. **if** $i = n$ **then**
            if $deg(1)$ is odd
                then *minsum* $= 2$
                else *minsum* $= 1$
        2. **if** $i < n$ **then**
            if $deg(i)$ is odd

> $\quad$ **then** $minsum = 1$
> $\quad$ **else** $minsum = 0$
> 3. **if** $i < n$ **and** $deg(i) = 1$ [vertex v_i is a leaf]
> $\quad$ **then** $childsum = 0$
> $\quad$ **else** $childsum = \Sigma_{j:parent[j]=i} f(j)$
> 4. **if** $childsum < minsum$
> $\quad$ **then**
> $\quad\quad$ **while** $childsum < minsum - 1$ **do**
> $\quad\quad\quad$ for a vertex j with $parent[j] = i$ and $f(j) = -1$, set $f(j) = 1$
> $\quad\quad$ **od**
> $\quad\quad$ $f(i) = 1$
> $\quad$ **else if** $\exists j$ $parent[j] = i$ **and** $0 \le sum(j) \le 1$
> $\quad\quad$ **then** $f(i) = 1$
> $\quad\quad$ **else** $f(i) = -1$
> 5. $sum(i) = childsum + f(i)$
> **od**
> **end**

We note in passing that the authors did not present an algorithm for computing the value of $\Gamma_s(T)$ for a tree T.

In 2001 [6] Damaschke provides the following three complexity results for signed domination.

Theorem 6 (Damaschke [6]) *SDF is NP-hard for planar graphs G with $\Delta(G) \le 3$.*

Theorem 7 (Damaschke [6]) *For some $\varepsilon > 0$ and for graphs G with $\Delta(G) \le 3$, $\gamma_s(G)$ cannot be approximated in polynomial time within a factor of $1 + \varepsilon$ unless $P = NP$.*

Proposition 8 (Damaschke [6]) *For every fixed k, there is a polynomial algorithm for deciding whether a graph G with $\Delta(G) \le 5$ satisfies $\gamma_s(G) \le k$.*

Damaschke points out that a simple algorithm can be constructed that runs in $O(n^{2k})$ time for deciding whether a graph G with $\Delta(G) \le 5$ satisfies $\gamma_s(G) \le k$.

In 2008 [26] Lee and Chang study variations of Y-domination such as $\{k\}$-domination, k-tuple domination, signed domination, and minus domination on a wide variety of graphs, including strongly chordal graphs, a class which includes as subclasses trees, block graphs, interval graphs, and directed path graphs. This paper also gives NP-completeness results for these problems on doubly chordal graphs, dually chordal graphs, bipartite planar graphs, chordal bipartite graphs, and planar graphs. Some definitions of these classes of graphs might be helpful here.

A graph G is an *undirected path graph* if it is the intersection graph of paths in a tree.

A graph G is a *directed path graph* if it is the intersection graph of directed paths in a rooted directed tree.

Recall that:

$$interval \subset directedpath \subset undirectedpath \subset chordal$$

A chord uv in a cycle C of even length $2k$ is called an *odd chord* if the shortest distance $d(u, v)$ between u and v in the cycle C is odd.

A graph G is *strongly chordal* if it is chordal and every even cycle in G of length at least 6 contains an odd chord.

A graph G is *chordal bipartite* if it is bipartite and every cycle of length at least 6 has a chord.

A vertex $u \in N[v]$ is called a *maximum neighbor* of v if for all vertices $w \in N[v]$, $N[w] \subseteq N[u]$, that is, the closed neighborhood of every neighbor of v is contained in the closed neighborhood of u.

Let $v_1, v_2, \ldots, v_n$ be an ordering of the vertex set V, and for $1 \le i \le n$, let G_i be the subgraph of G induced by the vertices $v_i, v_{i+1}, \ldots, v_n$.

A vertex ordering is called a *maximum neighborhood ordering* if for all $1 \le i \le n$, vertex v_i has a maximum neighbor u_i in G_i, that is, for every $w \in N_i[v_i]$, $N_i[w] \subseteq N_i[u_i]$.

A graph G is *dually chordal* if it has a maximum neighborhood ordering. It is worth noting that $\gamma(G)$ can be computed for dually chordal graphs in linear time [4].

A graph G is *doubly chordal* if it is both chordal and dually chordal.

In this paper, Lee and Chang [26] introduce the following variant of signed domination.

A *signed zero dominating function* $f : V \rightarrow \{-1, 1\}$ satisfies the condition that for all $v \in V$, $f(N[v]) \ge 0$. The *signed zero domination number* $\gamma_{s0}(G)$ equals the minimum weight of a signed zero dominating function on G.

SIGNED ZERO DOMINATING FUNCTION (SZDF))
Instance: Graph $G = (V, E)$, integer k.
Question: Does G have a signed zero dominating function of weight at most k?

Theorem 9 (Lee, Chang [26]) *SZDF is NP-complete for chordal graphs.*

Proof Sketch. Use a transformation from DOMINATING SET for chordal graphs. Given a chordal graph G, construct another chordal graph H by attaching to each vertex $v \in V(G)$, $d(v)$ paths of length three. One can then show that $\gamma_{s0}(H) = 2m - n + 2\gamma(G)$, where $n = |V|$, $m = |E|$, and $|V(H)| = n + 6m$. Thus, $\gamma(G) \le k$ if and only if $\gamma_{s0}(H) \le 2m - n + 2k$. $\square$

Theorem 10 (Lee, Chang [26]) *SDF is NP-complete on doubly chordal graphs.*

Proof Sketch. Use a transformation from SZDF for chordal graphs above. Given a chordal graph $G = (V, E)$ of order n, construct a graph H by adding a new vertex x adjacent to all vertices in G. Then add n new vertices as leaves attached to x. The graph H so constructed can be seen to be a doubly chordal graph, since it is chordal and the newly added vertex x serves as a maximum neighbor for all vertices. One can show that $\gamma_s(H) = \gamma_{s0}(G) + (n+1)$. Thus, $\gamma_{s0}(G) \leq k$ if and only if $\gamma_s(H) \leq k + (n+1)$. $\square$

In 2015 [28], Lin, Liu, and Poon show that SDF is W[2]-hard and that for graphs with $\Delta(G) \leq 6$, SDF is contained in APX-hard, the class of problems which can be approximated within some constant ratio.

This statement warrants some discussion for those readers not familiar with this terminology. For a basic discussion of W[2]-hardness, the reader is referred to [7] and [8] by Downey and Fellows. Most domination-related problems fall within one of the following complexity classes:

$$FPT \subseteq W[1] \subseteq W[2] \subseteq W[3] \ldots$$

The class of problems in FPT are called Fixed Parameter Tractable. These are problems like the following:

SPANNING TREE
Instance: Connected graph $G = (V, E)$.
Parameter: k.
Question: Does G have a spanning tree with at least k leaves?

Theorem 11 (Bodlaender [3]) *For every fixed k, it can be determined in $O(n)$ time whether a connected graph G of order n has a spanning tree with at least k leaves.*

Another example is the following:

K-CYCLE
Instance: Graph $G = (V, E)$.
Parameter: k.
Question: Does G have a cycle of length at least k?

Theorem 12 (Fellows, Langston [14]) *For every fixed k, it can be determined in $O(n)$ time whether a graph G of order n has a cycle of length at least k.*

For our purposes we can say that a problem with a parameter k is *fixed parameter tractable* if there exists a constant α and an algorithm which can solve any instance of size n in time $f(k)n^{\alpha}$. Problems which are not thought to be fixed parameter tractable are organized hierarchically in increasing classes of hardness. The computational problem used to define the class $W[1]$ is the following.

WEIGHTED 3CNF SATISFIABILITY
Instance: A Boolean expression X in conjunctive normal form with 3-literals per clause.
Parameter: k.
Question: Is there a satisfying truth assignment of weight at least k, that is, k literals are assigned the value *true*?

The following problem is an example of a problem in $W[1]$ [7].

IRREDUNDANT SET
Instance: Graph $G = (V, E)$.
Parameter: k.
Question: Does G have an irredundant set of cardinality at least k?

DOMINATING SET is known to be $W[2]$-hard, which means, in effect, that no algorithm is known that can do any better than try all possible sets of size k and then check each such set in $O(n)$ time to see if it is a dominating set; thus, no algorithm better than $O(n^{1+k})$ is known for DOMINATING SET for fixed k.

Examples of dominating problems which are $W[2]$-hard include TOTAL DOMINATING SET, CONNECTED DOMINATING SET, DOMINATING CLIQUE, and MAXIMAL IRREDUNDANT SET. DOMINATING SET and INDEPENDENT DOMINATING SET are, in fact, $W[2]$-complete, in the same way that decision problems can be shown to be NP-hard or NP-complete.

A problem that appears to be even more difficult than DOMINATING SET is the following which has been shown to be in $W[3]$.

DOMINATING THRESHOLD SET
Instance: Graph $G = (V, E)$.
PARAMETER: Positive integers k and r.
Question: Does there exist a subset $V' \subseteq V$ with $|V'| \leq k$ such that for every $v \in V$, $|N[v] \cap V'| \geq r$?

In 2015 [27] Lin and Poon present constant factor approximation algorithms for SDF on subgraphs of cubic graphs, graphs of maximum degree four, and graphs of maximum degree five. In addition, they prove the NP-completeness of SDF on sub-cubic, planar bipartite graphs. They also present an $O^*(5.1957^{\gamma_s})$-time FPT-algorithm for SDF on sub-cubic graphs. It follows that SDF on graphs with maximum degree three is NP-complete.

3 Minus Domination

The minus domination number, where $Y = \{-1, 0, 1\}$ and $\gamma_Y(G) = \gamma^-(G)$, was introduced in 1996 by Dunbar, Goddard, Hedetniemi, Henning, and McRae [9] and [11] and in 1999 by Dunbar, Hedetniemi, Henning, and McRae [12].

In [12], Dunbar et al. briefly discuss the possible applications of minus dominating functions, including assigning positive or negative electrical charges, or positive or negative spins to electrons, or positive, negative or neutral votes or preferences to people in a social network. The minus domination number indicates the minimum number of people whose positive votes can assure that all local (closed) neighborhoods have more positive than negative voters, even though the entire network (graph) has far more negative voters than positive voters. Similarly, the upper signed domination number indicates the largest number of positive voters necessary to offset a few negative voters so that all local neighborhoods have positive vote totals.

In [11] Dunbar et al. show that the following two decision problems are NP-complete, even when restricted to bipartite or chordal graphs.

MINUS DOMINATING FUNCTION (MDF)
Instance: Graph $G = (V, E)$, integer k.
Question: Does G have a minus dominating function of weight at most k?

UPPER MINUS DOMINATING FUNCTION (UMDF)
Instance: Graph $G = (V, E)$, integer k.
Question: Does G have a minimal minus dominating function of weight at least k?

Theorem 13 (Dunbar et al. [11]) *MDF is NP-complete for bipartite and chordal graphs.*

Proof Sketch. Use a transformation from DOMSET.

Given either a bipartite or a chordal instance $G = (V, E)$ of DOMSET, to each vertex $v \in V$, attach a path of length three, and denote the resulting graph by H. One can show that $\gamma^-(H) = \gamma(H) = \gamma(G) + |V(G)|$. Therefore, $\gamma(G) \leq k$ if and only if $\gamma^-(H) \leq k + |V(G)|$. Notice as well that if G is either bipartite or chordal, so is the constructed graph H. $\square$

Theorem 14 (Dunbar et al. [11]) *UMDF is NP-complete for bipartite graphs.*

Proof Sketch. Use a transformation from ONE-IN-THREE 3SAT.

ONE-IN-THREE 3SAT
Instance: Set $U = \{u_1, u_2, \ldots, u_n\}$ of n variables, a collection $C = \{c_1, c_2, \ldots, c_m\}$
   of 3-variable clauses over U, where no clause contains a negated variable.
Question: Is there a truth assignment for U such that each clause $c \in C$ has exactly
   one true variable?

Given an instance of ONE-IN-THREE 3SAT, with variables $U = \{u_1, u_2, \ldots, u_n\}$, and clauses $C = \{c_1, c_2, \ldots, c_m\}$, construct a bipartite graph in which each $u_i \in U$ is a vertex in an associated 4-cycle, and to each clause $c_j \in C$ there is a path of length two, the middle vertex of which is labeled c_j. Three edges are then added from each clause vertex $c_j = (v_{j1} \vee v_{j2} \vee v_{j3})$ to the three vertices v_{j1}, v_{j2}, and v_{j3}. One then shows that U, C has a One-in-Three 3SAT truth assignment if and only if the bipartite graph so constructed has a minimal minus dominating function of weight at least $k = 2n + 3m$.

For example, if the instance (U, C) of ONE-IN-THREE 3SAT in Figure 11 has a solution, e.g., $u_1 = u_5 = True$ and $u_2 = u_3 = u_4 = False$, then as indicated in Figure 11:

Assign 0 to vertices u_1 and $v_{1,2}$ and $+1$ to $v_{1,1}$ and $v_{1,3}$.
Assign 0 to vertices u_5 and $v_{5,2}$ and $+1$ to $v_{5,1}$ and $v_{5,3}$.
Assign -1 to vertices u_2 and $+1$ to $v_{2,1}$, $v_{2,2}$, and $v_{2,3}$.
Assign -1 to vertices u_3 and $+1$ to $v_{3,1}$, $v_{3,2}$, and $v_{3,3}$.
Assign -1 to vertices u_4 and $+1$ to $v_{4,1}$, $v_{4,2}$, and $v_{4,3}$.

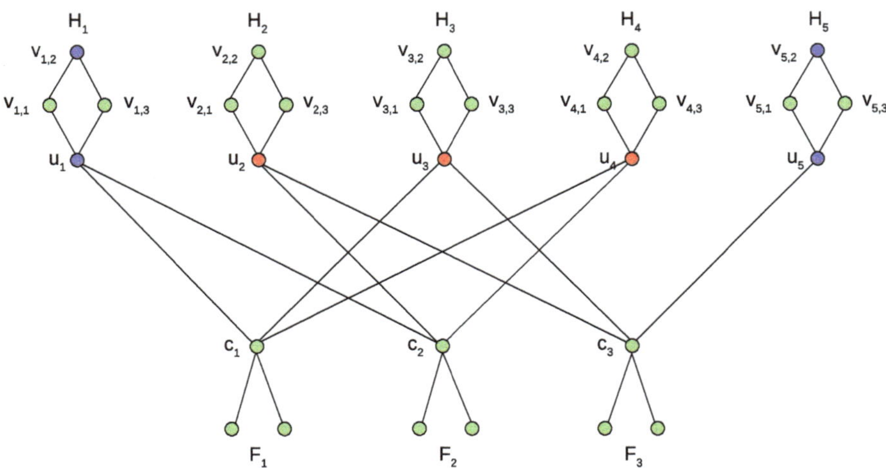

Fig. 11 The Graph G_1 for $(u_1 \vee u_3 \vee u_4) \wedge (u_1 \vee u_2 \vee u_4) \wedge (u_2 \vee u_3 \vee u_5)$ (figure reproduced from [11])

Then assign $+1$ to each of the nine vertices in F_1, F_2, and F_3.

It can be seen that this is a minimal minus dominating function of the bipartite graph $G(U, C)$ of weight $2n + 3m$. $\square$

The authors also prove the following.

Theorem 15 (Dunbar et al. [11]) *UMDF is NP-complete for chordal graphs.*

Proof Sketch. Use a transformation from ONE-IN-THREE 3SAT.

Given an instance of ONE-IN-THREE 3SAT, with variables $U = \{u_1, u_2, \ldots, u_n\}$ and clauses $C = \{c_1, c_2, \ldots, c_m\}$, construct a chordal graph in which each $u_i \in U$ is a single vertex labeled u_i, and to each clause $c_j \in C$, there is a path of length two, the middle vertex of which is labeled c_j. Three edges are then added from each clause vertex $c_j = (v_{j1} \vee v_{j2} \vee v_{j3})$ to the three vertices v_{j1}, v_{j2}, and v_{j3}. Finally, add n additional vertices and form a clique of size $2n$ with the n vertices $U = \{u_1, u_2, \ldots, u_n\}$.

One then shows that (U, C) has a One-in-Three 3SAT truth assignment if and only if the chordal graph so constructed has a minimal minus dominating function of weight at least $k = 3m + 1$. $\square$

The authors also present the following simple linear algorithm for computing $\gamma^-(T)$ for any tree T, in which *minsum(i)* denotes the minimum possible sum of values that may be assigned to vertex i $(f(i))$ and its children $f(j)$, i.e., vertices j with $parent[j] = i$.

Algorithm Tree MDF (Dunbar et al. [11]).

Input:

array $parent[1, \ldots, n]$ representing the vertices of a tree T rooted at vertex v_n, with vertices labeled $v_1, v_2, \ldots, v_n$ in such a way that the parent of vertex v_j is a vertex v_i with $i > j$, that is, if $parent[j] = i$, then $i > j$

array $deg[1, \ldots, n]$, where $deg[i] = deg(v_i)$ in T

array $minsum[1, \ldots, n]$

array $childsum[1, \ldots, n]$

array $sum[1, \ldots, n]$

array $f[1, \ldots, n]$

Output: a minus dominating function $f : V \rightarrow \{-1, 0, 1\}$ of minimum weight.

begin

for $i = 1$ **to** n **do**

    1. **if** $i = n$

        then $minsum(i) = 1$

        else $minsum(i) = 0$

    2. **if** $i < n$ and $deg(i) = 1$

        then $childsum(i) = 0$

        else $childsum(i) = \Sigma_{j:parent[j]=i} f(j)$

    3. **if** $childsum(i) < minsum(i)$

        then

            3.1 **while** $childsum(i) < minsum(i) - 1$ **do**

                for a vertex j with $parent[j] = i$ and $f(j) < 1$, set $f(j) = f(j) + 1$

            od

            3.2 $f(i) = 1$

            3.3 $childsum(i) = minsum(i) - 1$

    4. **if** $childsum(i) \geq minsum(i)$ **then**

        4.1 **if** $(\exists j)(parent[j] = i$ **and** $sum(j) = 0)$

            then $f(i) = 1$

        4.2 **else if** $(\exists j)(parent[j] = i$ **and** $sum(j) = 1)$

            then $f(i) = 0$

        4.3 **else if** $childsum(i) = minsum(i)$

            then $f(i) = 0$

        4.4 **else** $f(i) = -1$

    5. $sum(i) = childsum(i) + f(i)$

    od

end

Dunbar et al. conclude by noting that they have not constructed a linear algorithm for computing $\Gamma^-(T)$ for any tree T.

In 2001, Damaschke [6] established several algorithm and complexity results for minus domination in graphs with small maximum degrees. Among other things, he noted that in any graph G, there will always exist minus dominating functions f where $w(f)$ can be any value, $\gamma^-(G) \leq w(f) \leq n$.

Damaschke observed, as did Dunbar et al. [11], that if you construct a graph H from a graph G of order n by attaching a path of length three to every vertex $v \in V(G)$, say with vertices labeled v, x, y, z, then you can observe that without loss of generality, there is always a minus dominating function f of H of weight $\gamma^-(H)$ which assigns the values $f(z) = 0, f(y) = 1, f(x) = 0$, and in particular $f(v) \geq 0$. From this, it follows that $\gamma^-(H) = n + \gamma(G)$. Therefore, many NP-completeness results for domination apply equally to minus domination.

Damaschke shows that in graphs G with $\Delta(G) \leq 4$, $\gamma^-(G)$ can be approximated within some constant ratio, and a simple algorithm achieves an approximation ratio of 15. (However, the analysis is not very easy.)

Damaschke gives a simple proof of the following result, in which $\alpha(G)$ denotes the maximum cardinality of an independent set in a graph G.

Theorem 16 (Damaschke [6]) *MDF is NP-hard for planar graphs having maximum degree $\Delta \leq 4$.*

Proof. Construct a transformation from the following NP-complete problem (cf. Garey and Johnson [15]).

INDEPENDENT SET $\Delta \leq 3$
Instance: Planar graph $G = (V, E)$ with $\Delta(G) \leq 3$, positive integer k.
Question: Does G have an independent set of size at least k?

Let $G = (V, E)$ be a planar graph of order $n = |V|$, size $m = |E|$ and $\Delta(G) \leq 3$. We may assume without loss of generality that no vertex $v \in V$ has degree 1.

Construct a planar graph H with $\Delta(H) \leq 4$ by subdividing every edge $uv \in E(G)$ with a vertex w and attaching to w a leaf x and a path w, y, z of length two.

Let $f : V \to \{-1, 0, 1\}$ be a minus dominating function of H with $w(f) = \gamma^-(H)$. One can show that, without loss of generality, $f(x) = f(z) = 0$, and $f(y) = f(w) = 1$. It follows therefore, that even if a vertex $u \in V(G)$ has $f(u) = -1$, it will still be the case that $f(N[u]) \geq 1$; since we are assuming that there are no vertices of degree 1, every vertex $u \in V(G)$ will have either two or three w-neighbors with $f(w) = 1$. It only remains to show that $f(N[w]) \geq 1$, for every subdivision vertex w. This means that both $f(u) = -1$ and $f(v) = -1$ cannot hold, and thus, the set of vertices v with $f(v) = -1$ must form an independent set in G. From this it follows that $\gamma^-(H) = 2m - \alpha(G)$. $\qquad\square$

Damaschke shows that you can use the same transformation above (but without planarity) to show the following. The class MaxSNP (for strict NP) is a subclass of NP optimization problems consisting solely of constant factor approximable problems, for example, MAX-3SAT, in which you want to find a truth assignment satisfying as many 3-literal clauses as possible. Since there is a fixed-ratio approximation algorithm for solving any problem in MaxSNP, MaxSNP is contained in APX, the class of problems which can be approximated within some constant ratio.

Theorem 17 (Damaschke [6]) *For some $\varepsilon > 0$, $\gamma^-(G)$ in graphs with $\Delta(G) \leq 4$ cannot be approximated in polynomial time within a factor of $1 + \varepsilon$ unless $P = NP$.*

Proof. It has been shown by Berman and Fujito [2] that INDEPENDENT SET for graphs with $\Delta(G) \leq 3$ is MAX SNP-complete. So there exists a constant $\varepsilon > 0$ such that in the graph H constructed above, $\alpha(H)$ cannot be approximated for such G in polynomial time within, say, $1 + 12\varepsilon$, unless $P = NP$. If we assume that in graphs G with $\Delta(G) \leq 4$ $\gamma^-(G)$ can be approximated within factor $1 + \varepsilon$, then since $2e \leq 3n$ in G and $\alpha(H) \geq n/4$, this could be used to approximate $\alpha(H) = 2e - \gamma^-(G)$ within $1 + 12$, a contradiction. $\qquad\square$

Damaschke then shows the following.

Theorem 18 (Damaschke [6]) *In graphs G with $\Delta(G) \leq 4$, $\gamma^-(G)$ can be approximated in linear time within some constant ratio, at least* 15.

Theorem 19 (Damaschke [6]) *For every fixed k, there is a polynomial algorithm deciding whether a graph G with $\Delta(G) \leq 4$ satisfies $\gamma^-(G) \leq k$.*

In 2008 [26] Lee and Chang establish the NP-completeness of the minus domination problem MDF on the following two classes of graphs using the exact same transformation as used in 1996 by Dunbar et al. [11] and in 2001 by Damaschke [6].

Theorem 20 (Lee, Chang [26]) *MDF is NP-complete for chordal bipartite graphs and bipartite planar graphs.*

Proof Sketch. Use a transformation from DOMSET for chordal bipartite graphs and bipartite planar graphs. Given such a graph G, construct a graph H by adding a path of length 3 to each vertex of G. Then show that $\gamma^-(H) = \gamma(H) = \gamma(G) + n$. Thus, $\gamma(G) \leq k$ if and only if $\gamma^-(H) \leq k + n$. $\qquad\square$

Lee and Chang also present an $O(|V| + |E|)$ algorithm for computing $\gamma^-(G)$ for strongly chordal graphs. They also raise the question of the NP-completeness of minus domination on doubly chordal graphs.

In 2015 [28] Lin, Liu, and Poon show that MDF is W[2]-hard for general graphs and is NP-complete for sub-cubic, bipartite planar graphs. They also show that MDF is APX-hard for graphs of maximum degree seven and present the first fixed-parameter algorithm for MDF on sub-cubic graphs, which runs in $O*(2.3761^{5k})$ time, where $k = \gamma^-(G)$.

In 2016 [13] Faria, Hon, Kloks, Liu, Wang, and Wang establish the following theorems.

Theorem 21 (Faria et al. [13]) *MDF is fixed-parameter tractable, that is, there exists a linear algorithm for finding a minus dominating function of size at most k (at most k vertices are assigned the value $+1$) in d-degenerate graphs, the class of graphs each of whose induced subgraphs have a vertex of degree at most d.*

The authors note that since minus domination of bounded size can be formulated in monadic second-order logic, without quantification over subsets of edges, there is a linear algorithm for solving MDF on graphs of bounded treewidth or rankwidth or cliquewidth.

Cographs (complement reducible graphs) can be defined as the class of graphs that can be constructed recursively as follows: (i) K_1 is a cograph; (ii) if G and H are cographs, so are $G \cup H$ and $G + H$, where the *join* of graphs G and H is the graph $G + H = (V(G) \cup V(H), E(G) \cup E(H) \cup \{uv : u \in V(G), v \in V(H)\})$.

Theorem 22 (Faria et al. [13]) *There exists a polynomial algorithm for computing $\gamma^-(G)$ for any cograph G.*

Distance-hereditary graphs are graphs having the property that for every pair u, v of nonadjacent vertices, all chordless u-v paths have the same length.

Theorem 23 (Faria et al. [13]) *There exists a polynomial algorithm for computing $\gamma^-(G)$ for any distance-hereditary graph G.*

Theorem 24 (Faria et al. [13]) *For any fixed k, there exists a polynomial algorithm for computing $\gamma^-(G)$ for any graph of rankwidth k.*

A strongly chordal graph G is a graph in which every cycle C of even length at least 6 has an odd chord, that is, an edge between two non-consecutive vertices u and v on C, whose minimum distance $d_C(u, v)$ on the cycle is odd.

Theorem 25 (Faria et al. [13]) *There exists a polynomial algorithm for computing $\gamma^-(G)$ for any strongly chordal graph G.*

Corollary 26 (Faria et al. [13]) *There exists a linear algorithm for computing $\gamma^-(G)$ for any interval graph G.*

A *split graph* is a graph $G = (V, E)$ whose vertices can be partitioned $V = \{V_1, V_2\}$ such that V_1 is an independent set and the induced graph $G[V_2]$ is a complete subgraph, where either of V_1 or V_2 can be empty.

Theorem 27 (Faria et al. [13]) *MDF is NP-complete for split graphs.*

Proof Sketch. Use a transformation from the following NP-complete problem.

(3, 2)-HITTING SET
Instance: A set $U = \{u_1, u_2, \ldots, u_n\}$, a collection $C = \{c_1, c_2, \ldots, c_m\}$ of 3-element
subsets of U and a positive integer k.
Question: Does there exist a subset $U' \subseteq U$ with $|U'| \leq k$ such that for each $c_i \in C$,
$|c_i \cap U'| \geq 2$?

Given an instance (U, C) of (3, 2)-HITTING SET, construct the following split graph $G(U, C)(V_1, V_2, E)$: into the independent set V_1, place one vertex for each $c_i \in C$; into the clique V_2, place one vertex for each $u_j \in U$. Add to the clique V_2 a set $X = \{x_0, x_1, x_2, \ldots, x_{n+m}\}$, and add to the independent set V_1 a set $X' = \{x'_0, x'_1, x'_2, \ldots, x'_{n+m}\}$. The edges in $E(G)$ consist of (i) the edges in the clique V_2 of order $n + n + m + 1$, (ii) the edges of a matching between the vertices in X' and the vertices in X, and (iii) three edges from each vertex c_i to the three vertices corresponding to the 3-elements of c_i in U.

It remains then to show that a smallest (3, 2)-HITTING SET of U, C corresponds to a minimum minus dominating function on the graph $G(U, C)$. $\square$

Faria et al. then present the following variant of MDF, attributed to Hattingh, Henning, and Slater [18].

ZERO MINUS DOMINATING FUNCTION (ZMDF)
Instance: Graph $G = (V, E)$.
Question: Does G have a minus dominating function of weight at most 0?

They show that even for this fixed value of 0, ZMDF is NP-complete, as a result of which they conclude that MDF is not Fixed Parameter Tractable.

4   Signed and Minus Total Domination

Recall from Section 1 the following definitions:

4.1 $Y = \{-1, 1\}$ with $f(N(v)) \geq 1$

When $Y = \{-1, 1\}$, $\gamma_{\{-1,1\}t}(G) = \gamma_{st}(G)$, the *signed total domination number* of a graph G, and $\Gamma_{\{-1,1\}t}(G) = \Gamma_{st}(G)$, the *upper signed total domination number* of G.

4.2 $Y = \{-1, 0, 1\}$ with $f(N(v)) \geq 1$

When $Y = \{-1, 0, 1\}$, $\gamma_{\{-1,0,1\}t}(G) = \gamma_t^-(G)$, the *minus total domination number* of a graph G, and $\Gamma_{\{-1,0,1\}t}(G) = \Gamma_t^-(G)$, the *upper minus total domination number* of G.

Signed total domination was first introduced and studied by Zelinka in 2001 [35]. In 2004, Henning [20] was among the first to offer NP-completeness results for SIGNED TOTAL DOMINATING FUNCTION, using a transformation from TOTAL DOMINATING SET.

TOTAL DOMINATING SET (TDS)
Instance: Graph $G = (V, E)$, integer k.
Question: Does G have a total dominating set of cardinality at most k?

SIGNED TOTAL DOMINATING FUNCTION (STDF)
Instance: Graph $G = (V, E)$, integer k.
Question: Does G have a signed total dominating function of weight at most k?

Theorem 28 (Henning [20]) *STDF is NP-complete, even when restricted to bipartite or chordal graphs.*

Proof Outline. Use a transformation from TDS. Given an instance of TDS, a graph $G = (V, E)$ and a positive integer k, construct a graph H by adding, for each vertex $v \in V$, $deg(v)$ paths P_5 and joining v to the middle vertex of each of these $deg(v)$ paths. Note that if $n = |V(G)|$ and $m = |E|$, then $|V(H)| = n + 5 \Sigma_{v \in V} deg_G(v) = n + 10m$, and $|E(H)| = m + 5 \Sigma_{v \in V} deg_G(v) = m + 10m = 11m$.

It is easy to see that if the original graph G is either bipartite or chordal, then so is the constructed graph H.

It only remains to show the following;

$$\gamma_{st}(H) = |V(H)| + 2\gamma_t(G) - 2n.$$

This follows essentially because all of the vertices in $V(H) \setminus V(G)$ must be assigned the value $+1$ in any signed total dominating function f of H, and from this it follows that every vertex in $V(G)$ must have at least one neighbor in $V(G)$ which is also assigned the value $+1$. Thus, the vertices in $V(G)$ which are assigned the value $+1$ form a total dominating set of G.

Thus, one can show that $\gamma_t(G) \leq k$ if and only if $\gamma_{st}(G) \leq j = |V(H)| + 2k - 2n$. $\square$

Also in 2004, Harris and Hattingh [16] show that the decision problems STDF and MTDF are NP-complete when restricted to bipartite graphs or chordal graphs. They also present linear algorithms for computing $\gamma_t^-(G)$ and $\gamma_{st}(T)$ for an arbitrary tree T.

We next present the Harris-Hattingh linear algorithm for computing $\gamma_t^-(T)$ of any tree T.

Algorithm: Minus Total Domination (MTD) (Harris, Hattingh [16]).
Input A rooted tree $T = (V, E)$ with vertices $V = \{1, 2, \ldots, n\}$, where the root is vertex n, and (i) an array $parent[1, \ldots, n]$, such that $i < parent[i]$, (ii) an array $deg[1, \ldots, n]$ of the degrees of the vertices, and (iii) a listing $N(1, \ldots, n)$ of the vertices in the open neighborhood $N(i)$ of i in T.
Output A minimum weight minus total dominating function $f : V \to \{-1, 0, +1\}$
begin
    for $j = 1$ **to** n **do** $f(j) = -1$ **od**
    for $j = 1$ **to** n **do**
        1. **if** vertex j is a leaf **and** $j < n$
            then
                $OpenSum = 1$
                $f(parent[j]) = 1$
            else
                $OpenSum = f(N(j))$
        2. **if** $j < n$
            then while $OpenSum < 1$ **and** $f(parent[j]) < 1$ **do**
                $f(parent[j]) = f(parent[j]) + 1$
                $OpenSum = OpenSum + 1$
            od
        3. **while** $OpenSum < 1$ **do**
            Choose a child i of j, e.g. $parent[i] = j$, with $f(i) < 1$

$$\textbf{while } Opensum < 1 \textbf{ and } f(i) < 1 \textbf{ do}$$
$$f(i) = f(i) + 1$$
$$OpenSum = OpenSum + 1$$
od

od
end

Page limitations will not permit a detailed and complete proof of the correctness of this linear algorithm. But it is easy to see that when the algorithm is finished, the function $f : V \to \{-1, +1\}$ is a minus total dominating function. Note that initially all values are set to -1 and at no time in the algorithm is the value assigned to any vertex decreased. In addition, after every vertex j has been considered, either the value assigned to its $parent[j]$ or the values assigned to its children have been altered so that $f(N(i)) \geq 1$. It only remains, then, to show that the function f has minimum weight among all total minus dominating functions; this is found in [16].

For completeness, we next present the Harris and Hattingh, linear algorithm for computing the signed total domination number $\gamma_{st}(T)$ for any tree T, which is very similar to that above for computing $\gamma_t^-(T)$. It is based in part on the observation that in a signed total domination function $f : V \to \{-1, 1\}$, the minimum sum such that $f(N(i)) \geq 1$ if the degree of a vertex i is even is $f(N(i)) = 2$.

Algorithm: Signed Total Domination (TMD)(Harris, Hattingh [16]).
Input A rooted tree $T = (V, E)$ with vertices $V = \{1, 2, \ldots, n\}$, where the root is vertex n, (i) an array $parent[1, \ldots, n]$, such that $i < parent(i)$, (ii) a listing $N(1, \ldots, n)$ of the children of i in T, and array $deg[1, \ldots, n]$ of the degrees of the vertices in T.
Output A minimum weight signed total dominating function $f : V \to \{-1, +1\}$
begin
  for $j = 1$ **to** n **do** $f(i) = -1$ **od**
  for $j = 1$ **to** n **do**
    1. **if** $deg[j]$ is odd
      then $MinSum = 1$
      else $MinSum = 2$
    2. **if** $deg[j] = 1$ **and** $j < n$
      then
        $OpenSum = 1$
        $f(parent(j)) = 1$
      else
        $OpenSum = f(N(j))$
    3. **if** $OpenSum < MinSum$
      then
        if $j < n$ **and** $f(parent[j]) = -1$
          then
            $f(parent[j]) = 1$
            $OpenSum = OpenSum + 2$

> **while** $OpenSum < MinSum$ **do**
>   Choose a child i of j, $parent[i] = j$, with $f(i) < 1$
> $f(i) = f(i) + 2$
> $OpenSum = OpenSum + 2$
>
> **end**

In 2009 [23] Lee presents unified methods to compute $\gamma_t^-(G)$ in $O(n^2)$ time for chordal bipartite graphs – a class of graphs that includes bipartite permutation graphs, biconvex bipartite graphs, and convex bipartite graphs – and $\gamma_{st}(G)$ in $O(n+m)$ time for trees. Lee also proves that STDF is NP-complete for doubly chordal graphs.

Lee defines a *signed total zero-dominating function* $f : V \to \{-1, 1\}$ such that for every vertex $v \in V$, $f(N(v)) \geq 0$. The *signed total zero-domination number* $\gamma_{st}^0(G)$ equals the minimum weight of a signed total zero-dominating function on G.

SIGNED TOTAL ZERO-DOMINATING FUNCTION (STZDF)
Instance: Graph $G = (V, E)$, integer k.
Question: Does G have a signed total zero-dominating function of weight at most k?

Theorem 29 ([23]) *STZDF is NP-complete for chordal graphs.*

Proof Sketch. Use a transformation from TOTAL DOMINATING SET for chordal graphs. Given a chordal graph $G = (V, E)$ with $n = |V|$ and $m = |E|$, construct a chordal graph H by adding to each vertex $v \in V(G)$ a set of $deg(v) - 1$ paths, each of length three. One can show that $\gamma_{st}^0(G) = 2m - 2n + 2\gamma_t(G)$. Thus, $\gamma_t(G) \leq k$ if and only if $\gamma_{st}^0(G) \leq 2m - 2n + 2k$. $\qquad\qquad\square$

Theorem 30 ([23]) *STDF is NP-complete for doubly chordal graphs.*

Proof Sketch. Use a transformation from STZDF for chordal graphs. Given a chordal graph $G = (V, E)$ with $n = |V|$, construct a doubly chordal graph H by adding a new vertex x adjacent to every vertex $v \in V(G)$ and attaching to vertex x a set of $n + 1$ paths of length two. One can show that $\gamma_{st}(H) = \gamma_{st}^0(G) + 2n + 3$. Thus, $\gamma_{st}^0(G) \leq k$ if and only if $\gamma_{st}(H) \leq 2n + 3 + k$. $\qquad\qquad\square$

Lee [25] introduces the concept of *R-total domination* in an effort to develop a unified approach to the signed and minus total domination problems. Let

$$P = \{I_1, I_{1+d}, I_{1+2d}, \ldots, I_{1+(l-1)d}\},$$

where l, d, I_1 are fixed integers and l, $d > 0$. Let $G = (V, E)$ be a graph and R a labeling function which assigns an integer $R(v)$ to each $v \in V$. An *R-total dominating function* of G is a function $f : V \to P$ such that $\Sigma_{u \in N(v)} f(u) \geq R(v)$ for all vertices $v \in V$. The *R-total domination number* $\gamma_{t,R}(G)$ equals the minimum weight of an R-total dominating function on G. By design, R-total domination includes total domination, signed total domination and minus total domination as special cases.

Lee points out that this R-domination labeling is very similar to labeled domination of Lee and Chang in [26].

In 2011 [24] Lee studies the signed and minus total domination problems STDF and MTDF for two subclasses of bipartite graphs: biconvex bipartite graphs and planar bipartite graphs. He presents algorithms for computing $\gamma_t^-(G)$ and $\gamma_{st}(G)$ for biconvex bipartite graphs in $O(n+m)$ time, and in 2012 [25] Lee shows that the R-total domination problem can also be solved in $O(n+m)$ time for convex bipartite graphs. He also proves the following.

Theorem 31 (Lee [24]) *STDF is NP-complete for planar bipartite graphs of maximum degree 3.*

Theorem 32 (Lee [24]) *MTDF is NP-complete for planar bipartite graphs of maximum degree 4.*

In 2012 [31] Pradhan considers total domination, signed total domination and minus total domination on chordal bipartite graphs. He shows that given a chordal bipartite graph $G=(V,E)$ of order $n=|V|$ and size $m=|E|$ with a weak elimination ordering, $\gamma_t(G)$ can be computed in $O(n+m)$ time. This reduces the complexity of computing $\gamma_t(G)$ for chordal bipartite graphs from $O(n^2)$ to $O(n+m)$ time. He then shows that both $\gamma_t^-(G)$ and $\gamma_{st}(G)$ can be computed for chordal bipartite graphs in $O(n+m)$ time.

5 Majority Domination

An excellent discussion of properties of majority dominating functions discussed in this section, as of 1998, can be found in a chapter by Hattingh, entitled "Majority Domination and Its Generalizations" in [19].

In 1995 [5], Broere, Hattingh, Henning, and McRae modified the definition of signed domination as follows: a function $f:V \to \{-1,1\}$ is called a *majority dominating function* if $f(N[v]) \geq 1$ for at least half of the vertices $v \in V$. The *majority domination number* $\gamma_{maj}(G)$ equals the minimum value of $f(V)$ overall majority dominating functions on G. In applications, a majority of all closed neighborhoods could vote in favor of some proposition, even though the population of all people might be overwhelmingly opposed to it. In this paper, the authors attribute the following interesting theorem and proof to N. Alon. A *bipartition* of a graph $G=(V,E)$ is a vertex partition $\pi = \{V_1, V_2\}$. A bipartition of a graph G of order n is called *balanced* if $|V_1| = |V_2| = k$ when $n=2k$, and $|V_1| = k$ and $|V_2| = k+1$ when $n=2k+1$.

Theorem 33 (Alon) *For any connected graph G of order $n \geq 2$, $\gamma_{maj}(G) \leq 2$.*

Proof. Let $G=(V,E)$ be a graph of odd order $2k+1$. Among all balanced bipartitions $\pi = \{V_1, V_2\}$, where $|V_1| = k$ and $|V_2| = k+1$, let π^* have a minimum number of edges between V_1 and V_2. In such a partition, every vertex $v \in V_2$ must have at least as many neighbors in V_2 as it has in V_1; if a vertex $v \in V_2$ has more neighbors in V_1 than it has in V_2, then moving v to set V_1 will produce another balanced bipartition π' having even fewer edges between V_1 and V_2, a contradiction.

Therefore, if we assign the value $+1$ to all vertices in V_2 and the value -1 to all vertices in V_1, we will produce a majority dominating function. Every vertex in V_2 will have $f(N[v]) \geq 1$ and the majority of vertices in G are in V_2. In this case, $f(V) = 1$.

If the graph G has even order, then delete an arbitrary vertex x from G. Let π^* be a balanced bipartition of $G - x$ having a minimum number of vertices between V_1 and V_2 as in the argument above when a graph G has odd order. Once again, assign the value $+1$ to all vertices in V_2, -1 to all vertices in V_1 and $+1$ to vertex x. In this case, $f(V) = 2$. $\qquad\square$

The proof of this theorem raises the following question: given a connected graph G of order $n \geq 2$, can one construct a majority dominating function of weight at most 2 in polynomial time? Such a polynomial algorithm is indeed possible, since constructing a balanced bipartition with a minimum number of edges between V_1 and V_2 is not necessary. Assume that the graph G has odd order, as in Alon's proof, and in $O(n)$ time, construct an arbitrary balanced bipartition $\pi = \{V_1, V_2\}$ of $V(G)$. Repeat the following step:

While the larger set of V_1 and V_2 contains a vertex v having more neighbors in the smaller set than in the larger set,

do move vertex v to the smaller set **od**

Since every execution of this step decreases the number of edges between V_1 and V_2, at most $O(m)$ steps can be executed, where $m = |E|$.

Using a polynomial transformation from DOMINATING SET for 4-regular planar graphs, the authors show that MAJORITY DOMINATING FUNCTION is NP-complete.

MAJORITY DOMINATING FUNCTION (MDF)
Instance: Graph $G = (V, E)$, integer k.
Question: Does G have a majority dominating function of weight at most k?

Theorem 34 (Broere, Hattingh, Henning, McRae [5]) *MDF is NP-complete.*

Proof Sketch. Use a transformation from DOMINATING SET for planar 4-regular graphs [15]. Given an instance of this problem, a 4-regular planar graph $G = (V, E)$ of order n and a positive integer k, construct the following graph H. Let K_{n+8} be a complete graph of order $n + 8$, and let $\overline{K_8}$ be the graph consisting of 8 isolated vertices. Let H be formed from the disjoint union $G \cup K_{n+8} \cup \overline{K_8}$, by adding edges as follows: (i) from every vertex in $\overline{K_8}$ add an edge to every vertex of G, (ii) from each of four of the vertices in K_{n+8} add an edge to every vertex in G. Let this set of four vertices in K_{n+8} be denoted by K_4. Note that $|V(H)| = 2n + 16$.

One can then show that if $S \subseteq V(G)$ is a dominating set of G of cardinality at most k, then H has a majority dominating function $f : V \rightarrow \{-1, 1\}$ of weight at most $2k - 2n - 8$. This function can be defined as follows: Let S be a dominating set of

G of cardinality at most k. For each vertex $v \in S$, let $f(v) = 1$. For each vertex $x \in K_4$, let $f(x) = 1$. And for all other vertices $w \in V(H)$, let $f(w) = -1$.

Since each vertex $v \in \overline{K_8}$ has degree 4, having four neighbors in the set K_4, it follows that $f(N[v]) = 3 \geq 1$. Each vertex $v \in S$ has degree 8, four neighbors in $V(G)$ and four neighbors in K_4. Thus, $f(N[v]) \geq 1$, since $f(v) = 1$ and all four neighbors $w \in K_4$ have $f(w) = 1$. Finally, every vertex $u \in V(G) - S$ has four neighbors in $V(G)$ and four neighbors in K_4, but one neighbor of u in $V(G)$ is in S which is assigned the value 1. Thus, $f(N[u]) \geq 1$. Thus, in all, at least $n + 8$ vertices have $f(N[v]) \geq 1$ which is at least half of the total of $2n + 16$ vertices. This function assigns the value 1 to $k + 4$ vertices, and the value -1 to the remaining $n - k + n + 8 - 4 + 8 = 2n - k - 4$ vertices, for a total weight of $2k - 2n - 8$.

We will not provide the details of the converse: given a majority dominating function of weight at most $2k - 2n - 8$ show that there must exist a dominating set of G of size at most k. It is based on the following. Among all minimum majority dominating functions of H, let f be one that assigns the value $+1$ to the maximum number of vertices. Let P denote the set of vertices assigned the value $+1$. Claim 1. $|P| \leq k + 4$. Claim 2. $f(N[v]) \leq 0$ for all $v \in K_{n+8}$. Claim 3. $f(K_4) = 4$. $\square$

The authors close by asking: is MDF NP-complete when restricted to trees?

A similar question can be asked for the parameter $\Gamma^-_{maj}(G)$, which can be defined as the maximum weight of a minimal majority dominating function. To the best of our knowledge, this parameter has not been studied.

In 1997, Yeh and Chang [34] extend the NP-completeness result of Broere et al. for general graphs by showing that MAJORITY DOMINATING FUNCTION can be solved in $O(n^2) = O(n \sum_{v \in V} deg(v))$ time for trees, $O(n^3)$ time for cographs, and polynomial time for k-trees for any fixed k.

The class of k-trees is the class of graphs that can be defined recursively by: (i) the complete graph K_k is a k-tree, and (ii) any graph H obtained from a k-tree G by adding a new vertex joined to all vertices of a complete subgraph of order k in G is a k-tree.

The authors note that since it is possible to embed a *partial k-tree* (a subgraph of a k-tree) into a k-tree in polynomial time, there is also a polynomial-time algorithm for computing the majority domination number of a partial k-tree for any fixed k.

In 2001 [21] Holm proves the NP-completeness of the following restriction of MDF.

MAJORITY DOMINATING FOR COMPLETE GRAPH UNIONS

Instance: A disjoint union G of complete graphs $G = K_{n_1} \cup K_{n_2} \cup \ldots \cup K_{n_m}$ and a positive integer k.

Question: Does G have a majority dominating function f of weight $f(V) \leq k$?

The NP-completeness of this decision problem is shown using a transformation from the following well-known NP-complete problem.

PARTITION

Instance: A finite set A and a *size* $s(a) \in Z^+$, for all $a \in A$.

Question: Does there exist a subset $A' \subset A$ for which

$$\sum_{a \in A'} s(a) = \sum_{a \in A-A'} s(a)?$$

At the end of this paper, Holm asks if the following decision problem is NP-complete:

MAJORITY FOR COMPLETE MULTIPARTITE GRAPHS
Instance: A complete multipartite graph $G = K_{n_1,n_2,\ldots,n_m}$, a positive integer k.
Question: Does G have a majority dominating function f of weight $w(f) \le k$?

6 Efficient Y-Domination

In 1996 [1] Bange, Barkauskas, Host, and Slater generalize the definitions of signed and minus dominating functions to Y-valued dominating functions, as given in Section 1 above. They also introduce the generalized notion of Y-efficient domination for functions $f : V \to Y$. If the closed neighborhood sum $f(N[v]) = 1$ for every $v \in V$, then f is called an *efficient Y-dominating function* of G. The authors point out, however, that there are graphs that do not have efficient Y-dominating functions for any subset Y of real numbers.

They showed that if the closed neighborhood matrix of G is invertible, then G has an efficient Y-dominating function for some set Y, since if the closed neighborhood matrix N is invertible, then the matrix equation $NX = 1$ has a unique solution X, which determines the efficient Y-dominating function.

Two Y-dominating functions are *equivalent* if they have the same closed neighborhood sum at every vertex of G. Illustrations of this are given in Figure 12, with the following three equivalent functions defined on the vertices of the cycle C_6: (i) (0, 1, 0, 0, 1, 0), (ii) (1/3, 1/3, 1/3, 1/3, 1/3, 1/3), and (iii) (1, -1, 1, 1, -1, 1). It is proved that G has an efficient Y-dominating function if and only if all equivalent Y-dominating functions have the same weight or, equivalently, that if f_1 and f_2 are any two efficient Y-dominating functions, then they have the same weight, $f_1(V) = f_2(V)$.

Moreover, the problem of the existence of an efficient signed dominating function $f : V \to \{-1, 1\}$ is shown to be NP-complete for general graphs using a transformation from ONE-IN-THREE 3SAT given earlier.

EFFICIENT SIGNED DOMINATING FUNCTION (ESDF)
Instance: Graph $G = (V, E)$.
Question: Does G have an efficient signed dominating function?

The corresponding decision problem for efficient minus domination is the following.

EFFICIENT MINUS DOMINATING FUNCTION (EMDF)

Fig. 12 Efficient
y-dominating functions
$f_1(V)=f_2(V)=f_3(V)=2$

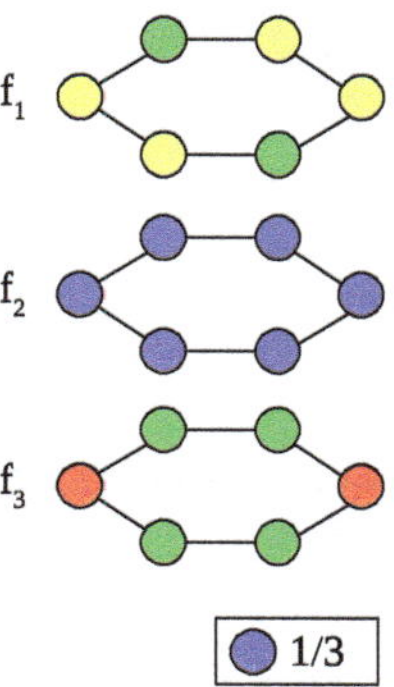

Instance: Graph $G=(V, E)$.
Question: Does G have an efficient minus dominating function?

EFFICIENT DOMINATING FUNCTION (EDF)
Instance: Graph $G=(V, E)$.
Question: Does G have an efficient dominating function?

In 2000 [29] and 2003 [30] Lu, Peng and Tang showed that EMDF is NP-complete for (i) chordal graphs, (ii) chordal bipartite graphs (in which every cycle of length greater than 4 has a chord), (iii) planar bipartite graphs, and (iv) planar graphs of maximum degree 4 and that ESDF is NP-complete for chordal graphs.

The NP-completeness of EMDF on these four classes of graphs follows from the fact that (i) EDF is NP-complete on chordal, chordal bipartite, and planar bipartite graphs, and the class of planar graphs of maximum degree 3 and (ii) the construction given by Dunbar et al. [11], whereby to a graph in any of these classes you can add to each vertex a path of length three and create a graph that is still chordal, chordal bipartite or planar bipartite, or planar of maximum degree 4.

The NP-completeness of ESDF for chordal graphs is shown using a transformation from One-in-Three 3SAT.

With regard to the class of trees, the authors show that if tree T has an efficient minus dominating function f, then it must be that for every vertex $v \in V$, $f(v) \geq 0$. Thus, the efficient minus domination problem is equivalent to the efficient domination problem on trees.

7 Signed Star Domination

In 2001 [32] and 2005 [33] Xu introduces the following variation of signed dominating functions. For a graph $G=(V, E)$ without isolated vertices, for every vertex $v \in V$, let $E(v)=\{uv \in E : u \in N(v)\}$ be the set of edges incident with v. A function $f : E(G) \to \{-1, 1\}$ is said to be a *signed star dominating function* of G if for every $v \in V$, $\Sigma_{e \in E(v)} f(e) \geq 1$. The minimum value of $\Sigma_{e \in E} f(e)$, taken over all

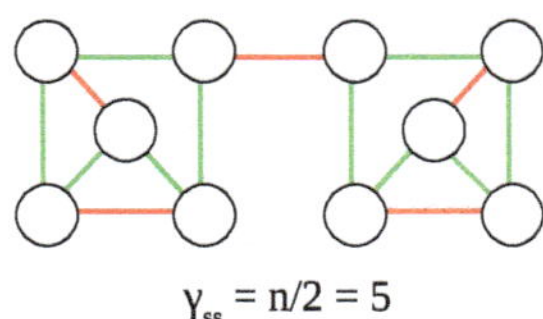

Fig. 13 Efficient signed star domination

signed star dominating functions f of G is called the *signed star domination number* of G and is denoted by $\gamma_{ss}(G)$. The cubic graph in Figure 13 illustrates a signed star dominating function in which $\Sigma_{e \in E(v)} f(e) = 1$ for every vertex; two of the three edges incident with every vertex are assigned the value $+1$ (green), and the third is assigned the value -1 (red). Indeed, this is always possible for any cubic graph having a perfect matching, for the set of red edges is a perfect matching in Figure 13.

In 2019 [36] Zhao, Shan, Miao, and Liang study bounds for the signed star domination numbers in arbitrary graphs, e.g., for any graph G of order n, $\lceil n/2 \rceil \leq \gamma_{ss}(G) \leq 2n - 4$. They also present a linear algorithm for computing $\gamma_{ss}(T)$ for any tree T.

8   Open Problems

1. In [18] Hattingh, Henning, and Slater presented a linear algorithm for computing $\gamma_s(T)$ for any tree T, but did not present a linear algorithm for computing $\Gamma_s(T)$ for any tree T.
2. In [11] Dunbar, Hedetniemi, Henning, and McRae presented a linear algorithm for computing $\gamma^-(T)$ for any tree T, but did not present a linear algorithm for computing $\Gamma^-(T)$ for any tree T.
3. Can you show that the decision problem for upper majority domination is NP-complete?
4. Is the Minimum Minus Domination Problem NP-complete on doubly chordal graphs? Recall that a graph is *doubly chordal* if it is both chordal and dually chordal. *Dually chordal graphs* are the clique graphs of chordal graphs.
5. The minus variant of majority dominating functions has apparently not been investigated. In this case you seek a minimum weight function $f : V \rightarrow \{-1, 0, 1\}$ such that the number of closed neighborhoods voting "yes," i.e., with $f(N[i]) \geq 1$, is greater than the number of closed neighborhoods voting "no," with $f(N[j]) \leq -1$. In applications, this permits voters to abstain from voting. In this model the number of neighborhoods voting "yes" need not be a majority of all closed neighborhoods, as in majority domination, it only needs to be greater in number than those neighborhoods voting "no." Recall that $\gamma_{maj}(G)$ is designed for signed functions $f : V \rightarrow \{-1, 1\}$.
6. Is MAJORITY DOMINATING FUNCTION NP-complete when restricted to complete multipartite graphs?
7. What can you say about upper signed star domination, $\Gamma_{ss}(G)$?

References

1. D.W. Bange, A.E. Barkauskas, L.H. Host, P.J. Slater, Generalized domination and efficient domination in graphs. Discrete Math. **159**(1–3), 1–11 (1996)
2. P. Berman, T. Fujito, On approximation properties of the independent set problem for low degree graphs. Theory Comput. Syst. **32**, 115–132 (1999)
3. H.L. Bodlaender, On linear time minor tests and depth-first search, in *Proceedings of the 1st Workshop on Algorithms and Data Structures*, ed. by F. Dehne, et al. Lecture Notes in Computer Science, vol. 382 (1987), pp. 577–590
4. A. Brandstädt, V.D. Chepoi, F.F. Dragan, The algorithmic use of hypertree structure and maximum neighbourhood orderings. Discrete Appl. Math. **82**, 43–77 (1998)
5. I. Broere, J.H. Hattingh, M.A. Henning, A.A. McRae, Majority domination in graphs. Discrete Math. **138**(1–3), 125–135 (1995)
6. P. Damaschke, Minus domination in small-degree graphs. Discrete Appl. Math. **108**(1–2), 53–64 (2001)
7. R.G. Downey, M.R. Fellows, Fixed-parameter tractability and completeness. Congr. Numer. **87**, 161–187 (1992)
8. R.G. Downey, M.R. Fellows, Fixed-parameter tractability and completeness I: basic results. SIAM J. Comput. **24**(4), 873–921 (1995)
9. J. Dunbar, W. Goddard, S.T. Hedetniemi, M.A. Henning, A. McRae, Minus domination in regular graphs. Discrete Math. **149**, 311–312 (1996)
10. J. Dunbar, S.T. Hedetniemi, M.A. Henning, P.J. Slater, Signed domination in graphs, in *Graph Theory, Combinatorics and Algorithms* (Kalamazoo, MI, 1992), vols. 1, 2 (Wiley-Intersci. Publ., Wiley, New York, 1995), pp. 311–321
11. J. Dunbar, W. Goddard, S.T. Hedetniemi, M.A. Henning, A. McRae, The algorithmic complexity of minus domination in graphs. Discrete Appl. Math. **68**, 73–84 (1996)
12. J. Dunbar, S.T. Hedetniemi, M.A. Henning, A. McRae, Minus domination in graphs. Discrete Math. **199**(1–3), 35-47 (1999)
13. L. Faria, W.-K. Hon, T. Kloks, H.-H. Liu, T.-M. Wang, Y.-L. Wang, On complexities of minus domination. Discret. Optim. **22**, 6–19 (2016)
14. M.R. Fellows, M.A. Langston, Nonconstructive tools for proving polynomial time decidability. J. Assoc. Comput. Mach. **35**, 727–739 (1988)
15. M.R. Garey, D.S. Johnson, *Computers and Intractability, A Guide to the Theory of NP-Completeness* (Freeman, New York, 1979)
16. L. Harris, J.H. Hattingh. The algorithmic complexity of certain functional variations of total domination in graphs. Australas. J. Combin. **29**, 143–156 (2004)
17. J.H. Hattingh, Majority domination and its generalizations, in *Domination in Graphs, Advanced Topics* (Marcel Dekker New York, 1998)
18. J.H. Hattingh, M.A. Henning, P.J. Slater, On the algorithmic complexity of signed domination in graphs. Australas. J. Combin. **12**, 101–112 (1995)
19. T.W. Haynes, S.T. Hedetniemi, P.J. Slater (eds.), *Domination in Graphs, Advanced Topics* (Marcel Dekker, New York, 1998)
20. M.A. Henning, Signed total domination in graphs. Discrete Math. **278**(1–3), 109–125 (2004)
21. T.S. Holm, On majority domination in graphs. Discrete Math. **239**(1–3), 1–12 (2001)
22. L. Kang, E. Shan, Signed and minus dominating functions in graphs, in *Topics in Domination in Graphs*, ed. by T.W. Haynes, S.T. Hedetniemi, M.A. Henning (Springer, Berlin, 2020), pp. 301–348
23. C.-M. Lee, On the complexity of signed and minus total domination in graphs. Inform. Process. Lett. **109**(20), 1177–1181 (2009)
24. C.-M. Lee, Signed and minus total domination on subclasses of bipartite graphs. Ars Combin. **100**, 129–149 (2011)
25. C.-M. Lee, R-total domination on convex bipartite graphs. J. Combin. Math. Combin. Comput. **81**, 209–224 (2012)

26. C.-M. Lee, M.-S. Chang, Variations of Y-dominating functions on graphs. Discrete Math. **308**(18), 4185–4204 (2008)
27. J.-Y. Lin, S.-H. Poon, Algorithms and hardness for signed domination. Theory Appl. Model. Comput. **9076**, 453–464 (2015)
28. J.-Y. Lin, C.-H. Liu, S.-H. Poon, Algorithmic aspect of minus domination on small-degree graphs, in *Computing and Combinatorics*. Lecture Notes in Computer Science, vol. 9198 (Springer, Cham, 2015), pp. 337–348
29. C.L. Lu, S.-L. Peng, C.Y. Tang, Efficient minus and signed domination in graphs, in *Algorithms and Computation (Taipei, 2000)*. Lecture Notes in Computer Science, vol. 1969 (Springer, Berlin, 2000), pp. 241–253
30. C.L. Lu, S.-L. Peng, C.Y. Tang, Efficient minus and signed domination in graphs. Theoret. Comput. Sci. **301**(1–3), 381–397 (2003)
31. D. Pradhan, Complexity of certain functional variants of total domination in chordal bipartite graphs. Discrete Math. Algorithm. Appl. **4**(3), 19 (2012), 1250045
32. B. Xu, On signed edge domination numbers of graphs. Discrete Math. **239**, 179–189 (2001)
33. B. Xu, Note on edge domination numbers of graphs. Discrete Math. **294**, 311–316 (2005)
34. H.-G. Yeh, G.J. Chang, Algorithmic aspects of majority domination. Taiwan. J. Math. **1**(3), 343–350 (1997)
35. B. Zelinka, Signed total domination number of a graph. Czechoslovak Math. J. **51**, 225–229 (2001)
36. Y.-C. Zhao, E.F. Shan, L.-Y. Miao, Z.-S. Liang, On signed star domination in graphs. ACTA Math. Appl. Sin. Engl. Ser. **35**(2), 452–457 (2019)

Algorithms and Complexity of Power Domination in Graphs

Stephen T. Hedetniemi, Alice A. McRae, and Raghuveer Mohan

1 Power Domination in Graphs

Based on the early research of Baldwin, Mili, Boisen and Adapa in 1993 [6], Brueni in 1993 [13] and Boisen, Baldwin and Mili in 2000 [10], Haynes, Hedetniemi, Hedetniemi and Henning introduced the concept of power domination in graphs in 2002 [24], as follows. Electric power companies continuously monitor the state of their electrical power lines with the use of *phasor measurement units* or *PMUs*, which estimate the magnitude and phase angles of electrical phasor quantities such as voltage or current. These are placed at *substation buses* where transmission lines, loads, and generators are connected and transmission lines connecting two electrical nodes are represented by edges in a graph G. For more information on power domination, please see the excellent chapter by Dorbec in [17].

In the corresponding graph theoretical model, a set S of vertices is initially selected, at which PMUs are to be located.

A set S is said to be a *power dominating set* of a graph $G = (V, E)$ if every vertex and every edge in the system are monitored by S in accordance with the following rules:

R1. If a vertex v is in the set S, then both v and all edges incident to v are called *observed*.

R2. Every vertex incident with an observed edge is observed.

R3. If both vertices u and v of an edge uv are observed, so is the edge uv.

S. T. Hedetniemi
School of Computing, Clemson University, Clemson, SC, USA
e-mail: hedet@clemson.edu

A. A. McRae · R. Mohan (✉)
Computer Science Department, Appalachian State University, Boone, NC 28608, USA
e-mail: mcraeaa@appstate.edu; moharr@appstate.edu

T. W. Haynes et al. (eds.), *Structures of Domination in Graphs*, Developments in Mathematics 66, https://doi.org/10.1007/978-3-030-58892-2_15

R4. If all of the edges but one incident to an observed vertex v are observed, then the one final edge incident to v is observed.

A *power dominating set* of a graph $G = (V, E)$ is a set $S \subseteq V$ having the property that all vertices $v \in V$ and all edges $uv \in E$ are observed by the vertices in S.

The minimum cardinality of a power dominating set of a graph is called the *power domination number*, denoted $\gamma_P(G)$.

This definition immediately gives rise to the following decision problem.

POWER DOMINATING SET (PDS)
Instance: Graph $G = (V, E)$, positive integer $k > 1$
Question: Does G have a power dominating set of cardinality at most k?

In 2002 this was solved as follows.

Theorem 1 (Haynes et al.) *PDS is NP-complete when restricted to bipartite or chordal graphs.*

Proof Sketch. For the bipartite case, use a transformation from the well-known NP-complete problem 3-SAT.

3-SAT
Instance: A set $U = \{u_1, u_2, \ldots, u_n\}$ of Boolean variables and a set $C = \{C_1, C_2, \ldots, C_m\}$ of three-variable clauses, where each clause contains three distinct occurrences of either a variable u_i or its complement $\overline{u_i}$.
Question: Does C have a satisfying truth assignment, that is, an assignment of TRUE or FALSE to each Boolean variable, such that in each clause C_i at least one variable (or its complement) is assigned the value TRUE?

Given an instance (U, C) of 3-SAT, construct a bipartite graph $G(U, C)$ instance of PDS, as follows. For each variable u_i, construct a 4-cycle C_4, with two non-adjacent, variable vertices, labeled u_i and $\overline{u_i}$.

For each clause, for example, $C_i = \{u_i, \overline{u_j}, u_k\}$, create two non-adjacent vertices $C_{i,1}$ and $C_{i,2}$, and join both of these two vertices to the three vertices in the 4-cycles corresponding to u_i, $\overline{u_j}$, and u_k. Thus, each of the two vertices $C_{i,1}$ and $C_{i,2}$ will have degree 3 in the bipartite graph $G(U, C)$ so constructed (cf. Figure 1).

One can then show that (U, C) has a satisfying truth assignment if and only if the constructed bipartite graph $G(U, C)$ has a power dominating set S of cardinality at most $k = n$. Note that if S contains one vertex in a 4-cycle, either u_i or $\overline{u_i}$, then all four vertices and all four edges in the 4-cycle will be observed, using rules R1, R2, R4, and then R2. Thus, from a satisfying truth assignment, one can choose one variable vertex from each 4-cycle and observe all vertices and all edges in all 4-cycles. This will also suffice to observe all edges between a 4-cycle and a clause vertex, and all clause vertices, using rules R1, R2, and R3.

Conversely, any power dominating set S for the constructed graph G must contain at least one vertex from each 4-cycle. One vertex per 4-cycle accounts for all k vertices in S. Since every clause vertex is duplicated, rule R4 cannot be used to observe clause vertices. Every clause vertex is observed through R1 by being

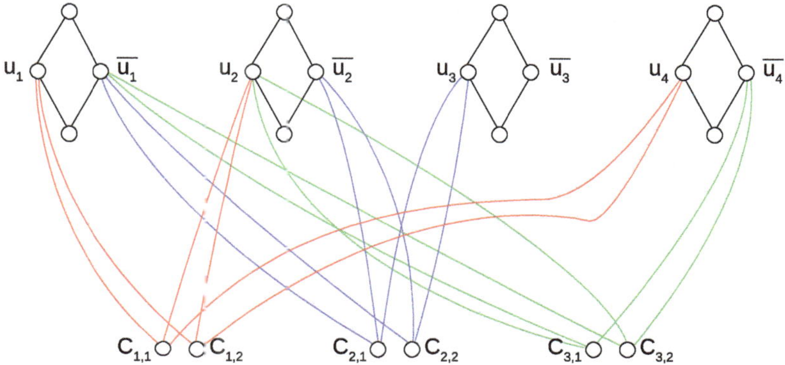

Fig. 1 Example of a constructed bipartite graph for proof of Theorem 1. The clauses are $C_1 = (u_1 \cup u_2 \cup u_4)$, $C_2 = (\bar{u}_1 \cup \bar{u}_2 \cup u_3)$, and $C_3 = (\bar{u}_1 \cup u_2 \cup \bar{u}_4)$

adjacent to vertices in S. A truth assignment for U exists by assigning a true value to the literals corresponding to the U vertices in S.

The fact that PDS remains NP-complete for chordal graphs can be shown by creating a clique of cardinality $2n$ among the $2n$ variable vertices, which creates a chordal graph. $\qquad\square$

Although Haynes et al. did not point this out, this same clique construction creates a split graph, that is, a graph whose vertices can be partitioned into two sets, $V = \{V_1, V_2\}$, such that V_1 induces a clique and V_2 is an independent set. Thus, we have as a corollary the following.

Corollary 1 *PDS is NP-complete when restricted to split graphs.*

When restricted to the family of trees, Haynes et al. develop a noteworthy result, as follows. Let T be a tree obtained from a single vertex x by attaching to x any number of paths, of any finite length. Such a tree is called a *spider*. The *spider number* $sp(T)$ of a tree T equals the minimum order k of a vertex partition $V = \{V_1, V_2, \ldots, V_k\}$ such that each subset V_i induces a spider.

For spiders the authors present the following two results.

Proposition 1 (Haynes et al.) *For any tree T, $\gamma_P(T) = 1$ if and only if T is a spider.*

Proof Sketch. Let $S = \{x\}$. Then by rules R1, R2, and R4, all vertices and edges of T will be observed. $\qquad\square$

Theorem 2 (Haynes et al.) *For any tree T, $\gamma_P(T) = sp(T)$.*

On the basis of these two results, Haynes et al. present a linear algorithm to compute the value $\gamma_P(T)$ for any tree T.

In 2005, [14] Brueni and Heath, following on the earlier master's thesis of Brueni [13], offered the following simpler vertex definition of power domination in graphs, which they prove is equivalent to the vertex-edge definition.

Rule B1. If a vertex $v \in S$, then v and all vertices in $N(v)$ are observed.

Rule B2. (Kirchhoff's Rule) If a vertex v is observed and there is a vertex $u \in N(v)$ that is the only unobserved vertex in $N(v)$, then vertex u is observed.

These two rules appear in several other papers in the following form.

For a connected graph G and a vertex set $S \subseteq V$, the set $M(S)$ of vertices *monitored* by S is defined recursively as follows:

(1) $M(S) \leftarrow S \cup N(S)$ (for every $v \in S$, v and all of its neighbors in $N(v)$ are monitored),
(2) While there is a vertex $v \in M(S)$ having exactly one unmonitored neighbor w, that is, $N(v) \cap (V(G) - M(S)) = \{w\}$, set $M(S) \leftarrow M(S) \cup \{w\}$.

A set S is called a *power dominating set* of G if $M(S) = V(G)$. The *power domination number* $\gamma_p(G)$ is the minimum cardinality of a power dominating set of G. This contrasts with dominating sets S, which can only observe, or dominate, vertices at distance 1 from vertices in S. Power dominating sets are not "local" in that they can monitor or observe vertices arbitrarily far from vertices in S.

Brueni and Heath provide an $O(|V| + |E|)$ algorithm for computing the set of vertices observed by any set $S \subseteq V$ in any graph G.

In [14] Brueni and Heath prove the following results.

Theorem 3 (Brueni-Heath) *For any connected graph G of order $n = |V|$, $\gamma_P(G) \leq n/3$, and this bound is tight.*

Theorem 4 (Brueni and Heath) *PDS is NP-complete for planar bipartite graphs.*

Proof Sketch. Use a transformation from what is called PLANAR 3-SAT. To each Boolean variable $U = \{u_1, u_2, \ldots, u_n\}$, construct a 4-cycle, two non-adjacent vertices of which are labeled u_i and $\overline{u_i}$. From each of these two variable vertices, attach a leaf. For each clause C_j in $C = \{C_1, C_2, \ldots, C_m\}$, construct a K_2, one vertex of which is labeled c_j. Then add an edge between c_j and the three variable vertices the clause contains. It is assumed in PLANAR 3-SAT that this graph is always planar and it is by construction bipartite. One can then show that U, C has a satisfying truth assignment if and only if the planar bipartite graph $G(U, C)$ has a power dominating set of size at most n. $\qquad \square$

In 2005 [23] and again in 2008 [24], Guo, Niedermeier, and Raible show that the power dominating set (PDS) problem can be solved by a dynamic programming algorithm for graphs of bounded treewidth. Moreover, they simplify and extend several NP-completeness results, by showing that PDS remains NP-complete for planar graphs, circle graphs (intersection graphs of chords of a circle, where two vertices are adjacent if and only if the corresponding chords cross each other), and split graphs.

Guo et al. show that PDS, when parameterized by $|S| = k$, the size of a power dominating set, is W[2]-hard for general graphs and, like DOMINATING SET, is only $\Theta(\log n)$-approximable, meaning that it cannot be approximated any better than DOMINATING SET.

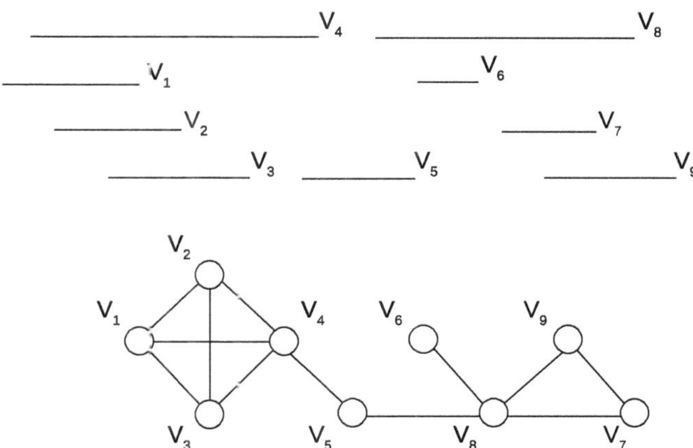

Fig. 2 Interval graphs

DOMINATING SET (DOMSET)
Instance: Graph $G = (V, E)$, positive integer k.
Question: Does G have a dominating set of size at most k?

Guo et al. were perhaps the first to discuss the non-locality of power domination. Whereas a dominating set can be verified by examining the neighborhoods of all vertices in a graph, the same check is not sufficient for verifying that a set is a connected dominating set. In this case one must verify that the subgraph induced by a dominating set is connected. However, verification that a set is a power dominating set is neither 1-local, as in the case of domination, nor a question of the subgraph induced by a power dominating set; rather, it is a question of the effect that a power dominating set has at arbitrary distances from vertices in the power dominating set.

In [23] Guo et al. present a linear algorithm for computing $\gamma_P(T)$ for any tree T that is simpler than the one given by Haynes et al. [25].

In 2008 [24] Guo, Niedermeier, and Raible construct a dynamic programming algorithm for computing $\gamma_P(G)$ for graphs of treewidth k. The running time of their algorithm is $O(c^{k^2} n)$, where c is a constant.

In 2005 [33] Liao and Lee show that the PDS decision problem is NP-complete for split graphs, which are a subclass of chordal graphs. In [33] the authors present a linear algorithm for computing $\gamma_P(G)$ for interval graphs G (cf. Figure 2), provided the interval ordering of the graph is provided, and they show that if the interval ordering is not given, the algorithm with $O(n \log n)$ time complexity is asymptotically optimal, where n is the number of intervals. They also show that the same results hold for the class of proper circular arc graphs, where *circular arc graphs* are the intersection graphs of sets of arcs of a circle, and a circular arc graph is *proper* if no arc properly contains another arc (cf. Figure 3).

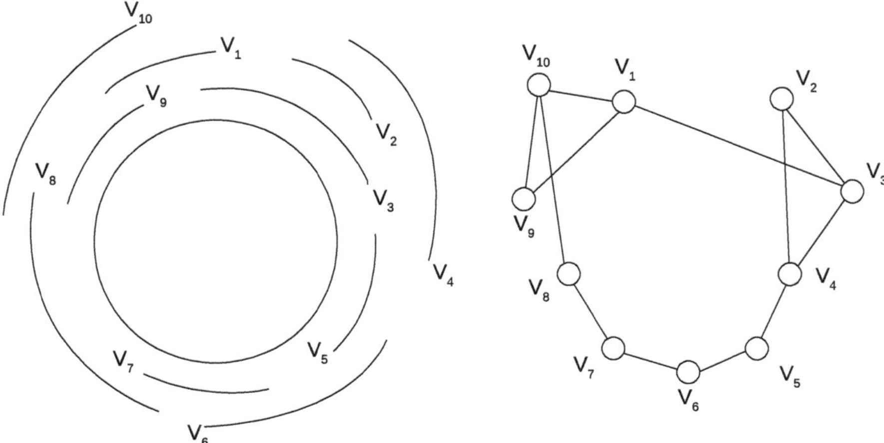

Fig. 3 Circular arc graphs

In 2006 [30] Kneis, Molle, Richter and Rossmanith present a linear-time algorithm for computing $\gamma_P(G)$ in graphs having bounded treewidth. They also prove the following.

Theorem 5 (Kneis et al.) *POWER DOMINATING SET is W[2]-hard for the parameter k, the number of PMUs, or the size of the power dominating set S.*

Proof. Use a reduction from DOMINATING SET, where the parameter is the size k of the dominating set. DOMINATING SET has been shown to be W[2]-complete by Downey and Fellows [20]. Given an instance of DOMINATING SET, a graph $G = (V, E)$ and a positive integer k, construct the corona $G' = G \circ K_1$ of G, by attaching a leaf v' to every vertex $v \in V$. Show that $\gamma(G) \le k$ if and only if $\gamma_P(G') \le k$.

We note at this point that the same corona reduction was independently used by Guo, Niedermeier, and Raible in [23] and in [24].

Let $S \subset V$ be a dominating set of G with $|S| \le k$. Clearly, S dominates all vertices in G' except possibly some leaves v' attached to vertices v. Since all neighbors of any such vertex v except v' are power dominated by S, it follows that vertex v' is power dominated by S using Kirchhoff's Rule. Thus, the same set S is a power dominating set of G' with $|S| \le k$.

Let $S \subseteq V'$ be a power dominating set of G' with $|S| \le k$. If a leaf $v' \in S$, then $S' = S - \{v'\} \cup \{v\}$ is also a power dominating set of G' with $|S'| = |S| \le k$. Thus, we may assume, without loss of generality, that $S \subseteq V$ and contains no leaves of G'. It remains to show that Kirchhoff's Rule never applies to any vertex $v \in V$, which means that S is also a dominating set of G.

Let $M(S) = N[S]$ be the set of vertices initially observed by the power dominating set S; thus, all vertices initially in $M(S)$ are either in S or are dominated by a vertex in S. Let $U = V(G') - M(S)$ be the set of vertices that are unobserved, and assume that $U \neq \varnothing$.

The claim is that the only vertices in U are leaves of G'. The only vertices that can be added to the set $M(S)$ of observed vertices are those which are observed using Kirchhoff's Rule. If there is a vertex $v \in V$ that is not in $M(S)$, then it has no neighbor in S. This means that vertex v has at least two unobserved neighbors, at least one unobserved neighbor in V and its leaf neighbor in G', since we are assuming that S contains no leaves of G'. Thus, v can never be observed, contradicting the assumption that S is a power dominating set of G'. $\qquad\square$

As observed by Guo et al., the corona operation preserves bipartiteness, planarity, and the property of being a circle graph (an intersection graph of chords of a circle). The corona operation also preserves the property of being a chordal bipartite graph. DOMINATING SET has been shown to be NP-complete for chordal bipartite graphs by H. Müller and A. Brandstädt in 1987 [36].

Corollary 2 *PDS is NP-complete for bipartite, planar, circle, and chordal bipartite graphs.*

Earlier we observed that a minor modification of the proof of Haynes et al. suffices to show that PDS is NP-complete for both chordal graphs and split graphs.

Guo, Niedermeier, and Raible have used a transformation from VERTEX COVER to show that PDS is NP-complete for split graphs [24].

In 2006, Xu, Kang, Shan, and Zhao [42] construct a linear algorithm for computing $\gamma_P(G)$ for any connected block graph G. A *block* of a graph G is a maximal 2-connected subgraph of G. A graph G is a *block graph* if and only if every block of G is a complete subgraph. Their algorithm is based in part on the following result.

Theorem 6 (Xu et al.) *If G is a block graph having at least one cut vertex, then G has a $\gamma_P(G)$-set in which every vertex is a cut vertex.*

The algorithm works by first constructing a decomposition tree of a block graph called the *refined cut tree*, as shown in Figure 4. This tree has one vertex for every block and one vertex for every cut vertex of G. An edge exists between a block vertex and a cut vertex if and only if the cut vertex is contained in that block. The *refined cut tree* is formed by re-labeling the blocks of the graph as *block vertices*. This tree can be constructed in linear time. The algorithm then roots the tree at a cut vertex and processes the cut vertices from the leaves up to the root. At each level, a power dominating set is constructed for that level by modifying the power dominating set constructed for the lower level.

Xu et al. also prove that the power domination number of a block graph of order n is no more than $n/3$, with equality if and only if the graph is the corona $G \circ K_2$ or $G \circ \overline{K_2}$ for some block graph G. Power domination in block graphs is also studied by Atkins, Haynes and Henning in 2006 [5].

In 2006 [43] Zhao, Kang, and Chang show that $\gamma_P(G) \leq n/3$ for any connected graph G of order $n \geq 3$ and $\gamma_P(G) \leq n/4$ for any connected claw-free cubic graph G

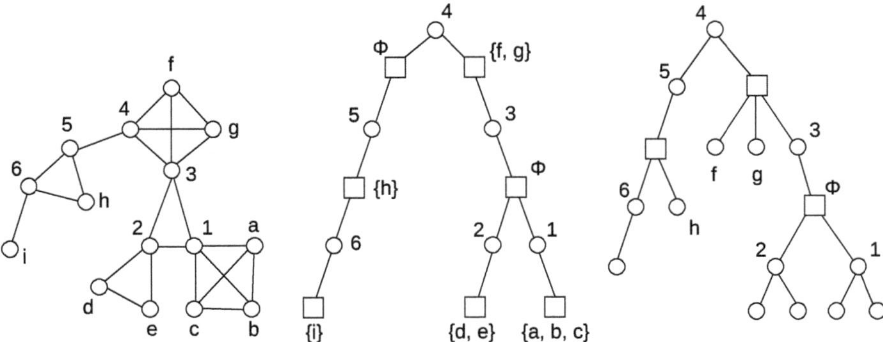

Fig. 4 Block graph cut trees (figure reproduced from [42])

of order n. It is interesting to note how many authors have noted the $n/3$ upper bound for various classes of graphs, for example, for trees [25] or block graphs [42].

In 2007 [26] Hon, Liu, Peng, and Tang present a linear algorithm for computing the power domination number of block-cactus graphs, which are defined as follows.

A *cut vertex* in a graph G is a vertex whose removal increases the number of connected components. A *cactus* is a graph in which every edge is a member of at most one cycle. Finally, a *block-cactus* graph is a graph in which every block is either a complete graph or a cycle. A linear algorithm for computing the power domination number of a block-cactus graph is based on the following observation.

Theorem 7 (Hon, Liu, Peng, Tang) *In any block-cactus graph, there exists a minimum power dominating set that contains only cut vertices.*

In 2006 [19] Dorfling and Henning determine the power domination numbers of $n \times m$ grid graphs.

Theorem 8 (Dorfling, Henning) *If $G = P_n \square P_m$ is an $m \times n$ grid graph, where $1 \le m \le n$, then*

(i) $\gamma_P(G) = \lceil \frac{m+1}{4} \rceil$ *if $m \equiv 4 \bmod 8$, and*
(ii) $\gamma_P(G) = \lceil \frac{m}{4} \rceil$ *otherwise.*

The authors show how to construct a minimum power dominating set S of a grid graph $G = P_m \square P_n$, as follows: let $m = 8k + j$, where $0 \le j \le 7$ and vertices are represented by their row and column numbers, such as the vertex $v = (2, 3)$ in row 2, column 3.

If $k = 0$, let $S' = \varnothing$; otherwise let $S' = \{(8i+3, 2), (8i+5, 3) : 0 \le i \le k-1\}$. Then,

if $j = 0$, let $S = S'$.
if $j \in \{1, 2\}$, let $S = S' \cup \{(m, 1)\}$.
if $j = 3$ let $S = S' \cup \{(m-1, 1)\}$.
if $j = 4$ let $S = S' \cup \{(m-2, 1), (m-1, 1)\}$.

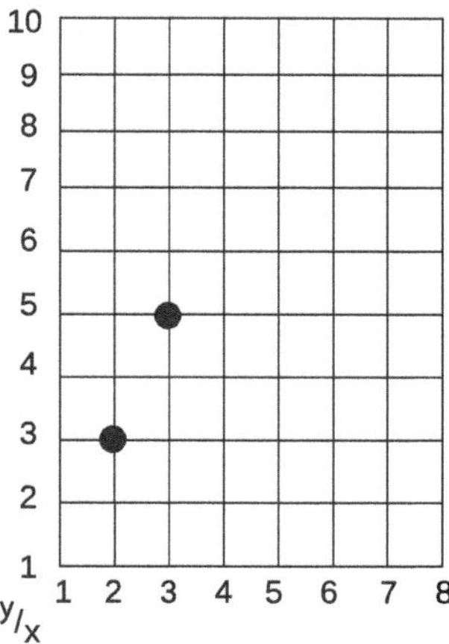

Fig. 5 A minimum power dominating set of an 8×10 grid graph

if $j \in \{5, 6, 7\}$ let $S = S' \cup \{(m + 3 - j, 2), (m + 5 - j, 3)\}$.

Thus, for example, a minimum power dominating set of an 8×10 grid graph needs only two vertices, such as $v = (2, 3)$ and $w = (3, 5)$ (cf. Figure 5).

In 2007 [37] Pai, Chang and Wang provide a somewhat simpler algorithm for placing a minimum power dominating set in a grid graph than that given above by Dorfling and Henning.

In 2008 [39] and in 2012 [9] Raible and Fernau show that PDS is NP-hard on planar cubic graphs and design an $O_*(1.7548^n)$ algorithm for computing $\gamma_P(G)$ for arbitrary graphs. Their NP-completeness result can be briefly described as follows.

Theorem 9 (Binkele-Raible, Fernau) *PDS is NP-hard on planar cubic graphs.*

Proof Sketch. Use a reduction from PLANAR CUBIC VERTEX COVER, as follows:

VERTEX COVER

Instance: Graph $G = (V, E)$, positive integer k.

Question: Does G have a vertex cover of cardinality at most k, that is, a set $S \subseteq V$ such that for every edge $uv \in E$, either $u \in S$ or $v \in S$, that is, $|\{u, v\} \cap S| \geq 1$?

Given a planar cubic instance of VERTEX COVER, with $V = \{v_1, v_2, \ldots, v_n\}$, replace each vertex $v \in V$ with the graph in Figure 6. Notice that if G is planar and cubic, then the constructed graph G' is planar and cubic. One can then show that G has a vertex cover of cardinality at most k if and only if G' has a power dominating set of cardinality at most k. In the gadget in Figure 6, if a set S is a vertex cover, then all edges incident to a vertex v will have a vertex in S. This means that the vertices in S, when viewed as a power dominating set, will observe vertices q_1, c_{v1}, q_2, c_{v2}, q_3, c_{v3}. Once these have been observed, then all remaining vertices a_{v1}, z_v, a_{v2}, a_{v3} will be observed. Thus, $\gamma_P(G') \leq k$.

It only remains to show that starting with a power dominating set S' of G' of cardinality at most k, there will be a vertex cover of G of cardinality at most k. This

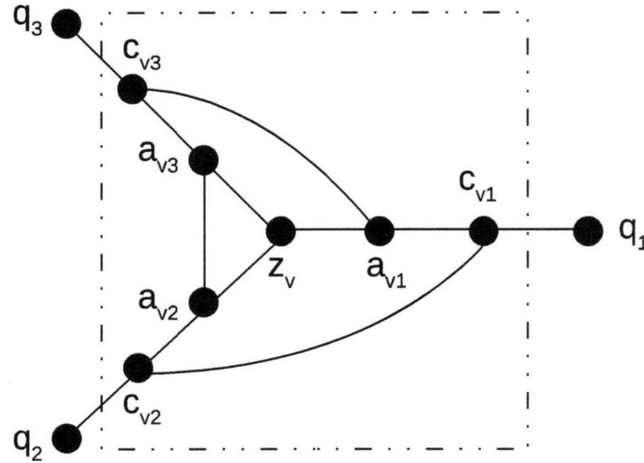

Fig. 6 Gadget for cubic planar graphs (figure reproduced from Binkele [9])

is based on the observations that (i) any minimum power dominating set of G' will contain at most one vertex from any gadget; (ii) if a gadget contains a vertex, then it must be the z_v vertex; and (iii) if a gadget T_v contains no vertices in S', then the a_{vi} vertices are indirectly observed by all three of the c_{vi} vertices and the z_v vertex is indirectly observed by the a_{v1} vertex.   □

Binkele-Raible and Fernau conclude their paper by presenting an 11-page, highly technical, exact algorithm, for computing the value of $\gamma_P(G)$ for any graph G, whose running time is $O*(1.7548^n)$ and which when applied to cubic graphs has a running time of $O*(1.6212^n)$. We refer the interested reader to [9] for the details.

In 2009 [3] Aazami and Stilp present several approximation results for PDS, in particular demonstrating for the first time a "gap" in the approximation guarantees between DOMINATING SET and PDS, especially since it has been shown that DOMINATING SET has an $O(\log n)$ approximation guarantee, and it has been shown that no polynomial algorithm can give a better approximation guarantee. Indeed, Feige [21] has shown that DOMINATING SET is even hard to approximate within a ratio of $(1 - \varepsilon)\, n$. Notable as well is their introduction of power domination applied to directed graphs.

Aazami and Stilp present a transformation from the MIN-REP decision problem to PDS which shows that PDS cannot be approximated within a factor of $2^{\log^{1-\epsilon} n}$, unless $NP \subseteq DTIME(n^{polylog(n)})$, since the same result holds for the MIN-REP problem, which can be stated as follows.

MIN-REP

Instance: (i) Bipartite graph $G = (A, B, E)$; (ii) a partition of the two partite sets $A = \{A_1, A_2, \ldots, A_j\}$ and $B = \{B_1, B_2, \ldots, B_j\}$ into sets of equal size, that is, the sets A_i and B_j all have the same size, say N; and (iii) a positive integer k.

Question: Does there exist a set $C \subset A \cup B$, with $|C| \le k$, such that whenever there exist vertices $u \in A_i$ and $v \in B_j$ where $uv \in E$, there exist a vertex $a \in C \cap A_i$ and a vertex $b \in C \cap B_j$ with $ab \in E$?

For undirected graphs, the authors introduce the notion of *strong regions*, meaning sets of vertices which contain vertices which must appear in a $\gamma_P(G)$-set, as a means of obtaining lower bounds on the size of an optimum solution for PDS. Using this idea, they develop an algorithm for finding power dominating sets that have an approximation guarantee of $O(k)$ for graphs of treewidth k. The algorithm requires as input the partial-k tree decomposition and runs in time $O(n^3)$, independent of k.

Since it is known that planar graphs have treewidth $O(\sqrt{n})$, the Aazami-Stilp algorithm provides an $O(\sqrt{n})$ approximation for PDS. The authors then show that their methods cannot improve on this $O(\sqrt{n})$ approximation guarantee.

Aazami and Stilp also describe a simple algorithm with an approximation guarantee of $O(\frac{n}{\log n})$ for the PDS problem. Their algorithm can be described as follows. Partition the vertices of a graph G into $k = \log n$ equal-sized sets $\{V_1, V_2, \ldots, V_k\}$. Then consider all possible ways of selecting some nonempty collection of these k sets. For each nonempty collection of these k sets, form the union of all vertices contained in these sets, and then, in polynomial time, check to see if this union is a power dominating set. Among all of these different collections of k sets, output one that power dominates G and has the minimum total number of vertices. Note that in this algorithm, we can consider at most $2^k = 2^{\log n} = n$ different collections of the k sets $V_1, V_2, \ldots, V_k$. Clearly, this algorithm runs in polynomial time, since it can be tested in $O(|V| + |E|)$ time whether any set of vertices is a power dominating set. Let S^* be a minimum power dominating set, of cardinality $\gamma_P(G) = |S^*|$. It is easy to see that the union of the subsets in $V_1, V_2, \ldots, V_k$ that intersect S^* is a power dominating set of G and that the number of vertices in this union of sets is at most $\frac{n}{\log n}|S^*|$. This establishes the approximation guarantee of $O(\frac{n}{\log n})$.

In 2010 [2] Aazami introduces a variation of the power domination problem [also introduced by Liao, but as yet unpublished in his 2009 PhD Thesis [32]], which involves an integer, which we will denote r, for the number of *rounds* of propagation permitted before all vertices must be observed. Thus, for a graph of order n, $1 \le r \le n - 1$. The DOMINATING SET problem corresponds to $r = 1$, while the POWER DOMINATING SET (PDS) problem corresponds to $r = n - 1$. In PDS the goal is to find a minimum cardinality set of vertices S that power dominates all vertices $v \in V$, where a node v is power dominated if (1) $v \in S$ or it has a neighbor in S or (2) v has a neighbor u such that u and all of its neighbors except v are power dominated. Rule (1) is the DOMINATING SET problem, and Rule (2) is a propagation rule that applies iteratively. The *r-round PDS problem*, or *r*PDS, has the same set of rules as PDS, except that Rule (2) is applied in parallel to all vertices that are newly observed in the current *round*. The requirement is to find a minimum cardinality set S such that all vertices can be observed in at most r rounds. The *r-round power domination number* $\gamma_{rP}(G)$ equals the minimum cardinality of an *r*-round power dominating set in G.

The parameter $\gamma_{rP}(G)$ is a power dominating version of the *distance-r domination number*, denoted $\gamma_{\le r}(G)$, which equals the minimum cardinality of a set $S \subseteq V$

such that for every vertex $w \in V - S$, $d(w, S) \le r$, that is, every vertex $w \in V - S$ is at distance at most r from at least one vertex in S.

Aazami provides a proof that $rPDS$ is NP-hard by means of a simple modification of the NP-hardness proof of PDS provided by both Guo et al. [23] and Kneis et al. [30], in which they attach a single leaf vertex to every vertex in a planar graph G.

Theorem 10 (Aazami) *For any $r \ge 1$, $rPDS$ is NP-hard for planar graphs.*

Proof Sketch. Use a transformation from PLANAR DOMINATING SET. Given a planar graph $G = (V, E)$, construct a planar graph G' by attaching a path of length $r - 1$ to every vertex $v \in V(G)$. One can show that G has a dominating set of cardinality at most k if and only if G' has an r-round power dominating set of cardinality at most k.

Given a dominating set $S \subseteq V(G)$ of cardinality at most k, after the first round, S will dominate, or observe, all vertices in $V(G)$. Then, in the next $r - 1$ rounds, all vertices on all paths attached to the vertices $v \in V(G)$ will be observed. Thus, the same set S is an r-round power dominating set of G' of cardinality at most k.

Conversely, if S' is an r-round power dominating set of G', it is easy to see that there is another r-round power dominating set S'' of the same cardinality as S' which contains no vertices on any path attached to the vertices $v \in V(G)$. Assume that this set S'' is not a dominating set of G, and let $w \in V(G)$ be any vertex not dominated by S'' in the first round. This means that at best v is power dominated in the second round. But this implies that it will take at least $r + 1$ rounds to power dominate the last vertex on the path of length $r - 1$ attached to vertex w, a contradiction.　　□

As shown previously with the Guo et al. [23] construction for power dominating set, the above construction holds for bipartite graphs, circular graphs, chordal graphs, and chordal bipartite graphs. Therefore we have the following corollary.

Corollary 3 *For any $r \ge 1$, $rPDS$ is NP-hard for bipartite, circular, chordal, and chordal bipartite graphs.*

Aazami shows that $rPDS$, or $\gamma_{rP}(G)$, cannot be approximated better than $2^{\log^{1-\epsilon} n}$ even for $r = 4$ in general graphs. He provides a dynamic programming algorithm to compute the value $\gamma_{rP}(G)$ in polynomial time for graphs of bounded treewidth. He also presents a PTAS (polynomial time approximation scheme) for $rPDS$ on planar graphs for $r = O(\frac{\log n}{\log \log n})$. Finally, he gives an integer programming formulation of r-round PDS.

In 2011 [7] Barrera and Ferrero provide upper bounds for the power domination numbers of cylinders $P_m \square C_n$, and exact values of the power domination numbers of toroidal grid graphs $C_m \square C_n$ and some generalized Petersen graphs.

In 2016 [31] Liao continues the development of r-round power domination by presenting linear algorithms for computing $\gamma_{rP}(G)$ for trees and for block graphs.

In 2010 [38] Pai, Chang, and Wang consider a variation of power domination in which PMUs may only be placed within a restricted subset of the vertices V of a graph, called a *forbidden zone* Z. Thus, the parameter $\gamma_P(G, Z)$ equals the minimum cardinality of a power dominating set S such that $S \cap Z = \varnothing$. This leaves open the possibility, of course, that for some restricted sets Z, no restricted power dominating set may exist. As an illustration, they present algorithms to solve this restricted type of power domination on grids, under the restriction that only certain consecutive rows or columns form a forbidden zone.

Pai et al. [38] also introduce another variation of power domination, as follows. Given a graph $G = (V, E)$ and an integer k, with $0 \le k \le |V|$, a set $S \subseteq V$ is called a *k-fault-tolerant power dominating set*, or *kFPDS*, of G if $S - F$ is still a PDS of G for any subset $F \subset S$ with $|F| \le k$. The *k-fault-tolerant power domination number* of G, denoted $\gamma_{kFP}(G)$, equals the minimum cardinality of a *kFPDS* of G. Notice that from this definition, it follows that:

(i)   for any $k \ge 0$, $\gamma_{kFP}(G) \le \gamma_{k+1FP}(G)$,
(ii) $\gamma_{0FP}(G) = \gamma_P(G)$,
(iii) $\gamma_P(G) + k \le \gamma_{kFP}(G)$, and
(iv) if G contains $k + 1$ mutually disjoint $\gamma_P(G)$-sets, then the union of these $k + 1$ sets can form a *kFPDS*, which implies that $\gamma_{kFP}(G) \le (k + 1)\gamma_P(G)$.

Pai et al. establish the following results for $1 \times n$, $2 \times n$, and $3 \times n$ grid graphs.

Proposition 2 (Pai, Chang, Wang) *For* $G_{1,n}$, $G_{2,n}$, $G_{3,n}$, $G_{4,n}$, *and* $G_{5,n}$,

(1) $\gamma_{1FP}(G_{1,n}) = 2$; *let* $S = \{(1, 1), (1, n)\}$,
(2) $\gamma_{1FP}(G_{2,n}) = 2$; *let* $S = \{(1, 1), (1, n)\}$,
(3) $\gamma_{1FP}(G_{3,n}) = 2$; *let* $S = \{(2, 1), (2, n)\}$,
(4) $\gamma_{1FP}(G_{4,n}) = 3$; *let* $S = \{(2, 1), (3, 3), (4, 1)\}$,
(5) $\gamma_{1FP}(G_5,)n = 3$; *let* $S = \{(2, 1), (3, 3), (4, 1)\}$.

Pai et al. conclude by presenting placement algorithms that do the following:

(i)   approximate $\gamma_{1FP}(G_{m,n})$ within a factor of 1.60 for $6 \le m \le n$,
(ii) approximate $\gamma_{2FP}(G_{m,n})$ within a factor of 2.34 for $7 \le mn$,
(iii) approximate $\gamma_{3FP}(G_{m,n})$ within a factor of 3.34 for $11 \le m \le n$.

Figure 7 presents two examples of one-fault power dominating sets, one in a $15 \times n$ grid graph and the other in a $17 \times n$ grid graph.

In 2012 [15] Chang, Dorbec, Montassier, and Raspaud introduce the concept of k-power domination, a direct generalization of power domination, by changing in a

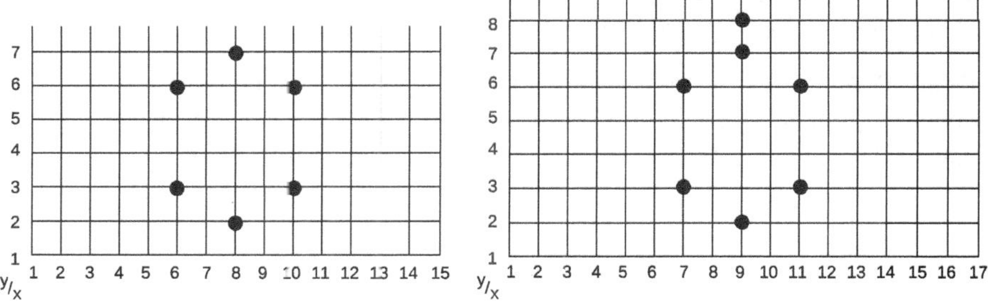

Fig. 7 One-fault power dominating sets (figure reproduced from [38])

natural way Rule 2 (Kirchhoff's Rule), sometimes called the *propagation rule*, as follows, for some fixed nonnegative integer k:

Rule B1. If a vertex $v \in S$, then v and all vertices in $N(v)$ are observed.

Rule B2. (Kirchhoff's Rule) If a vertex v is observed and there is a vertex $u \in N(v)$ that is the only unobserved vertex in $N(v)$, then vertex u is observed.

Rule B3. (Chang, Dorbec, Montassier, Raspaud) If a vertex v is observed and there are at most k vertices in $N(v)$ that are unobserved, then all vertices in $N(v)$ are observed.

This can also be stated as follows:

For a connected graph G and a vertex set $S \subseteq V$, the set $M_k(S)$ of vertices k-observed, or k-monitored, by S is defined recursively as follows:

(1) $M_k(S) \leftarrow S \cup N(S)$ (v and all of its neighbors in $N(v)$ are k-monitored),
(2) While there is a vertex $v \in M_k(S)$ having at most k unmonitored neighbors, that is, $|N(v) \cap (V(G) - M(S))| \leq k$, set $M_k(S) \leftarrow M_k(S) \cup N(v)$.

A set S is called a *k-power dominating set* of G if $M_k(S) = V(G)$. The *k-power domination number* $\gamma_{kP}(G)$ is the minimum cardinality of a k-power dominating set of G.

One can observe that when $k = 0$, $\gamma_{0P}(G) = \gamma(G)$, and when $k = 1$, $\gamma_{1P}(G) = \gamma_P(G)$.

The authors quickly establish an upper bound for $\gamma_{kP}(G)$ which generalizes known results for $k = 0$ and $k = 1$.

Theorem 11 (Chang, Dorbec, Montassier, Raspaud) *For any connected graph G of order $n \geq k + 2$,*

$$\gamma_{kP}(G) \leq n/(k+2)$$

and this bound is best possible.

Chang et al. provide one complexity result and one algorithm.

KPDS
Instance: Graph $G = (V, E)$, positive integer t.
Question: Does G have a k-power dominating set of cardinality at most t?

Theorem 12 (Chang, Dorbec, Montassier, Raspaud) *KPDS is NP-complete for chordal graphs and bipartite graphs.*

Proof Sketch. For any graph G and any nonnegative integer k, let G_k be the graph obtained from G by attaching k leaves to every vertex $v \in V(G)$. It is easy to show that $\gamma_{kP}(G_k) = \gamma(G)$. Furthermore, if G is chordal or bipartite, so is G_k. $\qquad \square$

Theorem 13 (Chang, Dorbec, Montassier, Raspaud) *There is a linear algorithm for computing the value $\gamma_{kP}(T)$ for any tree T.*

Proof Sketch. The algorithm is based on the method used to compute the value of $\gamma(T)$ for any tree T given by Mitchell, Cockayne, and Hedetniemi in [35] and Cockayne, Goodman, and Hedetniemi in [16]. Assign two values $L(v) = (a_v, b_v)$ to each vertex v in a rooted tree T, where $a_v \in \{B, F, R\}$ and $b_v \in \{0, 1, \ldots, k\}$. The value $a_v = R$ (required) means that the vertex will be a vertex in the minimum k-power dominating set; the value $a_v = F$ (free) means that the vertex has been monitored/observed; and the value $a_v = B$ (bound) means that the vertex has not been monitored. The second label b_v is only applied to vertices labeled F or B and records the number of neighbors of v, which is at most k, that may yet be monitored once v is monitored. The algorithm works on a rooted tree from the leaves up to the root and is based on the following five rules for a leaf vertex x and its parent vertex y, where $T' = T - x$ is the tree remaining after leaf x has been deleted and where initially all vertices v are assigned the label $L(v) = (B, k)$.

(1) if $a_x = R$, then $\gamma_{kP}(T) = 1 + \gamma_{kP}(T')$ and change $a_y = F$ if $a_y = B$.
(2) if $(a_y = R)$ or $(a_x = F$ and $b_z = 0)$, then $\gamma_{kP}(T) = \gamma_{kP}(T')$.
(3) if $a_x = B$ and $b_y > 0$, then $\gamma_{kP}(T) = \gamma_{kP}(T')$, where $b_y = b_y - 1$ in T'.
(4) if $a_x = B$ and $b_y = 0$, then $\gamma_{kP}(T) = \gamma_{kP}(T')$, where $a_y = R$ in T'.
(5) otherwise, if $a_x = F$, $b_x > 0$ and $a_y \neq R$, then $\gamma_{kP}(T) = \gamma_{kP}(T')$, where $a_y = F$ in T'.

Chang et al. illustrate their algorithm with the example in Figure 8.

In 2013 [18] Dorbec, Henning, Löwenstein, Montassier, and Raspaud continue the development of k-power domination by applying it to the consideration of

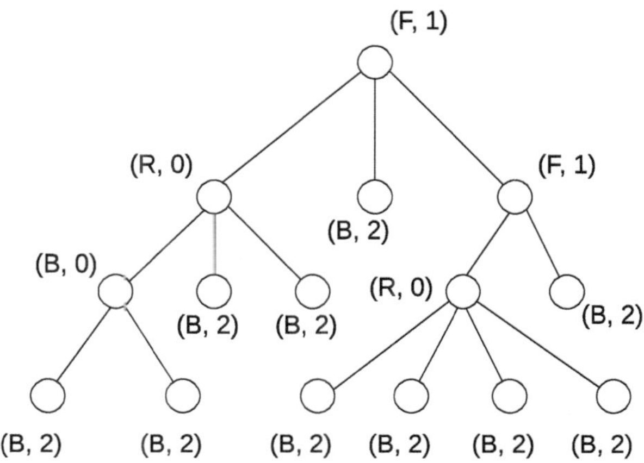

Fig. 8 2-power domination in trees (figure reproduced from [15])

regular graphs. The main result of their paper is a 12-page proof of the following theorem.

Theorem 14 (Dorbec, Henning, Löwenstein, Montassier, Raspaud) *For $k \geq 1$ and G a connected $(k+2)$-regular graph of order n, if $G \neq K_{k+2,k+2}$, then $\gamma_{kP}(G) \leq \frac{n}{(k+3)}$, and this bound is tight.*

They conclude with the following conjecture.

Conjecture 1 (Dorbec, Henning, Löwenstein, Montassier and Raspaud) *For $k \geq 1$ and $r \geq 3$, if $G \neq K_{r,r}$ is a connected r-regular graph of order n, then $\gamma_{kP}(G) \leq \frac{n}{(r+1)}$.*

In 2016 [41] Wang, Chen, and Lu generalize the linear, k-power domination algorithm by Chang, Dorbec, Montassier, and Raspaud [15] for computing $\gamma_{kP}(G)$ for trees, to an $O(|V| + |E|)$ algorithm for computing $\gamma_{kP}(G)$ for block graphs.

In 2013 [34] Liao and Lee give an 11-page presentation of a linear algorithm for computing $\gamma_P(G)$ for interval graphs, if the interval ordering of the graph is provided as input. In addition, they show that if the interval ordering is not given, their algorithm runs in $\theta(n \log n)$ time, where n is the number of intervals, and is asymptotically optimal.

Liao and Lee also give a 7-page presentation in which they extend their methods to the class of circular arc graphs. They construct a linear algorithm for computing $\gamma_P(G)$, first for proper circular arc graphs, in which no arc is properly contained within another arc, and then for general circular arc graphs, provided the circular arc endpoints are sorted.

In 2015 [40] Stephen, Rajan, Ryan, Grigorious, and William apply power domination to the graphs of the chemical compounds of polyphenylene, dendrimers (cf. Figure 9), rhenium trioxide (cf. Figure 10), and silicate networks (cf. Figure 11). The rhenium trioxide (ReO_3) graphs, cf. Figure 8, are essentially subdivision graphs of three-dimensional grid graphs, that is, the graphs $S(P_p \square P_q \square P_r)$, but denoted by the authors as $RO(p, q, r)$, where the subdivision vertices, the vertices of degree 2, represent oxygen atoms and the vertices of degree 3 represent rhenium atoms.

Although the authors do not solve the problem of computing the power domination number of rhenium trioxide lattices, they do provide several interesting observations.

Theorem 15 (Stephen, Rajan, Ryan, Gregorious, William) *For any rhenium trioxide graph $RO(p, q, r)$, the following must hold:*

(i) *Each unit cell in $RO(p, q, r)$ must contain at least three vertices of any power dominating set.*

(ii) *The three vertices in every unit cell of $RO(p, q, r)$ belonging to a power dominating set cannot all lie on the same face.*

(iii) *Every face in each unit cell must have at least one vertex in any power dominating set.*

(iv) *In any power dominating set of $RO(p, q, r)$, every unit cell must contain at least two rhenium atoms (vertices of degree 3).*

Fig. 9 Polyphenylene
dendrimers (figure
reproduced from Stephen et
al. [40])

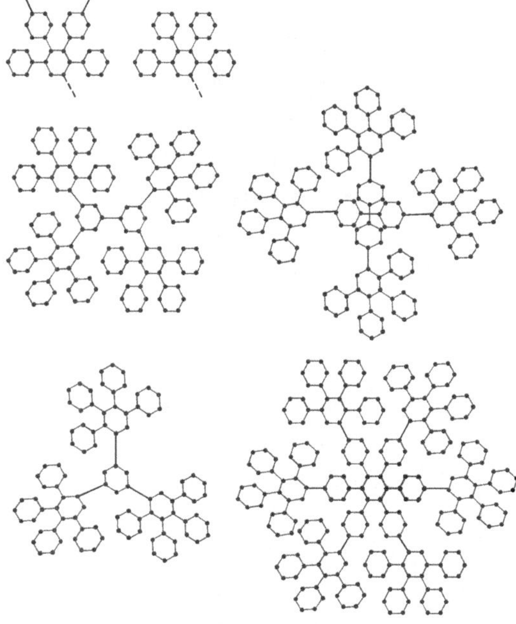

Fig. 10 Rhenium trioxide
(figure reproduced from
Stephen at el. [40])

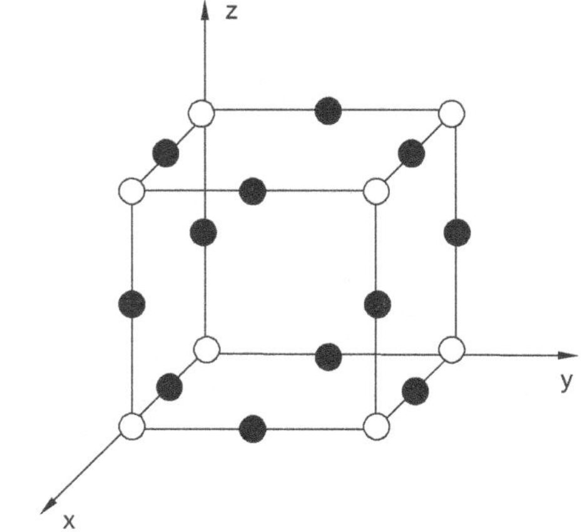

The graphs of *silicate networks* $SL(n)$ are obtained from the graphs of *honey-comb* networks $HC(n)$ as follows. The graphs $HC(n)$ are defined recursively: (i) $HC(1) = C_6$, (ii) $HC(2)$ is obtained from $HC(1)$ by attaching a layer of six hexagons to the outer edges of $HC(1)$, and (iii) $HC(n)$ is obtained from $HC(n-1)$ by attaching hexagons to the outer edges of $HC(n-1)$.

We should interject here that the power domination number of these honeycomb networks $HC(n)$ were determined by Ferrero and Varghese in 2011 [22] to be: $\gamma_P(HC(n)) = \lceil \frac{2n}{3} \rceil$.

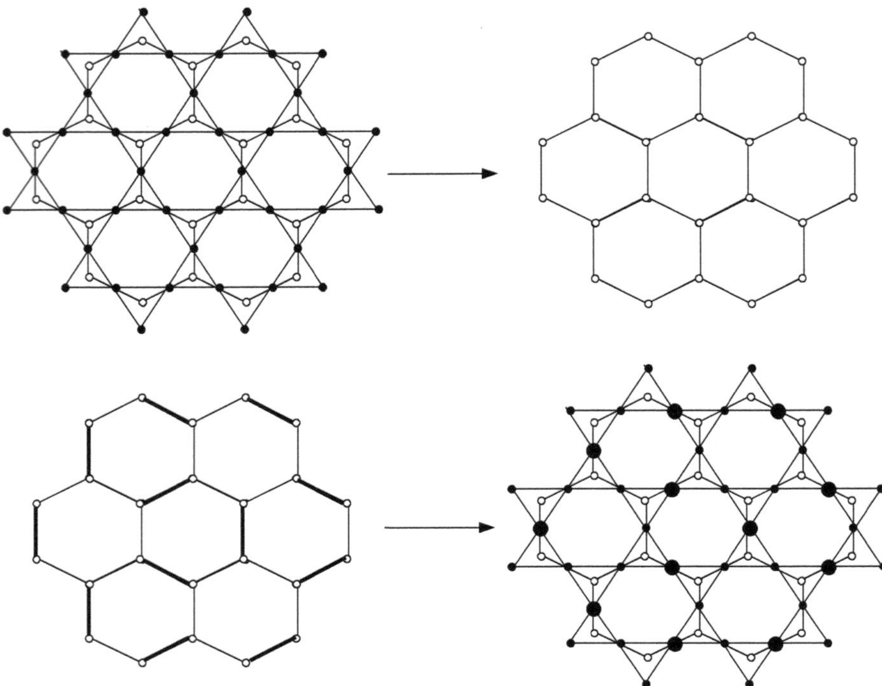

Fig. 11 Silicate networks (figure reproduced from Stephen et al. [40])

In order to construct a silicate network $SL(n)$, to each vertex in $HC(n)$, assign a silicon ion. Then subdivide every edge of $HC(n)$, and assign to each of these subdivision vertices an oxygen ion. To each of the $6n$ degree-2 vertices on the outer face of $HC(n)$, attach a leaf and assign an oxygen atom to these $6n$ leaves. Finally, add edges to form a triangle of oxygen ions surrounding each silicon ion, thereby creating a tetrahedron, the center of which is a silicate ion, cf. Figure 11. The resulting graph is the silicate network $SL(n)$, which has $15n^2 + 3n$ vertices, $36n^2$ edges, and diameter $4n$.

The algorithm for constructing a minimum power dominating set of a silicate network $SL(n)$ becomes very simple, as follows:

Step 1. In each bounding cycle of $HC(i)$, for $1 \le i \le n$, choose alternate edges. [In $HC(1)$ choose 3 alternating edges; in $HC(2)$ choose 12 alternating edges; etc.]

Step 2. For every $1 \le i \le n$, choose the oxygen ion vertices in $SL(i)$ that subdivide the edges chosen in the bounding cycle of $HC(i)$.

The authors then show that the $3n^2$ vertices so chosen form a minimum cardinality power dominating set of $SL(n)$.

Theorem 16 (Stephen et al.) *For every silicate network $SL(n)$, $\gamma_P(SL(n)) = 3n^2$.*

In 2017 [28] Kang and Wormald present two heuristics for finding a small power dominating set in a random cubic graph. They analyze the performance of these heuristics on random cubic graphs using differential equations. In this way, they prove that the proportion of vertices in a minimum power dominating set of a random cubic graph is asymptotically almost surely at most 0.067801. They also provide a corresponding lower bound of 1/29.7, which is approximately 0.03367, using known results on bisection width.

In this paper, lower and upper bounds for $\gamma_P(G)$ are given for a random cubic graph G.

For the lower bound, Kang and Wormald first prove that $\gamma_P(G) \geq bw(G) - 13$ for any cubic graph G, where $bw(G)$ denotes the bisection width of G, which is defined as $min\{|\partial S| : S \subset V(G)\}, |S| = |V(G)|/2\}$. Coupled with a result of Kostochka and Melnikov [29], this gives $\gamma_p(G) > 0.03367n$ asymptotically almost surely for an n-vertex random cubic graph G, as $n \rightarrow \infty$.

For upper bounds, Kang and Wormald present two greedy algorithms that find power dominating sets and analyze them, again using the differential equation method. The analysis of the second algorithm gives the main upper bound: asymptotically almost surely, $\gamma_p(G) \leq 0.067801n$ for a random cubic graph G of order n.

In 2018 [8] Benson, Ferrero, Flagg, Furst, Hogben, Vasilevska, and Wissman discuss the close connection between the power domination number, what is called the *zero forcing number* $Z(G)$ of a graph G, and the *maximum nullity* of G. We will need a few definitions.

The concept of *zero forcing* can be explained in terms of a process of coloring the vertices of a graph G. Initially some subset $S \subseteq V$ of vertices are colored, say blue, and all vertices in $V - S$ are colored white. Then, just as in the power domination propagation rule, if u is a blue vertex and exactly one neighbor $v \in N(u)$ is white, then the color of v changes to blue. In this way vertex u *forces* vertex v to change color. This can be denoted by $u \rightarrow v$ and is called a *zero forcing rule*. A set S is called a *zero forcing set* of G if after coloring all vertices in S blue, repeated applications of the zero forcing rule result in all vertices being colored blue. The *zero forcing number*, $Z(G)$, equals the cardinality of a zero forcing set in G.

It is easy to see that the power domination process in a graph G can be described as choosing a set $S \subseteq V$ and applying the zero forcing process to the closed neighborhood $N[S]$ of S. Thus, as first observed by Aazami [1], a set S is a power dominating set of a graph G if and only if $N[S]$ is a zero forcing set of G.

Notice also that for any graph G with minimum degree $\delta(G)$, $\delta(G) \leq Z(G)$.

Next, let $S_n(\mathcal{R})$ denote the set of all $n \times n$ real symmetric matrices. For $A = [a_{ij}] \in S_n(\mathcal{R})$, the graph of A, denoted by $G(A)$, is the graph with vertices $V = \{v_1, v_2, \ldots, v_n\}$ and edges $\{v_i v_j : a_{ij} \neq 0, 1 \leq i < j \leq n\}$. Note that the diagonal of A is ignored in defining $G(A)$.

The set of symmetric matrices described by a graph G of order n is defined as $S(G) = \{A \in S_n(\mathcal{R}) : G(A) = G\}$. The *maximum nullity* of G is $M(G) = max\{nullA : A \in S(G)\}$, where $nullA$ is the dimension of the null space of A, and the *minimum rank* of G is $mr(G) = min\{rankA : A \in S(G)\}$, where $rankA$ is the dimension of the column space of A. By definition, $M(G) + mr(G) = |V(G)|$.

The term *zero forcing* comes from the process of forcing zeros in a null vector of a matrix $A \in S(G)$. Thus, we have the basic inequality, as observed in [4]:

Proposition 3 (AIM) *For any graph G, $M(G) \leq Z(G)$.*

Starting from this basic inequality, Benson et al. prove the following.

Theorem 17 (Benson et al.) *For any non-empty graph G, $\frac{Z(G)}{\Delta(G)} \leq \gamma_P(G)$.*

Having established a connection between the power domination number and the zero forcing number, the authors mention an interesting variation of the power domination propagation rule, or the zero forcing rule, called a *skew zero forcing rule*, first introduced in [27] by the IMA-ISU research group on minimum rank.

In the skew zero forcing rule, a vertex u can force a neighbor $v \in N(u)$ to change color from white to blue if v is the only neighbor of u colored white. But it is permitted that the color of blue itself can be white, whereas in the zero forcing rule, the color of vertex u must be blue. This modified forcing rule then gives rise to the *skew zero forcing number*, $Z^-(G)$, which equals the minimum cardinality of a skew zero forcing set, that is, a set S of vertices that can force all vertices $v \in V$ to be colored blue using the skew zero forcing rule. This variation, essentially in power domination, seems worth studying.

In 2019 [11] Bozeman, Brimkov, Erickson, Ferrero, Flagg, and Hogben consider the problem of determining the minimum number of additional PMUs needed to observe a power network when the network is expanded, but the existing devices S remain in place. They also study the related problem of finding the smallest zero forcing set that must contain a given set of vertices S. The sizes of such sets in a graph G are, respectively, called the *restricted power domination number* and the *restricted zero forcing number* of G subject to S, which can be denoted $\gamma_P(G, S)$ and $Z(G, S)$.

Bozeman et al. present a linear algorithm for computing $\gamma_P(G, S)$ on graphs with bounded treewidth.

Theorem 18 *For any graph $G = (V, E)$ of order n and bounded treewidth, and any set $S \subseteq V$, a minimum power dominating set of G containing S can be computed in $O(n)$ time.*

In 2019 [12] Brimkov, Mikesell, and Smith consider the problem of finding a minimum power dominating set in which the subgraph $G[S]$ induced by the initial set S of vertices is connected. The minimum cardinality of such a power dominating set is called the *connected power domination number*, denoted $\gamma_{cP}(G)$. They show that the connected power domination problem (CPDS) is NP-hard for arbitrary graphs, but can be computed in linear time for trees, cactus graphs, and block graphs.

Tables 1 and 2 and Figure 12 summarize the complexity results for power domination and the several variations of power domination discussed in this chapter.

Table 1 Complexity results for power dominating set

NP-complete	Polynomial
Bipartite graphs	Trees
Haynes et al. 2002 [25], Guo et al. 2005, 2008 [23, 24]	Haynes et al. 2002 [25], Guo et al. 2005 [23]
Chordal graphs	Graphs with bounded treewidth
Haynes et al. 2002 [25]	Guo et al. 2008 [24], Kneis et al. 2006 [30]
Split graphs	Interval graphs
Guo et al. 2005, 2008 [23, 24], Liao and Lee 2005 [33]	Liao and Lee 2005, 2013 [33, 34]
Planar graphs	Proper circular arc graphs
Guo et al. 2005, 2008 [23, 24]	Liao and Lee 2005, 2013 [33, 34]
Planar bipartite graphs	Circular arc graphs
Brueni and Heath 2005 [14]	Liao and Lee 2013 [34]
Circle graphs	Block graphs
Guo et al. 2005, 2008 [23, 24]	Xu et al. 2006 [42]
Planar cubic graphs	Block-cactus graphs
Raible and Fernau 2008, 2012 [9, 39]	Hon et al. 2007 [26]
	Grid graphs
	Dorfling and Henning 2006 [19]
	Silicate networks
	Stephen et al. 2015 [40]

Table 2 Complexity results for variations of power domination

	NP-complete	Polynomial
r-PDS	Planar, bipartite, chordal, circle graphs	Bounded treewidth, trees, block graphs
	Aazami 2010 [2]	Liao 2016 [31]
k-PDS	Chordal, bipartite, circle, planar, chordal bipartite graphs	Trees
	Chang et al. 2012 [15]	Chang et al. 2012 [15]
		Block graphs
		Wang et al. 2016 [41]
PDS with forbidden zone		Bounded treewidth
		Bozeman et al. 2019 [11]
c-PDS	Arbitrary graphs	Trees
	Brimkov et al. 2019 [12]	Bozeman et al. 2019 [11]
		Cactus, block graphs
		Brimkov et al. 2019 [12]

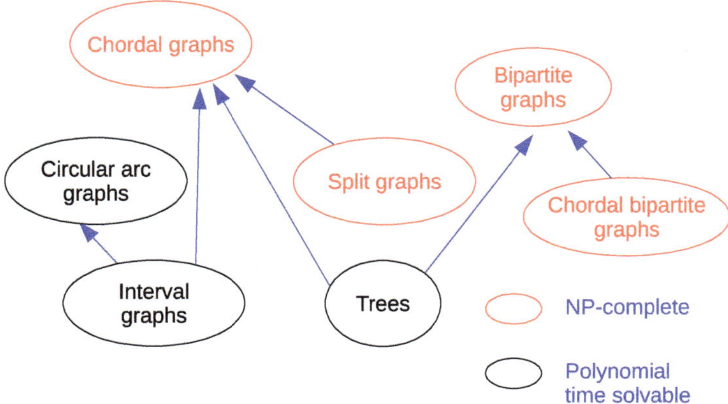

Fig. 12 Summary of power domination results

References

1. A. Aazami, Hardness results and approximation algorithms for some problems on graphs. PhD Thesis, University of Waterloo (2008)
2. A. Aazami, Domination in graphs with bounded propagation: algorithms, formulations and hardness results. J. Combin. Optim. **19**(4), 429–456 (2010)
3. A. Aazami, K. Stilp, Approximation algorithms and hardness for domination with propagation. SIAM J. Discrete Math. **23**(3), 1382–1399 (2009)
4. AIM Minimum Rank Special Graphs Work Group, F. Barioli, W. Barrett, S. Butler, S.M. Cioaba, D. Cvetkovic, S.M. Fallat, C. Godsil, W. Haemers, L. Hogben, R. Mikkelson, S. Narayan, O. Pryporova, I. Sciriha, W. So, D. Stevanovic, H. van der Holst, K. Vander Meulen, A. Wangsness, Zero forcing sets and the minimum rank of graphs. Linear Algebra Appl. **428**, 1628–1648 (2008)
5. D. Atkins, T.W. Haynes, M.A. Henning, Placing monitoring devices in electric power networks modeled by block graphs. Ars Combin. **79**, 129–143 (2006)
6. T.L. Baldwin, L. Mili, M.B. Boisen Jr., R. Adapa, Power system observability with minimal phasor measurement placement. IEEE Trans. Power Syst. **8**, 707–715 (1993)
7. R. Barrera, D. Ferrero, Power domination in cylinders, tori, and generalized Petersen graphs. Networks **58**, 43–49 (2011)
8. K.F. Benson, D. Ferrero, M. Flagg, V. Furst, L. Hogben, V. Vasilevska, B. Wissman, Zero forcing and power domination for graph products. Australas. J. Combin. **70**, 221–235 (2018)
9. D. Binkele-Raible, H. Fernau, An exact exponential time algorithm for power dominating set. Algorithmica **63**(1–2), 323–346 (2012)
10. M.B. Boisen, Jr., T.L. Baldwin, L. Mili, Simulated annealing and graph theory applied to electrical power networks. Manuscript (2000)
11. C. Bozeman, B. Brimkov, C. Erickson, D. Ferrero, M. Flagg, L. Hogben, Restricted power domination and zero forcing problems. J. Combin. Optim. **37**(3), 935–956 (2019)
12. B. Brimkov, D. Mikesell, L. Smith, Connected power domination in graphs. J. Combin. Optim. **38**(1), 292–315 (2019)
13. D.J. Brueni, Minimal PMU placement for graph observability: a decomposition approach. Master's thesis, Virginia Polytechnic Institute and State University (1993)

14. D.J. Brueni, L.S. Heath, The PMU placement problem. SIAM J. Discrete Math. **19**(3), 744–761 (2005)
15. G.J. Chang, P. Dorbec, M. Montassier, A. Raspaud, Generalized power domination of graphs. Discrete Appl. Math. **160**(12), 1691–1698 (2012)
16. E.J. Cockayne, S.E. Goodman, S.T. Hedetniemi, A linear algorithm for the domination number of a tree. Inform. Process Lett. **4**, 41–44 (1975)
17. P. Dorbec, Power domination in graphs, in *Topics in Domination*, ed. by T.W. Haynes, S.T. Hedetniemi, M.A. Henning (Springer, Berlin, 2020)
18. P. Dorbec, M.A. Henning, C. Löwenstein, M. Montassier, A. Raspaud, Generalized power domination in regular graphs. SIAM J. Discrete Math. **27**, 1559–1574 (2013)
19. M. Dorfling, M.A. Henning, A note on power domination in grid graphs. Discrete Appl. Math. **154**(6), 1023–1027 (2006)
20. R.G. Downey, M.R. Fellows, *Parameterized Complexity* (Springer, Berlin, 1999)
21. U. Feige, A threshold of lnn for approximating set cover. J. Assoc. Comput. Mach. **45**, 634–652 (1998)
22. D. Ferrero, S. Varghese, A. Vijayakumar, Power domination in honeycomb networks. J. Discrete Math. Sci. Cryptogr. **14**(6), 521–529 (2011)
23. J. Guo, R. Niedermeier, D. Raible, Improved algorithms and complexity results for power domination in graphs, in *Fundamentals of Computation Theory*. Lecture Notes in Computer Science, vol. 3595 (Springer, Berlin, 2005), pp. 172–184
24. J. Guo, R. Niedermeier, D. Raible, Improved algorithms and complexity results for power domination in graphs. Algorithmica **52**(2), 177–202 (2008)
25. T.W. Haynes, S.M. Hedetniemi, S.T. Hedetniemi, M.A. Henning, Domination in graphs applied to electric power networks. SIAM J. Discrete Math. **15**(4), 519–529 (2002)
26. W.-K. Hon, C.-S. Liu, S.-L. Peng, C.-Y. Tang, Power domination on block-cactus graphs, in *Proceedings of the 24th Workshop Combinatorial Mathematics and Computation Theory* (2007), pp. 280–284
27. IMA-ISU research group on minimum rank, M. Allison, E. Bodine, L.M. DeAlba, J. Debnath, L. DeLoss, C. Garnett, J. Grout, L. Hogben, B. Im, H. Kim, R. Nair, O. Pryporova, K. Savage, B. Shader, A. Wangsness Wehe, Minimum rank of skew-symmetric matrices described by a graph. Linear Algebra Appl. **432**, 2457–2472 (2010)
28. L. Kang, N. Wormald, Minimum Power dominating sets of random cubic graphs. J. Graph Theory **85**(1), 152–171 (2017)
29. A.V. Kostochka, L.S. Melnikov, On a lower bound for the isoperimetric number of cubic graphs, in *Probabilistic Methods in Discrete Mathematics* (1993), pp. 251–265
30. J. Kneis, D. Mölle, S. Richter, P. Rossmanith, Parameterized power domination complexity. Inform. Process. Lett. **98**, 145–149 (2006)
31. C.-S. Liao, Power domination with bounded time constraints. J. Combin. Optim. **31**(2), 725–742 (2016)
32. C.-S. Liao, Graph-theoretic domination and related problems with applications. Ph.D. Thesis, Dept. Computer Science and Information Engineering, National Taiwan University, Taiwan (2009)
33. C.-S. Liao, D.-T. Lee, Power domination problem in graphs, in *Computing and Combinatorics*. Lecture Notes in Computer Science, vol. 3595 (Springer, Berlin, 2005), pp. 818–828
34. C.-S. Liao, D.T. Lee, Power domination in circular-arc graphs. Algorithmica **65**(2), 443–466 (2013)
35. S.L. Mitchell, E.J. Cockayne, S.T. Hedetniemi, Linear algorithms on recursive representations of trees. J. Comput. Syst. Sci. **18**, 76–85 (1979)
36. H. Müller, A. Brandstädt, The NP-completeness of Steiner tree and dominating set for chordal bipartite graphs. Theoret. Comput. Sci. **53**, 257–265 (1987)
37. K.-J. Pai, J.-M. Chang, Y.-L. Wang, A simple algorithm for solving the power domination problem on grid graphs, in *Proceedings of the 24th Workshop Combinatorial Mathematics and Computation Theory* (2007), pp. 256–260

38. K.-J. Pai, J.-M. Chang and Y.-L. Wang, Restricted power domination and fault-tolerant power domination on grids. Discrete Appl. Math. **158**(10), 1079–1089 (2010)
39. D. Raible, H. Fernau, Power domination in $O(1.7548^n)$ using reference search trees, in *Algorithms and Computation, Lecture Notes in Computer Science*, vol. 5369 (Springer, Berlin, 2008), pp. 136–147
40. S. Stephen, B. Rajan, J. Ryan, C. Grigorious, A. William, Power domination in certain chemical structures. J. Discrete Algorithm. **33**, 10–18 (2015)
41. C. Wang, L. Chen, C. Lu, k-Power domination in block graphs. J. Combin. Optim. **31**(2), 865–873 (2016)
42. G. Xu, L. Kang, E. Shan, M. Zhao, Power domination in block graphs. Theoret. Comput. Sci. **359**(1–3), 299–305 (2006)
43. M. Zhao, L.Y. Kang, G.J. Chang, Power domination in graphs. Discrete Math. **306**, 1812–1816 (2006)

Self-Stabilizing Domination Algorithms

Stephen T. Hedetniemi

1 Introduction

In this chapter, we introduce the elegantly simple, self-stabilizing algorithm model to researchers having an interest in domination in graphs. In 1974 [12, 13], Dijkstra introduced the algorithm paradigm called *self-stabilizing algorithms* as a special case of distributed algorithms. But algorithms of this type were not studied and developed until the late 1980s, and it was not until the early 2000s that self-stabilizing domination algorithms began to appear.

In this chapter, we present the basic framework and definitions of self-stabilizing algorithms. An in-depth treatment of self-stabilizing algorithms is given in the book by Dolev [16]. We then present self-stabilizing algorithms for finding in an arbitrary connected graph: (i) a maximal independent set, (ii) a maximal matching, (iii) a minimal dominating set, (iv) a minimal total dominating set, and (v) two disjoint minimal dominating sets. It is important to note at the outset that these algorithms are not designed to find either minimum or maximum sets having some domination property, only minimal or maximal sets. We then discuss a variety of other domination-related, self-stabilizing algorithms that have been published. Finally, we present a list of domination-related self-stabilizing algorithms that have yet to be designed.

S. T. Hedetniemi (✉)
School of Computing, Clemson University, Clemson, SC, USA
e-mail: hedet@clemson.edu

T. W. Haynes et al. (eds.), *Structures of Domination in Graphs*, Developments in Mathematics 66, https://doi.org/10.1007/978-3-030-58892-2_16

2 Self-Stabilizing Framework

In the self-stabilizing algorithm paradigm, we assume that a distributed computing system or computer network is modeled by a connected, undirected graph $G = (V, E)$, of *order* $n = |V|$ nodes or processors, and *size* $m = |E|$ edges or bidirectional communication links $\{u, v\}$ between pairs of nodes. If $\{u, v\} \in E$, we say that u and v are *neighbors*, and $N(u) = \{v : \{u, v\} \in E\}$ is the set of neighbors of node u, or the *open neighborhood* of u, while $N[u] = N(u) \cup \{v\}$ is the *closed neighborhood* of u. If $S \subseteq V$ is a set of vertices in a graph G, we let $\overline{S}$ (sometimes denoted $V - S$ or $V \setminus S$) denote the vertices not in set S.

2.1 *Program and Computation*

Every node, at all times, continues to execute the same program or self-stabilizing algorithm and in doing so maintains a set of variables common to all nodes. A node can only change the value of its own variables. The *state* of a node is defined by the vector of current values of all of its variables. The union of the states of all nodes in the graph/system defines the *global state* and constitutes the *current configuration* of the whole system.

The algorithm, which is being executed independently and simultaneously at every node of the system, consists of the same finite list of *rules*, called *guarded commands*, of the form,

$$Rule : Guard \rightarrow Action$$

or

$$Rule : \textbf{if..}\ Guard\ \textbf{..then..}\ Action,$$

where *Guard* is a Boolean expression involving some or all of the variables of the nodes in the closed neighborhood of a node u; this is called the *shared-variable model*. If this expression (Guard) is evaluated to be true, then node u is said to be *enabled* or *privileged* to execute the corresponding Action. This gives rise to two types of execution. In what is sometimes called *coarse scheduling*, both reading/expression evaluation and writing/making a move are done in one step, while in what is called *read/write atomicity*, two steps are required. A *move* by node u consists of the execution of the designated Action, which consists of changing the values of the variables at node u as specified by the Action.

Normally at most one rule at a node will be enabled at any moment, but if several rules are simultaneously enabled, only the Action in the first enabled rule in the list will be executed.

2.2 Distance-k Knowledge

In 2004 [19], Gairing, Goddard, Hedetniemi, Kristiansen, and McRae introduce the idea that self-stabilizing algorithms can be designed in which the rules have guards whose Boolean expressions involve some or all of the variables of the nodes within distance-k of the given node. It is shown in 2004 [19] and subsequently in 2008 [28] by Goddard, Hedetniemi, Jacobs, and Trevisan how to convert a distance-k algorithm to one in the distance-1 model, but it comes with an increased cost in running time, for example, a self-stabilizing algorithm, which stabilizes in $O(n^2)$ moves in the distance-2 model, will stabilize in $O(n^5)$ moves in the distance-1 model; see also Turau in 2012 [64].

2.3 Anonymous Systems

In an *anonymous* system, or network, nodes do not have unique identifiers, e.g., $ID(u) = ID(v)$, for all $v \in N(u)$, which means that the same rule applies equally to all nodes. By contrast, in *non-anonymous* networks, a rule can compare the identifier of a node u with the identifiers of nodes in its neighborhood $N(u)$, in order to determine if node u is enabled, for example, if $ID(u) > ID(v)$, then the node u may become enabled, otherwise node v may become enabled.

2.4 Schedulers

If, at any time, several nodes are enabled to make a move, a mechanism, called a *scheduler*, or an *adversarial daemon*, is assumed to determine, decide, or choose which node or nodes make the next move(s). In the *central scheduler model*, also called the *serial model*, one node is adversarially selected to make its move. In the *distributed* model, any number of enabled nodes can be adversarially selected to make their moves simultaneously, while in the *synchronous* model, all enabled nodes must make their moves simultaneously.

A further distinction can also be made between *fair* and *unfair* schedulers. With a fair scheduler, every node that is continuously enabled is eventually selected to make a move. With an unfair scheduler, there is no such condition.

2.5 Self-Stabilization

A *computation* c is a finite sequence of global configurations $c = c_0, c_1, \ldots, c_k$, where configuration c_i results from configuration c_{i-1} after all enabled nodes selected by the scheduler have made their next moves.

A configuration is said to be *stable* if no node is enabled. A self-stabilizing algorithm is said to be *stabilizing* if, regardless of any initial configuration c_0, the system always reaches a stable state c_k after a finite number of moves.

The major objective of self-stabilization is for a system to always achieve a *desired* or *legitimate* stable state. An algorithm is called *self-stabilizing* if (i) when started in any initial illegitimate state, it always reaches a legitimate state after a finite number of moves, and (ii) for any legitimate state and for any move enabled by that state, the next state is always a legitimate state.

2.6 Running Times

The (worst-case) running time of a self-stabilizing algorithm under a *central* scheduler is defined to equal the maximum possible number of moves from any initial configuration to a stable configuration.

The running time of an algorithm under the *distributed* scheduler can be measured by the total number of moves, or the number of time steps, or rounds. A *round* as discussed by Dolev in [16] is a minimal sequence of time steps where every enabled node at the start of the round either makes a move or has its move disabled by the move of a neighbor; if the scheduler is *fair*, every round is guaranteed to finish.

For the *synchronous* scheduler, the number of time steps and the number of rounds are identical. In general, the number of moves is an upper bound on the number of time steps.

2.7 Rationale for Self-Stabilizing Algorithms

One of the most important requirements of modern distributed systems is that they should be fault tolerant, which means that a system should be able to function correctly in spite of intermittent or infrequent faults. Ideally, the global state of the system should be legitimate and should remain legitimate. But often enough, system malfunctions can put the system in some arbitrary illegitimate state. It is desirable, therefore, that some mechanism, other than a system-wide reset or external agent, is in place, which can automatically bring the system back to a legitimate global state.

The traditional approach to this type of fault tolerance is to assume worst-case scenarios and make significant efforts to protect the system against such eventualities at the cost of additional hardware and software. Such additional costs may not be an economic option, especially in cases when faults are only transient, subsequent repairs can be made, and short-term unavailability of system service can be tolerated while the system re-establishes a legitimate state.

Since the stabilization time must be small with respect to the frequency of faults, the speed of self-stabilization is important. A self-stabilizing system cannot guarantee that the system is able to operate properly when a node or link continuously injects faults into the system or when communication errors occur so frequently that a new legitimate state cannot be reached. But once the offending fault is removed or corrected, the system can once again provide its necessary services after a reasonable amount of self-stabilizing time.

3 Self-Stabilizing Maximal Independent Set Algorithms

In this section, we present what may well be the simplest and most elegant of all self-stabilizing algorithms, due to Skukla, Rosenkrantz, and Ravi in 1995 [59]. This algorithm only has two rules and in $O(n)$ time finds a maximal independent set of nodes, which of course is also a minimal dominating set. This algorithm assumes that there is a central scheduler, whereby only one, adversarially chosen node can make a move at a time. All nodes are anonymous and make no use of identifier information. Notice, before we get started, that this algorithm does not find either a minimum cardinality maximal independent set or a maximum cardinality independent set, only a maximal independent set, which is all that is required in many distributed system applications.

Recall that a set $S \subset V$ is *independent* if no two nodes in S are neighbors.

In this self-stabilizing algorithm, each node maintains only one Boolean variable x, such that $x(i) = 1$ if node i is in the maximal independent set S and $x(i) = 0$ if node i is not in S. Algorithm MIC in Figure 1 only has the following two, very simple rules.

Rule C1 says that if node i is not in S and has no neighbor in S, then it is enabled to enter S.

Rule C2 says that if node i is in S and has a neighbor in S, then it is enabled to leave the set S.

Given this algorithm, one must prove each of the following:

Algorithm MIC: Maximal Independent - Central
C1: if $(x(i) = 0) \wedge (\not\exists j \in N(i) : x(j) = 1)$
      then $x(i) := 1$ [enter set]
C2: if $(x(i) = 1) \wedge (\exists j \in N(i) : x(j) = 1)$
      then $x(i) := 0$ [leave set]

Fig. 1 Algorithm MIC: Central Model [59]

(i) Under the central scheduler, regardless of the initial global state, and regardless of the sequence of moves made, a stable state must be reached after a finite number of moves.

(ii) In every stable state, the set of nodes S, for which $x(i) = 1$, must always define a maximal independent set.

It is also important to ascertain the running time, i.e., worst-case performance, of this algorithm. We will show that it stabilizes after at most $O(n)$ moves for any graph of order n.

In order to prove that this algorithm stabilizes, we will use the following lemmas.

Lemma 1 *After a node executes Rule C1, it can never make another move.*

Proof After a node i executes Rule C1, $x(i) = 1$ and all of its neighbors $j \in N(i)$ have $x(j) = 0$. As long as $x(i) = 1$, node i cannot execute Rule C1, and it can only execute Rule C2 if $x(i) = 1$ and a neighbor $j \in N(i)$ has $x(j) = 1$. But as long as $x(i) = 1$, no neighbor $j \in N(i)$ can execute Rule C1, and therefore every neighbor j must remain in state $x(j) = 0$. $\square$

Lemma 2 *After a node executes Rule C2, it can only execute Rule C1.*

Proof After a node executes Rule C2, its value has changed from $x(i) = 1$ to $x(i) = 0$, and therefore it is no longer able to execute Rule C2, which requires $x(i) = 1$. $\square$

Theorem 1 (Shukla et al.) *Algorithm MIC stabilizes in at most $2n$ moves.*

Proof A node can only execute four possible move sequences: (i) no move at all, (ii) Rule C1, (iii) Rule C2, and (iv) Rule C2 followed by Rule C1. Thus, if there are n nodes, at most $2n$ moves can ever be executed. $\square$

Lemma 3 *If Algorithm MIC is stable, the set $S = \{i \mid x(i) = 1\}$ is an independent set.*

Proof Assume that Algorithm MIC is in a stable set and S is not an independent set. Then, by definition, there must be two adjacent nodes i and j, both of which have $x(i) = 1$ and $x(j) = 1$. But in this case both node i and node j are enabled to execute Rule C2, and hence Algorithm MIC is not stable: a contradiction. $\square$

Lemma 4 *If Algorithm MIC is stable, then the set S is a maximal independent set.*

Proof Assume that Algorithm MIC is in a stable state and S is an independent set but is not a maximal independent set. Then, by definition, there must exist a node i that is not in S and has no neighbors in S, which means that $x(i) = 0$ and for every $j \in N(i)$, $x(j) = 0$. But in this case node i is enabled to execute Rule C1 and, therefore, Algorithm MIC is not stable. $\square$

Thus, as desired, Algorithm MIC stabilizes and finds a maximal independent set in $O(n)$ time, in fact, in at most $2n$ moves. This is arguably the simplest of all self-stabilizing graph algorithms. It is worth pointing out that Algorithm MIC is general, in that it can stabilize with any possible maximal independent set, and can do so starting from the initial All-Zero configuration in which $x(i) = 0$, for all nodes i. For example, let $S = \{v_1, v_2, \ldots, v_k\}$ be any maximal independent set. Starting in the All-Zero configuration, Algorithm MIC, under the central scheduler, could select each of these nodes, in order, to execute Rule 1; at the end of k moves, the maximal independent set will be determined and the algorithm will be stable.

3.1 Distributed Model Maximal Independent Set Algorithm

While Algorithm MIC is designed to run under a central scheduler, Algorithm MID, shown in Figure 2, due to Ikeda, Kamei and Kakugama in 2002 [41], is designed to find a maximal independent set under a distributed scheduler. This means that at any time, any adversarially chosen subset of enabled nodes can simultaneously make a move. Such a set of moves constitutes one *round*.

We show here that Algorithm MID stabilizes in $O(n)$ rounds.

Again, each node maintains only one Boolean variable x such that $x(i) = 1$ if node i is in S and $x(i) = 0$ if node i is not in S. But notice the key difference between Algorithms MIC and MID; Algorithm MID, under the distributed scheduler, uses the ID, which is the integer index i of a node, to determine its eligibility to make a move, where we assume that all nodes have a unique ID. In fact, we only need to assume that no two nodes in any closed neighborhood have the same ID.

Note that while Rule D1 is the same as Rule C1 in Algorithm MIC, Rule D2 is slightly different than Rule C2 and asserts that a node in the set S can only be forced to leave S if it has a neighbor in S whose ID is larger.

Theorem 2 (Ikeda, Kamei, Kakugawa) *Starting from an arbitrary state, Algorithm MID stabilizes in at most $O(n^2)$ moves, and when stable, the set $S = \{i : x(i) = 1\}$ is a maximal independent set.*

In [41], Ikeda et al. construct an example where Algorithm MID takes $\Theta(n^2)$ time steps. In 2008 [26], Goddard, Hedetniemi, Jacobs, Srimani, and Xu determine the running time of Algorithm MID in terms of rounds, as follows.

Theorem 3 (Goddard et al.) *Starting from an arbitrary state, Algorithm MID stabilizes in at most n rounds.*

Proof We prove this by showing that in every round R there is a node v_R, which moves and, having moved, never moves again.

Algorithm MID: Maximal Independent - Distributed
D1: if $(x(i) = 0) \land (\nexists j \in N(i) : x(j) = 1)$
      then $x(i) := 1$ [enter set]
D2: if $(x(i) = 1) \land (\exists j \in N(i) : j > i \land x(j) = 1)$
      then $x(i) := 0$ [leave set]

Fig. 2 Algorithm MID: Distributed Model [41]

Case 1. Assume that some node executes Rule D1 in round R. Since no two nodes have the same ID value, let v_R be a node with maximum ID that executes Rule D1 in round R and sets $x(v_R) = 1$. Since v_R executed Rule D1, before this time step none of its neighbors were in S. By the choice of v_R, any neighbor of v_R that also executes Rule D1 in round R has a smaller ID value than node v_R. After this round, no other neighbor of v_R can execute Rule D1, since $x(v_R) = 1$. Furthermore, v_R will never leave the set S by executing Rule D2, since it has a larger ID than any of its neighbors also in S. Hence, v_R will never move again.

Case 2. If no node executes Rule D1 in round R, let v_R be a node in S, with $x(v_R) = 1$, that executes Rule D2 during this round, because it has a neighbor, say w_R with $x(w_R) = 1$ and $w_R > v_R$. Assume furthermore that over all such pairs of neighbors, v_R and w_R, where v_R executes Rule D2, w_R has the maximum ID.

By the choice of v_R and w_R, w_R is not enabled to execute Rule D2 in round R. Hence, w_R stays in S for the rest of the round. It follows that all neighbors v_R of w_R in S that execute Rule D2 in round R will ever move again.

Because of Cases 1 and 2, it follows that the number of rounds is at most the number of nodes. $\qquad\qquad\square$

As with Algorithm MIC, it is possible for Algorithm MID to stabilize with any possible maximal independent set and can do so starting from the initial All-Zero configuration.

3.2 *Synchronous Model Maximal Independent Set Algorithm*

In this section, we present a synchronous model, self-stabilizing Algorithm MIS, in Figure 3, for finding a maximal independent set, due to Goddard, Hedetniemi, Jacobs, and Srimani in 2003 [22].

We again assume that no two neighbors have the same ID and that every node can compare its ID with the IDs of all of its neighbors.

Rule S1 says that a node not in S may enter S provided it does not have a neighbor with larger ID already in S. If it enters S with a neighbor with smaller ID already in S, then subsequently that neighbor with a smaller ID will be forced to leave S.

Similarly, Rule S2 says that a node must leave set S if it has a neighbor in S, which has a larger ID.

<table>
<tr><td colspan="2">Algorithm MIS: Maximal Independent - Synchronous</td></tr>
<tr><td>S1: if $(x(i) = 0)(\nexists j \in N(i) : j > i \wedge x(j) = 1)$</td><td></td></tr>
<tr><td>   then $x(i) := 1$</td><td>[enter set]</td></tr>
<tr><td>S2: if $(x(i) = 1) \wedge (\exists j \in N(i) : j > i \wedge x(j) = 1)$</td><td></td></tr>
<tr><td>   then $x(i) := 0$</td><td>[leave set]</td></tr>
</table>

Fig. 3 Algorithm MIS: Synchronous Model [22]

The proofs of correctness and the running time of this algorithm are given by Goddard et al. [22] as follows.

Lemma 5 *If at any time t, the set S of nodes with $x(i) = 1$ does not form an independent set, then at least one node will execute Rule S2 during the next round.*

Proof Assume that at some time t there exists at least one pair of adjacent nodes, both of which are in S, that is, the set S is not independent. Among all nodes in S, which have neighbors also in S, let node v_R have the smallest ID. It follows that this node is enabled to execute Rule S2 and must do so during the next round. $\square$

Lemma 6 *If at any time t, the set S of nodes with $x(i) = 1$ forms an independent set but does not form a maximal independent set, then at least one node will execute Rule S1 during the next round.*

Proof Assume that at some time t the set S of nodes with $x(i) = 1$ forms an independent set but does not form a maximal independent set. Then there must exist a node v_R for which $x(v_R) = 0$ and all neighbors $w_R \in N(v_R)$ have $x(w_R) = 0$. Clearly, node v_R is enabled to execute Rule S1 during the next round. $\square$

Theorem 4 *If Algorithm MIS stabilizes, then the set S of nodes with $x(i) = 1$ forms a maximal independent set.*

Proof From Lemma 5, we know that if Algorithm MIS stabilizes then S must be an independent set, and from Lemma 6, we know that if S stabilizes then S must be a maximal independent set. $\square$

Theorem 5 *Algorithm MIS stabilizes in $O(n)$ rounds.*

Proof At time $t = 1$, after the first round, we know that all nodes v whose ID is larger than the IDs of all of their neighbors will have value $x(v) = 1$. If they have $x(v) = 1$ at time $t = 0$, then they are not enabled to execute Rule S2 and will remain after the first round with $x(v) = 1$. If they have $x(v) = 0$ at time $t = 0$ and will be enabled to execute Rule S1, then they have $x(v) = 1$ after the first round. Furthermore, none of these largest ID nodes will ever be enabled to execute rule S2. Since there is one largest ID node, call it v_1, it will be permanently set to $x(v_1) = 1$ after round one, and every neighbor $w \in N(v_1)$ will be permanently set to $x(w) = 0$ after round two.

By time $t = 3$, after the third round, the node, say v_3, with the largest ID among the nodes in $V - N[v_1]$ will be permanently set to $x(v_3) = 1$, and after time $t = 4$, all neighbors of v_3 will have their x-values set permanently to zero.

This process will continue until all nodes are stable after at most n rounds. $\square$

3.3   Other Self-Stabilizing Independent Set Algorithms

Several other self-stabilizing algorithms have appeared for finding maximal independent sets. For example, in 2007 [63] Turau presents such an algorithm using an unfair distributed scheduler, which stabilizes in at most $max\{3n - 5, 2n\}$ moves.

In 2013 [35], Hedetniemi, Jacobs, and Kennedy present several self-stabilizing algorithms for finding disjoint independent sets S_1 and S_2, where S_1 is a maximal independent set, and S_2 is a maximal independent set in the graph $G[\overline{S}_1]$ induced by $\overline{S}_1$.

Maximal k-Packings

An equivalent definition of an independent set is a set S of nodes having the property that for any $i, j \in S$, $d(i, j) > 1$, that is, no two nodes are adjacent. This immediately generalizes to a *k-packing*, which is a set S of nodes having the property that for any $i, j \in S$, $d(i, j) > k$. It should be noted that a maximal k-packing is also a *minimal distance-k dominating set*, which means that every node $i \in \overline{S}$ is within distance-k of at least one node in S.

Note, in this regard, that every maximal independent set is a minimal distance-1 dominating set. Self-stabilizing maximal k-packing algorithms have been developed by Kristiansen in his 2002 PhD thesis [49], by Gairing, Geist, Hedetniemi, and Kristiansen in 2004 [17], Goddard, Hedetniemi, Jacobs, and Srimani in 2005 [25], by Shi in 2012 [58], and by Trejo-Sánchez, Fernández-Zepeda, and Ramírez-Pacheco in 2017 [62].

Maximal k-Dependent Sets

Still another equivalent definition of an independent set is that it is a set S of nodes having the property that the maximum degree of a node in the subgraph $G[S]$ of G induced by S is zero. A *k-dependent set* is a set S of nodes having the property that the maximum degree of a node in the induced subgraph $G[S]$ is at most k, or equivalently if for every $i \in S$, $|N(i) \cap S| \le k$.

In 2004 [18], Gairing, Goddard, Hedetniemi, and Jacobs present the following simple, two-rule, self-stabilizing Algorithm MKD, in Figure 4, for finding a maximal k-dependent set, using a central scheduler; it stabilizes in at most $2kn + 3n$ moves. This algorithm uses only one, non-negative integer variable $f(i) \in \{0, 1\}$, where $f(i) = 1$ if node $i \in S$, $f(i) = 0$ if node $i \notin S$, and $f(N(v_i)) = \Sigma_{j \in N(i)} f(j)$.

Algorithm MKD: Maximal k-Dependent
KD1: if $(f(i) = 0) \wedge (f(N(i)) \le k)$
     then $f(i) := 1$
KD2: if $(f(i) = 1) \wedge (f(N(i)) > k)$
     then $f(i) := 0$

Fig. 4 Algorithm MKD: Central Model [18]

Minimal Vertex Covers

A *vertex cover* is a set S of nodes having the property that every edge $e = uv$ contains a vertex in S, that is, either $u \in S$ or $v \in S$ or both. It is well known and easily proved that the complement $\overline{S}$ of every (maximal) independent set S is a (minimal) vertex cover, and conversely, the complement of every (minimal) vertex cover is a (maximal) independent set. Given this, every self-stabilizing algorithm for finding a maximal independent set S also finds a minimal vertex cover $\overline{S}$, that is the nodes i with $x(i) = 0$.

Several papers have focused on finding minimal vertex covers within a constant factor of optimality, such as Turau in 2010 [65], Turau and Hauck in 2011 [69], and Delbot, Laforest, and Rovedakis in 2014 [11].

4 Self-Stabilizing Maximal Matching Algorithms

Given an undirected graph $G = (V, E)$, a *matching* is defined to be a set $M \subseteq E$ of pairwise disjoint edges. That is, no two edges in M are incident with the same node. A matching M is *maximal* if there does not exist another matching M' such that $M' \supset M$.

4.1 Central Model Maximal Matching Algorithm

In 1992 [38], Hsu and Huang present the first self-stabilizing algorithm for finding a *maximal matching* in a distributed network $G = (V, E)$ under a central scheduler. They show that their algorithm stabilizes in $O(n^3)$ moves. A further running time analysis of their algorithm is given in by Tel in 1994 [61], who shows that Algorithm Hsu–Huang stabilizes in $O(n^2)$ moves. A subsequent paper by Hedetniemi, Jacobs, and Srimani in 2001 [37] shows that, in fact, Algorithm Hsu–Huang, in Figure 5, stabilizes in $O(m)$ moves, where $m = |E|$ is the number of edges. We present next the Hsu–Huang algorithm.

Algorithm Hsu-Huang

M1: if $(i \rightarrow \text{null}) \wedge (\exists j \in N(i) : j \rightarrow i)$
     then $i \rightarrow j$ [accept proposal]

M2: if $(i \rightarrow \text{null}) \wedge (\forall k \in N(i) : \neg(k \rightarrow i)) \wedge (\exists j \in N(i) : j \rightarrow \text{null})$
     then $i \rightarrow j$ [make proposal]

M3: if $(i \rightarrow j) \wedge (j \rightarrow k) \wedge (k \neq i)$
     then $i \rightarrow \text{null}$ [withdraw proposal]

Fig. 5 Algorithm Hsu–Huang, Central scheduler [38]

Each node maintains just one variable, a pointer, which is either null, denoted $i \to null$, or points to a neighbor $j \in N(i)$, denoted $i \to j$. The algorithm has just three rules.

Rule M1 allows a node i to *accept* a proposed match with another node j, which is pointing to node i, provided $i \to null$.

Rule M2 allows a node i to *propose* matching with a neighbor j, which currently is not matched ($j \to null$), provided no other node k is currently proposing a match with node i by pointing to i.

Rule M3 allows a node i to *withdraw* a proposal if the node j to which it is pointing is currently pointing to some other node k.

An edge between two adjacent nodes i and j becomes a permanent edge of a maximal matching when each is pointing to the other, $i \to j$ and $j \to i$, in which case we say that nodes i and j are *matched*. The maximal matching M produced by the Algorithm Hsu–Huang is the set of edges $e = \{i, j\}$ such that $i \leftrightarrow j$.

We present the proof given in [37] that this algorithm stabilizes in at most $2m + n$ moves.

For every move made by a node i, there is a corresponding node j that enables the move; we will denote such a move by (i, j, Mk), for $1 \le k \le 3$, and say that it is an (i, j)-move. Let $c(i, j)$ denote the number of (i, j)-moves that has been executed, and let $c(e)$ denote the number $c(e) = c(i, j) + c(j, i)$.

After a move $(i, j, M1)$ or $(j, i, M1)$ has been executed, we will say that i and j are *matched*.

Lemma 7 *After nodes i and j have been matched, neither node can make another move.*

Proof After an $(i, j, M1)$ move, neither node i nor node j will have a null pointer and are therefore not enabled to make move M1 nor M2. Furthermore, since $i \leftrightarrow j$, neither node is enabled to execute Rule M3. $\qquad\square$

Lemma 8 *After an $(i, j, M2)$-move, at most one more (i, j)-move is possible, namely $(i, j, M3)$.*

Proof Let $m = (i, j, M2)$ be a move on the edge (i, j), and let $m' = (i, j, Mk)$ be the next move on the same edge. Clearly, it can only be $(i, j, M3)$. It then suffices to show that no further (i, j)-move can occur.

After move m, we must have $i \to j$ and $j \to null$, and prior to move m', we must have $i \to j$ and $j \to k$ for some $k \ne i$. Thus, sometime after move m and before move m', there must have been a move m'' of the form $m'' = (j, k, M1)$, which implies that node j is permanently matched with node k. Being permanently matched, there can be no more (i, j)-moves. $\qquad\square$

Lemma 9 *Following a move $(i, j, M2)$, there can be only one more move on the edge (i, j), either $(j, i, M1)$ or $(i, j, M3)$.*

Proof Once a proposal has been made with an $(i, j, M2)$ move, node j is enabled to make move $M1$. It must either accept a proposal from node i or another node k. If it chooses node i and makes the move $(j, i, M1)$, then it will become permanently matched with node i and no further move can be made on this edge. But if node j chooses another node k, it will become permanently matched with node k, forcing node i to execute $M3$, after which no further move can be made on edge (i, j). $\square$

Lemma 10 *Following a move $(i, j, M3)$, there are at most two more moves on the edge $e = (i, j)$.*

Proof If there is to be another move on edge (i, j) following a move $(i, j, M3)$, then node j will have to reset its pointer to null by executing $M3$. With both pointers set to null, the next move on the edge can only be a proposal, $(i, j, M2)$ or $(j, i, M2)$. But by Lemma 9, there can only be one more move on this edge. $\square$

Consider an arbitrary initial state of the system, and if initially $i \rightarrow j$, let edge (i, j) be called an *initial edge*. Let I denote the set of all initial edges. Note that initially there can only at most n initial edges, one for each node i, thus, $|I| \leq n$. Recall that we have defined $c(e) = c(i, j) + c(j, i)$ to count the number of moves made on the undirected edge $e = (i, j)$.

Lemma 11 *For each edge $e \in E$, $c(e) \leq 3$ and for at most n edges, $c(e) = 3$.*

Proof If $c(e) > 0$, then there is a first move m on this edge $e = (i, j)$, either $m = (i, j, M1)$, or $m = (i, j, M2)$ or $m = (i, j, M3)$. Lemmas 7, 9 and 10 prove that $c(e) \leq 3$.

In order to prove that for at most n edges, $c(e) = 3$, let $C_3 = \{e | c(e) = 3\}$. If some edge $e \in C_3$, then the first move on this edge must be of the form $(i, j, M3)$. But this implies that the initial state of node i is $i \rightarrow j$, and this means that $e \in I$ and so $C_3 \subseteq I$, and therefore $|C_3| \leq n$. $\square$

Theorem 6 *For any graph $G = (V, E)$ having order $n = |V|$ and size $m = |E|$, Algorithm Hsu–Huang stabilizes in at most $2m + n$ moves under the central scheduler.*

Proof This follows from Lemma 11. $\square$

4.2 Synchronous and Distributed Model Maximal Matching Algorithm

In 2003 [22], Goddard, Hedetniemi, Jacobs, and Srimani show that the following Algorithm MMDS, in Figure 6, finds a maximal matching and stabilizes for any graph of order n in at most $n + 1$ rounds, under the synchronous scheduler. Notice that in Rule DM2, a node i having a null pointer and no node k pointing to it may point to a neighbor j whose pointer is null, and thereby make a proposal of a match, provided that j has the minimum ID among the neighbors of node i whose pointer in null. The proof of correctness of this algorithm and its running time is considerably longer than that of Algorithm Hsu–Huang and is omitted.

In 2008 [26], Goddard, Hedetniemi, Jacobs, Srimani, and Xu proved that Algorithm MMDS also finds a maximal matching and stabilizes in at most $O(n)$

> **Algorithm MMDS: Maximal Matching - Distributed or Synchronous**
>
> **DM1: if** $(i \to null) \wedge (\exists j \in N(i) : j \to i)$
>
>   **then** $i \to j$ [accept proposal]
>
> **DM2: if** $(i \to null) \wedge (\forall k \in N(i) : k \not\to i) \wedge (\exists j \in N(i) : j \to null)$
>
>   **then** $i \to \min\{j \in N(i) : j \to null\}$ [make proposal]
>
> **DM3: if** $(i \to j) \wedge (j \to k \text{ where } k \neq i)$
>
>   **then** $i \to null$ [withdraw proposal]

Fig. 6 Algorithm MMDS: Maximal matching, distributed, and synchronous scheduler

rounds and at most $O(n^3)$ time steps under a distributed scheduler. We provide their proof here, as it is instructive. At any point in the execution of Algorithm MMDS, under the distributed scheduler, let $M = \{\{i, j\} : i \leftrightarrow j\}$ denote the set of matched edges.

Recall that a *round*, as discussed by Dolev in [16], is a minimal sequence of time steps where every enabled node at the start of the round either makes a move or has its move disabled by the move of a neighbor; if the scheduler is *fair*, every round is guaranteed to finish.

Theorem 7 *If Algorithm MMDS stabilizes, then the set M is a maximal matching in the graph G.*

Proof It is clear that the set M is a matching since a node can only be matched with one other node; thus, no two edges can have a node in common. Assume that Algorithm MMDS has stabilized but M is not a maximal matching. Since Algorithm MMDS is stable, no node is enabled to execute Rule DM3. Therefore, every node either has a null pointer or is matched. Since M is not maximal, there must be two adjacent nodes, both of which have null pointers. But in this case, both nodes are enabled to execute Rule DM2, and the algorithm is not stable, which is a contradiction. $\qquad\square$

Lemma 12 *After nodes i and j have been matched, neither node can make another move.*

Proof After nodes i and j have been matched, neither node i nor node j will have a null pointer and are therefore not enabled to execute Rule DM1 or DM2. Furthermore, since $i \leftrightarrow j$, neither node is enabled to execute Rule DM3. $\qquad\square$

Lemma 13 *Consider a time step where at least one node executes Rule DM2 and makes a proposal, but no new match occurs. Then there exists some node that is proposed to but does not make a move.*

Proof Suppose that during a time step no new match occurs, and some node i executes DM2 and proposes to node j. If during this time step, node j does not make a move, then the lemma is true. Suppose, therefore, that node j makes a move. Since no match occurs, it must execute DM2 and propose to some node k. If node

k does not make a move, the lemma is proved. One can then follow the sequence of proposals: node i proposes to node j, which proposes to node k, which proposes to still some other node, etc. Since the graph is finite, either a node is reached which does not make move or there must exist a cycle of proposals.

But consider the node in the cycle having the largest ID, say u. Some node, say v, must propose to u in this cycle. But, in turn, some node w must have proposed to node v and node w must have a smaller ID than node u. Therefore, node v should have proposed to w, a contradiction.

Therefore, if during some time step, no match occurs but some node proposes to some other node, then there must be a node that receives a proposal but does not make a move. $\square$

Lemma 14 *In the execution of Algorithm MMDS, there cannot be two consecutive rounds without a new match.*

Proof Let R be a round in which no new match occurs. If no more rounds are executed, then the algorithm is stable and the lemma is true, so assume that there is another round R'. We will show that a new match must occur.

Case 1. Assume that during round R no new match occurs, that is, no node executes DM1, but some node executes DM2. Then, by Lemma 13, at that time step, some node x is proposed, which does not make a move. It follows that x is enabled at the end of round R to execute DM1 by the end of the following round, creating a new match after round R', since every node pointed to must execute DM1 for some node pointing to it in any given round.

Case 2. Assume that during round R no node executes DM1 or DM2. That is, all moves in R are DM3. It follows that by the end of round R, every node is either matched, has a null pointer, or points to a neighbor that has a null pointer.

So, the first time step of the next round R' is an execution of DM1 or DM2. If a DM1 move is executed, then the lemma is proved. If no new match occurs, a DM2 move must be executed during R'. Then, by Lemma 13, some node x must be proposed to. But x was privileged at the start of the round and so must accept by the end of the round, creating a new match. $\square$

Theorem 8 *Starting from an arbitrary state, Algorithm MMDS stabilizes in at most n rounds.*

Proof By Lemma 12, all matched nodes remain matched. By Lemma 14, there cannot be two consecutive rounds without a new match. Since every new match matches two nodes, the theorem follows. $\square$

The following result shows that the number of time steps is a bit larger than for previous algorithms:

Lemma 15 *Algorithm MMDS stabilizes in at most $O(n^3)$ time steps under a distributed scheduler.*

Proof We know there are at most $O(n)$ time steps where a new match occurs.

We claim that there are at most n^2 time steps in which some node executes DM2 but no new match occurs. By Lemma 13, in each such time step there is some node that is pointed to but does not make a move. Each node can be pointed to only $n - 1$ times. Thus, the claim follows.

In between the above time steps, each node can execute DM3 at most once. Thus the total number of time steps between two matches is at most $O(n^2)$. Since there can be at most $n/2$ matches, it follows that the total number of time steps is at most $O(n^3)$. $\qquad\qquad\square$

4.3 Other Self-Stabilizing Matching Algorithms

In 2001 [4] Blair, Hedetniemi, Hedetniemi and Jacobs present a self-stabilizing algorithm for finding a maximum, rather than the typical maximal, matching in an arbitrary tree.

In 2006 [29] Goddard, Hedetniemi and Shi present an anonymous self-stabilizing algorithm for finding a 1-maximal matching in a tree, and ring of length not divisible by 3. Their algorithm converges in $O(n^4)$ moves under a central daemon.

In 2007 [52], Manne, Mjelde, Pilard, and Tixeuil present a self-stabilizing algorithm for finding a maximal matching, using a distributed scheduler, which stabilizes in $O(|E|)$ rounds, improving on previous bounds of $O(n^2)$ and $O(\Delta|E|)$. Their algorithm also has the same running time as previous self-stabilizing, maximal matching algorithms, using central, distributed, and synchronous schedulers.

In 2009 [53], Manne, Mjelde, Pilard, and Tixeuil present a self-stabilizing algorithm for the maximal matching problem that improves the running time of the previous best algorithm for a distributed scheduler and at the same time meets the bounds of the previous best algorithms for the sequential and distributed fair schedulers. Their algorithm requires unique IDs at distance two and uses a Boolean variable at each node, which enables neighbors to communicate whether this node is already matched.

In 2015 [2], Asada and Inoue present a self-stabilizing algorithm for finding a 1-maximal matching, which is guaranteed to stabilize, under the anonymous model, with a fair central scheduler, but only when restricted to graphs having no cycles of lengths a multiple of 3; this includes all bipartite graphs, including grid graphs and trees. Since it stabilizes in $O(|E|)$ moves, it stabilizes in $O(n)$ moves for trees and cycles C_n, for n not a multiple of 3.

In 2016 [8], Cohen, Lefévre, Maâmra, Pilard, and Sohier present a self-stabilizing algorithm for finding a maximal matching in an anonymous network. The running time is $O(n^2)$ moves with high probability, under the adversarial distributed scheduler. Among all self-stabilizing algorithms using a distributed scheduler and the anonymous model, their algorithm provides the best known running time. Moreover, the previous best known algorithm working under the same scheduler and using IDs has an $O(m)$ running time, leading to the same order of growth than their anonymous algorithm. Although their algorithm does not make the assumption that a node can determine whether one of its neighbors points to it or to another node, it still has the same asymptotic behavior.

In 2016 [42], Inoue, Ooshita, and Tixeil present a self-stabilizing 1-maximal matching algorithm, using the unfair distributed scheduler. Their algorithm is restricted to graphs having no cycles of length a multiple of 3 and stabilizes in $O(|E|)$ moves. It also provides a 2/3-approximation of a maximum matching in these graphs, which improves on the 1/2-approximation guaranteed by any maximal matching.

Generalized b-Matchings

Given a graph $G = (V, E)$, let $E_i = \{(i, j) \in E\}$ denote the set of edges incident to a node i and let $d(i) = |E_i|$ denote the *degree* of node i.

Let $b : V \rightarrow \{0, 1, \ldots, n - 1\}$ define a bound $b(i)$ on the number of edges that can be incident to node i. A subset $M \subseteq E$ is called a *b-matching* if for all $1 \leq i \leq n$, $b(i) \leq d(i)$. A b-matching M is called *maximal* if there does not exist a b-matching M' such that $M \subset M'$.

In 2003 [23], Goddard, Hedetniemi, Jacobs, and Srimani present a self-stabilizing maximal b-matching algorithm that stabilizes in $O(m)$ moves under an unfair central scheduler, independently of the particular b-values $b(i)$.

Self-Stabilizing Matching Approximation Algorithms

In 2011 [54], Manne, Mjelde, Pilard, and Tixeuil present the first self-stabilizing algorithm for finding a 2/3-approximation of a maximum matching in an arbitrary graph. Their algorithm stabilizes in at most $O(n^2)$ rounds, under a distributed scheduler. However, it might make an exponential number of moves.

In 2011 [68], Turau and Hauck present a more refined analysis of the running time of the first self-stabilizing algorithm for computing a 2-approximation of a maximum matching by Manne and Mjelde [51], who showed that their algorithm stabilizes in $O(2n)$ moves under a central scheduler, and in $O(3n)$ moves under a distributed scheduler. Turau and Hauck show that the Manne–Mjelde algorithm, in fact, stabilizes in $O(mn)$ moves under a central scheduler and, when modified, can stabilize in $O(mn)$ moves under a distributed scheduler.

In 2016 [10], Datta, Larmore, and Masuzawa present an anonymous-model, silent self-stabilizing algorithm for computing the maximum matching number of any tree. Their algorithm stabilizes in $O(n \cdot diam)$ moves, where *diam* is the diameter of the tree.

In 2017 [9], Cohen, Maâmra, Manoussakis, and Pilard present the first polynomial, self-stabilizing algorithm for finding a 2/3-approximation of a maximum matching in an arbitrary graph. The previous best known algorithm, by Manne et al. in 2011 [54], has a sub-exponential time running time under the distributed scheduler. The algorithm by Cohen et al. is an adaptation of the Manne et al. algorithm, works under the same scheduler, but stabilizes in $O(n^3)$ moves.

In 2017 [43], Inoue, Ooshita, and Tixeuil present an ID-based, self-stabilizing, 1-maximal matching algorithm that works under the distributed unfair scheduler for arbitrary graphs. It finds a 2/3-approximation of a maximum matching and stabilizes in $O(|E|)$ moves. The algorithm assumes that node IDs are distinct up to distance three.

The proposed algorithm closes the running time gap between two recent results: in 2016 [42], Inoue et al. present a 1-maximal matching algorithm that stabilizes in $O(|E|)$ moves but requires that the graph not contain a cycle of length a multiple of three; the algorithm of Cohen et al. in 2017 [9] stabilizes on arbitrary graphs but makes $O(n^3)$ moves. The Inoue–Ooshita–Tixeuil algorithm makes the same $O(|E|)$ moves but stabilizes on arbitrary graphs.

Strong Matchings

The definition of a maximal matching can be generalized to distance-k matchings. In particular, a *strong matching* is a matching $M \subseteq E$ having the property that no two edges in M are connected by an edge. This is equivalent to saying that for any two edges $e_1, e_2 \in M$, $d(e_1, e_2) > 1$. In 2005 [25], Goddard, Hedetniemi, Jacobs, and Srimani present an exponential running time, self-stabilizing algorithm for finding a maximal strong matching; this algorithm has only one rule; see also [24] in 2003 by the same authors.

5   Self-Stabilizing Dominating Set Algorithms

In this section, we present self-stabilizing algorithms for finding minimal dominating sets in arbitrary connected graphs $G = (V, E)$, first under a central scheduler, then under a synchronous scheduler, and finally under an unfair distributed scheduler. We conclude this section by presenting the first self-stabilizing algorithm for finding a minimal total dominating set.

A *dominating set* is a subset S of nodes such that $\forall i \in V : N[i] \cap S \neq \varnothing$, that is, every node i is either a member of S or is adjacent to a node j in S. A dominating set S is *minimal* if it does not contain a proper subset that is also a dominating set. It is important to know that a dominating set S is *minimal* if and only if every node $i \in S$ is either (i) not adjacent to any other vertex in S, in which case we say that node i is *its own private neighbor* or (ii) node i is the only vertex in S, which dominates some vertex j not in S, $j \in \overline{S}$, in which case we say that node j is an *external private neighbor* of node i.

5.1   *Central Model Minimal Dominating Set Algorithm*

The following Algorithm MDC, in Figure 7, is the first self-stabilizing algorithm for finding a minimal dominating set in an arbitrary graph, due to Hedetniemi, Hedetniemi, Jacobs, and Srimani in 2003 [32]; it assumes a central scheduler.

Algorithm MDC: Minimal Dominating - Central

D1: **if** $(x(i) = 0) \wedge (\forall j \in N(i))(x(j) = 0)$
     then $x(i) = 1$ [enter set]

D2: **if** $(x(i) = 1) \wedge (\not\exists j \in N(i))(j \rightarrow i) \wedge (\exists k \in N(i))(x(k) = 1)$
     then $x(i) = 0$ [leave set]

P1: **if** $(x(i) = 1) \wedge (i \not\rightarrow null)$
     then $i \rightarrow null$ [no private neighbor]

P2: **if** $(x(i) = 0) \wedge (\exists \text{ unique } j \in N(i))((x(j) = 1) \wedge (i \not\rightarrow j))$
     then $i \rightarrow j$ [point to private neighbor]

P3: **if** $(x(i) = 0) \wedge (\exists \text{ more than one } j \in N(i))((x(j) = 1) \wedge (i \not\rightarrow null))$
     then $i \rightarrow null$ [no private neighbor]

Fig. 7 Algorithm MDC: Central Model [32]

The first rule D1 says that if a node i is currently not a member of the dominating set S ($x(i) = 0$) and no neighbor is in S, then it is enabled to enter S (by setting $x(i) = 1$).

Rule D2 says that if a node i is currently in S ($x(i) = 1$) but is not a private neighbor of any vertex (no node is pointing to i) and node i has a neighbor in S, then node i can leave the set S (by setting $x(i) = 0$).

Algorithm MDC has three kinds of pointer moves.

Rule P1 says that if node i is in S, its pointer should be null.

Rule P2 says that if node i is not in S and has a private neighbor j in S, then it should point to j.

Rule P3 says that if node i is not in S and has two or more neighbors in S, then its pointer should be null.

The proof of correctness of Algorithm MDC proceeds as follows. We will omit some of the details. Let S_t denote the set of nodes i having $x(i) = 1$ at time t.

Lemma 16 *If at any time t, S_t is not a minimal dominating set, then Algorithm MDC is not stable.*

Proof Suppose that Algorithm MDC is stable but the set S is not a dominating set. If S is not a dominating set, then there exists a node i not in S ($x(i) = 0$) and no neighbor of i is in S. This means that node i is enabled to execute D1, and thus, Algorithm MDC is not stable.

Assume therefore that S is a dominating set but is not a minimal dominating set. Thus, there exists a node i in S such that $S - \{i\}$ is a dominating set. This implies that node i must have a neighbor, say k in S, since it is not its own private neighbor, and node i does not have an external private neighbor.

There must also be a neighbor of i, say j, with $j \rightarrow i$, for if not, then node i is enabled to execute D2. Furthermore, $j \notin S$, else node j is enabled to execute P1. In addition, node j must not have another neighbor than i in S, else it is enabled to execute P3. Therefore, j is not in S, has exactly one neighbor in S, namely i, and therefore, node i has a private neighbor, contradicting our assumption that it has no private neighbor. $\qquad\square$

Lemma 17 *If a node i executes D1, then it will never again make move D1 or D2.*

Proof If a node i executes D1 at some time t, then none of its neighbors are in S_t, meaning that for all neighbors $j \in N(i)$, $x(j) = 0$. As long as $x(i) = 1$, it could only execute D2, but it can only execute D2 if it has a neighbor j with $x(j) = 1$. $\qquad\square$

Lemma 18 *A node i can execute at most two D1 or D2 moves.*

Proof If a node i makes its first move D1, then by Lemma 17, it will never make another D1 or D2 move. If node i makes its first move D2, then it can only make move D1, after which it can make no further D1 or D2 moves. $\qquad\square$

Lemma 19 *There can be at most n consecutive pointer moves, P1, P2, or P3.*

Proof If a node i executes a pointer move P1, P2, or P3, and subsequently, there are no moves D1 or D2 made by any node, then node i is not enabled to execute P1, P2, or P3. Therefore, in any sequence of consecutive pointer moves, each node can only execute one pointer move. $\qquad\square$

Lemma 20 *Algorithm MDC can make at most $n^2 + n$ moves.*

Proof By Lemma 18, there can be at most $2n$ moves D1 and D2. By Lemma 19, there can be at most n consecutive pointer moves between successive D1 or D2 moves. $\qquad\square$

Theorem 9 *Algorithm MDC finds a minimal dominating set and stabilizes in $O(n^2)$ moves.*

Proof This follows from Lemmas 16 and 20. $\qquad\square$

Algorithm MDS: Minimal Dominating - Synchronous

$S \equiv \{ i : x(i) = 1 \}$

SD1: **if** $(x(i) = 0) \wedge c(i) \neq |\{j \in N(i) : x(j) = 1\}|$
      then $c(i) = |\{j \in N(i) : x(j) = 1\}|$ [correct S-neighbor count]

SD2: **if** $(x(i) = 0) \wedge (|N(i) \cap S| = 0) \wedge (c(i) = 0) \wedge (\nexists j \in N(i)((j < i) \wedge (c(j) = 0))$
      then $x(i) := 1$ [enter set S]

SD3: **if**$(x(i) = 1) \wedge (|N(i) \cap S| > 0) \wedge (\forall j \in N(i)(\textbf{if } x(j) = 0 \textbf{ then } c(j) := 2)$
      then $x(i) := 0$ and
$$c(i) := \begin{cases} 1 & \text{if } |N(i) \cap S| = 1 \\ 2 & \text{otherwise} \end{cases}$$
      [leave set S]

Fig. 8 Algorithm MDS: Synchronous Model [26]

5.2 Synchronous Model Minimal Dominating Set Algorithm

We next present Algorithm MDS, in Figure 8, which is the first, synchronous model, self-stabilizing algorithm for finding a minimal dominating set, due to Goddard, Hedetniemi, Jacobs, Srimani, and Xu in 2008 [26]; an earlier 2003 version of this algorithm, by Xu, Hedetniemi, Goddard, and Srimani [71]. As with previous synchronous model algorithms, this algorithm assumes that all nodes have unique ID values. Again, $x(i) = 0$ means that node i is not in the dominating set S, and $x(i) = 1$ means that node i is in S. The variable $c(i) \in \{0, 1, 2\}$ keeps count of the number of neighbors of node i in the set S, where $c(i) = 2$ means that node i has 2 or more neighbors in S. Thus, if $|\{j \in N(i) : x(j) = 1\}| \geq 2$, then we set $c(i) = 2$. The value of $c(i)$ is not used if node i is a member of S.

Rule SD1 makes sure that a node i not in S has the correct value of $c(i)$.

Rule SD2 says that a node i can enter S if it has no neighbor in S, its current value $c(i) = 0$ is correct, and its ID is smaller than any neighbor j with $c(j) = 0$.

Rule SD3 says that a node i is enabled to leave S, by setting $x(i) = 0$ and setting a correct value of $c(i)$, if it has at least one neighbor in S, and according to the $c(j)$-values of its neighbors, it has no private neighbors in $\overline{S}$.

We first show the correctness.

Theorem 10 *If Algorithm MDS stabilizes, then the set $S = \{i : x(i) = 1\}$ is a minimal dominating set.*

Proof Suppose that Algorithm MDS is stable but S is not a dominating set. Thus, there is a node i such that $S \cap N[i] = \varnothing$. Among all such undominated nodes, let i have the minimum ID. Then, $x(i) = 0$. Further, since Algorithm MDS is stable, node i is not enabled to execute Rule SD1, and therefore, $c(i) = 0$ is correct.

Consider any neighbor $j \in N(i)$ whose ID j is smaller than i, $j < i$. Then, $x(j) = 0$, but by the choice of i, j must be dominated by a node in S. So, since node j is not enabled to execute Rule SD1, $c(j) > 0$ must be true. It follows then that node i is enabled to execute Rule SD2, a contradiction. Therefore, S is dominating.

Suppose that S is a dominating set but is not minimal. Then there is a node $i \in S$ such that $S - \{i\}$ is a dominating set. It follows that, for each $j \in N[i]$, we have $|N[j] \cap S| > 1$. If $j \in N(i) - S$, then by Rule SD1, since $x(j) = 0$, $c(j) = 2$. Hence, node i must be enabled to execute Rule SD3, a contradiction. Thus, S is a minimal dominating set. $\square$

We next show that Algorithm MDS stabilizes.

Lemma 21 *If $x(i)$ changes from 0 to 1, then $x(i)$ will never again change.*

Proof If $x(i)$ changes from 0 to 1, then by Rule SD2, all nodes j in the neighborhood $N(i)$ must have $x(j) = 0$. By Rule SD2, only the node of i and j with smaller ID is enabled to execute Rule SD2 in the same time step according to the synchronous model, so $x(j)$ does not change in the same time step. Therefore, after this time step no neighbor of i is in S. After that, no neighbor j of i can enter S since there is at least one node (namely i) in $S \cap N(j)$, and i will not leave S since none of its neighbors are in S. $\square$

Theorem 11 *Starting from any arbitrary state, Algorithm MDS stabilizes in at most $4n + 1$ time steps under the synchronous scheduler.*

Proof By Lemma 21, each node can change its x-value at most twice. Therefore, there can be at most $2n$ changes of x-values on all n nodes. If there is no change in the x-value of any node during a time step, then the time step only involves corrections of c-values. The change in a c-value is determined only by x-values. Since we are working with the synchronous scheduler, there cannot be two consecutive time steps without a change in x-value. Therefore, the upper bound of execution time is $4n + 1$ time steps. $\square$

One can also show that Algorithm MDS converges under the distributed scheduler.

Theorem 12 *Algorithm MDS stabilizes with a minimal dominating set in at most $5n$ moves under the distributed scheduler.*

Proof We claim that every node can make at most 5 moves under a distributed scheduler.

Case 1. Assume that for a node i, $c(i)$ never changes to 0. By Lemma 21, if i executes Rule SD2 and changes from $x(i) = 0$ to $x(i) = 1$, then $x(i)$ will never change again. Thus, we may assume that after its first move, $x(i) = 0$. So, apart from possibly its first move being Rule SD3, node i makes only Rule SD1 moves. Each such move changes the value of $c(i)$, which must oscillate between 1 and 2. Each 1-to-2 move is due to a neighbor entering S; once two neighbors have entered, i has two neighbors in S until the end of the algorithm, and so cannot move again. It follows that the longest possible sequence of changes for $c(i)$ is ?–2–1–2–1–2.

Case 2. Assume that $c(i)$ changes to 0 at some point. No neighbor enters before $c(i)$ goes to 0. So before the move $c(i) = 0$, node i may make at most two moves (a leave move or a $c(i) = 2$ move, perhaps followed by a $c(i) = 1$ move). After $c(i)$ becomes 0, i may make either an enter move, or a $c(i) = 1$ move, perhaps followed by a $c(i) = 2$ move. $\square$

5.3 *Distributed Model Minimal Dominating Set Algorithm*

We next present a $4n$-move, self-stabilizing Algorithm MDD, in Figure 9, for finding a minimal dominating set using an unfair distributed scheduler, by Chiu, Chen, and Tsai in 2014 [7]. An earlier 2013 version of this algorithm by Chiu and Chen appears in [6]. For reasons of consistency with the notation used in our previous

---

Algorithm MDD: Minimal Dominating - Distributed

R1: **if** $(x(i) = 00) \wedge |N(i) \cap S| = 0 \wedge (\nexists j \in N(i))(x(j) = 00) \wedge (j < i))$
    then $x(i) := 1$ [enter S]

R2: **if** $(x(i) = 1) \wedge |N(i) \cap S| = 1 \wedge (\nexists j \in N(i))(x(j) = 01)$
    then $x(i) := 01$ [leave with unique private neighbor]

R3: **if** $(x(i) = 1) \wedge |N(i) \cap S| > 1 \wedge (\nexists j \in N(i))(x(j) = 01)$
    then $x(i) := 02$ [leave with private neighbors]

R4: **if** $(x(i) = 00) \wedge |N(i) \cap S| = 1$
    then $x(i) := 01$ [stay out with unique private neighbor]

R5: **if** $(x(i) = 01 \vee 00) \wedge |N(i) \cap S| > 1$
    then $x(i) := 02$ [stay out with private neighbors]

R6: **if** $(x(i) = 01 \vee 02) \wedge |N(i) \cap S| = 0$
    then $x(i) := 00$ [stay out with no S-neighbors]

Fig. 9 Algorithm MDD: Distributed Model [7]

self-stabilizing algorithms, we will change the notation used by Chiu, Chen, and Tsai to be similar to that used in this chapter.

Algorithm MDD assigns to each node i a four-valued variable $x(i)$, which defines the local state of node i, such that $x(i) \in \{1, 00, 01, 02\}$. As before, at any time, $S = \{i : x(i) = 1\}$ and all such nodes are called *S-nodes*. All other nodes, those in states 00, 01, or 02, are called *Out nodes*, nodes in $\overline{S}$.

A node in state 1 is a member of S.

A node in state 00 is not in S and has no neighbor in S.

A node in state 01 is not in S but has a unique neighbor in S.

A node in state 02 is not is S but has at least two neighbors in S

The correctness of Algorithm MDD can be proved as follows: we omit the details.

Lemma 22 *If Algorithm MDD is stable, then S is a minimal dominating set.*

Lemma 23 *If a node executes R1, it will never make another move.*

Lemma 24 *A node can execute R6 at most once.*

Theorem 13 *Algorithm MDD stabilizes under an unfair distributed scheduler in at most $4n - 2$ moves.*

5.4 Minimal Total Dominating Set Algorithm

In this section, we present Algorithm MTDC, in Figure 10, which is the first self-stabilizing, minimal total dominating set algorithm, due to Goddard, Hedetniemi, Jacobs, and Srimani in 2005 [25]; it assumes the central scheduler model (see also [21] in 2003 by the same authors). Recall that a total dominating set of a graph

---

Algorithm MTDC: Minimal Total Dominating - Central
R1: if $(x(i) \neq pointedto(i)) \vee (p(i) \neq q(i))$

   then $x(i) := pointedto(i)$ **and** $p(i) = q(i)$

Fig. 10 Algorithm MTDC: Central Model [25]

$G = (V, E)$ is a set $S \subseteq V$ having the property that $N(S) = V$, which means that every node in $\overline{S}$ is adjacent to at least one node in S, and every node $v \in S$ is adjacent to another node $w \in S$, where $v \neq w$. This means that a graph G does not have a total dominating set if it has an isolated node. Therefore, we assume that G is a nontrivial connected graph. This algorithm is based on the fact that in any minimal total dominating set S, every node $v \in S$ has an external private neighbor.

In this algorithm, each node i has two variables:

(i) a Boolean x, where $x(i) = true$ if node i is in the minimal total dominating set S, and $x(i) = false$ if node i is not in S;
(ii) a pointer variable $p(i)$ such that if $p(i) = j$ then $i \rightarrow j$.

We need the following three definitions:

Definition 1 $minbr(i) = min\{j : j \in N(i)\}$, the neighbor of i having the smallest ID.

Definition 2 Boolean: $pointedto(i) = (\exists j \in N(i))(j \rightarrow i)$

Definition 3 $q(i)$ is the following pointer expression:

$$q(i) := \begin{cases} minbr(i) & \text{if} N(i) \cap S = \emptyset \\ j & \text{if} N(i) \cap S = \{j\} \\ null & \text{if} |N(i) \cap S| \geq 2 \end{cases}$$

The minimal total dominating set algorithm has but one rule.

This one rule says that if there is a node i pointing to a node j, then node j should become a member of the minimal total dominating set by setting $x(j) = true$. It also says that if a node i is in S $(x(i) = true)$ and has no neighbors in S, then it should point to that node j in its neighborhood having the smallest ID, in which case node j must become a member of S by setting $node(j) = true$. A node having two or more neighbors in S sets its pointer to null, and if it has exactly one neighbor i in S, then it must point to that node, informing it that node i must remain in S.

The correctness of this algorithm can be shown as follows.

Lemma 25 *If Algorithm MTDC stabilizes, then the set* $S = \{i : x(i) = true\}$ *is a minimal total dominating set.*

Proof We first show that if Algorithm MTDC is stable, then S is a total dominating set. If S is not a total dominating set, then there must exist a node i such that $N(i) \cap S = \emptyset$. Since the algorithm is stable, it must be true that $p(i) = q(i) = minbr(i)$ and $minbr(i) \notin S$. But this implies that $pointedto(minbr(i))$ is true but $x(minbr(i)) = false$ so $minbr(i)$ is enabled to execute Rule R1, a contradiction.

Next, we must show that S is a minimal total dominating set. Assume that there is some node j such that $S - \{j\}$ is a total dominating set. Since $j \in S$, $x(j) = true$ and there must be some node $i \in N(j)$ for which $p(i) = j$. But since the algorithm is stable, it must be the case that since $p(i) = q(i)$, node j must be the unique neighbor of i in S. Thus, the removal of j from S will leave node i undominated, a contradiction. $\square$

We say that *node i invites node j* if at some time t, node i has no neighbor in S and then executes Rule R1, causing $p(i) = q(i) = j$. In order for a node j to become a member of S, it must either be pointed to from an initial erroneous state or be invited to be a member by being pointed to by a node i in S.

In order to show that Algorithm MTDC stabilizes, we note that if the set S does not change its membership, then every node can only execute at most once, to correct its pointer value. We say that an *in-move* is a move that causes $x(i)$ to become true.

Lemma 26 *If during some time interval, there is no in-move by a node having a larger ID than some node i, then during this time interval node i can make at most two moves.*

Proof The first in-move made by a node i maybe have been because a neighbor $j \in N(i)$ happened to be pointing to i. A second in-move made by node i must be by invitation. So suppose that node i is invited by a neighbor, node j. Then j must be the smallest node in $N(i)$ since $minbr(j) = i$ and at the time of the invitation, all other nodes in $N(i)$ are not in S.

By our assumption that during some time interval there is no in-move by a node having a larger ID than node i, their membership in S does not change, so node j remains pointing to i throughout the time interval, and node i remains in S for the remainder of the time interval. $\square$

Theorem 14 *Algorithm MTDC always stabilizes and finds a minimal total dominating set.*

Proof It suffices to show that every node makes only a finite number of in-moves. By Lemma 26, node n, which has the largest ID, makes at most two in-moves. During each of the three time intervals between such in-moves, using Lemma 26 again, node $n - 1$ can make at most two in-moves. By repeating this argument, it is easy to show that each node can make only finitely many in-moves during the intervals in which larger nodes are inactive. $\square$

It can be shown, although we will not do so here, that in the worst case, Algorithm MTDC can make an exponential number of moves. This is our first example of a worst-case exponential time self-stabilizing algorithm. In the field of self-stabilizing algorithms, this is often acceptable, since on average, these algorithms can stabilize fairly quickly.

In 2014 [3], Belhoul, Yahiaoui, and Kheddouci present the first polynomial, self-stabilizing algorithm for finding a minimal total dominating set in an arbitrary graph.

They also generalize their algorithm to find a minimal total k-dominating set. Both of their algorithms stabilize in $O(mn)$ moves.

6   Other Self-Stabilizing Domination Algorithms

The reader is referred to an excellent 2010 survey by Guellati and Kheddouci [31] on self-stabilizing algorithms for finding maximal independent sets and minimal dominating sets. Several other papers have been published, which present self-stabilizing, minimal dominating set algorithms.

In 2003 [71], Xu, Hedetniemi, Goddard, and Srimani present a synchronous, self-stabilizing algorithm for finding a minimal dominating set, which stabilizes in $4n$ rounds, starting from any arbitrary global state. A round is defined as the period of time during which every node receives messages from all of its neighbors. The algorithm is general in the sense that it can stabilize with every possible minimal dominating set, as distinct from other self-stabilizing minimal dominating set algorithms, which stabilize only with independent dominating sets.

In 2015 [15], Ding, Wang, and Srimani present a synchronous model, self-stabilizing algorithm for finding a minimal dominating set, which finds a dominating set in just two rounds, but then takes additional n rounds to obtain a minimal dominating set.

Distance-k Dominating Sets
A dominating set $S \subseteq V$ is called a *distance-k dominating set* if for every node $j \in \overline{S}$ there exists a node $i \in S$ such that $d(i, j) \leq k$.

In 2008 [50], Lin, Huang, Wang, and Chen present a self-stabilizing algorithm for finding a minimal distance-2 dominating set in an arbitrary graph.

Distance-k Independent Dominating Sets
Given a graph $G = (V, E)$, a *distance-k independent dominating set*, also called a *maximal distance-k independent set*, is both a distance-k independent set and a distance-k dominating set. That is, given any node $v \in S$, no other node $u \in S$ is at distance k or less from v, and any node $w \in \overline{S}$ is at distance k or less from some node in S.

In 2014 [44], Johnen presents a self-stabilizing algorithm for finding a distance-k independent dominating set, under the unfair distributed scheduler, which stabilizes in at most $4n + k$ rounds. This is further discussed in a subsequent paper by the author in 2015 [45].

Disjoint Dominating Sets
A well-known theorem of Ore [56] states that in any graph having no isolated nodes, the complement $\overline{S}$ of every minimal dominating set is a dominating set. This means that any self-stabilizing algorithm for finding a minimal dominating set in effect finds two disjoint dominating sets, although the complement $\overline{S}$ need not be a minimal dominating set.

Algorithm 2DSC: 2 Dominating Sets - Central
S1: if $(x(i) = 0)(\forall j \in N(i))(x(j) = 0)$
     then $x(i) := 1$ [enter set V_1]
S2: if $(x(i) = 1) \wedge (\forall j \in N(i))(x(j) = 1)$
     then $x(i) := 0$ [enter set V_0]

Fig. 11 Algorithm 2DS: Central Model [32]

Algorithm Unfriendly - Central
S1: if $(C(i) = Blue) \wedge (B(i) > R(i))$
     then $C(i) = Red$
S2: if $(C(i) = Red) \wedge (R(i) > B(i))$
     then $C(i) = Blue$

Fig. 12 Algorithm Unfriendly: Central Model [34]

A *dominating bipartition* is a bipartition $V = V_0 \cup V_1$ into two disjoint dominating sets, neither of which needs to be a minimal dominating set.

In 2003 [32], Hedetniemi, Hedetniemi, Jacobs, and Srimani present the following very simple, self-stabilizing Algorithm 2DSC, in Figure 11, under the central scheduler, for creating a dominating bipartition.

This algorithm stabilizes in at most $n - 1$ moves.

An *unfriendly partition* is a two-coloring of the nodes of a graph, say with colors Red and Blue, having the property that every node colored Red has at least as many Blue neighbors as it has Red neighbors, and every node colored Blue has at least as many Red neighbors as Blue neighbors. These partitions were originally defined and studied by Borodin and Koshtochka in 1977 [5], Aharoni, Milner, and Prikry in 1990 [1] and Shelah and Milner in 1990 [57]. They observed the following simple result.

Theorem 15 *Every finite connected graph G of order $n \geq 2$ has an unfriendly partition.*

It is immediate from the definition that every unfriendly partition $V = R \cup B$ is a bipartition into two dominating sets. In 2013 [34], Hedetniemi, Hedetniemi, Kennedy, and McRae present three self-stabilizing algorithms for finding an unfriendly partition, all using the central scheduler model. The first and simplest of these is Algorithm Unfriendly—Central, in Figure 12, where $C(i) \in \{Blue, Red\}$, $B(i) = |\{j : j \in N(i) \wedge C(j) = Blue\}|$ equals the number of Blue neighbors of node i and $R(i) = |\{j : j \in N(i) \wedge C(j) = Red\}|$ equals the number of Red neighbors of node i.

This algorithm stabilizes with an unfriendly partition in at most $m = |E|$ moves.

In 2015 [36], Hedetniemi, Jacobs, and Kennedy, using the distance-2 model, in which nodes can utilize state information of all nodes within distance-2 in making a move, present a self-stabilizing algorithm for finding one maximal independent set, and a second disjoint minimal dominating set. This algorithm stabilizes in $O(n^2)$ moves, which can be converted to a distance-1 model algorithm that makes $O(n^5)$

Algorithm Optimally Efficient - Central
S1: if $(x(i) = 0) \wedge (|N_0(i)| > |N_1(i)| + 1SNbr(i)$
      then $x(i) = 1$
S2: if $(x(i) = 1) \wedge (|N_2(i)| \geq |N_1(i)| - 1SNbr(i))$
      then $x(i) = 0$

Fig. 13 Algorithm Optimally Efficient: Central Model [66]

moves. They also present a distance-2, self-stabilizing algorithm for finding two disjoint minimal dominating sets, which also stabilizes in $O(n^2)$ moves. Two other self-stabilizing, unfriendly partition algorithms are also given, using the distance-2 model, where the objective is to increase the number of bicolored edges in the resulting unfriendly partition.

Optimally Efficient Sets

The *efficiency* of a set $S \subseteq V$ is defined as $\varepsilon(S) = |\{v \in \overline{S} : |N(v) \cap S| = 1\}|$, which equals the number of nodes not in S that are adjacent to exactly one node in S, or are dominated exactly once by the nodes in S. The *efficiency of a graph G* is defined to be $\varepsilon(G) = max\{\varepsilon(S) : S \subseteq V\}$. A set S is called *optimally efficient* if adding nodes cannot increase its efficiency, but deleting a node decreases its efficiency.

In 2012 [33], Hedetniemi, Hedetniemi, Jiang, Kennedy, and McRae present a self-stabilizing algorithm, under the central scheduler and the distance-2 model, to find a maximal optimally efficient set S in $O(n^2)$ distance-2 moves, or $O(n^5)$ distance-1 moves.

In 2013 [66], Turau presents two self-stabilizing algorithms, the first of which considerably improves on the algorithm mentioned above, by Hedetniemi et al. [33], for finding an optimally efficient set, and which stabilizes in $O(n^5)$ moves. Algorithm Optimally Efficient—Central, in Figure 13, by Turau, operating under the unfair distributed scheduler, stabilizes in just $O(nm)$ moves. Since this algorithm has just two rules, we present this algorithm. Once again we change the notation to be similar to that used throughout this chapter. This algorithm is designed for a central scheduler.

Let $N_k(i) = \{j \in N(i) : x(j) = 0 \wedge N_S(j) = k\}$ denote the neighbors of node i having $x(j) = 0$ and exactly k neighbors in S. Let $1SNbr(i) = 1$ if node i has exactly one neighbor in S, and $1SNbr(i) = 0$ otherwise.

The second algorithm in [66] is the first self-stabilizing algorithm, sequential or otherwise, which computes the exact value of the efficiency $\varepsilon(T)$ of a tree T.

7 Avenues for Further Study

As indicated in the introduction to this chapter, research on self-stabilizing domination algorithms has only been going on for about 20 years. Furthermore, researchers who design self-stabilizing, domination-related algorithms are relatively few in number. But given the relative ease of designing self-stabilizing domination algorithms, and their usefulness, it may be a fruitful area for both graph theorists and algorithms researchers.

In this closing section, we will list a number of areas of domination in graphs in which self-stabilizing algorithms have either not yet been designed or in which relatively little has been done. Keep in mind that for any one type of domination listed below, self-stabilizing algorithms can be designed with three types of schedulers: central, synchronous, and distributed; they can be ID-based or anonymous; and they can use distance-k knowledge for varying values of k. Thus, one always has a lot of design options. The reader is challenged to see if you can design a self-stabilizing algorithm for finding a minimal dominating set of any of the following types.

1. **Paired domination.** A dominating set $S \subseteq V$ is called a *paired dominating set* if the induced subgraph $G[S]$ has a perfect matching.
2. **Restrained domination.** A dominating set $S \subseteq V$ is called a *restrained dominating set* if the subgraph $G[\overline{S}]$ induced by $\overline{S}$ contains no isolated vertices, that is, every vertex in $\overline{S}$ has at least one neighbor in $\overline{S}$.
3. **Signed domination.** A function $f : V \rightarrow \{-1, 1\}$ is called a *signed dominating function* if for every vertex $v \in V$, $f(N[v]) = \Sigma_{w \in N[v]} f(w) \geq 1$.
4. **Minus domination.** A function $f : V \rightarrow \{-1, 0, 1\}$ is called a *minus dominating function* if for every vertex $v \in V$, $f(N[v]) = \Sigma_{w \in N[v]} f(w) \geq 1$.
5. **Odd domination.** A dominating set S is called an *odd dominating set* if for every vertex $v \in V$, $|N[v] \cap S|$ is an odd number.
6. **Secure domination.** A dominating set S is called a *secure dominating set* if for every vertex $v \in \overline{S}$ there exists an adjacent vertex $u \in S$ such that $S - \{u\} \cup \{v\}$ is a dominating set.
7. **Roman domination.** A function $f : V \rightarrow \{0, 1, 2\}$ is called a *Roman dominating function* if every vertex $v \in V$ with $f(v) = 0$ is adjacent to at least one vertex $w \in N(v)$ with $f(w) = 2$.
8. **Cost-effective domination.** A dominating set $S \subseteq V$ is called a *cost-effective dominating set* if every vertex $v \in S$ has at least as many neighbors in $\overline{S}$ as it has in S.
9. **Capacity-k domination.** A dominating set $S = \{v_1, v_2, \ldots, v_r\} \subseteq V$ is called a *capacity-k dominating set* if there exists a partition $V = \{V_1, V_2, \ldots, V_r\}$ such that for every $1 \leq i \leq r$, (i) $v_i \in V_i$, (ii) $V_i \subseteq N[v_i]$, and (iii) $|V_i| \leq k$.
10. **Connected domination.** This deserves some discussion. Given a graph $G = (V, E)$, a dominating set S is a *connected dominating set* if the subgraph $G[S]$ induced by S is connected. The problem of finding a minimal connected dominating set in a graph has been quite a challenge using the self-stabilizing paradigm. Indeed, is such an algorithm even possible, given that each node only has local knowledge of the graph?

In 2010 [46], Kamei and Kakugawa present a self-stabilizing algorithm that approximates the connected domination number $\gamma_c(G)$ within a factor of at most $7.6\gamma_c(G) + 1.4$. Their algorithm stabilizes in $O(k)$ rounds, where k is the depth of an input breadth-first-search spanning tree of G.

In 2010 [30], Goddard and Srimani present two self-stabilizing algorithms for finding reasonably minimal connected dominating sets, but not guaranteed to be minimal, the second of which constructs a breadth-first spanning tree and then discards the leaves. Their algorithms run with anonymous nodes and with a distributed scheduler.

In 2012 [47], Kamei and Kakugama improve on their previous result by presenting a self-stabilizing algorithm, which approximates the connected domination number within a factor of 6, when restricted to unit-disk graphs.

In 2013 [48], Kamei, Kakugawa, Devismes, and Tixeuil present a self-stabilizing algorithm that approximates the maximum number of leaves in any spanning tree of a graph G within a factor of 1/3, meaning that it is guaranteed to have at least 1/3 of the maximum possible number of leaves. Their algorithm stabilizes in at most $O(n^2)$ rounds.

11. **Weakly connected domination**. Given a graph $G = (V, E)$, a dominating set S is called *weakly connected* if the subgraph induced by the edges having at least one node in S is connected. While the problem of designing a self-stabilizing algorithm for finding a minimal connected dominating set has proven to be difficult, in 2009 [67] Turau and Hauck present a self-stabilizing algorithm for finding a weakly connected dominating set. In 2015 [14], Ding, Wang, and Srimani present another self-stabilizing algorithm for finding a weakly connected dominating set.

12. **$\{k\}$-domination**. Given a graph $G = (V, E)$, let $f : V \rightarrow \{0, 1, \ldots, k\}$ be a function from the node set V to the set of integers $\{0, 1, \ldots, k\}$. For any subset $S \subset V$, define $f(S) = \Sigma_{v \in S} f(v)$. Such a function f is called a $\{k\}$-*dominating function* if for every node $i \in V$, $f(N[i]) \geq k$.

In 2003 [20], Gairing, Hedetniemi, Kristiansen, and McRae present a five-rule, self-stabilizing algorithm for finding a minimal $\{k\}$-dominating function, which stabilizes in at most $(2n + 1)(n + 1)2^{n+2}$ moves. They also present a self-stabilizing algorithm for finding a $\{2\}$-dominating function, which stabilizes in at most $3n + 2m$ moves. This in turn provides a self-stabilizing algorithm, which stabilizes in $O(n)$ moves when restricted to planar graphs. A version of this self-stabilizing algorithm for $k = 2$ can be found in the 2002 PhD thesis of Kristiansen [49].

13. **Strong and weak domination**. Given a graph $G = (V, E)$, a dominating set S is called *strong* if for every node $j \in \overline{S}$ there exists a node $i \in N(j) \cap S$ whose degree $d(i)$ satisfies $d(i) \geq d(j)$. Similarly, a dominating set S is called *weak* if for every node $j \in \overline{S}$ there exists a node $i \in N(j) \cap S$ with $d(i) \leq d(j)$.

In 2015, Neggazi, Guellati, Haddad, and Kheddouci [55] present a self-stabilizing algorithm for finding an independent strong dominating set, which operates under the unfair distributed scheduler and stabilizes in at most $n + 1$ rounds. The authors show that using rules that choose nodes having larger degrees than their neighbors (strong domination) results empirically in smaller dominating sets than the maximal independent sets and minimal dominating sets found by previous self-stabilizing algorithms.

Algorithm Maximal Irredundant - Central
ENTER: if $(x(i) = 0) \wedge (i$ **is safe)**
   then $x(i) = 1$
LEAVE: if $(x(i) = 1) \wedge (i$ **has no private neighbor)**
   then $x(i) = 0$

Fig. 14 Algorithm Maximal Irredundant: Central Model [28]

14. **k-domination**. Given a graph $G = (V, E)$, a *k-dominating set* is a set $S \subseteq V$ having the property that for every node $j \in \overline{S}$, $|N(j) \cap S| \geq k$, that is, every node in $\overline{S}$ is dominated at least k-times. Self-stabilizing algorithms for finding a minimal 2-dominating set have been designed in 2007 [40] by Huang, Lin, Chen, and Wang, using a distributed scheduler, and in 2008 [39] by Huang, Chen, and Wang, using a central scheduler.

15. **Maximal irredundant sets**. Given a graph $G = (V, E)$, a set $S \subseteq V$ is called *irredundant* if for every node $i \in S$, either (i) there exists a node $j \in \overline{S}$ such that $N(j) \cap S = \{i\}$, in which case we say that node j is an *external private neighbor* of node i, with respect to the set S, or (ii) node i is not adjacent to any node in S, in which case we say that node i is *its own private neighbor*. An irredundant set S is *maximal* if for every node $j \in \overline{S}$, the set $S \cup \{j\}$ is not irredundant. This means that either node j does not have a private neighbor with respect to the set $S \cup \{j\}$ or there exists a node $i \in S$ such that i has a private neighbor with respect to S but does not have a private neighbor with respect to $S \cup \{j\}$. In this latter case, we say that adding node j to S *destroys* node i. We say that a node $j \in \overline{S}$ is *safe* with respect to a set S if adding it to S does not destroy any node in S and j has a private neighbor with respect to the set $S \cup \{j\}$.

It is well known that every minimal dominating set is maximal irredundant. Thus, every self-stabilizing algorithm for finding a minimal dominating set also finds a maximal irredundant set. But since there are maximal irredundant sets that are not dominating sets, a true self-stabilizing, maximal irredundant algorithm had not been designed until 2008, when Goddard, Hedetniemi, Jacobs, and Trevisan [28] found a way to design such an algorithm using distance-4 knowledge. Their Algorithm Maximal Irredundant—Central, in Figure 14, has only two rules.

The authors show that this algorithm finds a maximal irredundant set in $O(n^7)$ moves.

16. **Alliances in graphs**. Given a graph $G = (V, E)$, a set $S \subset V$ is a *global offensive alliance* if each node $i \in \overline{S}$ has $|N[i] \cap S| \geq |N[i] \cap \overline{S}|$, that is, every node in $\overline{S}$ has at least as many neighbors in S as it has in $\overline{S}$ plus itself. Similarly, a set $S \subset V$ is a *global defensive alliance* if each node $i \in S$ has $|N[i] \cap S| \geq |N[i] \cap \overline{S}|$, that is, every node in S has at least as many neighbors in S plus itself as it has in $\overline{S}$. A set S is a *global powerful alliance* if it is both a global defensive and global offensive alliance. Self-stabilizing global algorithms have

been designed in 2006 [70] by Xu, in 2013 [72] by Yahiaoui, Belhoul, Haddad, and Kheddouci, and in 2014 [60] by Srimani.

17. **Dominating sets with external private neighbors**. According to a well-known theorem of Bollobás and Cockayne, every graph G without isolated vertices has a minimum dominating set in which every vertex has an external private neighbor. So far, no self-stabilizing algorithm has appeared for a minimal dominating set having this property.

18. **Self-stabilizing algorithms on special classes of graphs**. In general, self-stabilizing algorithms are designed for arbitrary graphs. But it may well be possible to design algorithms that stabilize even faster on special classes of graphs, such as grid graphs, n-cubes, planar graphs, trees, and chordal graphs.

19. **Relative performance of self-stabilizing algorithms**. The algorithms we have presented in this chapter always have rules, whereby a node i enters the set S in question. But these in-moves can be modified in ways other than by comparing the IDs of nodes in the neighborhood of a node i, for example, you could give preference to allowing a node to make an in-move if its degree is either greater than or less than the degrees of nodes in its neighborhood. This would allow preference to be given to nodes of larger degree, or nodes of smaller degree. And this then means that you can get empirical data on the relative speeds and relative performance of self-stabilizing algorithms, namely, on average which algorithms stabilize more quickly, and when they stabilize, on average how large or how small are the maximal independent sets or minimal dominating sets that are found.

20. **Distance-k self-stabilizing algorithms**. Recent research has expanded the assumption that a node can only "see" the values of the variables of the nodes in its neighborhood $N(i)$; this is called the *shared-variable model*. But what if a node could see not only the values of the variables of its neighbors, but the neighbors of its neighbors? Several papers have been published on *distance-k knowledge* and how it is possible to convert a distance-k self-stabilizing algorithm to a standard distance-1 model algorithm, albeit at an increased cost in running time. These algorithms quickly become more sophisticated but enable other types minimal and maximal sets to be found, like maximal irredundant sets or k-packings (cf. Gairing et al. in 2004 [19], Goddard et al. in 2006 [27] and Goddard et al. in 2008 [28]).

Acknowledgment The author would like to thank the reviewers for the time it took them to carefully read this chapter and for their many suggestions to improve its presentation.

References

1. R. Aharoni, E.C. Milner, K. Prikry, Unfriendly partitions of a graph. J. Combin. Theory Ser. B **50**(1), 1–10 (1990)
2. Y. Asada, M. Inoue, An efficient silent self-stabilizing algorithm for 1-maximal matching in anonymous networks, in *WALCOM: Algorithms and Computation*. Lecture Notes in Computer Science, vol. 8973 (2015), pp. 187–198

3. Y. Belhoul, S. Yahiaoui, H. Kheddouci, Efficient self-stabilizing algorithms for minimal total k-dominating sets in graphs. Inform. Process. Lett. **114**(7), 339–343 (2014)

4. J.R.S. Blair, S.M. Hedetniemi, S.T. Hedetniemi, D.P. Jacobs, Self-stabilizing maximum matchings. Congr. Numer. **153**, 151–159 (2001)

5. A.V. Borodin, A.V. Kostochka, On an upper bound of a graph's chromatic number, depending on the graph's degree and density. J. Combin. Theory Ser. B **23**, 247–250 (1977)

6. W.Y. Chiu, C. Chen, Linear-time self-stabilizing algorithms for minimal domination in graph, in *Proceedings of the International Workshop on Combinatorial Algorithms, IWOCA 2013*. Lecture Notes in Computer Science, vol. 8288 (2013), pp. 115–126

7. W.Y. Chiu, C. Chen, S.-Y. Tsai, A $4n$-move self-stabilizing algorithm for the minimal dominating set problem using an unfair distributed demon. Inform. Process. Lett. **114**(10), 515–518 (2014)

8. J. Cohen, J. Lefévre, K. Maâmra, L. Pilard, D. Sohier, A self-stabilizing algorithm for maximal matching in anonymous networks. Parallel Process. Lett. **26**(4), 1650016 (2016)

9. J. Cohen, K. Maâmra, G. Manoussakis, L. Pilard, Polynomial self-stabilizing maximum matching algorithm with approximation ratio 2/3. *20th International Conference on Principles of Distributed Systems*, Art. No. 11, 17 pp., LIPIcs. Leibniz Int. Proc. Inform., 70, Schloss Dagstuhl. Leibniz-Zent. Inform., Wadern, 2017

10. A.K. Datta, L. Larmore, T. Masuzawa, Maximum matching for anonymous trees with constant space per process. *19th International Conference on Principles of Distributed Systems*, Art. No. 16, 16 pp., LIPIcs. Leibniz Int. Proc. Inform., 46, Schloss Dagstuhl. Leibniz-Zent. Inform., Wadern, 2016

11. F. Delbot, C. Laforest, S. Rovedakis, Self-stabilizing algorithms for connected vertex cover and clique decomposition problems, in *Principles of Distributed Systems*. Lecture Notes in Computer Science, vol. 8878 (2014), pp. 307–322

12. E.W. Dijkstra, Self-stabilizing systems in spite of distributed control. Commun. ACM **17**(11), 643–644 (1974)

13. E.W. Dijkstra, A belated proof of self-stabilization. J. Distributed Comput. **1**, 5–6 (1986)

14. Y. Ding, J.Z. Wang, P.K. Srimani, New self-stabilizing algorithms for minimal weakly connected dominating sets. Internat. J. Found. Comput. Sci. **26**(2), 229–240 (2015)

15. Y. Ding, J.Z. Wang, P.K. Srimani, Self-stabilizing algorithm for minimal dominating set with safe convergence in an arbitrary graph. Parallel Process. Lett. **25**(4), 1550011 (2015)

16. S. Dolev, *Self-Stabilization* (MIT Press, Cambridge, 2000)

17. M. Gairing, R.M. Geist, S.T. Hedetniemi, P. Kristiansen, A self-stabilizing algorithm for maximal 2-packing. Nordic J. Comput. **11**(1), 1–11 (2004)

18. M. Gairing, W. Goddard, S.T. Hedetniemi, D.P. Jacobs, Self-stabilizing maximal k-dependent sets in linear time. Parallel Process. Lett. **14**(1), 75–82 (2004)

21. M. Gairing, W. Goddard, S.T. Hedetniemi, P. Kristiansen, A.A. McRae, Distance-two information in self-stabilizing algorithms. Parallel Process. Lett. **14**, 387–398 (2004)

20. M. Gairing, S.T. Hedetniemi, P. Kristiansen, A.A. McRae, Self-stabilizing algorithms for $\{k\}$-domination, in *Proceedings of the Sixth Symposium on Self-Stabilization (SSS 2003)*, San Francisco. Lecture Notes in Computer Science, vol. 2074 (Springer, Berlin, 2003), pp. 49–60

21. W. Goddard, S.T. Hedetniemi, D.P. Jacobs, P.K. Srimani, A self-stabilizing distributed algorithm for minimal total domination in an arbitrary system graph, in *Proceedings of the Eighth IPDPS Workshop on Formal Methods for Parallel Programming: Theory and Applications*, Nice (2003)

22. W. Goddard, S.T. Hedetniemi, D.P. Jacobs, P.K. Srimani, Self-stabilizing protocols for maximal matching and maximal independent sets for ad hoc networks, in *Proceedings of the Fifth IPDPS Workshop on Advances in Parallel and Distributed Computational Models*, Nice (2003)

23. W. Goddard, S.T. Hedetniemi, D.P. Jacobs, P.K. Srimani, A robust distributed generalized matching protocol that stabilizes in linear time, in *Proceedings of the ICDCS International Workshop on Mobile Distributed Computing (MDC03)*, Rhode Island (2003)

24. W. Goddard, S.T. Hedetniemi, D.P. Jacobs, P.K. Srimani, Self-stabilizing distributed algorithm for strong matching in a system graph, in *HiPC 2003*, ed. by T.M. Pinkston, V.K. Prasanna, Lecture Notes in Computer Science, vol. 2913 (Springer, Berlin, 2003), pp. 66–73

25. W. Goddard, S.T. Hedetniemi, D.P. Jacobs, P.K. Srimani, Self-stabilizing global optimization algorithms for large network graphs. Internat. J. Dist. Sensor Netw. **1**, 329–344 (2005)
26. W. Goddard, S.T. Hedetniemi, D.P. Jacobs, P.K. Srimani, Z. Xu, Self-stabilizing graph protocols. Parallel Process. Lett. **18**(1), 189–199 (2008)
27. W. Goddard, S. Hedetniemi, D. Jacobs, V. Trevisan, Distance-k information in self-stabilizing algorithms, in *13th Colloquium on Structural Information and Communication Running Time (SIROCCO)*, (Chester, UK, 2006). Lecture Notes in Computer Science, vol. 4056 (2006), pp. 349–356
28. W. Goddard, S.T. Hedetniemi, D.P. Jacobs, V. Trevisan, Distance-k knowledge in self-stabilizing algorithms. Theoret. Comput. Sci. **399**, 118–127 (2008)
29. W. Goddard, S.T. Hedetniemi, Z. Shi, An anonymous self-stabilizing algorithm for 1-maximal matching in trees, in *Proceedings of the 2006 International Conference on Parallel and Distributed Processing Techniques and Applications (PDPTA"6)*, Las Vegas, vol. II (2006), pp. 797–803
30. W. Goddard, P.K. Srimani, Anonymous self-stabilizing distributed algorithms for connected dominating set in a network graph, in *Proceedings of the International Multi-Conference Complexity, Informatics and Cybernetics: IMCIC 2010*
31. N. Guellati, H. Kheddouci, A survey on self-stabilizing algorithms for independence, domination, coloring, and matching in graphs. J. Parallel Distrib. Comput. **70**(4), 406–415 (2009)
32. S.M. Hedetniemi, S.T. Hedetniemi, D.P. Jacobs, P.K. Srimani, Self-stabilizing algorithms for minimal dominating sets and maximal independent sets. Comput. Math. Appl. **46**, 805–811 (2003)
33. S.M. Hedetniemi, S.T. Hedetniemi, H. Jiang, K.E. Kennedy, A. McRae, A self-stabilizing algorithm for optimally efficient sets in graphs. Inform. Process. Lett. **112**(16), 621–623 (2012)
34. S.M. Hedetniemi, S.T. Hedetniemi, K.E. Kennedy, A. A. McRae, Self-stabilizing algorithms for unfriendly partitions into two disjoint dominating sets. Parallel Process. Lett. **23**(1), 1350001 (2013)
35. S.T. Hedetniemi, D.P. Jacobs, K.E. Kennedy, Linear-time self-stabilizing algorithms for disjoint independent sets. Comput. J. **56**(11), 1381–1387 (2013)
36. S.T. Hedetniemi, D.P. Jacobs, K.E. Kennedy, A theorem of Ore and self-stabilizing algorithms for disjoint minimal dominating sets. Theoret. Comput. Sci. **593**, 132–138 (2015)
37. S.T. Hedetniemi, D.P. Jacobs, P.K. Srimani, Maximal matching stabilizes in time $O(m)$. Inform. Process. Lett. **80**, 221–223 (2001)
38. S.-C. Hsu, S.-T. Huang, A self-stabilizing algorithm for maximal matching. Inform. Process. Lett. **43**, 77–81 (1992)
39. T.-C. Huang, C.-Y. Chen, C.-P. Wang, A linear-time self-stabilizing algorithm for the minimal 2-dominating set problem in general networks. JISE J. Inf. Sci. Eng. **24**(1), 175–187 (2008)
40. T.-C. Huang, J.-C. Lin, C.-Y. Chen, C.-P. Wang, A self-stabilizing algorithm for finding a minimal 2-dominating set assuming the distributed demon model. Comput. Math. Appl. **54**(3), 350–356 (2007)
41. M. Ikeda, S. Kamei, H. Kakugawa, A space-time optimal self-stabilizing algorithm for the maximal independent set problem, in *PDCAT Proceedings of the Third International Conference on Parallel and Distributed Computing, Applications and Technologies* (2002), pp. 70–74
42. M. Inoue, F. Ooshita, S. Tixeuil, An efficient silent self-stabilizing 1-maximal matching algorithm under distributed scheduler without global identifiers, in *Stabilization, Safety and Security of Distributed Systems*. Lecture Notes in Computer Science, vol. 10083 (2016), pp. 195–212
43. M. Inoue, F. Ooshita, S. Tixeuil, An efficient silent self-stabilizing 1-maximal matching algorithm under distributed scheduler for arbitrary networks, in *Stabilization, Safety and Security of Distributed Systems*. Lecture Notes in Computer Science, vol. 10616 (2017), pp. 93–108
44. C. Johnen, Fast, silent self-stabilizing distance-k independent dominating set construction. Inform. Process. Lett. **114**(10), 551–555 (2014)

45. C. Johnen, Memory efficient self-stabilizing distance-k independent dominating set construction, in *Networked Systems*. Lecture Notes in Computer Science, vol. 9466 (2015), pp. 354–366
46. S. Kamei, H. Kakugawa, A self-stabilizing distributed approximation algorithm for the minimum connected dominating set. Int. J. Found. Comput. Sci. **21**(3), 459–476 (2010)
47. S. Kamei, H. Kakugawa, A self-stabilizing 6-approximation for the minimum connected dominating set with safe convergence in unit disk graphs. Theoret. Comput. Sci. **428**, 80–90 (2012)
48. S. Kamei, H. Kakugawa, S. Devismes, S. Tixeuil, A self-stabilizing 3-approximation for the maximum leaf spanning tree problem in arbitrary networks. J. Comb. Optim. **25**(3), 430–459 (2013)
49. P. Kristiansen, *New Results on the Domination Chain, Graph Homomorphisms, Alliancesand Self-stabilizing Algorithms*. Ph.D Thesis, Department of Informatics, University of Bergen, 2002
50. J.-C. Lin, T.-C. Huang, C.-P. Wang, C.-Y. Chen, A self-stabilizing algorithm for finding a minimal distance-2 dominating set in distributed systems. JISE J. Inf. Sci. Eng. **24**(6), 1709–1718 (2008)
51. F. Manne, M. Mjelde, A memory efficient self-stabilizing algorithm for maximal k-packing, in *Eighth International Symposium Stabilization, Safety and Security of Distributed Systems, SSS* (November 2006, Dallas), ed. by A.K. Datta, M. Gradinariu. Lecture Notes in Computer Science, vol. 4280 (2006)
52. F. Manne, M. Mjelde, L. Pilard, S. Tixeuil, A new self-stabilizing maximal matching algorithm, in *Structural Information and Communication Complexity*. Lecture Notes in Computer Science, vol. 4474 (2007), pp. 96–108
53. F. Manne, M. Mjelde, L. Pilard, S. Tixeuil, A new self-stabilizing maximal matching algorithm. Theoret. Comput. Sci. **410**(14), 1336–1345 (2009)
54. F. Manne, M. Mjelde, L. Pilard, S. Tixeuil, A self-stabilizing 2/3-approximation algorithm for the maximum matching problem. Theoret. Comput. Sci. **412**(40), 5515–5526 (2011)
55. B. Neggazi, N. Guellati, M. Haddad, H. Kheddouci, Efficient self-stabilizing algorithm for independent strong dominating sets in arbitrary graphs. Int. J. Found. Comput. Sci. **26**(6), 751–768 (2015)
56. O. Ore, *Theory of Graphs*. American Mathematical Society Colloquium Publications (1962)
57. S. Shelah, E.C. Milner, Graphs with no unfriendly partitions, in *A Tribute to Paul Erdös*, ed. by A. Baker, B. Bollobás, A. Hajnal (Cambridge University Press, Cambridge, 1990), pp. 373–384
58. Z. Shi, A self-stabilizing algorithm to maximal 2-packing with improved complexity. Inform. Process. Lett. **112**(13), 525–531 (2012)
59. S. Shukla, D.J. Rosenkrantz, S.S. Ravi, Observations on self-stabilizing graph algorithms for anonymous networks, in *Proceedings of the Second Workshop on Self-Stabilizing Systems* (1995), pp. 7.1–7.15
60. P.K. Srimani, Self-stabilizing minimal global offensive alliance algorithm with safe convergence in an arbitrary graph, in *Theory and Applications of Models of Computation*, Lecture Notes in Computer Science, vol. 8402 (2014), pp. 366–377
61. G. Tel, Maximal matching stabilizes in quadratic time. Inform. Process. Lett. **49**, 271–272 (1994)
62. J.A.Trejo-Sánchez, J.A. Fernández-Zepeda, J.C. Ramírez-Pacheco, A self-stabilizing algorithm for a maximal 2-packing in a cactus graph under any scheduler. Int. J. Found. Comput. Sci. **28**(8), 1021–1045 (2017)
63. V. Turau, Linear self-stabilizing algorithms for the independent and dominating set problems using an unfair distributed scheduler. Inform. Process. Lett. **103**(3), 88–93 (2007)
64. V. Turau, Efficient transformation of distance-2 self-stabilizing algorithms. J. Parallel Distrib. Comput. **72**, 603–612 (2012)
65. V. Turau, Self-stabilizing vertex cover in anonymous networks with optimal approximation ratio. Parallel Process. Lett. **20**(2), 173–186 (2010)

66. V. Turau, Self-stabilizing algorithms for efficient sets of graphs and trees. Inform. Process. Lett. **113**(19–21), 771–776 (2013)
67. V. Turau, B. Hauck, A self-stabilizing algorithm for constructing weakly connected minimal dominating sets. Inform. Process. Lett. **109**(14), 763–767 (2009)
68. V. Turau, B. Hauck, A new analysis of a self-stabilizing maximum weight matching algorithm with approximation ratio 2. Theoret. Comput. Sci. **412**(40), 5527–5540 (2011)
69. V. Turau, B. Hauck, A fault-containing self-stabilizing $(3 - \frac{2}{\Delta+1})$-approximation algorithm for vertex cover in anonymous networks. Theoret. Comput. Sci. **412**(33), 4361–4371 (2011)
70. Z. Xu, *Self-stabilizing Protocols for Distributed Systems*. Ph.D. thesis, Clemson University, 2002
71. Z. Xu, S.T. Hedetniemi, W. Goddard, P.K. Srimani, A synchronous self-stabilizing minimal dominating protocol in an arbitrary network graph, in *Proceedings of the Fifth International Workshop on Distributed Computing (IWDC2003)*. Lecture Notes in Computer Science, vol. 2918 (2003), pp. 26–32
72. S. Yahiaoui, Y. Belhoul, M. Haddad, H. Kheddouci, Self-stabilizing algorithms for minimal global powerful alliance sets in graphs. Inform. Process. Lett. **113**(10–11), 365–370 (2013)

Algorithms and Complexity of Alliances in Graphs

Stephen T. Hedetniemi

1 Introduction

Throughout this chapter we will use the following terminology and notation. Given a graph $G = (V, E)$ of *order* $n = |V|$ and *size* $m = |E|$, let $N(v) = \{u | uv \in E\}$ be the set of *neighbors* of vertex v and let $d(v) = |N(v)|$ be the *degree* of v. For a set $S \subseteq V$ of vertices, let $N(S) = \cup_{v \in S} N(v)$ and let $\overline{S}$ denote the set of vertices not in S. If $d_S(v) = |N(v) \cap S|$ denotes the number of neighbors of v that are in S and $d_{\overline{S}}(v) = |N(v) \cap \overline{S}|$ denotes the number of neighbors of v in $\overline{S}$, then $d(v) = d_S(v) + d_{\overline{S}}(v)$. The *boundary* of a set S of vertices is the set $\partial(S) = N(S) \cap \overline{S}$.

A set S of vertices in a graph $G = (V, E)$ is called a *defensive alliance* if for every vertex in $v \in S$, $d_S(v) + 1 \geq d_{\overline{S}}(v)$, that is, every vertex in S has at least as many neighbors in S, including itself, as it has neighbors in $\overline{S}$. A defensive alliance S is called *strong* if the degree inequality is strict, $d_S(v) + 1 > d_{\overline{S}}(v)$. The *defensive alliance number* $a(G)$ equals the minimum cardinality of a defensive alliance in G, while the *strong defensive alliance number* $\hat{a}(G)$ equals the minimum cardinality of a strong defensive alliance in G.

For a vertex-weighted graph G, where each vertex $v \in V$ has a non-negative weight $w(v)$, a set S is called a *weighted defensive alliance* if for every vertex $v \in S$, $\Sigma_{u \in N[v] \cap S} w(v) \geq \Sigma_{u \in N(v) \cap \overline{S}} w(u)$.

This is generalized as follows: a set S is a *defensive k-alliance* if for every vertex $v \in S$, $d_S(v) \geq d_{\overline{S}}(v) + k$. A defensive k-alliance S is called a *global defensive alliance* if S is a dominating set, or equivalently, if for every vertex $w \in \overline{S}$, $N(w) \cap S \neq \varnothing$, every vertex in $\overline{S}$ has at least one neighbor in S.

S. T. Hedetniemi (✉)
School of Computing, Clemson University, Clemson, SC, USA
e-mail: hedet@clemson.edu

T. W. Haynes et al. (eds.), *Structures of Domination in Graphs*, Developments in Mathematics 66, https://doi.org/10.1007/978-3-030-58892-2_17

A set S of vertices in a graph $G = (V, E)$ is called an *offensive alliance* if for every vertex in $v \in \partial(S)$, $d_S(v) \geq d_{\overline{S}}(v) + 1$. That is, every vertex in $\partial(S)$ has at least as many neighbors in S as it has neighbors in $\overline{S} + 1$. This is generalized as follows: a set S is an *offensive k-alliance* if for every vertex $v \in \partial(S)$, $d_S(v) \geq d_{\overline{S}}(v) + k$. An offensive k-alliance S is called a *global offensive alliance* if S is a dominating set, or equivalently, if for every vertex $w \in \overline{S}$, $N(w) \cap S \neq \varnothing$, every vertex in $\overline{S}$ has at least one neighbor in S. An offensive alliance S is called *strong* if the degree inequality is strict, $d_S(v) > d_{\overline{S}}(v) + 1$. The *offensive alliance number $a_o(G)$* equals the minimum cardinality of an offensive alliance in G, while the *strong offensive alliance number $\hat{a}_o(G)$* equals the minimum cardinality of a strong offensive alliance in G.

Finally, a set S is called a *(global) powerful k-alliance* if S is both a (global) defensive k-alliance and a (global) offensive k-alliance.

2 Algorithms and Complexity of Alliances in Graphs

With respect to defensive, offensive, powerful, and global alliances, there are six basic decision problems as follows:

DEFENSIVE ALLIANCE (DA)
Instance: Graph $G = (V, E)$, positive integer k.
Question: Does G have a defensive alliance of cardinality at most k?

OFFENSIVE ALLIANCE (OA)
Instance: Graph $G = (V, E)$, positive integer k.
Question: Does G have an offensive alliance of cardinality at most k?

POWERFUL ALLIANCE (PA)
Instance: Graph $G = (V, E)$, positive integer k.
Question: Does G have a powerful alliance of cardinality at most k?

GLOBAL DEFENSIVE ALLIANCE (GDA)
Instance: Graph $G = (V, E)$, positive integer k.
Question: Does G have a global defensive alliance of cardinality at most k?

GLOBAL OFFENSIVE ALLIANCE (GOA)
Instance: Graph $G = (V, E)$, positive integer k.
Question: Does G have a global offensive alliance of cardinality at most k?

GLOBAL POWERFUL ALLIANCE (GPA)
Instance: Graph $G = (V, E)$, positive integer k.
Question: Does G have a global powerful alliance of cardinality at most k?

However, another six decision problems arise if one seeks *strong* alliances, in which the degree inequalities are strict. Even more, there are corresponding decision problems for defensive and offensive k-alliances.

The complexity of each of these basic six decision problems has been settled; all of these problems are NP-complete.

In his 2001 Ph.D. thesis, Shafique [18] proves that (STRONG) POWERFUL ALLIANCE is NP-complete.

In 2002 Favaron, Fricke, Goddard, Hedetniemi, Hedetniemi, Kristiansen, Laskar, and Skaggs [6] prove the following, where $\beta(G)$ is the *vertex covering number* of a graph G, that is, the minimum cardinality of a set S of vertices such that for every edge $uv \in E$, $\{u, v\} \cap S \neq \varnothing$.

Theorem 1 (Favaron et al. [6]) *If G is a cubic graph, then every vertex cover is a strong offensive alliance, and vice versa. Therefore, $\hat{a}_o(G) = \beta(G)$.*

Proof Let S be a vertex cover of a cubic graph G. Then the complement $\overline{S}$ of S is an independent set of vertices. Therefore, S is a strong offensive alliance. Conversely, let S be a strong offensive alliance in G, and let $w \in \overline{S}$. If $w \in \partial(S)$, then all neighbors of w must be in S. $\qquad\qquad\square$

Assume $w \notin \partial(S)$. Let $u \in \overline{S}$ be a vertex at the shortest distance from w that has a neighbor in S, and let $(w = u_1, u_2, \ldots, u_k = u)$ be a shortest u-w-path. Note that u_2 could be u. But by our previous argument, every neighbor, including u_{k-1}, of u is in S, contradicting our choice of u.

Since VERTEX COVER is known to be NP-hard, even for cubic graphs, it follows that (STRONG) OFFENSIVE ALLIANCE is NP-hard for cubic graphs.

In 2006 Cami, Balakrishnan, Deo, and Dutton [3] show that the decision problems for global defensive alliances (GDA), global offensive alliances (GOA), and global powerful alliances (GPA) are all NP-complete for general graphs, and thus, it follows that these problems are also NP-complete for weighted graphs. In proving these NP-completeness results, they use transformations from the following well-known NP-complete problem.

DOMINATING SET (DOMSET)
Instance: Graph $G = (V, E)$, positive integer k.
Question: Does G have a set of cardinality at most k?
We present one simple NP-completeness proof.

Theorem 2 (Cami, Balakrishnan, Deo, Dutton [3]) *GLOBAL DEFENSIVE ALLIANCE is NP-complete.*

Proof Sketch. Given an instance of DOMSET, that is, a graph $G = (V, E)$ and a positive integer k, construct a graph $G' = (V', E')$ by attaching to each non-isolated vertex $v \in V$, $d(v) - 1$ paths of length 2. Let A denote the set of all vertices of degree 2 on the paths of length 2 added to G to form G'. One can then show that if G has a dominating set S of cardinality at most k, then the set $S' = S \cup A$ is a global defensive alliance of G' of cardinality at most $k' = k + 2|E| - |V|$. Similarly, one can show that if the constructed graph G' has a global defensive alliance of cardinality at most $k' = k + 2|E| - |V|$, then G has a dominating set of cardinality at most k. $\qquad\square$

In 2007 Fernau and Raible [7] show that OFFENSIVE ALLIANCE is NP-complete, as follows.

Theorem 3 (Fernau, Raible [7]) *OFFENSIVE ALLIANCE is NP-complete.*

Proof Sketch. Use a transformation from STRONG OFFENSIVE ALLIANCE, which is shown to be NP-complete by Favaron et al. [6] above. Given an instance of STRONG OFFENSIVE ALLIANCE, namely, a graph $G = (V, E)$ and a positive integer k, construct a graph $G' = (V', E')$ as follows. Let $G_1 = (V_1, E_1)$ and $G_2 = (V_2, E_2)$ be two copies of G. Thus, for every vertex $v \in V$, there are now two copies $v_1 \in V_1$ and $v_2 \in V_2$. For every vertex $v \in V$, create a vertex v' which is adjacent to v_1 and v_2. Finally create a clique C of order $2k + 1$, and let x be an arbitrary vertex in C. Join x to each of the $|V|$ vertices v'. This, then, is the graph G'.

One can then show that G has a strong offensive alliance of cardinality at most k if and only if G' has an offensive alliance of cardinality at most $2k$. $\square$

Fernau and Raible also show that all of these alliance decision problems are fixed-parameter tractable (FPT) when parameterized by the size k of the alliance set. We illustrate these results with one of their theorems.

Theorem 4 (Fernau, Raible [7]) *DEFENSIVE ALLIANCE is in $\mathcal{FPT}$.*

Proof Given an instance (G, k) of DEFENSIVE ALLIANCE, consisting of a graph G and the parameter k, we show fixed-parameter tractability by constructing at most n search trees, whose sizes are exponential in the fixed parameter k.

Note that in any minimal defensive alliance S, we can assume that $G[S]$ is connected, since if not, then any connected component would necessarily be a smaller defensive alliance.

Let $V = \{v_1, v_2, \ldots, v_n\}$. Starting with $v_1 \in V$, assume that v_1 is a member of a defensive alliance S, with $|S| \leq k$, and assume that $G[S]$ is connected. Then, since $G[S]$ is connected, every other vertex in S can be reached from vertex v_1 in at most k steps. Therefore, we can branch at v_1 as follows: first, if $S = \{v_1\}$ is already a defensive alliance, then we are done. Otherwise, we have to add some neighbor of v_1 to our potential defensive alliance S. This leads to an initial branching of at most $2k - 1$, because if S is a defensive alliance of size at most k, then any vertex $v \in S$ has at most $k - 1$ neighbors in S. This means that vertex v can have at most k neighbors not in S. Thus, $d(v) \leq 2k - 1$, and if S is a defensive alliance of order at most k, then for all $v \in S$, $d(v) \leq 2k - 1$.

Now, after the first branching from vertex v_1, say to neighbor v_2 of v_1, we have a potential defensive alliance of order 2, from which we can branch over the union of the two neighborhoods of v_1 and v_2, resulting in a branch of size at most $2(2k - 1)$. If we continue in this way, we obtain a search tree of size bounded by $\Pi_{j=1}^{k} j(2k-1) = (2k - 1)k!$

Checking if we have found a defensive alliance can be done in $O(k^2)$ time. If we encounter a vertex v^* of degree greater than $2k - 1$, it cannot be part of the search for a defensive alliance of order at most k. And because we assume that S is connected, for any two vertices w_1, w_2, there is a path P connecting w_1 and w_2 with the property that $P \subseteq S$. This means that we do not have to branch over v^*, because we encounter every vertex of S by branching only over vertices with degree $\leq 2k - 1$.

Because our first assumption, that v_1 must be in a defensive alliance, does not have to be true, we might have to do an initial branch over all $v \in V$. Hence, we can decide DEFENSIVE ALLIANCE in $O(k^3 k! n)$ time. $\qquad\square$

In 2007 and 2009 Jamieson, Hedetniemi, and McRae [13, 14] prove that DEFENSIVE ALLIANCE is NP-complete, even when restricted to split graphs, chordal graphs, or bipartite graphs.

In 2007 Jamieson and Dean [12] consider weighted versions of alliance problems. Given a graph $G = (V, E)$, a non-negative integer weighting function $w : V \to \mathcal{N}$, and a set S, define $w(S) = \Sigma_{v \in S} w(v)$.

A set $S \subset V$ is called a *weighted defensive alliance* if for all $v \in S$, $w(N[v] \cap S) \geq w(N[v] \cap \overline{S})$, or equivalently if the ratio $\rho(v) = w(N[v] \cap S)/w(N[v]) \geq 1/2$.

They show that the problem of finding a minimum-cost, weighted alliance of any type (defensive, offensive, or powerful) is NP-hard, even on a star, that is, a tree of the form $K_{1,n}$. We present one of their results; the other results and their proofs are very similar.

Theorem 5 (Jamieson, Dean [12]) *The problem of computing a (global) minimum-cost, weighted powerful alliance is NP-hard, even when restricted to stars.*

Proof Use a transformation from the following, well-known NP-complete problem.

SUBSET SUM
Instance: Finite set A, integer weight $w(a) > 0$ for each $a \in A$, integer $B > 0$.
Question: Is there a subset $A' \subset A$ such that $w(A') = B$?

Consider an n-element instance (A, w, B) of SUBSET SUM, and assume that $B > w(A)/2$. Let $G = K_{1,n+1}$, where the central vertex x has $w(x) = w(A)$, one leaf y has $w(y) = w(A)$, and the remaining n leaves have weight $w(a)$, for each $a \in A$.

Show that G has a weighted powerful alliance of cost $w(A) + B$ if and only if (A, w, B) is a "yes" instance of SUBSET SUM.

Assume that there exists a subset $A' \subseteq A$ with $w(A') = B$. Form a powerful alliance S by taking the center vertex x along with the leaves corresponding to A'. We must show that S is both a weighted defensive alliance and a weighted offensive alliance.

It is easy to see that the ratio $\rho(x) = w(N[x] \cap S)/w(N[x]) = (w(A) + B)/3w(A) \geq 1/2$, and $\rho(v) = 1$ for all leaves $v \in S$. Thus, S is a weighted defensive alliance. Similarly, for every leaf $w \in \overline{S}$, $\rho(w) \geq 1/2$. Thus, S is a weighted offensive alliance.

Conversely, let S be a minimum-cost powerful alliance of G with cost $w(A) + B$. Clearly S must contain the center vertex x, or else S cannot be a weighted powerful alliance. But in this case, S cannot include the leaf y with $w(y) = w(A)$, or else $w(S) > w(A) + B$. Therefore, S includes vertex x together with leaves whose weight sums to exactly B. This gives a solution to the instance of SUBSET SUM. Since S is actually a global powerful alliance, this argument applies to both the global and non-global minimum-cost, weighted powerful alliance problems. $\qquad\square$

Using very similar arguments, Jamieson and Dean prove the following.

Theorem 6 (Jamieson, Dean [12]) *The problem of computing a (global) minimum-cost, weighted defensive alliance is NP-hard, even when restricted to stars.*

Theorem 7 (Jamieson, Dean [12]) *The problem of computing a minimum-cost, weighted offensive alliance is NP-hard, even when restricted to stars.*

The authors point out that it is trivial to compute a *global* minimum-cost weighted offensive alliance on a star, since an optimal solution must either consist of all leaves and not the center or the center vertex plus every leaf having weight greater than the center. But this global weighted problem remains NP-hard for trees.

Theorem 8 (Jamieson, Dean [12]) *The problem of computing a minimum-cost, global weighted offensive alliance is NP-hard, even when restricted to trees.*

The authors conclude their paper by presenting an $O(|V|^3)$ algorithm for computing a minimum *cardinality* weighted defensive alliance on a tree. Using similar approaches, one can also find a minimum cardinality weighted global defensive alliance, (global) offensive alliance, and (global) powerful alliance in a tree.

In 2008 Araujo-Pardo and Barrière [1] study defensive alliances in ℓ-regular graphs and show that for $\ell \leq 5$, there are simple algorithms for finding a minimum defensive alliance in any ℓ-regular graph.

In 2009 Fernau, Rodriquez, and Sigarreta [8] study the complexity of offensive r-alliances in graphs. The *global offensive r-alliance number* $\gamma_{or}(G)$ is the minimum cardinality of a global offensive r-alliance in G. The authors show that the problem of finding a minimum cardinality (global) offensive r-alliance is NP-complete.

Since Cami et al. [3] had previously proved the NP-completeness for $r = 1$, Fernau et al. were able to modify their proof to show NP-completeness of GLOBAL OFFENSIVE r-ALLIANCE for any fixed r.

GLOBAL OFFENSIVE r-ALLIANCE (GOr-A)
Instance: Graph $G = (V, E)$, positive integer $k \leq |V|$.
Question: Does G have a global offensive r-alliance of size at most k?

Theorem 9 (Fernau et al. [8]) *For every r, GOr-A is NP-complete.*

Proof Sketch. We present only the proof for $r \leq 1$, although the construction in Cami et al. [3] can be modified to work for any $r \geq 1$.

Let (G, k) be an instance of DOMSET with G having minimum degree $r + 1$. In order to create an instance of GOr-A, to every vertex $v \in V$, attach $d_G(v) + r - 1 \geq 0$ paths P_3 of length 2, yielding a new graph $G' = (V', E')$, having G as an induced subgraph. Let A denote the new neighbors of vertices in V, and let B denote the set of leaves of all of the attached paths.

If $D \subseteq V$ is a dominating set in G, then $S = D \cup A$ is a global offensive r-alliance. Clearly, S is a dominating set in G'. Now, consider a vertex $v \in B$. Obviously, $N(v) \subseteq A$, and therefore $|N_{G'}(v) \cap S| \geq |N_{G'}(v) \cap \overline{S}| + r$. Since any vertex $v \in V - D$ has at least one neighbor in D, it follows that $|N_{G'}(v) \cap \overline{S}| \leq d_G(v) - 1$, while

$|N_{G'}(v) \cap S| \geq (d_G(v) + r - 1) + 1 = d_G(v) + r$. Therefore, S is a valid global offensive r-alliance.

Conversely, let S be a global offensive r-alliance of G'. Since S is a dominating set, for each P_3 attached to G, either the corresponding A-vertex or the corresponding B-vertex must be in S.

Consider some $v \in \overline{S}$. It must be dominated. If no neighbor of $v \in V$ is in S, then $|N_{G'}(v) \cap S| \leq d_G(v) + r - 1$, while $|N_{G'}(v) \cap \overline{S}| \geq d_G(v)$, which leads to a contradiction. Hence, $S \cap V$ is a dominating set in G.

Combining these arguments, one can conclude that $G = (V, E)$ has a dominating set of size at most k if $G' = (V', E')$ has a global offensive r-alliance of size $k + \Sigma_v(d_G(v) + r - 1) = k + (r - 1)|V| + 2|E|$. $\qquad\square$

In 2012 Chang, Chia, Hsu, Kuo, Lai, and Wang [4] present an $O(n \log \Delta)$ algorithm to determine the global defensive alliance number of a tree.

In 2014, Dourado, Faria, Pizaña, Rautenbach, and Szwarcfiter [5] prove that it is NP-complete to decide for a given 6-regular graph G and a given integer k whether G contains a defensive alliance of order at most k. This completes the results of Araujo-Pardo and Barriére [1] for $r \leq 5$ regular graphs. They also prove that the problem of computing the strong, global offensive alliance number $\gamma_{\hat{o}}(G)$ is APX-hard, even for cubic graphs, and is NP-complete for chordal graphs.

Theorem 10 (Dourado, Faria, Pizaña, Rautenbach, Swarcfiter [5]) *DEFENSIVE ALLIANCE is NP-complete for 6-regular graphs.*

The authors' proof of this theorem uses a transformation from ONE-IN-THREE 3SAT, having only positive literals. Given an instance of this NP-complete problem, they construct a 6-regular graph G and a positive integer k such that the instance of ONE-IN-THREE 3SAT has a solution if and only if the 6-regular graph G has a defensive alliance of cardinality at most k. The details of this proof are too numerous to be included here.

Theorem 11 (Dourado, Faria, Pizaña, Rautenbach, Swarcfiter [5]) *GLOBAL STRONG OFFENSIVE ALLIANCE is NP-complete for chordal graphs.*

Proof Use a transformation from DOMSET, which is known to be NP-complete for chordal graphs. Let $G = (V, E)$ be a chordal graph of size $m = |E|$. Construct another chordal graph $G' = (V', E')$ by attaching $d(v)$ leaves to every vertex $v \in V$. Thus, the total number of added leaves is $2m$. Let L denote this set of added leaves. Furthermore, every global strong offensive alliance of G' must contain all leaves in L.

If S is a dominating set of G, let $S' = S \cup L$. Then clearly every vertex $v \in \overline{S'}$ has at least $d_G(v) + 1$ neighbors in S'. Therefore, S' is a global strong offensive alliance.

Conversely, if S' is a global strong offensive alliance of G', then $L \subseteq S'$. Let $S = S' - L$. Since every vertex $u \in \overline{S}$ has at least $d_G(u) + 1$ neighbors in S', every such vertex has at least one neighbor in S, that is, S is a dominating set of G. Since $|S'| - |S| = 2m(G)$ in both cases, the desired result follows. $\qquad\square$

Since it is known that DOMINATING SET can be solved in polynomial time for strongly chordal graphs, it is natural to ask the following question:

Question 1 *Is GLOBAL STRONG OFFENSIVE ALLIANCE NP-complete for strongly chordal graphs?*

Theorem 12 (Dourado, Faria, Pizaña, Rautenbach, Swarcfiter [5]) *There is some $\varepsilon > 0$ such that approximating GLOBAL STRONG OFFENSIVE ALLIANCE within a factor of $(1 + \varepsilon)$ is NP-hard for cubic graphs.*

In 2015 [9] Fernau, Rodriguez, and Sigarreta show that the problem of computing the minimum cardinality of a powerful r-alliance is NP-hard.

In 2015 [10] Harutyunyan and Legay present linear algorithms for computing the global offensive alliance number and the global powerful alliance number of a weighted tree. Recall that earlier Jamieson and Dean [12] presented $O(|V|^3)$ algorithms for computing these alliance numbers of trees. We present here the Harutyunyan-Legay algorithm for computing the global weighted offensive alliance number of a tree. In this algorithm, $L(v)$ denotes the set of leaves adjacent to vertex v; the parent of a leaf is called a *support vertex*; $S(T)$ denotes the set of support vertices in a rooted tree T; $r(T)$ denotes the root vertex of T; $p(v)$ denotes the parent of vertex v; $w(S)$ denotes the sum of the weights of all vertices in S; d denotes the *depth* of T with respect to the root r; and a vertex labeled "–" is not included in the global offensive alliance S.

Note that for a leaf v and its parent $p(v)$, if the weight of v is greater than the weight of $p(v)$, then v must be in every minimum weight global offensive alliance S. But if the weight of v is less than or equal to the weight of $p(v)$, then we can be guaranteed that we can put $p(v)$ into S.

Algorithm 1 Global weighted offensive alliance number in trees

---

1: **for** $v \in S(T)$ **do**
2:    **for** $\ell \in L(v)$ **do**
3:       **if** $w(\ell) > w(v)$ **then**
4:          put ℓ in S.
5:       **else**
6:          Label ℓ with "–"
7:       **if** there exists $\ell \in L(v)$ such that $w(\ell) \leq w(v)$ **then**
8:          put v in S.
9: **for** vertices v at depth $d - i$, $i = 1$ **to** d **do**
10:    **if** $v \notin S$ **and** v is not labeled "–" **and** all of v's children are labeled "–" **then**
11:       **if** $w(p(v)) \geq w(N[v] - p(v))$ **then**
12:          put $p(v)$ in S (if it already is not) and label v with "–".
13:       **else**
14:          put v in S.
15:    **else if** $v \notin S$ **and** v is not labeled "–" **and** v has at least one neighbor u already in S **then**

16:     **if** $w(N(v) \cap S) \geq w(N[v] - S)$ **then**
17:         label v with "–".
18:     **else if** $w((N(v) \cap S) \cup p(v)) \geq w(N[v] - S - p(v))$ **then**
19:         put $p(v)$ in S and label v with "–".
20:     **else**
21:         put v in S.

In 2016 Lewoń, Malafiejska and Malafiejski [16] introduce the concept of a *defensive set* in a graph G, as follows. Recall that a defensive alliance is a set S having the property that for every $v \in S$, $|N[v] \cap S| \geq |N[v] \cap \overline{S}|$. A set S is a *defensive set* if for every vertex $v \in S$, either $|N[v] \cap S| \geq |N[v] \cap \overline{S}|$, or there is a neighbor $u \in S$ of v, such that $|N[\{u, v\}] \cap S| \geq |N[\{u, v\}] \cap \overline{S}|$. Equivalently, a defensive set is a set S, such that every vertex $v \in S$ either satisfies the boundary condition by itself, or it has a neighbor u in S such that $\{u, v\}$ together satisfy the boundary condition.

A defensive alliance or a defensive set S is *total* if S is a total dominating set.

This paper presents (i) an $O(n \log \Delta)$ algorithm for finding a minimum total defensive alliance, (ii) an $O(n \, \Delta^2 \log \Delta)$-time algorithm for finding a minimum global defensive set, and (iii) an $O(n \, \Delta^2 \log \Delta)$-time algorithm for finding a minimum total defensive set, all in trees.

The authors also establish the NP-completeness of the decision problems for (i) global defensive alliances, (ii) global defensive sets, (iii) total defensive alliances, and (iv) total defensive sets, even when restricted to planar bipartite, subcubic graphs. It is known that TOTAL DOMINATING SET is NP-complete for subcubic, bipartite planar graphs [15].

Proposition 1 (Lewoń et al. [16]) *If G is a subcubic graph, then S is a total dominating set if and only if S is a total defensive alliance.*

Proposition 2 (Lewoń et al. [16]) *If G is a subcubic graph, then S is a defensive alliance if and only if S is a defensive set.*

Theorem 13 (Lewoń et al. [16]) *The problems TOTAL DEFENSIVE ALLIANCE and TOTAL DEFENSIVE SET are NP-complete for subcubic, bipartite planar graphs.*

In 2018 [2] Bliem and Woltran consider the question of whether DEFENSIVE ALLIANCE is fixed-parameter tractable when parameterized by the *treewidth $tw(G)$* of G.

The authors show that DEFENSIVE ALLIANCE is $W[1]$-hard when parameterized by the treewidth $tw(G)$ of G. Thus, under a standard complexity-theoretic assumption, this problem has no $(f(tw(G))|V|^{O(1)})$-time algorithm for any function $f : \mathcal{Z}^+ \to \mathcal{Z}^+$.

For the parameter of treewidth, the question of whether DEFENSIVE ALLIANCE is FPT had remained open. This $W[1]$-hard result is surprising since subset problems which are fixed-parameter tractable when parameterized by solution size k usually are fixed-parameter tractable for treewidth as well.

2.1 Self-Stabilizing Alliance Algorithms

In 2006 Xu [20], in his Ph.D. thesis, and in 2007 Srimani and Xu [19] initiate the study of self-stabilizing alliance algorithms. They design two self-stabilizing algorithms for finding (i) a minimal global offensive alliance and (ii) a minimal global defensive alliance. It is important to note at the outset that alliances found with self-stabilizing algorithms are only required to be minimal in cardinality, not necessarily of minimum cardinality.

A global offensive (or defensive) alliance S is called *minimal* if no proper subset of S is also a global offensive (or defensive) alliance.

A global offensive (or defensive) alliance S is called *1-minimal* if there does not exist a vertex $v \in S$ such that $S' = S - \{v\}$ is also a global offensive (or defensive) alliance.

It is easy to see that a global offensive alliance is minimal if and only if it is 1-minimal. However, it is possible that a 1-minimal global defensive alliance may not be minimal.

To those readers unfamiliar with self-stabilizing algorithms, we present this simple algorithm below, along with some explanation.

Algorithm MGOA: Minimal global offensive alliance

$v.diff := |\{x \in N[v] : x.state = IN\}| - |\{x \in N[v] : x.state = OUT\}|$

R1:if$v.state = OUT \wedge (v.diff < 0)$ **then**$v.state := IN$

R2:if$v.state = IN \wedge (v.diff \geq 1)$ **then**$v.state := OUT$

Each vertex $v \in V$ in a graph G maintains at all times two variables. Its state $v.state = IN$ means that v is a member of the minimal global offensive alliance S being formed, that is, $v \in S$, while $v.state = OUT$ means that $v \in \overline{S}$.

The second variable maintained by vertex v is $v.diff$, which at any time records the difference between $|N[v] \cap S|$ and $|N[v] \cap \overline{S}|$. It simply counts the number of its neighbors u (including itself) with $u.state = IN$ and the number of neighbors with $u.state = OUT$ and computes the difference $v.diff$.

At any time t while the self-stabilizing algorithm is executing, all vertices v know their current $v.state$ and $v.diff$ values.

Rule R1, above, says that if a vertex v currently has $v.state = OUT$ and has a negative difference ($v.diff < 0$), then vertex v is *enabled* to make a move, which means set $v.state = IN$, that is, enter the set S.

Rule R2 says that if vertex v is currently in the set S ($v.state = IN$) and has a positive difference of at least 1 ($v.diff \geq 1$), then vertex v is *enabled* to make a move, which means set $v.state = OUT$, that is, go into $\overline{S}$.

At any time t, only one of the vertices which are currently enabled to make a move, according to either Rule 1 or Rule 2, is chosen by what is called a *central scheduler*, or sometimes called an *adversarial daemon*, to execute their move and change the value of its $v.state$.

A scheduler is said to be *fair* if a continuously enabled vertex will be eventually chosen to make a move. Otherwise, it is called *unfair*.

A fundamental idea in self-stabilizing algorithms is that the algorithm can be started when the system (the vertices in the graph G) may be started in any random global state. But after a finite amount of time, the system must reach a correct global state and one which is stable, meaning that after that, it does not change state. Such an algorithm is *self-stabilizing* if (i) regardless of the initial global state in which the algorithm is started, it always reaches a correct global state after a finite number of moves (local state changes) and (ii) once a correct global state has been achieved, any subsequent states must also be correct global states. Such a self-stabilizing algorithm is called *silent* if whenever a correct global state is reached, no vertices are enabled to make another move.

One has then to show that no matter what sequence of enabled vertices the central scheduler chooses to make a move, sooner or later, a state must be reached in which no vertex is enabled to make a move, at which point the algorithm has stabilized. When stable, one must then be able to prove that if no vertex is eligible to make a move, then the set S must be a minimal global offensive alliance, as desired.

Algorithm MGOA must be proved to be correct. This is usually done by a series of lemmas, such as the following.

Lemma 1 (Xu [20]) *If Algorithm MGOA stabilizes, then the set $S = \{v \in V : v.state := IN\}$ is a global offensive alliance.*

Proof If Algorithm MGOA stabilizes, any vertex v not in S must have $v.diff \geq 0$; otherwise, it would be enabled to make move Rule 1; hence, Algorithm MGOA would not be stable, a contradiction. Thus, S must be a global offensive alliance. $\square$

Lemma 2 (Xu [20]) *If at any time the set S is a global offensive alliance, but not a minimal global offensive alliance, then there must exist a vertex v which is enabled to make move Rule 2.*

Proof Suppose that at any time during the execution of Algorithm MGOA there is a subset $S' \subset S$ which is a global offensive alliance. Let v be a vertex in S but not in S'. By the definition of a global offensive alliance, $|N(v) \cap S'| \geq |N[v] \cap \overline{S'}|$. This can be restated as $|N(v) \cap S'| \geq |N(v) \cap \overline{S'}| + 1$.

But since $S' \subset S$, we have $|N(v) \cap S| \geq |N(v) \cap S'|$ and $|N(v) \cap \overline{S}| \leq |N(v) \cap \overline{S'}|$. Thus, $|N(v) \cap S| \geq |N(v) \cap \overline{S}| + 1$. Thus, vertex v is enabled to make move Rule 2. $\square$

Theorem 14 (Xu [20]) *When Algorithm MGOA stabilizes, the set S is a minimal global offensive alliance.*

Proof By Lemma 1, when MGOA stabilizes, S is a global offensive alliance. If S is not a minimal global offensive alliance, by Lemma 2 there must be a vertex which is enabled to execute Rule 2. Thus, if Algorithm MGOA has stabilized, there can be no such vertex, and S must be a minimal global offensive alliance. $\square$

Theorem 15 (Xu [20]) *Algorithm MGOA stabilizes in at most* $2|E| + |V| = 2m + n$ *moves.*

Proof At any time during the execution of Algorithm MGOA, let $X = \{uv : u.state := IN$ and $v.state := OUT\}$. Thus, X defines the set of edges whose two vertices u and v are neither both IN nor both OUT of S.

If Rule 1 is executed on vertex v, then before the move, there are $|N(v) \cap S|$ edges in X incident to v, and after the move, there are $|N(v) \cap \overline{S}|$ edges in X incident to v. Since $|N(v) \cap S| < |N(v) \cap \overline{S}|$ before vertex v makes this move, the number of edges in X cannot decrease after a Rule 1 move.

Similarly, if vertex v executes Rule 2, there are $N(v) \cap \overline{S}$ edges in X incident to v before the move and $N(v) \cap S$ edges in X incident to v after the move. But since $|N(v) \cap \overline{S}| \geq |N(v) \cap \overline{S}| + 1$, executing Rule 2 will always increase the number of X edges.

Since $0 \leq |X| \leq m$, Rule 2 can be executed at most m times.

Each execution of Rule 2 will decrease $|S|$ by 1, and each execution of Rule 1 will increase S by 1. Since $0 \leq |S| \leq n$, there can be at most $m + n$ executions of Rule 1.

Therefore, the total number of moves on all vertices is at most $2m + n$. $\qquad\square$

In addition to the self-stabilizing Algorithm MGOA, Xu [20] also constructs a much more complex, self-stabilizing algorithm for finding a 1-minimal global defensive alliance.

In 2013 Yahiaoui, Belhoul, Haddad, and Kheddouci [21] present a very simple self-stabilizing algorithm for finding a minimal global powerful alliance in an arbitrary graph, using what is called a *distributed scheduler*. Unlike a central scheduler, which can only select one enabled vertex to make a move at a time, a distributed scheduler is permitted to select any subset of currently enabled vertices to execute their moves simultaneously at any given time. The authors also present self-stabilizing algorithms for some generalizations of this problem. Using an unfair distributed scheduler, their proposed algorithms converge in $O(mn)$ moves starting from an arbitrary state. We next present their self-stabilizing, global powerful alliance algorithm, which still uses the central scheduler.

In this algorithm a vertex is said to be *satisfied* (satisfies the powerful alliance property that for every vertex $v \in V$, $|N[v] \cap S| \geq |N[v] \cap \overline{S}|$) if its closed neighborhood has at least as many neighbors in S as it has in $\overline{S}$. Thus, a vertex is satisfied if $v.diff \geq 0$. In addition, $v.diff \geq 2$ means that v remains satisfied even if a neighbor of v leaves the set S. The value $0 \leq v.diff < 2$ means that v is satisfied but will not remain satisfied if a neighbor leaves S.

Algorithm MGPA: Minimal global powerful alliance (expression model)

$\varepsilon = \{diff := |\{x \in N[v] : x.state = IN\}| - |\{x \in N[v] : x.state = OUT\}|\}$

R1: $v.state = OUT \wedge (v.diff < 0 \vee (\exists u \in N(v) : u.state = IN \wedge u.diff < 0)) \rightarrow v.state := IN$

R2: $v.state = IN \wedge v.diff \geq 2 \wedge (\forall u \in N(v) : u.diff \geq 2) \rightarrow v.state := OUT$

For this algorithm the authors prove each of the following lemmas.

Lemma 3 *When no vertex is enabled to make a move using MGPA, then the set S forms a minimal powerful alliance in G.*

Lemma 4 *Once a vertex has $v.diff \geq 0$, it will always have $v.diff \geq 0$.*

Lemma 5 *Once a vertex executes Rule 2, it will never execute another move.*

Lemma 6 *Algorithm MGPA stabilizes after at most 2n moves using the unfair central scheduler.*

Theorem 16 (Yahiaoui, Belhoul, Haddad, Kheddouci [21]) *Algorithm MGPA is a silent, self-stabilizing algorithm in the expression model using the unfair central scheduler and stabilizes after O(n) moves with a minimal global powerful alliance set S.*

In 2013 Hedetniemi, Hedetniemi, Kennedy, and McRae [11] present three self-stabilizing algorithms for producing an *unfriendly partition* of vertices of a graph. An *unfriendly partition* is a partition $V = \{V_1, V_2\}$, such that the vertices in each set have at least as many neighbors in the set they are not in as in the set they are in. This is equivalent to saying that $V = \{V_1, V_2\}$ is a partition into two disjoint global offensive alliances. The simplest of these three self-stabilizing algorithms uses the central scheduler. Associated with each vertex $v \in V$ are three variables: (i) a color $C(v) \in \{Red, Blue\}$, the current color of vertex v; (ii) $B(v) = |\{u \in N(v) : C(u) = Blue\}|$, the number of neighbors of v whose color is Blue; and (iii) $R(v) = |\{u \in N(v) : C(u) = Red\}|$, the number of neighbors of v whose color is Red.

Algorithm Unfriendly
R1:if $C(v) = Blue \wedge B(v) > R(v)$ **then** $C(v) = Red$.
R2:if $C(v) = Red \wedge R(v) > B(v)$ **then** $C(v) = Blue$

Theorem 17 (Hedetniemi, Hedetniemi, Kennedy, McRae [11]) *Algorithm UNFRIENDLY stabilizes to an unfriendly partition $V = \{R, B\}$ in at most $m = |E|$ moves.*

Proof It is easy to see that each execution of either Rule 1 or Rule 2 strictly increases the number of bicolored edges. Thus, at most m rules can be executed, since the maximum number of bicolored edges is m. When Algorithm Unfriendly stabilizes, any vertex not meeting the unfriendly condition, that it have at least as many neighbors with the opposite color as its own color, will be enabled to make a move. Thus, the algorithm could not be stable. $\square$

Algorithm Unfriendly assumes that every vertex can see the current color of all of its neighbors. Since a vertex cannot see the colors of the neighbors of its neighbors, this algorithm runs under what is called the *distance-1 model*. More sophisticated self-stabilizing algorithms can be designed if one can assume that all vertices can see the current states of all vertices at distance-2. In this paper the authors present two more self-stabilizing unfriendly partition algorithms, Algorithm Unfriendlier and Algorithm More Unfriendly, using the distance-2 model. The motivation for these two algorithms was to make decisions, based on greater local knowledge, that would increase, on average, the number of bicolored edges when the algorithms stabilize. We leave the details of these two algorithms to the interested reader.

3 Open Problems

Along with the concepts of defensive, offensive, powerful, and global alliances, the existence of variations of alliances like the following suggest that many combinations of the four basic types of alliances with these variations have yet to be studied:

(i) strong alliances, in which the degree inequality, $d_S(v) + 1 > d_{\overline{S}}(v)$, is strict;

(ii) k-alliances, in which $d_S(v) \geq d_{\overline{S}}(v) + k$;

(iii) weighted alliances, in which for every vertex $v \in S$, $\Sigma_{u \in N[v] \cap S} w(v) \geq \Sigma_{u \in N(v) \cap \overline{S}} w(u)$;

(iv) minimum-cost weight alliances, in which the cost $w(S) = \Sigma_{v \in S} w(v)$ is minimized;

(v) α-alliances, in which the ratio $\rho(v) = \frac{w(N[v] \cap S)}{w(N[v])} \geq \alpha$ must be satisfied for some $0 < \alpha \leq 1$; this variation is proposed by Jamieson and Dean in [12];

(vi) defensive sets, in which for every vertex $v \in S$ either $|N[v] \cap S| \geq |N[v] \cap \overline{S}|$, or there is a neighbor $u \in S$ of v, such that $|N[\{u, v\}] \cap S| \geq |N[\{u, v\}] \cap \overline{S}|$; this variation is proposed by Lewoń, Malafiejska, and Malafiejski in [16];

(vii) total alliances, which are global alliances, but the set S in addition to being a dominating set is a total dominating set; this was also proposed by Lewoń, Malafiejska, and Malafiejski in [16].

In [13] Jamieson makes the following statement, which suggests that more polynomial alliance algorithms can be constructed.

"Using straightforward dynamic programming techniques, we can also construct a polynomial-time (in fact, linear-time) algorithm for computing a minimum-cost (global) weighted alliance on a cycle, a tree of bounded degree, or a graph of bounded path-width. On trees, since a straightforward dynamic programming algorithm runs in pseudo-polynomial time, we can obtain a polynomial running time for the special case where vertex weights are bounded by a polynomial function of n."

In [5] Dourado, Faria, Pizaña, Rautenbach, and Swarcfiter note that DOMINAT-ING SET can be solved in polynomial time for strongly chordal graphs. Given the NP-completeness result for GLOBAL STRONG OFFENSIVE ALLIANCE, using a transformation from DOMINATING SET, it is natural to ask if GLOBAL STRONG OFFENSIVE ALLIANCE can be solved in polynomial time for strongly chordal graphs.

Finally, the following theorem is well-known in the theory of domination in graphs.

Theorem 18 (Ore [17]) *For any graph G having no isolated vertices, the complement $\overline{S}$ of every minimal dominating set S is a dominating set.*

Corollary 1 *The vertices of any graph G having no isolated vertices can be partitioned into two dominating sets.*

Ore's Theorem raises the following general question: Under which conditions can the vertices of a graph G be partitioned into two sets, $V = \{V_1, V_2\}$, such that V_1 and V_2 are alliances of some given type(s)? As noted above by Hedetniemi, Hedetniemi, Kennedy, and McRae in [11], the vertices of every nontrivial connected graph can be partitioned into two global offensive alliances.

References

1. G. Araujo-Pardo, L. Barriére, Defensive alliances in regular graphs and circulant graphs. Preprint 2008. http://upcommons.upc.edu/e-prints/bitstream/2117/2284/1/aliances1002.pdf
2. B. Bliem, S. Woltran, Defensive alliances in graphs of bounded treewidth. Discrete Appl. Math. **251**, 334–339 (2018)
3. A. Cami, H. Balakrishnan, N. Deo, R. Dutton, On the complexity of finding optimal global alliances. J. Combin. Math. Combin. Comput. **58**, 23–31 (2006)
4. C.-W. Chang, M.-L. Chia, C.-J. Hsu, D. Kuo, L.-L. Lai, F.-H. Wang, Global defensive alliances of trees and Cartesian product of paths and cycles. Discrete Appl. Math. **160**(4), 479–487 (2012)
5. M. C. Dourado, L. Faria, M.A. Pizaña, D. Rautenbach, J. L. Szwarcfiter, On defensive alliances and strong global offensive alliances. Discrete Appl. Math. **163**(2), 136–141 (2014)
6. O. Favaron, G. Fricke, W. Goddard, S.M. Hedetniemi, S.T. Hedetniemi, P. Kristiansen, R.C. Laskar, D. Skaggs, Offensive alliances in graphs. Discuss. Math. Graph Theory **24**, 263–275 (2002)
7. H. Fernau, D. Raible, Alliances in graphs: a complexity-theoretic study, in *Proceedings of the SIFSEM 2007*, ed. by J. van Leeuwen, G.F. Italiano, W. Van der Hoek, C. Meinel, H. Sack, F. Plá—vsil, M. Bieliková, vol. II (Institute of Computer Science ASCR, Prague, 2007), pp. 61–70
8. H. Fernau, J.A. Rodriguez, J.M. Sigarreta, Offensive k-alliances in graphs. Discrete Appl. Math. **157**(1), 177–182 (2009)
9. H. Fernau, J.A. Rodriguez, J.M. Sigarreta, Global powerful r-alliances and total k-domination in graphs. Util. Math. **98**, 127–147 (2015)
10. A. Harutyunyan, S. Legay, Linear time algorithms for weighted offensive and powerful alliances in trees. Theoret. Comput. Sci. **582**, 17–26 (2015)
11. S.M. Hedetniemi, S.T. Hedetniemi, K.E. Kennedy, A.A. McRae, Self-stabilizing algorithms for unfriendly partitions into two disjoint dominating sets. Parallel Process. Lett. **23**(1), 1350001 (2013). https://doi.org/10.1142/S0129626413500011
12. L.H. Jamieson, B.C. Dean, Weighted alliances in graphs. Congr. Numer. **187**, 76–82 (2007)
13. L.H. Jamieson, *Algorithms and Complexity for Alliances and Weighted Alliances of Various Types*. Ph.D. thesis, Clemson University, 2007
14. L.H. Jamieson, S.T. Hedetniemi, A.A. McRae, The algorithmic complexity of alliances in graphs. J. Combin. Math. Combin. Comput. **68**, 137–150 (2009)
15. A. Kosowski, M. Malafiejski, P. Żyliński, Cooperative mobile guards in grids. Comput. Geom. **37**(2), 59–71 (2007)

16. R. Lewoń, A. Malafiejska, M. Malafiejski, Global defensive sets in graphs. Discrete Math. **339**(7), 1861–1870 (2016)
17. O. Ore, *Theory of Graphs*. American Mathematical Society Colloquium Publications, vol. 38 (American Mathematical Society, Providence, 1962)
18. K.H. Shafique, *Partitioning a Graph in Alliances and Its Application to Data Clustering*, Ph.D. thesis, School of Computer Science, University of Central Florida, 2001
19. P.K. Srimani, Z. Xu, Distributed protocols for defensive and offensive alliances in network graphs using self-stabilization, in *Proceedings of the International Conference on Computing: Theory and Applications, Kolkata, India*, 2007, pp. 27–31
20. Z. Xu, *Self-Stabilizing Protocols for Distributed Systems*. Ph.D. thesis, School of Computing, Clemson University, August 2006
21. S. Yahiaoui, Y. Belhoul, M. Haddad and H. Kheddouci, Self-stabilizing algorithms for minimal global powerful alliance sets in graphs. Inform. Process. Lett. **113**(10–11), 365–370 (2013)